PLANT PHYSIOLOGY

PLANT PHYSIOLOGY

THIRD EDITION

Frank B. Salisbury
Utah State University

Cleon W. Ross
Colorado State University

Wadsworth Publishing Company
Belmont, California
A Division of Wadsworth, Inc.

Biology Editor: Jack C. Carey
Production Editor: Harold Humphrey
Designer: MaryEllen Podgorski
Copy Editor: Yvonne Howell
Technical Illustrators: Darwen Hennings, Vally Hennings,
Marilyn Krieger
Cover Photograph: Harold Dorr
Section Opening Drawings: Vally Hennings
Print Buyer: Karen Hunt

Printed in the United States of America

3 4 5 6 7 8 9 10——89 88 87 86

ISBN 0-534-04482-4

Library of Congress Cataloging in Publication Data

Salisbury, Frank B.
  Plant physiology.

  Bibliography: p.
  Includes indexes.
  1. Plant physiology.   I. Ross, Cleon W., 1934-
II. Title.
QK711.2.S23   1985   581.1   84-20953
ISBN 0-534-04482-4

To
Herman Henry Wiebe
1921–1984

This edition of *Plant Physiology* is dedicated to the memory of Herman H. Wiebe, who died on March 14, 1984. Professor Wiebe made major contributions to this book and to the two previous editions. He reviewed every chapter and was especially helpful with the chapters in his own speciality, which included most of the section on plant–water relations. Professor Wiebe was a doctoral student of Paul Kramer at Duke University and taught at Utah State University since 1954. He had a deep concern for students, and this concern is reflected in many ways in this book and in other books that he reviewed for Wadsworth Publishing Company. As it happens, Cleon Ross was a doctoral student of Herman Wiebe, and Frank Salisbury and Herman Wiebe were close friends and colleagues for the eighteen years since Salisbury came to Utah State University. We will miss our dear friend and advisor.

Preface

Much new information has been gained in the field of plant physiology since 1978 when the second edition of our textbook appeared. It would have been relatively easy to double the size of our third edition just by adding new information to our second. Yet as teachers of plant physiology, we realize that time always limits what can be presented in a plant physiology course. Hence, it has been our assignment to choose and summarize those new findings that are most representative of the status of the science in the mid-1980s. We hope that you as students and teachers will agree with many of our choices. We also realize that our book contains more information than can be presented in the typical one-quarter or one-semester course in plant physiology. Thus teachers will also have to choose topics most relevant for their classes, but we hope that all students will gain some impression of the breadth of our science by surveying those topics in our text that are not assigned in class.

The organization of this text is basically the same as in previous editions. We realize that beginning plant physiology students have studied the plant cell in previous courses, but we have greatly expanded the discussion of the cell in our prologue. Knowledge of the cell is so fundamental that a complete review seems justified—and was requested by many reviewers. The rest of our text is divided into four sections: The first seven chapters (Section 1) deal with physical processes, the middle seven chapters (Section 2) with metabolism and biochemistry, and the eight chapters of Section 3 with growth and development. The last two chapters (Section 4) are built upon the other twenty-two to relate plant function to the environment that so strongly influences it. We assume that students have some knowledge of organic chemistry as background for our biochemistry section. We also assume that students have studied DNA replication and protein synthesis (molecular biology), topics usually covered in basic biology courses. Nevertheless, in Appendix C we present several topics about molecular biology for quick review. Of course, students will also have covered many other topics that we describe; we hope our presentation will be appropriate review and reinforcement.

In preparing this edition and in teaching plant physiology, it is apparent that biochemical information is becoming increasingly important in understanding the so-called physical processes of plant function (mostly plant–water relations), although physics is still of central importance to water relations. We have retained the previous order (physical processes before biochemical), because a survey of teachers by our publisher revealed that a significant majority of users still prefer this organization. Nevertheless, we suggest that some teachers might like to discuss Section 2 before Section 1, since little background in physical processes is required to understand the biochemistry chapters. We might reverse the order of these two sections in our fourth edition, and we would greatly appreciate comments from teachers about such an approach.

We have retained most of the features (teaching and learning aids) of our second edition, since they were well received by many users. In addition, we have made a few changes, partially as a result of our own teaching experiences and partially in response to reviewers suggestions.

The Format   Our publisher designed a new format for this edition, one that is a bit more conventional and should be easier for readers to handle. This format provides fewer words on a page than in our previous editions, so the number of pages has increased significantly. Part of the page increase is because of new material added in this edition, although we compensated by condensing our writing and by eliminating some older material. We also added many new illustrations, deleted a few of the older ones, and had some of our earlier illustrations redrawn. Many of the new illustrations were prepared by Darwen and Vally Hennings, a pair of talented botanical illustrators.

Guest Essays and Boxed Essays   For the 1978 edition, we requested that several plant physiologists write guest essays that often related some of their own personal experiences and impressions of our science and that sometimes presented facts and applications of topics in the text. In this third edition we kept many of these guest essays and added some new ones. We find them enjoyable and informative, and we hope that you will too. We retained and added a number of small boxed essays as well. (In Chapter 7 on phloem transport, it was necessary to add a rather large boxed essay on the chemistry of carbohydrates, because numerous carbohydrates and related compounds occur in the phloem translocation stream.)

Appendices In our second edition we presented a number of special topics at the end of the book. Most reviewers suggested that these either be deleted or incorporated into appropriate chapters. We have followed this suggestion, but three important topics are not amenable to incorporation in other chapters. These are now Appendices. The first presents tables and some discussion describing the Système Internationale (SI units). The use of SI units has expanded considerably in all sciences during recent years. Our goal in this edition has been to use SI units whenever they are being used by current journals that publish major advances in plant physiology. We were thus able to use SI units almost exclusively, although a few of the older units are still in wide use and will probably continue to be used (liter for volume; molality and molarity for solution concentrations; and minute, hour, and day for time). The second appendix is an updated version of our previous Special Topic on radiant energy. It seems logical to consider this topic as a unit in an appendix rather than discussing several aspects of it in various chapters in the text or to give it undue emphasis by including it with other material in a single chapter. As noted, our third appendix concerns molecular biology.

References to the Literature of Plant Physiology Users of our previous editions have commented favorably on the references in our text. Thus, on the one hand we were tempted to document most of our important conclusions. We did this by referring to authors by last name and year of publication. On the other hand, we realized that frequent references distract some students. In the prologue and the first few chapters, not many references to the literature are given in the text, although the bibliography at the end of the book contains many reviews and other references for these chapters. The interested user can scan the titles in these lists to find specific sources of information. After the first few chapters, we provided many more references. Students should realize that they must become accustomed to such references, because they are an essential part of the scientific literature.

The increased use of references in this edition has led to some other problems. In the first (1969) edition of our book, we provided first names of authors if we knew them and otherwise provided initials when scientists were first mentioned in the text. Frequently, we also mentioned where their work was performed. We felt that this was one way to make it apparent to students that plant physiology is the product of many individuals working in diverse places all over the world. That is, we hoped that we could personalize our science somewhat. We realize, however, that some students may become frustrated by feeling that they should remember all about who

did what and when. In this edition, we have often retained the use of first names (and some locations), but we have not made an effort to use first names and intials of everyone whose name is mentioned. Be admonished that full names are there mainly to personalize our science, but references are given for scientific documentation of important discoveries and recent facts and to point out recent reviews.

Some Specific Chapter Changes All chapters were thoroughly revised for this edition, although the first three chapters (water potential, osmosis, and transpiration) have less new material than the others. The first three were revised to make them easier for students to follow. The order of Chapters 5 and 6 was reversed, so that students would learn about the elemental composition of plants before they investigated how such elements are absorbed. Chapter 7 (phloem transport) was substantially reorganized and expanded in response to suggestions from teachers and students. We reorganized Chapters 8 to 11 less than Chapter 7, although we placed photorespiration in Chapter 10 instead of Chapter 11. We also deleted a section on cell-wall chemistry from Chapter 10 to save space; some of that material (as related to growth) is now in Chapter 15. Chapter 14 was expanded almost as much as Chapter 7. This chapter describes numerous chemical reactions and products common to most organisms plus reactions that occur only in plants; it also describes ecological functions of lipids and several other natural products. This chapter and others in Section 2 (except for Chapter 12 on respiration) emphasize the unique biochemistry of green plants. The chapters in Section 3 on growth and development have all been thoroughly revised (Chapter 20 on the biological clock perhaps the least). Stress physiology has been advancing rapidly in recent years, so Chapter 24 was extensively revised and expanded.

The Reviewers As in our previous editions, we are indebted to the reviewers retained by our publisher or requested by us to examine individual chapters or the entire manuscript. These reviewers, along with many others who volunteered comments, provided many valuable suggestions that we followed. We are grateful to them, but of course we accept complete responsibility for the text in its present form.

Final Observations Preparation of any text involves frustrations in checking details and reconciling points of view, yet we have learned much about our science in writing for you. Although we each write somewhat differently, we hope that our writing displays organization and readability as well as facts and major principles. Frequently, we emphasize or imply that many problems remain unsolved. If that were not true, our science could not advance. Current and future researchers (many of whom are now students) must solve many present and future problems, so that years from now plant physiology texts will be somewhat less speculative and contain even more principles.

We hope that our enthusiasm and love for the science of plant physiology is apparent to you and that you will come to share these feelings with us. These are the feelings that motivate the rapid advances now being made in virtually all scientific disciplines.

Frank B. Salisbury, Logan, Utah
Cleon W. Ross, Fort Collins, Colorado
December 1984

Reviewers

Luke S. Albert University of Rhode Island
Betty D. Allamong Ball State University
James A. Bassham University of California, Berkeley
J. Clair Batty Utah State University
Clanton Black, Jr. University of Georgia
Caroline Bledsoe University of Washington
John S. Boyer University of Illinois
C. H. Brueske Mount Union College
Michael J. Burke Oregon State University
Clyde L. Calvin Portland State University
William V. Dashek West Virginia University
R. Dean Decker University of Richmond
Gerald F. Deitzer Smithsonian Institution
Stanley Duke University of Wisconsin, Madison
Frank Einhellig University of South Dakota
Emanuel Epstein University of California, Davis
Donald B. Fisher University of Georgia
Douglas C. Fratianne Ohio State University
D. J. Friend University of Hawaii
Donald R. Geiger University of Dayton
Benz Z. Ginzburg Hebrew University of Jerusalem
Shirley Gould Lewis and Clark College
Richard H. Hageman University of Illinois
John Hendrix Colorado State University
Theodore Hsiao University of California, Davis
Ray C. Huffaker University of California, Davis
William T. Jackson Dartmouth College
Andre Jagendorf Cornell University
Peter B. Kaufman University of Michigan, Ann Arbor

George G. Laties University of California, Los Angeles
Edwin H. Liu University of South Carolina
James A. Lockhart University of Massachusetts
Abraham Marcus The Institute for Cancer Research
Maurice M. Margulies Smithsonian Institution
Ann G. Matthysse University of North Carolina, Chapel Hill
Jerry McClure Miami University
Linda McMahan Utica College
Robert Mellor University of Arizona
Joseph E. Miller Argonne National Laboratory
Terence M. Murphy University of California, Davis
Murray Nabors Colorado State University
Dave Newman Miami University, Ohio
Park S. Nobel University of California, Los Angeles
James O'Leary University of Arizona
Herbert D. Papenfuss Boise State University
Wayne S. Pierce California State College
Donald W. Rains University of California, Davis
Walter Shropshire, Jr. Smithsonian Institution & George Washington University
Wendy Kuhn Silk University of California, Davis
George Spomer University of Idaho
Cecil R. Stewart Iowa State University
Larry L. Tieszen Augustana College
George Welkie Utah State University
Herman Wiebe Utah State University
Conrad J. Weiser Oregon State University
Jan A. D. Zeevaart Michigan State University

Contents in Brief

Contents in Detail

Prologue

Plant Physiology and Plant Cells

Plant physiology is the science that studies plant function: what is going on in plants that accounts for their being alive. The promise of plant physiology is to give you some insight into these functions.

Of course you know that plants are not really as inanimate as they appear. (It is often difficult to tell a plastic plant from its real counterpart.) But studying plant physiology should greatly broaden your appreciation for the many things that are happening inside plants. Water and dissolved materials are moving through special transport pathways: water from soil through roots, stems, and leaves to the atmosphere; and inorganic salts and organic molecules in many directions within the plant. Thousands of kinds of chemical reactions are underway in every living cell, transforming water, mineral salts, and gases from the environment into organized plant tissues and organs. And from the moment of conception, when a new plant begins as a zygote, until the plant's death, which could be thousands of years later, organized processes of development are enlarging the plant, increasing its complexity, and initiating such qualitative changes in its growth as the formation of flowers in season and the loss of leaves in autumn.

Plant physiology studies all these things.

P.1 Some Basic Postulates

Plant physiology is one of many branches of biological science. Like the other branches, it studies life processes, which are often similar or identical in many organisms, including plants. In this prologue, we shall state eleven postulates, or generalizations, about science in general and plant physiology in particular. These postulates are presented with a minimum of discussion, but information on cells is so fundamental to plant physiology that after the list we provide a review of plant cells as the main body of this prologue. Here are the postulates:

1. *Plant function can ultimately be understood on the basis of the principles of physics and chemistry.* Plant physiologists accept the philosophical statement called the Law of the Uniformity of Nature, which states that the same circumstances or causes will produce the same effects or responses. This concept of cause and effect must be accepted as a working hypothesis (i.e., on faith). Although there is no way to prove that the principle always applies everywhere in the universe, there is also no reason to doubt that it does. It is possible that life depends on a spirit or entelchy not subject to scientific investigation; but if that is presumed, then by definition science cannot be applied to anything. The assumption that plants are mechanistic leads to fruitful research; the contrary assumption, called **vitalism,** has been completely unproductive in science. For example, convictions (yours or ours) about the existence of a Creator may help or hinder your appreciation of plant physiology, but they cannot play a direct role in the science itself.

2. *Modern plant physiology in particular and biology in general depend upon the physical sciences: upon physics and chemistry, with their reliance on mathematics.* Plant physiology is essentially an application of modern physics and chemistry to the understanding of plants. Indeed, progress in plant physiology has been almost completely dependent upon progress in the physical sciences. Today, the technology of applied physical science provides the instrumentation upon which research in plant physiology depends.

3. *Botanists and plant physiologists study members of four of the five kingdoms of organisms presently recognized by many biologists (Table P-1); but much discussion in this book is concerned with the true plants and, indeed, with a relatively few species of gymnosperms and angiosperms.*

Table P-1 A Simplified Outline of the Classification of Organisms.

VIRUSES: Exhibit properties of life only when present in cells of other organisms; considered by most biologists not to be alive when isolated.

1. MONERA:* prokaryotic organisms (no organized nucleus or cellular organelles); include bacteria, blue-green algae (sometimes called cyanobacteria), and mycoplasms.

2. PROTISTA: Eukaryotic (true organelles and nucleus), mostly single-celled organisms; include protozoa (single-celled "animals"), some algae,* and slime molds.* (Some authors include all the eukaryotic algae, even multicellular forms.)

3. FUNGI:* The true fungi

4. PLANTAE:* Most algae and all green plants; include the following plus some minor groups not mentioned:
    Brown algae
    Red algae
    Green algae**
    Mosses and liverworts*
    Vascular plants (higher plants)
       Ferns and relatives*
       Cycads and rare gymnosperms*
       Conifers (common gymnosperms)**
       Flowering plants (angiosperms)**
          Monocotyledons (monocots)
          Dicotyledons (dicots)

5. ANIMALIA: Multicellular animals

*Studied by plant physiologists
**Emphasized by plant physiologists

Modern biologists feel that a five-kingdom approach to a classification of living organisms is far superior to the previous attempts to classify all organisms as either plants or animals. Nevertheless, there is still much controversy about the placement of certain groups such as the slime molds and some of the algae. Suffice it to say that plant physiologists study the blue-green algae (or cyanobacteria) and other prokaryotes studied by bacteriologists, various groups of algae, slime molds, true fungi, and representatives of all major groups in the plant kingdom. Nevertheless, our discussions here will strongly emphasize gymnosperms and the flowering plants, only occasionally referring to the other groups.

4. *The cell is the fundamental unit of life; all living organisms consist of cells, which contain either membrane-bound nuclei or comparable structures without membranes. Life does not exist in units smaller than cells. Cells arise only from the division of pre-existing cells.* Collectively, these statements are known as the **cell theory.** Coenocytic organisms (certain algae, fungi, and slime molds) do not have their organelles (mitochondria, nuclei, etc.) partitioned into cells by cell membranes and walls, but other typical structures of cells are present.

5. *Eukaryotic cells contain such membrane-bound organelles as chloroplasts, mitochondria, nuclei, and vacuoles, whereas prokaryotic cells (by definition) contain no membrane-bound organelles.*

6. *Cells are characterized by special macromolecules, such as starch and cellulose, which consist of hundreds to thousands of identical sugar or other molecules, or repeating groups of molecules.*

7. *Cells are also characterized by such macromolecules as proteins and nucleic acids (RNA and DNA), which consist of chains of hundreds to thousands of simpler molecules of various kinds (20 or more amino acids in protein and 4 or 5 nucleotides in nucleic acids). These chains include long segments of nonrepeating sequences that are preserved and duplicated when the molecules are reproduced.* These molecules, so typical of life, contain *information,* much as the sequence of letters in this sentence accounts for the information in the sentence. Information is transferred from cell generation to generation through DNA and from DNA to protein via RNA. The information in a protein bestows upon the molecule certain physical characteristics and the ability to catalyze (speed up) chemical reactions in cells; proteins that catalyze reactions are called **enzymes** and are fundamental to life function.

8. *In multicellular organisms, cells are organized into tissues and organs; different cells in a multicellular organism often have different structures and functions.* The tissue–organ concept is more difficult to apply to plants than to animals; but typical plant-stem tissues include, for example, epidermis, cortex, vascular tissues, and pith. The principal organs of plants include roots, stems, leaves, flowers, and fruits.

9. *Living organisms are self-generating structures.* Through the process called **development,** which includes cell divisions, cell enlargement (especially elongation in stems and roots), and cell specialization, a multicellular organism begins as a single cell (fertilized egg, the **zygote**) and eventually becomes a mature organism. Though much descriptive information is available, development is probably the least understood phenomenon of contemporary biology (about as mysterious as the functioning of the human brain).

10. *Organisms grow and develop within environments and interact with these environments and with each other in many ways.* For example, plant development is influenced by temperature, light, gravity, wind, and humidity.

11. *In living organisms, as in other machines, structure and function are intimately wedded.* Clearly, there could be no life functions without the structures of genes, enzymes, organelles, cells, and often tissues and or-

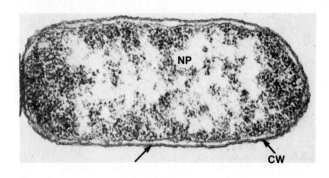

a

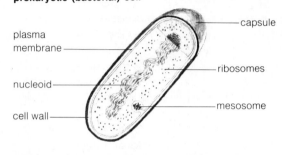

prokaryotic (bacterial) cell

plasma membrane —
nucleoid —
cell wall —

— capsule
— ribosomes
— mesosome

b

Figure P-1 (**a**) A prokaryotic cell, the bacterium *Escherichia coli*, magnified 21,500 times. The nucleoid (NP), the prokaryote equivalent of a nucleus, occupies the center of the cell. The cytoplasm surrounding the nucleus is packed with ribosomes. The cell is surrounded by the cell wall (CW): the plasma membrane (arrow) lies just beneath the cell wall. (Micrograph courtesy of William A. Jensen.) (**b**) An interpretation of a generalized prokaryotic cell. (W. A. Jensen and F. B. Salisbury, 1984, *Botany*, p. 47.)

gans. Yet the functions of growth and development create the structures. Studies in plant physiology depend strongly upon plant anatomy and **cytology** (study of cells) and also upon structural and functional chemistry. At the same time, such structural sciences as anatomy become more meaningful because of plant physiology.

P.2 Prokaryotic Cells: Bacteria and Blue-green Algae[*]

Membranes are the extremely thin layers of material, consisting mostly of lipids and protein, that separate cells, and most cell parts, from their surroundings. We discuss their nature in later chapters, especially

---

[*]The rest of this prologue has been condensed and updated from W. A. Jensen and F. B. Salisbury, 1984, *Botany*, Second Edition. Wadsworth Publishing Company, Belmont, California. Chapter 3 (originally prepared by Salisbury).

Chapter 6. **Prokaryotic cells,** those of bacteria, blue-green algae (cyanobacteria), and mycoplasms, have only the surface membrane that surrounds each cell. Any membranous material found inside such cells is likely to be an inward extension of the cell membrane. **Eukaryotic cells,** on the other hand, contain several kinds of **organelles** ("little organs"), each surrounded by a single or a double membrane system.

The **nucleus** of the eukaryotic cell is surrounded by a double membrane; but the prokaryotic cell has only a central body called a **nucleoid,** which is surrounded by **cytoplasm** (all the substance enclosed by the plasma membrane and outside the nucleoid) but not by a membrane. In bacteria, the nucleoid consists of a single piece of DNA about 1 millimeter (mm) long, closed into a circle. This is the essential genetic material.

The term *prokaryotic* means "*before* a nucleus" (from the Greek) and not *without* a nucleus. Indeed, fossil prokaryotes as old as 3.3 billion years are known, whereas the oldest eukaryotic fossils are less than 1 billion years old. (*Eukaryote,* also from the Greek, means *true nucleus.*)

Prokaryotic cells are comparatively small, seldom more than a few micrometers (μm) long and only about 1 μm thick (Fig. P-1). (Metric and SI units are summarized in Appendix A.) Blue-green algae cells are usually much larger than bacterial cells. Also, all blue-green algae carry out photosynthesis with chlorophyll *a*, not found in bacteria, and by metabolic pathways common to plants and algae but not bacteria. Thus, the term *cyanobacteria*, which implies that blue-green algae are just another form of bacteria, is unfortunate.

Prokaryotic cells are surrounded by **cell walls,** usually lacking cellulose and, hence, chemically different from the typical walls of higher plants. The wall may be anywhere from 10 to 20 nanometers (nm) thick and sometimes is coated with a relatively thick, jellylike **capsule** or slime of proteinaceous material. Inside the wall, and tightly pressed against it, is the outer membrane of the prokaryotic cell, the **plasma membrane** or **plasmalemma,** which may be smooth or have infoldings that extend into the cell, forming structures called **mesosomes.** Besides controlling what enters and leaves the cells, membranes have other important functions. Many enzymatic reactions including photosynthesis and respiration take place on the proteins contained in membranes, and the plasma membranes of prokaryotes are thought to play a role in cell replication.

Small spherical bodies about 15 nm in diameter, the **ribosomes,** crowd the cytoplasm and are the sites of protein synthesis. The cytoplasm of the more complex prokaryotes may also contain **vacuoles** (saclike structures), **vesicles** (small vacuoles), and reserve

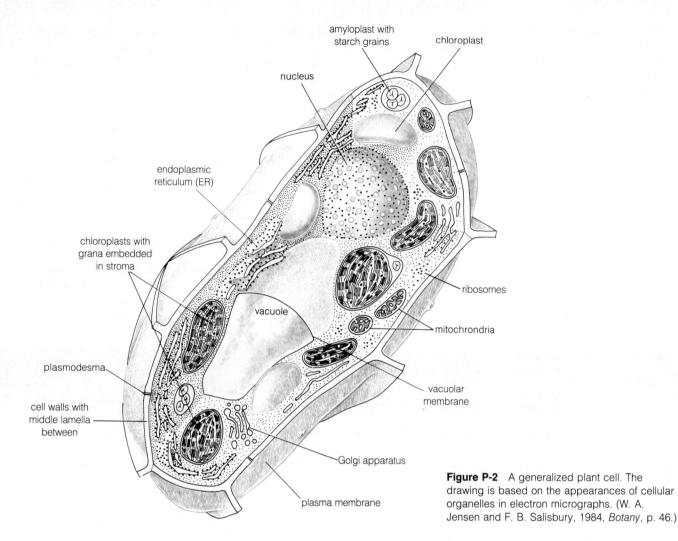

amyloplast with
starch grains

chloroplast

nucleus

endoplasmic
reticulum (ER)

chloroplasts with
grana embedded
in stroma

vacuole

ribosomes

mitochrondria

plasmodesma

cell walls with
middle lamella
between

vacuolar
membrane

Golgi apparatus

plasma membrane

Figure P-2   A generalized plant cell. The drawing is based on the appearances of cellular organelles in electron micrographs. (W. A. Jensen and F. B. Salisbury, 1984, *Botany*, p. 46.)

Table P-2   The Components of a Prokaryotic Cell.

I.   CELL WALL (with or without a capsule)

II. PLASMA MEMBRANE or PLASMALEMMA (sometimes with infoldings called mesosomes)

III. NUCLEOID (single circular strand of DNA—the genetic material)

IV. CYTOPLASM (all the substance enclosed by the plasma membrane except the nucleoid)
  1. RIBOSOMES (sites of protein synthesis)
  2. VACUOLES (saclike structures)
  3. VESICLES (small vacuoles)
  4. RESERVE DEPOSITS (complex sugars and other materials)

V.   FLAGELLA (threadlike structures protruding from cell surfaces; capable of beating to cause cell movement; consist of single protein fibers)

Source: W. A. Jensen and F. B. Salisbury, 1984, *Botany*, p. 46.

---

deposits of complex sugars or inorganic materials. In some rare blue-green algae, the vacuoles are filled with nitrogen gas.

Many bacteria are capable of relatively rapid movement, generated by the action of threadlike structures, the **flagella,** that protrude from the cell surface. Table P-2 summarizes the structures of prokaryotic cells.

P.3 Eukaryotic Cells: Protist, Fungal, and Plant

The principal structures of prokaryotic cells are also present in eukaryotic cells, but the latter have several additional structures as well, most of them bound by membranes. A useful fiction in studying eukaryotic cells is the "typical" plant cell, illustrated in Fig. P-2 and summarized in Table P-3. There is, of course, no such thing as the typical cell or the "average teenager." Both are statistical creations, composites of features characteristic of a class but seldom found together in one individual. Nevertheless, the **parenchyma cells** (thin-walled, often isodiametric living cells) found in pith, cortex, root and shoot tips, and so forth, have most of the features of the typical plant cell. Let's begin from the outside of our typical cell and work toward the cellular inclusions. The exercise will provide a basic idea of how plants, protistans, and fungi differ at the cellular level from each other and from animals.

The Cell Wall Many protistans and virtually all fungal and plant cells are surrounded by a cell wall. Indeed no other feature is more characteristic of fungal and plant cells than the wall. All cells have membranes that enclose their contents, but animal and some protist cells have no walls—*only* membranes. Wall structures and other cell features are illustrated in Fig. P-3. Young growing cells, storage cells, the photosynthesizing cells of leaves, and some other cell types have only a **primary wall,** a wall characterized by being thin and formed while the cell is undergoing rapid growth and elongation. The cell wall surrounds the **protoplast,** which includes the plasma membrane and all that it encloses. This membrane is usually pressed tightly against the wall because of the pressure of the fluids inside. Many mature plant cells, especially those that have finished growing, have laid down a **secondary wall** between the primary wall and the cell membrane. Between the primary walls of adjacent cells is the **middle lamella** that cements the two cell walls together.

The primary cell wall   Compared with an entire cell, or even with the secondary wall, the primary wall is thin, on the order of 1 to 3 μm thick. It consists of about 9 to 25 percent **cellulose.** About 30 to 40 pairs

---

Table P-3   The Components of a Eukaryotic Plant Cell.

I. CELL WALL*
   A. Primary wall (about ¼ cellulose); about 1 to 3 μm thick
   B. Secondary wall (½ cellulose + ¼ lignin); may be 4 μm thick or more
   C. Middle lamella (pectin-cementing layer between cells)
   D. Plasmodesmata (strands of cytoplasm penetrating wall); 40 to 100 nm thick
   E. Simple and bordered pits

II. PROTOPLAST (contents of the cell exclusive of the cell wall); 10 to 100 μm diameter
   A. Cytoplasm (cytoplasm + nucleus = protoplasm)
      1. Plasma membrane (plasmalemma); 0.0075 μm (7.5 nm) thick
      2. Vacuolar membrane (tonoplast); 7.5 nm thick
      3. Microtubules; 18 to 27 nm thick
      4. Microfilaments; 5 to 7 nm thick
      5. Endoplasmic reticulum (ER); 7.5 nm thick (each membrane)
      6. Ribosomes; about 15 to 25 nm diameter
      7. Golgi bodies (dictyosomes); 0.5 to 2.0 μm diameter
      8. Mitochondria; 0.5 to 1.0 by 1 to 4 μm
      9. Plastids†
         a. Proplastids (immature plastids)
         b. Leucoplasts (colorless plastids)
         c. Amyloplasts (contain starch grains) and other food-storage plastids
         d. Chloroplasts; 2 to 4 μm thick by 5 to 10 μm diameter (may also contain starch grains)
         e. Chromoplasts (colored plastids other than chloroplasts; often red, orange, yellow, etc.)
     10. Microbodies; 0.3 to 1.5 μm diameter
     11. Spherosomes; 0.5 to 2.0 μm diameter
     12. Cytosol
   B. Nucleus (cytoplasm + nucleus = protoplasm)
      1. Nuclear envelope (double membrane); 20 to 50 nm thick
      2. Nucleoplasm (granular and fibrillar substance of nucleus)
      3. Chromatin (chromosomes become apparent during cell division)
      4. Nucleolus; 3 to 5 μm
   C. Vacuoles (up to 95% of cell volume or more)
   D. Ergastic substances (inclusions of relatively pure materials, often in plastids)*
      1. Crystals (such as calcium oxalate)
      2. Tannins †
      3. Fats and oils
      4. Starch grains (in amyloplasts and chloroplasts, see above)†
      5. Protein bodies
   E. Flagella and cilia; 0.2 μm thick, 2 to 150 μm long

---

*Occur in fungal, plant, and some protistan cells but seldom in animals.
†Occur only in plant cells and some protistans.

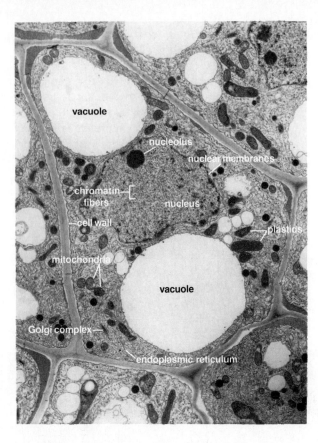

Figure P-3 An electron micrograph of a cell from the developing fruit of cotton, showing cell walls and other cell parts. The cell wall is conspicuous in this photograph, and the middle lamella can be seen joining the walls of adjacent cells (especially at the corners). The nucleus shows some DNA-protein (genetic material or chromatin) just inside the nuclear membranes, and a prominent nucleolus. Two large vacuoles take up much of the cell volume, but not as much as in more mature plant cells, especially those in stems and roots. Also visible in the cell are plastids, mitochondria, parts of the endoplasmic reticulum, and at least one faint Golgi body. (Micrograph by William A. Jensen and Paula Stetler.)

of long, unbranched cellulose molecules form a long cylindrical fiber called a **microfibril.** Recent data suggest that microfibrils are about 3.5 nm thick (Fig. P-4). Because of the parallel arrangement of the cellulose molecules, microfibrils behave like crystals and have as much tensile strength, for their weight, as steel wires in a cable.

The microfibrils are laid down roughly at right angles to the long axis of the cell. They are like hoops around the *inside* of a barrel. As the cell elongates, the microfibrils slip past each other and are pulled into the long axis of the cell (discussed in Chapter 15).

Since the most recent microfibrils to be deposited are still laid down predominantly parallel to the circumference, the microfibrils cross each other almost as threads in cloth (but do not go over and under each other as do woven threads).

The cellulose microfibrils are embedded in a **matrix** of other materials, which are chemically much more complex. Principal among these are the **hemicelluloses,** which form a branching, molecular network filled with water. A typical primary wall may contain 25 to 50 percent hemicelluloses. Closely related are the **pectic substances,** which make up 10 to 35 percent of a primary wall. Primary walls also usually have about 10 percent protein. The matrix materials of the primary wall are not crystalline like cellulose, so the matrix cannot be seen in Fig. P-5; it was dissolved away to make the microfibrils more visible (also see Fig. 6-11). The primary wall is admirably adapted to growth. In response to growth-regulating chemicals, it softens in some way so the microfibrils can slide past each other in the water-filled matrix (Chapter 16). This sliding occurs as the protoplast absorbs water, expanding like a balloon and creating pressure against the wall. Thus the wall stretches **plastically** (irreversibly, like bubble gum) rather than **elastically** (like a rubber balloon) as the cell grows. Some primary walls increase their area as much as 20 times during growth.

When the cell is not growing, even the primary wall resists stretching, thanks to the high tensile strength of its cellulose microfibrils, which can't slip past each other. Yet the wall is porous enough to allow the free passage of water and materials dissolved in the water. The pore diameters are about 3.5 to 5.2 nm (Carpita et al., 1979; Carpita, 1982) compared with about 0.3 nm diameter for a water molecule and about 1 nm for a sugar molecule.

Imagine a cotton (mostly cellulose) cloth bag with a water-filled balloon inside. The cloth is porous, freely allowing the passage of water and materials dissolved in the water. It also has tensile strength, resisting stretching as one tries to force more water into the balloon. Yet it collapses when the water is released from the balloon. Likewise, if the cell loses water and hence its hydraulic pressure, the primary cell wall collapses (although not as much as a cotton bag). Leaves and young stems are made of cells that have mostly primary walls. They are rigid while the fluid in their cells pushes against their walls, but they wilt when enough water is lost to decrease the internal pressure (Fig. P-6).

The secondary cell wall In many plant cells, especially those that will provide support for the plant or will be involved in the conduction of fluids under tension (negative pressure), the protoplast begins to secrete a

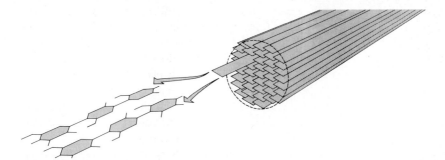

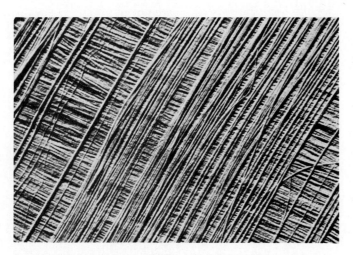

Figure P-4 (left) A schematic drawing of how cellulose molecules are arranged to form a microfibril of cellulose. Pairs of molecules are held together by hydrogen bonding to form the sheetlike strips, of which there are about 40 in each microfibril. Each strip is held to the ones above, below, and to its sides by hydrogen bonding. The hexagons represent glucose molecules in the long-chain cellulose molecules. (W. A. Jensen and F. B. Salisbury, 1984, *Botany*, p. 49.)

Figure P-5 Ordered pattern of cellulose microfibril deposition in a maturing cell wall of a green alga. × 21,000. Courtesy of E. Frei and R. D. Preston, from *Proceedings of the Royal Society*, Series B, 154 (1961): 70.

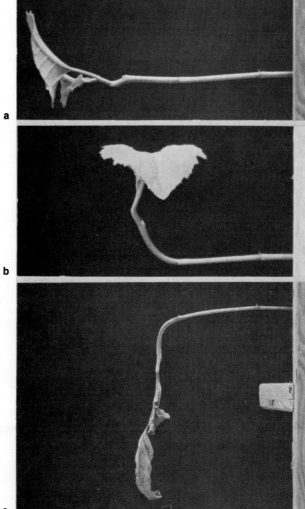

a

b

c

Figure P-6 How turgor pressure in cells determines the form of nonlignified tissues. (**a**) A cocklebur plant that has just been laid on its side. (**b**) The same plant 18 h later. The upward bend has occurred in the actively growing part of the stem where lignin has not yet been deposited in the cell walls. (**c**) The same plant, 8 days later, after it has dried out and wilted. Note that the lignified part of the stem that did not bend upwards in response to gravity also did not wilt as the plant dried out. (Photographs by Frank B. Salisbury.)

secondary wall after the cell has stopped enlarging. Often, after the secondary wall has been secreted, the protoplast dies, and its constituents are removed from the cell so only the wall remains. This is the case for all but a small portion of the cells in wood and cork.

Secondary walls are usually much thicker than primary walls; some are on the order of several micrometers thick. Secondary walls consist of about 41 to 45 percent cellulose, 30 percent hemicelluloses, and 22 to 28 percent **lignin** (Chapter 14), which is not easily compressed and resists changes in form; lignin is much more rigid than cellulose. But the combination of stretched cellulose microfibrils embedded in lignin, like the steel rods under tension that are embedded in concrete to form prestressed concrete, gives wood its strength. For its weight, wood is one of the strongest materials known. It certainly does not wilt when it loses water (Fig. P-6)!

When the cell that is to have a secondary wall stops enlarging, lignin is first deposited in the already-formed middle lamella, then in the existing primary wall, and finally in the secondary wall, as it forms.

The middle lamella The pectic substances that cement adjacent cells together and form the middle lamella are ideally suited to this role, since they exist as gels. Indeed, we extract them from unripe fruits, where they are plentiful, and use them in making jams and jellies. Pectins can be broken down by certain enzymes, as happens when many fruits ripen. A green peach, for example, is rock hard, but as it ripens the tissues become mushy or pulpy.

Pits, plasmodesmata, and other cell-wall features Primary walls usually have thin areas called primary **pit-fields.** Extremely thin strands of cytoplasm, called **plasmodesmata** (singular: plasmodesma), extend through the walls of adjacent cells, often in the primary pit-fields, connecting the protoplasts of adjacent cells. Plasmodesmata have been seen for decades, but their detailed structures could only be understood after development of the electron microscope (discussed in Section 6.4). They appear as channels lined by the cellular membranes of adjacent cells and filled with a strand about 40 nm in diameter of the endoplasmic reticulum, a special membrane system we shall discuss below. Plasmodesmata are thought to be of considerable importance, since they unite the many cells of a tissue or plant into one functional whole. We can calculate that substances like glucose can pass from cell to cell through the plasmodesmata some thousand times faster than across membranes and cell walls.

Secondary wall material is not laid down over the primary pit-fields, so the secondary wall includes characteristic depressions called **pits** (Fig. P-7). Sometimes pits also appear in secondary walls where there are no primary pit-fields. A pit in a cell wall usually occurs opposite a pit in the wall of the adjacent cell, and the two primary walls and middle lamella between the pits form the **pit "membrane."** There are two kinds of pits, simple and bordered (Fig. P-7). In bordered pits, the secondary wall arches over the pit cavity.

In conducting cells with fluids under tension (and sometimes in other cells), secondary wall deposition often produces rings, spirals, or networks (Fig. 4-6). These beautiful structures prevent collapse from pressures produced by adjacent cells filled with pressurized fluids.

Eukaryotic Protoplasts; the Components of Cytoplasm As you can see from Table P-3, protoplasts consist of four main parts: cytoplasm, nucleus, va-

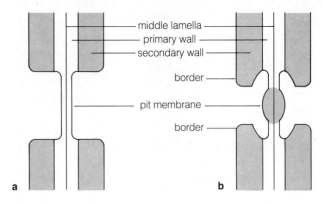

Figure P-7 The main characteristics of simple (**a**) and bordered (**b**) pits. (W. A. Jensen and F. B. Salisbury, 1984, *Botany*, p. 51.)

cuole(s), and ergastic substances. All eukaryotic cells have cytoplasm and, when young, at least one nucleus. (The nucleus disappears in cells such as sieve elements, as they mature.) As in the prokaryotes, the **cytoplasm of eukaryotes** is a complex, watery matrix containing many molecular substances, some in colloidal suspension; membrane-bound organelles are also present. Originally, the term cytoplasm was used to designate the matrix surrounding the nucleus; but because of advances in electron microscopy and the discovery of organelles, the concept of cytoplasm is evolving and imprecise. Some now use the term **cytosol** for the matrix in which cytoplasmic organelles are suspended.[*] In any case, the cytoplasm and nucleus combined may be called **protoplasm,** a term used less often now than formerly. Since most of the chemically functioning parts of the cell occur in the protoplasm, we might think of it as the "living" part of the cell, but chemical changes also occur in the cell wall (e.g., the softening that allows growth) and even in the **vacuole,** which is a volume of water and dissolved materials surrounded by a membrane, often occupying 90 percent or more of a mature cell. Most mature, living plant and fungal cells have a large central vacuole. Small vacuoles (1.0 μm in diameter) are present in certain animal and protist cells, but these seldom resemble the typical vacuoles of plant and fungal cells. A few plant cells also have accumulations of relatively pure nonliving substances such as calcium oxalate, protein bodies, gums, oils, and resins, collectively called **ergastic substances.**

Each kind of organelle in the cytoplasm is the site

---

[*]Originally the term was defined as "that portion of the cell which is found in the supernatant fraction after centrifuging the homogenate at 105 000 × g for 1 hour." (See Clegg, 1983.) It has become common practice to use the term in reference to intact cells, however, and we shall follow this practice.

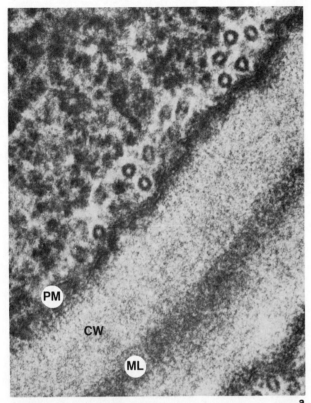

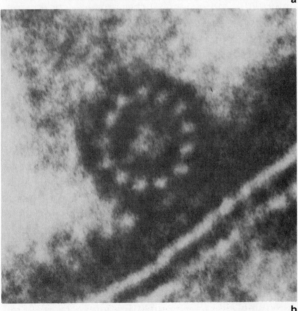

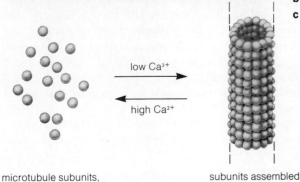

microtubule subunits, disassembled

subunits assembled into microtubule

of specific chemical processes, as we learn when we separate them by ultracentrifugation and study their biochemical activities. This segregation of processes makes the complex chemistry of cells possible by increasing efficiency, much as an assembly line does for industrial production. It also allows seemingly incompatible activities, such as the synthesis and breakdown of the same kinds of molecules, to go on within the same cell at the same time.

Actually, dynamic action is a living cell's normal condition. Organelles grow, divide, change shape, contain the enzymes that catalyze thousands of metabolic reactions, and secrete substances through the membrane to the outside wall or the outside world. They take part in growth and specialization of cells and are involved in a myriad of vital activities. Indeed, in many cells, the cytoplasm can be observed with the light microscope to stream around the cell periphery. The cell is the unit of life, and like all living things it is a changing, vibrant entity.

The membranes of eukaryotic cells The membranes of eukaryotic and prokaryotic cells are similar. In both cases, the plasma membrane, or plasmalemma, regulates the flow of dissolved substances in and out of cells, and osmosis regulates the flow of water. Furthermore, both the chloroplasts, where photosynthesis occurs, and the mitochondria, where respiration takes place, are surrounded by double membranes. The inner membrane is highly elaborate in form and is the membrane in or on which much of the metabolic activity takes place and that regulates molecular traffic into and out of the organelle. A double membrane also surrounds the nuclear material. A single membrane of great importance in plant and fungal cells is the one that surrounds the vacuole, known as the **vacuolar membrane,** or **tonoplast.** We shall discuss the detailed structures of membranes and some of their functions in Chapter 6.

Figure P-8 Microtubules. (**a**) Part of a cell from a *Juniperus chinensis* root tip showing microtubules in cross section (small circles) next to the plasma membrane (PM). The primary cell wall (CW) and middle lamella (ML) are especially prominent. ×51,000. The microtubules are most abundant in a zone about 0.1 μm thick adjacent to the plasmalemma and are arranged circumferentially around meristematic cells, much as hoops around the inside of a barrel. Along both the side and the end walls of these cells, the arrangement of these microtubules is very similar to that of the cellulose microfibrils (Fig. P-4 and P-5) in the adjacent wall. (**b**) Higher magnification (×650,000) of a microtubule from the juniper root cell. Note the 13 subunits that make up the microtubule, which is about 25 nm in diameter. (**c**) The assembly and disassembly of microtubule subunits in high and low concentrations of calcium ions. (Micrographs courtesy of Myron C. Ledbetter; see Ledbetter, 1965. Used by permission.)

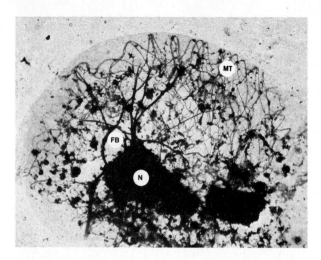

Figure P-9 Whole mount of a negatively stained cytoskeleton from a carrot cell protoplast grown in suspension culture. When such cells are extracted in an iso-osmotic, microtubule-stabilizing buffer that contains detergent, cytoskeletons remain that are insoluble in the detergent. When negatively stained, these are seen to consist of cortical microtubules (MT) seen as looping beneath the cell periphery, and of bundles of 7-nm fibrils (FB) that appear to link the nucleus (N) and the cell cortex. The latter cytoskeleton elements were found not to consist of the protein, actin. (Micrograph courtesy of Clive Lloyd; see Powell et al., 1982.)

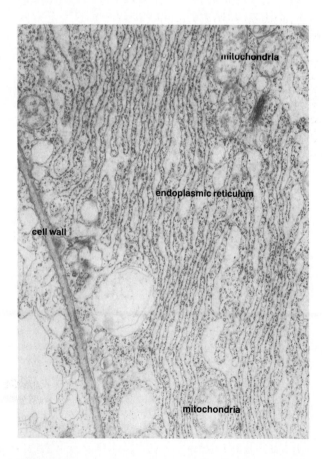

Figure P-10 A section through two adjacent cells of the developing fruit of shepherd's purse (*Capsella*) showing extensive rough endoplasmic reticulum (ER) in the upper cell. (By Patricia Schulz.)

Microtubules and microfilaments The cytoplasm of eukaryotic cells contains **microtubules,** which are long hollow cylinders about 25 nm in diameter and varying in length from a few nanometers to many micrometers. They consist of spherical molecules of a protein called **tubulin,** which spontaneously assemble under certain conditions to form the long hollow cylinders (Fig P-8). Microtubules are often located near the plasmalemma of nondividing but growing cells; they are probably involved in wall formation. Their orientation seems to determine the orientation of the cellulose microfibrils of both primary and secondary walls. They are especially prevalent, for example, near isolated regions of secondary wall thickening. They play important roles other than cell wall formation, however, especially in cell division and movements of or within cells. For example, the spindle fibers that separate the chromosomes during cell division consist of microtubules.

Microfilaments are smaller, solid structures (5 to 7 nm in diameter) that act alone or in coordination with microtubules to produce cellular movements. They also consist of protein, specifically the protein **actin,** which is a significant constituent of animal muscle tissues. Activities of microfilaments apparently cause movements such as muscle contraction, **cytoplasmic streaming** (a flow of cytoplasm around the inside of plant cells), and **ameboid movements** (movements of single protist, fungal, and animal cells in which protoplasm flows out from the cell, forming a kind of "false foot," followed by the rest of the cell flowing toward the false foot, and resulting in movement of the cell along a surface). Slime molds, in particular, exhibit ameboid movements.

Activities of the two filamentous structures are coordinated in animal cell division: The chromosomes are separated by microtubules in the spindle, and the cytoplasm is divided by microfilaments that form a furrow around the circumference of the cell, shrinking toward the center until the original cell has been pinched into two new cells.

With animal cells, it has been shown that actin microfibrils, microtubules, and intermediate filaments form three different **cytoskeletal systems,** each with characteristic distribution in the cell and each helping to determine the form of the cell. Do plant cells also have filamentous cytoskeletons? The cell wall supports plant protoplasts, so a cytoskeleton might not be necessary; and cell walls have hindered the search for plant cytoskeletons. Nevertheless, cytoskeletonlike structures have been found in isolated protoplasts from several dicot and monocot species (Powell et al., 1982). In these studies, cell walls are dissolved away with enzymes, leaving the protoplasts, and then strong detergent solutions are used to remove the membranes and other cytoplasmic materials, leaving the fibrillar cytoskeleton (Fig. P-9). Several bundles are found, each having a few hundred 7-nm microfibrils, that seem to emerge from the nucleus and subdivide into finer bundles as they approach the periphery of the cell, where microtubules are found. The 7-nm fibrils do not consist of actin, but they could be related to the P-protein (phloem protein) discussed in Chapter 7. The role of the plant cytoskeleton remains to be determined, but the fibril bundles could (among other things) position the nucleus in the cell.

Endoplasmic reticulum (ER) and ribosomes   In very thin sections, prepared for transmission electron microscopy, the entire cytoplasm of many eukaryotic cells appears to be filled with a double-membrane system that looks something like a collapsed sack folded onto itself again and again; the **endoplasmic reticulum,** or **ER** (Fig. P-10). Recently, high-voltage transmission electron micrographs of much thicker sections have clarified the three-dimensional aspect of the ER. In many cells, at least, the ER is less like a collapsed sack than it is like an assemblage of thousands of tiny tubes called **tubules.** The tubules are connected to pillowcaselike pairs of membranes called **cisternae,** with hundreds of tubules extending outward from each cisterna. The ER forms a system of transport for various molecules within the cell and even between cells via plasmodesmata. As will be explained in subsequent chapters, many chemical activities are associated with the ER.

One of these activities is protein synthesis, which occurs on the numerous ribosomes (about 15 to 25 nm in diameter) that are often associated with the ER, always on the cytosol side of the membrane. ER with attached ribosomes is called **rough ER; smooth ER** lacks ribosomes. Identical ribosomes are also attached to the cytosol side of the outer membrane that surrounds the nucleus. Furthermore, thousands of free ribosomes occur in the cytosol, unbound to any membrane. Often, ribosomes form a chain like a string of beads; these structures are called **polyribosomes,** or **polysomes,** and each is held together by a strand of **messenger-RNA (mRNA),** the genetic information of which is being translated into a protein. Smaller ribosomes (15 nm), the size of those in prokaryotes, are present in mitochondria and chloroplasts, where they synthesize some of the protein found in these organelles; other chloroplast and mitochondrial proteins are formed on cytoplasmic ribosomes and transported into the organelles. Nuclei contain no true ribosomes, and modern biochemical studies indicate that nuclei import all of their proteins from the cytoplasm.

In addition to holding ribosomes, the ER synthesizes sterols and phospholipids, essential parts of all membranes. The ER also forms microbodies, small organelles to be discussed later. The ER transports certain enzymes and other proteins across the plasma membrane and out of the cytoplasm: This is secre-

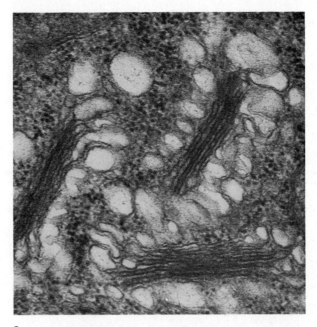

a

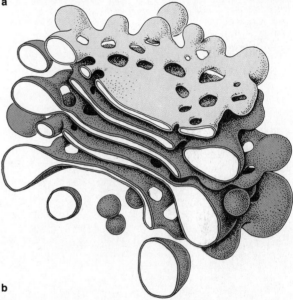

b

Figure P-11   (**a**) Elements of the Golgi apparatus in a root-cap cell of maize (corn) fixed with potassium permanganate. Fragmentation of the flattened sacs into separate vesicles is clearly visible. ×28,000. (Micrograph by William A. Jensen.) (**b**) A three dimensional drawing of a generalized Golgi body from a plant cell. (W. A. Jensen and F. B. Salisbury, 1984, *Botany*, p. 54.)

tion. Some secreted proteins are essential for function of the cell wall, for example, allowing it to stretch plastically during growth.

Golgi bodies   Electron micrographs of suitably prepared tissues show stacks of flattened hollow disks with convoluted margins that are surrounded by spherical bodies (Fig. P-11). The stacks are called **Golgi bodies** after the Italian, Camillo Golgi, who

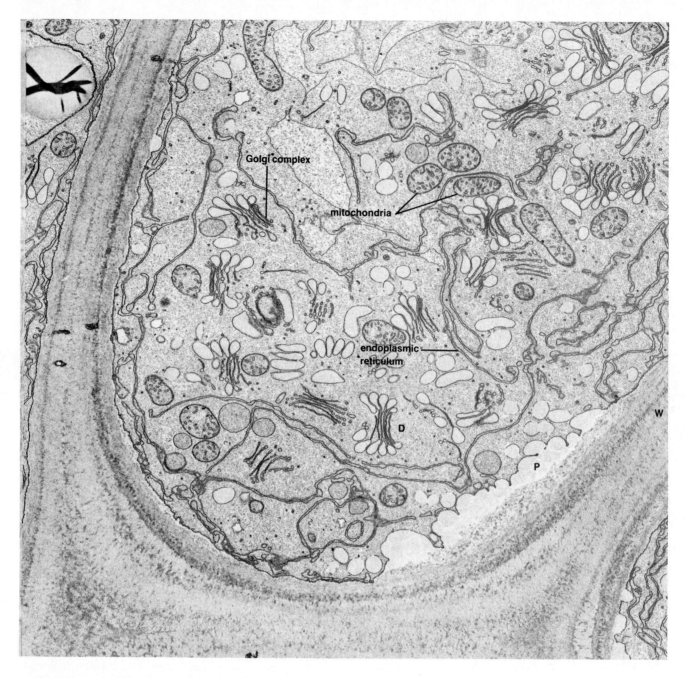

Figure P-12 Electron micrograph of a corn root-cap cell showing fusion of secretory products (P), pinched off dictyosomes (D)(Golgi apparatus). The plasma membrane and cell wall (W) are also shown. (From D. J. Morre, D. D. Jones, and H. H. Mollenhauer, 1967.)

discovered them with a light microscope in 1898; or **dictyosomes** (used especially by many botanists), and the spherical bodies are called **Golgi, or dictyosome, vesicles.** The structures of Golgi bodies constantly change, because some of the flattened hollow disks (**cisternae**) grow while others shrink and disappear. This growth and disappearance of the cisternae help explain both the origin and major function of Golgi bodies. One side grows on its forming

face, as tiny vesicles from the ER fuse with the cisterna on that face. Such fusion increases the size of each cisterna, and its contents are changed slightly because membranes of the ER are not identical to those of Golgi bodies. Inside the cisterna cavity, the newly absorbed compounds are changed into others.

At the opposite side of a Golgi body (the maturing or releasing face), other vesicles with new chemical properties are released. Some of these vesicles are close to the plasma membrane adjacent to the cell wall. Indeed, we know that fusion of Golgi vesicles with the plasma membrane increases its surface area during growth. Furthermore, each vesicle contains polysaccharides that contribute to the growth of the wall, although cellulose is not present in Golgi bodies and is probably formed very near the plasma membrane just before being transported into the wall. Golgi bodies probably have functions other than contributing to growth of the plasma membrane and transporting materials into the cell wall. For example, the slime on the outside of a root cap, which lubricates the root tip as it grows through the soil, is probably secreted when Golgi vesicles fuse with the plasma membrane as just described (Fig. P-12).

The outer membrane of the nuclear envelope, the ER, Golgi bodies, the plasma membrane, and even the tonoplast membrane surrounding the vacuole seem to be related in origin. The outer membrane of the nuclear envelope apparently forms young parts of the ER, which then grows much more on its own as it synthesizes its own lipids and proteins. As we noted, vesicles from the ER contribute to the Golgi body cisternae, and Golgi vesicles contribute to growth of the plasma membrane and the tonoplast.

Mitochondria In the light microscope, **mitochondria** appear as small spheres, rods, or filaments varying in shape and size, usually from about 0.5 to 1.0 μm in diameter and 1 to 4 μm in length. They were first seen in about 1900. The electron microscope shows them to have a rather elaborate internal structure and frequently an oval shape (see Fig. P-2). There is a smooth outer membrane and an inner membrane that is intricately folded into several forms, including shelves or tubular protuberances called **cristae.** The tubular forms are common in plant mitochondria. As already noted, small ribosomes are also present. Mitochondria are the sites of much chemical activity in cells, perhaps over half of the cell's metabolism. They are discussed in Chapter 12.

Chloroplasts and other plastids **Plastids** are special structures that occur only in plants and some protistans and are bound by a double-membrane system. Their inner membrane is not folded as in mitochondria, but other membranes arranged in various ways often occur within the plastids. Many plastids contain pigments. If they contain chlorophyll, they are **chloroplasts** (see Fig. P-2); if they contain other pigments, they are called **chromoplasts** (e.g., the plastids that contain the red pigments of tomato skins). Colorless plastids are called **leucoplasts** (from the Greek: *leucos* = white, *plastos* = formed) and may contain storage proteins or, most commonly, two or more grains of starch. In the latter case, they are

called **amyloplasts.** Most plastids, including chloroplasts, are thought to develop from **proplastids,** which are small colorless bodies found in plants growing in the dark as well as in the light and difficult to distinguish from mitochondria. Chloroplasts are discussed in Chapter 9.

Microbodies and spherosomes **Microbodies** are spherical organelles bounded by a single membrane. They range in diameter from 0.5 to 1.5 μm and have a granular interior, sometimes with crystalline inclusions of protein. There are two kinds, **peroxisomes** and **glyoxysomes,** each of which plays a special role in the chemical activities of cells, especially peroxisomes in photosynthesis (Chapter 10) and the conversion of fats to carbohydrates by glyoxysomes during seed germination (Chapter 14).

Spherosomes are also spherical and are bounded by one-half of a single membrane. They range in diameter from 0.5 to 2.0 μm. Many spherosomes mostly contain fatty materials and may well be centers of fat synthesis and storage. Cells may also contain storage bodies of fat, called **oleosomes** (*ole* = oil), or of protein, called **protein bodies.**

The Nucleus The **nucleus** is the control center of the eukaryotic cell—and often is its most conspicuous organelle. It is spherical or elongated and often 5 to 15 μm or more in diameter. It exercises its control over cell functions by determining the kinds of enzymes made in the cell, and these determine the chemical reactions that take place and thus the structures and functions of cells. The control lies in the same structure as the genetic or hereditary information and is contained in long fibers of DNA combined with protein, which forms a material called **chromatin.** This material is duplicated before cell division by chemical processes; then, during division of the nucleus, the chromatin fibers condense by coiling into elongated, dark-staining bodies called **chromosomes,** which are visible with the light microscope. Between divisions the coiling relaxes, and chromosomes cannot be observed in the nucleus. The nucleus also contains a watery, enzyme-filled solution known as **nucleoplasm** in which the chromatin, or chromosomes, and nucleoli are suspended.

The nucleus contains one or more roughly spherical bodies, the **nucleoli** (singular: nucleolus), each about 3 to 5 μm (up to 10 μm) in diameter. Nucleoli are dense, irregularly shaped masses of fibers and granules suspended in the nucleoplasm. The resemblance of these granules to cytoplasmic ribosomes is more than coincidence, for the subunits of ribosomes, composed mostly of RNA and protein, are made in the nucleolus.

Surrounding the nucleus are two parallel surface

Figure P-13 A replica of a fractured onion root-tip cell prepared by freezing the tissue and then causing it to fracture, after which metal is deposited onto the fractured surfaces to make the replica. The fracture lines often split lipid bilayers of membranes. In this splendid photograph, the nuclear envelope is exposed, showing many pores. Fractures in the endoplasmic reticulum are visible in the cytoplasm outside the nucleus. ×40,000. (Micrograph by Daniel Branton.)

membranes, together called the **nuclear envelope** (Fig P-13). The inner membrane lies 20 to 30 nm inside the outer membrane. The nuclear envelope is perforated by many **pores,** which are thought to allow communication between nucleus and cytoplasm. In surface view, the pores are octagonal and about 70 nm in diameter. The inner and outer membranes are joined to each other to form the margin of the pores, which appear to be lined with material, giving rise to a structure known as the **annulus** (Latin, *ring*), which fills the pore except for a narrow central channel. Sometimes these channels seem to be filled with particles about the right size to be ribosomal subunits caught in transit from nucleus to cytoplasm. As noted above, the outer membrane of the nuclear envelope often seems to be continuous with the ER.

Vacuole As characteristic of plant cells as the cell wall and plastids is the vacuole. The form and rigidity

of tissues (e.g., in leaves and young stems—see Fig. P-5) made of cells that have only primary walls is caused by the water with its dissolved materials exerting pressure in the vacuole. The pressure develops by osmosis, which we discuss in Chapter 2.

There is another important aspect of vacuoles that make plants what they are (Wiebe, 1978a, 1978b). To survive, a plant must absorb relatively large amounts of water, mineral elements, carbon dioxide, and sunlight. Each of these, even sunlight, may be and often is relatively scarce or dilute in the environment. Large surface areas greatly facilitate the absorption of each of these by plants: the surfaces of finely divided roots penetrating large volumes of soil and the surfaces of leaves capturing sunlight and absorbing carbon dioxide from the atmosphere. One way to achieve large surfaces is to begin with relatively large volumes and spread them out into thin layers, such as most leaves, or into long narrow structures, such as roots or conifer needles. Plants have relatively large volumes because their vacuoles are filled with water, which is much more abundant in the environment than any other constituent of protoplasm. If plant cells were of solid protoplasm without vacuoles, as are most animal cells, they would be able to expose only a fraction of their needed surface area. It is just as important for animals to be compact, with the smallest possible surface area and their protoplasm concentrated to produce energy and reduce inertia for motion.

The concentration of dissolved materials in vacuoles is high, approximately as high as the concentration of salt in sea water and as high as that of the cytosol. There are hundreds of dissolved materials, including salts, small organic molecules, and some proteins (enzymes) and other molecules. Some vacuoles have high concentrations of pigments, which produce the colors of many flowers or the red of red maple leaves (so concentrated that they mask the green of the chloroplasts). In some plant parts, the vacuole contains materials that would be harmful to the cytoplasm. The sharp taste of oranges and lemons comes from the high concentration of citric acid in their vacuoles. Such vacuoles may have a *p*H as low as 3.0, with the *p*H of surrounding cytoplasm between 7.0 and 7.5 (close to neutral). Most vacuoles are slightly acidic (*p*H of 5 to 6). Sometimes, too, the vacuole contains crystals; calcium oxalate crystals are especially common in some species. Such crystals may remove excessive calcium from the protoplasm.

Probably, the enzymes in the vacuole digest various materials absorbed into it or may even digest much of the cytoplasm when the cell dies and the tonoplast (vacuolar membrane, actually part of the vacuole) breaks down. This probably happens when the protoplasts of wood cells break down and die, for example. In this sense, the vacuole is like a **lysosome,** a cellular organelle common in animal and some fungal and protist cells. Lysosomes contain digestive enzymes that break down the materials they absorb, or the enzymes digest much of the protoplasm following cell death and breakdown of lysosome membranes.

Young, dividing cells in the growing tips of stems and roots have only minute vacuoles, many in each cell, formed from the ER. These grow with the cell, taking up water by osmosis and coalescing with each other until mature cells often have vacuoles that occupy 95 percent or more of their volume, with the protoplasm spread in a thin layer between the tonoplast and the plasmalemma. But not all dividing cells lack large vacuoles. Cambial cells between the bark and wood of a growing stem have large vacuoles and divide to produce both phloem and xylem cells, likewise with large vacuoles. The vacuole is one of the most variable features of plant cells, varying from being almost nonexistent to filling most of the volume of most mature living plant cells. In that respect the "typical" cell of Fig. P-2 is not at all typical. Actually, considering the large quantities of wood on earth, the typical plant cell is dead!

Table P-3 includes ergastic substances (which we have already noted), flagella, and cilia. The latter two features of cells are found mostly in algae, fungi, protozoa, very small (microscopic) animals, specialized cells of other animals, and sex cells of certain gymnosperms. Since we emphasize the higher plants (mostly conifers and angiosperms) in this book, these structures will not be further treated in our discussions.

It is worth remembering that three features are especially characteristic of plant cells as compared to animal cells: the cell wall with its cellulose, the vacuole (which provides pressure and a large volume and surface area with a minimum of protoplasm), and plastids, especially chloroplasts. Animal cells never have walls, nor do many protist cells, and the walls of prokaryotes and fungi differ in significant ways from those of plants. Vacuoles can be found in all five kingdoms, but the large, central vacuoles we have been discussing are present in practically all cells of plants, fungi, and some protists. Chloroplasts occur only in plants and in some protists (depending on how they are classified); in a sense, blue-green algae (cyanobacteria) are structures comparable to chloroplasts. Animals and fungi never produce chloroplasts.

P.4 A Definition of Life

Based upon the postulates in the early part of this chapter and our discussion of cells, we can summarize this chapter by attempting to define life. Of course, there are difficulties. Special macromolecular structures exist after an organism has died and might even exist on the surface of a dead world somewhere in the universe, so in that sense they are not characteristic of life. Viruses display many of the properties of life but only when they are associated with living organisms. Nevertheless, consider the following statement:

Life is a peculiar series of functions associated with a peculiar series of organized structures in which certain macromolecules, having building blocks arranged in nonrepeating but reproducing sequences, are capable of reproduction, transfer and utilization of information, and catalysis of metabolic reactions. All this is organized at least to the level of a cell with its surrounding membrane, which level allows the functions of growth, metabolism, irritability or response to environment, and reproduction.

one

Water, Solutions, and Surfaces

1

Diffusion
and Water Potential

 The science of plant physiology can be broadly outlined in a few simple questions about cells:

1. How do things get in and out of cells?

2. What goes on inside a cell?

3. How do cells reproduce and change to form multicellular organisms?

4. How do cells interact with their environments?

This book summarizes in an introductory way the present status of answers to these questions. During the past quarter century it has become fashionable to replace the more formal terms *hypothesis*, *theory*, and *law* with the more general term **model,** which is truly an excellent word to describe the products of a scientist's creative thought. This book is, to a great extent, a description of the models that have been developed to explain and predict plant function. For example, how does the water "run up hill" in moving to the top of a tall tree? A physical analogue and a conceptual model, continually being modified by new information and new ideas, exist to help us understand.

The chapters in this first section describe models relating to the first question: movement of water and other substances across membranes, throughout the plant, and between the plant and its environment. This question overlaps somewhat with the question of what goes on inside cells, since cellular chemistry influences **plant–water relations** (as the combined topics of this first section are called) in interesting ways. Even growth depends on water uptake, and much of plant–water relations depends upon cells interacting with their environment. The remaining three questions are discussed in the other three sections of this book—again with much overlap among sections.

1.1 Plants and Water

Plant physiology is, to a surprising degree, the study of water. Many plant functions depend quite directly upon the properties of water and of substances dissolved in the water. Thus a brief review of water's properties is a good way to begin our study of plant physiology.

The Hydrogen Bond, Key to Water's Properties
Most of the unique properties of water can be ascribed to the interesting fact that line segments connecting the centers of the two hydrogen atoms with the center of the oxygen atom do not form a straight line. Instead, they form an angle of about 105°, closer to a right angle than to a straight line (Fig. 1-1). The angle is exact in ice but only an average in liquid water. The two electrons that fill the first shell of the hydrogen atom (one belonging to the hydrogen atom and the other borrowed from the oxygen) are usually closer to the oxygen nucleus. Thus, in a water molecule the hydrogen atoms approximate naked protons on the surface of the oxygen atom. While the net charge for the molecule as a whole is, of course, neutral, the protons, distributed 105° apart on the surface of the oxygen atom, provide a slight positive charge on one side of the molecule. This is balanced by an equal negative charge on the other side of the molecule. Such a molecule is said to be **polar.** The result is that the positive side of one water molecule is attracted to the negative side of another (Fig. 1-1). The bond formed this way between the two molecules is called a **hydrogen bond.**

Compared to ionic and covalent bonds, hydrogen bonds are weak. They occur in many substances besides water, those between oxygen and nitrogen being especially important in plants. The strength of the bond between a hydrogen atom in one molecule and some negative part in another molecule varies,

WATER

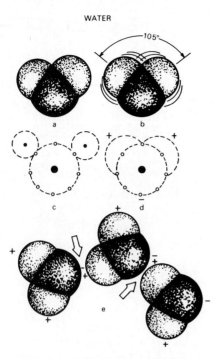

Figure 1-1 Water molecules. The angle between the two hydrogen atoms (lighter) attached to the oxygen atom (darker) averages 105° (**a** and **b**). The angle is not absolutely stable (**b**) but represents an average sharing of electrons and distribution of charge (**c** and **d**). Attraction of the negative side of one water molecule to the positive side of another produces the hydrogen bond (**e**). Arrows indicate hydrogen bonds.

depending upon the other molecule, from about 8 to 42 kilojoules/mole ($kJ\ mol^{-1}$). **Ionic bonds,** in which electrons move from one atom and become attached to another, have strengths that vary from about 582 $kJ\ mol^{-1}$ for CsI to 1004 $kJ\ mol^{-1}$ for LiF; NaCl is intermediate at 766 $kJ\ mol^{-1}$. The strengths of **covalent bonds,** in which electrons are shared by two atoms, overlap ionic-bond strengths but are usually weaker. Important examples include (all given in $kJ\ mol^{-1}$) 138 for O—O, 293 for C—N, 347 for C—C, 351 for C—O, 414 for C—H, 460 for O—H (as in water), 607 for C=C, and 828 for C≡C.

Van der Waals attractive forces are even weaker than hydrogen bonds, having bond strengths of about 4.2 $kJ\ mol^{-1}$. In neutral, nonpolar molecules, these forces result from the fact that electrons are continually in motion, so that the molecule's center of negative charges does not always correspond to its center of positive charges. Thus, as two like molecules approach each other very closely, they may induce slight polarizations in each other, with the regions of unlike charge attracting each other. Such forces hold molecules together in liquid hydrocarbons, for example, and also in membranes and the internal parts of proteins. All these molecular bonds are electrical in nature.

Liquid at Room Temperature The higher the molecular weight of an element or a compound, the greater the likelihood that it will be a solid or a liquid at a given temperature, such as room temperature. The lower the molecular weight, the greater the likelihood of its being a liquid or a gas. The larger the molecule, the more energy (heat) required to cause it to break the forces that bind it to surrounding molecules and become a liquid or a gas. For example, the low-molecular-weight hydrocarbons (methane, ethane, propane, and butane) are all gases at room temperature. Their respective molecular weights are 16, 30, 44, and 58 grams per mole. Ammonia (mol. wt. = 17) and carbon dioxide (mol. wt. = 44) are also gases at room temperature; they must be cooled to very low temperatures before they become liquids or solids. Yet water, with a molecular weight of only 18, is a liquid at room temperature. The explanation for this difference is that hydrogen bonds provide a disproportionately high attractive force among water molecules, inhibiting their separation and escape as vapor. The hydrocarbons, on the other hand, have only the relatively weak van der Waals forces among their molecules in the liquid state.

It is interesting that other liquids with low molecular weights are also polar molecules with hydrogen bonding between them. Good examples are the lower alcohols (methyl = CH_3OH) or the lower organic acids (formic = $CHOOH$, or acetic = CH_3COOH). The presence of oxygen and hydrogen atoms makes hydrogen bonding possible in these compounds.

Life as we know it is unthinkable without liquid water. This becomes increasingly evident as we examine other properties of this crucial liquid.

Nearly Constant Volume For all practical purposes liquid water is incompressible. This means that the laws of hydraulics apply to organisms, since they consist largely of water. That a young growing plant is a hydraulic system becomes strikingly evident when the plant wilts (Fig. 1-2). The normal form of such a plant is maintained by the pressure of water within the protoplasts pressing against the cell walls. Furthermore, plants grow as they absorb water, causing their cells to expand. Some petals and leaves, such as those of the sensitive plant (*Mimosa* sp.), move or fold up as water moves in or out of special cells at their bases. Stomates on leaf surfaces open as their guard cells take up water and close when water moves out of them. Substances are transported in moving fluids in both plants and animals.

Specific Heat Almost exactly one small **calorie** (cal) is required to raise 1 g of pure water 1°C. A kilocalorie, or large Calorie (spelled with a capital C), equals 1000 small calories. The SI unit of heat (en-

Figure 1-2 Normal (left) and wilted (right) *Fittonia* plants. The normal appearance of a plant is dependent on having sufficient water in the cells to provide turgidity. (Photographs by Ginny Mickelson.)

ergy) is the **joule** (see Appendix A), and the small calorie is now defined as 4.184 joules (exactly). The amount of energy required to raise the temperature of a unit mass of a substance 1°C is called its **specific heat.** The specific heat of water varies only slightly over the entire range of temperatures at which water is a liquid, and it is higher than that of any other substance except liquid ammonia. The specific heat of liquid water is caused by the arrangement of molecules, which allows the hydrogen and oxygen atoms to vibrate freely, almost as if they were free ions. Thus they can absorb large quantities of energy without much temperature increase. Plants (think of a large succulent cactus) and animals consist largely of water and thus have relatively high temperature stability, even when gaining or losing energy.

Latent Heats of Vaporization and Fusion Some 2452 joules (586 cal) are required to convert 1 g of water to 1 g of water vapor at 20°C. This unusually high **latent heat of vaporization** can again be ascribed to the tenacity of the hydrogen bond. It is important for cooling leaves by transpiration.

To melt 1 g of ice at 0°C, 335 J (80 cal) must be supplied. This is also a high **latent heat of fusion,** caused again by the hydrogen bonds, although ice has fewer per molecule than water. Each H_2O molecule in ice is surrounded by *four others*, forming a tetrahedral structure. (Each oxygen atom attracts two extra hydrogen atoms.) The tetrahedrons are arranged in such a manner that the ice crystal is basically hexagonal, as demonstrated in the pattern of snowflakes. As is usual during conversion from the solid to the liquid state, molecules of water move slightly farther apart during melting. Yet water is

extremely unusual because its total volume *decreases* during melting. This is because the molecules are packed more efficiently in the liquid than in the solid. Each molecule in the liquid is surrounded by five or more other molecules.

The result of this packing difference is that water expands as it freezes, and ice has a lower density than water. Thus ice floats on the tops of lakes in the winter rather than going to the bottom where it might remain without thawing through the following summer. The expansion is also a potential source of harm to plant and animal tissue in freezing weather. Because water expands upon freezing, increased pressure will make ice melt at a lower-than-normal temperature. (That is, increased pressure lowers the melting point.) With other substances, increased pressure usually raises the melting point.

Viscosity Since hydrogen bonds must be broken for water to flow, one might expect water **viscosity,** or resistance to flow, to be considerably higher than it is. But in liquid water each hydrogen bond is shared on the average by two other molecules, and thus the bonds are somewhat weakened and fairly easily broken. Water can flow readily through plants. In ice there are fewer bonds per oxygen atom; hence, each is stronger. Viscosity of water decreases markedly with increasing temperature (Table 1-1).

Adhesive and Cohesive Forces of Water Because of its polar nature, water is attracted to many other substances; that is, it wets them. Molecules of proteins and cell-wall polysaccharides are excellent examples. This attraction between unlike molecules (water and other molecules in this case) is called **adhesion.** In the

Table 1-1 Viscosities of Fluids.

Fluid	Temperature (°C)	Coefficient of Viscosity[a] (centipoises)	Percent of H_2O Viscosity at 20°C
Water	0	1.787	177
	10	1.307	130
	20	1.002	100
	30	.7975	80
	40	.6529	65
	60	.4665	47
	80	.3547	35
	100	.2818	28
Ethyl alcohol	20	1.20	120
Benzene	20	.65	65
Glycerine	20	830.0	83,000
Mercury	20	1.60	160
Machine oil	19	120.0	12,000

[a]In centipoise = poise multiplied by 100. One poise represents a force of 1 dyne per cm^2 required to displace a large plane surface in contact with the upper surface of a layer of liquid 1 cm thick over a distance of 1 cm in 1 sec. A more convenient method of measurement notes the time required for a given volume of liquid to flow by gravity through a tube of given dimensions and then applies a suitable equation.

case of water, it involves hydrogen bonding between the water and the other molecules. The attraction of like molecules for each other (because of hydrogen bonding) is called **cohesion.** Cohesion bestows upon water an unusually high tensile strength. In a thin, confined column of water such as that in the xylem elements of a stem, this tensile strength can reach high values, allowing water to be pulled to the tops of tall trees (Chapter 4).

Cohesion between water molecules thus accounts for **surface tension:** The molecules at the surface of a liquid are continually being pulled into the liquid by the cohesive (hydrogen-bond) forces. The result is that a drop of water acts as if it were covered by a tight elastic skin. It is the surface tension that makes a falling drop spherical. The surface tension of water is higher than that of most other liquids.* Surface tension often plays a role in the physiology of plants. For example, under normal pressures, the passage of air bubbles through pores and pits in cell walls is prevented because the surface tension of the

*Hydrazine and most metals (including mercury) in the liquid state have higher surface tensions.

water that surrounds an air bubble is too great to allow deformation of the bubble.

Water as a Solvent Water will dissolve more substances than any other common liquid. This is partially because it has one of the highest known **dielectric constants,** which is a measure of the capacity to neutralize the attraction between electrical charges. Because of this property, water is an especially powerful solvent for electrolytes. The positive side of the water molecule is attracted to the negative ion and the negative side to the positive ion. Water molecules thus form a "cage" around ions so that the ions are often unable to unite with each other and precipitate or crystallize.

If water contains dissolved electrolytes, then these will carry a charge, and water becomes a good conductor of electricity. If water is absolutely pure, however (and pure water is extremely difficult to obtain), then it is a very poor conductor. Hydrogen bonding makes it too rigid to carry a charge readily.

The importance of water as a solvent in living organisms will become quite evident in this first section. The process of osmosis, for example, which forms such an important part of the discussion in these chapters, depends upon the presence of dissolved materials in the water of living cells. We shall also be concerned with the movement of dissolved materials by diffusion and other processes in plants.

Protoplasm itself is an expression of the properties of water. Its protein and nucleic acid components owe their molecular structures, and hence their biological activities, to their close association with water molecules. Indeed, virtually all the molecules of protoplasm owe their specific chemical activities to the water milieu in which they exist. Exceptions are the molecules contained in cellular oil bodies (oleosomes) or the liquid (fatty) portions of membranes, but the oleosomes and the membranes are themselves strongly influenced by the surrounding water.

Water molecules actively enter into the chemistry that is so important to life. Along with carbon dioxide molecules, water is an essential raw material for photosynthesis, for example. Few processes of metabolism would be possible without the utilization or production of water molecules. Nevertheless, water is relatively inert chemically. On balance, it is much more important as the environment for chemical reactions than as a chemical reactant or product.

Ionization of Water and the _p_H scale Some of the molecules in water separate into hydrogen [H^+] and hydroxyl [OH^-] ions. The tendency for these ions to recombine is a function of the chances for collisions between them; that is, recombination depends upon the relative number of ions in the solution. This **mass**

law relationship may be expressed mathematically by saying that the product of the molal concentrations equals a constant: $[H^+] \cdot [OH^-] = K$. Near room temperature, $K = 10^{-14}$, so in pure water both $[OH^-]$ and $[H^+] = 10^{-7}$ molar (M). (To multiply 10^{-7} by 10^{-7}, exponents are added to give 10^{-14}.) Water is seldom pure enough to contain equal numbers of hydrogen and hydroxyl ions. Presence of dissolved carbon dioxide, as in distilled water or in cells, may raise the hydrogen ion concentration as high as 10^{-4} M ($pH = 4$). The hydroxyl ion concentration is then 10^{-10} M. If tap water contains much dissolved limestone (calcium carbonate), the hydrogen ion concentration can be close to neutral or even slightly basic ($pH = 7$ to 8).*

With this discussion of water as a background, it is time to return to the question of how water and other materials move in and out of cells. One of the simplest transfer processes in living organisms is diffusion, but even diffusion has enough interesting implications to occupy us for the rest of this chapter.

1.2 Bulk Flow and Diffusion

The contents of plant cells are under considerable pressure (at least as much as the air in an automobile tire). Say that we make a small hole through the cell wall and membrane. The cell contents will flow out through the hole until the pressure inside the cell is equal to the pressure outside (atmospheric pressure, probably). **Fluids** are substances, such as liquids and gases, that flow or conform to the shape of their container. When the flow occurs in response to differences in pressure and involves groups of atoms or molecules moving together, it is called **bulk flow.** Sometimes the differences in pressure are established by gravity (the weight of fluid), in which case we speak of **hydrostatic pressures.** In other cases, the pressure is produced by some mechanical compression applied to all or part of the system. In animals, the pump called the heart is a good example. In plants, fluids flow through the vascular tissues by bulk flow in response to pressure differences that are created in ways we shall discuss in later chapters.

Usually, water and the substances dissolved in it

move in and out of cells not by bulk flow but one molecule at a time. The net movement from one point to another because of the random kinetic activities of molecules or ions is called **diffusion.** Because diffusion in liquids is slow over macroscopic distances and because the bulk flow of gases and liquids is so common (convection currents), diffusion is not something that we readily notice. Nevertheless, it is easy to observe diffusion: Carefully place a crystal of dye into a beaker of still water. As the dye dissolves, you can observe it slowly spreading from its source (diffusing) throughout the liquid. Diffusion in air is much more rapid than in water, as you can observe when a bottle of some strong smelling substance such as ammonia is opened a meter or two away. Yet the transfer of odor from the bottle to your nose by diffusion is often aided by air currents (bulk flow).

As we explain below, diffusion often occurs in response to differences in the concentration of substances between one point and another. (As the dye begins to dissolve, it is highly concentrated in the water close to the crystal but absent some distance away.) Concentration differences are extremely common in living cells in particular and in organisms in general. For example, as sugar in the cytosol is taken in and metabolized by a mitochondrion, its concentration close to the mitochondrion is kept lower than its concentration close to a photosynthesizing (sugar-producing) chloroplast in the same cell. At the microlevel in cells, diffusion of many substances including water occurs constantly

Table 1-2   Some Molecular Values for Three Gases.

	H_2	O_2	CO_2
Molecular weight of gas	2.01	32.0	44.0
Average velocity at 0°C, meters/second (m s^{-1})	1695	425	362
Average velocity at 30°C, m s^{-1}	1787	448	381
Mean-free-path between collisions with other molecules, 0°C, 1 atmosphere (atm) pressure, nanometers	112	63	39
Number of collisions of 1 molecule/sec, in billions, which is the number shown × 10^9	15.1	6.8	9.4
Diameter of each molecule, in nanometers	0.272	0.364	0.462
Number of molecules (× 10^{19}), 1 atm pressure, in 1 cm^3	2.70	2.71	2.72

---

*The hydrogen ion concentration is indicated by the **pH scale,** on which $pH = -\log[H^+]$. Another way to state this is that pH equals the absolute value of the hydrogen ion concentration, expressed as a negative exponent of 10. For example, when $[H^+] = 10^{-4}$ M, then $pH = 4$. Neutrality is expressed by $pH = 7$; decreasing values below 7 indicate increasing acidity; increasing values above 7, increasing alkalinity. The pH units are multiples of 10 on a logarithmic scale and should therefore not be added together or averaged. Only a tenth as many H^+ need to be added to an unbuffered solution to change the pH from 7 to 6 as from 6 to 5.

and virtually everywhere. Thus, to understand cells it is imperative that we understand diffusion.

The beginning is to understand one of the most fundamental principles of physics: kinetic theory. The basic ideas of kinetic theory are taught early and often reviewed, but the details are easily forgotten. You may never have been taught some of the quantitative aspects of the theory. We shall present a brief review.

1.3 Kinetic Theory

Kinetic theory states that the elementary particles (atoms, ions, and molecules) are in constant motion at temperatures above absolute zero. The average energy of a particle in a homogeneous substance rises as temperature increases but is constant for various substances at a given temperature. It is instructive when using this model to consider some of the actual velocities and masses of the moving particles. Velocities can be easily calculated for particles in gases, but it is much more difficult to obtain values for liquids and solids. The average velocity (V_{ave}) of particles in a gas is calculated by the following formula (see modern texts on statistical thermodynamics):

$$V_{ave} = \left(\frac{8RT}{\pi M}\right)^{1/2} \tag{1.1}$$

where

V_{ave} = average velocity in centimeters/second (cm s^{-1})

R = molar gas constant (8.31×10^7 J mol^{-1} K^{-1})

T = absolute temperature in kelvins (K)

M = molecular weight in grams/mole (g mol^{-1})

π = 3.1416

As this equation shows, the average velocity is proportional to the square root of the absolute temperature; that is, the higher the temperature the faster the motion of the particles. At the same time, the average velocity is inversely proportional to the square root of the mass; so the smaller the particle the faster it moves at a given temperature.

Applying this equation and others produces some impressive numbers, as illustrated in Table 1-2. Average velocities are surprisingly high. The average hydrogen molecule, near room temperature, is moving close to 2 km s^{-1}, which is 6433 km h^{-1} (3997 mph)! Even the much heavier CO_2 molecule has an average speed of 1372 km h^{-1}. However, at atmospheric pressure, particles do not move far between collisions, only 150 to 400 times their own diameters. With such high velocities and short pathways between collisions, the number of collisions of each molecule is enormous: on the order of billions each

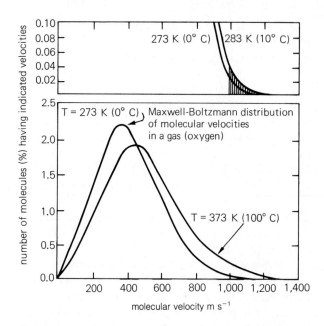

Figure 1-3 The Maxwell–Boltzmann distribution of molecular velocities in a gas at two temperatures 100°C apart. The curves at the top show the high-velocity portion for a gas at two temperatures 10°C apart. The area under the curves indicated by the vertical lines represents the number of highly energetic particles and approximately doubles in going from the lower to the higher temperature.

second. In liquids, for which no one has yet written satisfactory equations (i.e., models), the velocities are of the same order of magnitude at room temperatures, but the mean-free-paths are much shorter; thus the number of collisions is much greater. In solids, the particles are more or less held in place but vibrate against each other. Realizing the astronomical number of collisions possible in such short time intervals helps us to understand how chemical reactions can be so rapid. Table 1-2 shows that raising the temperature from 0° to 30°C, which is a large part of the temperature range of life functions, increases the average particle velocities only about 5 percent.

It is important to realize that the actual velocities of individual particles in a homogeneous substance vary widely from the average velocity. The velocities are distributed according to the Maxwell–Boltzmann equation, which produces curves such as those shown in Fig. 1-3. We suspect that this distribution would be similar for liquids and the solutes dissolved in them. It is easy to understand why particle speeds vary so widely if we think of the random nature of particle collisions. Unless the collision is perfectly symmetrical, one of the participants gains energy and the other loses it; hence, the speed of a particle will probably change billions of times per second, at virtually every collision. Yet statistically, the particle velocities will be distributed at any instant as in Fig. 1-3.

High-speed (high-energy) particles are the ones most likely to cause melting, evaporation, and chemical reactions. As the water molecules with the highest speeds enter the vapor state during evaporation, the average energy of the remaining molecules is lowered, which is the same as saying that the remaining liquid water is cooled. This is why evaporation is a cooling process. Particles with lower energies (left side of curves in Fig. 1-3) are the first to condense from vapor to the liquid state or to solidify from liquid to solid (freeze). All these processes are important in living plants.

Note in Fig. 1-3 and Table 1-2 that *average* particle velocities do not change much even with an increase of 100 K (equivalent to 100°C), yet the number of *high-velocity* particles (under the right-hand tails of the curves) increases considerably even with a 10-K rise in temperature. This is a feature of the shapes of the curves. If the number of *high-energy* particles doubles with a small temperature increase, and if it is these particles that take part in chemical reactions, then we can understand why the reaction rates of many chemical reactions double with an increase of just a few degrees. The factor by which a reaction rate increases with a 10°C increase in temperature is called the Q_{10}.*

1.4 A Model of Diffusion

Let's apply the concept of molecular motion (kinetic theory) to gain an understanding of diffusion. Imagine a model: two rooms connected by an opening; one room containing white balls in free motion, the other containing black balls also in motion. Our imaginary balls lose no energy as heat when they bounce off the walls or each other; like molecules, they are "perfect bouncers." The chances of a black ball going through the opening into the other room in an interval of time depends on the speed and concentration (number per unit volume) of black balls and on the size of the opening. At the beginning, the concentration of black balls is higher in one room than the other, but as black balls go through the opening, their concentration builds up in the other room. Gradually, a condition of **dynamic equilibrium** is approached in which the concentration of black balls is the same in both rooms. Now balls are still passing through the opening, but the chance that a black ball will go in one direction through the opening is the same as the chance for its going in the opposite direction. The same is true for the white balls. Indeed, the direction of diffusion before equilibrium of each type of ball will be independent of the other, provided the two do not stick to each other (i.e., interact chemically). The equilibrium is *dynamic*, because balls are still passing through the opening, but there is no *net* movement.

The model just described is nothing more than an expansion in size of what really happens (say, in two compartments, each containing a different gas, and the two connected by an opening). The value of the model in this case is that the size and velocities of the "perfect bouncing balls" might be easier to visualize than real molecular size particles. With the model balls in mind, it is not difficult to mentally accelerate their velocities and shrink their sizes to those of molecules given in Table 1-2.

Diffusion as just described occurs in response to a concentration gradient. **Concentration** is the amount of substance or number of particles per unit volume. A **gradient** occurs when something such as concentration changes gradually from one point in space to another. (A temperature gradient is another example easy to visualize; it could be expressed as degrees of temperature difference per centimeter.)

Gradients in properties other than concentration can also lead to diffusion. This is especially important when we consider the diffusion of water. As noted in the previous discussion, liquid water is virtually incompressible. Since a given amount of liquid water always occupies essentially the same volume, its concentration remains nearly constant at 55.2 to 55.5 moles/liter (*M*). Slight changes occur when substances are dissolved in water and when the temperature of the water changes, but these changes in concentration of water have little effect upon the diffusion of the water. Addition of many substances, such as sugar and various salts, causes the water volume to expand (as does increasing temperature) so that the water concentration is diminished slightly. Addition of certain other substances actually causes the water volume to shrink slightly, increasing the water concentration. In no cases can these changes in water concentration account in a quantitative way for the observed diffusion of water, which is so important in plants. So our diffusion model based upon concentration gradients is too simple.

The science of thermodynamics provides concepts that make it possible to refine the model so that

---

*Q_{10} values may be calculated at any two temperatures when reaction rates are known:

$$Q_{10} = \left(\frac{k_2}{k_1}\right)^{10/(T_2 - T_1)}$$

or

$$\log Q_{10} = \left(\frac{10}{T_2 - T_1}\right)\log \frac{k_2}{k_1},$$

where

T_1 = lower temperature

T_2 = higher temperature

k_1 = rate at lower temperature

k_2 = rate at higher temperature

it explains the observed phenomena much more accurately. We cannot teach thermodynamics in the few paragraphs available here, but a brief overview of some important thermodynamic principles will provide an intuitive feeling for what happens.

1.5 Thermodynamics

Matter has mass and occupies space, but what is energy? Energy is something that occupies no space and has no mass but can transform or act on matter. We observe energy only by observing its effects on matter. So energy is not easy to deal with, but **thermodynamics** (the science of energy transformation) is a system of thinking devised during the last century to help us understand heat and machines, especially steam engines. The principles of thermodynamics now apply to energy in general (not just heat) and are widely used in virtually every field of science.

In thermodynamics the word **system** means that region of space or quantity of matter on which we have focused our attention. A system might be a chlorophyll molecule, a beaker of sugar solution, a photosynthesizing leaf, the Canadian Rockies, or the Milky Way Galaxy. Everything not in the system is called the **surroundings** (environment). The system is separated from the surroundings by the **boundary,** which must usually be imagined. Thermodynamics is often concerned with the energy transfers or **interactions** that take place across the boundary. Nevertheless, it is important to realize that thermodynamics always operates within the limits of a specific known portion of the universe. That is, the system and its surroundings constitute an even larger system that has no interactions with *its* surroundings. In this sense, thermodynamics deals only with closed systems. (Special approaches that we shall not consider here are required to deal with open systems.) Much of thermodynamics is based upon two fundamental laws with which you might already be familiar.

Entropy (S) and the Laws of Thermodynamics The **First Law of Thermodynamics** can be stated in several ways: *In all chemical and physical changes, energy is neither created nor destroyed but is only transformed from one form to another.* Or, *in any process the total energy of the system plus its surroundings remains constant.* Or, *you can't get something for nothing.*

The First Law puts some important limitations on what can and cannot be done. For example, the energy trapped in organic molecules by photosynthesis cannot exceed the light energy absorbed.

The **Second Law of Thermodynamics** is difficult to put into words, since it is very abstract. Nevertheless, it can be stated in several ways: *Any system, plus its surroundings, tends spontaneously toward increasing disorder.* Or, *heat cannot be completely converted into work without changing some part of the system.* Or, *in any energy conversion some energy is transferred to the surroundings as heat.* Or, *no real process can be 100 percent efficient.* Or, *there can never be a perpetual motion machine.* Or, *you can't even break even.*

The consequences of the Second Law are extremely important. For example, photosynthesis will never be 100 percent efficient because some of the light energy driving the process will be converted into heat. Because some of the energy driving *any* process will be converted to or will remain as heat, there will never be a perpetual motion machine.

The statement that randomness or disorder must always increase for the system and its surroundings is especially significant. The measure of this randomness is called **entropy** (S). Entropy is defined as $S = k \ln w$, where k is a constant of proportionality (Boltzmann's constant), and $\ln w$ is the natural logarithm of w, which is the **thermodynamic probability.** The thermodynamic probability is the number of ways in which the system can be microscopically arranged without changing its macroscopic features. (How many arrangements can the molecules of a petunia assume and still make a petunia?)

Imagine a classroom with 10 chairs and 10 students. We ask ourselves in how many ways can those 10 people arrange themselves on those 10 chairs? The first person to sit down has 10 choices, the second has 9 choices, the third has 8 choices, and so on. We see that there are $10 \times 9 \times 8 \times 7 \times 6 \times 5 \times 4 \times 3 \times 2 \times 1 = 3,628,800$ ways of arranging 10 students on 10 chairs. If we have no additional information about the seating, we must conclude that all arrangements are equally likely, and the probability of finding the students in any one particular arrangement is 1 divided by 3,628,800. Obviously, as the number of possible arrangements increases, the probability of observing a particular arrangement decreases. Thus entropy is a measure of our uncertainty or lack of information about the microscopic arrangement of a system, and the entropy associated with this example could be computed as: $S = k \ln 3,628,800 = 15.1\ k$.*

As the number of possible arrangements increases, the value of entropy in absolute terms becomes so large that it is meaningless, and often it is impossible to calculate. One approach (not taken in this text) is to base calculations of absolute entropy on the so-called **Third Law of Thermodynamics,** which assumes that the entropy of any substance is zero at absolute zero. We shall follow the common practice of those who apply thermodynamic concepts to

---

*Part of the discussion in this section is taken from an essay by J. Clair Batty, published in the second edition of this textbook on pages 402–404.

plant-water relations, metabolism, and most other processes. We shall calculate the *entropy changes (ΔS)* that take place in a system and its surroundings during physical and chemical processes.

If the entropy of a system plus its surroundings *increases* during a process, then it is a **spontaneous process** (which is one way of defining such a process). As a rule, if entropy for the system alone *decreases*, it is because the process is being driven by an input of external energy; it is not spontaneous. A simple spontaneous cooling process is an exception, however.

Spontaneous processes are **irreversible** without an input of energy into the system that increases entropy in the surroundings. Perfume molecules do not spontaneously condense back into an open bottle, for example. When the entropy reaches its maximum level for a system and its surroundings, then the system is said to be at **equilibrium.** At equilibrium there is no net change in entropy or any other property of the system. We shall see that there are other ways of defining equilibrium, but one way is to define it as the state of maximum entropy.

Is life as a system an exception to the Second Law of Thermodynamics? As a zygote grows into a mature organism, the order and complexity of the substances it takes from the environment increase greatly. But, if the organism is the system, large amounts of both energy and raw materials must cross the boundary into the system from the surroundings. The Second Law says that the overall entropy of an organism and its surroundings increases as organisms grow. So most biologists do not think life is an exception to the Second Law—although it may be impossible to *prove* that it is not. Certainly there is something marvelous about living organisms that enables them to gather surrounding raw materials and energy and to decrease their own entropy as they grow. What are the mechanisms that control this organizing of materials and energy? How did they come about? These are the unsolved problems in the study of organism development.

Entropy is a valuable concept, because it leads us to think about the degree of orderliness in the universe and how the changes that we observe around us are spontaneously driven by the overall increase in disorder. The concept applies nicely, for example, to our model of diffusion. Think of the energy (and intelligence) required to create the high degree of order of the white and the black balls at the beginning of our thought experiment. Then think about how that order was destroyed—converted to random disorder—by spontaneous mixing as the balls passed through the window. Equilibrium was reached at maximum disorder (entropy). Our musings about plant function are often aided by the entropy concept.

The Gibbs Free Energy (G) We need to know more about a system than its entropy. Basically, an energy transfer or interaction across the boundary between a system and its surroundings is only a pushing, pulling, or striking of some kind. If the pushing or pulling across the boundary is highly organized, we call that a **work interaction.** A work interaction usually involves many, many particles that are organized to move together in the same direction and at the same time, such as a spinning shaft, a hammer striking a nail, or a stream of electrons flowing in a wire. The thermodynamic definition of work might be stated: *A work interaction is one in which what happens at the interaction boundary could be repeated in such a way that its sole effect would be the raising or lowering of a weight in the surroundings*. Notice that the change in the height of a weight necessitates the marshalling of many particles into motion in the same direction and at the same time.

If the pushing, pulling, or striking at the interaction boundary is random, chaotic, and on the scale of atoms and molecules, we call that a **heat interaction.** A lighted match held against the bottom of a metal table will not lift the table because, even though, on the average, the molecules in the hot combustion gases are striking the molecules in the table surface more often and with greater vigor than the molecules of the table strike back, they lack organization. With organization (e.g., as when hot gases drive a piston in a hydraulic system), the table could be raised. Some interactions are apparently neither all heat nor all work. An example is a beam of light (photons) striking a system. The photons certainly have some degree of organization because of direction and uniform speed, but much of that organization is lost upon striking the surface of the system.

J. Willard Gibbs in the 1870s developed a thermodynamic measure that allows us to think of the energy passing across the boundary between a system and its surroundings that is available to do work—available to be involved in a work interaction. This measure is now called the **Gibbs free energy,** which is *a measure of the maximum energy available for conversion to work (at constant temperature and pressure)*. At this point, it would be highly desirable to insert one or more chapters telling how Gibbs derived the free-energy concept and thereby defining it as it should be defined. Since space is limited, we will have to be content with a word description of free energy, an equation that defines it, and applications of the concept that eventually lead us to a definition of water potential. The water-potential concept is fundamental to the first chapters of this book and is used occasionally in later chapters.

To begin with, free energy includes all components of the internal energy (E) in a substance.

Internal energy includes the velocity or translational kinetic motions of the particles discussed previously, as well as their rotations and vibrations. Internal energy also includes the electron energies (discussed in Appendix B), which involve energy levels of electrons in molecules and the effects of absorption of radiant energy, and the molecule–electron configurations that we refer to collectively as chemical bonds. Part of the internal energy of gasoline, for example, is the energy locked in the bonds that hold its atoms of hydrogen and carbon together. Internal energy, like entropy, is usually not calculated; again, *changes* in internal energy (ΔE) are more important than exact amounts.

We can think of internal energy as existing in two forms. For example, the chemical-bond energy in gasoline is **potential energy.** When the gasoline burns, some of the bonds break, releasing energy that causes the translation, rotation, and vibration of atoms and molecules (which we observe as an increase in temperature); this atomic and molecular motion is called **kinetic energy.** In an engine, some of this random, disorganized energy is organized and transmitted to the surroundings as work by means of moving pistons. Many of the processes in plants involve conversions from potential to kinetic energy, or vice versa. Photosynthesis converts kinetic light energy (moving photons) to potential bond energy, and respiration releases it again as kinetic heat or mechanical or other forms of energy. We can think of potential energy as being a function of position or condition and kinetic energy as being due to motion (of objects, molecules, photons, electrons, and so on). So internal potential and kinetic energy is an important part of free energy.

The basic equation for the Gibbs free energy (G) has an enthalpy factor and an entropy factor. The **enthalpy** (H) of a system consists of the internal energy (E), plus the absolute pressure (P) multiplied by the volume (V). In the basic equation, the entropy factor, which is the entropy (S) multiplied by the absolute temperature (T), is subtracted from the enthalpy to give the Gibbs free energy:

$$G = E + PV - TS = H - TS \qquad (1.2)$$

where

G = Gibbs free energy

E = internal energy

PV = pressure–volume product

H = enthalpy ($E + PV$)

TS = entropy or disorder factor

Since S and E cannot be known exactly, we cannot calculate G; but the *change* in G between two energy states of a system ($\Delta G = G_2 - G_1$) can be calculated. For example, a standard free-energy change (ΔG^0) of a chemical reaction can be determined by the following equation:

$$\Delta G^0 = -RT \ln K_{eq} \qquad (1.3)$$

where

ΔG^0 = standard free energy change in joules (J) or calories (cal)

R = the ideal gas constant (8.304 J mol^{-1} K^{-1}; 1.987 cal mol^{-1} K^{-1})

T = absolute temperature (in kelvins, K)

$\ln$ = natural logarithm

K_{eq} = equilibrium constant

$\quad = \dfrac{\text{activities of products multiplied together}}{\text{activities of reactants multiplied together}}$
(Activities are "corrected concentrations"; discussed below.)

The results of the calculation are the same as would be obtained if it were possible to subtract the absolute free energies of the products from those of the reactants. This standard free-energy change is the maximum useful work that can be obtained when one mole of each reactant is converted to one mole of each product. Thousands of experiments have shown that free energy decreases for spontaneous reactions and increases for nonspontaneous reactions. For spontaneous processes, then, ΔG will be *negative*, which is another way to define such processes. When a reaction is spontaneous, at equilibrium the activities of the products are higher than those of the reactants; the equilibrium constant is greater than 1. So in equation 1.3, $\ln K_{eq}$ is positive, and ΔG is negative. If there are more reactants than products at equilibrium, then the equilibrium constant is a fraction, and ΔG is positive.

It is important in plant physiology to have some feeling for the components of G, especially those incorporated in E. Further, it is important to know that, though the entropy *increases* in spontaneous processes, the free energy that is lost from the system ($-\Delta G$) is an indication of the maximum work that could be done by the process if it were 100 percent efficient (which it never can be). If the process is not spontaneous (if ΔG is positive), then ΔG indicates the minimum energy that must be added to the system to make the process go. When equilibrium is reached, all macroscopic properties cease changing, so there is no longer any change in free energy (that is, $\Delta G = 0$).

The Chemical Potential and Water Potential We can speak of the free-energy change of a total system, or of any one of its components. We note, however, that a large volume of water has more free energy than a smaller one, under otherwise identical conditions. Hence, it is convenient to consider the free energy of a substance in relation to some unit quantity of the substance. The free energy per unit quantity of substance, specifically per gram molecular weight (i.e., the free energy mol^{-1}), is called the **chemical potential.** Chemical potential, like solute concentration and temperature, is independent of the quantity of substance being considered.

For a **solute** (dissolved substance) in a **solvent** (the liquid in which the solute is dissolved—in plants, mostly water), the chemical potential is approximately proportional to the concentration of solute. (Actually, concentration is usually corrected by some factor that depends upon concentration and other parameters to produce a corrected concentration called the **activity.**) A diffusing solute tends to move from areas of high chemical potential (free energy per mole) to areas of low chemical potential.

The chemical potential of water is an extremely valuable concept in plant physiology. In 1960, Ralph O. Slatyer in Canberra, Australia and Sterling A. Taylor at Utah State University proposed that the chemical potential of water be used as the basis for a property of water in plant–soil–air systems. They defined the **water potential** of any system or part of a system that contains water, or could contain water, as being equivalent to the chemical potential of the water in that system or system part compared with the chemical potential of pure water at atmospheric pressure and the same temperature; and they suggested that the water potential of the reference pure water be considered to be zero. In the next chapter we shall apply these concepts; but for now suffice it to say that the water potential is negative if the chemical potential of water in the system under consideration is lower than that of the reference pure water, and it is positive if the chemical potential of the water in the system is greater than that of the reference water.

In thermodynamics, the chemical potential of any substance, including water, has the units of energy, as in equation 1.3. Appropriate SI units are joules per kilogram (J kg^{-1}) or joules per mole (J mol^{-1}), though in the past, calories per mole or per kilogram have also been used. In 1962, Taylor and Slatyer recommended that the energy terms of chemical potential be divided by the partial molar volume of water, which would give the water potential in units of *pressure*. Plant physiologists had long been discussing the diffusion of water in terms of pressure, so this suggestion was accepted and is now almost universally applied. It is still valid to use energy units when speaking of water potential (e.g., Campbell, 1977), but most plant physiologists and soil scientists now use the following definition of **water potential:** *The water potential (ψ) is the chemical potential of water in a system or part of a system, expressed in units of pressure and compared to the chemical potential (also in pressure units) of pure water at atmospheric pressure and at the same temperature; the chemical potential of pure water is arbitrarily set at zero.*

As solutes move in response to differences in solute chemical potential, so water diffuses in response to differences in water potential. When water potential is higher in one part of a system than in another, and nothing (e.g., some impermeable membrane) prevents the movement of water, it moves from the high point of water potential to the low point. The process is spontaneous: Free energy is released to the surroundings and the system's free energy decreases. This released energy has the potential to do work, such as osmotically lifting water upward in stems in the phenomenon known as root pressure. The maximum possible work is equivalent to the free energy released, but sometimes no work is done. Then the free energy simply appears in the system and its surroundings as heat or increased entropy. In any case, it is important to remember that equilibrium is reached when the change in free energy (ΔG) or the water potential difference ($\Delta \psi$) is equal to zero. At this point, entropy for the system and its surroundings will be at a maximum, but entropy change (ΔS) will equal zero.

1.6 Chemical and Water Potential Gradients

Because gradients in chemical potential or water potential produce the **driving force** for diffusion, it is important to understand the five factors that most commonly produce chemical-potential or water-potential gradients in the soil–plant–air continuum.

Concentration or Activity For solute particles (mineral ions, sugars, and so on), the activity (effective concentration) is by far the most common and important factor in establishing the chemical-potential gradients that drive diffusion. In this text, when we discuss the movement of solute particles in and out of cells and throughout the plant, we shall be thinking almost exclusively of activity differences: Particles diffuse from points of high to points of low activity.

Water is the common solvent in plants. As we have seen (Section 1.1), water is virtually incompressible, so its concentration remains nearly constant, changing only slightly with addition of solutes and changes in temperature. Thus a model based strictly upon water concentration will not explain dif-

Factors Affecting Diffusion

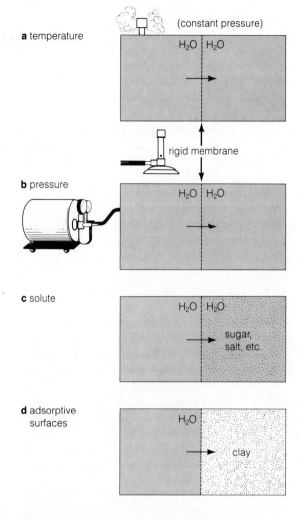

a temperature

(constant pressure)

H_2O | H_2O

rigid membrane

b pressure

H_2O | H_2O

c solute

H_2O | H_2O

sugar, salt, etc.

d adsorptive surfaces

H_2O

clay

Figure 1-4 Models of diffusional systems.

fusion of water in plants. Indeed, differences in water concentration can readily be ignored.

Temperature   We all know that water vapor diffuses from cold food to the even colder freezer coil in a refrigerator. By the same process, *liquid water* or *water vapor* often diffuses from deep in the soil to the surface when the surface is cooled at night, and deeper into the soil during the day. (These processes often involve evaporation, condensation, or freezing and thawing.) But consider a gas at two different temperatures, separated by a barrier that permits diffusion: The cooler gas diffuses into the warmer gas, opposite to the diffusion in the examples just mentioned. This is because the cooler gas is the more concentrated gas, when pressures are equal, and the concentration difference is more important to diffusion than the slightly higher velocities of the molecules in the warmer gas. (On page 68, John Dacey

explains the diffusion of gases into water lilies.) So temperature effects on diffusion are complex.

Actually, temperature differences are usually ignored in discussions of plant–soil water relations for a very good reason: *The thermodynamic equations that we have been considering assume constant temperature throughout the system and its surroundings.* However, there are ways to estimate the effects of temperature changes, as we shall see in the next chapter. It is important to consider temperature effects, because strong temperature gradients may exist in plants. Consider, for example, the plants of the arctic or alpine tundras, sometimes only a few centimeters high, whose roots are in soil close to freezing while their leaves are warmed by the sun to over 20°C. Or think of temperate-zone plants in springtime, when their buds are beginning to become active though their roots are still covered by snow.

Pressure   Increasing the pressure will increase the free energy and hence the chemical potential in a system. Imagine a closed container (Fig. 1-4) separated into two parts by a rigid membrane that permits passage only of solvent molecules (water, we assume). If pressure is applied to the solution on one side of the membrane but not to the other, the water potential of the pressurized side will be increased; water molecules will then diffuse more rapidly through the membrane into the compartment of lower pressure. This will occur even when membrane pores are too small to permit bulk flow. This effect of pressure is an extremely important consideration in studying plants, because the contents of most plant cells are under pressure compared with their surroundings, and because fluids in the xylem can be under tension (negative pressure).

Effects of Solutes on Solvent Chemical Potential   It has been observed that solute particles decrease the chemical potential of the solvent molecules. This decrease is independent of any effect on solvent (water) concentration, which may be decreased, increased, or not changed, according to the kind of solute.*

---

*To see this for yourself, study the many tables by A. V. Wolf, Morden G. Brown, and Phoebe G. Prentiss entitled "Concentrative Properties of Aqueous Solutions: Conversion Tables" in the *Handbook of Chemistry and Physics*, CRC Press, Boca Raton, Florida. In the 1983-1984 edition, the tables are on pages D-223 to D-272. For a range of molar and other concentrations of 99 solutions, including those of many salts and organic compounds, not to mention sea water, human plasma, and urine from humans and various other animals, the tables show such parameters as densities, total water concentration, freezing point depression, osmolality, viscosity, and conductance. Osmolality allows computation of osmotic potentials at any temperature with the Van't Hoff equation. (These concepts are discussed in Chapter 2; the Van't Hoff equation is equation 2.2.)

Rather, it depends primarily on the **mole fraction,** which is the number of solvent particles (ions or molecules) compared with the total number of particles:

$$\text{mole fraction of solvent} = \frac{\text{moles of solvent}}{\text{moles of solute} + \text{moles of solvent}}$$

(1.4)

Consider again the closed container in Figure 1-4. If there is pure water on one side of the membrane and a solution on the other side, a water potential gradient will exist, with water potential being lower on the solution side. Water will diffuse through the membrane from the pure water side into the solution. This special case of diffusion is **osmosis,** which is often the process by which water moves from the soil into the plant and from one living plant cell to another.

Of course, a steep chemical-potential gradient exists for the solute particles in our container. If they can penetrate the membrane, they will move from the solution side (where their chemical potential is high) to the other side, where initially their chemical potential is infinitely low. When they become equally concentrated on both sides, there will be no more net movement. Differences in the chemical potential of solutes across cell membranes is a major factor in the movement of ions from the soil into the plant and for transport of ions and un-ionized solutes in and out of plant cells. We shall see that solutes can also move across membranes *against* chemical potential gradients, but metabolic energy must be used to do so.

It is fundamental that the chemical potential of solutes not be confused with the chemical potential of water.

Matrix   Many charged surfaces, such as those of clay particles, proteins, or cell-wall polysaccharides, have a great affinity for water molecules. These surfaces usually have a net negative charge that attracts the slightly positive sides of the polar water molecules. However, because of hydrogen bonding, even surfaces that have no net charge, such as those of starch, also bind water. A material with surfaces that bind water is called a **matrix** (one of several uses of the word). Binding is a spontaneous process that releases free energy (ΔG is negative). In our imaginary double compartment (Fig. 1-4), we might have pure water on one side of the membrane and dry protein or clay particles on the other. This condition would establish a steep gradient in water potential, with the water potential being high in the pure water and extremely low in the dry protein or clay. Water molecules would diffuse down the water-potential gradient into the other compartment and become bound to the protein or clay. (Of course, in this case the membrane is required only to keep the water from *flowing* into the

protein or clay.) This process of water becoming adsorbed on a matrix is called **hydration.** It is primarily responsible for the first phase of water uptake by a seed prior to germination.

1.7 Vapor Density, Vapor Pressure, and Water Potential

If a sample of water or solution is exposed to an evacuated container or to one filled with gas (as in Fig. 1-5), water molecules will evaporate into the container until they reach an equilibrium vapor concentration. At equilibrium, water molecules will be condensing back into the liquid phase at the same rate as they are evaporating into the gas phase. A convenient way to express the concentration of water molecules in the **vapor phase** (the gas phase) is as grams per cubic meter (g m^{-3}). Such an expression is called the **vapor density.** The vapor molecules, colliding with the liquid surface and the walls of the container, exert a pressure called the **vapor pressure.** Note that *vapor density* and *vapor pressure* are two different ways of expressing the amount of vapor that is in equilibrium with the liquid. This amount is independent of the presence of other gases. It will be the same if the volume above the liquid contains air at atmospheric pressure or if the volume was originally a vacuum, in which case the liquid would boil until the **saturation vapor density** or **saturation vapor pressure** had been achieved.

Applying the rule that water potentials are equal in all parts of a system when the system is at equilibrium (i.e., the differences in water potential are 0; $\Delta\psi = 0$), we find that factors affecting water potential in the liquid phase will also affect saturation vapor density or pressure. Figure 1-5 illustrates the three most important effects and their magnitudes. Note that vapor density almost doubles when the temperature changes from 20 to 30°C. This effect is much greater than the effect of adding solutes (which lowers vapor density slightly) or the effect of raising pressure (which raises vapor density slightly).

The effect of adding solutes can be calculated by **Raoult's Law,** formulated in 1885. This law states that the vapor pressure over perfect solutions is proportional to the mole fraction of solvent:

$$p = X_1 p^0$$

(1.5)

where:

p = the vapor density or pressure of the solution

X_1 = the mole fraction of solvent as defined in equation 1.4

p^0 = saturation vapor density or pressure of pure solvent

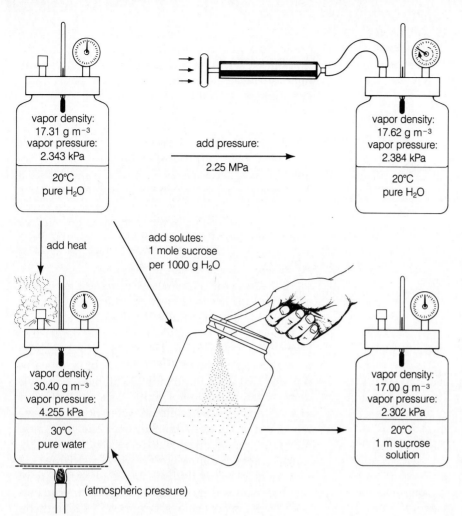

Figure 1-5 Illustrating effects of pressure, temperature, and solutes on vapor density and vapor pressure (and thus indirectly on water potential). Units of vapor density are grams per cubic meter (g m^{-3}); vapor pressure units are kilopascals (kPa). These two ways of expressing the amount of vapor are interconvertible with a suitable equation. The equation for calculating water potentials from vapor pressures is given in Chapter 2 (equations 2.4 and 2.5). Figures for vapor density and pressure of pure water in a closed volume at different temperatures are given in standard tables. Note the relatively large effect of temperature on vapor density or pressure compared with the much smaller effects of pressure and added solutes.

Perfect solutions (those that exactly follow such laws) are seldom, if ever, encountered in nature, but Raoult's law provides a close approximation for real solutions. Furthermore, it is beautifully simple.

Since the solvent is usually present in quantities much greater than the solute, the actual lowering of the vapor pressure is usually not very great. Consider 1 gram-molecular-weight (1 mol) of sucrose (342.3 grams) dissolved in 1000 g (55.508 mol) of water. The mole fraction of water in this solution is calculated as follows:

$$X_{H_2O} = \frac{55.508}{55.508 + 1.000} = 0.9823 \ (98.23\%)$$

At equilibrium, the vapor density or pressure above pure water in a closed container is by definition equal to 100 percent **relative humidity.** At 20°C, this vapor density equals 17.31 g m^{-3}, so by Raoult's law the vapor density above a 1.00 molal solution of sucrose equals $17.31 \times 0.9823 = 17.00$ g m^{-3}. Reversing the calculation, you see that 17.00 is 98.23 percent of 17.31, so the relative humidity above a 1.00 molal sucrose

solution is 98.23 percent. Thus, relative humidity above a solution (at equilibrium) is equal to the mole fraction of solvent expressed as percent.

With modern methods, to be discussed in Chapter 2, it is possible to measure relative humidities to a small fraction of a percent, so the vapor pressure above a solution in a closed container can be accurately determined. From this, as we shall see, the water potential can be calculated. Water potentials of leaves or other plant materials placed in closed containers are determined in this way.

1.8 The Rate of Diffusion

If the chemical potential of a given solute is different in different parts of a system, there will be a tendency for the solute to diffuse from the point of high chemical potential toward the point of low chemical potential. The same is true, of course, for water molecules, which diffuse in response to gradients in water potential. But how fast does diffusion take place? The rate of diffusion is given by a simple but highly significant equation called **Fick's First Law,** formulated in 1855. Fick's law can be written in various ways, but the following integrated form is easy to understand and to apply:

$$J = \frac{\psi_1 - \psi_2}{r}$$

or

$$J = \frac{\Delta\psi}{r} \qquad (1.6)$$

where:

J = the water flux density (g m^{-2} s^{-1})

ψ_1 = water potential at its highest point

ψ_2 = water potential at its lowest point

r = resistance to diffusion

The difference in water potential ($\psi_1 - \psi_2$) can be expressed simply as $\Delta\psi$. This is the water-potential gradient, the driving force for diffusion. The rate of diffusion expressed as a **flux,** the movement of a given quantity of water molecules or solute particles across a unit cross-sectional area in a unit amount of time, is proportional to the magnitude of the driving force ($\Delta\psi$). The flux is also inversely proportional to the resistance to diffusion encountered between ψ_1 and ψ_2.

Some workers prefer to think of permeability rather than resistance to a diffusing substance. The permeability (P_j) is simply the inverse of the resistance:

$$P_j = \frac{1}{r} \qquad (1.7)$$

Thus a medium (e.g., a membrane) that has a high resistance to diffusion of water or a solute has a low permeability. We can write

$$J = P_j\Delta\psi \qquad (1.8)$$

The application of Fick's law can be further refined by considering the difference in water potential (or chemical potential of a diffusing solute) over some unit distance such as a centimeter. This gives an indication of the steepness of the gradient in chemical or water potential. The steeper the gradient the more rapid the diffusion. Resistance or permeability is measured over the same unit distance.

Increasing the temperature increases the average velocity of all the ions and molecules in a solution and thus increases the rate of diffusion; but as we saw in Table 1-2, this effect is not great over the narrow range of Kelvin temperatures in which organisms are normally active. The Q_{10} for diffusion of many gases is about 1.03 (meaning that a gas diffuses only 1.03 times faster when temperature is raised 10°C; see footnote on page 24). Solutes in water, however, have Q_{10} values of 1.2 to 1.4 for diffusion. This is because increased temperature breaks hydrogen bonds in water, so solutes diffuse more rapidly; the viscosity of water (Table 1-2) is decreased, and permeability of water to solutes is increased. Since less-massive particles have higher average velocities at a given temperature, they will diffuse more rapidly than larger particles (equation 1.1), all other factors being equal.

2

Osmosis

It is an everyday experience to turn on a water faucet or flush a toilet. Thus we are perfectly familiar with water movement as a bulk flow phenomenon—our plumbing systems see to that! But in the world around us, vast quantities of water are moving, usually invisibly, by diffusion or in bulk in response to pressure gradients established by diffusion.

It takes some mental effort to visualize this more unfamiliar aspect of the real world. With our mind's eye (there is no other way) we must see those water molecules, flying and bouncing billions of times each second in the vapor state, holding each other in the liquid state with their hydrogen bonding—positive side of one to negative of another—even while their kinetic motions tend to make them fly apart. We must somehow conceptualize the entropy, free energies, and chemical potentials and how these properties can drive the molecules to diffuse down a gradient. We must realize that pressure increases these quantities even while solute particles and matric surfaces decrease them.

With these models in mind, we are ready to extend our concepts to the cells of plants. We are ready to discuss osmosis and related matters.

2.1 An Osmotic System

A device that measures osmosis is an **osmometer.** This is usually a laboratory device, but a living cell may be thought of as an osmotic system (Fig. 2-1). In both cases, two things are usually present: First, two or more volumes of solutions or pure water are always isolated from each other by a membrane that restricts the movement of solute particles more than it restricts the movement of solvent molecules. Second, there is usually some means of allowing pressure to build up in at least one of the volumes. In the laboratory osmometer, pressures usually build up

hydrostatically by raising the solution in the tube against gravity, but other means are also used, such as a piston for automatically increasing the pressure in the system as soon as the volume of liquid begins to expand by the first small increment. In the cell, the rigidity of the plant cell wall is responsible for the increase in pressure.

It is important to note that the cell wall and the cell membrane are different structures. The membrane allows water molecules to pass more rapidly than solute particles; the primary cell wall is usually highly permeable to both. It is the plant cell membrane that makes osmosis possible, but it is the cell wall that provides the rigidity to allow a buildup in pressure. Animal cells do not have walls, so if pressures build up in them, they often burst, as happens when red blood cells are placed in water. Turgid cells provide much of the rigidity of nonwoody plant

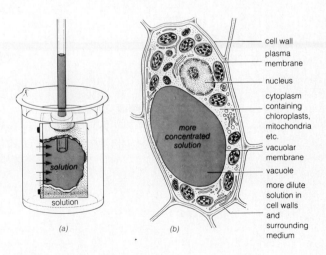

Figure 2-1 (**a**) A mechanical osmometer in a beaker. (**b**) A cell as an osmotic system.

a Possible Levels of Water Potential

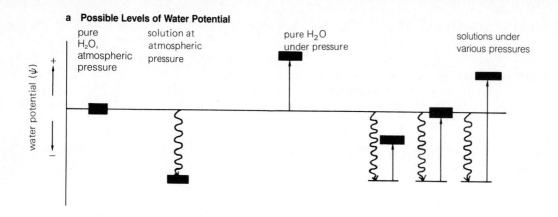

pure
H₂O,
atmospheric
pressure

solution at
atmospheric
pressure

pure H₂O
under pressure

solutions under
various pressures

water potential (ψ)

+

−

b Diffusion of Water in Response to △ψ

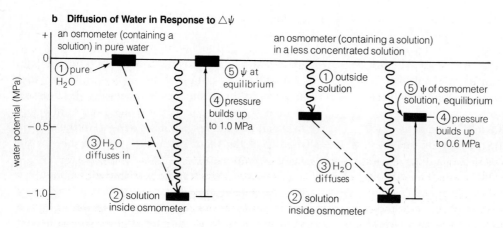

an osmometer (containing a
solution) in pure water

an osmometer (containing a solution)
in a less concentrated solution

water potential (MPa)

+

0

−0.5

−1.0

① pure
H₂O

⑤ ψ at
equilibrium

④ pressure
builds up
to 1.0 MPa

③ H₂O
diffuses in

② solution
inside osmometer

① outside
solution

⑤ ψ of osmometer
solution, equilibrium

④ pressure
builds up
to 0.6 MPa

③ H₂O
diffuses

② solution
inside osmometer

c Two Water-Potential Gradients

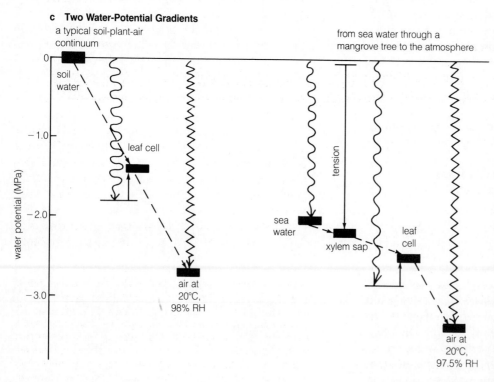

a typical soil-plant-air
continuum

from sea water through a
mangrove tree to the atmosphere

water potential (MPa)

0

−1.0

−2.0

−3.0

soil
water

leaf cell

air at
20°C,
98% RH

tension

sea
water

xylem sap

leaf
cell

air at
20°C,
97.5% RH

Figure 2-2 A schematic
illustration of various effects on
water potential. Black rectangles
show the water potential (scale on
the left). Wavy lines indicate
depression in water potential
caused by solutes (the osmotic or
solute component of water
potential). Unbroken lines suggest
the effect of pressure on water
potential. (Arrows point up for
positive pressure, down for
tension.) Zig-zag lines indicate
effects of relative humidities below
100 percent on water potential of
the atmosphere (discussed later in
the chapter and in Chapter 4). (**a**)
The basic effects of solutes and
pressures, alone and in
combination, on water potential of
open liquid water. (**b**) Diffusion
of liquid water in response to
gradients in water potential,
showing how water potential
changes as diffusion occurs into
an osmometer. (**c**) Water diffusion
down a water potential gradient
from the soil through a plant into
the atmosphere and from sea
water through a plant into the
atmosphere. The sea-water
example (note *tension* in the
xylem sap) is discussed in
Chapter 4 (see Figure 4-13).

same solution at atmospheric pressure) can be expressed in energy terms as -2.5 kJ kg^{-1} and in pressure terms as -2.5 MPa. In the SI system, kJ kg^{-1} have the same numerical value as megapascals. This transformation was suggested by Taylor and Slatyer in their 1962 paper.

In 1887, J. H. van't Hoff discovered an empirical relationship that allows calculation of an approximate osmotic potential from the molal concentration of a solution. He plotted osmotic potentials from direct osmometer readings as functions of molal concentration, obtaining the following relationship, which has the exact form as the law for perfect gases:

$$\pi = -miRT \qquad (2.2)$$

where

π = osmotic potential

m = molality of the solution (moles of solute/1000 g H_2O)

i = a constant that accounts for ionization of the solute and other deviations from perfect solutions

R = the gas constant (0.00831 liter MPa mol^{-1} K^{-1}, or 0.00831 liter kJ mol^{-1} K^{-1}, or 0.0831 liter bars mol^{-1}K^{-1}, or 0.080205 liter atm mol^{-1}K^{-1}, or 0.0357 liter cal mol^{-1} K^{-1})

T = absolute temperature (K) = degree C + 273

If m, i, and T are known for a solution, then osmotic potential can be easily calculated (in the absence of highly active colloids). For nonionized molecules, such as glucose and mannitol in dilute solutions, i is 1.0, but in other cases i varies with concentration. This is partly because activity varies with concentration and partly because the extent to which a salt or acid ionizes depends upon its concentration. Total π for a complex solution such as cell sap is the sum of all osmotic potentials caused by all solutes. It may be expressed as **osmolality.**

Consider some examples of simple solutions. For a 1.0 m glucose solution at 30°C,

$$\pi = -\left(1.0 \ \frac{mol}{liter \ H_2O}\right)(1.0)\left(0.00831 \ \frac{liter \ MPa}{mol \ K}\right)(303 \ K)$$

$$\pi = -2.518 \text{ MPa (at 30°C = 303 K)}$$

For the same solution at 0°C,

$$\pi = -(1.0 \ m)(1.0)\left(0.00831 \ \frac{liter \ MPa}{mol \ K}\right)(273 \ K)$$

$$\pi = -2.269 \text{ MPa (at 0°C = 273 K)}$$

Note that the osmotic potential is less negative at 0°C than at 30°C, which seems to imply that water would diffuse from the cold toward the warm liquid. That is contrary to our experience (and to the discussion in Chapter 1), and we are incredulous only if we forget that the calculated osmotic potentials are compared with pure water at atmospheric pressure *and at the same temperature.* The fact that the warmer solution has a more negative osmotic potential than the colder one means that pure warm water, at one atmosphere pressure *and at the same temperature* as the solution in the osmometer, would produce a higher equilibrium pressure in a warmer solution than cooler water would produce in the cooler solution. Since the thermodynamic equations we have presented depend on constant temperature, they cannot be used directly to calculate the forces that might be developed by **thermoosmosis:** a situation in which warm water might be separated from cold water by a membrane that permits pressure buildup in the cold water as warm water diffuses in (see Fig. 1-4). The effects of solutes, pressure, and temperature on vapor pressure, however, suggest that high pressures might be built up on the cold side if a perfect thermoosmometer could be built. Figure 1-5 shows the relative magnitudes of solute pressure and temperature effects on vapor pressure and thus, by implication, on other properties of solutions related to water potential. The temperature effect is much greater than the solute or pressure effects.

For an application of the van't Hoff equation to an example in which ionization causes i to have a value different from one, consider a 1.0 m NaCl solution at 20°C. Sodium chloride ionizes 100 percent in solution, but the value of i is not quite what one would predict from this. Two ions for each formula (NaCl) would suggest that $i = 2$; actual measurements show that $i = 1.8$:

$$\pi = -(1.0 \ m)(1.8)\left(0.00831 \ \frac{liter \ MPa}{mol \ K}\right)(293 \ K)$$

$$\pi = -4.38 \text{ MPa}$$

Since the van't Hoff equation has the same form as the equation for perfect gases, the **osmotic pressure** (the actual pressure developed in an osmometer rather than the osmotic potential) is the same as would be exerted on the wall of the container if the solute particles existed as a gas in an equivalent volume. At 0°C, the pressure of 1 mole of a perfect gas in a volume of 1 liter is 2.27 MPa (2.52 MPa at 30°C), and 2.27 MPa is the osmotic pressure developed by a 1-molal solution of nonionizing solute at 0°C in a perfect osmometer. This interesting relationship, which once caused considerable confusion, is now considered coincidental; it is certainly incorrect to think of solute particles exerting pressure on the walls of a container as though they were a gas. Pressures on the walls of an open beaker of solution or an ordinary osmometer at equilibrium are hydrostatic pressures only, caused by the weight of the solution.

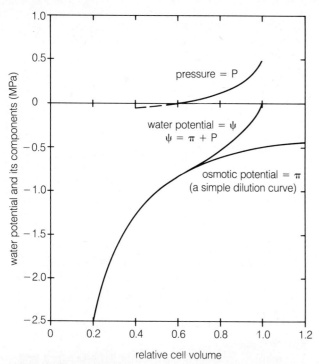

Figure 2-3 The Höfler Diagram. The components of water potential are shown as they change with cell volume. Osmotic potential is a dilution curve calculated by the relationship: $\pi_1 V_1 = \pi_2 V_2$, as described in the text. The pressure curve is arbitrary but expresses the fact that as a cell with zero pressure takes up water, pressure increases slowly at first and then more rapidly. The water potential curve is the algebraic sum of the pressure and osmotic potential curves.

Incidentally, the gas law applies only to "perfect" gases, and the van't Hoff equation is at best only an approximation for ideal solutes.

The van't Hoff equation is not merely coincidence, however, because it can be derived from Raoult's law (equation 1.5) and other thermodynamic considerations. Its derivation was an early confirmation of the developing thermodynamic principles.

2.4 Dilution

We have neglected one factor that is usually important in a real osmotic system as contrasted to a perfect osmometer. As water diffuses across a membrane in a real system, it not only causes an increase in pressure but also dilutes the solution. This increases the osmotic potential in the solution (makes it less negative), so that the pressure required to reach equilibrium is less than would have been predicted from the original osmotic potential.

The relationship between water potential and its two primary components during dilution is well illustrated by the **Höfler diagram** (Fig. 2-3). The concept of this diagram was devised by K. Höfler in Austria in 1920. It describes the changing magnitudes of water potential, pressure potential, and osmotic potential

as volume changes, assuming that the system expands only by taking up water and that no solutes move out or in. The curve for osmotic potential is derived from the simple dilution relationship, which is a close approximation for dilute molal solutions,

$$\pi_1 V_1 = \pi_2 V_2 \qquad (2.3)$$

where

π_1 = osmotic potential before dilution

V_1 = volume before dilution

π_2 = osmotic potential after dilution

V_2 = volume after dilution

The curve for pressure potential, on the other hand, is more hypothetical. Its shape depends on the tube diameter of the osmometer, or the stretching properties of the cell wall, being steep if the tube is narrow or the wall is rigid, and less steep if the tube is wide or the wall is less rigid. Actually, cell walls stretch easily at first; as pressure grows, their resistance increases somewhat and then becomes rather constant (McClendon, 1982). This is suggested by the pressure line in Fig. 2-3. The water potential curve is the algebraic sum of the pressure potential and the osmotic potential curves, as given by equation 2.1.

The Höfler diagram provides us with a good way to visualize the principles of equation 2.1 and the complications of dilution. It describes what would happen if mature cells were placed in solutions of different osmotic potentials so that water is either gained or lost from the cells, *but the total quantity of solutes within remains constant.* We shall discuss some of these approaches later in this chapter. The Höfler diagram may also describe what happens in some plant cells under normal conditions, at least over short time intervals, but growing cells do not act as suggested by the Höfler diagram. For one thing, the osmotic potential usually remains rather constant in growing cells as they absorb and/or produce solutes within (see Chapter 15). Furthermore, when cells grow, their walls become softened (Chapter 16), so that they stretch irreversibly (plastically), and pressures in them often decrease instead of increasing (Ross and Rayle, 1982).

2.5 The Membrane

Membranes exist in a wide variety, but osmosis will occur regardless of how the membrane functions, as long as solute movement is restricted compared to water movement (Fig. 2-4). The membrane could consist of a layer of material in which the solvent is more soluble than the solute, which would allow more solvent molecules than solute particles to pass through. A layer of air between two water solutions provides a barrier that completely restricts movement of nonvolatile solutes while allowing water vapor to pass. A third membrane model we can visualize is a sieve with holes of such a size that water molecules could pass but the larger solute particles could not. We shall see in Chapter 6 that water passes rapidly through cell membranes and thus may be slightly soluble in them, and that cell membranes also act like they have pores. It has been suggested that water sometimes passes in the vapor state from soil particles to the root in dry soils.

In 1960, Peter Ray brought an interesting problem to the attention of plant physiologists. Calculations of the thicknesses of certain membranes and the rates of osmotic water movement across them showed that this movement cannot occur by diffusion alone; the rates are too high. Ray suggested that the zone of diffusion might be very thin: an interface, say, between the water that is in the pores of the membrane and the solution that is inside the osmotic system. At this interface, the water-potential gradient would be extremely steep, resulting in rapid diffusion. This rapid diffusion of water across the interface into the solution would create tension in the water remaining in the pore, pulling it along in a bulk flow (Fig. 2-4d). This fourth membrane mechanism once again illustrates the complexities of nature. Note that the thermodynamic relations (direction and equilibrium) still hold.

The vapor-membrane model is an example of a truly semipermeable membrane, but probably all membranes that occur in plants allow some solute to pass with the water. In such cases, we are dealing with **differentially permeable membranes** rather than truly semipermeable ones. Though these membranes are permeable to both solvent and solute, they are usually much more permeable to solvent. The permeability of membranes to solutes introduces a complication into our consideration of osmosis: It determines the rate at which the equilibrium, established by solute concentration and pressure, gradually shifts as osmotic potentials on either side of a membrane change in response to the passage of solute particles.

2.6 Measuring the Components of Water Potential

Soon after the water potential concept was formulated by Otto Renner in 1915, methods were developed for measuring water potential and its components. Newer methods have been introduced since, but the older methods can help us understand plant–water relations. The newer methods are more useful in agriculture and research. We shall summarize both methods, not so much as a "cookbook" of useful techniques as an illustration and application of the principles that we have been discussing.

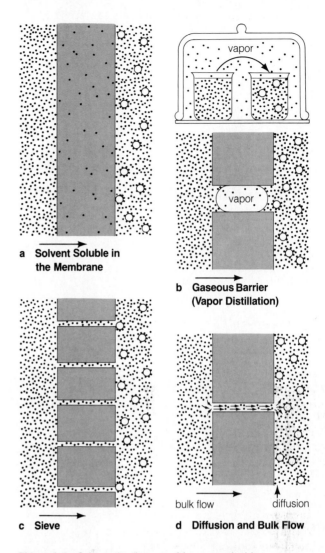

a Solvent Soluble in the Membrane

b Gaseous Barrier (Vapor Distillation)

c Sieve

d Diffusion and Bulk Flow

bulk flow   diffusion

Figure 2-4 Schematic diagram of four conceivable membrane mechanisms. The black dots represent water molecules about 0.3 nm in diameter, open circles represent sucrose molecules about 1.0 nm in diameter, and the membranes are drawn to the scale of most cellular membranes at a thickness of 7.5 nm. Note that water concentration is about the same on both sides of the membranes. Water molecules move rapidly through membranes in cells, possibly by mechanisms similar to those represented by models (**a**) and (**c**). This is discussed in Chapter 6. Model (**d**) is a refinement of model (**c**), as discussed in the text. The vapor model (**b**) may apply in plants and soils in various situations, but no membrane is known to have gas-filled pores.

Water Potential Probably the most meaningful property that we can measure in the soil–plant–air system is the water potential. It is not only the final determinant of diffusional water movement, but also it is the indirect determinant of bulk water movement, which occurs in response to the pressure gradients that are set up by diffusional movement. Furthermore, in principle and in practice, water potential is probably the simplest component of an osmotic system to measure. Remember that at equi-

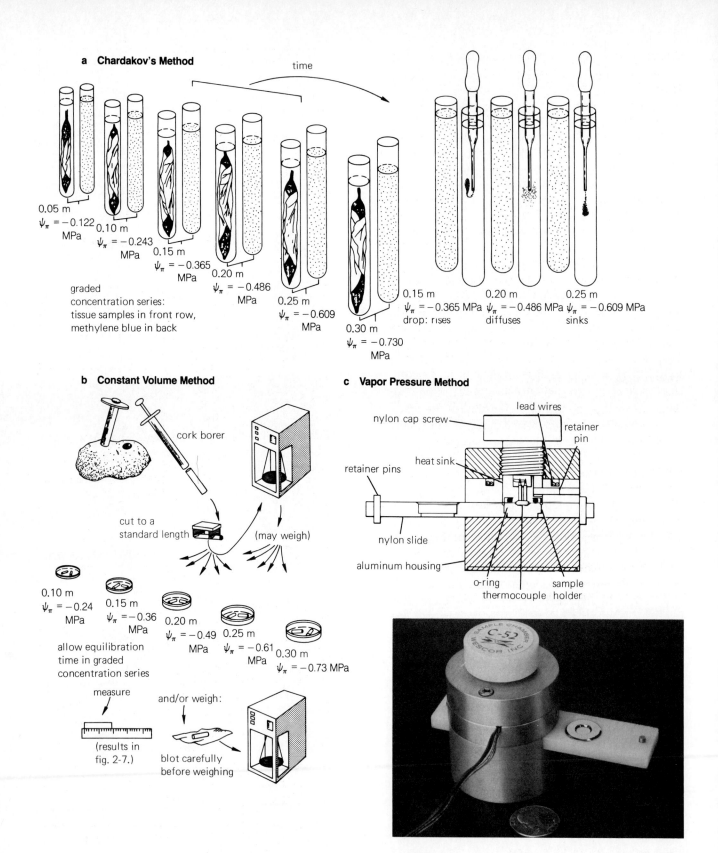

a Chardakov's Method

time

0.05 m
$\psi_\pi = -0.122$ MPa

0.10 m
$\psi_\pi = -0.243$ MPa

0.15 m
$\psi_\pi = -0.365$ MPa

0.20 m
$\psi_\pi = -0.486$ MPa

0.25 m
$\psi_\pi = -0.609$ MPa

0.30 m
$\psi_\pi = -0.730$ MPa

graded
concentration series:
tissue samples in front row,
methylene blue in back

0.15 m
$\psi_\pi = -0.365$ MPa
drop: rises

0.20 m
$\psi_\pi = -0.486$ MPa
diffuses

0.25 m
$\psi_\pi = -0.609$ MPa
sinks

b Constant Volume Method

cork borer

cut to a
standard length

(may weigh)

0.10 m
$\psi_\pi = -0.24$ MPa

0.15 m
$\psi_\pi = -0.36$ MPa

0.20 m
$\psi_\pi = -0.49$ MPa

0.25 m
$\psi_\pi = -0.61$ MPa

0.30 m
$\psi_\pi = -0.73$ MPa

allow equilibration
time in graded
concentration series

measure

(results in
fig. 2-7.)

and/or weigh:

blot carefully
before weighing

c Vapor Pressure Method

nylon cap screw

lead wires

retainer pin

heat sink

retainer pins

nylon slide

aluminum housing

o-ring
thermocouple

sample
holder

Figure 2-5 Three different ways to measure water potential.
Osmotic potentials are calculated at 20°C by equation 2.2.
The vapor-pressure device is made by Wescor, Inc., Logan,
Utah. (Drawing and photograph used by permission.)

librium, $\Delta\psi = 0$; that is, ψ is equal in all parts of the system. Thus, a plant part can be introduced into a closed system; and after equilibrium has been achieved, ψ can be known or determined for any part of the system, and therefore for the plant part. There are several possibilities for applying this principle, of which three general approaches are most widely used (Fig. 2-5).

In the **tissue-volume method** (Fig. 2-5b), a sample of the tissue in question is placed in each of a series of solutions of varying but known concentrations (usually sucrose, sorbitol, mannitol, or, best, polyethylene glycol, PEG). The best solute for such measurements is one that does not easily pass through membranes or harm the tissue. *The object is to find the solution in which the tissue volume does not change*, indicating no gain or loss in water, which means that the tissue and the solution were in equilibrium to begin with. So the water potential of the tissue must be equal to that of the solution. At atmospheric pressure, when $P = 0$, $\psi = \pi$; and π can be calculated from equation 2.2. (Figure 2-6 shows curves for π as a function of PEG concentration.)

In practice, there are several ways to determine changes in tissue volume. One is to measure the volume of each piece of tissue before placing it in the solution (usually standard volumes are cut) and then to measure the volume (or only the length) again after time has been allowed for exchange of water (Fig. 2-5b). Volume change is plotted as a function of solution concentration, which indicates an increase in volume in relatively dilute solutions and a decrease in volume in relatively concentrated ones. On such a plot (Fig. 2-7), the point at which the volume curve crosses the zero line indicates the solution that had the same water potential as that of the tissue at the start of the experiment.

In a second approach (not illustrated) tissue samples are allowed to equilibrate in small closed containers with the vapor over solutions of known concentration rather than with the solutions themselves. Thus, the solutions are not contaminated with solutes from the tissue. Weights rather than volumes are usually measured in this method.

Rather than measuring changes in the tissue, one might measure the *concentration of the test solution.* If it becomes less concentrated, the tissue will have lost water. This is a better approach than measuring tissue volumes, which often gives ψ values that are too negative because solutes are absorbed from the test solution by the tissue samples.

In 1948 a Russian scientist, V. S. Chardakov, devised a simple, efficient method of determining the test solution in which no change in concentration occurs (Fig. 2-5a). **Chardakov's method** can often be used in the field. Test tubes that contain solutions of

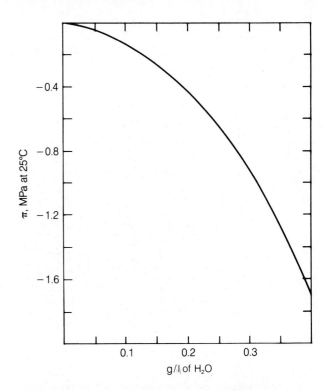

Figure 2-6 Osmotic potentials for PEG 4000 for various concentrations. PEG 6000 gave almost identical results. (Unpublished data obtained by Cleon Ross with a thermocouple psychrometer; π values of PEG solutions cannot be measured correctly by freezing-point depression.)

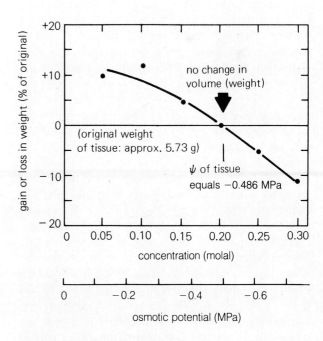

Figure 2-7 Weight of plant tissue samples as a function of the concentration of solutes with which the samples have been allowed to come into equilibrium. (Data from a student plant physiology laboratory report, Colorado State University.)

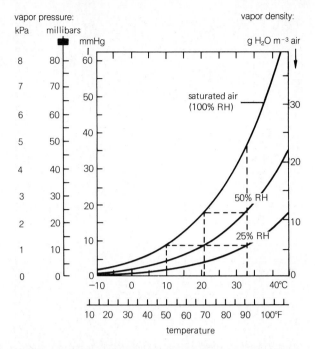

Figure 2-8 Relationship between moisture in the atmosphere (expressed as both vapor pressure and vapor density) and temperature for saturated air and for air at 50 percent relative humidity. Dashed lines indicate, for example, that air at 10°C, 100 percent RH, has the same amount of moisture as air at 21°C, 50 percent RH, and as air at 33°C, 25 percent RH.

graded concentrations are colored slightly by dissolving in them a small crystal of a dye such as methylene blue. (Addition of the dye does not change osmotic potential significantly.) Tissue samples are placed in test tubes that contain solutions of equivalent concentration but no dye. Time is allowed for exchange of a certain amount of water, but it is not essential that the tissue reach equilibrium with the solution. This may happen in as little as 5 to 15 min. Then the tissue is removed, and a small drop of the equivalent colored solution is added to the test tube. If the colored drop rises, the solution in which the tissue was incubated has become more dense, indicating that the tissue has taken up water; in this case, the tissue had a lower (more negative) water potential than the original solution. If the drop sinks, the solution has become less dense, having absorbed water from the tissue; the solution, then, had a lower water potential than the original tissue. If the drop diffuses evenly into the solution without rising or sinking, then no change in concentration has occurred, and the water potential of the solution equals that of the tissue. When several solutions with different concentrations are used, there is usually one in which the drop neither sinks nor floats. With equation 2.2 we calculate its π (or we determine π empirically). At $P = 0$, π of the solution is equal to ψ and thus the average ψ of the tissue.

In the **vapor-pressure method,** tissue is placed in a small closed volume of air. The water potential of the air comes into equilibrium with the water potential of the tissue, which changes only insignificantly in the process (Fig. 2-5c). The water potential of the air is then determined by measuring its vapor density (humidity) at a known temperature.

Vapor pressure and vapor density were defined in Chapter 1 (page 30). **Relative humidity** (RH) is the amount of water vapor in air at a particular temperature compared with the amount of water vapor that the air *could* hold at that temperature (the **saturation vapor pressure** or **density**). The saturation vapor density approximately doubles for each 10°C (or 20°F) increase in temperature. If a volume of air that is saturated with water vapor (100 percent RH) is warmed by 10°C, it can hold about twice as much water as before; so its relative humidity is about 50 percent. This relationship is illustrated in Fig. 2-8.

If we know the absolute temperature (T in kelvins), the saturation vapor density or pressure of pure water (p^0) at that temperature, the vapor density or pressure in the test chamber (p), and the molar volume of water (V_1 in liters mol^{-1}), we can calculate the water potential from the following formula, which is derived from Raoult's law (equation 1.5):

$$\psi = -\frac{RT}{V_1} \ln \frac{p^0}{p} \qquad (2.4)$$

The ratio of 100 p/p^0 is the relative humidity. If we convert to common logarithms and use numerical values for R and V_1, equation 2.4 simplifies to

$$\psi(\text{in MPa}) = -1.06 \, T \log_{10}\left(\frac{100}{RH}\right) \qquad (2.5)$$

Note that equation 2.5 gives the water potential of the air when the temperature and the relative humidity are known.

The measurement of water potential in tissue samples by this method is simple in principle, but in practice a number of difficulties are involved. These difficulties were solved only in the 1950s and since then, but now this is one of the most commonly used methods. To begin with, the temperature must be uniform, at least within a hundredth of a degree celsius, if the method is to be sufficiently accurate. This is because slight changes in temperature result in large changes in RH and water potential at constant vapor density.*

Another problem involves the measurement of the humidity inside the test chamber. An ingenious

---

*Air at 100 percent RH always has a water potential of zero. If air at 100 percent RH and 20°C is warmed only to 21°C, its RH drops to 94.02 percent, and $\psi = -8.34$ MPa, a value more negative than that found in almost any cell!

method was developed in 1951 by D. C. Spanner in England and has since been improved. Two thermocouple junctions are built into the test chamber. One of these has a relatively large mass and thus remains close to the temperature of the air in the chamber. The second thermocouple is very small. When a weak current is passed through the two junctions (in the right direction), the small one cools rapidly by the Peltier effect.* As it cools, a minute amount of moisture condenses on it from the air inside the chamber (Fig. 2-9). This point of moisture is then cooled by evaporation and thus acts as a "wet bulb," and the difference between the temperature of the "wet bulb" and that of the dry thermocouple indicates RH and thus the water potential of the air in the chamber. Uniform air temperatures between the two thermocouples are maintained by immersing the chamber in a water bath or by placing it in an aluminum block (or a smaller silver block since silver is the better conductor of heat). In practice, the drop evaporates so rapidly that the actual temperatures cannot be measured. Rather, the system is arbitrarily calibrated with solutions of known concentrations. Typically, measurements (which require less than a minute) are made at regular intervals until they stabilize after an hour or two, indicating that the tissue has reached equilibrium with the air in the chamber.

Still another method for the measurement of water potential in plant stems and leaves (petioles) involves use of a pressure bomb. This is discussed later in this section and again in Chapter 4.

Osmotic Potential Since the absolute value of osmotic potential (which is negative) is equivalent to the real pressure (positive) in a "perfect" osmometer in pure water at equilibrium, the osmotic potential of a solution can be measured directly. Many measurements of this property were made, particularly in 1877 by Wilhelm F. P. Pfeffer, who made nearly perfect, rigid, semipermeable membranes by soaking a porous clay cup in potassium ferrocyanide and then in cupric sulfate, precipitating cupric ferrocyanide in the pores. Columns of mercury were used to determine pressure. Van't Hoff used Pfeffer's data in

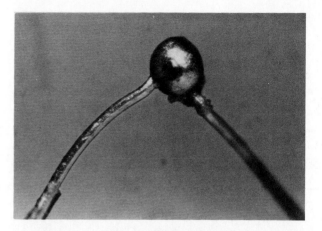

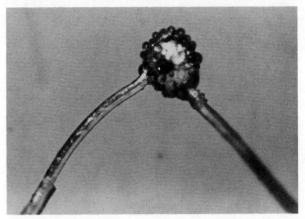

Figure 2-9 The thermocouple junction (spherical metal object—about 100 μm in diameter) in a psychrometer used to measure atmospheric humidity, and thus water potential. Top: the dry junction before the test. Bottom: the wet junction just after Peltier cooling. Drops of water have condensed from the air. (Photographs courtesy of Herman Wiebe.)

working out equation 2.2. With increased understanding of the properties of solutions, it became apparent that other, simpler measurements could be made and the data converted to osmotic potential. Excellent ways of doing this with free liquids have been developed, but no completely satisfactory method is yet available for measuring the osmotic potentials of the liquids in plant cells: attempts to measure it almost invariably change it.

The vapor-pressure method described earlier for the measurement of water potential applies equally well to the measurement of osmotic potential of a free liquid. This measurement is now performed routinely on human body fluids with an instrument similar to that shown in Figure 2-5c. But to use this method for plants, the pressure in the plant cells must first be reduced to zero so that only the osmotic component influences the vapor pressure. This is accomplished by rapid freezing in the laboratory, which produces ice crystals that rupture all the membranes; so pressure within the cells becomes zero, and $\psi = \pi$. The treatment results in mixing of cytoplasm, vacuolar sap, and water contained in the cell

---

*A thermocouple operates on the principle of the **thermoelectric effect.** When a circuit consists of two wires made of different metals (e.g., copper and the alloy constantan), and the two junctions between the two wires are at different temperatures, a current flows around the circuit. When the temperature of one junction is known and the current is measured, the temperature of the other junction (the thermocouple) can be calculated. The **Peltier effect** is the opposite of the thermoelectric effect. When a current is passed through a circuit of two different metals, the two junctions will have different temperatures. One will heat and one will cool, depending on the direction of the current.

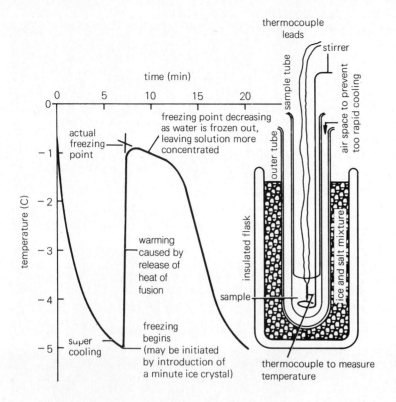

time (min)

actual freezing point

freezing point decreasing as water is frozen out, leaving solution more concentrated

warming caused by release of heat of fusion

freezing begins (may be initiated by introduction of a minute ice crystal)

super cooling

thermocouple leads

stirrer

sample tube

air space to prevent too rapid cooling

outer tube

insulated flask

ice and salt mixture

sample

thermocouple to measure temperature

Figure 2-10 Determination of the freezing point of a solution.

walls. This often leads to a change in osmotic potential, so values differ from those of an intact cell, but the method is nevertheless in frequent use.

The freezing point, as well as the vapor pressure, is a function of the mole fraction and hence of the osmotic potential. Properties of solutions that are functions of the mole fraction are called **colligative properties.** They include freezing point, boiling point, vapor pressure, and osmotic potential. Osmotic potential can be calculated from any of the first three values. Measuring freezing points to calculate osmotic potentials is called the **cryoscopic** or **freezing point method.** As mentioned, the osmotic potential of a molal nonionized solution at 0°C is ideally −2.27 MPa. Its freezing point (Δ_f) proves to be −1.86°C. The effect of osmotic potential on freezing points holds even for nonideal solutions such as plant saps, so the osmotic potential (osmolality) of an unknown dilute solution can be estimated from its freezing point by the following relationship:

$$\frac{\pi}{\Delta_f} = \frac{-2.27 \text{ MPa}}{-1.86°C} \qquad (2.6)$$

or

$$\pi(\text{in MPa}) = 1.22\Delta_f \qquad (\Delta_f \text{ in degrees C}) \qquad (2.7)$$

Determination of the freezing point proves to be relatively simple; highly accurate mercury thermometers (to 0.01°C) and thermocouples are com-

mercially available for this purpose, as are complete instruments for accurately determining the freezing points of solutions. A time–temperature curve is determined (Fig. 2-10) to account for supercooling and the subsequent warming upon freezing that is caused by release of the heat of fusion. Results are compared with those of pure water.

Obtaining a pure plant sap is far more difficult. One can squeeze out the sap with a press, freeze the tissue to rupture the cells and then squeeze out the sap, or homogenize the tissue in a blender and filter the sap. All these methods, applied to the same tissue, typically give different values of π; and the difference may be as high as 50 percent. Blender values are usually the most negative (most concentrated) and hand-squeezing sap through cheesecloth, the least negative. A principal problem is that the various methods involve different degrees of mixing of cytoplasmic contents, cell-wall water, and other substances with the vacuolar sap, which is usually much larger in volume than the cytoplasm and is thus primarily responsible for the osmotic behavior of plant tissues. (The cytoplasm and vacuole are in osmotic equilibrium, because there is no rigid barrier between them.) The problem with pressing living cells is that almost pure water is obtained because of osmotic filtration. In spite of these limitations, the cryoscopic method has been widely used for many decades.

If the sap in any plant tissue is in osmotic equilibrium with some outside surrounding solution ($\Delta\psi$

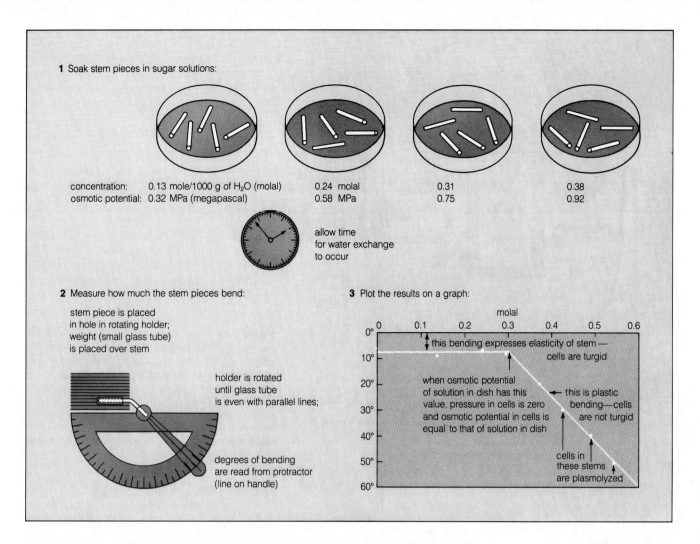

1 Soak stem pieces in sugar solutions:

concentration: 0.13 mole/1000 g of H$_2$O (molal) 0.24 molal 0.31 0.38
osmotic potential: 0.32 MPa (megapascal) 0.58 MPa 0.75 0.92

allow time
for water exchange
to occur

2 Measure how much the stem pieces bend:

stem piece is placed
in hole in rotating holder;
weight (small glass tube)
is placed over stem

holder is rotated
until glass tube
is even with parallel lines;

degrees of bending
are read from protractor
(line on handle)

3 Plot the results on a graph:

molal

this bending expresses elasticity of stem —
cells are turgid

when osmotic potential
of solution in dish has this
value, pressure in cells is zero
and osmotic potential in cells is
equal to that of solution in dish

this is plastic
bending—cells
are not turgid

cells in
these stems
are plasmolyzed

Figure 2-11 How to detect when cells are just beginning to plasmolyze without actually looking at them. At the moment when the turgor pressure in the cells reaches zero, the tissues (stem, in this case) become much more flexible and much less elastic. (Jensen and Salisbury, *Botany*, 1984, p. 66; see Lockhart, 1959.)

= 0) and *no pressure or tension existed within the tissue*, then the osmotic potential of the sap would be equal to the osmotic potential of the surrounding solution. The problem with such a measurement is to obtain zero pressure within the tissue without changing the other osmotic properties any more than necessary (which may be too much). This is the method of measuring osmotic potential by observing **incipient plasmolysis.** Samples of a tissue are placed in a graded-solution series of known osmotic potentials. (Sucrose or mannitol solutions can be used, but substances such as PEG have the advantage of not penetrating the tissue or being changed metabolically by the tissue as easily.) After an equilibration period (usually 30 min to 1 h), the tissue is examined under a microscope. It has been assumed by plant physiologists that incipient plasmolysis occurs in tissue in which about half the cells are just beginning to **plasmolyze** (protoplasts are just beginning to pull away

from the cell wall), and that this represents an internal pressure of zero. If this assumption is true, then the osmotic potential of the solution producing incipient plasmolysis is equivalent to the osmotic potential of the cells within the tissue, *after they have come to equilibrium with the solution.*

If this is true for the tissue at equilibrium (and its being true depends entirely on the truth of the assumption that incipient plasmolysis represents zero pressure), then we must ask how the tissue has changed as incipient plasmolysis developed. The pulling away of the protoplasts from the wall is caused by shrinkage or decrease in volume, so the sap solution inside the protoplasts has become more concentrated and therefore has developed a more negative osmotic potential. (Remember the Höfler diagram, Fig. 2-3.) If careful volume measurements of the original tissue and the tissue at incipient plasmolysis are made (either the overall volume of the

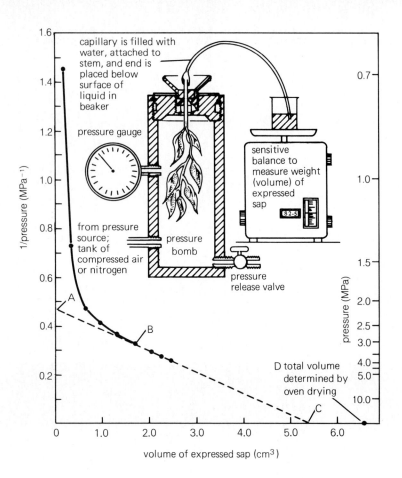

Figure 2-12 Illustrating the pressure bomb as a method for measuring several parameters concerning the water relations of plants. Point A: the water potential of hydrated tissue, about 2.1 MPa. Point B: incipient plasmolysis, indicating an osmotic potential of about 3.1 MPa. Point C: volume of free water in the tissue, about 5.35 cm³. Point D: total volume of tissue water, about 6.72 cm³. Bound water equals D minus C, or 1.37 cm³ (see Tyree and Hammel, 1972, and Tyree et al., 1973).

tissue or, better, the dimensions of a fairly large sample of protoplasts), then the change in osmotic potential caused by change in volume can be calculated. When this correction is not made, the values of osmotic potential obtained by the plasmolytic method are too negative, often by a value of 0.1 or more MPa (5 to 10 percent or more).

Figure 2-11 illustrates another way in which incipient plasmolysis in a tissue such as a stem can be estimated. Stems are placed in a graded series of solutions as in the method just discussed, and after some time has elapsed for water to be exchanged, the elasticity and plasticity of the tissue are measured in the simple manner illustrated in the figure. At the onset of incipient plasmolysis, the plastic (irreversible) bending of the stem becomes noticeable; and it increases as more and more water is lost from the cells in solutions of increasingly negative osmotic potential. The method nicely illustrates a point that has already been made several times: The rigidity and form of young tissues that consist of cells with non-lignified primary walls depend completely upon the pressures in the cells produced by osmotic water uptake (see also Fig. 1-2).

During recent years, another somewhat sophisti-cated approach has been coming into wide use. This method uses a **pressure bomb** and is capable of providing several kinds of data relating to the water status of a plant. To measure its osmotic potential, a leaf or branch is removed from a plant and hydrated. This can be done by placing the cut end in pure water for several hours or overnight, covered with a plastic bag to ensure 100 percent RH. Studies indicate that osmotic potential changes only slightly during this time. Another way is to force pure water into the specimen under pressure to effect rapid hydration. The hydrated branch is placed in the pressure bomb (Fig. 2-12) with the cut end protruding. Pressure is applied, and sap begins to exude from the cut end. This sap is almost pure water, because it is exuded by reverse osmosis (pressures in the bomb are increased to higher positive values than osmotic potentials are negative, so water diffuses out of the cells).

The reciprocal of the balancing pressure ($1/P$) is plotted as a function of the reciprocal of the volume of exuded sap ($1/V$) to obtain the characteristic curve, as shown in Fig. 2-12. With increasing pressure, the result is at first a curved line; but later (at point B) it becomes linear. The cells at this point are at incipient plasmolysis, and the negative value of the pressure in the bomb is equal to the osmotic potential in the cells (see papers of Hellkvist et al., 1973; Tyree and Dainty, 1973; and Tyree and Hammel, 1972). Extrapolating the line back to an expressed volume of zero (point A) gives the reciprocal of the osmotic potential of the cells of the original hydrated tissue, and a Höfler diagram can be constructed to calculate the volume of water in the tissue at that pressure or any other. Extrapolating the line ahead to infinite pressure (point C) gives the reciprocal of the volume of liquid originally present in the cells. The water that remains in the sample after the highest pressure has been applied can be measured by weighing it before and after drying in an oven (point D). The excess over the amount calculated at point C is assumed to be **bound water** that is held with great force to hydrophilic surfaces.

Measurements of osmotic potentials by these methods (especially the use of freezing points) have yielded results varying from values of about -0.1 MPa in aquatic plants to extremely negative values, -20 MPa or lower, in **halophytes** (plants that grow in salty soil). These values are subject to all the problems just described, plus others. The extreme halophyte values, for example, are made far too negative by salt crystals on the leaf surfaces. The true osmotic potentials of halophyte cells are probably -5 to -8 MPa, and the osmotic potentials of most other plant saps lie between -0.4 and -2.0 MPa (Table 2-1). As might be expected, values of osmotic potential vary within a single growing plant: Those of the youngest leaves are most negative, those of root cells change in response to drying soils, and those of many other plant cells also change in response to changing environmental conditions (e.g., salt or water stress). This process of **osmoregulation** in response to stress is discussed in Chapter 24.

Pressure In a laboratory osmometer, pressure is measured directly; but direct measurement of pressure in plant cells is difficult. Usually, the pressure is calculated after water potential and osmotic potential have been determined:

$$P = \psi - \pi \qquad (2.8)$$

Paul B. Green and Frederick W. Stanton (1967) described a method for the direct measurement of **turgor pressure** (cell pressure, equal to the pressure of water in the cell) in large cells, such as those of the alga *Nitella axillaris*. The method has since been greatly elaborated with sophisticated equipment and applied to higher plant cells (Hüsken et al., 1978).

Green and Stanton made a minute manometer by fusing closed one end of a capillary tube (diameter 40 μm) and fashioning the other end into a tip like

Table 2-1 Some Examples of Empirically Determined Osmotic Potentials of Leaves.

Species	Osmotic Potential ψ_π (MPa)
Shadscale (*Atriplex confertifolia*)[a]	-2.4 to -20.5
(The most negative values are highly questionable.)	
Pickleweed (*Allenrolfea occidentalis*)[b]	-8.9
Sagebrush (*Artemisia tridentata*)[a]	-1.4 to -7.4
Salicornia (*Salicornia rubra*)[a]	-3.2 to -7.3
Blue spruce (*Picea pungens*)[b]	-5.2
Mandarin orange (*Citrus reticulata*)[b]	-4.8
Willow (*Salix babylonica*)[b]	-3.6
Cottonwood (*Populus deltoides*)[c]	-2.1
White oak (*Quercus alba*)[c]	-2.0
Sunflower (*Helianthus annuus*)[c]	-1.9
Red maple (*Acer rubrum*)[c]	-1.7
Waterlily (*Nymphaea odorata*)[c]	-1.5
Bluegrass (*Poa pratensis*)[c]	-1.4
Dandelion (*Taraxacum officinale*)[c]	-1.4
Cocklebur (*Xanthium spp.*)[c]	-1.2
Chickweed (*Stellaria media*)[c]	-0.74
Wandering Jew (*Zebrina pendula*)[c]	-0.49
White pine (*Pinus monticola*)[d], dry site, August, exposed to sun	-2.5
White pine (*Pinus monticola*)[d], moist site, April, more shaded	-2.0
Big sage (*Artemisia tridentata*)[d], March to June, 1973	-1.4 to -2.3
Big sage (*Artemisia tridentata*)[d], July to August, 1973	-3.8 to -5.9
Herbs of moist forests[e]	-0.6 to -1.4
Herbs of dry forests[e]	-1.1 to -3.0
Deciduous trees and shrubs[e]	-1.4 to -2.5
Evergreen conifers and Ericaceous plants[e]	-1.6 to -3.1
Herbs of the alpine zone[e]	-0.7 to -1.7

[a]Harris, 1934, pp. 65, 70, 110. Harris made thousands of measurements by observing freezing point depressions of expressed sap. The ranges shown indicate the variability he encountered within a single species, in a single state (i.e., Utah), and in a single year. Obviously, little certainty can be attached to the specific values shown for individual species. Presence of salt crystals on leaf surfaces must account for the extremely negative values in some cases.

[b]Student reports, Plant Physiology class, Colorado State University, Fort Collins.

[c]Meyer and Anderson, 1939.

[d]Cline and Campbell, 1976; Campbell and Harris, 1975.

[e]Pisek, 1956.

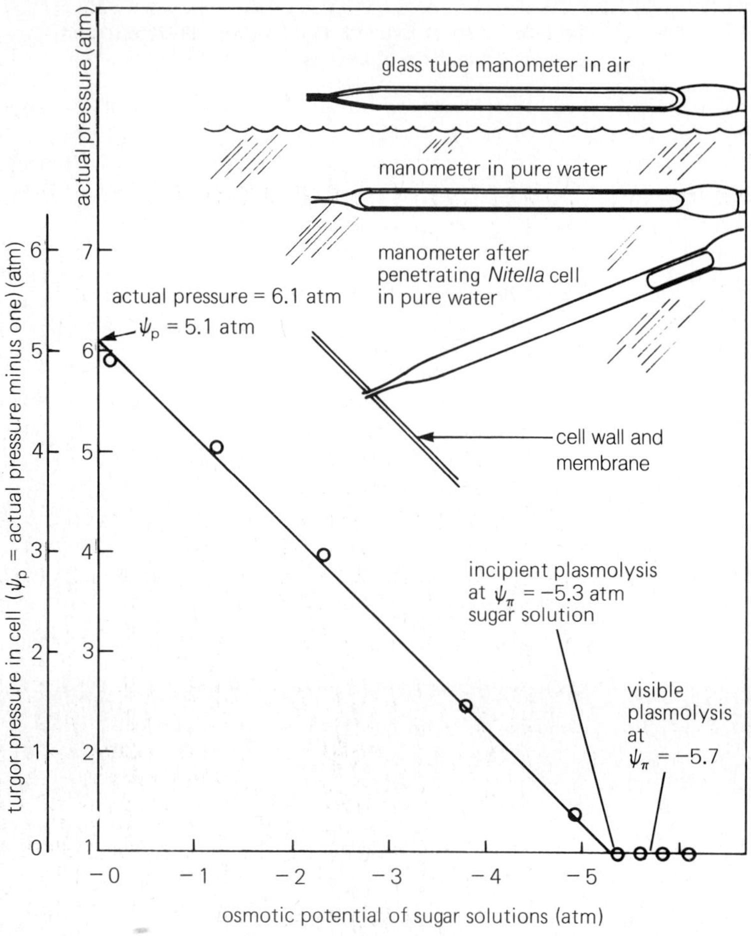

glass tube manometer in air
manometer in pure water
manometer after penetrating *Nitella* cell in pure water
actual pressure = 6.1 atm
ψ_p = 5.1 atm
cell wall and membrane
incipient plasmolysis at ψ_π = −5.3 atm sugar solution
visible plasmolysis at ψ_π = −5.7
actual pressure (atm)
turgor pressure in cell (ψ_p = actual pressure minus one) (atm)
osmotic potential of sugar solutions (atm)

Figure 2-13 Green and Stanton's experiment. After equilibration in sugar solutions of various osmotic potentials, *Nitella* cells were penetrated by capillary tubes (closed at one end) as illustrated. In the original experiment, atmospheres were the unit of measurement; they are retained here. (1 atm = 0.1013 MPa) (Data from Green and Stanton, 1967.)

that of a syringe needle. If such a tube, with its open end placed in water, is observed with a microscope, water is seen to enter the tube by capillarity, somewhat compressing the air inside the tube. The position of the meniscus inside the tube is noted and the volume of air calculated. When the open end of the tube punctures a cell, pressure in the cell is transferred to the air in the tube, compressing it further, to an extent indicated by the movement of the meniscus (Fig. 2-13). The final pressure in the tube is always equal to its pressure before penetration of the cell, multiplied by the ratio of the original to the final volume (according to Boyle's law). The pressure in the cell before penetration can be approximated by multiplying atmospheric pressure by the ratio of the original volume in the tube to the volume after entrance of water by capillarity. There will be a slight change in pressure within the cell upon penetration by the tube, but even this can be determined by penetrating the cell with a second tube while observing the change in pressure in the first. The method measures actual pressure in the cell, but according to convention this will be close to 1 atm (about 0.1 MPa) higher than the turgor pressure.

Matric Potential Hydrophilic surfaces (e.g., of such colloids as protein, starch, and clay; see box essay) adsorb water, and the tenacity with which the molecules of water are adsorbed is not only a function of the nature of the surface but also of the distance between the surface and the adsorbed water molecules. Those located directly on the adsorbing surface will be held extremely tightly; those at some distance from the surface will be held less tightly. The adsorption of water by hydrophilic surfaces is called **hydration,** or **imbibition.**

The **matric potential** (τ) is a measure at atmospheric pressure of the tendency for a matrix to adsorb additional water molecules. This tendency is equal to the average tenacity with which the least tightly held (most distant) layer of water molecules is adsorbed. Matric potential is expressed in units of water potential. A dry colloid or hydrophilic surface, such as filter paper, wood, soil, gelatin, or the stipe of a brown alga, often has an extremely negative matric potential (as low as −300 MPa), while the same colloid in a large volume of pure water at atmospheric pressure has a matric potential of zero (since it is saturated and therefore in equilibrium with the

Colloids: Characteristic Components of Protoplasm

Protoplasm is unique not only because it consists of highly complex and special molecules but also because of its physical nature. At first glance this physical nature seems to be just a kind of high viscosity that makes protoplasm a bit like gelatin pudding or sometimes like glue. But the physical nature of protoplasm is determined by vast areas of interface between some of those special molecules, especially the large ones called enzymes, and the protoplasmic solutions in which they are suspended. The reactions of life are catalyzed at these interfaces, and much else that is going on in cells is influenced by the microscopic surfaces in protoplasm. Soils are also characterized by the huge interfaces between clay and humus particles and their surroundings. Technology takes advantage of such systems in its water softeners, catalytic converters, and numerous other applications.

The special physical nature of protoplasm is caused by membranes and by particles that are too small to settle out by gravity but are larger than the atoms, small molecules, and ions that are true solute particles. When these larger particles are suspended in water, they sometimes form a glue; so they have been termed **colloids,** from the Greek word *kolla,* or glue.

Why do colloids not settle out? Because they are surrounded by much smaller water molecules in rapid motion, as discussed in Chapter 1. The result of constant collisions is that a colloidal particle is never at rest. It is small enough that the random velocities of the impacting water molecules do not average out: At any given moment there is a high probability that such a particle is bombarded more strongly on one side than on the opposite side. When colloidal particles are observed in a light microscope by strong illumination from one side, they appear as points of light (the **Tyndall effect,** first noticed by John Tyndall, 1820–1893). They seem to dance around with many random hops (impacts) per second. The largest (brightest) particles dance less than the smaller (dimmer) particles.

This is **Brownian movement,** discovered by the Scottish botanist Robert Brown in 1827. It is a beautiful, even spectacular, confirmation of kinetic theory. It is this erratic and continuous motion, caused by unequal kinetic bombardment of different sides of colloidal particles by water molecules, that keeps colloids from settling. Indeed, we might define a colloidal particle (not a true solute) as one that is small enough to remain in suspension because of its

Brownian movement. With slightly larger particles, there is a much greater chance that the random bombardment of water molecules on any side of the particle will approach an average value for the system as a whole. Thus larger particles are less affected by the random bombardments of the molecules that make up their milieu, so gravity wins and they settle out.

The upper size limit (diameter) for particles exhibiting Brownian movement is about 100 to 2000 nm, depending on shapes and densities. Since light waves in the visible part of the spectrum are 385 to 776 nm long (Appendix B), only the larger colloidal particles are large enough to cast shadows. The smaller ones refract light waves causing the Tyndall effect, though they are not themselves actually visible in a light microscope. The electron microscope, which operates with electron beams having wavelengths of less than 0.1 nm, easily resolves most colloidal particles, the smallest of which have diameters of about 10 nm. (Particles in solution that are smaller than colloids are true solutes, but the distinction is not precise). Many of the particles in a cell, including the ribosomes and all the single protein molecules that are enzymes, are in the colloidal size range.

Most colloidal particles pass through filter paper, but they cannot pass through cellophane as can true solute particles. Suspension particles are too large to pass through filter paper.

Although colloidal particles are very small, each is large enough to present a surface (a layer of atoms) to the surrounding water molecules and solute particles. Another consequence of the small size of colloidal particles is that the total amount of surface presented by the particles in a given volume is relatively huge. Imagine a solid cube of material 1 cm on each of its edges. There are six faces, so it has a surface area of 6 cm^2. Cut it once. You expose 2 cm^2 more of surface. Continue slicing the cube until you have reduced each of its parts to a cube 10 nm long on its edges. Now the total surface area is 6,000,000 cm^2 (600 m^2). A single cube with the same surface area would be 10 m high with a volume of 1000 m^3! Protein molecules are not cubes, but they are of comparable size.

The reactions of life occur on surfaces, and it is easy to see how relatively large surface areas can exist in a single cell. It is also easy to see how hydration, matric forces, can influence the water milieu of cells and soils.

water). In general, when any colloid at atmospheric pressure is in equilibrium with its surroundings, the least tightly held water molecules have the same free energy as the water molecules in the surroundings, so the matric potential of the colloid is equal to the water potential of the surroundings.

In modern discussions, the radius of curvature of the water surfaces between colloidal particles (the **meniscus;** plural, menisci) is often mentioned. The meniscus is the basis of the phenomenon of capillarity, discussed in Chapter 4 (see Fig. 4-4). The smaller the radius of curvature of the meniscus, the

Figure 2-14 A pressure-plate (or pressure-membrane) apparatus used to measure matric potentials of soils and other materials. The wet samples are placed in the circular holders on the pressure plate, a porous plate that allows diffusion of water. After the top has been secured in place with wing nuts and against a rubber O-ring, pressure is introduced and water begins to diffuse through the plate and out through the small tube into the beaker (center of photograph). When water has stopped coming out the tube (often after 24 h), the matric potential (a negative value) of the sample is numerically equal to the pressure (a positive value) in the apparatus, as read on the gauges at the bottom of the picture. A higher pressure might then be introduced to expel more water, producing a moisture-release curve such as that of Figure 2-15. (Pressure plate in the laboratory of Ray W. Brown, Forestry Sciences Laboratory, Logan, Utah; photograph by Frank B. Salisbury.)

tighter the water is held to the colloidal or other hydrophilic surface by hydration and the more negative the matric potential.

A commonly used modern way of measuring matric potential is instructive. A hydrated colloid is enclosed in a pressure chamber by a membrane filter that is supported by a screen to withstand high pressure (Fig. 2-14). The pores of the membrane are 2 to 5 nm in diameter, large enough to allow passage of solutes and water but not the colloid. Surface tension prevents the passage of air through the wet membrane. Assume (the simplest case) that the colloid is wet with pure water (no solutes). Compressed gas is introduced into the pressure filter. Increasing the pressure raises the water potential of the water that is adsorbed on the colloid toward, and finally above, zero; so the water molecules begin to diffuse out through the membrane. (You can also think of the air as pressing on the menisci and making their radii of curvature smaller.) Further increases in pressure on the colloid result in additional but smaller increments of water movement from the colloid out through the membrane.

When all water movement stops, the water on the colloid under pressure will be in equilibrium, through the membrane, with the pure water that is at atmospheric pressure and the same temperature on the outside of the membrane. This means that the water potential of the least tightly adsorbed water

molecules on the colloid under pressure is equal to zero. If pressure and the matric potential are the only components that are influencing water potential on the colloid side of the membrane, as indicated by

$$\psi = P + \tau \qquad (2.9)$$

then when water potential is zero, the absolute numerical value of the negative matric potential will be equal to the positive pressure potential that is produced by the compressed gas ($-\tau = P$). Since the pressure is known and positive, the matric potential is also known, but negative.

Usually, after equilibrium has been reached at a given pressure, the water content of the colloid is determined by weighing before and after drying in an oven. The matric potential of the colloid, plotted as a function of its water content, provides the **moisture release curve** (Fig. 2-15).

A test of the assumption that the final pressure in the pressurized membrane apparatus is a measurement of matric potential is to measure the water potential of the colloid at atmospheric pressure by the vapor method described in Section 2.6. The two measurements agree closely, indicating the validity of the approach.

It is evident from this definition of matric potential that, at equilibrium, the matric potentials of all hydrophilic surfaces in a system are equal. Thus the

Pursuing the Questions of Soil-Plant-Atmosphere Water Relations

F.B.S. 1971

Ralph O. Slatyer

Ralph O. Slatyer was educated in Australia in the 1950s. Since 1967, he has been a Professor of Biology at the Australian National University in Canberra City. His essay well illustrates the international nature of plant physiology; it also reinforces and expands several topics that we have been considering.

To me, the most exciting and stimulating part of scientific research is when you make an observation or generate a hypothesis that you think is original and may, in addition, be at variance with accepted phenomena or attitudes.

Of course, in most cases, more careful observation, more thorough reading of the literature, or a critical discussion with colleagues leads to an awareness either that your ideas or observations won't hold up, or that someone else has anticipated you. But occasionally there is the breakthrough, and scientific knowledge and understanding moves forward.

In my own case, the first such heady moments came in the mid-1950s when I was investigating the effects of progressive and prolonged water stress on physiological plant responses. At that time, it was widely accepted that the permanent wilting percentage was a soil constant, the soil water content below which no plant growth would occur and further transpiration would cease. This concept had strong empirical support, mainly through painstaking work by Veihmeyer and Hendrickson at the University of California, Davis, on irrigated fruit crops, but dated back to previous works of Briggs and Shantz in the early part of the century. Associated with the concept was the notion that soil water was equally freely available to the plant at water contents down to the permanent wilting percentage, although this had come under challenge, particularly from scientists at the United States Salinity Laboratory at Riverside, California.

My initial research work, associated with my master's and doctoral studies, was concerned with crop and native species responses in arid and semiarid regions, and I repeatedly found evidence that consistently contradicted the established dogma. Accordingly, I set off to spend what was essentially a brief postdoctoral period with Professor Paul Kramer at Duke University. I found him receptive to my ideas and full of encouragement.

Basically, all I had done was to propose that as water stress was progressively imposed by soil water depletion, the turgor pressure of the leaf cells decreased until it reached zero, when the leaf water potential equaled the osmotic potential. I argued that at this point the leaves would be permanently wilted and that growth might be reasonably expected to have ceased. Even with stomatal closure, however, continued transpiration would be expected to continue to deplete soil water until desiccation of the plant itself reached lethal levels. From this argument it followed that the permanent wilting percentage should not be a soil constant, but merely the soil water content at which the soil water potential and the plant water potential were balanced, at a level equal to the leaf cell osmotic potential, so that zero turgor pressure existed.

Somewhat surprisingly, these views, published in both experimental and review papers, were rapidly accepted by the scientific community and have since been reconfirmed in general aspects by numerous investigators. In the process, of course, some of the more specific assertions have required qualification, but overall the dynamic nature of the soil-plant-water interaction and the permanent wilting percentage was clearly established.

This approach also seemed to provide a better basis for interaction between plant and soil scientists interested in plant-environment interactions, and it led to a requirement for a more integrated term to describe the state of water in plants and soils. Through the 1950s, "diffusion pressure deficit," "total soil moisture stress," and related terms were being used by these two groups of scientists who were basically talking about the same thing. This matter finally came to a head, informally, over dinner in a restaurant in Madrid during a UNESCO Plant Water Relations conference, attended by, among others, Sterling Taylor, Wilford Gardner, Robert Hagan, Fred Milthorpe, and myself. We proposed the term "water potential" (already suggested many years earlier by soil scientists), based thermodynamically on the chemical potential of water, as a single term for both soil and plant scientists, to be divided into component potentials, as appropriate. The meeting asked Sterling Taylor and me to draft a letter to *Nature* and a more definitive paper on the subject, and from this rather informal and personal beginning, the new terminology was launched. As far as I know, it is now used almost universally, although it too has been improved by modification and qualification.

I began this brief essay by referring to the excitement of scientific discovery. I conclude by referring to the spirit of cooperation that has existed in that part of the scientific community with which I have been associated. While it has always been a challenge to be first with a new piece of work, my life has been enriched by the warm personal relationships that have been developed between both my immediate colleagues and those further afield, whom one first meets by correspondence, or at conferences, and then by sharing space and facilities in a common laboratory.

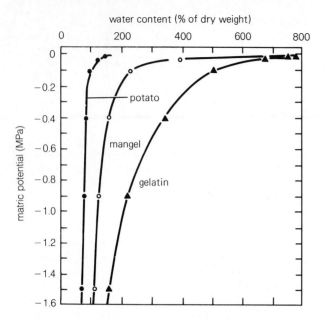

water content (% of dry weight)

Figure 2-15 Moisture release curves for two plant materials and gelatin. (From Wiebe, 1966.)

water potential of the layer of least tightly adsorbed molecules on any surface at equilibrium is equal to the water potential of the system as a whole. If it is more negative, water is adsorbed; if less negative, water is lost.

The term *matric potential* has been used in a different sense, however. In a complex system, such as a cell that contains molecules of colloidal protein and other hydrophilic surfaces, as well as simple solute ions and molecules, the final water potential will be determined not only by the solute particles and the pressure but also by the matric effects of the proteins and other surfaces.

The complexities of natural systems are not well understood, but we can imagine a number of them. One is that some of the polar solutes, especially ions, are adsorbed on the hydrophilic surfaces, influencing their water-adsorbing properties. Another is that the adsorption of water molecules on these surfaces decreases the quantity of water in the part of the system not influenced by the matrix, so that the solution in this part is more concentrated than it otherwise would be. It is possible that no water molecule in the system will be far enough away from a hydrating surface (e.g., a protein molecule) to be completely unaffected by such surfaces. This is unlikely in the cytosol, as it turns out (Passioura, 1980, 1982); but it is often true of soil water with its dissolved solutes or of water in the cell wall. At a certain level of dryness, virtually no water is beyond the hydrating influence of the clay particles in the soil or the polysaccharides in the cell wall, as the case may be.

Some modern authors (for example, Passioura,

1980, 1982; Tyree and Garvis, 1982) suggest that matric effects can be accounted for solely in terms of solute and pressure effects. Consider what happens at the molecular level when a hydrophilic surface adsorbs water. In the simplest case, the adsorption may be only a matter of London–van der Waals forces or hydrogen bonds, which probably extend for only one or two diameters of a water molecule (0.3 to 0.6 nm). But many hydrophilic surfaces in nature, especially those on clays, proteins, and the microstructures of cell walls, have a large unbalanced negative charge. The positive sides of water molecules are strongly attracted to such negatively charged surfaces, and several layers of water molecules can build up on such a surface (Fig. 2-16). These forces are great enough to produce pressures of tens to hundreds of megapascals in the molecules closest to the surfaces. This effect is much like the pressure produced by gravity in deep water, except that the pressure gradient occurs over nanometers instead of hundreds of meters. This all makes good sense in terms of solute and pressure effects, because the water molecules that are attracted to a hydrophilic surface lose much of their free energy. That is, extending our thermodynamic discussion to the microscale (which may not be completely valid), we would say that the matric forces, acting like solutes, would produce an extremely negative water potential close to the adsorbing surfaces if this effect were not balanced by the very high positive pressures that exist there. The water potential throughout the layer of adsorbed water molecules is thus constant, as it must be at equilibrium.

If solutes, especially cations, are present, they are strongly attracted to the hydrophilic surfaces, and their concentration there is extremely high (calculated to be at least 2 *M*). This produces a negative osmotic potential, which forms a gradient from extremely negative close to the charged surface to much less negative some distance away. Again, the water molecules strongly diffuse toward the charged surface where solute concentrations are so high, producing a gradient in pressure opposite to the gradient in osmotic potential and keeping the water potential constant throughout (Fig 2-16).

Clearly, the effects of hydrophilic surfaces on water molecules play an important role in plants. Membranes, proteins, ribosomes, and several components of the cell wall can become hydrated and thus are part of the matric component that combines with the osmotic component and pressure to determine water potential. There are two important things to remember, however: First, *at equivalent weights*, small inorganic ions reduce the free energy (and thus the water potential) of an amount of water far more (hundreds to millions of times) than a hydrophilic substance such as protein. Thus, in the cytosol, the

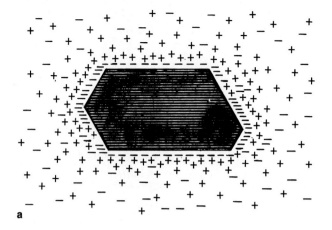

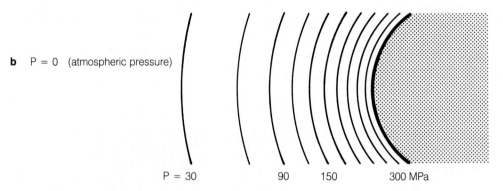

b   P = 0   (atmospheric pressure)

P = 30   90   150   300 MPa

Figure 2-16 (**a**) The electrical double layer of ionic distribution around a charged colloidal particle. Charges next to the surface of the particle are meant to indicate surface charges. These are mostly negative, but the negative charge is interrupted at three points with positive charges. Distribution of ionic charge is statistically opposite to the distribution of surface charge, but a few irregularities are apparent. (**b**) Pressure increase in water as it approaches a hydrated, charged surface. The lines represent isobars of pressure increasing by 30-MPa intervals as the surface is approached. The pressure increase is caused by attraction between the charged surface and water molecules (by hydrogen bonding) and by the steeply increasing solute concentration, which is caused by mutual attraction of dissolved solutes (ions) and the charged surface.

contribution of the matric component to final water potential is probably small and could be negligible compared to that of solute ions and molecules. Yet the matric component is quite significant in the cell wall. Second, at equilibrium (and, in this respect, cells must always be close to equilibrium) the matric potential of matric surfaces in cells will be equal to the water potential of the system as a whole, so it is not valid to measure pressure, osmotic potential, and matric potential of matric surfaces and algebraically sum the three values to obtain the water potential. At equilibrium, the pressure, osmotic, and matric components will determine water potential in some complex way, and then the water potential will determine the matric potential of hydrophilic surfaces (the free energy of the water molecules attached to such surfaces).

In dry seeds and in the soil, matric effects are especially important. Water uptake into seeds is at first largely a function of matric effects; the surfaces of proteins and some cell-wall polysaccharides must become hydrated before germination can begin. Dry wood is also hydrophilic, and for centuries rocks have been split by a technique that makes use of this property: Holes are drilled along a desired fracture line in the rock, dry wooden pegs are driven into the holes, and water is run over the pegs, which then hydrate, expand, and crack the rock. Growing fungi, and perhaps roots and other systems, also develop powerful expansion force that may depend on hydration.

Perhaps someday our understanding will advance to the point at which we can calculate the relative contributions of solutes (osmotic potential), pressure, and adsorption of water on hydrophilic surfaces (matric effects) to the true water status of plants. In the meantime, it is important to realize that all three factors play a role and that it is not valid (because of the complications just discussed) to consider matric potential as a component of water potential in the same sense that osmotic potential and pressure are thought of as components.

3

The
Photosynthesis–Transpiration
Compromise

During the summer of 1974, John Hanks, a soil scientist at Utah State University, kept careful track of the amount of water required to grow a crop of maize (corn) on the college Greenville farm. To mature the crop, water equivalent to 600 mm (23.6 in.) of rain was added to the field. A fourth of this evaporated from the soil, but most of the remaining 450 mm passed through the plants into the atmosphere. This evaporation of water from plants (and animals, according to most dictionaries) is called **transpiration.** In plants, it normally refers to internal water lost through stomates, cuticle, or lenticels. Continuing the calculations, Hanks showed that 600 kg of water were transpired by the maize plants for each 1 kg of dry maize (grain) produced. As a matter of fact, 225 kg of water were transpired to produce 1 kg of dried plant material, including leaves, stems, roots, cobs, and seeds. This large amount of transpired water is typical, although there are substantial differences among species. (For reviews see Hanks, 1982, 1983.)

Why is so much water lost by transpiration to mature a crop? Because an essential part of those dry corn seeds and all other plant parts are the carbon atoms that form the skeletons of the organic molecules of which those plant parts consist, and virtually all of this carbon must come from the atmosphere.

It enters the plant as carbon dioxide (CO_2) through the stomatal pores, mostly on leaf surfaces, and water exits by diffusion through these same pores as long as they are open. This is the dilemma faced by the plant: How to get as much CO_2 as possible from an atmosphere in which it is extremely dilute (about 0.03 percent by volume), and at the same time retain as much water as possible, water that must fill and keep turgid all the cells and provide the medium in which the CO_2 can be photosynthetically combined with water to form the molecules of life. The agriculturalist faces a similar challenge: How to achieve a maximum crop yield with a mini-

mum of irrigation water, an important natural resource (Sinclair et al., 1984).

Understanding the environmental factors and how they influence transpiration from and CO_2 absorption into a leaf in the field at various times turns out to be a difficult assignment, because the factors interact in so many ways. Environmental factors influence not only the physical processes of evaporation and diffusion but also the stomates on the leaf's surfaces and their apertures, through which over 90 percent of the transpired water and the CO_2 pass. Increased leaf temperature, for example, promotes evaporation considerably and diffusion slightly but may cause the stomates to close, or to open wider, depending on species. At dawn, stomates open in response to the increasing light, and the light increases the temperature of the leaf. The increasing temperature allows the air to hold more moisture, so evaporation is promoted, and perhaps stomatal aperture is affected. Wind brings more CO_2 and blows away the water vapor, causing an increase in evaporation and CO_2 uptake. But if the leaf is warmed above air temperature by sunlight, wind will lower its temperature, causing a decrease in transpiration. When soil moisture becomes limiting, transpiration and CO_2 uptake are inhibited because stomates close. So if one were to visit the Greenville farm hoping to learn about the extent of transpiration or CO_2 uptake at any odd moment, one would have to be armed with a formidable array of environmental measuring devices, not to mention a pocket calculator.

3.1 Measurement of Transpiration

In this chapter, we shall be more concerned with transpiration than with CO_2 uptake. It is perhaps more challenging, since CO_2 concentration in the atmosphere is fairly constant, while transpiration studies must consider evaporation as well as diffusion.

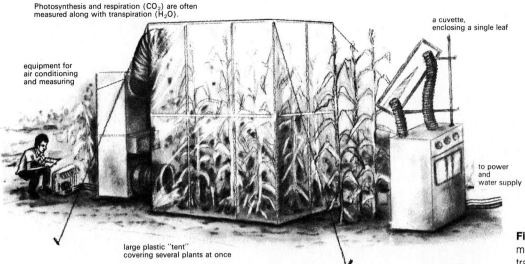

Photosynthesis and respiration (CO_2) are often measured along with transpiration (H_2O).

equipment for air conditioning and measuring

a cuvette, enclosing a single leaf

to power and water supply

large plastic "tent" covering several plants at once

Figure 3-1 Tent and cuvette methods for measuring transpiration in the field.

Anyway, understanding transpiration provides the foundation for understanding CO_2 uptake, but note that water molecules diffuse 1.56 times as fast as CO_2 molecules (because of their lower molecular weight) and that the atmosphere normally contains much more water vapor than CO_2 (at 35°C, H_2O = 0.3 to 1.4 mol m^{-3}; CO_2 = 0.0156 mol m^{-3}).

How could we measure transpiration at the Greenville farm to see if our developing models and calculations were valid? In one elaborate method (Fig. 3-1), a tent of transparent plastic is placed over a number of plants, and humidities, temperatures, and carbon dioxide levels are measured for air entering and leaving the tent. From these data, the amount of transpiration and photosynthesis can be calculated. But the problem is obvious: How can we be sure that the environment surrounding the plants is not influenced by the tent? With elaborate instrumentation, it is possible to control temperatures, humidities, and gas concentrations within the tent, but this is no simple assignment for a basic plant physiology student laboratory. Tens of thousands of dollars or equivalent must be invested in such equipment.

A much simpler approach is to weigh at intervals on a sensitive balance a potted plant with its soil sealed against water loss. Stephen Hales did this 200 years ago; it is known as the **lysimeter method.** Since the amount of water used in plant growth is less than 1 percent of the final dry weight of the plant (note Hanks' figures), virtually all the change in weight can be ascribed to transpiration. The problem here is being sure that the potted plant is really representative of other plants in the field, after being moved from its location in the field to the balance and back, and with its roots confined.

Hanks and others have considerably expanded this simple approach to arrive at trustworthy figures. His lysimeters on the Greenville farm are large con-

tainers (several cubic meters) full of soil and buried so their top surfaces are level with the field's surface. The container is placed on a large plastic bag buried beneath it and filled with fluid (water and antifreeze) that extends into a standpipe above the surface (Fig. 3-2). The level of liquid in the pipe is a measure of the weight of the lysimeter, so it changes with the water content of the soil in the lysimeter and the growing plants, but their weight is small compared to the soil. Water in the soil depends only upon evaporation from the soil surface and transpiration from plants. Evaporation from soil and transpiration combined are called **evapotranspiration.** Lysimeters provide the most reliable field method for studying evapotranspiration, although they are obviously expensive and involved and cannot be moved around on the spur of the moment. Though not universally available, lysimeters are widely used.

A more convenient method is to remove a leaf from a plant in the field and put it immediately on a sensitive balance. The loss of weight during the first minute or two has often been used as an indication of transpiration. Since the very act of detaching the leaf from the plant often changes the rate of transpiration, this method has only limited usefulness in comparative studies.

The potted-plant method works nicely in the laboratory. Another laboratory approach is to detach a plant and insert the cut stem into a device that allows easy measurement of the water absorbed by the plant. Sometimes the stem is attached to a burette, or water absorption might be measured as an air bubble moves through a capillary tube connected to the closed water reservoir from which the plant is absorbing water (a **potometer**). This can be useful when relative transpiration rates must be studied for brief time intervals, but of course no one is comfortable with the idea that a detached plant acts like a plant

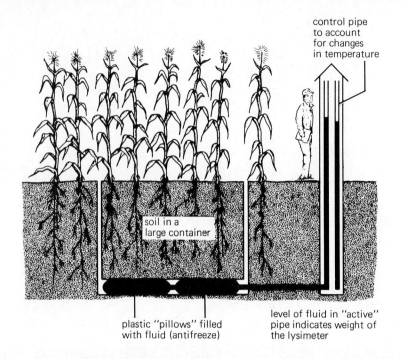

control pipe
to account
for changes
in temperature

soil in a
large container

plastic "pillows" filled
with fluid (antifreeze)

level of fluid in "active"
pipe indicates weight of
the lysimeter

Figure 3-2 Diagram of a large field lysimeter, operating on a hydraulic principle.

with its roots intact. (Actually, whole plants with intact roots can also be used in potometers.)

The **cuvette** (illustrated in field use in Fig. 3-1) is widely used in the laboratory for study of single leaves. As in the tent method, moisture or CO_2 entering and exiting is measured. In the laboratory, where purposes are special, cuvettes and potometers may be highly appropriate. Instead of trying to follow natural transpiration, laboratory studies may be aimed at understanding the effects of individual environmental factors upon transpiration. Factors can be standardized and measured under steady-state conditions.

3.2 The Paradox of Pores

Nature often proves to be more complex than we expect. Suppose we compare the evaporation rate from a beaker of water and from an identical beaker that is half covered with metal strips or a similar material. We would expect evaporation from the second beaker to be about half of that from the first. Now let's cover all but about 1 percent of the second beaker. We will use a thin piece of foil with small holes making up about 1 percent of the total area. Will we measure about 1 percent as much evaporation? Not if the holes have about the same size and spacing as the stomates found in the epidermis of a leaf. We will in fact measure about half as much evaporation (50 percent) as from the open surface.

How can this be? Why isn't evaporation directly proportional to area? It certainly seems paradoxical that stomatal openings on the leaf make up only

about 1 percent of the surface area, whereas the leaf sometimes transpires half as much water as would evaporate from an equivalent area of wet filter paper. We resolve this apparent paradox by realizing that evaporation is a diffusion process from water surface to atmosphere, and by applying the integrated form of Fick's law introduced in Chapter 1 (equation 1.6). Simply stated, the law says that diffusion is proportional to the driving force and inversely proportional to the resistance. In our example, the driving force is the same for both beakers: the difference in vapor density between the water surface (where the atmosphere is saturated with vapor) and the atmosphere some distance away (where it must be below saturation if evaporation is to occur).

The different evaporation rates depend on different resistances to diffusion. Part of the resistance is an inverse function of the area, and this part is much higher above the beaker covered with porous foil, which is what we expected. But the other part of the resistance depends on the distance in the atmosphere through which water molecules must diffuse before their concentration reaches that of the atmosphere as a whole. The greater the distance, the higher the resistance. This distance can be called the **boundary layer,** and it is much shorter above the pores in the foil than above the free water surface. Molecules evaporating from the free water will be part of a relatively dense column of molecules extending some distance above the surface, while molecules diffusing through a pore can go in any direction within an imaginary hemisphere centered above the pore. In the hemisphere, the concentration drops very rapidly with distance from the pore, which is to say that the

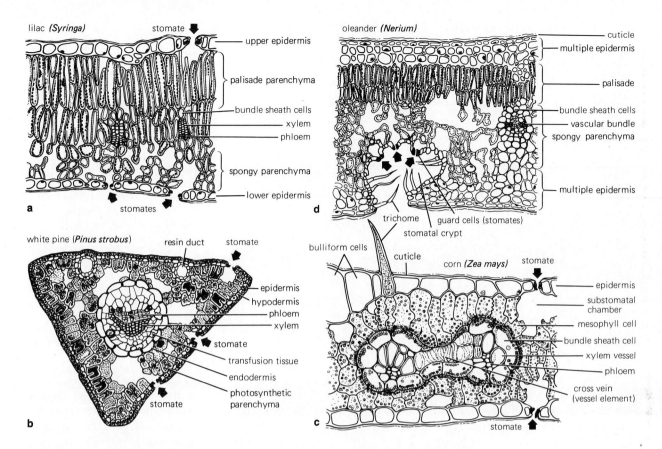

Figure 3-3 Cross sections through four representative leaves, one with "normal" stomates (**a**), one with slightly sunken stomates (**b**), one a grass with stomates about equal on both surfaces (**c**), and one with stomates deeply sunken in a substomatal cavity (**d**). Note other details of differing leaf anatomy. Pine leaves do not have a palisade layer, for example.

concentration gradient is very steep because the boundary layer is very thin. If pores are closer together than the thickness of their boundary layers, these hemispheres overlap and merge into the boundary layer.

Many empirical studies were made several decades ago to determine the effects of pore size, shape, and distribution on diffusion rates. The stomates of typical plants proved to be nearly optimal for maximum gas or vapor diffusion. Thus, plants are ideally adapted for CO_2 absorption from the atmosphere, but also for loss of water by transpiration. The stomates can close, however, and in most plants they are adapted to close when CO_2 absorption is unnecessary because photosynthesis is not possible (in darkness or with disease or pollution injury).

3.3 Stomatal Mechanics

Stomates come in considerable variety. Figure 3-3 shows drawings of cross sections through four kinds of leaves. The waxy **cuticle** on leaf surfaces restricts diffusion so that most water vapor and other gases must pass through the openings between the **guard cells.** Technically, the term **stomate*** refers only to this opening, but often the term is applied to the entire **stomatal apparatus,** including the guard cells. Adjacent to each guard cell are usually one or two other modified epidermal cells called **accessory** or **subsidiary cells.** Water evaporates inside the leaf from the **palisade parenchyma** and **spongy parenchyma** cell walls into the **intercellular spaces,** which are continuous with the outside air when the stomates are open. Carbon dioxide follows the reverse diffusional path into the leaf. Many of the cell walls of both palisade and spongy parenchyma cells (collectively called **mesophyll** cells) are exposed to the internal leaf atmosphere, although this is seldom evident for palisade cells in drawings and photomicrographs of leaf cross sections. It becomes much more apparent in sections through the palisade, parallel to the leaf surface, and is also strikingly apparent in scanning electron micrographs such as that of Fig. 3-4.

---

*Some anatomists prefer **stoma** (singular) and **stomata** (plural).

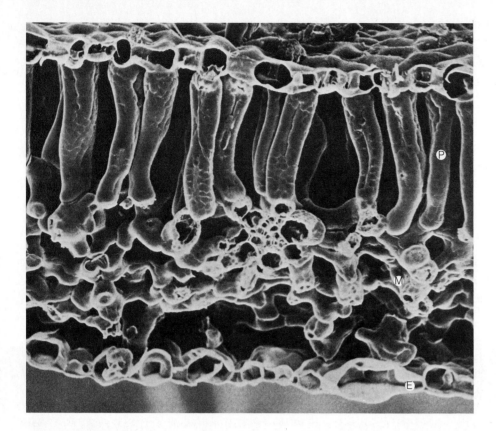

Figure 3-4 Transverse view of a mature broadbean leaf. The internal organization of cells in this leaf is characteristic of many plants, consisting of a layer of palisade cells (P) in the upper half of the leaf, spongy mesophyll cells (M) in the lower half (palisade and spongy mesophyll are collectively called mesophyll), and bounded on both sides by the epidermis (E). Note the large air gaps between the palisade cells as well as the spongy mesophyll cells. Most of the surfaces of the cells is exposed to the air; the area of cell wall exposed is about 10 to 40 times the surface area of the leaf (Nobel, 1980). The proportion of air volume to cell volume in a leaf can vary from 10 to 80 percent among different types of plants. ×420. (Scanning electron micrograph courtesy of John Troughton; see Troughton and Donaldson, 1972.)

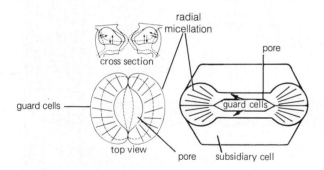

Figure 3-6 Schematic drawing of two stomates, showing radial micellation. Left, a dicot leaf (dashed lines show guard cells after loss of water, hence closing); right, a monocot (grass) stomate.

Stomates sometimes occur only on the bottoms of leaves but often are found on both tops and bottoms (with perhaps more on the bottoms). Waterlily pads have stomates only on top, and submerged plants have none at all. Grasses usually have about equal numbers on both sides. Sometimes, as in the oleander or pine (Fig. 3-3), the stomates occur in a substomatal crypt. Such **sunken stomates** are apparently an adaptation to reduce transpiration.

Figure 3-5 shows scanning electron micrographs of stomates from four species. Typical stomates of dicots consist of two kidney-shaped guard cells; grass guard cells tend to be more elongate (dumbbell shaped). Guard cells contain a few chloroplasts, whereas their neighboring epidermal cells seldom do. In some species, at least, there seem to be fewer, or no, plasmodesmata connecting the protoplasts of guard cells and accessory cells, although there may be plasmodesmata between guard cells and the mesophyll cells below.

Stomates open because the guard cells take up water and swell. At first, this is puzzling. One might imagine that the swelling guard cells would force the inside walls of the stomate together. Stomates function the way they do because of special features in the submicroscopic anatomy of their cell walls. The cellulose microfibrils, or micelles, that make up the plant cell walls are arranged around the circumference of the elongated guard cells as though they were radiating from a region at the center of the stomate (Fig. 3-6). The result of this arrangement of microfibrils, called **radial micellation,** is that when a guard cell expands by taking up water, it cannot increase much in diameter, because the microfibrils do not stretch much along their length. But the guard cell can increase in length; therefore, because two guard cells are attached to each other at both ends, they bend outward when they swell, which opens the stomate.

It was noticed as long ago as 1856 that the guard cells of some species were slightly more thickened along the concave wall adjacent to the stomatal opening. Since then, most authors have suggested that the thickening was responsible for the opening when the

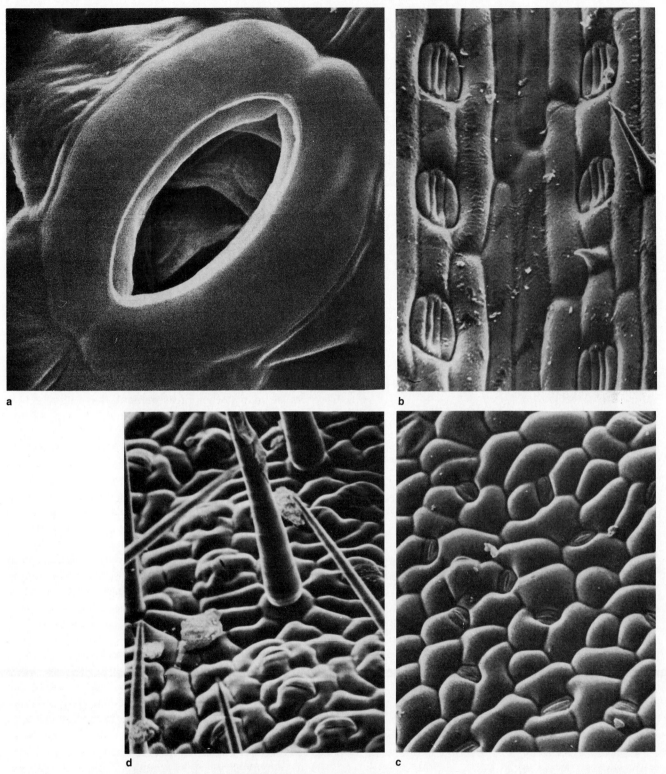

Figure 3-5 (**a**) Spongy mesophyll cells, as seen through a stomate on the lower surface of a cucumber leaf. ×7900. (Courtesy of John Troughton; see Troughton and Donaldson, 1972.) (**b**) Upper surface of wheat (*Triticum* sp.). Note the characteristic monocot stomates. (**c**) Upper surface of lambsquarter (*Chenopodium rubrum*). (**d**) Upper surface of velvet leaf (*Limnocharis flava*). Note the hairs. (Scanning electron micrographs courtesy of Dan Hess.)

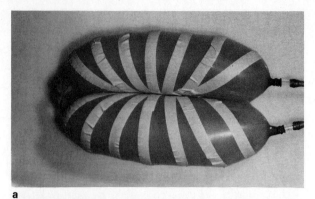

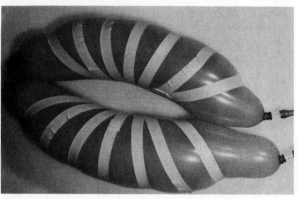

Figure 3-7 Two balloons representing a guard-cell pair. **(a)** Balloons in their "relaxed" state with masking tape applied to represent both the "radial micellation" and the thickening along part of the ventral walls. **(b)** The balloon pair in an inflated state. Balloons were glued together at the ends with rubber cement before inflating (which weakened the rubber and caused eight pairs to burst when inflated before achieving success with the pair shown!).

guard cell takes up water. Donald E. Aylor, Jean–Yves Parlange, and Abraham D. Krikorian (1973) at the Connecticut Agricultural Experiment Station re-investigated a 1938 discovery by H. Ziegenspeck. They showed with balloon models (Fig. 3-7) and mathematical modeling that the radial micellation is much more important in stomate opening than the thickening of the inside walls and that radial micellation is as important in grass as in dicot guard cells.

3.4 Stomatal Mechanisms

What is it that causes guard cells to take up water so that the stomates open? This classical problem of plant physiology has been discussed and studied for many decades. A prime suspect for the cause of stomatal opening is some osmotic relationship. There are several possibilities. If the osmotic potentials of the guard cell protoplasts become more negative than those of the surrounding cells, water should move into the guard cells by osmosis, causing an increase in pressure and swelling. Another possibility is a decrease in the resistance of the guard cell wall to stretching, which would decrease pressure inside and thus allow the uptake of more water, causing swelling. Or the surrounding subsidiary cells might shrink, again decreasing pressure on the guard cells.

By inserting microhypodermic needles filled with silicon oil into individual cells, Mary Edwards and Hans Meidner (1979) measured the pressures in epidermal, subsidiary, and guard cells directly. When they punctured a cell in a way that caused no leakage, the pressure inside forced some cell sap into the needle, pushing back the silicon oil. This activated a highly sensitive pressure transducer that applied pressure to the oil until the meniscus between it and the cell sap was pushed back to the cell surface (as revealed in a microscope). The pressure transducer gave an electrical signal that could be calibrated in terms of real pressures. Edwards and Meidner found that pressure drop in subsidiary cells did indeed contribute to the swelling of guard cells and the opening of stomates.

This is only a contributing effect, however, because many other measurements have shown that the osmotic potential of guard cells does become more negative when stomates open. For example, G. D. Humble and Klaus Raschke (1971) measured values of -1.9 MPa for the osmotic potential of broadbean (*Vicia faba*) guard cells with closed stomates and -3.5 MPa when the stomates were open. Since the guard cells nearly doubled in volume during opening, this increase in solute concentration occurs in spite of dilution. The increased solute concentration results in an osmotic transfer of water from the accessory cells to the guard cells. We can now restate the question more precisely: *What causes the change in the osmotic potential in guard cells that results in stomatal opening?*

Environmental Effects on Stomates Many changes in environmental factors influence stomatal apertures. Any theory purporting to explain guard cell action must account for the known environmental effects, so it is appropriate to review these effects before considering how osmotic potentials change in the guard cells.

Stomates of most plants open at sunrise and close in darkness, allowing entry of the CO_2 needed for photosynthesis during the daytime. Opening generally requires about an hour, and closing is gradual throughout the afternoon (Fig. 3-8). Stomates close faster if plants are suddenly placed in darkness. Certain succulents that are native to hot, dry conditions (e.g., cacti, *Kalanchoe*, and *Bryophyllum*) act in an opposite manner: They open their stomates at night, fix carbon dioxide into organic acids in the dark, and close their stomates during the day (see CAM photosynthesis in Chapter 10). This is an appropriate way to absorb CO_2 through open stomates at night, when transpiration stress is low, and conserve water during the heat of the day. The minimum light level for the opening of stomates in other plants is about 1/1000 to 1/30 of full sunlight, just enough to cause some net photosynthesis, which reduces the CO_2 concentration in the leaf. Light level influences not only the rate of opening but also the final aperture size, bright light causing a wider aperture.

Low concentrations of CO_2 in the leaves cause stomates to open, and removal of CO_2 by mesophyll cells during photosynthesis is the main reason that stomates of most species open in light. Succulents fix CO_2 into organic acids at night, diminishing CO_2 and causing stomatal opening. If CO_2-free air is blown across leaves even in darkness, then their slightly open stomates open wider. Conversely, high CO_2 concentration in the leaves causes the stomates to close partially, and this occurs in the light as well as the dark. When the stomates are completely closed, which is unusual, external CO_2-free air has no effect. The environmental factors that influence photosynthesis and respiration probably affect stomatal opening and closing by acting indirectly on the internal CO_2 concentration. Such coupling of stomatal action to photosynthesis has obvious survival value.

The water potential within a leaf also has a powerful effect on stomatal opening and closing. As water potential decreases (water stress increases), the stomates close. This effect can override low CO_2 levels and bright light. Its protective value during drought is obvious.

High temperatures (30 to 35°C) usually cause stomatal closing. This might be an indirect response to water stress, or else a rise in respiration rate might cause an increase in CO_2 within the leaf. High CO_2 concentration in the leaf is probably the correct explanation for high-temperature stomatal closing in some species, because it can be prevented by flushing the leaf continuously with CO_2-free air. In some plants, however, high temperatures cause stomatal opening instead of closing. As a result, the increased transpiration removes heat from the leaf.

Sometimes stomates partially close when the leaf is exposed to gentle breezes, probably because more CO_2 is brought close to the stomates, increasing its diffusion into the leaf. Wind can also increase transpiration, leading to water stress and stomatal closing.

Guard-Cell Uptake of Potassium Ions Since stomates open because the guard cells take up water, and water uptake is caused by more solute and hence a more negative osmotic potential, what is the solute and where does it come from? Since about 1968 (al-

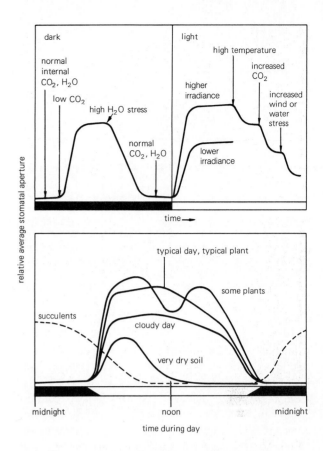

Figure 3-8 A summary diagram of stomatal response to several environmental conditions. In the top graph, arrows point to times when some environmental parameter was changed as indicated by the label.

though the first reports appeared as early as 1943), accumulated experimental evidence has made it clear that potassium ions (K^+) move from the surrounding cells into the guard cells when stomates open. Representative data are shown in Fig. 3-9.

The quantities of K^+ that build up in the vacuoles of guard cells during stomatal opening are sufficient to account for the opening, assuming that the K^+ is associated with a suitable anion to maintain electrical neutrality. Increases of up to 0.5 M in K^+ concentration are observed, enough to decrease the osmotic potential by about 2.0 MPa. Stomatal opening and K^+ movement into the guard cells are closely correlated in nearly every case investigated. Light causes a buildup of K^+ in guard cells, as does CO_2-free air. When leaves are transferred to the dark, K^+ moves out of guard cells into the surrounding cells, and stomates close. This has been observed in numerous species from all levels of the plant kingdom (e.g., mosses, ferns, conifers, monocots, and dicots) and for the stomates occurring on various plant organs (e.g., moss sporophytes, leaves, stems, and sepals).

When strips of epidermal tissue are removed

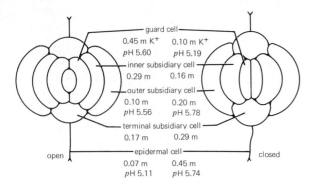

Figure 3-9 Quantitative changes in K^+ concentrations and pH values of the vacuoles in several cells making up the stomatal complex of *Commelina communis*. Values are given for the open and closed conditions of the stomatal pore. (Data of Penny and Bowling, 1974 and 1975.)

from the leaves of broadbean (*Vicia faba*), most epidermal cells are broken, but the guard cells remain intact. When these strips are floated on solutions, stomates will not open unless the solutions contain K^+. So guard cells must normally obtain K^+ ions from adjacent epidermal cells. Stomates also close in response to the application of abscisic acid, a plant growth regulator (see later), which causes loss of K^+ from the guard cells. So nearly all evidences agree that K^+ transport from accessory cells to guard cells is the cause of the more negative osmotic potentials and hence stomatal opening, and that reverse transport is the cause of closing. (Members of the orchid genus, *Paphio-pedilum*, seem to be exceptions. Their guard cells do not contain chlorophyll and do not take up sufficient K^+ ions when stomates open. See Outlaw, Manchester, and Zenger, 1982.)

Having learned about the K^+ flux, we can ask the next question: *What is the mechanism of K^+ movement?* Consideration of this question forces us to realize that the science of plant physiology has become increasingly dependent on biochemistry. Now we must add biochemical considerations to our thermodynamic concepts of water relations.

To begin, stomatal opening is not a simple pumping of K^+ into guard cells with energy provided by light. Such a mechanism does not account for dark opening in CO_2-free air (also accompanied by K^+ influx). Further, it has been observed that increasing the pH of guard cells by exposing leaves to ammonium vapor also causes opening and that stomates sometimes open in response to low oxygen levels. Neither of these observations is accounted for by a light-driven uptake of K^+ into guard cells.

In some species, Cl^- ions or other anions accompany K^+ into and out of guard cells, but Klaus Raschke and G. D. Humble (1973) observed that no anion accompanies the movement of K^+ into guard

cells of *Vicia faba* leaves. Instead, as potassium ions move into guard cells, an equal number of hydrogen ions come out (Fig. 3-9). Where does the H^+ come from? Organic acids are synthesized in the guard cells in response to factors that cause stomatal opening. Malic acid is the most common product under normal conditions. Since hydrogen ions are present in organic acids, the pH in the guard cells would drop (they would become more acidic) if H^+ were not exchanged for incoming K^+. The increased ions (K^+ and organic acid anions) make the osmotic potential more negative.

The organic acids are made largely from starch and sugars stored in guard cells. It has long been known that starch disappears in the guard cells as stomates open, although careful measurements could never demonstrate that sugars appeared in its place. Thus the increasingly negative osmotic potential is caused by an uptake of K^+ and sometimes Cl^- and/or a buildup of potassium salts of organic acids, mostly malate. Which of the two processes predominates apparently depends on the species and perhaps on the availability of Cl^-. But how? Much has been learned and postulated about the biochemical steps involved, steps that we shall discuss in later chapters. Starch is apparently broken down to produce a 3-carbon compound (phosphoenol pyruvate or PEP); this step is promoted by blue light. The PEP is then combined with CO_2 to produce oxaloacetic acid and then the 4-carbon malic acid. Finally, H^+ ions from the malic acid leave the cell to balance the K^+ ions that are entering. (For reviews, see Outlaw, 1982, and Permadasa, 1981).

In the succulents with stomates that open in the dark, CO_2 is combined with PEP to produce malic acid (in the dark) in many cells besides guard cells. Extrusion of H^+ ions from the malic acid in guard cells could lead to K^+ uptake and stomatal opening.

The Abscisic Acid Effect on Stomates Another observation of the early 1970s was almost as revolutionary as that of K^+ uptake. When the growth regulator, abscisic acid (ABA; see Chapter 17), is applied at extremely low concentrations (e.g., $10^{-6} M$), it causes stomates to close. Furthermore, when leaves are subjected to water stress, the ABA in their tissues builds up. When leaves dry at normally slow rates, ABA builds up before stomates close, suggesting that stomatal closure in response to leaf water stress is mediated by ABA.

There is one observation, however, that seems counter to this simple picture. When water stress develops rapidly, as when a leaf is removed and subjected to dry air at warm temperatures, stomates close *before* ABA begins to build up in the leaf tissue. This has caused concern during recent years, but the explanation may be quite simple: When water stress develops rapidly, water may evaporate from the guard cells themselves, causing them to lose turgor and thus causing stomates to close. Study of the problem has led plant physiologists to appreciate the fact that ABA in leaves can occur in at least three **pools** (a term used when evidence suggests that a substance that occurs in different parts of a tissue or a cell affects some process differently depending on where it is located). The three ABA pools in leaf tissue might be the cytosol of the cells where ABA is apparently synthesized, the chloroplasts where it is accumulated (or synthesized, according to some authors), and the cell walls outside of the protoplasts. The ABA pool that leads to stomatal closure must be in the cell walls and must be replenished from the leaf mesophyll cells where ABA is synthesized. (For recent papers and reviews, see: Cowan et al., 1982; Henson, 1981; Lancaster and Mann, 1977; Outlaw, 1982, 1983; Outlaw et al., 1982; Permadasa, 1981; Radin and Ackerson, 1982; Wilson et al., 1978.)

Assuming that ABA is indeed the messenger that causes stomates to close under slowly developing water stress (in which case direct loss of water from guard cells is not rapid enough to cause closure), it appears that there are at least two **feedback loops** that control stomatal opening and closing. When CO_2 decreases in the intercellular spaces and thus in the guard cells, K^+ moves into guard cells and stomates open, allowing CO_2 to diffuse in, thus completing the first loop. This meets the needs of photosynthesis. (In nonsucculents, it also leads to transpiration.) If water stress develops, ABA appears in the water that moves to the guard cells, so the stomates close, completing the second loop. The two loops interact: The degree of stomatal response to ABA depends upon CO_2 concentration in the guard cells, and response to CO_2 depends upon ABA. One feedback loop provides CO_2 for photosynthesis; the other protects against excessive water loss (Fig. 3-10). Stomates, as Raschke (1976) has said, have been "delegated the task of providing food while preventing thirst."

Light Quality and Stomatal Response Since stomates open in the dark in response to CO_2-free air, low oxygen levels, and other factors, light is not directly essential for their opening. When light is effective, it is partially because photosynthesis in the leaf mesophyll cells lowers the CO_2 in the intercellular spaces, and the guard cells respond by taking up more K^+ ions. We would expect, then, that the same wavelengths of light (colors) that are effective in photosynthesis would be effective in opening stomates. At fairly high light levels, this is exactly what is observed: Blue and red light are effective in both photosynthesis and stomatal opening. When the experiment is done quantitatively, however, it is seen that blue light is more effective (relative to red light) in causing stomatal opening than it is in photosynthesis. Furthermore, at low light levels, blue light

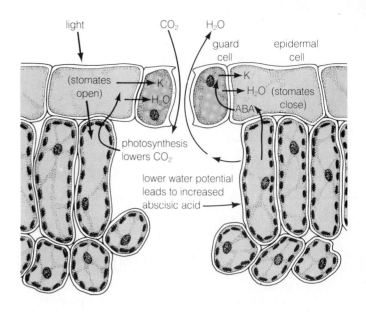

Figure 3-10 Two important feedback loops, one for CO_2 and one for H_2O, that control stomatal action. The left part of the drawing illustrates the effects of light: Light promotes photosynthesis, which lowers CO_2 levels in the leaf; the leaf's response is to cause more K^+ to move into guard cells, and water follows osmotically, causing stomates to open. (There is also a direct effect of light on stomatal opening, independent of CO_2 levels.) The right-hand side shows the effects of water stress: When more water exits than can enter from the roots, abscisic acid (ABA) is released or produced from mesophyll cells, which leads to movement of K^+ out of guard cells; water follows osmotically, so stomates close. If the rate of drying is extremely rapid, water is lost from the guard cells directly, bypassing the ABA step but still leading to closure. (From W. A. Jensen and F. B. Salisbury, *Botany*, 1984, p. 260.)

may cause stomatal opening when red light has no effect at all. Thus, there is an enhancement of stomatal opening by blue light acting directly on guard cells, in addition to the effects on photosynthesis in mesophyll cells and perhaps in guard cells. The blue light causes movement of K^+ independently of the effects of CO_2 or ABA. It has even been shown that isolated guard cell protoplasts respond to low levels of blue light by absorbing K^+ and then swelling osmotically (Zeiger and Hepler, 1977). It is now thought, as suggested above, that blue light promotes the breakdown of starch into the PEP molecules that can accept CO_2, producing malic acid. But much remains to be learned.

3.5 The Role of Transpiration: Water Stress

What Good Is Transpiration? Many philosophers of science would object to such a question, labeling it **teleological.** Such a question assumes that all things in the universe have purpose, an assumption that cannot be scientifically demonstrated. The biologist

Calculating CO₂ Fluxes into Leaves

Bruce G. Bugbee

Bruce Bugbee is a Research Associate Professor at Utah State University, where he teaches a course in crop physiology and works with Frank Salisbury on a project supported by the National Aeronautics and Space Administration to achieve maximum yields of wheat in the controlled environments of a space station or space craft. Plants in such a nearly closed system would not only provide food for astronauts but would remove CO_2, produce O_2, and purify water (when astronauts condensed the transpired water). Bruce grew up in Minnesota and obtained the doctoral degree in horticulture from Pennsylvania State University.

In this chapter, we stated that problems of CO_2 exchange are relatively simple to solve once the principles of transpiration have been worked out. This essay shows how that is done.

Carbon dioxide, diffusing in a direction opposite to water vapor, must pass across the resistance of the boundary layer (r_a), through the stomatal pore resistance (r_s), and finally diffuse through the mesophyll resistance (r_m) to the chloroplast. A few measurements and a few equations allow us to calculate the magnitude of each of these resistances. The first step is to calculate the total resistance ($r_{total} = r_a + r_s + r_m$). Diffusion of gases across a concentration gradient is analogous to the flow of electricity across a resistance. We can therefore use an equation analogous to Ohm's law and shown as an integrated form of Fick's first law in equation 1.6:

$$\text{total } CO_2 \text{ flux} = \frac{\Delta C}{\Sigma R}$$

Where ΔC = the carbon dioxide concentration gradient between the chloroplast and the external atmosphere, ΣR = total resistance to CO_2 (r_{total}), and total CO_2 flux = the photosynthetic rate per unit area of leaf. The carbon dioxide concentration in the chloroplast is assumed to be equal to the compensation point of photosynthesis (Section 11.2; Coombs and Hall, 1981).

If we rearrange the equation,

$$\Sigma R = \frac{\Delta C}{CO_2 \text{ flux}}$$

and measure the CO_2 flux by measuring the leaf photosynthetic rate in a gas exchange system, we can substitute units and calculate total resistance.

Assume:

CO_2 flux = 0.02 mmol m⁻² s⁻¹ (a typical photosynthetic rate)

CO_2 in atmosphere = 15 mmol m⁻³ (350 ppm)

CO_2 compensation point = 1.5 mmol m⁻³

$$\Sigma R = \frac{15 - 1.5 \text{ mmol m}^{-3}}{0.02 \text{ mmol m}^{-2} \text{ s}^{-1}}$$

$$= \frac{13.5}{0.02} \frac{\text{m}^{-1}}{\text{s}^{-1}}$$

$$= 675 \text{ s m}^{-1}$$

$$\Sigma R = r_a + r_s + r_m = 675 \text{ s m}^{-1}$$

We can now proceed to calculate the components of this resistance. The most direct method for measuring boundary layer resistance is to put a moist filter paper replica of a leaf in a closed system that can provide air flow over the leaf (Jarvis, 1971). The filter paper must be connected to a water reservoir outside the chamber so that it stays wet by capillary wicking of water. The only resistance to the vaporization of water is then the boundary layer resistance, which is determined by:

$$r_a = \frac{\Delta \text{ vapor density}}{\text{rate of H}_2\text{O addition to system}}$$

where Δ vapor density = vapor density gradient = difference between the saturation water vapor density of the air at the leaf temperature and the ambient water vapor density.

Assume:

Filter paper at 25°C. Saturation vapor density at 25°C = 23 g m⁻³

Air at 24°C, 40% relative humidity

Vapor density at 40% RH, 24°C = 0.4 × 22 g m⁻³ = 8.8 g m⁻³

Based on the simulated leaf area of wet filter paper, say that

Rate of H₂O addition to the

System = 0.8 g m⁻² s⁻¹

$$r_a = \frac{23 - 8.8 \text{ g m}^{-3}}{0.8 \text{ g m}^{-2} \text{ s}^{-1}}$$

$$= \frac{14.2}{0.8} \frac{\text{m}^{-1}}{\text{s}^{-1}}$$

$$= 17.75 \text{ s m}^{-1}$$

This is the estimated boundary layer resistance to water vapor loss from the leaf. Carbon dioxide molecules have a greater mass than water molecules, however, so CO_2 molecules diffuse more slowly. The ratio of the diffusivities of CO_2 and H_2O is proportional to their average velocities. Their average velocities are predicted by kinetic theory to be (see equation 1.1):

$$\frac{V_{H_2O}}{V_{CO_2}} = \frac{\text{Constant} \cdot \dfrac{1}{\sqrt{\text{Mass}}}}{\text{Constant} \cdot \dfrac{1}{\sqrt{\text{Mass}}}}$$

$$= \frac{\dfrac{1}{\sqrt{18}}}{\dfrac{1}{\sqrt{44}}}$$

$$= 1.56$$

Water vapor diffuses 1.56 times as fast as carbon dioxide. The boundary layer resistance to CO_2 is thus larger than H_2O. It would be 1.56 times 17.75 s m^{-1}, but transfer in the boundary layer is by both diffusion and by turbulent mixing. A ratio of 1.387 has been derived empirically to account for the turbulence (Thom, 1968).

The boundary layer resistance to CO_2 is therefore:

$$r_a^{CO_2} = 17.75 \times 1.387$$

$$= 24.6 \text{ s m}^{-1}$$

If we now replace our wet filter paper replica of the leaf with a real leaf we can use the same equation:

$$r_s^{H_2O} + r_a^{H_2O} = \frac{\Delta \text{ vapor density}}{\text{rate of } H_2O \text{ addition to the system}}$$

to calculate a new resistance value. This will include both the stomatal (r_s) and the boundary layer (r_a) components. By subtraction we can determine r_s.

Assume:

—measured rate of H_2O addition to system = 0.6 g m^{-2} s^{-1}

—leaf temperature = 25°C, air temperature = 24°C

—relative humidity in interstomatal cavity = 100% × 23 g m^{-3} = 23 g m^{-3}

—40% ambient relative humidity

—absolute vapor density at 40% RH, 24°C = 0.4 × 22 = 8.8 g m^{-3}

$$r_s^{H_2O} + r_a^{H_2O} = \frac{23 - 8.8 \text{ g m}^{-3}}{0.06 \text{ g m}^{-2} \text{ s}^{-1}}$$

$$= \frac{14.1}{0.06} \frac{\text{m}^{-1}}{\text{s}^{-1}}$$

$$r_s + r_a = 236.7 \text{ s m}^{-1}$$

$$r_s^{H_2O} = 236.7 - r_a^{H_2O}$$

$$r_s = 236.7 - 17.75$$

$$r_s^{H_2O} = 218.9 \text{ s m}^{-1}$$

This value for water vapor, multiplied by 1.56, gives us the resistance to CO_2 diffusion:

$$\begin{array}{r} r_s^{H_2O} = 218.9 \text{ s m}^{-1} \\ \times 1.56 \\ \hline r_s^{CO_2} = 341.5 \text{ s m}^{-1} \end{array}$$

The stomata clearly provide a much greater resistance to CO_2 flow than does the boundary layer.

Mesophyll resistance can now be easily determined by subtraction.

$$\Sigma R^{CO_2} = r_a^{CO_2} + r_s^{CO_2} + r_m^{CO_2}$$

$$r_m = \Sigma R - (r_a + r_s)$$

$$= 675 - (23.9 + 341.5)$$

$$r_m^{CO_2} = 309.6 \text{ s m}^{-1}$$

These values are typical for mesophytic C_3 plants when the water supply is ample for growth. The values for C_4 plants (see Chapter 10) are similar except that their mesophyll resistance is often much lower than C_3 plants. Quantitative measurements of CO_2 resistances allow us to analyze the barriers to CO_2 movement to the chloroplast and photosynthesis. Even under the optimum conditions of this example, stomatal resistance is about half of the total resistance to CO_2 diffusion. When water stress conditions occur, leading to stomatal closure, stomatal resistance quickly increases and accounts for nearly all the limitation to CO_2 passage to the chloroplast.

avoids teleology by rewording the question: *What is the selective advantage of transpiration?* Evolutionary theory states that features of living organisms must be either advantageous to their possessors or innocuous. A harmful feature will be eliminated by natural selection, if there is some less harmful feature that can be selected in its place. So what is the advantage of transpiration to a plant?

It is conceivable that the advantage might be a sort of "win by default." It is not only advantageous but essential to the life of a land plant to be able to absorb carbon dioxide from the atmosphere; it seems that the stomatal mechanism has evolved because of this need, and the disadvantageous consequence is transpiration. What about oxygen? Since stomates are typically closed at night, they obviously don't need to be open to absorb the oxygen used by the plant in respiration. This is probably because of the vast amounts of oxygen, compared with CO_2, in the atmosphere. During the day when leaves are photosynthesizing and stomates are open, oxygen diffuses out through them. But the stomates' real reason for existence is CO_2 absorption.

It can be argued that transpiration itself is neither essential nor of any advantage to the plant, since many plants can be grown through their life cycles in terraria at 100 percent RH, where transpiration is greatly reduced. It is a common observation that many plants grow better in atmospheres of high relative humidity. Certain alpine land plants (e.g., *Caltha leptosepala*) grow for days to weeks completely submerged in water. Obviously, no submerged aquatic plant transpires.

Nevertheless, careful investigation and thought have revealed several situations in which transpiration *per se* does seem to be beneficial to the plant. An unavoidable by-product of necessity may have been turned into an advantage. In most of these cases, it is possible for the plant to grow without transpiration, but when it occurs, it seems to confer some benefit.

Mineral Transport Minerals that are absorbed into the roots typically move up through the plant in the **transpiration stream,** the flow of water through the xylem caused by transpiration. But the transpiration stream is not essential for this movement, since minerals move upward in stems in spring before leaves appear. The young stems may transpire slightly but not much compared with transpiration after leaves expand. There is a circulation in plants (see Chapter 7): Solutions move through the phloem tissue from assimilating organs to utilizing organs. Even when there is no transpiration, the water in these solutions will return to the assimilating organs through the xylem tissue. Such a circulation has been demonstrated with radioactive tracers. Thus, transpiration is not essential for the movement of minerals within the

plant. Actually, the rate at which minerals arrive in the leaves is a function only of the rate of their movement into the xylem tissue, providing there is any xylem flow at all. The rate at which goods are delivered by an endless moving belt is a function only of the rate of loading. Studies have shown that the rate at which root cells transfer minerals from the soil to the xylem is usually much slower than the rate at which the xylem stream (transpiration stream) carries them away, even when transpiration is slow.

Nevertheless, when transpiration occurs, it may aid mineral absorption from the soil and transport in the plant. In one study, for example, tomato plants were grown in a greenhouse with high humidity and air enriched with CO_2. The abnormal CO_2 concentration partly closed the stomates while increasing photosynthesis. With partly closed stomates and high relative humidity, transpiration was reduced considerably, although not completely, because leaves in sunlight remain warmer than the air. The plants exhibited some calcium deficiency in their young leaves, however, as though calcium transport required a faster transpiration stream.

Optimum Turgidity If, as sometimes reported, plants do not grow as well at 100 percent RH as they do when some transpiration is allowed to occur, it could be because cells function best when there is some water deficit. There may be an optimum turgidity for cells, so that the various plant-cell functions are less efficient above and below this level. If transpiration is not allowed to occur, plants become overly turgid and do not grow as well as when there is some water stress. There is little conclusive evidence for this hypothesis, which needs more testing.

The main concern of plant physiologists and agriculturists is that of not enough water: **water stress,** or a too negative water potential. Plant responses to water stress are studied extensively in plant physiology partially because of the inhibitory effects of water stress on plant yield in natural and agricultural ecosystems. The topic of water-stress effects on plants is discussed in Chapter 24. Perhaps the most important point is that cell growth, which depends on absorption of water by cells, is one of the first processes to be affected by water stress. This frequently reduces yield. Other processes such as photosynthesis and synthesis of proteins and cell walls are also adversely affected by water stress.

3.6 The Role of Transpiration: Energy Exchange

For years, plant physiologists argued about whether transpiration was necessary to cool a leaf being warmed by the sun. On one hand, transpiration is certainly a cooling process; on the other, it was argued, if transpiration does not cool the leaf, other physical processes will—although in the absence of transpiration, it was conceded, leaves might be a few degrees warmer than otherwise. Growth of plants in atmospheres of 100 percent RH, in which transpiration is considerably reduced, were cited to support this view. As we have come to understand more about how the leaf exchanges energy with its environment, the argument has lost much of its interest. Truly, if transpiration doesn't remove the heat from a leaf, other processes will; but emphasizing this idea can obscure the important fact that in nature transpiration often plays an extremely important role in leaf cooling.

Evaporation of water is a powerful cooling process. Recall the Maxwell distribution of molecular velocities (see Fig. 1-3). It is the water molecules with high velocities that evaporate, and as they leave the liquid, the average velocity of the remaining molecules is lower, which is the same as saying that the liquid is cooler. Greenhouses in dry climates are often cooled by evaporative cooling: Air is drawn through a wet fibrous pad. When 1 g of water at 20°C evaporates, it absorbs 2.45 kJ (586 cal) from its environment; at 30°C the **latent heat of vaporization** is 2.43 kJ g^{-1} (580 cal g^{-1}). Plants evaporate tremendous amounts of water into their environments, and each gram of water transpired absorbs 2.4 to 2.5 kJ from the leaf and its environment.

Sometimes transpiration is the only important means of dissipating heat to the environment. Consider the large fan-shaped leaf of a native palm tree (*Washingtonia fillifera*) growing at a southern California oasis. Such a leaf is often cooler than the surrounding air, even though it is in full sunlight. When it is cooler than the surrounding air, it is absorbing some heat from the air, and it is absorbing more radiant energy from sunlight than it is radiating into its environment. It is cooler than air only because it is evaporating large quantities of water.

Investigations of the energy exchange between a plant and its environment provide an interesting example of plant biophysics. In the remainder of this chapter we shall consider some of the principles involved, without applying mathematics. (The mathematics are not difficult; see, for example, Gates, 1968, 1971, and Nobel, 1983.)

Leaf Temperature Consider the various factors that influence the temperature of a leaf. Transpiration cools the leaf, as we have seen. Condensation of moisture or ice on the leaf (dew or frost) releases the **latent heat of condensation** of water to the environment, and some of this heat is absorbed by the leaf. Incoming radiation absorbed by the leaf warms it. The leaf is itself radiating energy to its environment, however, and this is a cooling process. If the leaf is at a different temperature from that of the surrounding

air, it will exchange heat directly, first by **conduction** (in which the energies of molecules on the leaf surface are exchanged with those of the contacting molecules of air) and then by **convection** (in which a quantity of air that has been warmed expands, becomes lighter, and thereby rises—or sinks, if it has been cooled). In our discussion we shall refer to the combination of conduction and convection as convection.

If leaf temperature is changing, as is usually the case, the leaf is storing or losing heat. If a thin leaf stores a given amount of heat, its temperature rises rapidly; the same amount stored in a cactus raises the temperature much less, but the cactus stays warm longer. For simplicity, we shall consider only a leaf in a steady state of equilibrium with its environment, that is, at constant temperature. Some of the light energy absorbed by the leaf is converted to chemical energy by photosynthesis. This usually uses only a small percent of total light energy, however, so we can ignore it. Some energy is being produced by respiration and other metabolic processes; this is also small enough to ignore. Under steady-state conditions, three factors primarily influence leaf temperature: radiation, convection, and transpiration. Each of these is worthy of our consideration.

Radiation Appendix B discusses the principles of radiant energy, including those that apply to heat transfer studies. The thing to remember from the standpoint of leaf temperature is that it is the **net radiation** that is important. A leaf absorbs visible (light) and invisible (infrared) radiation from its surroundings and radiates infrared energy. If the leaf is absorbing more radiant energy than it radiates, then the excess will have to be dissipated by convection or transpiration, or both (or the temperature will rise). At night, leaves often radiate more energy than they absorb. Then they will probably absorb heat from the air and possibly from the water that condenses as dew on their surfaces. There are three important things to keep in mind when discussing the net radiation of a leaf: the wavelengths that are absorbed by the leaf, the total spectrum of the incoming radiation, and the amount of energy radiated by the leaf.

First is the **absorption spectrum** of the leaf. Of the energy incident on a leaf, some will be transmitted, some reflected, and some absorbed. The energy absorbed will depend on its wavelength. Leaves irradiated with white light absorb most of the blue and red wavelengths and a large portion of the green. Much of the green is reflected and transmitted, however, which is why leaves appear green. Leaves absorb very little of the near-infrared part of the spectrum; most is either transmitted or reflected. Thus, if our eyes could see in this part of the spectrum, vegetation would appear very bright to us. We can experience this impression by photographing vegetation

Ventilation in Waterlilies: A Biological Steam Engine

John Dacey

Graduate studies can provide some of the most exciting times in a person's life. This was the case for John Dacey as he unraveled the secrets of ventilation in waterlilies, as he describes in this personal essay. At the time, he was working on a doctoral degree in the Zoology Department at Michigan State University (completed in 1979). Before that, he had grown up in Kingston, Ontario, and he had obtained a combined undergraduate degree in biology and chemistry at King's College and Dalhousie University (neighboring institutions). He is now continuing his work on subsurface hydrology and root morphology of aquatic plants. This time his physiological ecology studies are centered in the salt marsh near the Woods Hole Oceanographic Institution, Massachusetts. (For reviews of this work, see Dacey, 1980, 1981; Dacey and Klug, 1982a, 1982b.)

Scientific research as an evolutionary process can be most exciting when it leads to questions that diverge from the expected course. In this essay I recount how my research unfolded; it is an effort to show how successful research depends on a combination of labor, logic, and luck. What began as a study of the role played by sediment gases in generating gas flows in water plants ultimately led to the discovery of a flowthrough ventilation system.

Oxygen is a basic requirement in the metabolism of all plants. Its absence around the roots of aquatic plants inhibits aerobic respiration in roots and allows potentially harmful materials to accumulate. It is generally accepted that the extensive network of internal gas spaces (**lacunae**) in these plants represents an adaptation to this environment, serving primarily to transport O_2 to buried roots and rhizomes.

In earlier studies, botanists measured gradients in the concentrations of O_2 and CO_2 in the lacunae. They found the highest levels of O_2 in photosynthesizing leaves and the lowest O_2 levels in the roots; vice versa for CO_2. They concluded that O_2 *diffused* to the roots from photosynthesizing leaves, and CO_2 *diffused* from the roots toward the leaves.

That idea is not unreasonable, but it is incomplete. A gas mixture is not only a mixture of randomly moving molecules; it is also a fluid. Just as a gradient in *partial* pressure drives the net diffusion of an individual component in a gas mixture, a gradient in *total* pressure drives a mass flow of the whole gas mixture (i.e., producing a wind). Under a gradient in total pressure, the movements of individual gases are not independent.

In my research I was interested in the relative contribution of diffusion and mass flow in the transport of gases in plants. My approach differed from previous research in two important ways: I measured the total pressure of lacunar gases by manometer, and I measured the concentrations of all principal gases by gas chromatography. This physical perspective made me especially conscious of potential mechanisms for generating pressure gradients.

I was studying a different plant when I saw gas streaming from a submerged waterlily leaf early in spring. The convenience of working with waterlilies quickly became apparent. They are large plants, with petioles up to 2 m long reaching from the sediment to the lake surface. Gas occupies at least half of the total volume of the plant, and I could withdraw samples by syringe. My first waterlily samples showed that methane (CH_4) was a significant constituent in lacunar gas, and early measurements confirmed that pressure could be developed in the root lacunae by methane from sediments. This confirmed my original hypothesis, but it soon became apparent that there was a more important process to investigate. Methane might be used as a tracer for gas movement!

Methane is produced in sediments by anaerobic bacteria and diffuses into lacunae in buried roots and rhizomes. As I studied its distribution in the waterlilies, I noticed some surprising patterns. For example, during nighttime in the summer months, CH_4 occurred throughout the plant in concentrations one would expect if gases moved principally by diffusion. During daylight, however, CH_4 was present in the petioles of older leaves but was absent in the petioles of young floating leaves. I could not explain its disappearance on the basis of diffusion, so I used experimental tracers to investigate the possibility of mass flow. By injecting small quantities of a tracer gas I monitored mass flow in the plant and found that during daylight gas moves from young floating leaves down their petioles to the rhizome. It subsequently moves from the rhizome up the petioles of older leaves to the atmosphere (Fig. 3-11). The mass flow down young petioles carried away any CH_4 tending to diffuse up from the rhizome. When ventilation stopped at night, CH_4 could accumulate in the petioles by diffusing from the rhizome.

Mass flow requires a gradient in total gas pressure. Using a sensitive manometer, I found that pressures in young leaves were slightly higher than the surrounding atmosphere (by less than 0.2 kPa). These small pressure differentials were sufficient to move air down the petioles at speeds up to 50 cm/min.

The next problem was to determine what caused the elevated pressures in young leaves. Ventilation stopped in darkness, so it seemed that the phenomenon was light-dependent. I confirmed that by showing that pressures in the lacunae of influx leaves were directly related to the levels of incident light. When I shaded a pressurized leaf, its lacunar gas pressure dropped immediately.

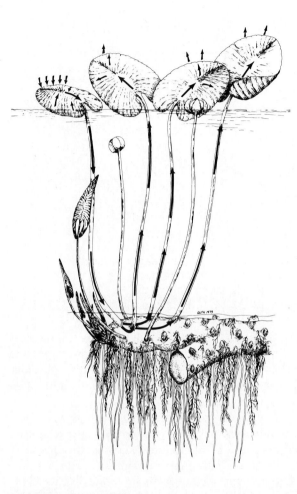

Figure 3-11 Waterlilly.

I soon found that photosynthesis did not play a role in pressurization. The pressures were not due to *de novo* gas production by the plant but resulted from movement of air from the surroundings into the lacunae. When I enclosed a pressurized leaf in a transparent bag, the bag tended to collapse, showing that the leaves drew air from the atmosphere against a pressure gradient.

I began to investigate the mechanism of this "pump." Temporarily forgetting the physical perspective I had used up to this point, I hypothesized that the pressurization must be a metabolic process, somehow using light energy to pump O_2 into the leaf. Any metabolic pump ought to be influenced by the composition of ambient gas. By varying the composition of gases in the plastic bag, I found that pressurization occurred regardless of the gas composition.

In an effort to understand the mechanism, I looked for differences in gas composition inside and outside the pumping leaves. Not surprisingly, the H_2O vapor pressure was higher inside the leaf. More significantly, the absolute amounts of both N_2 and O_2 were lower than ambient inside the leaf. There was, therefore, a diffusion gradient for N_2 and O_2 into the leaf even though the total gas pressure in the leaf exceeded ambient!

Another key discovery came on the heels of this observation. I found that lacunar gases pressurized in darkness when the leaf was held near a hot object. Pressurization was not dependent on light *per se* but on heat.

It was apparent that the answer lay in physics, not in biochemistry, so I went back to physical principles. Physical theory of gas flow through pores predicts that flow occurs only by diffusion when pores are very small (less than $0.1\mu m$ at atmospheric pressure). Larger pores allow both diffusion flow and mass flow. The fact that lacunar gas pressures were higher than ambient means that there was no significant mass flow between the lacunae and the atmosphere. Pores separating the lacunae from the atmosphere must have been very small, and diffusion must have been the dominant means of gas exchange.

At diffusive equilibrium, the partial pressures of all gases would be identical inside and outside a leaf with very small pores. There would also be no gradient in the total gas pressure (the sum of partial pressures). A leaf, however, is not in equilibrium with the ambient atmosphere, *since its H_2O vapor pressure almost always exceeds ambient*. There would be net diffusion of H_2O from the leaf. If the pores in the leaf were small enough to prohibit mass flow, the elevated partial pressure of H_2O inside the leaf would increase the total gas pressure in the leaf without influencing the partial pressures of N_2 and O_2. The total pressure in the leaf would remain elevated as long as water was available inside the leaf to keep its vapor pressure higher than ambient. In daylight, pressure inside the leaf increased as leaves were warmed by the sun. Warming caused more evaporation inside and thus an elevated H_2O vapor pressure. This phenomenon is known in physics as hygrometric pressurization.

As I described earlier, the pressurized gases in young leaves force a mass flow down the petiole to the rhizome. When flow occurs, it tends to lower the total pressure in the lacunae of pressurized leaves. As a result, the partial pressures of all the gases drop proportionately. The partial pressures of N_2 and O_2 drop below their respective levels in the atmosphere, setting up a gradient that causes those gases to diffuse into the leaf. This diffusion of N_2 and O_2 moves along a gradient in partial pressure (and incidentally against a gradient in total pressure). Simultaneously, water continues to evaporate into the lacunae, keeping its partial pressure near saturation. The continuous entry of air into the leaf sustains pressurization, thereby allowing it to drive a continuous mass flow.

(continued)

There is another relevant pressurization mechanism that has its basis in temperature differences between the lacunar gas and the atmosphere. Very briefly, physical theory predicts that gases diffuse through small pores in a partition more rapidly from the cool side (where gases are more compressed and thus concentrated) than from the warm side (where expansion causes lower concentration). The result is that the warm side tends to be slightly pressurized relative to the cool side (thermal transpiration). This mechanism, along with hygrometric pressurization, tends to elevate the pressure in warm young leaves.

Flowthrough in waterlilies depends on two other factors. First, pores in the older leaves are larger than in younger leaves, so older leaves allow mass flow to the atmosphere. Second, the lacunar system between the young and old leaves is continuous, so pressurized gas in young leaves moves freely by mass flow through the lacunae to the older leaves, where it escapes to the atmosphere.

This system is a flowthrough ventilation system. As much as 22 liters of air a day enter a single floating leaf and flow to the rhizome. This significantly accelerates O_2 transport to the roots over what would occur by diffusion alone. The system has the added advantage of also transporting CO_2 from the rhizome to the older leaves for use in photosynthesis.

Questions remain about the system: For example, are the flow-limiting pores in the stomatal apertures or are they in the palisade tissue beneath the stomates? Do similar mechanisms operate in other plants?

It is an essential characteristic of nature to be economical. Waterlily ventilation is a good example of simple design, using heat and the physics of gas behavior to operate a "biological steam engine."

---

with infrared-sensitive film (Fig. 3-12). Virtually all the far-infrared or thermal part of the spectrum is absorbed. If our eyes were sensitive only to that part of the spectrum, vegetation would appear as black as black velvet, if it were not for the fact that leaves emit thermal radiation and would appear to glow. Figure 3-13, showing an absorption spectrum of a leaf, presents these ideas quantitatively.

Second, the quality of radiation falling on a typical leaf varies considerably. Figure B-3, in Appendix B, illustrates the emission spectra for several radiation sources. The sun and the filament of an incandescent lamp emit **light** (the visible portion of the electromagnetic spectrum) because of their high temperatures. The higher the temperature, the more the peak of the emission spectrum is shifted toward the blue (see Wien's law in Appendix B). The temperature of the sun's surface is considerably above that of an incandescent filament in a light bulb, and thus sunlight is richer in blue and green wavelengths than light from an incandescent lamp.

The sun's radiation is further modified by passing through the atmosphere. Much of the ultraviolet is removed, and radiant energy is also absorbed by the atmosphere at several discrete wavelengths in the **far-red** (longer than 700 nm, but visible) and infrared parts of the spectrum. Most of the ultraviolet is absorbed by ozone in the upper atmosphere, and the infrared absorption bands are caused primarily by water and carbon dioxide.

Today, many plants used in physiological research are grown under artificial light sources, including fluorescent, incandescent, and such high-intensity discharge (HID) lamps as mercury vapor, low-pressure and high-pressure sodium, and metal halide lamps. These each have individual emission spectra (see Fig. B-3) that are absorbed in different ways by plants.

All objects at temperatures above absolute zero emit radiation. For objects at ordinary temperatures, most of this is in the far-infrared part of the spectrum, so plants are receiving this radiation from all their surroundings, including air molecules in the atmosphere. The quantity can be a sizable portion (e.g., 50 percent) of the total radiation environment, including sunlight.

The radiation absorbed by a plant is a function of the leaf's absorption spectrum and the emission spectrum of the sources irradiating the plant. The actual percentage of radiation absorbed varies considerably, therefore, since both absorption and emission spectra vary; but about 44 to 88 percent may be absorbed in common situations. Absorption is high when plants are irradiated with fluorescent light, since the leaf strongly absorbs most of the wavelengths emitted by fluorescent tubes. Absorption is much less when plants are irradiated with incandescent light, which is rich in the near-infrared part of the spectrum that is least absorbed by plants.

Third, plants emit radiant energy in the far-infrared part of the spectrum. The quantity of energy emitted can be calculated by application of the **Stephan–Boltzmann law** (Appendix B). This law states that the energy emitted is a function of the *fourth* power of the absolute temperature. Thus, as a leaf's temperature increases in sunlight, the radiant

a

b

Figure 3-12 Plants showing high reflectivity in the infrared portion of the spectrum. (**a**) Photograph taken with ordinary panchromatic film, which gives gray tones similar to the color intensities experienced by the human eye. (**b**) Photograph taken on infrared-sensitive film through a dark red filter that excludes most of the visible radiation. Note the brilliant white appearance of the vegetation and compare in particular the clumps of grass on the right. (Photographs by F. B. Salisbury.)

energy it emits increases sharply, so it heats up less than would otherwise be the case. Nevertheless, on the absolute scale, the normal temperature range of plants (from about 273 K to 310 K) is not very large. Energy emitted varies only by about 50 percent, but this can be significant. Even when a plant is illuminated by sunlight and is also receiving far-infrared radiation from its environment (e.g., from atmosphere, clouds, trees, rocks, and soil), the radiant energy emitted from the leaf is usually more than 50 percent of that absorbed and can reach 80 percent or more.

Convection   Heat is conducted–convected from the leaf to the atmosphere in response to a temperature difference between leaf and atmosphere. If the leaf is warmer, heat will move from the leaf to the atmosphere. *The temperature difference is the driving force;* the greater the difference the greater the driving force for convection.

With a given temperature difference, the rate of convective heat transfer is inversely proportional to the resistance to convection. The situation is exactly analogous to the integrated form of Fick's law introduced in Chapter 1 (equation 1.6) and applied in our discussion of diffusion through pores. It is also exactly analogous to **Ohm's law,** which says that the amount of current flowing through a wire is proportional to the driving force (the voltage difference)

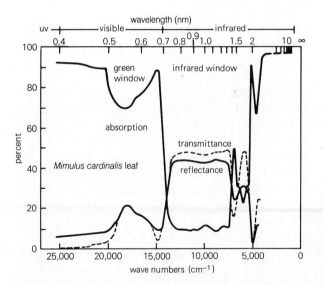

Figure 3-13 Absorption, transmission, and reflection spectra of a leaf. Note especially the "windows" in the green and in the near-infrared portions of the spectra. The leaves are thin and light green. (From Gates et al., 1965; used by permission.)

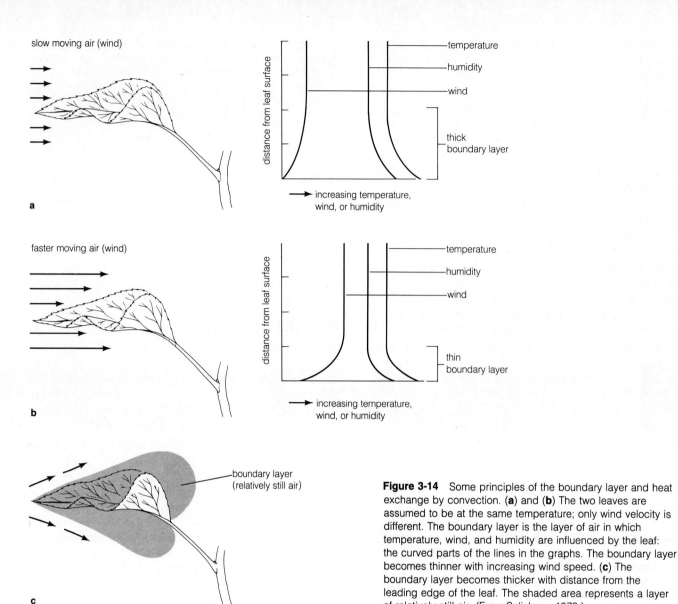

Slow moving air (wind)

a

distance from leaf surface

temperature
humidity
wind

thick boundary layer

→ increasing temperature, wind, or humidity

faster moving air (wind)

b

distance from leaf surface

temperature
humidity
wind

thin boundary layer

→ increasing temperature, wind, or humidity

c

boundary layer (relatively still air)

Figure 3-14 Some principles of the boundary layer and heat exchange by convection. (**a**) and (**b**) The two leaves are assumed to be at the same temperature; only wind velocity is different. The boundary layer is the layer of air in which temperature, wind, and humidity are influenced by the leaf: the curved parts of the lines in the graphs. The boundary layer becomes thinner with increasing wind speed. (**c**) The boundary layer becomes thicker with distance from the leading edge of the leaf. The shaded area represents a layer of relatively still air. (From Salisbury, 1979.)

and inversely proportional to the resistance encountered by the current in the wire. With convective heat transfer, the flow of heat is proportional to the temperature difference between the leaf and the atmosphere and inversely proportional to the resistance to heat flow encountered in the atmosphere.

The resistance to convective heat transfer is expressed by the thickness of the **boundary layer** (sometimes called the **unstirred layer**), which was introduced earlier in this chapter. This is the transfer zone of fluid (gas or liquid) in contact with an object (in this case, the leaf) in which the temperature, vapor density, or velocity of the fluid is influenced by the object (Fig. 3-14). Beyond the boundary layer, there is no influence of the object upon the medium. For a given temperature difference between the leaf and the air beyond the boundary layer (a given driving force), the convective transfer of heat is more rapid when the boundary layer is thin and slower

when it is thicker. In such a situation, the thickness of the boundary layer is inversely proportional to the steepness of the temperature gradient.

Usually there is air movement around a leaf. The more rapid the air movement the thinner the boundary layer (the steeper the temperature gradient). The boundary layer is thinnest next to a leaf's leading edge (the edge that faces into the wind). If the leaf's surface is parallel to the direction of wind movement, the boundary layer thickens from the leading edge toward the trailing edge of the leaf. Small leaves, especially conifer needles, have the thinnest boundary layers and are the most affected by convection. Large leaves, such as those of the desert fan palms, have the thickest boundary layers.

To summarize: The boundary layer is thinnest and offers the least resistance to convective heat transfer for small leaves and high wind velocities. Convective heat transfer is most efficient under such conditions, so smaller leaves have temperatures closer to air temperature than larger leaves, especially if there is a wind.

Transpiration In some ways, transpiration is closely analogous to convective heat transfer; in other ways, it is different. *The driving force for transpiration is the difference in water vapor density* within the leaf and in the atmosphere beyond the boundary layer.* The resistance to transpiration is partially the resistance of the boundary layer. To this extent, convection and transpiration are analogous, but there is an additional resistance to transpiration: the stomates. If the stomates are closed or nearly closed, resistance to transpiration can be very high; if they are open, resistance is relatively low. There are other resistances within the leaves besides that of the stomates, but these usually remain fairly constant. The resistance of the cuticle to passage of water depends upon atmospheric humidity, temperature, and perhaps light or other factors; but it is always relatively high, so it is seldom considered. The important thing to realize is that *there is always some leaf resistance;* that is, the leaf is never simply like a piece of wet paper. It is also important to know that the leaf resistance to transpiration can vary over a wide range as environmental factors influence stomatal apertures.

The vapor density gradient is influenced primarily by two factors: atmospheric humidity and leaf temperature. Usually, as a first approximation to the gradient, we imagine that the RH in the internal spaces of the leaf approaches 100 percent. Actually, it is somewhat less, because at equilibrium the water potential of the internal leaf atmosphere is equal to the water potential of the surfaces from which the water evaporates, and this is usually -0.05 to -3.0 MPa. (If equilibrium is not achieved, the water potential of the leaf atmosphere will be even lower.) Nevertheless, the water potential of the internal leaf is equivalent to an RH of about 98 percent or higher (see calculations following equations 2.4 and 2.5). Such high RHs seldom exist in the atmosphere beyond the boundary layer; therefore, *even if the leaf is at exactly the same temperature as the atmosphere beyond the boundary layer, there is under most conditions a vapor density gradient with higher values inside the leaf.*

Temperature can greatly accentuate the vapor density gradient, since vapor density is strongly a function of temperature (see Fig. 2-8). Warm air can

Table 3-1 Vapor-density gradients between leaves and the atmosphere when leaf and air temperatures are the same or different and when atmospheric humidity is different.

Conditions	Leaf	Difference	Air Beyond Boundary Layer
Temperature	20°C	none	20°C
Relative humidity	near 100%	near 90%	10%
Vapor density	10.9 g m^{-3}	9.8 g m^{-3}	1.1 g m^{-3}
Temperature	30°C	10°C	20°C
Relative humidity	near 100%	near 10%	90%
Vapor density	20.3 g m^{-3}	10.5 g m^{-3}	9.8 g m^{-3}

hold more water than cold air. An examination of Fig. 2-8 (summarized in Table 3-1) shows, for example, that air at 20°C and an atmospheric humidity of 10 percent establishes a vapor density difference of about 9.8 g m^{-3} between the leaf and the air, if they are at the same temperature and if the atmosphere within the leaf approaches 100 percent relative humidity. (At 20°C, saturated vapor pressure is about 10.9 g m^{-3}, and 10 percent of this is about 1.1 g m^{-3}.) If the leaf is at 30°C, however, and atmospheric humidity is as high as 90 percent (at 20°C), there still is a vapor density difference of about 10.5 g m^{-3}. (At 30°C, vapor density is 20.3 g m^{-3}; 90 percent of 10.9 g m^{-3} is 9.8 g m^{-3} which, when subtracted from 20.3 g m^{-3}, leaves a gradient of 10.5 g m^{-3}.) Thus, if the leaf is warmer than the air (a common phenomenon in sunlight), transpiration can occur into an atmosphere with an RH of 100 percent. As the vapor goes beyond the boundary layer, it may condense, as it does when a forest steams under the sun after a rainstorm, but this is of no consequence to the plant, which has lost water.

3.7 Plant Energy Exchange in Ecosystems

Application of heat transfer principles in field situations has provided considerable insight into the function of plant communities. To illustrate this, we shall consider a few of the principles discussed in previous sections as they apply to plants in the desert, the alpine tundra, and other situations.

The Desert In the desert, high temperatures are combined with high radiation, low humidity, and little available water. It is to the plant's advantage to conserve moisture and to maintain a relatively low leaf temperature. Many desert plants, perhaps most,

---

*Vapor density (expressed in grams or moles per cubic meter), discussed in Chapter 1 (page 30), can also be called **vapor concentration.** As also indicated in Chapter 1, we can speak of vapor pressure. Vapor density (or concentration) is somewhat easier to grasp intuitively than pressure, especially when diffusion is being discussed.

have small leaves, providing thin boundary layers with consequent efficient convective heat transfer. Their temperatures are thus closely coupled to air temperature, so at least they are not heated much above air temperature by sunlight. But such a situation might also produce a high rate of transpiration unless leaf resistance to evaporation is high. Some desert plants have sunken stomates and other mechanisms that result in high leaf resistance and retard transpiration. Some are also light gray and reflect much of the sun's radiation. We have seen that desert succulents close their stomates during the day and fix CO_2 into organic acids at night, thereby conserving water. But succulents often have large leaves (large surface area) that get much warmer than the air. Their protoplasm can evidently tolerate high temperatures.

Desert air temperatures are often extreme, so the ideal solution might be air conditioning by evaporative cooling, a technique applied widely by people who live in desert cities. A few desert species with roots that reach the water table achieve evaporative cooling. Typically, they have large leaves, as does the palm in an oasis, and their high rates of transpiration result in leaf temperatures several degrees below air temperature. This is possible because the low ambient humidity provides an ample vapor density difference even when the leaf is 10°C cooler than the air, and because the large leaf has a thick boundary layer, resulting in a low rate of convective heating of the leaf from the surrounding, warmer air. Leaf temperatures of large cocklebur (*Xanthium strumarium*) leaves (not native to the desert) growing along a ditch bank in southern Arizona were as much as 9 to 11°C cooler than the surrounding air, which was 36°C under a clear, midsummer sky. On the other hand, leaf temperatures of this same species measured in Oregon on a cool summer morning were several degrees warmer than the surrounding air. Wind tunnel studies (Drake et al., 1970) showed that, to a considerable extent, this plant does regulate its leaf temperature by controlling transpiration, staying warmer than cold air and cooler than hot air.

The Alpine Tundra Alpine plants are faced with a different situation. Cool, fairly humid, gusty air combines with solar radiation levels that can be extremely high. Water is seldom limiting, but evaporative cooling is of no advantage in a situation in which ambient temperatures are often well below the range considered optimum for metabolic processes. Several hundred measurements of leaf temperatures in the alpine tundra by Salisbury and George Spomer (1963) showed that when the sun is irradiating the leaves, their temperatures can often be as high as 30°C, which can be as much as 20°C above the temperature of the air just a few centimeters away from the plants. Alpine plants typically have small or finely divided leaves, but they grow in a layer only about 10 cm above the ground, where wind velocities are greatly reduced. Many have a cushion or rosette form that results in a thick boundary layer that is determined by the entire plant and nearby soil instead of by individual leaves. High leaf temperatures clearly indicate that transpiration does not provide much cooling under these conditions. Not all the reasons for this are apparent, so integrated field and laboratory research would be in order.

Other Situations: Effects of Wind The ecological problems of transpiration and heat transfer are less challenging in less extreme environments, but careful study of heat transfer can provide interesting insights into them. The great variety of leaf shapes in moderate environments is itself suggestive.

Data obtained during the first half of this century often seemed contradictory. Some indicated that wind increased transpiration (as it always increases evaporation from a free surface); others indicated that wind decreased transpiration. When radiation loads are relatively low and leaf resistance is also low, transpiration is certainly increased by wind; if leaf temperature is below air temperature, increasing wind velocity always tends to increase transpiration. But it is now clear that transpiration can be decreased by wind when the radiation heat load is high, particularly if leaf resistance is also high (i.e., stomates are closed). Under such conditions, the leaf temperature can be far above the air temperature, which would cause a high transpiration rate if stomates are open; but the wind cools the leaf by convection, and this cooling is more effective in reducing transpiration than is the wind in decreasing the boundary layer and thus increasing evaporation.

4

The Ascent of Sap

According to the *Guinness Book of World Records* (1983), the tallest tree in the world is the "Tallest Tree," a *Sequoia sempervirens* in Redwood Creek Grove, Redwood National Park, California (Fig. 4-1). The height in 1970 was estimated to be 111.6 m (366.2 ft). The book goes on to say that, in 1880, a *Eucalyptus regnans* (mountain ash) in Victoria, Australia was measured at 114.4 m by a qualified surveyor. The tallest non-sequoia conifer, about 95 m, is a *Pseudotsuga Menziesii* (Douglas fir) at Quinault Lake Park Trail, Washington. The tallest living broadleafed tree is an *E. regnans* in the Styx Valley, Tasmania, measured at 99.2 m. A Douglas fir felled by George Carey in British Columbia in 1895 was claimed to be 127 m tall with a circumference of 23.5 m. This claim has been obscured, though not necessarily invalidated, by a falsified photograph. In any case, water has probably moved in some of the tallest trees, from the roots to the uppermost leaves, a vertical distance of well over 120 m. What, asks the plant physiologist, is the mechanism of this movement?

4.1 The Problem

Although we tend to take tall trees for granted, the more we cogitate on how water moves uphill at rapid rates to their tops, the more we appreciate the challenge of the problem. A suction pump can lift water only to the **barometric height,** which is the height that is supported by atmospheric pressure (1.0 atm) from below (about 10.3 m or 34 ft at one **atmosphere,** the normal air pressure at sea level). If a long pipe is

Figure 4-1 The Tallest Tree (*Sequoia sempervirens*) (arrow), located in Redwood Creek Grove, Redwood National Park, California. The tree to the left is much shorter but closer to the camera. (Photograph by Frank B. Salisbury.)

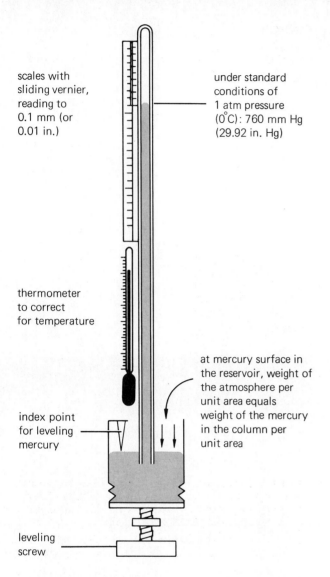

scales with sliding vernier, reading to 0.1 mm (or 0.01 in.)

under standard conditions of 1 atm pressure (0°C): 760 mm Hg (29.92 in. Hg)

thermometer to correct for temperature

index point for leveling mercury

at mercury surface in the reservoir, weight of the atmosphere per unit area equals weight of the mercury in the column per unit area

leveling screw

Figure 4-2 A mercury barometer.

sealed at one end and filled with water at some temperature, and then placed in an upright position with the open end down and in water, atmospheric pressure will support the water column to about 10.3 m. At this height the pressure equals the vapor pressure of water at its temperature (17.5 mm of Hg or 2.3 kPa = 0.0023 MPa at 20°C); above this height the water turns to vapor. At zero pressure, water will normally **vacuum boil** even at 0°C. (Actually, it boils at 0.61 kPa, its vapor pressure at 0°C.) When the pressure is reduced in a column of water so that vapor forms or air bubbles appear (the air coming out of solution), the column is said to **cavitate** (cavitation, noun). A laboratory barometer (Fig. 4-2) is analogous to our water-filled pipe but contains mercury instead of water. One atmosphere of pressure supports a column of mercury 760 mm high; 0.1 MPa or 1.0 bar supports 10.2 m of water or 750 mm of mercury.

To raise water from the ground level to the top of the Tallest Tree, a pressure at the base of 10.83 atm would be required, plus additional pressure to overcome resistance in the water's pathway and thus to maintain a flow. If overcoming the resistance requires a pressure about equal to that required to raise the water, a total of about 22 atm (11 plus 11) is necessary. To raise water to the top of the tallest tree that ever lived (say 150 m), a total of about 30 atm or bars (15 plus 15) might be required. (To be consistent we shall use 3.0 MPa instead of 30 atm or bars.) Clearly, water is not pushed to the tops of tall trees by atmospheric pressure.

Root pressures have been observed in several species. If the stem is cut from a grapevine, for example, and a tube with a mercury-manometer is attached, it can be seen that water is indeed sometimes forced from the roots under considerable pressure. Pressures of 0.5 or 0.6 MPa have been recorded, although in most species values do not exceed 0.1 MPa. Root pressures appear in most plants but only when ample moisture is present in the soil and when humidities are high, that is, when transpiration is exceptionally low. It is possible to see droplets of water exuded from openings in the tips or edges of grass or strawberry leaves, for example, a phenomenon called **guttation.** When plants are exposed to relatively dry atmospheres and low soil moisture, root pressures don't occur because water in their stems is under tension rather than pressure. Root pressures are not found in conifer trees (including the sequoia and the Douglas fir) under any conditions, although slight pressures have been observed in excised conifer roots. Furthermore, rates of movement by root pressure are too slow to account for the total water movement in trees. So we must reject root pressure as the means of moving water to the tops of tall trees, although it does operate in some plants sometimes.

How about capillarity? Most people who are unacquainted with the problem think that this is the mechanism that pulls water to the tops of trees. **Capillarity** is the interaction between contacting surfaces of a solid and a liquid that distorts the liquid surface from a planar shape. It causes the rise of liquids in small tubes. It occurs because the liquid wets the side of the tube (by adhesion) and is pulled up, which is evident from the curved **meniscus** at the top of the liquid column. As Fig. 4-3 shows, it is simple to calculate that liquids rise higher in tubes of smaller diameter. By the same token, it is easy to calculate that water will rise less than half a meter by capillarity in the xylem elements of plant stems, falling short by a factor of over 300 of accounting for the rise of sap in tall trees! Furthermore, a little reflection shows that capillarity cannot raise water in plants at all. Water rises in a small capillary tube because of the open meniscus at the top of the water's surface, but the xylem cells in plants are filled with water; they do not have open menisci. Menisci do exist at various locations in plants, and capillarity plays various roles in plant-water relations. For example, the submicroscopic menisci in the cell walls of plant leaves and other tissues are the *holding points* for xylem water. But they are not the source of movement.

During the 19th century, it was suggested that water moved up tree trunks in response to some living function or pumping action of stem cells. We can lift water to any height by pumping it over successive intervals, each one less than the barometric height. But careful anatomical study has failed to reveal any pumping cells. Indeed, most water moves in the dead xylem elements, which has been clearly demonstrated by the use of radioactive (tritiated) water as a tracer. Furthermore, in a paper published in 1893, Eduard A. Strasburger (a pioneer investigator of mitosis and meiosis in plants) tells how he sawed off trees 20 m tall but left them suspended upright in buckets of copper sulfate, picric acid, or other poisons. The fluids ascended all the way to the leaves, killing the bark as well as the scattered living cells (rays) in the wood. Water continued to move up through the trunk until the leaves died and transpiration stopped. He also scalded long sections of a wisteria vine, but sap continued to rise above 10 m.

The importance of living cells for sap flow in the wood cannot be overemphasized, however. The dead xylem in a plant was built by living cells, and new wood is laid down each year by the living, water-filled cambial cells. Yet the idea that water is pumped up through the trunk by cells along the way must be rejected. So how does water "flow uphill" to the tops of tall trees?

4.2 The Cohesion Mechanism of the Ascent of Sap

Around the turn of the century, a model was formulated to account for the rise of sap in tall trees. One of its elements, the cohesion of water, was not familiar from everyday experience, so the model was somewhat controversial. As a good hypothesis should, however, it suggested several consequences by which it could be tested. Now, after three quarters of a century, the numerous data that have accumulated support the model. Most difficulties and criticisms have been laid to rest, but we must still accept a visualization of reality that is not a familiar part of everyday experience—although entirely consistent with physics (reviews by Pickard, 1981; Zimmermann, 1983).

There are three basic elements of the **cohesion theory** for the ascent of sap: the *driving force, hydration* in the pathway, and the *cohesion of water.* The driving

Capillarity

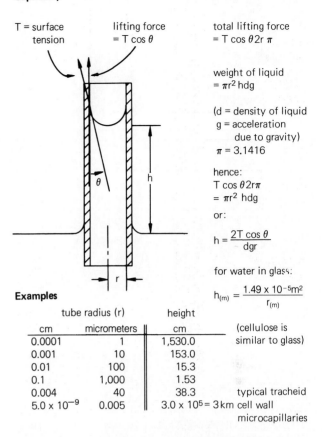

T = surface tension

lifting force = T cos θ

total lifting force = T cos θ2r π

weight of liquid = πr² hdg

(d = density of liquid
g = acceleration due to gravity)
π = 3.1416

hence:
T cos θ2rπ = πr² hdg

or:

$$h = \frac{2T \cos \theta}{dgr}$$

for water in glass:

$$h_{(m)} = \frac{1.49 \times 10^{-5} m^2}{r_{(m)}}$$

(cellulose is similar to glass)

Examples

tube radius (r)		height	
cm	micrometers	cm	
0.0001	1	1,530.0	
0.001	10	153.0	
0.01	100	15.3	
0.1	1,000	1.53	
0.004	40	38.3	typical tracheid
5.0 x 10⁻⁹	0.005	3.0 x 10⁵ = 3 km	cell wall microcapillaries

Figure 4-3 The principle of capillarity and the mathematics to predict the heights a liquid may be expected to reach in a tube of capillary dimensions. Examples of calculated heights are also shown. (See Nobel, 1983.)

force is the gradient in decreasing (more negative) water potentials from the soil through the plant to the atmosphere. Water moves in the pathway from the soil, through the epidermis, cortex, and endodermis, into the vascular tissues of the root, up through the xylem elements in the wood, into the leaves, and finally, by transpiration, through the stomates into the atmosphere. It is the special structure of this pathway (the relatively small diameter and thick walls that prevent collapse of the tubes), the osmotic potential of living leaf and stem cells, and the hydration properties of leaf parenchyma cell walls that make the system function. The hydration force between water molecules and the cell walls is caused by hydrogen bonding and is called **adhesion,** which is an attractive force between unlike molecules.

Cohesion is the key. This is the mutual force of attraction (also caused by hydrogen bonding, see Chapter 1) between the water molecules in the pathway. In that special environment, these cohesive forces are so great (water has such high tensile strength) that when water is pulled, by osmosis and

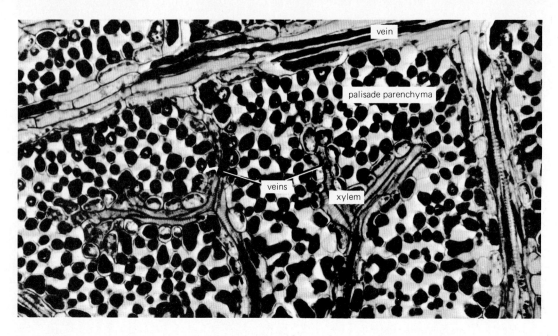

vein

palisade parenchyma

veins

xylem

Figure 4-4 A section cut parallel to the surface of a dicot leaf, showing the close proximity of vascular tissues to the mesophyll cells. (Photograph by William A. Jensen.)

evaporation, from its holding points in the cell walls at the top of a tall tree, the pull extends all the way down through the trunk and the roots into the soil. Whereas a water column in a vertical pipe of macro-dimensions would normally cavitate as just described, cavitation does not occur in the plant because of its highly specialized anatomy.

With this brief preview in mind, let us now examine each of the key points, beginning with the pathway and continuing with the nature of the driving force and the role of cohesion. (A thorough book on the topic, highly readable and often original, was authored by the late Martin H. Zimmermann of the Harvard Forest, Petersham, Massachusetts, 1983.)

4.3 The Anatomy of the Pathway

In the preceding chapter on transpiration, we considered the anatomy of the leaf (Figs. 3-3 and 3-4). Important features include the cuticle, the stomatal apparatus, and the considerable intercellular space with exposed moist cell surfaces (the mesophyll cells). Figure 4-4 shows the relationship between leaf cells and nearby vascular elements. It is obvious that no leaf cell is far from a vascular element. The following discussion and figures provide a review of stem and root anatomy. More details are in elementary botany or plant anatomy books.

The Vascular Tissues: Stem Anatomy In this chapter we are concerned primarily with the movement of water including dilute solutes through the xylem

tissues of stems and roots. In herbaceous monocot and dicot stems these tissues occur in vascular bundles associated with phloem tissues; these bundles are "open" in dicots, and often "closed" (meaning surrounded with a sheath of thick-walled fiber cells) in monocots. In woody tissues the xylem constitutes the wood, which is separated from the bark by a layer of cambium cells. The bark includes phloem and other tissues, such as cortex and cork. Figure 4-5 shows cross sections of a monocot and a dicot stem and a three-dimensional drawing of a woody stem showing xylem, cambium, and phloem. The phloem tissue is the site of transport for sugars and other products of assimilation, which are highly concentrated in the phloem sap. (Phloem transport is the topic of Chapter 7.) Xylem sap is a highly dilute solution, almost pure water. (Compositions of xylem and phloem saps are discussed in Chapter 7; see Table 7-1.)

Xylem tissue consists of four kinds of cells: **tracheids, vessel elements, fibers,** and **xylem parenchyma.** Fiber and parenchyma cells also occur in phloem. In xylem, especially in woody plants, only the parenchyma cells are living. These occur most abundantly in the **rays** that run radially through the wood of the tree, but some parenchyma cells are scattered throughout the xylem. It is the vertically arranged tracheids and vessel elements that are involved in transport of xylem sap. As a rule, gymnosperms (including conifers and their relatives) have only tracheids, whereas most angiosperms (flowering plants) have both vessel elements and tracheids. Both are elongate cells, but tracheids are

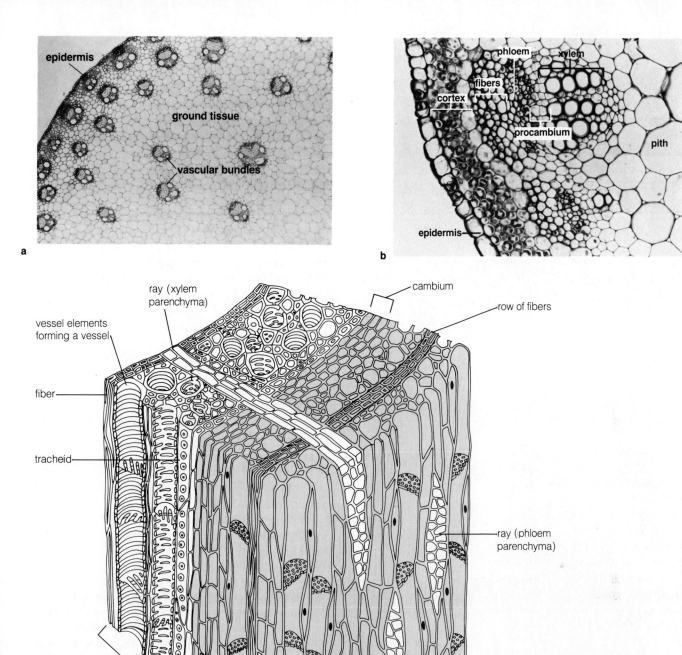

a epidermis · ground tissue · vascular bundles

b phloem · xylem · fibers · cortex · procambium · pith · epidermis

c ray (xylem parenchyma) · cambium · row of fibers · vessel elements forming a vessel · fiber · tracheid · ray (phloem parenchyma) · parenchyma cell · xylem · phloem · sieve-tube cells with sieve plates, companion cells

Figure 4-5 (**a**) A cross section through a typical monocot stem. Note the vascular bundles scattered in a ground tissue of pith. (**b**) A cross section through a typical herbaceous dicot stem. The vascular bundles form a ring with pith to the inside and cortex (typically with collenchyma cells—thickened corners in cross section) to the outside below the epidermis. In both types, xylem is usually (but not always) to the inside of the phloem. (**c**) A three-dimensional drawing of a woody dicot stem showing xylem (wood) to the inside of a layer of cambium, and phloem (part of bark) to the outside. (Micrographs courtesy of William Jensen.)

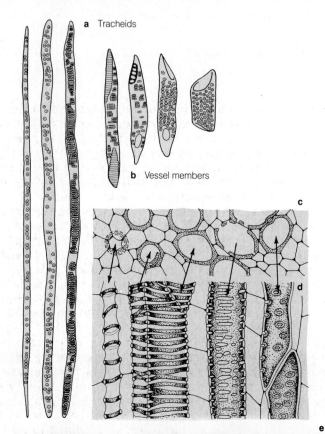

a Tracheids

b Vessel members

c

d

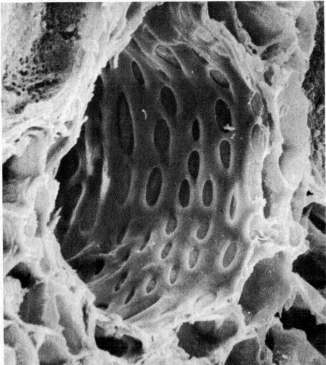

e

J. Troughton and L. A. Donaldson

Figure 4-6 (**a**) and (**b**) Tracheids and vessel members shown isolated from the tissue. Vessel members shown in cross section (**c**) and in longitudinal section (**d**). (**e**) Tracheid in pine as seen with a scanning electron microscope. Note that the vessels are shorter and thicker than tracheids, and note the pits in tracheids and vessels and the wall sculpturing in some of the vessels. (**a-d**, W. A. Jensen and F. B. Salisbury, 1984, *Botany*, p. 171. **e**, micrograph courtesy of John Troughton; see Troughton and Donaldson, 1972.)

longer and narrower than vessel elements (Fig. 4-6). They function as dead elements, that is, after they have been produced by growth and differentiation of meristematic cells, they die, and their protoplasts are absorbed by other cells. Before death, however, some changes in the walls occur that are important for water flow through them. One change is the formation of a **secondary wall,** consisting largely of cellulose, lignin, and hemicelluloses, that covers most of the **primary wall** (see discussion in the Prologue). This wall gives considerable compression strength to the cells and prevents them from collapsing under the extreme tensions that often exist in them. The lignified secondary walls are not as permeable to water as the primary walls, but in forming they leave **pits,** which are round thin places where the cells are separated only by the primary walls, which are quite permeable to water. Often the pits are **simple** (just a round opening); but sometimes, in both vessel elements and tracheids, they are somewhat more complex structures called **bordered pits,** in which the secondary walls extend over the center of the pit and the primary walls are swollen in the center of the pit to form a **torus** (Fig. 4-7). In electron micrographs, the primary wall around the torus appears porous. The figure shows that the torus can act as a valve, closing when pressure on one side is greater than pressure on the other.

Tracheid cells have tapered ends that overlap, as shown in Fig. 4-5. Pits in the tapered part allow water to move upward from one tracheid cell into the next; the tracheids thus form files of cells. Also there are numerous pits along the sides of tracheids allowing passage of water between adjacent cells. Vessel elements are usually strengthened by rings, spirals, or other thickenings like those that give rigidity to the hoses of vacuum cleaners (Fig. 4-6). They also have **perforation plates** at their ends; these plates have openings at which the secondary wall fails to form, and the primary wall and middle lamella dissolve away. These openings allow rapid movement of water, and vessel elements are aligned so that they form long tubes called **vessels** that extend from a few centimeters to several meters in some tall trees. Partially because of the perforation plates, resistance to water flow is usually considerably less in angiosperms than in the less porous tracheids of gymnosperms, but the relative diameters of the cells are especially important.

As we have noted, tracheids are relatively narrow, often in the range of 10 to 25 μm in diameter. Vessel elements are usually wider (say, 40 to 80 μm) and can be much wider (up to 500 μm = 0.5 mm, probably an upper limit). Rates of nonturbulent flow through small capillaries were studied independently in 1839 and 1840 by Gottfried H. L. Hagen in Germany and by Jean L. M. Poiseuille in France. The rate of flow is reduced by friction (adhesion) between the

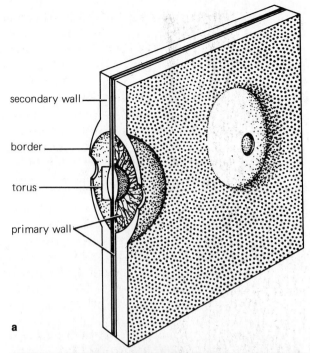

secondary wall

border

torus

primary wall

a

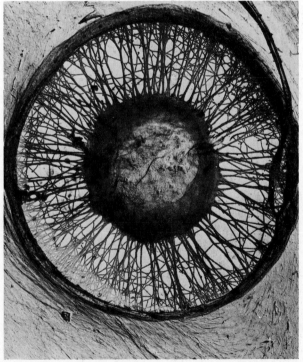

b

Figure 4-7 (**a**) Diagram of a bordered pit from a pine tracheid. If pressure on one side exceeds pressure on the other side, the torus will plug the hole, shutting off flow. (**b**) Transmission electron micrograph of a pit from Douglas fir (*Pseudotsuga*). The wood sample was dried by solvent exchange wih a low-boiling-point organic solvent that evaporated away in such a manner that pressures on opposite sides of the pit remained equal. The torus is the round structure in the center, and the pit membrane is seen to be highly porous and fibrous. The light circular area on the torus is the area of the pit aperature of the border behind the torus; the dark area is caused by the border, which is itself beyond the plane of focus. X7800 (Micrograph supplied by Wilfred A. Côté, Jr.; previously unpublished. See Comstock and Côté, 1968, for a description of the methods used.)

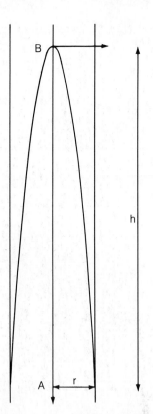

Figure 4-8 The flow paraboloid in a capillary of radius *r*, as calculated by Hagen and Poiseuille in the mid-1800s. If, at zero time, we could label all water molecules on a cross section of the tube at *A*, we would find them lined up on the surface of a paraboloid at time *t*. The fastest ones, in the center of the capillary, would have covered the distance *h* and reached point *B*. (From Zimmerman, 1983. Used by permission.)

fluid and the sides of the capillary tube; the molecules of fluid touching the capillary wall do not move at all, and the molecules in the center of the tube move the farthest in a given time. Thus, if at time zero all the molecules in a plane cross section of the capillary could be marked, and their positions along a longitudinal section could be noted at some later time, they would be found to form a parabola with the peak in the center (Fig. 4-8). Clearly, those molecules farthest from the walls of the tube flow the fastest. Indeed, Hagen and Poiseuille worked out an empirical equation for flow rate through capillaries as a function of capillary size and found the rate to be proportional to the *fourth* power of the radius of the capillary (details in Zimmermann, 1983).

Applying the Hagen–Poiseuille equation to tracheids and vessels produces some highly significant results. Consider a tracheid 20 μm in diameter and two vessels, one 40 μm and the other 80 μm in diameter. The relative diameters of the three tubes are 1, 2, and 4, but the relative flow rates will be 1, 16,

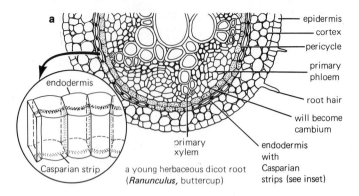

epidermis
cortex
pericycle
primary phloem
root hair
will become cambium
endodermis with Casparian strips (see inset)

endodermis

Casparian strip

a young herbaceous dicot root (*Ranunculus,* buttercup)

primary xylem

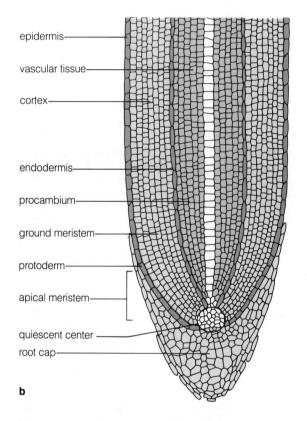

epidermis

vascular tissue

cortex

endodermis

procambium

ground meristem

protoderm

apical meristem

quiescent center

root cap

b

c

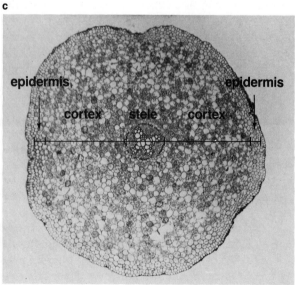

epidermis   epidermis

cortex   stele   cortex

and 256 (because of the fourth-power function). This means, all other factors being equal, that 256 times as much sap would flow through the 80-μm vessel as through the 20-μm tracheid.

Of course there are several complications. Applying the Hagen–Poiseuille equation to both situations assumes that friction between the sap and the inner walls of the tracheids and vessels is similar, but this would surely be influenced by the pitting and inner-wall sculpturing of the two kinds of transport elements and by the perforation plates of the vessels. The resistance as sap passes from one tracheid to another or from one *vessel* (not vessel *element*) to another also strongly influences flow rates. Tracheids probably have higher resistance to water flow because transfer occurs only through pits in the overlapping pointed ends, whereas transfer from one vessel to another often occurs over a considerable distance through *lateral* pitted walls as the two vessels lie in contact next to each other. (Transfer does not occur through vessel ends, which are not aligned.) In spite of the complications, it has been substantiated by measurements and calculations that flow through large vessels is much more rapid than it is through tracheids and small vessels. Indeed, in ring-porous wood, with large vessels laid down in early spring, often before leaves form, virtually all of the sap flow occurs through these vessels and only a small amount through the tracheids and smaller vessels.

Rapid flow through large vessels comes at a price, however, and the price is safety. There is a much greater chance that cavitation will occur in a large vessel than in a small one or in a tracheid. Indeed, in many species (e.g., *Castanea, Fraxinus, Quercus,* and *Ulmus*) cavitation occurs in the large vessels by the end of the year in which they are formed, so water transport is mostly dependent on large vessels that are less than a year old and on the smaller tracheids and vessels that form a backup system. No species with really large vessels occur in regions where freezing causes the expulsion of large bubbles of air, a situation that always leads to irre-

Figure 4-9 Anatomy of a young primary root. (**a**) A young dicot root in cross section. The inset shows the endodermal layer and the position of the Casparian strips in the radial walls. (**b**) A longitudinal section of a typical root. The cells surrounding the quiescent center divide to form the tissues of the root. (**c**) A light micrograph of a mature buttercup (*Ranunculus*) root at low magnification to show basic tissue patterns. There can be many variations. For example, there can be more or less cortex compared to stele (compare **a** and **c**), with pith in the center instead of xylem (many monocots such as maize), different arrangements of xylem, etc. (Micrograph courtesy of William A. Jensen; see Jensen and Salisbury, 1984.)

versible cavitation. (Small air bubbles may go back into solution.)

Some vessel elements with spiral thickenings exhibit the ability to elongate and grow while conducting water under tension. They grow as the surrounding cells (with contents under pressure) grow and pull them along, their spiral thickenings expanding like springs. Such growth occurs in the stems of most grasses, which support full-grown photosynthesizing and transpiring leaves.

In discussing anatomy of the pathway, we must also mention the cells in the **apical meristems** that produce the **primary xylem tissues,** and the **cambium,** consisting of living cells that divide to produce the **spring** and **summer wood** (both **secondary xylem tissues**). These are important features, because both meristems and cambium produce tracheids and vessels filled with sap.

Root Anatomy   Water enters the plant through its roots (Fig. 4-9). The xylem tissue in the root center is continuous with the xylem tissue in the stem. It is also closely associated with phloem tissue. As the root grows in diameter, cells between the xylem and the phloem form a vascular cambium that produces xylem tissue on the inside and phloem tissue on the outside.

The xylem and phloem elements are surrounded by a layer of living cells called the **pericycle.** The vascular tissue and the pericycle form a tube of conducting cells called the **stele.** Just outside the stele is a layer of cells called the **endodermis.** The endodermal cells are especially interesting and important from the standpoint of water movement in the plant, because their radial and transverse cell walls include thickenings called **Casparian strips** that are impregnated with **suberin,** which, like lignin and the cutin in the cuticle, is quite impermeable to water. The endodermal tangential walls (the inside and outside walls parallel to the surface of the root) are not so impregnated. Thus, water with its dissolved substances cannot pass around the endodermal cells via their walls but must pass directly through the cell itself (through the **protoplast** which is the cell contents, excluding the wall).

Outside the endodermis are several layers of relatively large, thin-walled living cells with intercellular air spaces at their edges. This is the **cortex.** The air spaces form interconnected air channels that seem essential for internal aeration. The cortical cell walls (primary) are highly permeable to water and its dissolved solutes, and thus it is quite likely that water can enter a root, moving via the cell-wall pathway in the cortex until it encounters the endodermal layer; it must pass directly through the endodermal cells (see Fig. 6-8).

The layer of somewhat flattened cells on the outside of the cortex is the **epidermis.** Some epidermal cells develop long projections called **root hairs** (see Figs. 6-3 and 6-8) that extend out among the soil particles around the root, greatly increasing the soil–root contact and enhancing water absorption and the volume of soil penetrated.

The root tips are continually growing through the soil, encountering new regions of moisture. A **root cap** protects the dividing meristematic cells and is continually sloughed off at the forefront and replenished by divisions of these meristematic cells. Since the tissues of the stele and the endodermis are formed as the cells in the meristem divide, enlarge, elongate, and differentiate, the stele will be open at the end where it is being formed. Could water enter through that end, bypassing the endodermal layer? Studies using dyes and radioactive (tritiated) water indicate that such movement is insignificant. Perhaps the cells in the meristematic region are so small and dense and their walls so thin that resistance to water movement is too high. Much of the water enters through the root hairs and their associated epidermal cells in the region of a young root where xylem vessels are mature and resistance is low.

The Apoplast–Symplast Concept   In 1930, E. Münch in Germany introduced a concept and terminology that are valuable in our discussion of the pathway of water and solute movement through plants. He suggested that the interconnecting walls and the water-filled xylem elements should be considered as a single system and called the **apoplast.** This is, in a sense, the "dead" part of the plant. It would consist of all cell walls in the root cortex including the tangential walls of the endodermis, but excluding the Casparian strip portion of the endodermal cells. All the tracheids and vessels in the xylem tissue would be part of the apoplast as would the cell walls in the rest of the plant, including those in the leaves, the phloem, and other cells in the bark. Except for the Casparian strips in the endodermal cells, the ascent of sap in a plant could take place entirely in the apoplast, particularly the xylem portion, but perhaps including the cell walls of the cortex and even the walls of living cells in the leaves.

The rest of the plant, the "living" part, Münch called the **symplast.** This includes the cytoplasm of all the cells in the plant, though some authors would exclude the large central vacuoles. The symplast is a unit, because the cytoplasm of adjoining cells is connected through plasmodesmata in the cell walls (see Figs. P-2 and 6-9).

Root Pressure   Based upon the apoplast–symplast concept, Alden S. Crafts and Theodore C. Broyer in 1938 proposed a mechanism to account for root pressure. A slightly modified version of their model still

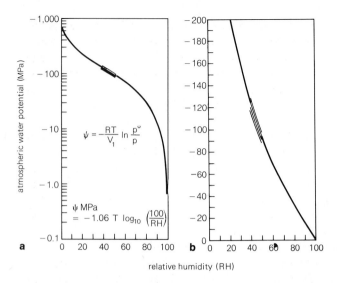

$$\psi = -\frac{RT}{V_1} \ln \frac{p^o}{p}$$

$$\psi\ \text{MPa} = -1.06\ T\ \log_{10} \left(\frac{100}{RH}\right)$$

relative humidity (RH)

Figure 4-10 The relationship of atmospheric water potential (20°C) to relative humidity, plotted on a logarithmic scale (**a**) and on a linear scale (**b**). The four thin lines are for different temperatures: 0° (bottom line), 10°, 20°, and 30°C (upper line). The curves were calculated with the equations shown, which are equations 2.4 and 2.5 on page 42.

seems reasonable. Assume that the root is in contact with a soil solution. Ions diffuse into the root via the apoplast (i.e., cell walls) across the epidermis, through the cortex, and up to the endodermal layer. Along the way, ions pass across cell membranes from the apoplast into the symplast in an active process that requires respiration. The result is an increase in the concentration of ions inside the cells (within the symplast) to levels higher than those outside (in the apoplast). Since the symplast is continuous across the endodermal layer, ions move freely into the pericycle and other living cells within the stele (Fig. 4-9). This might occur by movement through the plasmodesmata, and the velocity of movement inward might be increased by **cytoplasmic streaming,** the circular flowing of the cytoplasm that is often observed within such cells.

There is less oxygen within the stele than in more external cells. Crafts and Broyer suggested that this could result in a less efficient retention, so ions might leak into the apoplast (xylem) inside the stele. We now know that ions are actively pumped into the stele. In either case, the result is a buildup in concentration of solutes within the apoplast in the stele to a level higher than that of the soil solution (but still far below that of the cytoplasm and vacuoles), so the osmotic potential (π) in the stele is more negative than π in the soil. Because water must pass through the protoplasts of the endodermal layer, this layer acts as a differentially permeable membrane, and the root becomes an osmotic system. The buildup in pressure in this osmotic system must be the cause of root pressure. (Remember that root pressure does not commonly occur in most plants.)

4.4 The Driving Force: A Water-Potential Gradient

With this picture of the anatomy of the pathway in mind, we return to the mechanism of the ascent of sap. The first question concerns the driving force: Is there a water-potential gradient from the soil through the plant to the atmosphere sufficient to pull water along the pathway?

Atmospheric Water Potential The key to understanding is to realize the great capacity that dry air has for water vapor. As the RH of air drops below 100 percent, its affinity for water increases dramatically. This is shown by the rapid drop in the water potential (ψ)* of increasingly dry air (Fig. 4-10 and equations 2.4 and 2.5). At 100 percent RH (at all temperatures), ψ of air equals zero. At 20°C, ψ of air at 98 percent RH has dropped to −2.72 MPa (enough to hold a column of water 277 m high!); at 90 percent RH, ψ = −14.2 MPa; at 50 percent RH, ψ = −93.5 MPa; and at 10 percent RH, ψ = −311 MPa. Since soil water that is available to plants seldom has a water potential more negative than −1.5 MPa, air doesn't have to be very dry to establish a steep water-potential gradient from the soil, through the plant, and into the atmosphere. If the soil is fairly wet, a drop of only about 1 percent below 100 percent RH will establish the water-potential gradient.

The Role of Shoot-Cell Osmosis To grow, and usually even to remain alive, living cells in plants must be able to obtain water from the apoplast. This means that the water potential in living cells has to be slightly more negative than that in the surrounding walls and in the tracheids and vessels of the xylem. As we saw in Chapter 2 (Table 2-1), osmotic potentials of living plant cells are usually on the order of

---

*In Chapter 3, we noted that the rate of transpiration was proportional to the vapor-density difference between the inside of the leaf and the atmosphere beyond the boundary layer. This has often been experimentally determined (e.g., Wiebe, 1981). Yet the water potential of the air is a logarithmic function of the vapor density. So the rate of transpiration cannot be directly proportional to the water-potential difference between leaf and air. Nevertheless, the tendency for air to hold water vapor, the tendency that establishes the driving force for the ascent of sap, seems to be better indicated by the water potential of the atmosphere than by its vapor density. More work may be needed to clarify what presently appears to be an inconsistency in our application of the thermodynamic, water-potential concept to the study of plant water relations.

−0.5 to −3.0 MPa. Of course, osmotic potentials can be more negative than this: −4.0 to −8.0 MPa in halophytes. Pressures within living cells are usually on the order of 0.1 to 1.0 MPa, meaning that cell-water potentials will range from perhaps −0.4 MPa to more negative than −4.0 MPa. As noted already, sap in the apoplast, including the tracheids and vessels, is usually quite dilute, having an osmotic potential on the order of −0.1 MPa, or even less negative. But the cohesion theory says that water in the apoplast, particularly in the xylem, is under tension. This is expressed as negative pressure. Since we have reasoned that the osmotic potential of water in the apoplast must be less negative (higher) than that of water in the living cells, we can deduce that tensions in the apoplast will not be more negative than −0.3 to perhaps −3.5 MPa, which is a fairly wide range.

This line of reasoning leads to another important conclusion: In thinking of how sap gets to the leaves and stems of young plants and to the tops of tall trees, we could begin by ignoring the extremely negative atmospheric water potentials that we have just discussed. Even if the atmosphere were always at 100 percent RH throughout the boundary layer around the leaf, so that transpiration did not occur, sap could still be lifted to the tops of small plants and tall trees. The driving force would be the negative water potentials within living cells, these being established by the low (negative) osmotic potentials and low (positive) pressures in cells. If a plant could grow from a seedling to a large tree without ever losing any water by transpiration, water would still be entering the dividing and elongating cells in the meristems and stems and the expanding cells of the leaves osmotically, and this movement into living cells would pull columns of water up through the pathway into the plant. As it is, growing cells must compete with transpiration for the water they get.

The Role of Cell-Wall Hydration Having seen the importance of water movement into living cells by osmosis as a driving force that could raise water in the xylem even if there were no transpiration, now consider the usual situation in which transpiration does play an important role. Suppose, for example, that the atmosphere is quite dry and soil water begins to be depleted. Suppose that transpiration causes the water potential in the apoplast to be much more negative than that in the living cells; then water diffuses out of the cells, they lose pressure, and the soft tissues of the plant wilt. Will the columns of water in the xylem elements then fall back toward the soil, leaving empty tracheids and vessels? No, because in addition to the holding power of negative water potentials in living cells, there is the much greater holding power of hydration within the cell walls of the apoplast itself.

We can arrive at this conclusion in two ways. First, we can imagine that as the amount of water diminishes in the cell walls or the xylem elements of leaves, curved menisci begin to form between the cell-wall polysaccharides and in the intercellular spaces. The numbers in the table accompanying Fig. 4-3 show that the radii of curvature in these microcapillary pores would be so small that the water would be held with tremendous force, force strong enough to hold a column of water 3000 m high against gravity. Second, it is not even necessary to think of curved menisci to arrive at these figures. We have discussed matric potential (Chapter 1) and the forces of hydration or attraction of water molecules to hydrophilic surfaces (Chapter 2). Expressed in units of pressure comparable to those we have been discussing, water can be held by hydrophilic surfaces with tensions on the order of −100 MPa to −300 MPa. Clearly gravity (the weight of water in the xylem columns) could not remove water against such powerful forces unless the water columns were as tall as mountains. Yet a dry atmosphere can remove water even against the forces of hydration: If the RH is 1 percent at 20°C, the water potential of the air is −621 MPa.

Actually, such extreme tensions do not develop in plants. When the water potential in a drying plant drops to some critical level, cavitation occurs in the transporting elements. In 1966, H. A. Milburn and R. P. C. Johnson published a method of studying such cavitations. They attached an extremely sensitive microphone to a detached petiole and leaf of castor bean (*Ricinus communis*). As the leaf dried, the cavitation of water in a single vessel caused a distinct click, which was recorded. The clicks could be stopped by adding a drop of water to the cut end of the petiole or placing the leaf in a polyethylene bag. Placing the leaf in water for 24 h or (especially) vacuum infiltrating it led to recovery, from which one could infer that the vessels were again filled with water. The total number of clicks for a leaf (about 3000) were reduced by about 10 percent if the leaf was allowed to recover and then to wilt again, from which one could infer that 10 percent of the vessels were permanently vapor locked, probably with air.

Zimmermann (1983) has suggested that plants are constructed in such a way that cavitation occurs when water potential reaches a critical low level in a manner that allows refilling of the tracheid or vessel if conditions are right. Zimmermann calls the phenomenon **designed leakage** brought about by **air seeding.** According to the equation of Fig. 4-3, the meniscus in a capillary tube of 0.2 μm diameter exerts a force on the water below it that is equivalent to a pressure of about 1.4 MPa. Thus, if the pores of the cell walls (e.g., in the porous part of the pit membranes) of tracheids or vessels are about 0.2 μm in functional

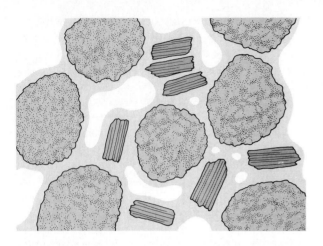

Figure 4-11 Diagram illustrating the concept of capillary water in soil, in which the large particles represent silt or sand and the smaller particles represent clay. Water is adsorbed to the particle surfaces by hydrogen bonding, hydrating the particles. Forces of hydration extend farther from the more highly charged clay surfaces. Curved surfaces are the menisci that appear in the capillary pores of the soil; they result from surface tension in the water. (Compare Figure 4-3.) (From W. A. Jensen and F. B. Salisbury, 1984, *Botany*, p. 253.)

diameter, a meniscus could not pass through them until the pressure in the cell was more negative than -1.4 MPa relative to outside pressure (assuming nearly atmospheric pressure in the intercellular air spaces). When the tension did become more negative, the meniscus would be pulled through the pore, allowing a minute quantity of air to enter. This is the *air seeding.* The air would lead to vacuum boiling, and the air and water vapor would immediately and explosively expand to fill the tracheid or vessel (accounting for the cavitation clicks of Milburn and Johnson, 1966). Note that the vessel would not be filled with air but mostly with water vapor at its vapor pressure for the prevailing temperature. The pressure in the cell would thus be positive but quite low (about 2 to 3 percent of atmospheric pressure at room temperature). Such pressures would immediately seal all the pores such as the one through which the air seeding occurred, since a water meniscus could never be pulled through such a pore by such small pressure differences on each side of the cell wall. Furthermore, when conditions were such as to increase the vapor pressure in the xylem above that of the vapor pressure of water for the particular temperature (as during a rain or by root pressure at night), the cells might refill by condensation of vapor and solution of the minute amount of air present. The situation would be analogous to the restoration of conduction in cells by application of water in the experiments of Milburn and Johnson.

Soil Water Consider a soil recently wet by a saturating rainstorm. Before the water has a chance to drain from the soil in response to gravity, the soil is said to be at **saturation.** Much of the water (called **gravitational water**) moves down through the pore spaces between the soil particles. If the concentration of ions is dilute (true of most soils), the water potential of this water is close to zero. After several hours, or even a day or so, the only water left in the soil is that which can be held against the force of gravity by the adhesion between the water molecules and the soil particles. This remaining water exists in capillary pores that are often wedge-shaped and seldom cylindrical (Fig. 4-11). Soil that contains all the **capillary water** it can retain against gravity is said to be wet to **field capacity.**

As water is removed from the soil by evaporation and plant roots, the water that remains is held in ever thinner water films, so its water potential becomes increasingly negative. Finally, the water potential in the soil and in the plant roots is nearly the same, so the roots no longer remove water from the soil. Water continues to transpire from the shoots for awhile, so the plant wilts. Even if the plant is placed in an atmosphere with 100 percent RH, so that it no longer loses water by transpiration, it still is not able to obtain enough water to overcome the wilting. At this point the amount of soil water is said to be at the **permanent wilting percentage.** Soil scientists have agreed to consider soil with a water potential of -1.5 MPa to be at the permanent wilting percentage, although most plants, especially those of the desert and those that can grow in salty soils, can remove water when the soil water potential is more negative than -1.5 MPa (see the personal essay of Ralph Slatyer in Chapter 2). At such low water potentials, however, water moves very slowly in soils (thousands of times slower than at field capacity), so most plants cannot absorb water rapidly enough to supply their needs. Their roots must grow to the water—but there is not enough water to support much root growth!

The amount of water in the soil at field capacity and at the permanent wilting percentage depends strongly upon soil texture and the amount and kind of soil organic matter. The **clay** fraction of soil consists of extremely small particles in the colloidal size range and with much hydrophilic surface. Water below the field capacity is thus held largely by the attractive forces between water molecules and the surfaces of clay particles and **humus** (soil organic matter decayed to the point that plant structures are not recognizable). Thus, by hydration, water molecules can be held against forces of tens to hundreds of megapascals in the soil as well as in the apoplasts of plants. The apoplasts might be able to compete with the dry soil for water, but if the plant is to live and grow, water must move into living cells. Their water potentials can never be as negative as the water potentials of soil much dryer than the permanent

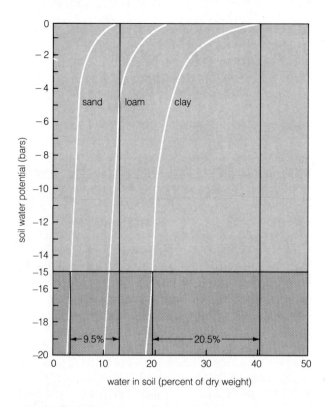
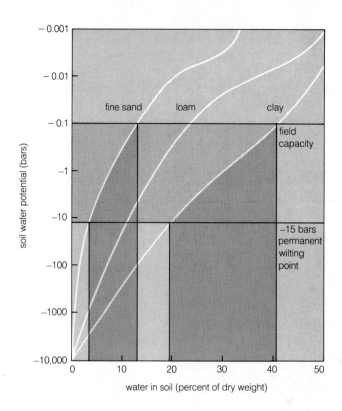

Figure 4-12 Soil water potential is a function of the amount of water in clay, loam, and sandy soils. Soil water between field capacity (ψ = -0.001 MPa) and the permanent wilting point (ψ = -1.5 MPa, arbitrarily) is considered to be available to plants. (**a**) Curves plotted on a linear scale. As water potential becomes more negative (downward on the ordinate), water is held more tightly by the soil particles; that is, it becomes less available to plants. Note that the water available to plants in the fine sand is only about 9.5 percent of the dry weight of the sand, while available water in the clay amounts to about 20.5 percent of the clay's dry weight. (**b**) Curves plotted on a logarithmic scale. The curves represent exactly the same data as in (a), but the logarithmic scale allows a much wider range of water potentials to be shown. (From W. A. Jensen and F. B. Salisbury, 1984 *Botany*, p. 255.)

wilting percentage, so such soils not only cause plants to wilt but also limit their growth and ultimately lead to the deaths of plant cells.

Although water in a soil that is below the permanent wilting percentage has an extremely low tendency to diffuse, it can be driven off by high temperatures—that is, by supplying a lot of energy. That is how we measure water in a soil: First the sample of soil is weighed, then it is held at temperatures just above the boiling point for several hours, and finally it is weighed again. The difference in

Table 4-1   Classifying Soil Particles by Diameter.

Particle	USDA System	World System
Coarse sand		2.0 to 0.2 mm
Fine sand	2.0 to 0.05 mm	0.2 to 0.02 mm
Silt	0.05 to 0.002 mm	0.02 to 0.002 mm
Clay	0.002 mm and smaller	0.002 mm and smaller

Source: W. A. Jensen and F. B. Salisbury, 1984, *Botany*, p. 253.

weight is that of the water driven off at high temperature.

Soil texture refers to the size of the particles that make up a soil. A soil with a given texture has a specific distribution of particle sizes. The various sizes are named according to two systems, as shown in Table 4-1. Practically all soils are mixtures of sand, silt, and clay. Soils with about 10 to 25 percent clay and the rest about equal parts of sand and silt are called **loams.**

Soils rich in clay and humus can hold the most water, but sometimes this means that there is less air space between the particles. Such soil may lack oxygen and thus may not be ideal for plant growth. Since plants require oxygen to support their root respiration, root penetration may be restricted in heavy clay soils. Sandy loams and clay loams, which make up many of our good agricultural soils, hold adequate water, have ample air space, and are easily penetrated by roots. Figure 4-12 shows soil water potential as a function of the amount of water in clay, loam, and sandy soils. Note that the soils with more clay can hold more water that is available to plants.

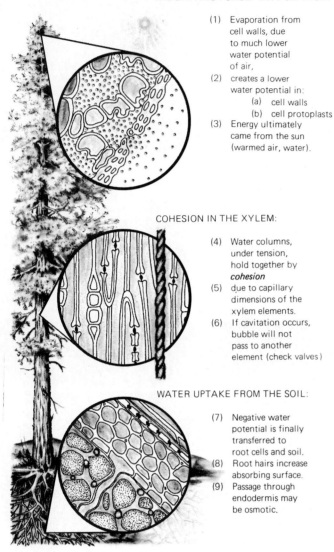

(1) Evaporation from cell walls, due to much lower water potential of air,

(2) creates a lower water potential in:
    (a) cell walls
    (b) cell protoplasts

(3) Energy ultimately came from the sun (warmed air, water).

COHESION IN THE XYLEM:

(4) Water columns, under tension, hold together by *cohesion*

(5) due to capillary dimensions of the xylem elements.

(6) If cavitation occurs, bubble will not pass to another element (check valves)

WATER UPTAKE FROM THE SOIL:

(7) Negative water potential is finally transferred to root cells and soil.

(8) Root hairs increase absorbing surface.

(9) Passage through endodermis may be osmotic.

Figure 4-13 A summary of the cohesion theory of the ascent of sap.

4.5 Tension in the Xylem: Cohesion

The second question concerns cohesion. Will it hold the water columns together? An Irish botanist, Henry H. Dixon (1914), carefully formulated the hypothesis that the tensions created by transpiration, osmotic uptake of water by living cells, and the hydration of the cell walls (all these drawing water from the pathway within the plant) were relieved by an upward movement of water from below, the columns of water being held together by cohesion (Fig. 4-13). Dixon and John Joley, a physicist colleague, began to develop the idea as early as 1894, as had E. Askenasy in 1895 (see 1897 reference) and Otto R. Renner in 1911. But it was published by Dixon in book form with a vast body of supporting data in 1914, so it is often called the **Dixon cohesion theory.**

The cohesion question breaks down into sub-questions, five of which form the subheadings of the following discussion. The cohesion hypothesis succeeds well at suggesting experimental approaches to its investigation. By now, enough data have accumulated to place the cohesion hypothesis on a solid footing.

Does water have a high enough tensile strength? The problem is whether water can sustain tensions of up to −3.0 MPa without cavitating. Determination of the tensile strength of water was a difficult task. Three approaches are worth discussing, and the third seems conclusive enough to provide a final solution.

First, knowledge about hydrogen bonding in water suggests that potential cohesive strength under ideal conditions is extremely high, enough to resist a tension of several hundred megapascals under ideal conditions. But our theories of liquids remain imperfect, and some assumptions must be made.

Second, several experimental measurements suggest high tensile strengths. The force required to separate steel plates held together by a water film has been measured. Fern annuli that are pulled apart by water under tension have been studied. Glass tubes were sealed while full of hot expanded water, and then cavitation was observed as the water cooled and contracted. These and other methods have produced values for the tensile strength of water, and even of tree sap with dissolved solutes and gases, on the order of −10 to −30 MPa, though a few workers measured only lower values, on the order of −0.1 to −3.0 MPa. These methods, designed to measure the tensile strength of water, also of necessity measure the ability of the solid surface in contact with the water to act as a seeding source for cavitation. The slightest bubble of air on the surface, for example, rapidly expands to produce cavitation, or air might be pulled through pores, as in Zimmermann's designed leakage. Hence, results probably reflect interface conditions instead of tensile strengths of water. The system must work as well as it does in plants because these interface characteristics allow water tensions on the order of −1.5 to −3.0 MPa.

Third, Lyman Briggs, former director of the National Bureau of Standards, published some especially impressive experimental results in 1950. He used capillary glass tubes bent in the form of a Z (Fig. 4-14). Centrifuging these tubes causes a tension on the water at the center, and the tension present when the water column breaks can easily be calculated. Briggs found that the greatest tensions appeared in the tubes of smallest diameter. With rather fine capillaries, values as negative as −26.4 MPa were measured, but even with capillary tubes of 0.5 mm diameter (the size of the largest vessels), air-saturated tap water did not cavitate under tensions of −2.0 MPa. Cavitation did occur when the center of the Z-tube was frozen with dry ice; since air is virtually insoluble in ice, freezing forced the dissolved air out of the water.

It is not obvious why the most negative pressures appeared in the narrowest tubes, but it could simply be that the larger tubes had the most surface and the most volume and therefore were statistically most likely to have points where cavitation could begin. More subtle reasons might involve the degree of curvature of internal tube surfaces; could flatter surfaces be more likely to harbor points of cavitation? Such possibilities need investigation, but Briggs' results support the observation that cavitation is more likely to occur in large vessels than in small tracheids.

What are flow velocities? Are the water columns really continuous? Results of the best and most recent experiments clearly indicate that the water columns are continuous. The observed flow velocities in stems clearly imply that the xylem elements are filled and continuous. The most elegant method to measure flow velocities was introduced by H. Rein in 1928 for measurement of blood flow velocities in animals. B. Huber and E. Schmidt (1936) made many measurements of sap velocities with the method, which consists of heating the liquid briefly at a point along the stem and then measuring (with a sensitive thermocouple) the time of arrival of the heat pulse a few centimeters downstream. There are many problems. We've seen (Fig. 4-8) that velocities depend on distance from the wall of the capillary, for example, but heat-pulse methods can at least give comparative information. As expected, velocities in wide-vessel trees such as ash are much greater than those in narrow-vessel trees such as birch. When velocities are measured at different points along the trunk, it can be seen that the rates increase first at higher points along the trunk in the morning as transpiration begins, strongly suggesting that transpiration pulls the sap up the trunk. Some reported peak velocities (at noon) range from 1 to 6 m h^{-1} for narrow-vessel trees to 16 to 45 m h^{-1} for trees with wide vessels. Slowest velocities occur in conifers.

Are the columns really under tension? In 1965, Per Scholander and his colleagues at the Scripps Institute of Oceanography in California published the elegantly simple and satisfactory **pressure bomb method** (see Fig. 2-13), which provided a way to measure tension in stems.* Scholander reasoned that if the water in the xylem of a stem is under tension, then the pressure outside compresses the cell walls of

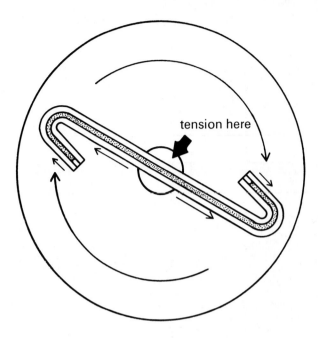

Figure 4-14 Method of measuring the cohesive properties of water utilizing a centrifuged Z-tube. Small arrows indicate the direction of centrifugal force and the principle of balancing. The shape of the Z-tube prevents water from flying out either end of the tube.

the xylem. Therefore, when the stem is cut, allowing pressure inside to equal pressure outside, the cell walls should expand, and the water columns in the xylem elements should recede from the cut surface. If the pressure difference is reestablished by increasing the pressure on the outside of the stem until the pressure difference is the same as it was before the stem was cut, the water should move back exactly to the cut. The method of testing this consists of cutting a twig from a tree, placing it in a pressure bomb, and increasing the gas pressure on the branch until the water in the xylem can be observed with a hand lens to return to the cut surface. The pressure in the bomb should then be equivalent to the absolute value of the tension in the stem before the cut.

Scholander and his colleagues measured tensions in stems under a variety of conditions. Results for different environments are shown in Fig. 4-15. Tensions were always observed, and they varied from a few tenths of a megapascal below zero to more negative than −8.0 MPa, the limit of the instrument. (The highest tensions were in a creosote bush, a shrub less than 2 or 3 m tall.) Measurements varied considerably for each plant, but trends are clear: Forest and freshwater species have the least negative tensions, and desert and seashore plants, growing in soils that are likely to be salty, have more negative tensions. As might be expected, tensions at night are less negative (less transpiration). Root pressures were never observed under these conditions, even at night.

---

*It is an interesting historical footnote that Henry Dixon, who developed the cohesion theory, understood the principle of the pressure bomb and constructed glass models (see pp. 142–154 in his book). After two rather serious explosions, he abandoned this approach!

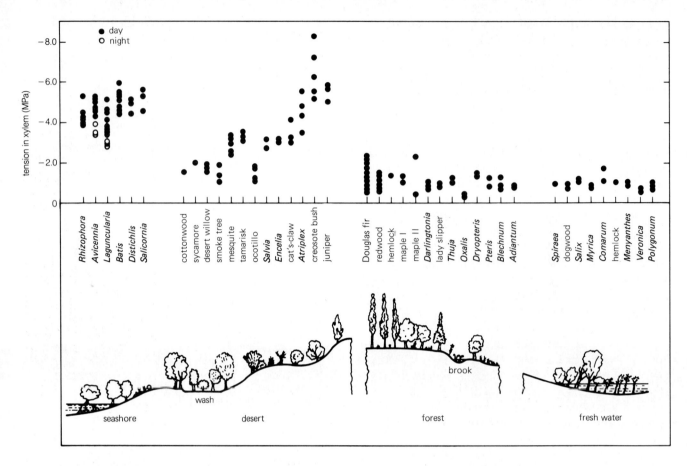

Figure 4-15 Negative sap pressures in a variety of flowering plants, conifers, and ferns. Most measurements were taken with a pressure bomb during the daytime in strong sunlight. Night values in all cases are likely to be several tenths of megapascals higher (less negative). (From Scholander et al., 1965; used by permission.)

In another interesting study (Fig. 4-16), a high-powered rifle was used to knock twigs off a tall Douglas fir tree at 27 and 79 m above the ground. Twigs were quickly placed in a pressure bomb. As might be expected, tensions varied with time of day, being most negative around noon when light levels were highest and humidities were low. The difference in heights between the two samples was about 52 m. The hydrostatic pressure difference for 52 m would be 0.51 MPa, which is close to the observed value of about 0.5 MPa. If water is to flow, the gradient must be greater than that established by gravity, so it is surprising that the measured gradient was so close to the hydrostatic gradient. Zimmermann (1983) suggests that this may be because twigs had to be used, and they probably (according to many studies cited by Zimmermann) had much more negative water potentials than xylem in the trunk. Sometimes, especially in small herbaceous plants, pressure gradients are much steeper (e.g., 0.08 MPa m^{-1} in tobacco; Begg and Turner, 1970). Reverse gradients in a tall eucalyptus during a rain, indicating downward water flow in the xylem, were reported by Legge (1980).

You can observe tensions in plant stems simply by immersing the stem of a transpiring plant in a dye solution and cutting through the stem. The dye instantly moves a considerable distance both up and down the stem inside the xylem elements and then stops. As with the pressure bomb experiments, the cut suddenly releases the tension on the walls of the xylem tubes, allowing them to expand and pull in dye solution.

Renner in 1911 performed an elegantly simple experiment. He attached a leafy branch to a burette to measure water uptake. He constricted the stem with a clamp to produce a high resistance. After measuring uptake under these conditions, he cut off the leafy end and applied suction with a pump that produced a tension of about −0.1 MPa. Only about one tenth as much water moved in response to the pump as in response to the leafy branch. We are compelled to conclude that the leaves exert a pull of some −1.0 MPa.

Except for plants showing guttation water, suggesting that root pressures are producing positive pressures in the xylem, water in the xylem of land

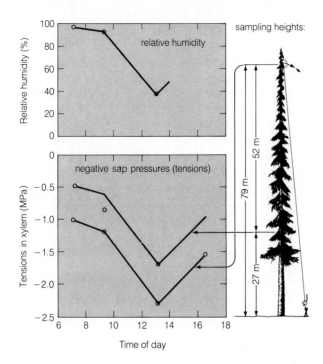

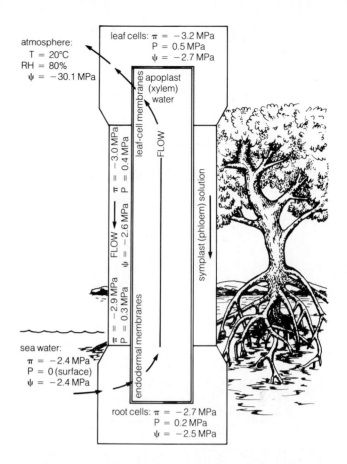

Figure 4-16 Negative sap pressures (tensions) showing hydrostatic gradients in a tall Douglas fir tree as a function of time of day and height. Pressures were measured by shooting down twigs and placing them in a pressure bomb (see Figure 2-12). The circles represent measured values. The weight of a column of water 52 m high (the distance between the two sampling heights) produces a pressure at the base of 0.503 MPa. This is the expected hydrostatic gradient. Thus the points taken from the highest sampling level are connected by lines (bottom lines in lower graph), and the lines above (for the lower sampling level) are drawn parallel and 0.5 MPa above (0.5 MPa less negative). Two of the three points fall almost on this calculated curve and the third point is not far away, demonstrating that the calculated hydrostatic gradient did exist in the tree. But why is no resistance component apparent? See text for discussion. Note that pressures become more negative as relative humidity drops (top graph). (Data from Scholander et al., 1965; used by permission. The data are typical of several measurements that were made. Figure from W. A. Jensen and F. B. Salisbury, 1984, *Botany*, p. 279.)

Figure 4-17 Water relations of a mangrove tree growing with its roots immersed in sea water. The diagram indicates the "essential" parts of the mangrove tree in this context. The endodermal membrane keeps salt out of the xylem (except for negligible amounts), and the leaf-cell membranes maintain a high solute concentration in the cells. The result is that water in the xylem must be under considerable tension both day and night to remain in equilibrium with sea water, and leaf cells have such a negative osmotic potential that they absorb water from the xylem in spite of its tension and low water potential. Only osmosis keeps the leaf cells from collapsing. (Data based on Scholander et al., 1965, but the hypothetical numbers have been modified to better match the discussions in this chapter and in Chapter 7.)

plants in summer must nearly always be under tension. This is certainly what pressure-bomb measurements have shown (e.g., see Fig. 4-15). Scholander (1968) studied a situation in which tensions in the xylem must always, day and night and in all weather conditions, be at least as negative as −3.0 MPa. Tropical mangrove trees grow with their roots bathed in sea water, which has an osmotic potential of about −3.0 MPa. As water enters the tree through the roots, salts are excluded, probably by the endodermal layer, so that the xylem sap is almost pure water with an osmotic potential of nearly zero. (Several other halophytes also have the ability to exclude salt.) The reason that water in the xylem sap remains in the tree and does not move out through the roots into the sea

water is that it has a water potential of −3.0 MPa or lower (Fig. 4-17). This is achieved by constant xylem tensions of −3.0 MPa or below. Scholander measured the predicted tensions with his pressure bomb. By this ultrafiltration mechanism, the mangroves avoid a lethal buildup of salt in their leaves.

Obviously, if the xylem sap in mangroves has a water potential of −3.0 MPa, the protoplasts of leaf cells must also have water potentials at least as negative as −3.0 MPa if they are to remain turgid. If they have a turgor pressure of about 0.5 MPa, then their osmotic potentials have to be about −3.5 MPa or lower if water is to enter. Osmotic potentials of −3.5 MPa have been measured in mangrove leaf cells. Ordinarily, trees that don't have their roots in salt

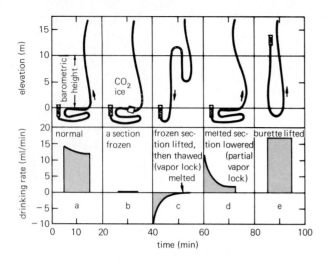

Figure 4-18 Scholander's experiments with tropical rattan vines (*Calamus* sp.). (**a**) The vine is cut off under water and a burette is attached, allowing measurement of the rate of water uptake. If the burette is stoppered, water continues to be taken up anyway until a vacuum is created in the burette, and the water boils. (**b**) To freeze the water in the vine, the burette first had to be taken off so that air entered all the xylem elements, vapor-locking the system. Then after freezing, the vapor locked portion (about 2 m) was cut off under water and the burette attached again. There was still no water uptake, indicating that freezing had indeed blocked the system. (**c**) If the vapor-locked portion was hoisted above the barometric height and allowed to thaw, some water ran out, but there was no uptake, indicating that the system was now vapor-locked. (**d**) If the vapor-locked portion was lowered to the ground, there was a rapid initial uptake as vapor condensed to water, breaking the vapor lock, but then uptake was slower than originally because some air had been excluded by freezing. (**e**) When the burette was elevated 11 m, rate of water uptake returned to the original level, indicating that the vapor lock had now been completely eliminated. (From Scholander et al., 1961.)

water have similar, if somewhat less spectacular, water relations. Tensions normally exist in their conductive xylem tissues, but their turgid leaf and other living cells have osmotic potentials even less negative than the water potentials of the xylem sap with its negative pressures. In any case, the importance of osmosis in the plant is clearly demonstrated. Without the differentially permeable membranes around the living cells and the highly negative osmotic potentials of the sap inside, the high tensions in the xylem system would lead to collapse of the living tissue. Without osmosis, plants would collapse!

What if the columns cavitate? Perhaps the most important question raised by plant physiologists about the cohesion hypothesis concerns what would happen if the continuous columns of water should in some way be broken. Say, for example, that a strong wind bends the stem, further stretching the columns of water and causing them to cavitate, or that the water in the tree trunk freezes, forming bubbles of gas. Or

what would happen if some columns were broken by sawing part way through the trunk? Investigators have taken many approaches to solve these and other problems. Figure 4-18 illustrates other elegant experiments of Scholander and his co-workers. They nicely support the cohesion theory.

Parts of the answers to our questions are found in the anatomy of xylem. Pathways of water have been studied by the use of dyes. A hole can be drilled in the trunk of a tree, for example, and dye inserted (summarized by Zimmermann, 1983). Although one must be careful to account for lateral movements caused by the relief of pressure when the hole is drilled, much can be learned about the patterns of vessels by following the dye above and below its insertion.

Another approach to understanding the anatomy is to make micrographs on movie film of consecutive cross sections through a stem. Upon projection, the longitudinal dimension of the stem is represented by time, and movements of individual vessels or vascular bundles (especially in monocots) can be followed. Their positions on various frames can also be measured so that individual vessels can be plotted in three-dimensional drawings. The conclusion of these and the dye studies is that vessels seldom if ever simply lie parallel to each other. When dye is inserted at a point in a trunk, for example, it spreads circumferentially around the trunk as it moves up and down, mostly within a single growth ring (except for the complications of radial transport in the rays). The spread amounts to about 1 to 2°, which equals about 17 to 35 mm per meter of trunk. Studies of individual vessels or vascular bundles show the same thing: a twisting and moving apart with progression up (or down) the stem. Usually the anatomy is highly complex. The conclusion is that water entering the xylem of any root is spread throughout the entire trunk, so that it reaches virtually all the branches that form the crown. Thus, if a given root is damaged, no part of the crown suffers from lack of water. (This is much less the case for small herbs.)

With this information about the twisting and spreading of vessels in the xylem, the results of saw-cut experiments seem far less mysterious. At least as long ago as 1806 (Cotta, cited in Hartig, 1878 and Zimmermann, 1983), researchers made sawcuts halfway through a tree trunk at different levels but from opposite sides. Often, one finds that water transport is not interrupted and that the tree survives (see more recent experiments of Preston, 1952; Greenridge, 1959; and Scholander et al., 1961), although the rate of water uptake decreases, and there is a much steeper tension gradient across the section of stem with the cuts. With xylem anatomy in mind, it is not too difficult to imagine how the path of water might continue uninterrupted around such sawcuts.

Studying Water, Minerals, and Roots

Paul J. Kramer

Paul J. Kramer, one of the deans of plant-water relations, has pondered the roles of water and minerals in plants since 1931 in his laboratory at Duke University, where he is Professor of Botany, producing many significant data and ideas, not to mention graduate students who are now carrying on the work all over the world.

My entrance into the field of plant water relations research was somewhat accidental. In 1928, when I was a young graduate student at Ohio State University searching for a thesis topic, Professor E. N. Transeau showed me a paper by Burton E. Livingston in which it was claimed that osmosis plays a negligible role in water absorption. Transeau suggested that, as Livingston presented little evidence for his view, I might investigate the problem, and I have been working in that field ever since.

Most textbook writers of those days assumed that osmosis was somehow involved in water absorption, but their discussions were so vague that it was quite impossible to understand how absorption really occurred. My early research was done very simply with T-tubes, pipettes, rubber tubing, a vacuum pump, a pressure chamber built from pipe fittings, tomato and sunflower root systems, and papaya petioles. It was intended to test ideas of Atkins, Renner, and other early workers, and the results indicated that root systems can function as osmometers, and transpiring shoots can absorb water through dead root systems. This research led me to support the neglected view of Renner that two mechanisms are involved in water absorption. According to this view, the root systems of plants growing in moist, well-aerated soil function as osmometers when transpiration is slow, resulting in the development of root pressure and the occurrence of guttation. When rapid transpiration lowers the pressure or produces tension in the xylem sap, however, water is pulled in through the roots, and osmotic movement is negligible. Thus, in transpiring plants, most or all of the water enters passively, and the roots act merely as absorbing surfaces. Further research demonstrated that factors such as cold soil and deficient aeration reduce absorption by increasing the resistance to water flow through roots, rather than by inhibiting some mysterious kind of active absorption mechanism. This early research contributed to the development of a relatively clear explanation of how water is absorbed and of how certain environmental factors affect the rate of absorption. Looking back, those early experiments seem prosaic, but at the time they were very exciting.

During the 1950s, I began to realize that many of the contradictory reports in the literature concerning the relationship between soil moisture and plant growth resulted from the fact that one cannot accurately predict plant water stress from measurement of soil water stress. Holger Brix, John Boyer, and others began working as students in my laboratory with the recently introduced thermocouple psychrometers, and I started a campaign to educate plant scientists about the necessity of measuring plant water potential. As they were finally beginning to realize the importance of water in relation to plant growth, this idea was generally accepted. I notice, however, that even today workers in water relations sometimes reduce the value of their research by failing to measure the degree of water stress to which their plants are subjected.

My work on water absorption naturally led to an interest in roots as absorbing organs. With the aid of a zoologist colleague, Karl Wilbur, I measured uptake of phosphorus by mycorrhizal roots. Then Herman Wiebe, as a graduate student in my laboratory, did some interesting experiments with radioactive tracers, which showed that the region of most rapid salt absorption and translocation to shoots often is several centimeters behind the root tip, rather than near it, as previously claimed. This conclusion was viewed with considerable skepticism at first, but we have had the pleasure of seeing it verified by more recent research, including work in R. Scott Russell's laboratory at Letcombe, England. I also concluded that considerable salt and water are absorbed by mass flow through suberized roots. This view has not yet been generally accepted, though evidence supporting it has been available for over 30 years. Perhaps this illustrates the difficulty of getting well-entrenched old ideas replaced by new ones.

Recently, Dr. Edwin Fiscus and other investigators in my laboratory have been studying various anomalies in water conduction, especially the reports of decrease in resistance with increase in rate of flow. Most of these reports involve the strict application of Ohm's law (flow being proportional to driving force and inversely proportional to resistance) without taking into account such anatomical and physiological facts as the role of phyllotaxy (leaf arrangement on the stem) in controlling the water supply to individual leaves or the fact that, as the rate of water flow through roots increases, the driving force changes from primarily osmotic to primarily mass flow. Thus, a simple application of Ohm's law is inadequate to explain water absorption without consideration of the driving forces involved.

I am glad that I entered plant physiology at a time when it was still possible to be a generalist. Thus, in addition to my work on water relations, salt absorption, and root systems, I was also able to carry on research on the physiology of woody plants. To me, research in whole plant physiology on the borderline between physiology and ecology, in what perhaps can be called environmental physiology, has been extremely interesting because of the variety of problems that it has presented. I also think that it has enabled me to contribute more to a general understanding of how plants live and grow than I could have contributed by concentration in a narrow area of research. In any event, there has never been any danger of boredom!

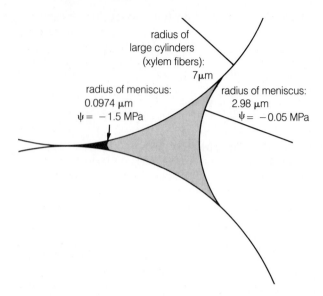

radius of
large cylinders
(xylem fibers):
7μm

radius of meniscus:
0.0974 μm
ψ = −1.5 MPa

radius of meniscus:
2.98 μm
ψ = −0.05 MPa

Figure 4-19 Capillary storage of water in the air spaces between cells. The large portions of circles represent wood fibers or other xylem elements. If these circles have radii of 7 μm, then the black area represents water stored when water potential in the xylem is about −1.5 MPa, and the hatched area represents water stored when xylem water potential is about −0.05 MPa. The menisci have curvatures in equilibrium with the water potentials (ψ) as shown.

Actually, if one is dealing with ring-porous wood with long and wide vessels, it is necessary to consider the rest of the transport system, because all the large long vessels are interrupted by the sawcuts. Yet there are always much shorter vessels in such trees, and there are also the tracheids that are shorter still. Water can always move circumferentially in the part of the trunk between cuts by zigzagging up and down from vessel to vessel or tracheid to tracheid.

What about air excluded from solution in the stem by freezing? Microscopic observation shows that air blockage occurs when some trees are frozen, just as when water was frozen in the spinning Z-tube. Inability to restore the water columns in the spring may be the factor that excludes certain trees and especially vines with large vessels from cold climates. How do trees that grow in such regions manage? Observations have shown that the blocked pathways are replaced or restored in such trees. But how?

Several explanations have been proposed. Ring-porous trees apparently use the "throw-away" method. Such species have such large and efficient vessels that a single growth increment of the trunk is sufficient to provide the crown with water. In this case, vessels are made before leaves emerge in the spring, and the vessels of previous years are not used. A second method involves refilling by root pressure in the spring, which is most clearly seen in the grapevine.

An ingenious explanation was proposed by E. Sucoff (1969), who studied the mathematics of bubbles in liquid. He showed that large bubbles expand more easily than small bubbles, especially if the liquid is under tension. Imagine a northern tree thawing in the spring. As the ice melts, the tracheids become filled with liquid containing the many bubbles of air that had been forced out by freezing. As melting continues and transpiration begins, tension begins to develop in the xylem. Sucoff showed that a critical point is reached for a large bubble, at which it expands explosively under tension as water turns to vapor in a fraction of a second (similar to the cavitation clicks already mentioned). This is confined to the tracheid in which it occurs, because of its anatomy, but it sends a shock wave to surrounding tracheids, driving their small air bubbles back into solution. What about the tracheid in which bubble expansion occurred? It would be vapor-locked and forever lost to sap movement. Study of wood in the spring indicated that about 10 percent of the tracheids were indeed filled with vapor, but the remaining 90 percent are ample to handle sap movement.

4.6 Xylem Anatomy:
A Fail-safe System

By now it should be apparent that water transport in response to gradients in negative pressures in plants, caused by negative water-potential gradients from soil through the plant to the atmosphere, depends upon the anatomy and physical characteristics of xylem tissue. Plants, especially tall trees, are apparently designed to allow sufficient flow in response to the pressure gradients to prevent (or at least usually to avoid) cavitation in the transport elements (but allowing it under some circumstances as in Zimmermann's "designed leakage"), to bypass cells that do become vapor locked, and sometimes even to restore (and often to replace) such vapor-locked cells. There are other safety features in the system worth mentioning but that can't be discussed in detail here (see Zimmermann, 1983).

For example, the wall sculpturing in vessel elements and the various kinds of perforation plates might keep bubbles that do form (as in winter freezing) from coalescing; small bubbles are much easier to dissolve than large ones. Furthermore, in many plants it can be shown that the highest resistance to water flow occurs at the leaf trace or in the base of the petiole. This means that as water stress builds during drought, cavitation occurs first in the leaves, so they may wilt and die, but the water system in the trunk remains relatively intact. It may be possible to produce new leaves but not a new trunk, especially in palm trees, for example, in which there is no secondary growth of xylem in the trunk.

We could discuss the special cells called **tyloses** that grow into vapor-locked tracheids and vessels in some species. These act as especially effective seals against water loss and pathogen invasion. Sometimes gums or resins are secreted into nonfunctional xylem cells. In many trees, these changes account for the heartwood that often develops in the center of the trunk. Usually, the heartwood contains many cells filled with air or vapor, but sometimes solutes build up in the heartwood, and these draw in water osmotically, so that solution occurs under pressure in the heartwood (called wetwood) at the same time that xylem sap is under tension in the rest of the trunk.

It seems obvious that some water is stored in the trunk of a tree because of the elasticity of the xylem cells. When transpiration is reduced at night or during rain, tension in the xylem is relaxed, but water may continue to be absorbed to relieve the tension as the cells expand elastically. This has often been observed experimentally. Zimmermann (1983) points out a less obvious role of capillarity in water storage. Remember that much of the trunk is filled with air (perhaps 10 to 15 percent in actively conducting wood and more in heartwood, although much of the "air" may be water vapor). The air must be essential for respiration of the living ray and other parenchyma cells in the wood. As we have seen, the air does not penetrate into the vessels or tracheids because of the small pore sizes in the cell walls. But the air spaces must contain capillary water with a water potential that is in equilibrium through the water-filled pores in the walls with the water in the conducting cells. When the water in the conducting system is under considerable tension (say $P = -1.5$ to -3.0 MPa), the menisci of the water in the air spaces have small radii of curvature; that is, there is a minimum of such water present. As tension in the conducting elements relaxes, however, water can move into the air spaces where menisci will then have larger radii of curvature (Fig. 4-19). Thus even more water might be stored in the wood than its elasticity would suggest.

All in all, our understanding of xylem anatomy combined with our application of the cohesion theory of the ascent of sap has greatly increased our appreciation for land plants as water conducting systems. In Chapter 7, we shall see that this is only part of the story: Plants are also anatomically constructed for highly efficient transport of the solutes produced in photosynthesis and other metabolic processes (this time, under positive pressure). But first we must learn more about the minerals required by plants and about the membranes that so strongly influence their movement.

5

Mineral Nutrition

What elements must a plant absorb to live and grow? Can a plant grow when provided only with elements in inorganic form (mineral salts)? Or do plants, like animals, require vitamins? If only minerals are required, then which ones and in what amounts? How can we know when a plant is lacking some essential element? How can we provide the limiting element to overcome its deficiency? What are the functions of these elements in the plant?

These are some questions of mineral nutrition, an important subscience of plant physiology. Since we must properly "feed" plants before we can feed ourselves, these are important questions. Answers obtained so far have greatly improved agriculture during the past century and a half, but still more improvements are needed. Answers to the questions of mineral nutrition also add to our basic understanding of plants. Plant growth requires the incorporation of elements into materials of which plants are made, and 15 to 20 percent of nonwoody plants is made from such elements, the rest being water.

5.1 The Elements in Plant Dry Matter

One method to determine which elements are essential to plants is to determine which elements healthy plants contain and in what proportions. When freshly harvested plants or plant parts are heated to 70 to 80°C for a day or two, nearly all the water is driven off; the remaining material is the so-called **dry matter**. Principal components of dry matter are cell-wall polysaccharides and lignin, plus such protoplasmic components as proteins, lipids, amino acids, organic acids, and certain elements, such as potassium that form no essential part of any compound.

Principal elements in a dry shoot system of maize (corn; reported in 1924) are listed in Table 5-1 (left column). Oxygen and carbon are by far the most abundant elements on a weight basis (about 44 percent of each), and hydrogen ranks third. This is approximately the same distribution of elements as in carbohydrates, including cellulose, the most abundant compound in wood. Smaller amounts of nitrogen are found, followed by several other elements in even lower concentrations. Also included are two elements, aluminum and silicon, that are believed nonessential for higher plants. More will be said of these elements later, but note that plants will absorb and accumulate numerous nonessential elements from soil solutions. At least 60 elements have been found in plants, including gold, lead, mercury, arsenic, and uranium. Had a complete elemental analysis of the maize plant been made, trace amounts of numerous other elements could have been found, some of which are essential to this and other plants.

A modern analysis of the leaf closest to the young maize cob shows concentrations of three additional elements essential for plants: zinc, copper, and boron (Table 5-1, right column). This table was prepared by analyzing leaves from a well fertilized and highly productive maize field. It emphasizes that leaves generally contain significantly more nitrogen, phosphorus, and potassium than do whole shoot systems, especially shoots of less-productive varieties grown more than a half-century ago.

Soils are composed largely of aluminum, oxygen, silicon, and iron, yet plants by no means reflect this composition, partly because they absorb carbon and most of their oxygen from the air. Other reasons are that most of the above-mentioned soil elements are not readily soluble in the soil water and that roots exhibit considerable selection over the rates at which elements are absorbed (see Chapter 6).

Table 5-1 Elemental Analysis of Whole Maize Shoot System and a Selected Maize Leaf. The shoot system included leaves, stem, cob, and grains.

Element	Maize Shoot[a] (% of dry weight)	Maize Leaf[b] (% of dry weight)
Oxygen	44.4	—
Carbon	43.6	—
Hydrogen	6.2	—
Nitrogen	1.5	3.2
Potassium	0.92	2.1
Phosphorus	0.20	0.31
Sulfur	0.17	0.17
Calcium	0.23	0.52
Magnesium	0.18	0.32
Chlorine	0.14	—
Silicon	1.2	—
Sodium	—	—
Iron	0.08	0.012
Manganese	0.04	0.009
Copper	—	0.0009
Boron	—	0.0016
Molybdenum	—	—
Zinc	—	0.003
Aluminum	0.89	—
Undetermined	7.8	—

[a]Data of Latshaw and Miller, *J. Agric. Research*, 27:854, 1924.
[b]Unpublished 1982 data of P. Soltanpour and S. Workman, Colo. State Univ. Soil Testing Laboratory.

5.2 Methods of Studying Plant Nutrition: Solution Cultures

Beginning in about 1804, scientists began to appreciate that plants require calcium, potassium, sulfur, phosphorus, and iron. Then, about 1860, three German plant physiologists—W. Pfeffer, Julius Sachs, and W. Knop—recognized the problems of determining the kinds and amounts of elements essential to plants growing in a medium as complex as a soil. They therefore grew plants with their roots in a solution of mineral salts (a **nutrient solution**), the chemical composition of which was controlled and limited only by the purity of chemicals then available. Growing plants in this way is referred to as **hydroponics,** or **solution culture** (Fig. 5-1). Although Pfeffer, Sachs, and Knop apparently did not realize it, other investigators later showed that many plants grew much better if the roots are aerated, as shown in Fig. 5-1a. (A history of hydroponics is given by Jones,

1982, and an evaluation of its potential future is given by Wilcox, 1982).

As techniques became available to purify salts and the water for hydroponic cultures, more exact control over the elements available to plants became possible. This proved especially important for several elements required only in very small (trace) amounts. Furthermore, requirements for molybdenum, copper, zinc, and boron are often difficult to demonstrate for species with large seeds, because large seeds sometimes contain enough of these elements to grow mature plants. In such cases, deficiency symptoms are more easily observed in the second generation grown from seeds taken from parents that have been grown without the added elements. Techniques to measure concentrations of elements in plants, soils, and nutrient solutions have also improved tremendously in the past twenty years. **Atomic absorption spectrometers** are now used to measure metals and some nonmetals. Even more valuable (and expensive) are **optical emission spectrometers,** in which elements are vaporized at temperatures above 5000 K. Such high temperatures temporarily excite electrons from their ground-state orbits into higher orbits, and as these electrons move back to their original ground states, electromagnetic energy is emitted at wavelengths different for each element. These wavelengths are measured and the energy quantified by the spectrometer. Concentrations of more than 20 elements in a single solution can be measured with great sensitivity in less than 1 minute (Soltanpour et al., 1982; Alexander and McAnulty, 1983).

In spite of solution culture advantages for mineral nutrition studies, the technique has disadvantages. One is the need for root aeration. Another is the need to replace the solution every day or two for maximum growth; this is because the solution composition changes continuously as certain ions are absorbed more rapidly than others. This selective uptake not only depletes certain ions but also causes undesirable pH changes. Owners of commercial greenhouses sometimes use recirculating solutions that flow in a thin layer through troughs around the roots of valuable crops, including lettuce and tomatoes. Such solutions are pumped from tanks in which the pH and solution composition can be monitored and adjusted automatically (Fig. 5-1c). The pump forces solution across the roots; then, when the pump is temporarily turned off, the solution is drained downhill, leaving a well-aerated film of nutrient solution on the root surfaces (the **nutrient-film-technique,** Cooper, 1979; Graves, 1983).

To avoid some problems of liquid cultures, many physiologists use washed white quartz sand or a mineral called perlite (expanded pumice) as a medium for the roots. Nutrient solutions are simply

a hydroponic culture

a hydroponic culture

glass jar
normally
covered
with
aluminum
foil to
exclude
light (and
thus algae)

aerator

b slop culture
(nutrients applied
by hand)

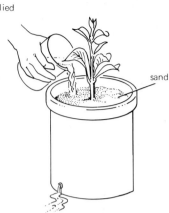

b slop culture
(nutrients applied
by hand)

sand

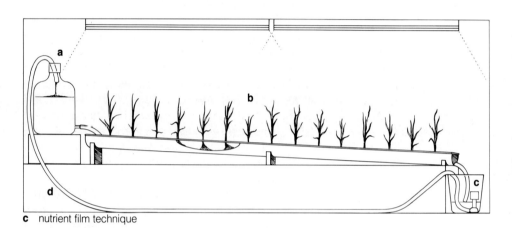

c nutrient film technique

Figure 5-1 Three methods for growing plants with nutrient solutions: (**a**) hydroponic culture (note aerated roots), (**b**) slop culture using sand, and (**c**) nutrient film technique. Reservoir A contains nutrient solution that drains down trough containing plants B. Plants can be supported in many ways. Unabsorbed solution flows into container C that has a pump to force solution through tube D back to reservoir. (Method **c** from M. W. Nabors, 1983; for practical use of the nutrient-film technique, see Cooper, 1979.)

poured or allowed to drip onto these media at suitable intervals in excess amounts to insure leaching of the old solution through holes in each pot (Fig. 5-1b). The technique is convenient but not suitable for detailed studies with certain elements needed only in trace amounts, because the sand and perlite contribute unknown amounts of certain essential elements. Furthermore, in solid media various roots are necessarily exposed at different times to varying amounts of essential elements and water, and there is no way to wash off all the sand or perlite without injuring small roots and root hairs. For some studies in which ion uptake measurements by roots are necessary, solution cultures provide the only adequate method. This method is described in Chapter 6 (see Fig. 6-15).

Numerous useful formulations of nutrient solutions were devised from studies in which various concentrations of elements were used. Hewitt (1966) described many of these and how to prepare and use them effectively. Two such recipes by pioneers of mineral nutrition in the United States are listed in Table 5-2, one by Dennis R. Hoagland and the second a modification of one by John Shive (modified by

H. A. Evans). Both have necessary elements in amounts to allow good growth of many higher plants, but a solution ideal for one species is seldom ideal for another. Note that the Evans solution in Table 5-2 contains all its nitrogen as nitrate, but nitrate is often absorbed so fast that it causes rapid rises in the nutrient solution pH, because absorption of nitrate (and other anions) is accompanied by absorption of H^+ or excretion of OH^- to maintain charge balances. The pH problem can be minimized by supplying part of the nitrogen as an ammonium salt (e.g. $NH_4H_2PO_4$, as in the Hoagland solution), because absorption of NH_4^+ and other cations occurs simultaneously with transfer of H^+ from the root to the surrounding solution (Section 6.8).

Nearly all nutrient solutions are more concentrated than soil solutions. For example, the phosphorus concentration of the Evans solution in Table 5-2 is $500 \mu M$ (micromolar), whereas about three-fourths of the determinations in one survey of 149 soils gave phosphorus levels less than 1.5 μM (Reisenauer, 1966). Over half of these soil solutions had potassium concentrations less than 1.25 mM (millimolar), whereas the Evans solution is 5.5 mM in

K$^+$. Many minerals in the soil are not in solution but are adsorbed on negatively charged surfaces of clay and organic matter or precipitated as insoluble salts. They dissolve only slowly as they are removed by plants from solution or as they are lost by leaching. Most plants grow well in solutions having concentrations of essential elements as low as those dissolved in the soil solution in which they normally grow, provided that the solutions are replenished often enough to maintain their concentrations or that dilute solutions in large volumes flow over their roots from a recirculating tank (Asher and Edwards, 1983). The higher concentrations typically employed in nonflowing solutions avoid the need to change solution more than once a day or once every few days, depending on growth rates. Of course, the concentration must be low enough to prevent plasmolysis of the root cells. Most solutions have osmotic potentials no more negative than -0.1 MPa, so this is not a problem.

5.3 The Essential Elements

The nutrient solutions of Table 5-2 contain 13 elements believed to be essential for all angiosperms and gymnosperms, although in fact the nutrient requirements of only certain cultivated species have been investigated well. Adding O, H, and C (from O_2, H_2O, and CO_2), 16 elements are considered essential. With these elements and sunlight, most plants can synthesize all the compounds they require. But is it possible that plants require some organic molecules synthesized by microorganisms that normally grow on them? Apparently not, since some plants have now been grown under sterile conditions in plastic or glass enclosures from which all microorganisms were excluded. Plants really are autotrophic, even though some associated microbes are beneficial (as in mycorrhizae, Chapter 6, and root nodules, Chapter 13). Often, these microbes are essential to plants in nature, though not in solution cultures, because they perform roles that allow plant survival in the face of competition and harsh environmental conditions (Rovira et al., 1983; Quispel, 1983).

There are two principal criteria by which an element can be judged essential or nonessential to any plant (Epstein, 1972): *First, an element is essential if the plant cannot complete its life cycle (i.e., form viable seeds) in the absence of that element. Second, an element is essential if it forms part of any molecule or constituent of the plant that is itself essential in the plant* (for example, N in proteins and Mg in chlorophyll). Either criterion is sufficient to demonstrate essentiality, and most elements in our list of 16 have met both. Historically, the first criterion was the main one employed, but improved chemical analyses have generally led to agreement on both criteria.

Although these two criteria are widely accepted by mineral nutrition experts, other criteria are often considered. Daniel Arnon and Perry Stout, at the University of California in Berkeley, suggested (1939) that a third criterion should be used. This is that, if an element is essential, it must be acting directly inside

Table 5-2   Two Nutrient Solutions for Hydroponic Culture.

Hoagland's Solution[a]			Evans' Modified Shive's Solution[b]		
Salt	Molarity	mg/l (ppm)	Salt	Molarity	mg/l (ppm)
KNO$_3$	0.010		Ca(NO$_3$)$_2$·4H$_2$O	0.005	
Ca (NO$_3$)$_2$	0.003		K$_2$SO$_4$	0.0025	
NH$_4$H$_2$PO$_4$	0.230		KH$_2$PO$_4$	0.0005	
MgSO$_4$·7H$_2$O	0.490		MgSO$_4$·7H$_2$O	0.002	
Mixture of 0.5% FeSO$_4$ and 0.4% tartaric acid:			Fe-versenate		0.5 Fe
0.6 ml/l added 3 times/week			KCl		9.0 Cl
MnCl$_2$·4H$_2$O		0.5 Mn; 6.5 Cl	MnSO$_4$		0.25 Mn
H$_3$BO$_3$		0.5 B	H$_3$BO$_3$		0.25 B
ZnSO$_4$·7H$_2$O		0.05 Zn	ZnSO$_4$		0.25 Zn
CuSO$_4$·5H$_2$O		0.02 Cu	CuSO$_4$		0.02 Cu
H$_2$MoO$_4$·H$_2$O		0.01 Mo	Na$_2$MoO$_4$		0.02 Mo

[a]From D. R. Hoagland and D. I. Arnon (1938). University of California Agricultural Experimental Station Circular # 347.
[b]From H. J. Evans and A. Nason (1953). Plant Physiology 28:233–254.

Figure 5-2 Comparisons among plants of *Echinochloa utilis* (left), *Amaranthus tricolor* (center), and *Kochia childsii* (right), which received either no addition (−Na) or 0.10 meq/liter sodium chloride (+Na). (From P. F. Brownell and C. J. Crossland, 1972, Plant Physiology 49:794–797.)

the plant and not causing some other element to be more readily available or antagonizing the effect of another element. This criterion has not been nearly as useful as the other two, but there are a few cases in which it has been applicable. One case, described in the boxed essay "Selenium" in this chapter, concerns the initial conclusion that selenium is essential for plants. It was later found that the growth-promoting effects of selenium resulted from the ability of the selenate ion to inhibit the absorption of phosphate, which was otherwise absorbed by the plants in toxic amounts.

We should also emphasize that many investigators consider that an element is essential if deficiency symptoms appear on plants when they are grown without addition of the element to the nutrient solution, even though such plants form viable seeds. The assumption, which seems reasonable, is that if the plants contained none of the element (if the element were absent from seeds, dust in the air, and nutrient solutions), these plants would develop deficiency symptoms so severe that they would die before they formed viable seeds. Use of this criterion has led to the evidence (mentioned later) that sodium and silicon are essential for certain plants.

It is usually easier to show that an element is essential than that it is not. Experimenters therefore often state that if an element in question is necessary, it is required only in concentrations less than the sensitivity limits of their detecting instruments. For example, it was reported that if vanadium is essential for lettuce or tomato plants, the amount needed is less than 0.02 μg per gram of dry tissue (0.02 ppm by weight). Because of such problems, it is likely that a few more nutrient elements needed in barely detectable amounts will eventually be added to our list.

Table 5-3 lists the 16 elements presently believed essential to all higher plants, the chemical (ionic) form most readily available, the approximate adequate concentration in the plant, and the approxi-

mate number of atoms of each element needed compared to molybdenum. About 60 million times as many atoms of hydrogen than of molybdenum are required, a dramatic difference. This difference reflects the importance of H in thousands of essential compounds, whereas Mo acts catalytically and in only one or a few compounds (enzymes). The first seven elements listed are often called **trace elements** or **micronutrients** (needed in tissue concentrations equal to or less than 100 μg g^{-1} of dry matter), and the last nine are the **macronutrients** (needed in concentrations of 1000 μg g^{-1} of dry matter, or more). The internal concentrations listed adequate should be considered only as useful guidelines because of variability among species and with age. Many data are summarized by Shear and Faust (1983) for several horticultural fruit and nut crops.

Besides these essential elements, some species require others. Evidence has existed for many years that sodium is required by or at least beneficial to certain desert species such as *Atriplex vesicaria*, common to dry inland pastures of Australia, and *Halogeton glomeratus*, a common introduced weed of salty arid soils in the western United States. More recently, Peter F. Brownell and C. J. Crossland (1972, also see Brownell, 1979) investigated and reviewed the sodium nutrition of 32 species and concluded that those having the C-4 photosynthetic pathway (Section 10.3) probably do require Na$^+$. Only *Atriplex vesicaria* failed to complete its life cycle without added sodium, but other C-4 species developed **chlorosis** (lack of chlorophyll) in leaves, and sometimes **necrosis** (dead tissues) developed in the leaf margins and tips. The assumption is that if tissue levels were reduced even more, the C-4 plants would show more pronounced deficiency symptoms and death would soon follow. On this basis we could say that sodium is essential for these species. Growth enhancement by Na$^+$ is shown in Fig. 5-2 for three C-4 pathway plants. Furthermore, certain species that fix CO$_2$ in photosynthesis via the crassulacean acid metabolism pathway (Section 10.6) also grow faster with sodium, and for them Na^+ is probably also essential (Brownell, 1979; Flowers and Läuchli, 1983). As explained in Chapter 10, there are several similarities between chemical reactions of the crassulacean acid metabolism pathway and the C-4 pathway, but what Na^+ has to do with this is not yet known.

Silicon is another element that increases growth of some plants, and the possibility that it is essential has been studied by many investigators (see reviews of Lewis and Reimann, 1969, and Werner and Roth, 1983). Maize (Table 5-1) and numerous other grasses accumulate silicon to the extent of 1 to 4 percent of their dry weight, whereas rice and *Equisetum arvense* (a horsetail or scouring-rush) contain up to 16 percent silicon. The contents of most dicots are much lower than in grasses or *E. arvense*.

As is so often the case in studies of essentiality, it is difficult to completely remove the element from the environment of plants to determine if they can grow without it. But with silicon the problems are especially serious, because it is present in glass and many nutrient salts, and it also exists as particulate SiO_2 in the atmosphere. In several species the amount of silicon has been reduced enough to create deficiency symptoms. In rice, for example, overall growth was retarded, transpiration was increased about 30 percent, and the older leaves died. In tomato, growth rates were lowered about half, new leaves were deformed, and many plants failed to set fruits. Other angiosperms also show deficiency symptoms without applied silicon, and Werner and Roth (1983) and others believe that it is generally an essential element. For certain algae (diatoms and some flagellate Chrysophyceae) that are surrounded by a silica-rich sheath, the element is certainly essential.

Silicon exists in soil solutions as silicic acid, H_4SiO_4 [or $Si(OH)_4$] and is absorbed in this form. It accumulates largely as hydrated amorphous silica ($SiO_2 \cdot nH_2O$), most abundantly in walls of epidermal cells but also in primary and secondary walls of other cells of roots, stems, and leaves and in grass inflorescences. When sheep and cattle eat grasses abundant in silica, the silica is largely excreted in the urine, but sometimes it forms kidney stones. Silica is also blamed for causing excessive wear on sheep's teeth (Jones and Handreck, 1967) and is implicated in throat cancers of people from northern China and Iran who eat bracts of inflorescences from a foxtail millet and of *Phalaris minor*, respectively (Hodson et al., 1982).

Cobalt is essential for many bacteria, including cyanobacteria. It is required for nitrogen fixation by bacteria in root nodules on legumes (Section 13.2). Figure 5-3 illustrates growth of soybeans with and

Figure 5-3 Specific cobalt requirement for nitrogen fixation in soybean. (From S. Ahmed and H. J. Evans, 1960, Soil Science 90:205, with the permission of the publisher. Copyright © 1960 by Williams & Wilkins Company, Baltimore, Maryland.)

without cobalt and with only atmospheric nitrogen, which was fixed in root nodules. Cobalt concentrations as low as $0.1 \ \mu g \ l^{-1}$ were high enough for rapid growth, and neither vanadium, germanium, nickel, nor aluminum could substitute for cobalt. Free-living bacteria that fix nitrogen apart from any symbiotic relationship with plants also require cobalt. Organisms requiring cobalt, including many animals, need it principally because it is a component of the vitamin B_{12} they require. Higher plants and algae have been thought to contain no vitamin B_{12} and to have no cobalt requirement, but there are scattered reports indicating that vitamin B_{12} is present in plants, along with enzymes that need it to function (Poston, 1978). There is also a claim that cobalt is essential for certain plants (Wilson and Nicholas, 1967) and a report that an enzyme needed to form aromatic compounds (Chapter 14) is nonfunctional without cobalt (Rubin et al., 1982). More research is certainly needed.

In addition to the 16 elements essential for plants, higher animals require sodium, iodine, cobalt, selenium, and apparently also nickel, silicon, chromium, tin, vanadium, and fluorine, but not boron (Miller and Neathery, 1977; Mertz, 1981). Some plants probably require nickel, because Ni^{2+} is an essential part of the enzyme **urease,** which catalyzes hydrolysis (breakdown with H_2O) of urea to CO_2 and NH_4^+. Legumes of tropical origin, including soybean (Section 13.2), form ureides in root nodules during nitrogen fixation, and utilization of such ureides in leaves, stems, and developing seeds likely requires urease (Kerr et al., 1983). A nickel requirement for certain algae has been demonstrated (Welch, 1981; Rees and Beckheet, 1982), and again the requirement seems to be for the function of urease. Another element that was thought to be essential for a few species is selenium (see boxed essay, "Selenium").

Selenium

Selenium is absorbed and accumulated in fairly high concentrations (up to at least 0.5 percent; i.e., 5 mg per g dry weight) by certain "accumulator species" of *Astragalus*. Interestingly, although this genus contains about 500 North American species, about 475 of these are nonaccumulators. Certain species of the genera *Stanleya*, *Haplopappus*, and *Xylorhiza* are also notable selenium accumulators.

Because *Astragalus* accumulators live only on seleniferous soils, workers in the 1930s sought to determine whether these plants require selenium. Although it was found that they grew much better on nutrient solutions to which up to nearly 30 parts per million of selenate (SeO_4^{2-}) were added, more recent studies indicate that this growth enhancement occurred because selenate reduced the toxic effects of phosphate; such species are unusually sensitive to phosphate (Bollard, 1983). The dramatic difference in ability to accumulate selenium is illustrated by results from *A. racemosus* and *A. missouriensis* growing side-by-side on a Nebraska soil containing 5 μg Se/g soil. *A. racemosus* contained 5560 μg Se/g dry weight, while *A. missouriensis* contained only 25 μg. This genus has apparently evolved in different directions with respect to selenium accumulation.

Such accumulators frequently poison livestock with an often fatal sickness called the alkali disease or blind staggers. This disease is noted occasionally in certain regions of the western Great Plains of North America, yet seleniferous soils and selenium accumulators are much more widespread (Brown and Shrift, 1982). Toxic forms of Se are certain amino acids in which Se has replaced the sulfur normally present, especially in selenomethionine. Why does the selenium in accumulators not also poison those plants in which it replaces sulfur? We do not know all the answers, but a major one is that accumulators form mainly seleno-amino acids that are not toxic themselves and are not incorporated into certain proteins that could become toxic (Brown and Shrift, 1982; Bollard, 1983).

In bacteria and animals, both of which require selenium, a few essential proteins have been found that contain selenium (Stadtman, 1980). Most such proteins are enzymes that catalyze oxidation-reduction reactions, and the selenium present is essential to their activity. Perhaps similar enzymes occur in selenium-accumulators, but so far there is no evidence that this is true. Bollard (1983) concluded that if selenium is essential for *Astragalus* species, it would have to function at tissue concentrations at or less than 0.08 ppm of dry matter, levels slightly lower than those of molybdenum.

5.4 Quantitative Requirements and Tissue Analysis

Table 5-3 lists concentrations of essential elements in tissues that appear necessary to prevent deficiency symptoms. Such values provide useful guides to physiologists, foresters, orchard managers, and farmers, because concentrations of elements in the tissues (especially in selected leaves) indicate more reliably than soil analysis whether plants will grow faster if more of a given element is provided (Bouma, 1983; Davidseau and Davidseau, 1982; Wolf, 1982). Figure 5-4 shows an idealized plot of growth rate as a function of the concentration of any given element in the plant. In the range of low concentrations called the **deficient zone,** growth increases dramatically as more of the element is provided and its concentration in the plant increases (Moorby and Besford, 1983). Above the **critical concentration** (minimal tissue concentration giving almost maximal growth), increases in concentration (fertilizations) do not appreciably affect growth (**adequate zone**). The adequate zone represents **luxury consumption** of the element, during which storage in vacuoles occurs. This zone is fairly wide for macronutrients but is narrower for micronutrients such as boron, zinc, iron, copper, manganese, and molybdenum. Continued increases of any element lead to toxicities and reduced growth (**toxic zone**).

Increasing environmental pollution is bringing much more attention to toxic effects of both essential and nonessential elements (see boxed essay, "Metal Toxicity"). Increased accumulation of salts is also a problem in many soils, and many approaches, including genetic tolerance, are being sought to relieve the toxicity (San Pietro, 1982; Epstein and Läuchli, 1983; Nabors, 1983—see Nabors' personal essay near the end of this chapter).

Figure 5-5 shows the average growth responses to calcium for 18 dicots and 11 monocots. The separate curves for these two groups show that critical calcium levels for dicots (about 0.2 percent, dry weight basis) are higher than for monocots (less than 0.1 percent) and emphasize the approximate nature of the tissue concentrations listed as adequate in Table 5-3. In general, grasses absorb more potassium and less calcium than do members of the legume family (Leguminosae or Fabaceae) and probably other dicots. However, fewer comparative data for trees and shrubs exist.

Table 5-3 Essential Elements for Most Higher Plants and Internal Concentrations Considered Adequate.[a]

Element	Chemical Symbol	Form Available to Plants[b]	Atomic Wt	Concentration in Dry Tissue (ppm)	Concentration in Dry Tissue (%)	Relative No. of Atoms Compared to Molybdenum
Molybdenum	Mo	$MoO_4^=$	95.95	0.1	0.00001	1
Copper	Cu	Cu^+, $\mathbf{Cu^{2+}}$	63.54	6	0.0006	100
Zinc	Zn	Zn^{2+}	65.38	20	0.0020	300
Manganese	Mn	Mn^{2+}	54.94	50	0.0050	1,000
Boron	B	H_3BO_3	10.82	20	0.002	2,000
Iron	Fe	Fe^{3+}, $\mathbf{Fe^{2+}}$	55.85	100	0.010	2,000
Chlorine	Cl	Cl^-	35.46	100	0.010	3,000
Sulfur	S	$SO_4^=$	32.07	1,000	0.1	30,000
Phosphorus	P	$\mathbf{H_2PO_4^-}$, $HPO_4^=$	30.98	2,000	0.2	60,000
Magnesium	Mg	Mg^{2+}	24.32	2,000	0.2	80,000
Calcium	Ca	Ca^{2+}	40.08	5,000	0.5	125,000
Potassium	K	K^+	39.10	10,000	1.0	250,000
Nitrogen	N	$\mathbf{NO_3^-}$, NH_4^+	14.01	15,000	1.5	1,000,000
Oxygen	O	O_2, H_2O	16.00	450,000	45	30,000,000
Carbon	C	CO_2	12.01	450,000	45	35,000,000
Hydrogen	H	H_2O	1.01	60,000	6	60,000,000

[a]Modified after P. R. Stout (1961). Proceedings of the Ninth Annual California Fertilizer Conference, pp. 21–23.
[b]If one of two forms is more common than the other, this is indicated by boldface type.

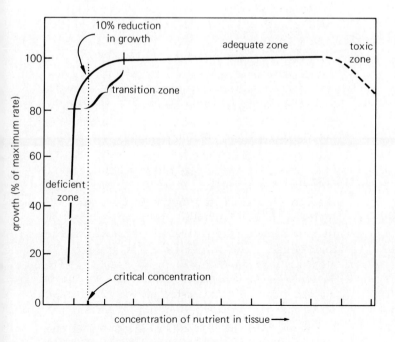

Figure 5-4 Generalized plot of growth as a function of the concentration of a nutrient in plant tissue. (After Epstein, 1972.)

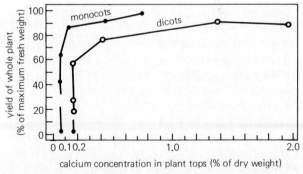

Figure 5-5 The relation between calcium concentration in tops and relative yields of 18 dicots and 11 monocots after 17 to 19 days' growth in constant calcium concentrations in solution. Each point represents the average values for all species in each plant group given a single Ca^{2+} treatment (0.3, 0.8, 2.5, 10, 100, or 1000 μM). (From J. F. Loneragan, 1968, Nature 220:1307–1308.)

Metal Toxicity

There is considerable genetic variation in the abilities of various species to tolerate otherwise toxic amounts of nonessential lead, cadmium, aluminum, silver, and of several essential micronutrient cations (Baker, 1981; Woolhouse, 1983). In some species, the elements are absorbed only to a limited extent; so this represents a case more of avoidance than true tolerance. In other species, the elements accumulate in roots with little transport to shoots. In still others, both roots and shoots contain much higher amounts of such elements than nontolerant species could live with. Mechanisms of true tolerance have not been studied much, but formation of stable nontoxic chelates (Section 5.5) and storage of elements in vacuoles are frequent suggestions. Selection of tolerant agricultural crops has not yet progressed far, with the important exception of wheat varieties in southern Brazil that are resistant to high aluminum concentrations in solutions of the acidic soils present there (Epstein and Läuchli, 1983). Of course, we should also be aware of plants that accumulate relatively high concentrations of elements in parts we eat. There is present concern, for example, that certain vegetables grown on soils formerly used as mining areas in Colorado contain excess lead and cadmium.

Data in Fig. 5-5 emphasize the differences in nutrient requirements among species. Furthermore, graphs similar to those in Figs. 5-4 and 5-5 have been effectively used to plan efficient uses of fertilizers for crop plants and forest trees. In the past, soils were not fertilized with N, P, or K beyond the critical tissue concentrations because of the diminishing yield return (cost ratio), but now we know that excess nitrate and some phosphate not absorbed by plants are leached from soils and ultimately appear in lakes and streams. Here they cause excessive growth of algae, which leads to the problems of **eutrophication** (nutrient enrichment leading to growth of algae and other plants; upon death, the decomposition of these plants by microorganisms uses so much dissolved oxygen that fish and other animals die). Furthermore, the manufacture of nitrogen fertilizers is one of the most energy-expensive aspects of modern intensive agriculture. Therefore, users of fertilizers must consider not only high yields but also water pollution and world energy requirements.

The abilities of plants to obtain essential nutrients from soils are important in determining where they grow. Although we know much about the mineral nutrition of crop plants, far too little is known about wild species, including forest trees (Christiansen and Lewis, 1982; Chapin, 1980; Plant and Soil, 1983). Except for carefully fertilized orchards, most trees and native grasses grow on rather unfertile soils, and their nutrient requirements are lower than those of crops bred to respond to fertilizers. These lower nutrient requirements result mainly from the ability of native trees, grasses, and herbaceous dicots to absorb nutrients faster than selected crops at low but not at high concentrations. Such species are therefore good competitors in their natural environments where growth is usually slow, but they could not compete in modern agriculture. Nevertheless, thousands of acres of genetically-selected forest trees in the northwestern United States are now fertilized with nitrogen. Of course, leaf fall from deciduous trees in autumn returns some absorbed nutrients to soils, especially N, P, and K. Furthermore, significant amounts of N, P, K, and Mg move out of tree leaves into twigs and branches before leaf fall (Ryan and Bormann, 1982; Titus and Kang, 1982). These are used in new growth the next season. Perennial range grasses similarly conserve minerals by translocation to roots and to lower stem tissues making up the crown in late summer.

5.5 Chelating Agents

The micronutrient cations iron, and to a lesser extent zinc, manganese, and copper, are relatively insoluble in nutrient solutions when provided as common inorganic salts, and they are also nearly insoluble in most soil solutions. This insolubility is especially marked if the pH is above 5, as it is in nearly all soils of the western United States and many other regions with low rainfall (Clark, 1982; Vose, 1982). Under these conditions, the micronutrient cations react with hydroxyl ions eventually precipitating insoluble hydrous metal oxides. An example in which the ferric form of iron yields the reddish-brown oxide (rust) is shown in reaction R5-1:

$$2Fe^{3+} + 6\,OH^- \rightarrow 2Fe(OH)_3 \rightarrow Fe_2O_3 \cdot 3H_2O \qquad \text{(R5-1)}$$

Because of this and other reactions that contribute to insolubility, certain plants, especially cereal grains and other grasses, cannot absorb enough iron and zinc. One way to overcome this deficiency is to fertilize with metal **chelates** (from the Greek, claw-like). A chelate is the soluble product formed when certain atoms in an organic molecule called a **chelating agent** or **ligand** donate electrons to the metal cation. Negatively charged carboxyl groups and nitrogen atoms possess electrons that can be shared in this way. One of the best known synthetic ligands provides both carboxyl groups and nitrogen atoms. This is **ethylene-diaminetetraacetic acid,** abbreviated **EDTA** and sold in garden shops under the trade name Versene. Extensive work by Willard L. Lindsay and his students at Colorado State University has greatly clarified the solubilities of numerous forms of iron, zinc, and other metals in soils (Lindsay, 1979; Schwab and Lindsay, 1983). Their work also explained quantitatively how a reducing atmosphere in soils keeps more iron in the soluble (reduced) Fe^{2+} form than in the far less soluble Fe^{3+} form, which is unavailable to roots unless chelated. Long-term studies by John C. Brown at the USDA Station in Beltsville, Maryland, and by others emphasized the ability of root surfaces of some dicots (though not of several grasses) to reduce Fe^{3+} to Fe^{2+} when iron was limiting and thus increase its solubility and speed its absorption (Brown, 1978, 1979; Olsen et al., 1982). Interestingly, roots of several species that absorb iron effectively even as it becomes deficient do so largely because they release more reductants and acidify the soil faster by releasing H^+ when iron-stressed than when nonstressed. Both the reductants and the H^+ facilitate iron absorption, thereby countering the iron deficient conditions.

Ligands such as EDTA are now commonly used to prevent or correct deficiency symptoms of iron in many parts of the western United States where the high pH of the calcareous soils renders iron unavailable to some species. The same problem occurs with zinc in certain areas, and iron or zinc chelates can often correct these deficiencies when applied to the foliage or soil, or even when injected into tree trunks. Iron chelates added to the soils of house plants often give improved green color and faster growth.

Good chelating agents used in soils have at least two important properties: They should be resistant to microbial attack, and they should form stable chelates with micronutrient cations but not with far more abundant competitive cations such as calcium or magnesium. EDTA has a high affinity for calcium and so is a poor chelating agent for calcareous soils, in which it has been replaced with **Fe-EDDHA, ethylenediamine di(o-hydroxyphenyl) acetic acid** (trade name is Sequestrene), a much better source of iron for calcareous soils. Still other synthetic ligands are used for certain purposes.

Naturally occurring chelates of micronutrient cations also exist in soils, maintaining far higher avail-

ability of these elements than would otherwise occur. Soil chelating agents have not been well characterized (Romheld and Marschner, 1983), but there is evidence that phenolic compounds, proteins, amino acids, and organic acids all contribute. Much more important for iron nutrition might be **siderophores,** dozens of kinds of which are produced and released into soils by many fungi and some bacteria (Emery, 1982; Powell et al., 1982). Siderophores form extremely stable chelates with iron in the Fe^{3+} form (but not with Fe^{2+} or other metals) and are transported into microbial cells where the Fe^{3+} is reduced to Fe^{2+} and released. Chemically, soil siderophores exist largely as two types, **catechols** (phenolic compounds with two adjacent hydroxyl groups) produced mainly by bacteria, and **hydroxamates** (peptides) produced mainly by fungi, including those of mycorrhizae (Chapter 6). Some hydroxamates carrying Fe^{3+} are absorbed by plants (Powell et al., 1982), but how important they are to overall plant iron nutrition remains to be learned. The effectiveness of commercial chelates such as Fe-EDTA and Fe-EDDHA is increased greatly because they transport tightly bound Fe^{3+} to the root surface, become reduced from the Fe^{3+} to Fe^{2+} form at the root surface, release Fe^{2+} to be absorbed by roots, and then diffuse back to soil particles to again pick up Fe^{3+} (Fig. 5-6).

Once absorbed, divalent metals are kept soluble partly by chelation with certain cellular constituents. Anions of organic acids, especially citric acid, appear most important as chelating agents for transport of iron through the xylem, but amino acids probably also participate. Ultimately, much iron, zinc, manganese, and copper are bound to proteins. In this form they speed electron transport processes of photosynthesis and respiration and increase the catalytic activity of enzymes. Monovalent cations such as K^+ and Na^+ do not form stable chelates, but even they are associated loosely by ionic attractions with both inorganic and organic acid anions, including proteins.

5.6 Functions of Essential Elements: Some Principles

Essential elements have sometimes been classified functionally into two groups, those having a role in the structure of an important compound, and those having an enzyme-activating role. There is no sharp distinction between these functions, because several elements form structural parts of essential enzymes and help catalyze the chemical reaction in which the enzyme participates. Carbon, oxygen, and hydrogen are the most obvious elements performing both functions, although nitrogen and sulfur, also found in enzymes, are equally important. Another example

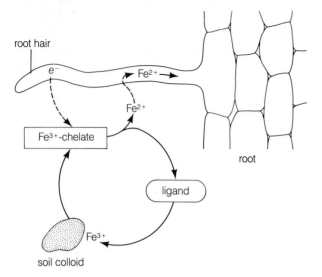

Figure 5-6 Action of chelates in solubilizing and transporting iron from soil particle to root-hair surface. (Redrawn from models of W. L. Lindsay, 1974, and W. L. Lindsay and P. Schwab, 1982.)

of an element with both structural and enzyme-activating roles is magnesium; it is a structural part of chlorophyll molecules and also activates many enzymes.

All elements, free or bound structurally to essential compounds, perform another function by contributing to osmotic potentials, thus aiding buildup of the turgor pressure necessary to maintain form, speed growth, and allow certain pressure-dependent movements (e.g., stomatal opening, Chapter 3; and "sleep" movements of leaves, Chapters 18 and 20). Abundant, nonbound potassium ions are dominant in this regard, but all ions contribute somewhat to osmotic potentials and, therefore, to turgor pressure. Potassium and perhaps chloride, both monovalent ions, are also elements necessary because they temporarily combine with and activate certain enzymes; no permanent structural roles that would make these elements essential are known, yet they perform transient structural roles.

5.7 Nutrient Deficiency Symptoms and Some Functions of Essential Elements

Plants respond to an inadequate supply of an essential element by forming characteristic **deficiency symptoms.** Such visually observed symptoms include stunted growth of roots, stems, or leaves and chlorosis or necrosis of various organs. Characteristic symptoms often help determine the necessary functions of the element in the plant, and knowledge of symptoms helps agriculturists and foresters determine how and when to fertilize crops. Several symp-

toms are described below and are illustrated in books by Sprague (1964), Gauch (1972), Hewitt and Smith (1975), and, for horticultural trees, by Sheer and Faust (1980).

Most of the symptoms described are those appearing on the plant's shoot system and are easily observed. Unless plants are grown hydroponically, root symptoms cannot be seen without removing the roots from the soil, so root deficiency symptoms have been less well described. Furthermore, all symptoms differ to some extent according to the species, the severity of the problem, the growth stage, and as you might suspect, complexities resulting from deficiencies of two or more elements.

The deficiency symptoms for any element depend primarily on two factors:

1. The function or functions of that element

2. Whether or not the element is readily translocated from old leaves to younger leaves

A good example emphasizing both factors is the chlorosis (caused by restricted chlorophyll synthesis) that results from magnesium deficiency. Because magnesium is an essential part of chlorophyll molecules, no chlorophyll is formed in its absence, and only limited amounts are formed when it is present in too low a concentration. Furthermore, chlorosis of lower, older leaves becomes more severe than that of younger leaves. This difference illustrates an important principle: *young parts of the plant have a pronounced ability to withdraw mobile nutrients from older parts, and reproductive organs, flowers and seeds, are especially good at withdrawing,* as mentioned in Chapter 7 and as you might predict if a species is to perpetuate itself. We

Table 5-4 A Guide to Plant Nutrient Deficiency Symptoms.

Symptoms	Deficient Element	Symptoms	Deficient Element
Older or lower leaves of plant mostly affected; effects localized or generalized.		Newer or bud leaves affected; symptoms localized.	
Effects mostly generalized over whole plant; more or less drying or firing of lower leaves; plant light or dark green.		Terminal bud dies, following appearance of distortions at tips or bases of young leaves.	
Plant light green; lower leaves yellow, drying to light brown color; stalks short and slender if element is deficient in later stages of growth.	Nitrogen	Young leaves of terminal bud at first typically hooked, finally dying back at tips and margins, so that later growth is characterized by a cut-out appearance at these points; stalk finally dies at terminal bud.	Calcium
Plant dark green, often developing red and purple colors; stalks short and slender if element is deficient in later stages of growth.	Phosphorus	Young leaves of terminal bud becoming light green at bases, with final breakdown here; in later growth, leaves become twisted; stalk finally dies back at terminal bud.	Boron
Effects mostly localized; mottling or chlorosis with or without spots of dead tissue on lower leaves; little or no drying up of lower leaves.		Terminal bud commonly remains alive; wilting or chlorosis of younger or bud leaves with or without spots of dead tissue; veins light or dark green.	
Mottled or chlorotic leaves; typically may redden, as with cotton; sometimes with dead spots; tips and margins turned or cupped upward; stalks slender.	Magnesium	Young leaves permanently wilted (wither-tip effect) without spotting or marked chlorosis; twig or stalk just below tip and seedhead often unable to stand erect in later stages when shortage is acute.	Copper
Mottled or chlorotic leaves with large or small spots of dead tissue.		Young leaves not wilted; chlorosis present with or without spots of dead tissue scattered over the leaf.	
Spots of dead tissue small, usually at tips and between veins, more marked at margins of leaves, stalks slender.	Potassium	Spots of dead tissue scattered over the leaf; smallest veins tend to remain green, producing a checkered or reticulating effect.	Manganese
Spots generalized, rapidly enlarging, generally involving areas between veins and eventually involving secondary and even primary veins; leaves thick; stalks with shortened internodes.	Zinc	Dead spots not commonly present; chlorosis may or may not involve veins, making them light or dark green in color.	
		Young leaves with veins and tissue between veins light green in color.	Sulfur
		Young leaves chlorotic, principal veins typically green; stalks short and slender.	Iron

From McMurtrey, 1950, *Diagnostic Techniques for Soils and Crops*. American Potash Institute.

do not understand this withdrawing power, but hormonal relations are involved (Chapters 15–17).

Whether withdrawal of an element from a leaf is successful, as with magnesium, depends on the element's mobility in the phloem of vascular tissues. This mobility is determined partly by the solubility of the chemical form of the element in the tissue and partly by how well it can enter the sieve tubes of the phloem. As discussed in Chapter 7, some elements readily move through the phloem from old leaves to younger ones and then to storage organs. These elements include nitrogen, phosphorus, potassium, magnesium, chlorine, and to some extent sulfur. Others such as boron, iron, and calcium are much less mobile, and the mobility of zinc, manganese, copper, and molybdenum is usually intermediate. If the element is soluble and loaded into the sieve tubes, its deficiency symptoms appear earliest and most pronounced in older leaves, whereas symptoms resulting from lack of a relatively immobile element such as calcium or iron appear first in younger leaves. A general guide to deficiency symptoms, partly emphasizing the phloem mobility principle, is given in Table 5-4.

Nitrogen Soils are more commonly deficient in nitrogen than any other element, although phosphorus deficiency is also widespread. Two major ionic forms of nitrogen are absorbed from soils: nitrate (NO_3^-) and ammonium (NH_4^+), as described in Chapter 13. Because nitrogen is present in so many essential compounds, it is not surprising that growth without added nitrogen is slow. Plants containing enough nitrogen to attain limited growth exhibit deficiency symptoms consisting of a general chlorosis, especially in older leaves. In severe cases these leaves become completely yellow and then light tan as they die. They frequently fall off the plant in the yellow or tan stage. Younger leaves remain green longer, because they receive soluble forms of nitrogen transported from older leaves. Some plants, including tomato and certain varieties of maize, exhibit a purplish coloration in stems, petioles, and lower leaf surfaces caused by accumulation of anthocyanin pigments.

Plants grown with excess nitrogen usually have dark green leaves and show an abundance of foliage, usually with a poorly developed root system and therefore a high shoot-to-root ratio. (A reverse ratio often occurs when nitrogen is deficient.) Potato plants grown with superabundant nitrogen show excess shoot growth with only small underground tubers. Reasons for this relatively high shoot growth are unknown, but undoubtedly sugar translocation to roots or tubers is affected in some way, due perhaps to a hormone imbalance. Excess nitrogen also causes tomato fruits to split as they ripen. Flowering and formation of seeds of several agricultural crops are retarded by excess nitrogen. However, flowering is not affected in some species, and still others that flower only in favorable daylength conditions (especially short days) do so faster with abundant nitrogen.

Phosphorus Second to nitrogen, phosphorus is most often the limiting element in soils. It is absorbed primarily as the monovalent phosphate anion ($H_2PO_4^-$) and less rapidly as the divalent anion (HPO_4^{2-}). The soil pH controls the relative abundance of these two forms, $H_2PO_4^-$ being favored below pH 7 and HPO_4^{2-} above pH 7. Much phosphate is converted into organic forms upon entry into the root or after transport through the xylem into the shoot. In contrast to nitrogen and sulfur, phosphorus never undergoes reduction in plants and remains as phosphate, either free or bound to organic forms as esters. Phosphorus-deficient plants are stunted and, in contrast to those lacking nitrogen, are often dark green in color. Anthocyanin pigments sometimes accumulate. Oldest leaves become dark brown as they die.

Maturity is often delayed compared to plants containing abundant phosphate. In many species there is a close interaction between phosphorus and nitrogen regarding maturity, excess nitrogen delaying and abundant phosphorus speeding maturity. If excess phosphorus is provided, root growth is often increased relative to shoot growth. This, in contrast to effects with excess nitrogen, causes low shoot-to-root ratios.

Phosphate is easily redistributed in most plants from one organ to another and is lost from older leaves, accumulating in younger leaves and in developing flowers and seeds. As a result, deficiency symptoms occur first in more mature leaves.

Phosphorus is an essential part of many sugar phosphates involved in photosynthesis, respiration, and other metabolic processes, and it is also part of nucleotides, as in RNA and DNA, and of the phospholipids present in membranes.

Potassium After nitrogen and phosphorus, soils are usually most deficient in potassium. Because of the importance of these three elements, commercial fertilizers list the percentages of N, P, and K they contain (but the last two are actually expressed as equivalent percents of P_2O_5 and K_2O). As with nitrogen and phosphorus, K^+ is easily redistributed from mature to younger organs, so deficiency symptoms first appear on older leaves. In dicots, these leaves initially become slightly chlorotic, especially close to dark **necrotic lesions** (dead or dying spots) that soon develop. In many monocots, such as cereal crops, cells at the tips and margins of leaves die first, and the necrosis spreads basipetally toward the younger, lower parts of leaf bases. Potassium deficient maize and other cereal grains develop weak stalks, and their roots become more easily infected with root rotting organisms. These two factors cause the plants to be rather easily bent to the ground **(lodged)** by wind, rain, or early snow storms.

Potassium is an activator of many enzymes that are essential for photosynthesis and respiration, and it also activates enzymes needed to form starch and proteins. This element is also so abundant that it is a major contributor to the osmotic potential of cells and therefore to their turgor pressure.

Sulfur Sulfur is absorbed from soils as divalent sulfate anions (SO_4^{2-}). It is metabolized by roots only to the extent that they require it, and most sulfate is translocated unchanged to the shoots in the xylem. Because enough sulfate is present in most soils, sulfur-deficient plants are fairly uncommon. Nevertheless, they have been observed in several parts of Australia, some regions of Scandinavia, southwestern grain-producing regions of Canada, and in scattered parts of the northwestern United States. Deficiency symptoms consist of a general chlorosis throughout the entire leaf, including vascular bundles (veins). Sulfur is not easily redistributed from mature tissues in some species, so deficiencies are usually noted first in younger leaves. In other species, however, most leaves become chlorotic at about the same time, or even first in older leaves. Many crop plants, including the roots, contain about one-fifteenth as much total sulfur as nitrogen, and this appears to be a useful guide to evaluate nutritional needs.

Most of the sulfur in plants occurs in proteins, specifically in the amino acids cysteine and methionine that are building blocks for proteins. Other essential compounds that contain sulfur are the vitamins thiamine and biotin and coenzyme A, a compound essential for respiration and for synthesis and breakdown of fatty acids.

Sulfur can also be absorbed by leaves through stomates as gaseous sulfur dioxide (SO_2), an environmental pollutant released primarily from burning coal and wood. SO_2 is converted to bisulfite (HSO_3^-) when it reacts with water in the cells, and in this form it both inhibits photosynthesis and causes chlorophyll destruction. Bisulfite is oxidized further to H_2SO_4; this acid has been blamed for toxic effects of acidic rainfall ("acid rains") in the northeastern United States and adjacent Canadian regions and in many Scandinavian regions.

Magnesium Magnesium is absorbed as divalent Mg^{2+}. In its absence, chlorosis of the older leaves is the first symptom, as already mentioned. This chlorosis is usually interveinal, because the mesophyll cells next to the vascular bundles retain chlorophyll for longer periods than the parenchyma cells between them. Magnesium is almost never limiting to plant growth in soils. Besides its presence in chlorophyll, magnesium is essential because it combines with ATP (thereby allowing ATP to function in many reactions) and because it activates many enzymes needed in photosynthesis, respiration, and formation of DNA and RNA.

Calcium Calcium is absorbed as divalent Ca^{2+}. Most soils contain enough Ca^{2+} for adequate plant growth, but acidic soils where high rainfall occurs are often fertilized with lime (a mixture of CaO and $CaCO_3$) to raise the pH. In contrast to Mg^{2+}, Ca^{2+} is almost immobile in phloem and as a result deficiency symptoms are always more pronounced in young tissues. Meristematic zones of roots, stems, and leaves, where cell divisions are occurring, are most susceptible, perhaps because calcium is required to bind pectate polysaccharides that form a new middle lamella in the cell plate that arises between daughter cells, or because calcium is needed to form microtubules of the mitotic spindle apparatus. Twisted and deformed tissues result from calcium deficiency, and

the meristematic areas die early. Calcium is also essential for normal membrane functions in all cells, probably as a binder of phospholipids to each other or to membrane proteins (Section 6.5).

Calcium is receiving renewed attention, because it is now recognized that all organisms maintain unexpectedly low concentrations of free Ca^{2+} in their cytosol, usually less than 1 μM (Clarkson and Hanson, 1980; Marmé, 1983). This is true even though calcium is as abundant in many plants, especially legumes, as phosphorus, sulfur, and magnesium. Most calcium is in central vacuoles and bound in cell walls to pectate polysaccharides. In vacuoles, calcium is frequently precipitated as insoluble crystals of oxalates, and in some species as carbonates, phosphates, or sulfates. Low cytosol concentrations of Ca^{2+} must be maintained partly to prevent formation of insoluble calcium salts of ATP and other organic phosphates. Few enzymes are activated by Ca^{2+}, and many are inhibited, the inhibition providing further requirement for cells to maintain unusually low Ca^{2+} concentrations in the cytosol where many enzymes exist.

Much calcium within the cytosol becomes reversibly bound to a small, soluble protein called **calmodulin** (Cheung, 1982). This binding activates calmodulin in such a way that it then activates several enzymes. The relation of calcium and calmodulin to enzyme activity in plants will require much more research, but for now we emphasize that an enzyme-activating role for Ca^{2+} likely exists mainly when the ion is bound to calmodulin or closely related proteins.

Iron Iron-deficient plants are characterized by development of a pronounced interveinal chlorosis similar to that caused by magnesium deficiency but occurring first on the youngest leaves. Interveinal chlorosis is sometimes followed by chlorosis of the veins, so the whole leaf then becomes yellow. In severe cases, the young leaves even become white with necrotic lesions. The reason that iron deficiency results in a rapid inhibition of chlorophyll formation is incompletely known, even though this problem has been studied for many years. Iron accumulated in older leaves is relatively immobile in the phloem, as in the soil, perhaps because it is internally precipitated in leaf cells as an insoluble oxide or in the form of inorganic or organic ferric phosphate compounds. Direct evidence that such precipitates are formed is weak, and perhaps other unknown but similarly insoluble compounds are formed. One abundant and stable form of iron in leaves is stored in chloroplasts as an iron-protein complex called **phytoferritin** (Seckback, 1982). Entry of iron into the phloem transport stream is probably minimized by formation of such insoluble compounds, although

Salinity and Agriculture: Problems and Solutions

Murray Nabors

Increasing the efficiency and quantity of food production is an important goal of agriculture and plant breeding. Researchers in plant physiology and development sometimes focus on the practical application of knowledge about plant cell form and function. In this essay, Dr. Murray Nabors, a colleague of Cleon Ross at Colorado State University, tells about the problem of excess salt in soils, its effect on agriculture, and possible solutions resulting from the study of plant tissue culture. Nabors went to Yale College and Michigan State University, where he worked on seed dormancy. Now he concentrates his efforts on developing methods to produce stress-tolerant plants from cultured plant cells and on teaching freshman biology.

I became interested in problems of soil stress and increasing agricultural production by an indirect route. Bugs, rocks, plants, and animals interested me as they do many children. Originally I intended to become a veterinarian, but through the unfortunate habit of petting all stray dogs I discovered that I was allergic to rabies vaccine. Soon after that my interest turned to growing plants. I grew them in the garden, in windows, all over the house, and ultimately in a small greenhouse built in our backyard.

By college I had decided to major in biology, concentrating on botany. One of the best decisions I made as an undergraduate was to become involved in faculty research projects for credit. I worked in Dr. Bruce Stowe's laboratory as a minor participant in his studies of indole compounds, and I worked for Dr. Ian Sussex on a small tissue culture problem. In this way I discovered not only that I liked textbook science, but also that I enjoyed laboratory research.

In graduate school at Michigan State, Dr. Anton Lang introduced me to the problem of seed dormancy by handing me a bag of seeds and asking me what I thought I could do with them. It turned out that quite a lot could be done. Studies on seed dormancy and particularly on phytochrome involvement in this process became my Ph.D. dissertation and have remained a major interest.

In recent years I have returned to plant tissue culture and to the many fascinating implications of **totipotency,** the notion that most plant cells contain a complete set of

genes and can, with sufficient prompting, produce complete plants. The power of this is simply that one small vial can contain a culture of 50,000 or more cells and therefore potential plants. Around 1970 it occurred to Dr. Peter Carlson, also a student of Sussex's, that the possibility of isolating agriculturally useful mutants was enhanced. After all, it is far easier to select salt tolerance from a million cells than from a million plants. The remainder of this essay is devoted to a discussion of why excess salts are a problem for agriculture and what tissue culture, as well as other techniques, can offer as solutions.

One of the characteristics of all living organisms is that they utilize only some of the many types of molecules available in the environment. Semipermeable membranes and various sorts of permease proteins are mechanisms by which desirable molecules are obtained and undesirable ones kept out or removed. For crop production, salinity problems occur chiefly from excesses of aluminum and sodium. The generic term salinity usually refers to Na^+, whereas acidity refers to Al^{3+}.

In regions of high rainfall, dissolved CO_2 in the form of carbonic acid enters the soil and ionizes into bicarbonate and H^+. Thus the negatively charged clay particles become acidic, and the aluminum in them, which is precipitated at normal soil pH, comes into solution and also binds to the clay. Both Al^{3+} and H^+ are near the top of the lyotropic series, so they displace other cations from negatively charged soil particles. These cations as well as bicarbonate and other less strongly bound anions stay in the soil solution and drain into the ground water. Thus acidic soil not only contains toxic aluminum, which itself makes the soil more acidic by binding hydroxide ions, but also is nutrient poor. Soluble aluminum enters plant cells where it lowers cell pH and disturbs normal metabolism. Approximately 25 percent of the world's arable land suffers from excess acidity and its accompanying problems. Soil is frequently limed in such regions to prevent aluminum damage, but this is expensive, which makes it impractical for developing countries. Crops with increased tolerance are also needed.

Sodium ions are generally toxic to living cells, which typically exclude, remove, or compartmentalize them. The active exclusion of Na^+ by animal cells is, in evolutionary terms, the *raison d'être* for the nerve impulse and ultimately the brain.

The earth contains considerable quantities of sodium salts. As snow melts in the mountains and rain falls, the streams become rivers and move down to the sea. Rivers become loaded with increasing quantities of dissolved salts as they pass through the land, and thus the ocean is salty. The Colorado River, for example, has salinity levels of around 0.87 mM (50 ppm) in the Rocky Mountains and around 17 mM (1000 ppm) as the river nears the Gulf of California. The salinity of rivers is increased beyond natural salt loading by municipal sewage and sewage-treatment plants and by irrigation, which leaches additional ground salts as well as applied fertilizer salts into the ground water and ultimately the rivers. In the case of the Colorado River Basin, approximately 50 percent of the river's salt comes from these human sources.

In addition to increasing the salinity of rivers, irrigation gradually adds salts to the soil, because any irrigation source, no matter how pure, contains dissolved salts. After each cycle of irrigation, the water evaporates or is transpired by the plants, leaving these salts behind. Eventually the salt concentration in soils can increase to the point at which crop production is lowered. Although plant species differ considerably in salinity tolerance, a good rule of thumb is that agricultural damage occurs when the soil salinity level reaches 14 mM (800 ppm). A typical safe drinking water standard is 500 ppm. Excess soil salinity is a problem in the vast agricultural valleys of California. Such damage in the Colorado River Basin alone is already estimated at over \$100 million per year (1984 dollars). Worldwide it may exceed one trillion dollars per year. Archaeological studies indicate that salinity in irrigated agriculture may have contributed to the downfall of ancient Sumer in Mesopotamia.

Irrigation practices that control the level of groundwater and limit the amount of water applied, such as trickle irrigation, can reduce overall agricultural losses and slow the accumulation of salts. At present about 14 percent of cultivated land is irrigated, and this land accounts for 50 percent of crop production. Yields could be increased on sizeable portions of cultivated land and could be sustained on currently uncultivated land if the extent of irrigation were increased. The major problems preventing this are the cost of installation and operation of irrigation systems and a lack of available water: About 80 percent of readily available fresh water is already utilized in agriculture.

Improving the salt and drought tolerance of crop plants must be seen as a major factor in overcoming these limitations to plant growth. These stresses usually occur together since dry land has not had natural salts leached from it and can accumulate additional salts if irrigated. Salinity damage is also a major problem in low-lying coastal areas. Around 25 percent of the earth's arable land suffers from excess salinity. Even rice, commonly thought of as a plant with ample water supplies, suffers from drought in about half of its acreage.

Plants can be grouped into two general categories based on their susceptibility to salt damage. **Halophytes** are "salt lovers" and tolerate relatively high concentrations of salts; **glycophytes** are "sugar lovers" and tolerate relatively low concentrations. Most commonly grown crop plants are glycophytes.

Salt and acid tolerance are not well understood physiologically and can be caused by mechanisms acting at the cellular level or the physiological or anatomical alterations involving the plant as a whole. Increases in the salt or acid tolerance of crop cultivars can be achieved in any of several ways.

1. Utilization of existing germplasm and transfer, by traditional breeding, of tolerance from resistant to sensitive cultivars.

2. Cross breeding of cultivars to a wide variety of wild tolerant plants in the same or in different species or genera.

3. Tissue-culture selection of cell lines, then regenerated plants, with increased levels of tolerance.

4. Direct movement of genes for increased tolerance from one plant to a crop plant.

At present the first three methods are practical and are being utilized by various groups of researchers. The fourth method requires advances in basic research before practical utilization can be achieved.

Our laboratory has been able to demonstrate for both tobacco and oats that selected, Na$^+$-tolerant cell cultivars can produce regenerated plants that are also tolerant and that pass stable tolerance on to subsequent generations. We are now working to increase stress tolerance of all major cereals including wheat, rice, maize, oats, and millet. Until recently, tobacco has been one of the few "white rats" of tissue culture because both cell culture and plant regeneration procedures were well established. Initially, working with cereals required the development of long-term, high-frequency plant regeneration methods. Our laboratory has concentrated on finding these methods for easily obtainable starting materials such as seeds. When we started three years ago, it was difficult to obtain more than a few plants from a rice tissue culture that originated from a single seed. Now, from one seed, we can produce over 110,000 plants in six months.

In addition to allowing production of salt-tolerant rice (now in the greenhouse testing stage), this technique means that a rare disease- or stress-resistant plant found in the field can be rapidly cloned for use by plant breeders and growers.

Figure 5-7 Iron deficiency in apple leaves: O, normal; A–D various levels of deficiency, with D the most severe. (Courtesy of M. Faust.)

in the plant without becoming a structural part of organic molecules. Most species absorb 10 to 100 times as much chloride than they require, so it represents a common example of luxury consumption. One function of chloride is to stimulate the split of H_2O during photosynthesis (Chapter 9), but it is also essential for roots and for cell division in leaves (Terry, 1977). Leaf deficiency symptoms consist of reduced growth, wilting, and development of chlorotic and necrotic spots. Leaves often eventually attain a bronze color. Roots become stunted in length but thickened, or club-shaped, near the tips. Chloride is rarely if ever deficient in nature, because of its high solubility and availability in soils and because it is also transported in dust or in tiny moisture droplets by wind and rain to leaves where absorption occurs. Because of its presence in human skin, it was necessary for investigators demonstrating its essentiality to wear rubber gloves.

Manganese Manganese exists in various oxidation states as insoluble oxides in soils, but it is absorbed largely as the divalent manganous cation (Mn^{2+}) after reduction of Mn in these oxides at the root surface (Uren, 1981). Deficiencies of manganese are not common, although various disorders such as "gray speck" of oats, "marsh spot" of peas, and "speckled yellows" of sugar beets result when inadequate amounts are present. Initial symptoms are often an interveinal chlorosis on younger or older leaves, depending on the species, followed by or associated with necrotic lesions. Electron microscopy of chloroplasts from spinach leaves showed that the absence of manganese causes disorganization of thylakoid membranes but has little effect on the structure of nuclei and mitochondria. This and much biochemical work show that the element plays a structural role in the chloroplast membrane system and that one of its important roles is, like that of chloride, in the photosynthetic split of H_2O (Section 9.4). The Mn^{2+} ion also activates numerous enzymes.

Boron Boron is almost entirely absorbed from soils as undissociated boric acid (H_3BO_3, more accurately shown as $B(OH)_3$). It is only slowly translocated out of organs in the phloem once it arrives there in the xylem (Raven, 1980). Deficiencies are not common in most areas, yet several disorders related to disintegration of internal tissues, such as "heart rot" of beets, "stem crack" of celery, "water core" of turnip, and "drought spot" of apples result from an inadequate boron supply. Plants deficient in boron show a wide variety of symptoms, depending upon the species and plant age, but one of the earliest symptoms is failure of root tips to elongate normally, accompanied by inhibited synthesis of DNA and RNA.

phytoferritin seems to represent a storehouse of iron. Another suggestion to explain poor mobility of iron is that it can enter the phloem in moderate amounts but then leaks into the xylem and moves back to the same leaf again. In any case, once it is taken into an organ from the xylem, its redistribution to younger tissues and to seeds is limited.

Iron deficiencies are often found in particularly sensitive species in the rose family, including shrubs and fruit trees (Fig. 5-7), and in cereal grains such as maize. In soils of the western United States, the high pH and the presence of bicarbonates contribute to iron deficiency, whereas in acidic soils soluble aluminum is more abundant and restricts iron absorption.

Iron is essential because it forms part of certain enzymes and part of numerous proteins that carry electrons during photosynthesis and respiration. It undergoes alternative oxidation and reduction between the Fe^{2+} and Fe^{3+} states as it acts as an electron carrier in proteins. The importance of iron, zinc, copper, and manganese in electron transport processes in plants was reviewed by Sandman and Böger (1983).

Chlorine Chlorine is absorbed from soils as the chloride ion (Cl^-) and probably remains in this form

Cell division in the shoot apex is also inhibited, as is that in young leaves. Boron plays an undetermined but essential role in elongation of pollen tubes.

Biochemical functions of boron remain unclear in spite of much study, partly because we don't know to what extent $B(OH)_3$ is modified in cells and partly because there might be several functions. Probably, much of this weak acid becomes bound as cis-diol borate complexes with adjacent hydroxyl groups of mannose and certain other sugars in cell-wall polysaccharides (but not with glucose, fructose, galactose and sucrose, which have no cis-diol arrangements of hydroxyl groups). The biochemical and physiological functions proposed for boron were reviewed by Dugger (1983), Pilbeam and Kirkby (1983), and Lovatt and Dugger (1984). No specific function is yet certain, but evidence favors special involvement of boron in nucleic acid synthesis and some unclear functions in membranes.

Zinc   Zinc is absorbed as divalent Zn^{2+}, probably often from zinc chelates. Disorders caused by zinc deficiency include "little leaf" and "rosette" of apples, peaches, and pecans, resulting from growth reduction of young leaves and stem internodes. Leaf margins are often distorted and puckered in appearance. Interveinal chloroses often occur in leaves of maize, sorghum, beans, and fruit trees, suggesting that zinc participates in chlorophyll formation or prevents chlorophyll destruction. The retardation of stem growth in its absence might result partly from its apparent requirement to produce a growth hormone, indoleacetic acid (auxin). Many enzymes contain tightly bound zinc essential for their function; considering all organisms, more than eighty such enzymes are known (Vallee, 1976).

Copper   Plants are rarely deficient in copper, mainly because they need so little of it (Table 5-3). Because it is sufficiently available in nearly all soils, deficiency symptoms are largely known from solution culture studies. Without copper, young leaves often become dark green in color and are twisted or otherwise mis-

shapen, often exhibiting necrotic spots. Citrus orchards are occasionally deficient, in which case the dying young leaves led to the name "die back" disease. Copper is absorbed both as the divalent cupric (Cu^{2+}) ion in aerated soils or as the monovalent cuprous ion in wet soils with little oxygen. Divalent Cu^{2+} is chelated in various soil compounds (generally unidentified), and these likely provide most copper to root surfaces. Partly because such small amounts are needed by plants (Table 5-3), it readily becomes toxic in solution culture unless its amounts are carefully controlled.

Copper is present in several enzymes or proteins involved in oxidation and reduction. Two notable examples are cytochrome oxidase, a respiratory enzyme in mitochondria (Section 12.7) and plastocyanin, a chloroplast protein (Section 9.4).

Molybdenum   Molybdenum exists to a large extent in soils as molybdate (MoO_4^{2-}) salts and also as MoS_2. In the former, molybdenum exists in the redox (valence) state of Mo^{6+} but in sulfide salts as Mo^{4+}. Probably because only trace amounts are required by plants, essentially nothing is known about the forms in which it is absorbed and how it is changed in plant cells. Most plants require less molybdenum than any other element, so molybdenum deficiencies are rare. Nevertheless, they are geographically widespread. Examples of disorders caused by inadequate molybdenum include "whiptail" of cauliflower and broccoli, found, for example, in certain areas of the eastern United States. Symptoms often consist of an interveinal chlorosis occurring first on the older or mid-stem leaves, then progressing to the youngest leaves. Sometimes, as in the "whiptail" disease, plants do not become chlorotic but develop severely twisted young leaves, which eventually die. In acidic soils, adding lime increases availability of molybdenum and eliminates or reduces the severity of its deficiency. The only well-documented function of molybdenum in plants is as part of the enzyme nitrate reductase, which reduces nitrate ions to nitrite ions (Chapter 13).

6

Absorption
of Mineral Salts

In previous chapters, we explained how water moves into, through, and out of plants and how osmosis is essential to water movement. We found it advantageous to ignore the much slower movement of solutes across membranes, because osmosis could not occur unless movement of water was much faster than that of solutes. Yet solutes do move from cell to cell and from one cellular organelle to another, and this is essential for life. Carbon, oxygen, and hydrogen are provided by H_2O and atmospheric CO_2 and O_2, but the 13 other elements essential to all plants are absorbed as ions from the soil by what has aptly been called a "solution mining" process. Just as leaves must absorb their carbon from a low concentration of CO_2 in the atmosphere, roots must absorb essential mineral salts from low concentrations in the soil solution. This chapter concerns the morphological and anatomical properties of roots that allow them to effectively absorb these mineral salts. Properties of membranes, which control absorption rates, are also described. Finally, some general mechanisms governing transport of all solutes across membranes are discussed.

6.1 Roots as Absorbing Surfaces

Plants solve the problem of absorbing frequently scarce water and mineral elements from soils by producing surprisingly large root systems (Fig. 6-1). The overall shapes of such systems are controlled mainly by genetic rather than environmental mechanisms. Thus grasses have fibrous and highly branched root systems near the soil surface, although the roots of perennial grasses usually extend more deeply than those of related annual species. Many herbaceous (nonwoody) dicots have a dominant taproot that can extend several meters downward (e.g., alfalfa), although the dominant taproot is shorter in most species (e.g., carrot, dandelion, and Canada thistle).

Other common herbaceous dicots such as soybean and tomato have root systems with a taproot difficult to distinguish from branch roots. The same is true for many trees and shrubs, both angiosperms and gymnosperms, although tree root morphology can be complex, especially in pines (Kramer and Kozlowsky, 1979). Roots commonly extend outwardly from tree trunks much farther than do aboveground branches.

Even though root morphology is genetically controlled, soil environments have influences. Branching patterns of roots are more flexible than those of shoots, perhaps because soil environments vary more than those of air. If a topsoil is only a thin layer covering a hardpan of clay or rocks, roots spread laterally near the surface because they cannot grow deeply. Essentially, roots grow where they can, and mechanical impedance, temperature, aeration, and availability of water and mineral salts are all important factors. In unusually moist or fertile regions, roots proliferate extensively (Fig. 6-2) until water or nutrients are exhausted; then they grow into new regions by formation of more branch or feeder roots. If water is more available deep in the ground, roots are generally also located deeply. Nevertheless, plants adapted to dry regions are not necessarily deep rooted, because shallow root systems take better advantage of brief rains. In fact, root systems of some species proliferate both near soil surfaces and at substantial depths, with a few long and relatively unbranched connections between; such systems probably represent adaptations to variable climates.

Far too little is known about root properties in soils, because roots are so difficult to observe. Yet careful studies show that branch roots of annual crops elongate for only a few days, and that those of perennial species live a year or more before decay occurs. Some desert shrubs replace up to a fourth of their root systems each year, absorbing water and mineral salts from new locations by the new roots

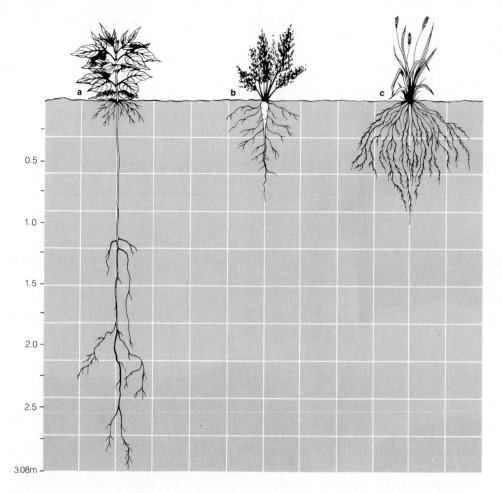

Figure 6-1 Types of root systems: (**a**) In poison ivy, the taproot is adapted to reach moisture deep in the ground. (**b**) The carrot taproot system is adapted to store food. (**c**) The fibrous root system of blue grama grass is adapted to absorb surface water. (From W. A. Jensen, et al., *Biology,* 1979.)

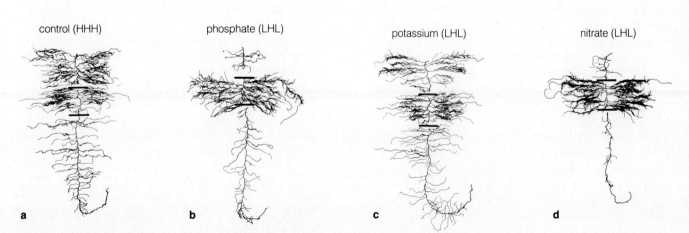

Figure 6-2 Root proliferation of barley in localized zones of sand fertilized with phosphate, potassium, or nitrate. Portions of root systems (shown separated by line-bars) were grown 21 days in sand compartments separated into three layers by wax barriers through which roots could grow but solution did not flow. Layers were fertilized with nutrient solution containing high (H) or low (L) levels of the particular element. Controls (HHH) received high levels of elements in all three layers. Plants exposed to varying potassium showed little proliferation in the well-fertilized central layer, but the acid-washed sand was found to contribute K^+. (From M. C. Drew, 1975.)

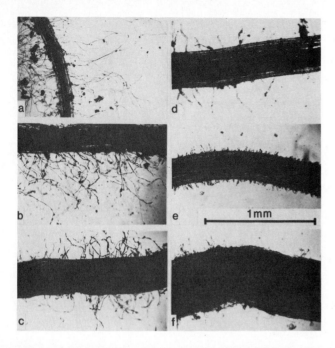

Figure 6-3 Root hairs of (**a**) Russian thistle, (**b**) tomato, (**c**) lettuce, (**d**) wheat, (**e**) carrot, and (**f**) onion. (From S. Itoh and S. A. Barber, *Agronomy Journal, 1983,* by permission of the American Society of Agronomy.)

(Caldwell and Camp, 1974). Yearly root losses from perennial grasses are slower than from perennial shrubs (Weaver and Zink, 1946; Troughton, 1981), and long retention of old roots contributes to the pronounced ability of grasses to prevent soil erosion. Not much is known about death and replacement of tree roots, but reports summarized by Sutton (1980) indicating that a 100-year-old Scots pine (*Pinus sylvestris*) tree had 5 million root tips, and a mature red oak (*Quercus rubrum*) tree had 500 million live tips indicate the vastness of such root systems (also see Perry, 1981).

The cylindrical and filamentous (long and narrow) form of roots proves unexpectedly important for absorption of water and solutes from soils. A cylinder has more strength per unit cross-sectional area than other shapes, and this shape (with a protective root cap) probably helps growing roots force soil particles aside without breaking the roots. The filamentous form of roots allows exploration of much more soil volume per unit root volume than if roots were shaped as spheres or discs (Wiebe, 1978). Exploration of large soil volumes is important for roots to contact water and ions. When soils are moist or wet, diffusion toward roots is reasonably rapid, but when they dry to a ψ near -1.5 MPa (a common permanent wilting point) water diffusion can decrease 1000 fold. Plants then have difficulty obtaining water and mineral ions dissolved in it. (See discussion on page 86.)

Besides filamentous roots, **root hairs** also contribute to absorption of ions and water. Root hairs are modified epidermal cells that have thin filaments up to 1.5 mm long (Dittmer, 1949). They develop just behind the short region of root elongation close to the tip, and the zone containing them is often less than 1 cm long. In the absence of soil but under conditions of adequate moisture and aeration, some species form an unusually extensive system of root hairs. Nevertheless, root hair formation in soils is minimized by microbes and by other soil conditions (Mosse, 1975; Reynolds, 1975). Figure 6-3 illustrates root hairs of various angiosperms growing in soil. In general, root hairs are more frequent and extend over a longer region of the root when soils are moderately dry than wet, but if soils are too dry, root hairs desiccate and die. Sutton's review (1980) indicates that although some conifers have root hairs, others probably have few or none. The presence of mycorrhizae, especially the ecto-type, minimizes root hair formation in conifers and other species.

6.2 Mycorrhizae

We usually learn about root structures from plants grown in a greenhouse. But in nature young roots of most species (perhaps 97 percent) look somewhat different, because they become infected with fungi present in native soils and form **mycorrhizae.** A mycorrhiza (fungus-root) is a **symbiotic** (intimate) and **mutualistic** (mutually beneficial) association between a nonpathogenic or weakly pathogenic fungus and living root cells, primarily cortical and epidermal cells. The fungi receive organic nutrients from the plant and, in turn, improve the mineral salt and water-absorbing properties of roots. Generally, only the tender young roots become infected by the fungus, perhaps because on the older parts the epidermis and cortex are lost and a protective layer of suberin develops in cork cells. Root-hair production either slows or ceases upon infection, so mycorrhizae often have few such hairs. This would greatly decrease the absorbing surface, except that the soil volume penetrated is greatly increased by the slender fungal hyphae extending from the mycorrhizae.

Two main groups of mycorrhizae are recognized: the **ectomycorrhizae** (Marks and Kozlowski, 1973) and the **endomycorrhizae** (Gerdemann, 1975; Sanders et al., 1976), although a more rare group with intermediate properties, the **ectendotrophic,** is sometimes encountered. In the ectomycorrhizae, the fungal hyphae form a mantle outside the root and also inside the root in the intercellular spaces of the epidermis and cortex. No intracellular penetration into epidermal or cortical cells occurs, but an extensive network called the **Hartig net** is formed between these cells (Fig. 6-4). Ectomycorrhizae are common on trees, including members of the Pinaceae family (pine, fir, spruce, larch, hemlock), the Fagaceae (oak, beech, chestnut), Betulaceae (birch, alder), Salicaceae (willow, poplar), and a few other families. Figure 6-5 shows scanning electron micrographs of two *Pinus contorta* roots, one noninfected root with root hairs and one infected with an ectomycorrhiza fungus.

Endomycorrhizae consist of three subgroups, but by far the most common are the **vesicular arbuscular mycorrhizae (VAM).** The fungi present in VAM are members of the Endogonacae, and they produce an internal network of hyphae between cortical cells that extends out into the soil where they absorb mineral salts and water. Although VAM fungi seem to penetrate directly into the cytosol of cortical cells (where they form structures called vesicles and arbuscules, giving them their name), that part of the fungus is surrounded by an invaginated plasma membrane of the cortex cell. The VAM are present on most species of herbaceous angiosperms, monocot and dicot, annual and perennial crops and native species; they also occur on the gymnosperm genera *Cupressus, Thuja, Taxodium, Juniperus,* and *Sequoia* (Gerdemann, 1975; Safir, 1980).

The fungal partner of both kinds of mycorrhizae receives sugars from the host plant, and plants that are grown in shade and deficient in sugars predictably have poor mycorrhizal development. Also, plants growing on fertile soils often have mycorrhizae that are less developed than those of wild plants growing on nonfertile soils. The most well-documented advantage of mycorrhizae to plants is increased phosphate absorption, although absorp-

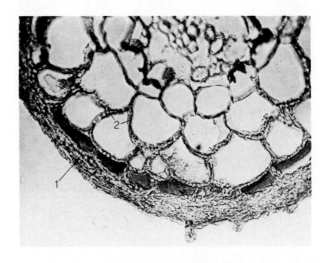

Figure 6-4 Ectomycorrhiza formed between *Pinus taeda* and the fungus *Thelephora terrestris*. Note the external mantle (1) and the Hartig net (2) between the cells. (Courtesy C. P. P. Reid.)

tion of other nutrients and water is often facilitated. The greatest benefit of mycorrhizae is probably to increase absorption of essential ions that normally diffuse slowly toward roots yet are needed in high demand, especially phosphate, NH_4^+, K^+, and NO_3^- (Chapin, 1980). Mycorrhizae offer great advantages to trees growing on unfertile soils (Fig. 6-6). In fact, without the nutrient-absorbing properties of mycorrhizae, many communities of trees could not exist. For example, some European pines introduced to the United States grew poorly until they were inoculated with mycorrhizal fungi from their native soils. A considerable potential exists for population of certain mine-waste areas, landfills, roadsides, and other in-

Figure 6-5 (**a**) Scanning electron micrograph of dichotomous roots and root hairs of *Pinus contorta*. Note the absence of a fungal mantle. (**b**) Scanning electron micrograph of ectomycorrhiza of *Pinus contorta* inoculated with *Cenococcum graniforme*. (Courtesy John G. Mexal, Edwin L. Burke, and C. P. P. Reid.)

Figure 6-6 Growth promotion of 6-month-old juniper (*Juniperus osteosperma*) plants by mycorrhizal formation. Plants were grown under identical conditions in a growth chamber. Only the three plants at right had mycorrhizae. (Courtesy of F. B. Reeves.)

fertile soils through the introduction of plants inoculated with fungi capable of forming mycorrhizae (Marx and Schenck, 1983). Greater contributions of mycorrhizae to agriculture and forestry should occur as we understand them better, and such understanding is developing rapidly.

6.3 Soils and Their Mineral Elements

Soils are a heterogenous and variable mixture of inorganic mineral particles, decaying organic matter, and living microorganisms, along with air and various inorganic salts and organic molecules dissolved in water. The mineral particles are present as sand, smaller silt, or still smaller clay, all composed mainly of silicon, oxygen, and aluminum. The clay mineral fraction is the most abundant in many soils, and the high surface-to-volume ratio of clay minerals and organic colloids is important to soil fertility because it allows considerable **adsorption** of cations to these particles. The surfaces of clay minerals become negatively charged because some atoms of Si^{4+} are replaced by Al^{3+}, and some atoms of Al^{3+} are replaced by Mg^{2+} or Fe^{2+}. Wherever such replacement occurs, a negative site becomes available, resulting in the attraction and adsorption of a dissolved cation such as $H^+ > Ca^{2+} > Mg^{2+} > K^+ = NH_4^+ > Na^+$ (the **lyotropic series;** ones to the left adsorbed most tightly). These adsorbed cations are not easily lost by leaching when water moves down through the soil, so they remain potentially available for absorption by roots.

The organic-matter fraction of soils also has an excess of negative charges resulting from ionized carboxyl groups ($-COOH \rightarrow -COO^- + H^+$), and also

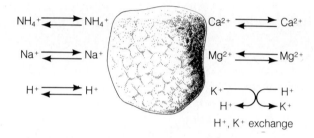

Figure 6-7 Cation exchange on a clay particle. The lower right reaction illustrates replacement of K^+ by H^+.

from ionized OH groups ($-OH \rightarrow -O^- + H^+$) of phenolic compounds present in decaying lignin from wood. Cations are adsorbed onto organic matter surfaces as well as to clay surfaces. In either case, these cations can exchange reversibly with identical or different cations dissolved in the soil solution, as illustrated in Fig. 6-7. Potassium replacement by H^+ (lower right) is an example of exchange between different cations and is called **cation exchange.** Exchange with H^+ is especially important, because roots release H^+ from malic acid and other organic acids into the soil, and H^+ ions are also released from H_2CO_3 formed when respiratory CO_2 reacts with water. These H^+ can replace any cation adsorbed to a soil colloid and make it available for the roots.

Among the anions, phosphate ($H_2PO_4^-$ or HPO_4^{2-}, depending upon pH) also occurs in solution in low concentrations, although most phosphate is reversibly precipitated as salts of aluminum, calcium, or iron. Nitrate (NO_3^-), sulfate (SO_4^{2-}) and chloride (Cl^-) are more soluble, and all are repelled by negatively charged colloids. Because of this repulsion, their salts are leached from soils as water passes

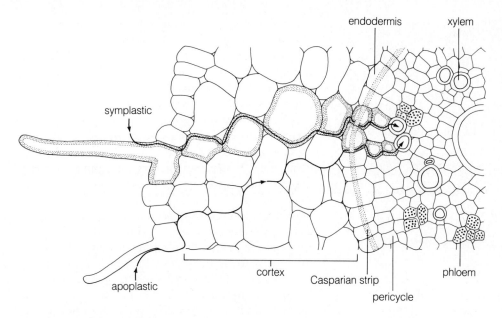

Figure 6-8 Anatomical aspects of symplastic and apoplastic pathways of ion absorption in the root-hair region. The symplastic pathway involves transport through the cytosol (stippled) of each cell all the way to nonliving xylem. The apoplastic pathway involves movement through the cell wall network as far as the Casparian strip, then movement through the symplasm. Casparian strip of endodermis is shown only as it would appear in end walls. (Redrawn from K. Esau, 1971.)

through it. Another form of nitrogen, NH_4^+, is adsorbed onto soil colloids, but most soils contain little NH_4^+ because it is so rapidly oxidized to NO_3^- by bacteria. Because of the high requirement of plants for nitrogen and the loss of NO_3^- by leaching, most crops (except legumes) require relatively large amounts of nitrogen fertilizers. If nitrogen can be added in organic matter such as animal manure or dead vegetation (compost), it is released only by decay and thus is available to plants for a longer time. The organic matter also improves the physical characteristics of the soil, including its cation exchange capacity, and provides many other essential elements.

6.4 Ion Traffic into the Root

The anions and cations most readily available for roots are those freely dissolved in the soil solution, even though their concentrations are usually low. In one survey of more than 100 agricultural soils at or near field capacity (with as much water present as the soil particles could hold), more than half had dissolved NO_3^- concentrations of less than 2 mM, phosphate of less than 0.001 mM, K^+ of less than 1.2 mM, and SO_4^{2-} of less than 0.5 mM (Reisenauer, 1966). Soils on which crops are grown are more fertile than range or forest soils. Even so, concentrations of such elements in crop plants can reach 10 to 1000

times these values. Such nutrients reach roots in three ways: by diffusing through the soil solution, by being passively carried along as water moves by bulk flow into the roots, and by roots growing toward them.

Mineral salts can be absorbed by and transported upward from regions of roots containing root hairs and also by much older regions many centimeters from the root tip (Clarkson and Hanson, 1980; Clarkson, 1981). Mycorrhizae have not been investigated nearly as much as nonmycorrhizal roots, but they almost surely absorb rapidly near the tips where fungal hyphae are concentrated and somewhat less rapidly in older regions. Root tips are frequently exposed to higher concentrations of dissolved mineral salts than older regions, because older regions exist in parts of the soil already explored and partially depleted by growing root tips.

In Section 4.3, we examined the pathway of water movement into young regions of nonmycorrhizal roots. There are three paths through which water and dissolved ions might move into the xylem cells of such roots (Fig. 6-8): through the cell walls (apoplast) of epidermal and cortical cells, through the cytoplasmic (symplast) system, moving from cell to cell, and from vacuole to vacuole of the living root cells (where the cytosol of each cell would necessarily form part of the pathway). Ions that are absorbed into root cell vacuoles usually move out

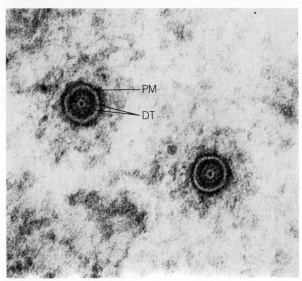

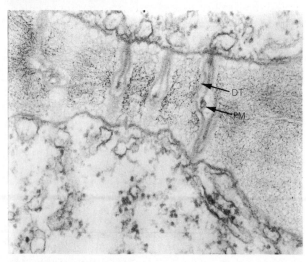

Figure 6-9 The structure of plasmodesmata. (**a**) Hundreds of plasmodesmata (small black dots) in a primary pit field in the end wall of a young barley endodermal cell. (**b**) High magnification of two such plasmodesmata, showing their tubular nature. (**c**) Longitudinal view of plasmodesmata across two adjacent young endodermal cells. PM, plasma membranes; DT, desmotubule. (Courtesy A. W. Robards, 1976.)

slowly, so the third pathway is less important than the other two. Most experts have assumed that both the symplastic and the apoplastic pathways contribute substantially to transport of most ions (Läuchli, 1976; Spanswick, 1976; Pitman, 1977). Nevertheless, Ca^{2+} is transported mainly by the apoplastic pathway (Clarkson, 1981). Apparently, both plant and animal cells have mechanisms to keep the concentration of Ca^{2+} in their cytosol at low concentrations, partly because Ca^{2+} precipitates both inorganic and organic phosphate compounds. It is important to note that the primary cell wall has holes between its polysaccharides that are large enough to easily pass dissolved salts (Carpita et al., 1979). Thus the wall represents little barrier to ion traffic whatever the pathway.

As described in Section 4.3, movement of water and ions via the apoplastic pathway can occur through walls of cortex cells until restricted by the impermeable Casparian strips of endodermal cells (see Fig. 4-10). The Casparian strip is present in all endodermal cells except those within a few millimeters of the root tip. It is usually formed near the end of the elongation phase of such cells. At the Casparian strip, further progress of solutes toward the xylem is controlled by the plasma membranes of the endodermal cells or of nearby cortical cells. These membranes help control the rates of ion absorption and the kinds of solutes absorbed. Furthermore, as dissolved ions move along in the walls of the epidermal and cortical cells, a certain percentage of each kind is absorbed by each of these cells into the cytosol and thus enters the symplast pathway. Some of these ions are transported into the vacuole, where they contribute greatly to the negative osmotic potentials of roots, facilitating water uptake, turgor pressure, and growth of roots through soils.

An ion that is absorbed by an epidermal cell and moves toward the xylem in the symplastic pathway must cross the epidermis, several cortical cells, the endodermis, and the pericycle. This movement could involve transport directly through each of the two primary walls, middle lamella, and plasma membranes between the cytosol of adjacent cells. Alternatively, the ion could move through **plasmodesmata,** which are tubular structures formed through the adjacent cell walls of nearly all living plant cells (Robards, 1975; Gunning and Overall, 1983). Frequencies of plasmodesmata are commonly greater than 1 million per square millimeter. Electron micrographs of plasmodesmata are shown in Fig. 6-9. Plasmodesmata are surrounded by a tube of plasma membrane that is continuous between the two adjacent cells. In their central region lies another tubular structure called a **desmotubule** that is apparently a compressed part of the endoplasmic reticulum extending from one cell to the other. Flow of solutes through plasmodesmata has been demonstrated in various ways, for example, by precipitation of Cl⁻ with Ag⁺ salts. Apparently, solutes cannot pass through the desmotubule but only between it and the plasma membrane. Even though plasmodesmata probably contribute to solute movement across cells, direct movement through other membrane regions is also involved.

Regardless of the pathway across the root, ions transported to the shoot must somehow get into dead conducting cells of the xylem. This involves transfer either from living pericycle cells or still-living xylem cells. Evidence obtained with inhibitors of respiration (especially those that block ATP formation) indicate that transfer into the conducting xylem requires metabolic energy and ATP formation. If so, this apparently means that pericycle or living xylem cells can absorb ions from other living cells on one side and secrete into dead xylem cells at the other side.

Most studies on pathways of ion absorption by roots have involved young and nonmycorrhizal roots, but our views of apoplastic and symplastic pathways for such roots must be modified for older roots and for mycorrhizae. Little attention has been given to how hyphae of mycorrhizal fungi transport ions to mycorrhizal roots and how these hyphae alter the apoplastic and symplastic pathways. But for older roots and nonmycorrhizal roots, information indicating importance of both pathways for absorption is accumulating (Clarkson, 1981). Mature parts of some dicot and grass roots where branch roots originate develop a suberized, water-repellent layer over the entire outer tangential wall of the endodermis. This layer is similar in composition to the Casparian strip already present on the radial walls of these cells, and the two often fuse. Plasmodesmata permeate this layer, and symplastic movement from the cortex into the endodermis is not slowed nearly as much by its presence as was thought before.

Roots of trees and other perennials having secondary growth (thickening due to division of vascular cambium) eventually lose their epidermal and most of their cortical cells, and the endodermis is also crushed and destroyed (Esau, 1971). These roots become covered with a thin layer of dead, corky tissue, frequently containing numerous specialized holes called **lenticels** through which oxygen can penetrate. Mineral salts and water move into such roots, even though suberin nearly covers the outer surface of the corky cells. The suberin layer is discontinuous in certain places, allowing some water and salts to enter, while other entry paths include lenticels and holes left where some of the branch roots have died. The pathway of ion transport into older regions of the roots of herbaceous dicots and grasses also becomes complicated by loss of the epidermis and cortex. This leaves the endodermis or pericycle as the outermost

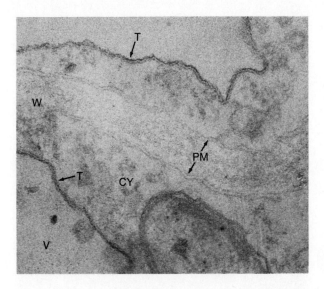

Figure 6-10 The three-layered appearance of the tonoplast (T) and plasmalemma (PM) in two root tip cells of potato. The cell wall (W), cytoplasm (CY), and part of the vacuole (V) are also shown. (Courtesy Paul Grun, 1963.)

layer of cells (Drew, 1979). These cells become brown because of tannin accumulation, and the tannins are apparently accompanied by lignin and suberin which act as barriers for transport of water and ions into the root (for description of tannins, lignin, and suberin, see Chapter 14).

Even though the pathways for ion traffic into the root are complex, it is clear that the plasma membranes of roots control the rates of movement and, therefore, to some extent even the kinds of ions that get in. Other membranes around organelles within root cells and membranes of other organs also control solute movement. In fact, the most important function of a membrane around any cell or organelle is to control the composition inside so that life's processes can occur normally even when changes in the surroundings occur. To understand such control, we need to understand membranes.

6.5 The Nature of Membranes

Electron micrographs show that most biological membranes are similar, regardless of their location. They are generally 7.5 nm to more than 10 nm thick and usually appear in cross-sectional high-resolution electron micrographs as two dark (electron dense) lines separated by a lighter (electron-transparent) layer. Figure 6-10 illustrates this three-layered appearance of both the plasmalemma and the tonoplast in two root-tip cells of the potato.

Membranes consist largely of proteins and lipids, the proteins usually representing about one-half

Figure 6-11 Scanning electron micrograph showing part of plasma membrane of pea root cell in surface view. Root section was rapidly frozen by a freeze-fracture technique that splits the membrane between the bilayers. As a result, the half of the bilayer facing the cytosol is absent, and only the portion adjacent to the cell wall is visible. Tiny bumps in the membrane half represent protein molecules (p). In the upper portion of the photograph, cellulose microfibrils (c) in the primary wall are shown. (Courtesy Dan Hess.)

to two-thirds of the membrane dry weight. Figure 6-11 shows an electron micrograph of a pea root membrane in surface view. When the tissue was cut, the knife tore away part of the membrane and exposed the microfibrils of cellulose in the wall below. The tiny bumps on the membrane represent protein molecules that extend out of the membrane into the wall. The kinds and proportions of proteins and lipids vary with the kind of membrane and the physiological state of the cell in which it occurs, so some differences exist between plasmalemma, tonoplast, endoplasmic reticulum, and membranes of dictyosomes, chloroplasts, nuclei, mitochondria, and microbodies (peroxisomes and glyoxysomes). Important composition differences in a given membrane such as the plasma membrane also exist among various species. Nevertheless, the principal lipids of most plant membranes are **phospholipids, glycolipids** (sugar-lipids), and **sterols.**

The four most abundant phospholipids are usually phosphatidyl choline, phosphatidyl ethanolamine, phosphatidyl glycerol, and phosphatidyl inositol (Fig. 6-12). The two most abundant glycolipids are monogalactosyldiglyceride, with one galactose sugar, and digalactosyldiglyceride, with two galactoses (Fig. 6-12). The glycolipids are found mainly in chloroplast membranes, in which phospholipids are

much less abundant. The structures of all these lipids have some common features that are important to membrane structure. First, they have a 3-carbon glycerol backbone (shown at the left of each structure) to which two long-chain fatty acids are esterified. These fatty acids usually have 16 or 18 carbons, often with double bonds between them (see Table 14.1 for a description of important fatty acids). The fatty acids are **hydrophobic** (water fearing), whereas the glycerol backbone with its oxygen atoms is more **hydrophilic** (water loving), because the oxygens can form hydrogen bonds with water. The final part of these lipids, shown at the bottom of each structure, is also a hydrophilic portion, because it is either electrically charged or has numerous oxygens, with which water can associate by hydrogen bonding. Molecules with distinct hydrophobic and hydrophilic regions are said to be amphipathic molecules.

In all membranes, the hydrophilic parts of lipids dissolve in water at either surface, and the fatty acids repelled by water extend toward the interior part of the membrane from each side and associate with each other by van der Waals attractive forces. This causes the membrane to become a lipid bilayer, as shown in Fig. 6-13. Sterols also have a long hydrophobic part rich in carbon and hydrogen and a small hydrophilic part (one hydroxyl group). Structures of major sterols are in Chapter 14 (Fig. 14-7), but Fig. 6-13 shows how they probably fit into the membrane structure. There are almost one-fifth as many sterols as phospholipids in the plasma membrane and tonoplast (Marty and Branton, 1980), although few other plant membranes have been purified enough from other cell contaminants for accurate measurement (Yoshida et al., 1983). The main function of sterols in membranes is apparently to stabilize the hydrophobic interior.

The proteins in membranes are of three types:

1. Catalytic proteins (enzymes), the most abundant of which are enzymes that catalyze hydrolysis of ATP to ADP and $H_2PO_4^-$. These ATP-hydrolyzing enzymes are called **ATPases,** and essentially all membranes of all organisms have at least one kind of ATPase. There are many other enzymes and proteins in the membranes of chloroplasts, mitochondria, dictyosomes, and the endoplasmic reticulum.

2. Several kinds of proteins (discussed later) variously called **carriers, permeases,** or **transporters,** each of which combines with and somehow transports a different ion or molecule across the membrane. (All evidence for the presence of these in plants is circumstantial, and none has been isolated and characterized.)

3. Structural proteins with no enzymatic or carrier activity, although it would be difficult to prove that such proteins contribute only to structure.

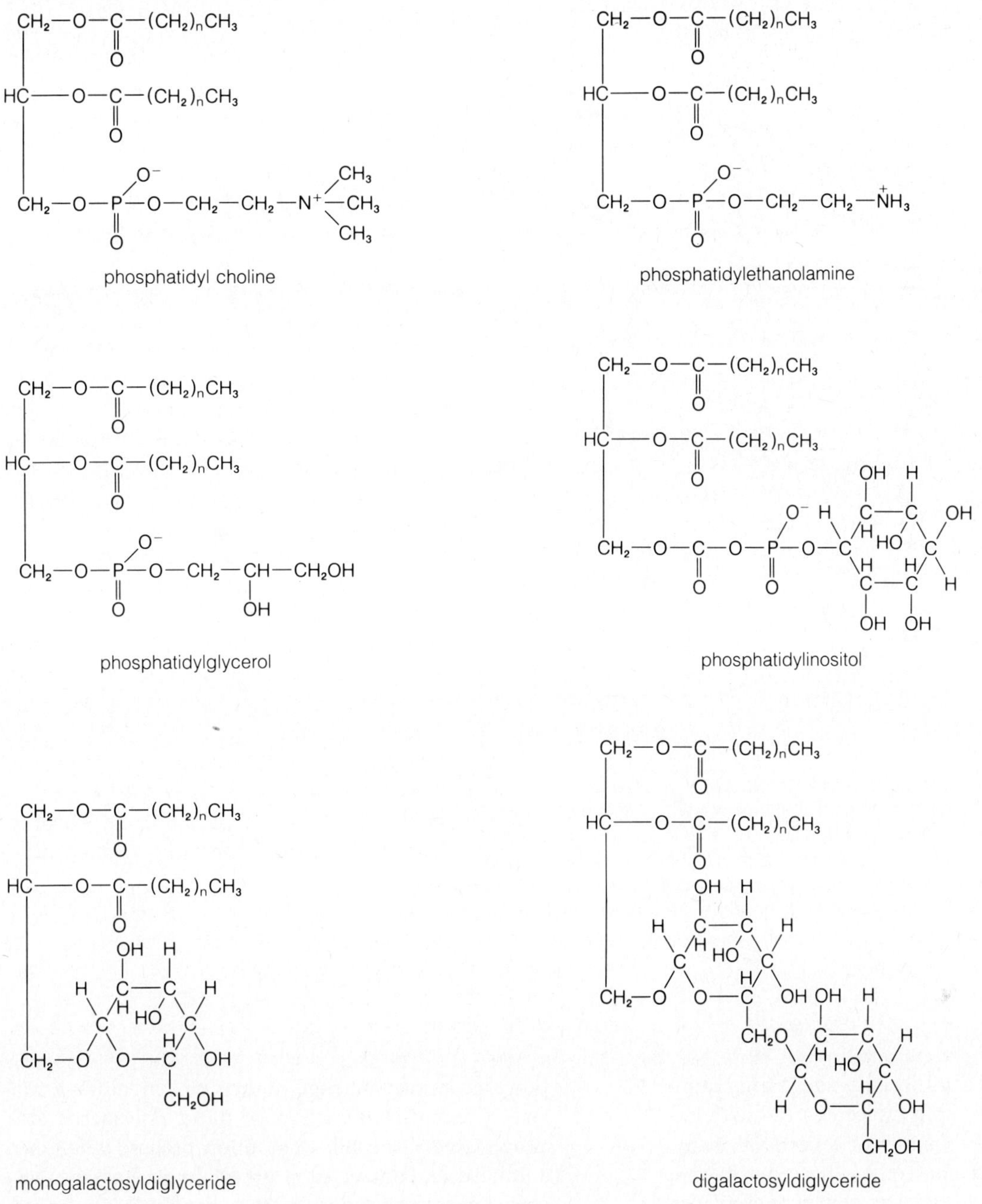

Figure 6-12 Structures of abundant membrane phospholipids and glycolipids. The subscript *n* in fatty acids represents a variable number of CH_2 groups or groups containing unsaturated carbons (usually 14 or 16).

Another essential membrane component is Ca^{2+}, without which membranes lose their ability to transport solutes inwardly and also become leaky for solutes they already contain. The function of Ca^{2+} is not well understood, but it probably bonds hydrophilic portions of phospholipids to each other and to negatively charged groups of proteins on the membrane surfaces.

The arrangement of proteins, lipids, and Ca^{2+} in membranes is a continuing problem receiving much attention from biologists, chemists, and physicists. In 1972, S. J. Singer and G. N. Nicholson proposed a now popular model called the **fluid mosaic model** (Fig.6-13). It has received great support, even though it is general and does not attempt to accurately describe any particular membrane. This model indicates that some protein molecules are imbedded in various places in the fluid bilayer of lipids such that "the

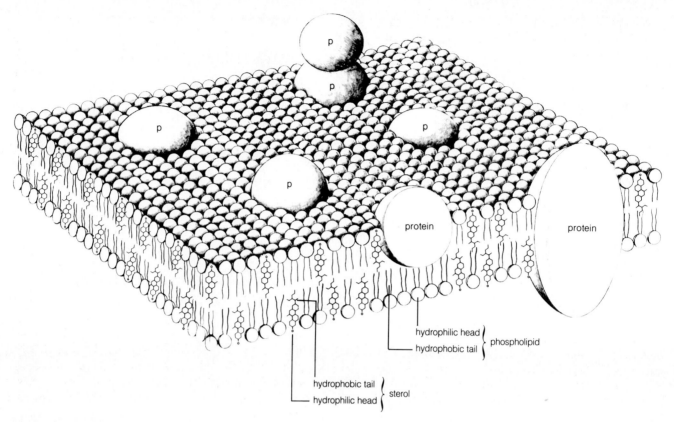

Figure 6-13 The fluid mosaic model of membrane structure. A bilayer of lipids such as those of Figure 6-12 gives stability to the membrane through association of the lipids' hydrophilic heads with H_2O on either side and association of hydrophobic fatty acids in the interior of the membrane. Sterols also have a relatively small hydrophilic head (made of one hydroxyl group) and a longer hydrophobic portion. Proteins (p) are either continuous across the membrane or lie preferentially toward either side. Most membranes contain substantially more protein than this drawing suggests.

proteins float in a lipid sea." The lipids are truly quite fluid; in fact, a phospholipid molecule can move laterally in one-half of the bilayer of a bacterial membrane from one end of the cell to the other in one second. (Flip-flop of a lipid between halves of the bilayer is rare.) Some of the hydrophobic proteins or their hydrophobic parts (those parts containing hydrophobic amino acids) penetrate deeply into the lipid-rich interior. Furthermore, some proteins extend all the way through the bilayer, as shown. These two types, called **integral proteins,** are bound tightly within the membrane and can be removed only with certain detergent solutions that break hydrogen bonds between all components of the membrane. Others, the **peripheral proteins,** are more loosely bound to one or the other side of the membrane surface and can be removed with dilute salt solutions or with detergents. The two sides (faces) of membranes are different, because the proteins or protein parts on each side differ.

In plasma membranes of animal cells, some peripheral proteins called **glycoproteins** contain short

polysaccharides attached at the outer membrane surface. Chemical interaction of those polysaccharides allows a very specific recognition process when two animal cells contact each other. In pathogenic bacteria, polysaccharides bound to lipids (**lipopolysaccharides**) in the outer membrane are the means by which these bacteria can successfully recognize and attach to a certain host plant in which they cause disease. These glycoproteins and lipopolysaccharides, then, provide a cell-to-cell recognition process. We wish to know if the outer surface of plant plasma membranes contains either glycoproteins or lipopolysaccharides that allow them to recognize bacteria (as in disease susceptibility or resistance) or each other (as in grafts or in pollen-stigma contacts). Thus far, limited evidence indicates that one or both kinds of molecules are present (Berkowitz and Travis, 1982; Larkin, 1982; Brummer and Parish, 1983). On the other hand, there is evidence that lipopolysaccharides of various bacteria recognize specific polysaccharides in the cell wall, not in the plasma membrane (Whatley and Sequeira, 1981). There is presently no reason to think that either glycoproteins or lipopolysaccharides participate in solute transport.

6.6 Some Principles of Solute Absorption

A primary goal in studying membranes with electron microscopes or by chemical methods is to understand how they selectively control the movement of solutes across them. Long before electron microscopes were developed or anyone knew how to isolate membranes for chemical analysis, people could study the absorption properties of membranes. Some properties gave important clues as to the nature of membranes. We shall mention four principles of solute absorption discovered by pioneers in the field. However, keep in mind that these principles say nothing about any mechanism of solute absorption. Later, we distinguish between passive and active mechanisms, but for now just remember that total transport by both mechanisms was being measured.

1. *If cells are not alive and metabolizing, their membranes become much more permeable to solutes.* If a cell is killed by high temperatures or poisons or its metabolism is inhibited by low temperatures, nonlethal high temperatures, or specific inhibitors, many solutes in the cell leak out.

2. *Water molecules and dissolved gases, such as N_2, O_2, and CO_2, penetrate all membranes rapidly.* No one knows how water can penetrate membranes far more rapidly than most of the solutes dissolved in it, but that it does is essential for the occurrence of osmosis. Interestingly, water even moves rapidly across artificial membranes formed only from phospholipids. This probably means that water normally moves through the lipids of membranes, not through protein channels as was long assumed.

For N_2, rapid movement seems innocuous for most cells. Dissolved nitrogen simply moves into and out of cells and their organelles at equal rates and without any effect. However, for certain freeliving prokaryotic cells and for legumes and other plants that contain such prokaryotes, the ability to trap N_2 and fix it into NH_4^+ after it moves inwardly from the air is important (see N-fixation in Chapter 13). For O_2, the rapid inward movement allows respiration to occur and is important for all nonphotosynthetic cells. For photosynthetic cells, the rapid movement of O_2 out into surrounding air is a normal process, and it is also important to minimize photorespiration (Section 10.4). For CO_2, rapid movement into photosynthetic cells is crucial, but for nonphotosynthetic cells the rapidity of entry or exit is probably less important. Movement of CO_2, however, is complicated by its ability to react with H_2O and form H_2CO_3 (car-

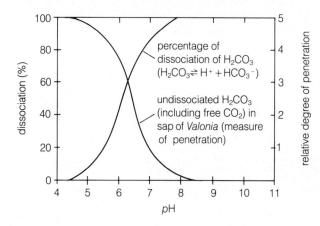

Figure 6-14 Influence of *p*H upon absorption of a weak electrolyte in the alga *Valonia*. After Osterhout and Dorcas, in A. C. Giese, 1962, *Cell Physiology, 2nd ed.*, W. B. Saunders Company, Philadelphia, p.233. Used by permission.)

bonic acid) and by the ability of H_2CO_3 to further ionize into H^+, HCO_3^- (bicarbonate), and CO_3^{2-} (carbonate). Regardless of those complications, CO_2 and all compounds or ions formed from it collectively move across membranes rapidly. For reasons mentioned below, CO_2 and H_2CO_3 move much faster than HCO_3^- or CO_3^{2-} ions.

3. *Hydrophobic solutes penetrate at rates positively related to their lipid solubility.* More hydrophobic, less hydrophilic solutes penetrate membranes more rapidly than those with opposite properties. Consider a few examples. Methyl alcohol (CH_3OH) is not much smaller than urea (H_2N—C—NH_2), but methyl

$$\parallel$$
$$O$$

alcohol is about 30 times as lipid soluble and moves into giant cells of the alga *Chara ceratophylla* about 300 times as rapidly as urea. Valeramide (5 carbons) is even larger than lactamide (3 carbons), and valeramide is about 40 times as lipid soluble and moves into Chara cells about 35 times as fast as lactamide. Urea, methyl alcohol, valeramide, and lactamide are presumed to move across the plasma membrane into cells by diffusing passively through the lipid bilayer toward a region of lower concentration. Such data initially gave evidence that membranes are rich in lipids, even before a bilayer was known to be present.

A practical problem of lipid solubility relates to whether a solute can become charged when dissolved in water, because any such charge (positive or negative) greatly decreases lipid solubility, increases water solubility, and usually decreases the permeability of cells to the solute. An important example (Fig. 6-14) concerns dissolved CO_2, its hydrated form H_2CO_3, the major ionic species HCO_3^-, and the further-ionized species CO_3^{2-} formed at *p*Hs above 8 (uncommon in plant environments). For cells of the

Table 6-1   Concentrations of Major Ions in Sea Water Compared with Their Concentrations in Vacuoles of Algae Living There.

| Ion | *Nitella obtusa*[a]–Baltic Sea | | *Halicystis ovalis*[b] | |
	Vacuole Concentration	Sea Water Concentration	Vacuole Concentration	Sea Water Concentration
Na^+	54 mM	30 mM	257 mM	488 mM
K^+	113 mM	0.65 mM	337 mM	12 mM
Cl^-	206 mM	35 mM	543 mM	523 mM

[a]Data for *Nitella* are from J. Dainty, 1962, Annual Review of Plant Physiology 13:379.
[b]Data for *Halicystis* are from R. W. Blount and B. H. Levedahl, 1960, Acta Physiologia Scandinavia 49:1.

alga *Valonia*, there is excellent correlation between the lack of charge and the form of carbon absorbed. Thus at low *p*H values, far more carbon is absorbed from dissolved CO_2 than at high *p*H values at which ionic and lipid insoluble HCO_3^- and CO_3^{2-} predominate.

A related example concerns absorption of an herbicide (weed-killer), 2,4-D (2,4-dichlorophen-oxyacetic acid). This herbicide has an acetic acid group that releases an H^+ to form negatively-charged 2,4-D at neutral or high *p*H, but at low *p*H little ionization occurs, and the uncharged 2,4-D molecule remains much more lipid soluble than the anionic form. Leaves absorb the herbicide far more effectively at *p*H 5 than at *p*H 8. Positive charges are also important. Contrary to the behavior of acidic compounds such as H_2CO_3 and 2,4-D, nitrogenous bases (in which the nitrogen attracts an H^+ and becomes positively charged at low *p*H) are usually absorbed more rapidly from neutral or slightly basic solutions in which no charge exists. Again, the major reason is their increased lipid solubility when noncharged.

4. *Hydrophilic molecules and ions with similar lipid solubilities penetrate at rates inversely related to their size.* For noncharged molecules, few data are available for plant cells, but our generalization about size is certainly valid for some bacteria (Giese, 1968) and might also be general for plants. The generalization was considered by investigators even in the 1800s to indicate that membranes contain tiny holes in the size range of the molecules being studied. Unfortunately, modern knowledge of membrane structure does not tell us where such holes are. We have come far, but there is still far to go.

For ions, the size relevant to penetration is that attained after *water of hydration* is attached (Clarkson, 1974). Each ion attracts a different (average) number of rather firmly bound water molecules to itself, depending upon the net charge density at its surface. For example, Li^+ (atomic mass 6.9), which has only one full shell of electrons around its nucleus, is 0.12 nm in diameter when unhydrated and binds about

five H_2O molecules. Alternatively, K^+ (atomic mass 39.1) has several shells of electrons, an unhydrated diameter of 0.27 nm, but binds only about four H_2O. Hydrated Li^+ is thus larger than hydrated K^+ and diffuses across membranes less rapidly than K^+. Divalent cations such as Mg^{2+} and Ca^{2+} have higher charge densities than either Li^+ or K^+, bind about a dozen H_2O, and are absorbed far more slowly than monovalent cations. Nevertheless, divalent cations are absorbed more rapidly than trivalent cations such as Fe^{3+}. Similar results occur with anions; thus, Cl^- and NO_3^- are absorbed far faster than SO_4^{2-} (Cram, 1983). Furthermore, monovalent $H_2PO_4^-$ is absorbed faster than divalent HPO_4^{2-} and much faster than trivalent PO_4^{3-}. At *p*H 7 (the approximate *p*H of the cytosol), ionization of $H_2PO_4^-$ into HPO_4^{2-} and H^+ is about half complete, so nearly equal amounts of monovalent and divalent forms of phosphate ions exist, with essentially no PO_4^{3-}. At a *p*H less than 6 (as in cell walls, vacuoles, and acidic soils), monovalent $H_2PO_4^-$ is dominant. Considering that the *p*H of many cell walls is about 5, even in neutral or slightly alkaline soils, phosphate transport into the cytosol across the plasma membrane usually involves mainly $H_2PO_4^-$. Transport from cytosol (*p*H about 7) into vacuoles at *p*H 5.5 or lower could involve either $H_2PO_4^-$ or HPO_4^{2-} based upon their roughly equal cytosol concentrations, but the former is probably again the main form involved.

6.7 Characteristics of Solute Absorption

Accumulation A remarkable fact about all cells is that they can absorb certain essential solutes so fast and over such long periods that their concentrations become much higher within the cells than in the external solution. We call such absorption **accumulation**. The extent to which the concentration becomes greater internally than externally is called the **accumulation ratio**. For example, storage tissues

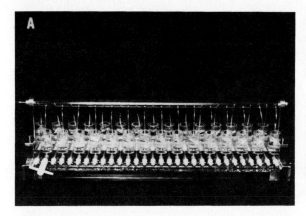

Figure 6-15 (**a**) and (**b**): An apparatus used for studies of solute absorption by root segments. The root tissues were placed in Plexiglass vials (**c**) which were then placed into aerated absorption-wells, as shown in (**b**). To terminate an experiment the Teflon valve is rotated and the solution drawn by vacuum into the chamber below the vials. (From L. V. Kochian and W. J. Lucas, 1982.)

such as slices of potato tubers placed in nutrient solutions often deplete the concentration of external ions to nearly zero within a day or two. During this time some ions (especially K^+) attain internal concentrations over a thousandfold higher than those finally present in the surrounding solution. Plant tissues usually contain at least 1 percent K^+ on a dry weight basis. When alive, such tissues typically contain 80 to 90 percent water, so the living plant contains about 0.1 percent or 25 mM K^+. This value is often attained by plants with roots growing in soils in which dissolved K^+ is no more concentrated than 0.1 mM, indicating overall whole plant accumulation ratios of about 250 to 1. Thermodynamic laws explained in Chapter 1 show that free diffusion not involving expenditure of metabolic energy could not be responsible for such great accumulation.

The accumulation of certain solutes is not restricted to roots of higher plants but is a universal phenomenon of living cells. As explained in Chapter 7, sieve tube elements or phloem sieve cells (in conifers) accumulate sucrose or other carbohydrates in high concentrations and transport them elsewhere. This transport is generally essential for photosynthetic leaf cells to supply other parts of the plant. Algae living in salty water must accumulate solutes to prevent plasmolysis, and one of their adaptation mechanisms to a saline environment is to increase their internal salt concentrations. Table 6-1 shows accumulation data for two species of algae living in different sea waters. Cells of both species have large central vacuoles from which solution was obtained for analysis. We conclude from these and other data that whether an ion is accumulated depends upon the ion and the species of plant. Note that both algae accumulated K^+, Nitella also accumulated Cl^- and to a lesser extent Na^+, while Halicystis appeared

to be in equilibrium with respect to Cl^- and actually restricted entry of Na^+.

The restriction of Na^+ is widespread in both plant and animal cells and in animals depends on an ATP-dependent "pump" (**sodium pump**) in the membrane to remove Na^+ that slowly but continuously diffuses in. Most plant cells apparently have similar pumping mechanisms to remove Na^+ from their cytoplasm, although such pumps have only occasionally been clearly demonstrated. A pump in the plasma membrane causes efflux of the Na^+ that diffuses in, as in animal cells, and a tonoplast pump is suspected to transport some of the cytoplasmic Na^+ into the vacuole. In roots of beans and some other species, much of the Na^+ in the vacuoles is never transported to the shoots. In some salt-tolerant species, however, transport of Na^+ to the shoot does occur; here it is likely that the translocated Na^+ ions are those in the cytosol, not in vacuoles, because transport of most ions out of vacuoles toward the shoot is slow.

Selectivity of Ion Absorption by Roots   That solutes are absorbed and accumulated by selective processes was further indicated by studies in which roots were excised from seedlings and allowed to accumulate ions. Seedlings were often grown in a solution that was allowed to become largely depleted of nutrients as growth occurred. Such "low-salt" roots have a high capacity for subsequent absorption of several ions, and this capacity is maintained for several hours even if the roots are cut off from the shoots (Epstein, 1972). Comparisons can be made with "high-salt" roots taken from seedlings provided abundant nutrients. Figure 6-15 shows a method by which such excised roots can be used in ion-uptake studies. Aeration of excised roots and those attached to seedlings

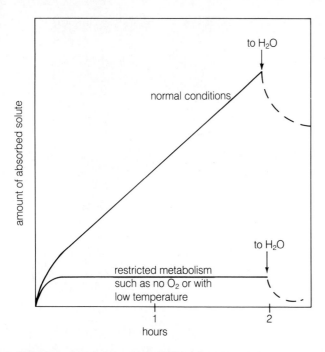

Figure 6-16 Progress of ion uptake with time under various conditions. For explanation, see text.

is essential to allow respiration needed for normal ion accumulation in most species (note tubes for forcing air through the solutions).

If such roots are provided with a solution containing dilute (about 0.2 mM) KCl and with about 0.2 mM Ca^{2+} to maintain normal membrane functions, the rate of absorption of K$^+$ is unaffected by similar concentrations of Na$^+$ salts. This is true even though Na$^+$ is chemically similar to K$^+$. The process of K$^+$ accumulation is therefore selective and uninfluenced by a related ion under these conditions. As expected, several other monovalent and less related divalent ions also have no influence on K$^+$ uptake. In similar experiments, absorption of chloride is unaffected by the related halides fluoride and iodide, as well as NO$_3^-$, SO$_4^{2-}$, or H$_2$PO$_4^-$. Calcium ions are essential for this selectivity, because without them K$^+$ absorption, for example, becomes inhibited by similar low concentrations of Na$^+$.

In spite of this apparent high selectivity, uptake mechanisms can sometimes be "fooled." Potassium absorption is inhibited competitively by Rb$^+$, and penetration of membranes by these two ions that have very similar charge densities apparently is governed by the same mechanism in all organisms. Similar competitive results are often obtained with Cl$^-$ and Br$^-$, with Ca^{2+}, Sr^{2+}, and sometimes Mg^{2+}, and with SO$_4^{2-}$ and selenate (SeO$_4^{2-}$). This selectivity of ion transport by roots also applies to organic compounds such as amino acids and sugars, and it occurs in all parts of the plant. It is an important evidence supporting the theory that proteinaceous carriers in the membranes move solutes into cells, because en-

zymes (which are proteins) are well-known to recognize selectively and be activated or inactivated by certain ions or molecules.

Failure of Absorbed Ions to Leak Out   Once ions are absorbed into cytoplasm or vacuoles of root cells, they do not readily leak out (efflux is slow). The same is true of many plant cells and of essential organic solutes in addition to ions. Rapid leakage can be induced by damaging the membrane with heat, poisons, lack of O$_2$, and to some extent by removing Ca^{2+}, but these are abnormal situations that eventually cause cell death. Slow leakage shows that absorption, especially in low-salt roots, is primarily a unidirectional influx. Efflux is much more extensive in roots that contain relatively high ion concentrations.

If the concentration of an ion in a low-salt tissue is measured during a time in which cells are exposed to that ion, graphs such as that of Fig. 6-16 are usually obtained. Under normal conditions of temperature and aeration (upper curve), there is an initial rapid influx of ions, although most of this simply represents diffusion into the cell walls, not actual movement across the plasma membrane. Subsequently, the absorption rate becomes essentially constant, often for as long as several hours. Now suppose those tissues are removed from solution at the time indicated by arrows in Fig. 6-16, placed in water at room temperature, and the amount of ion retained inside measured with time. Only a small fraction of the ions present leak out (top dashed line of Fig. 6-16). Ion loss occurs mainly from cell walls, not from inside the cells. Ions present in the cytoplasm and vacuole remain there much longer; and as the figure shows, they represent by far the major fraction in such experiments. Only if leakage experiments are continued for several hours can efflux first from the cytoplasm and then from the vacuole be detected in careful studies. The major point here is that efflux from the cytoplasm or vacuole is slow. Clearly, ions can be transported across membranes into cells much faster than they diffuse out; this is further evidence that carriers speed their influx.

Figure 6-16 (lower curve) also shows the time course of accumulation when respiration is inhibited by lack of O$_2$, cold temperature, or numerous respiratory poisons. The initial rapid absorption phase is not greatly altered, but the absorption rate then quickly falls to nearly zero (i.e., the curve becomes horizontal). When these tissues are transferred to water, nearly all the apparently absorbed ions diffuse outwardly into the water. This efflux is of the same magnitude as that from the healthy tissues, and efflux from both tissue types occurs mostly from the cell wall by simple diffusion. Such results illustrate the failure of ions absorbed across the controlling membrane to leak out. They also show the inhibition of absorption and accumulation by conditions that

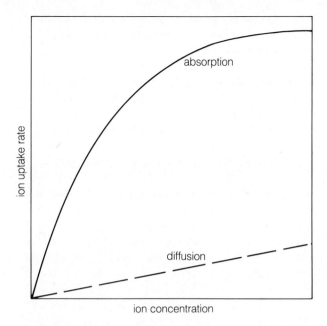

Figure 6-17 Influence of ion concentration surrounding plant cells on the rate of ion uptake. If free diffusion were responsible for uptake, the rate would be low and essentially proportional to concentration, but actual rates are considerably higher and show saturation kinetics.

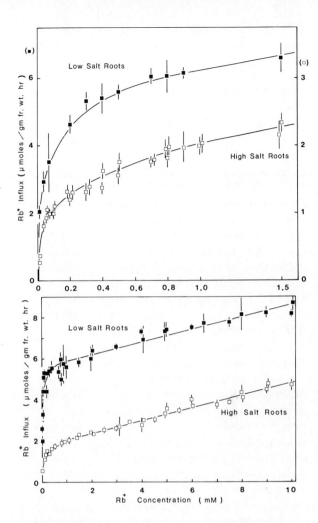

Figure 6-18 Kinetics of rubidium absorption by detached corn root segments. ^{86}Rb was used as a radioactive tracer, and its absorption is nearly identical to that of potassium. In the upper graph, data for the Mechanism 1 concentration range are expanded for clarity. Absorption period was 30 min. (From L. V. Kochian and W. J. Lucas, 1982.)

prevent normal respiration. More will be said of the relation of respiration and solute uptake later, because respiration provides ATP needed for solute uptake.

Solute Absorption at Various Solute Concentrations

To learn the mechanisms of solute absorption and to understand relations between fertilizer absorption and fertilizer application rates, hundreds of studies relating absorption rates to external concentrations have been made. A major conclusion for most plants in nature is that for nutrients used in abundance (nitrate, ammonium, phosphate, and potassium) diffusion to the root surface is the limiting factor, so absorption properties of the roots are only of limited importance for plant nutrition. For crop plants well fertilized with N, P, and K, and for other essential solutes, absorption through membranes is often rate-limiting. An increase in the external solute concentration around roots or other cells speeds solute absorption, but rates at both low and high concentrations are generally faster than if the solute simply diffused inward across the membrane. The lower line of Fig. 6-17 indicates that diffusion across the membrane is directly proportional to the external concentration of solute and that it is much slower than actual absorption rates (top line of Fig. 6-17). Actual absorption does not rise linearly as concentration increases but approaches a maximum at concentrations near those normally existing in soils. Apparently, absorption in this concentration range, often called

Mechanism 1, is sped mainly by a carrier. Such carriers become nearly saturated at low solute concentrations (1 mM). Thus as the curve levels off, it is assumed that the carrier is transporting as fast as it can, so additional solute cannot increase overall transport. Investigators in the 1960s and early 1970s frequently observed that the curve leveling represented by Mechanism 1 was followed by another rise and leveling as the solute concentration was raised much higher. These kinetics indicated that a second mechanism became functional at high concentrations, and this was called Mechanism 2. Epstein (1976), who was a pioneer in such studies, listed 61 studies of 12 ions in which so-called dual kinetics were apparently observed.

In the late 1970s and early 1980s, it was more common to observe kinetic curves that were smooth and nonsaturating, especially when many data points were obtained (Fig. 6-18) (Polley and Hopkins,

1979; Kochian and Lucas, 1982). A saturable Mechanism 2 seems nonexistent in these more recent studies, so the curves at high concentrations for both low-salt and high-salt roots are almost straight lines with positive slope. The uptake at high concentrations could represent a process controlled only by the rate of diffusion, but it seems much too fast and is slowed by certain metabolic inhibitors. These results suggest that uptake at high concentrations is also mediated by some carrier system (Kochian and Lucas, 1982, 1984). Reasons for the differences between earlier and more modern results are not clear, but certainly the improved methods by Leon Kochian and William J. Lucas at the University of California in Davis allow generation of many more results (data points on graphs) in a single experiment. With more numerous results, curves that show Mechanism 1 followed by a strong linear component seem to better account for absorption kinetics than do earlier explanations based on two saturating mechanisms.

It has long been evident (as in Fig. 6-18) that overall absorption of many ions is faster for cells with low rather than high internal ion concentrations (Lee, 1982), consistent with observations that roots become increasingly efficient in absorbing certain ions as these become depleted in the soil. Also, high internal concentrations of ions somehow "signal" the cell to slow further absorption (Glass, 1983). When the shoot system is growing rapidly, ions are quickly removed from the roots through the xylem. Processes such as these help various cells of various species to regulate and optimize their internal solute concentrations in various environments.

6.8 General Mechanisms of Solute Absorption

In the preceding section, we mentioned that solute uptake occurs by diffusion across the membrane and, more rapidly, by carriers. We also mentioned ATP-dependent pumps in relation to solute accumulation. How are carriers and pumps related and how do these relate to membrane structure and the need of ATP for accumulation? Briefly, although both carriers and pumps transport solutes, carriers are not linked *directly* to an energy source such as ATP, whereas pumps are energy linked. Some interesting facts and theories described below clarify that distinction and further expand on absorption mechanisms.

Diffusion and Facilitated Diffusion Consider first the relatively simple case of diffusion. If the concentration (actually, the chemical potential—see Section 1.5) of a solute is lower inside than outside a membrane, the solute will diffuse inward until equal

concentrations are attained. For neutral molecules such as sugars, their concentration in the cytoplasm is often kept low by their conversion to sugar phosphates and storage products such as starch and cell-wall polysaccharides and by their breakdown to CO_2 and H_2O during respiration. These reactions maintain inward diffusion. Cytoplasmic reactions also keep the concentrations of many ions low. For NO_3^-, its reduction to NH_4^+ and the conversion of NH_4^+ into amino acids and proteins favors NO_3^- uptake by diffusion. Sulfate is also converted rapidly into amino acids and proteins. For $H_2PO_4^-$, conversion into sugar phosphates, nucleotides, RNA, and DNA keeps its free cytoplasmic concentration low. For Mg^{2+} and Ca^{2+}, combination with negative sites on proteins is important, and for Mg^{2+} combination with negatively charged ATP also occurs. Divalent iron, zinc, manganese, and copper ions are also combined tightly with proteins. All of these chemical reactions in the cytoplasm favor inward diffusion of ions. Only the monovalent K^+, Na^+, Cl^-, and perhaps some boron as $H_2BO_3^-$ remain largely unchanged and uncombined, although each is attracted temporarily to charged sites on proteins.

When ions penetrate membranes by diffusion they do not move through the lipid bilayer as water does. Apparently, they are absorbed largely through channels in proteins by a process called **facilitated diffusion.** In this process they combine with membrane proteins, which then facilitate (speed) their movement across the membrane. Various polypeptide (small protein) antibiotics produced by certain bacteria greatly facilitate transport of several cations and seem to provide model systems for membrane proteins. These antibiotics are called **ionophores** because of their ability to transport cations across natural membranes or even artificial phospholipid bilayer membranes (Pressman, 1976). The ionophores gramicidin, nigericin, and valinomycin have been important for plant physiology research. Ionophores have donut shapes with hydrophilic groups near the hole and hydrophobic groups on the outside. Valinomycin functions by combining with a hydrated cation (which loses its water of hydration as it enters the hydrophilic hole), transporting the cation across the membrane and releasing it into the cytosol, and then diffusing back and picking up another cation. Each valinomycin can transfer about 1000 K^+ per second across a membrane, but Na^+ is transported only about one ten-thousandth as fast because it is too small to fit the hole properly. Nigericin can also carry ions across membranes by diffusion; it, too, is much more active with K^+ than Na^+. Gramicidin and certain other ionophores seem not to diffuse across the membrane; instead they stack to each other so as to provide a continuous channel through which the cation can diffuse in its hydrated form. Channel-forming ionophores allow even faster diffusion of cations than do the diffusible type (up to 10^7 ions per second).

Each ionophore is especially selective for a certain cation, consistent with the selective properties of membranes. Furthermore, each can be saturated with that cation at high concentrations comparable to the upper line of Fig. 6-17. We wish to know whether proteins in plant membranes also act as ionophores, as many scientists believe. If so, how do they act, and are there comparable ionophores for anions and comparable protein channels for noncharged sugar molecules? Such proteins could account for the inward transport of solutes, which is more rapid than expected from free diffusion across the lipid bilayer, and thus they could be true carriers in membranes. Assuming that membrane proteins are carriers that act like ionophores, it is probable that those involved are the integral proteins that span the membrane and that they provide continuous channels like those of gramicidin.

Passive and Active Transport, Electropotentials, and the Role of ATP Our discussion above emphasized the importance of concentration differences across membranes to solute uptake by diffusion and the possible importance of membrane proteins to facilitated diffusion. Yet no form of diffusion can account for accumulation against a concentration gradient. (However, complications for ions due to electrical charge differences will be described shortly.) *Many studies indicate that solute transport into cells is strongly dependent upon ATP and the ability of cells to respire aerobically and produce ATP.* Let us see how ATP provides such an important function. As described in Chapter 12, nearly all ATP is provided to all membranes of nonphotosynthetic cells by respiratory reactions in mitochondria that depend upon rapid absorption of oxygen and its conversion to H_2O. In photosynthetic cells, ATP that is produced in chloroplasts with the aid of energy in light (see Chapter 9) also helps drive solute absorption. How does ATP act?

The major function of ATP is to react with membrane ATPases. In animal plasma membranes, three different ATPases are known, and all are integral proteins that span the membrane (Kyte, 1981). The same is probably true of plant plasma membrane ATPases. The reaction they catalyze is shown in R6-1, although it is likely that both ATP and ADP are present as chelates of Mg^{2+}:

$$ATP^{3-} + H_2O \rightarrow ADP^{2-} + H_2PO_4^- \quad \text{(R6-1)}$$

The reaction is shown as irreversible, because the free energy released is so large (approximately 31.8 kJ or

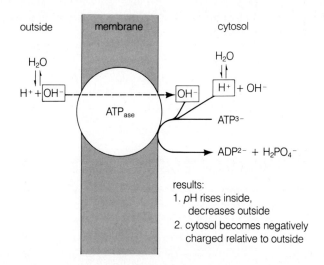

Figure 6-19 Simplified model to explain how ATPase hydrolyzes ATP using OH^- from outside the membrane and H^+ from the cytosol.

7.6 kcal per mol), even though chloroplasts and mitochondria rapidly form ATP by the reverse reaction (see Chapters 9 and 12). The energy released when ATP is hydrolyzed has two important effects. First, it causes the inside of the membrane (i.e., the cytosol) to become slightly negatively charged relative to the outside (i.e., the cell wall or soil solution). Second, it causes the pH of the cytosol to rise and that of the external solution to decrease. We do not understand how these effects occur, but it is assumed from the classic experiments and deductions of Peter Mitchell, in England, that the ATPase on the inside of the membrane obtains H_2O needed for hydrolysis by obtaining OH^- from the outside of the membrane and H^+ from the inside. An oversimplified model is shown in Fig. 6-19. A detailed description of the role of ATPases in causing solute transport across membranes is given by Tanford (1983).

These electropotential differences have no direct influence on the absorption of neutral molecules such as sugars, but they strongly attract cations and repel anions. Thus they complicate our simple interpretation above that only a lower solute concentration in the cytosol causes diffusion of ions into it. We now must evaluate the concentration effect and the electropotential effect to learn whether ion absorption is favored energetically. To do this we express each effect mathematically and add them to obtain the overall electrochemical potential difference, $\Delta\mu$,

Table 6-2   Use of the Nernst Equation to Predict whether Ions Are Absorbed Passively or Actively.[a]

	Ion Concentrations in Tissues (μ equiv./g water content)			
	Pea Roots		Oat Roots	
Ion	Predicted conc.	Measured conc.	Predicted conc.	Measured conc.
K^+	74	75	27	66
Na^+	74	8	27	3
Mg^{++}	2,700	3	350	17
Ca^{++}	10,800	2	1,400	3
NO_3^-	0.0272	28	0.0756	56
Cl^-	0.0136	7	0.0378	3
$H_2PO_4^-$	0.0136	21	0.0378	17
$SO_4^=$	0.000094	19	0.00071	4

[a]Ion concentrations in water extracts of tissues were measured after 24 h of absorption from solutions having known ion concentrations. Predicted values were obtained by assuming that electrochemical equilibrium existed and using the Nernst equation to calculate C_i/C_o. Knowing C_o, C_i was predicted and tabulated above for both pea roots and oat roots. (Data of N. Higinbotham et al., 1967.) Example for K^+ in pea roots:

$$\log(C_i/C_o) = \frac{(0.110\ V)(1.0\ equiv)(96{,}490\ J/V\ equiv)}{(2.3)(8.31\ J/mol\ deg)(298\ deg)} = 1.86$$

$$C_i/C_o = 10^{1.86} = 72.$$

Because C_o for K^+ was 1.0 $\mu mol/ml$ = 1.0 $\mu equiv/g$, C_i = 72 $\mu equiv/g$, compared with 74 predicted.

across the membrane. A slightly modified **Nernst equation** shows how this is done:

$$\Delta\mu = \Delta(RT\ln C) + \Delta(zFE) \qquad (6\text{-}1)$$

Here, $\Delta(RT\ln C)$ represents the chemical potential difference due to the concentration effect and $\Delta(zFE)$ the electropotential difference. R is the gas constant ($8.314\ J\ mol^{-1}\ deg^{-1}$); T is the absolute temperature; C is the ion concentration inside and outside of the membrane (the Δ sign preceding this term means that the difference in concentration must be measured); z is the number of charges on the ion (e.g., +1 for K^+ and -2 for SO_4^{2-}); F is the **Faraday constant,** 96,490 joules per electron volt equivalent; and E is the electropotential in volts on each side of the membrane (as for C, a difference in volts, ΔE, must be measured).

If we assume the temperature to be the same on both sides of the membrane, as is usually true, and convert to log base 10 values, equation 6-1 can be simplified for easy use to produce equation 6-2:

$$\log\frac{Ci}{Co} = \frac{-zF\Delta E}{2.3RT} \qquad (6\text{-}2)$$

Here Ci/Co is the *predicted* ratio of ion concentrations

inside (Ci) and outside (Co) the membrane *at equilibrium*[*] for any measured electropotential difference (ΔE) across the membrane. If the *actual* ratio measured experimentally is greater than the predicted ratio, the cell is performing work (using energy) to absorb the ion. This is called **active absorption,** as originally defined in 1949 by H. H. Ussing. If the actual ratio is equal to or less than predicted, **passive absorption** is occurring. Note that these definitions of active and passive absorption say nothing about mechanisms, that is, whether free or facilitated diffusion is occurring or whether energy is being used to create an electropotential difference across the membrane. This will become more clear from the example below.

Table 6-2 lists several representative data for excised roots. Pea and oat roots were allowed to absorb ions from balanced nutrient solutions for periods up to 48 h. The predicted concentration in the cells, Ci, was calculated from equation 6-2, and the measured concentrations represent actual data. After absorption, the ΔE across the pea root cells was -110 mV,

[*]At equilibrium, $\Delta\mu = 0$.

and for oat the ΔE was -84 millivolts. Nearly all ions were accumulated in both species so that C_i/C_o was greater than one (data not shown in the table), but the interesting comparisons are between predicted and measured concentrations. For K^+ and pea roots, predicted and measured concentrations are similar, indicating that the ion was near equilibrium and was probably absorbed passively; for oats, the measured concentration is greater than that predicted, indicating active transport. For Na^+, Ca^{2+}, and Mg^{2+}, the internal concentration is always less than predicted, so these ions apparently moved in passively by diffusion, even though a carrier might have facilitated their inward diffusion. Similar results in other experiments for Mn^{2+}, Fe^{2+}, and Zn^{2+} indicate that their absorption is also passive, even though the passive process depends on the previous energy-dependent production and hydrolysis of ATP to cause a negative charge inside the cytosol. For all anions, the measured internal concentrations are far higher than predicted for both species, showing that anions were absorbed actively. Clearly, the cells must use energy to accumulate anions because the negative charge inside the membrane strongly repels them.

Cotransport and Countertransport   We now face the problem of how energy from ATP hydrolysis at the inside surface of the plasma membrane can drive absorption of cations and anions when the negative charge resulting from that hydrolysis favors only cation absorption. To understand this phenomenon, consider the second effect of ATP hydrolysis, the difference in pH across the membrane, because this H^+ gradient represents potential energy used to drive absorption of anions. The process seems to work as follows: H^+ ions outside the membrane diffuse inward by a passive process, because their concentration is 10 to 100 times lower inside. Such diffusion can occur rapidly only if they combine with a carrier, but that carrier only transports H^+ if it simultaneously combines with and transports an anion in the same direction. This is an example of **cotransport,** or **symport** (Fig. 6-20). Based on lack of competition for absorption, there are presumably different carriers for nitrate, chloride, phosphate, and sulfate, yet each carrier simultaneously cotransports H^+. Furthermore, amino acids and sugars move into cells largely by cotransport with H^+ (Baker, 1978; Giaquinta, 1983), but just how many specific carriers there are is still unknown. In all these cotransport examples, H^+ moves in passively down an electrochemical potential gradient, whereas the anion or neutral molecule is transported actively. Passive H^+ absorption can also be used to transport cations simultaneously out of cells. Here the carrier seems to combine with H^+

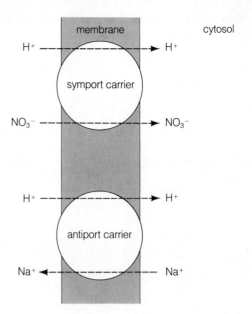

Figure 6-20   Simplified models showing how separate carriers in membranes could cause symport (cotransport) and antiport (countertransport). In each case, the energy released during transport of H^+ down its concentration gradient into the cytosol is coupled to movement of another solute against its concentration gradient.

on the outside and, frequently, Na^+ on the inside, somehow transporting them in opposite directions. This is an example of **countertransport,** or **antiport** (Fig. 6-20).

Relations among ATPases, Ion Pumps, and Active Transport   In animals, three ATP-dependent pumps have been characterized as closely related plasma membrane ATPases that also transport cations (Kyte, 1981). However, another ATPase that is sensitive to anions might also be present in such plasma membranes where it transports Cl^- or HCO_3^- (Schuurmans Stekhoven and Bonting, 1981). These ATPases are called pumps because they transport actively, using energy obtained from ATP hydrolysis. One is a Na^+/K^+-dependent ATPase that transports two K^+ inward at the same time that it moves three Na^+ outward. Another is a calcium pump that transports two Ca^{2+} outward for each ATP hydrolyzed, with no accompanying movement of other ions. The third is a H^+/K^+ pump that transports one K^+ out and one H^+ in each time ATP is hydrolyzed. It is very likely that plant plasma membranes contain this H^+/K^+ pump and that it is the most active one (Hodges, 1976; Briskin and Leonard, 1982), but it still has not been proven that the ATPase actually transports K^+ (as opposed to causing a negative cytosolic electropotential that favors its passive uptake by another carrier protein). Plant cells also appear to pump Na^+ and Ca^{2+} (and perhaps Mg^{2+}) from their cytosol both outward into the cell wall and inward into the central vacuole (Jeschke et al., 1983), but this could result only indirectly from ATPase action via antiport carriers. Certain ATPases that are believed to be present in the plasma membrane and the tonoplast of plants are also activated by Cl^-, and it is likely that they pump that anion into the cytosol and into the vacuole. Most ions absorbed by plant cells (except those transported into the xylem by roots) go into the large vacuole, so we need to learn much more about properties of the tonoplast. Progress in that area is going much faster than before, because techniques to iso-

late intact vacuoles were discovered in the mid-1970s. Those vacuoles can be analyzed to learn what solutes they contain; they can be provided with ATP to learn how they respond with respect to solute transport; and their tonoplasts can be stripped off without contamination by other membranes to learn their composition and ATPase characteristics (Wagner, 1982; Leigh, 1983). New techniques to isolate less contaminated plasma membranes have also been developed (Yoshida et al., 1983), and these should allow rapid increases in our knowledge of their properties.

7

Transport
in the Phloem

The plant, thrusting its roots into the soil and its leaves into the atmosphere and sunlight, is a complex of specialized organs. To function properly, a balanced and integrated transfer of materials must take place within this complex structure. Some of the materials absorbed by the roots must be moved to the leaves for assimilation. We have considered the movement of water and inorganic salts in this direction in the xylem, but minerals are redistributed from the leaves in the phloem. As photosynthesis proceeds, its products and other metabolic products must move out of the leaves to other parts of the plant. Roots, stems, and young leaves will require these materials, and developing fruits and flowers will either metabolize or store them. Certain hormones are synthesized only in specific parts of the plant and then are moved to other parts.

Plant physiologists have devoted considerable effort to the study of this movement or **translocation** of dissolved materials throughout the plant. To begin with, we need descriptive information. In which tissues does the movement take place? How fast? In what form? Then we can ask: What are the mechanisms of translocation and what are the controls that coordinate supply with demand?

The concept of blood circulation has hardly been in doubt since William Harvey lectured on it in the early 1600s, but the mechanisms of solute translocation in plants have only begun to yield to our probing during recent years, although the problem has been studied by scientific methods for well over a century. Why has elucidation of the transport mechanisms in plants been so difficult? Partly because of the size of the transport vessels, which are essentially all microscopic compared with the easy-to-comprehend heart with its macroscopic valves and associated plumbing. (And the stumbling block in understanding animal circulatory systems was the microscopic capillary system that returns arterial blood to the veins.) Perhaps even more important than the dimensions of the plant vascular system is the fact that fluids in phloem sieve tubes are under high pressure—many times higher than that in animal circulatory systems. When cells in the phloem are cut, the pressure is released, and the instantaneous surging of phloem contents alters or destroys the cellular structures that existed before cutting. Only in recent years have rapid freezing, followed by freeze drying and other techniques, provided help with this problem.

During the late 1970s and the early 1980s, most plant physiologists arrived at a consensus, agreeing that a mechanism of phloem transport that was first suggested in 1926 is indeed basically correct. Of course, the transport model has been modified somewhat since its initial presentation, but in its contemporary form it is supported by a large and convincing body of evidence. Virtually all of the once disturbing counter-evidences have now been satisfactorily explained and thus laid to rest.

7.1 Transport of Organic Solutes

In 1675, the Italian anatomist and microscopist Marcello Malphighi **girdled** a tree by removing a strip of bark from around its trunk. The experiment was repeated by Stephen Hales in 1727. This was one of the first experiments in the field of plant physiology. Such experiments are still used as demonstrations, and in their most sophisticated development, they have been combined with radioactive tracers. Bark can be surgically separated from wood; or wood can be removed, leaving the bark virtually intact.

It was long the aim of plant physiologists to measure translocation directly by following the movement of marked materials in the transport system.

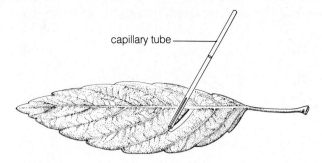

capillary tube

Figure 7-1 The reverse-flap technique for applying material to leaves. Note how the reverse feature makes it unlikely that solution will be pulled from the leaf to relieve tension in the xylem, since there is no direct connection between the flap and xylem in the stem. (Courtesy John Hendrix.)

Early investigators used dyes; indeed, the dye fluorescein moves readily in phloem cells and is still used as an effective tracer. Viruses and herbicides have also been used, but by far the most important tracers are the radioactive nuclides that became available after World War II. Radioactive phosphorus, sulfur, chlorine, calcium, strontium, rubidium, potassium, hydrogen (tritium), and carbon have been used in these studies. Heavy stable isotopes such as those of oxygen (^{18}O) or nitrogen (^{15}N) are also used.

Tracers can be applied by the **reverse-flap technique,** in which a flap including a vein is cut from a leaf, as shown in Fig. 7-1. Another widely used approach is to abrade the epidermis of a leaf so that the cuticle is removed, sometimes breaking open a few epidermal cells. Then solutions are applied, which readily penetrate to the leaf mesophyll cells, the leaf veins, or both. Another common approach exposes the leaf in a closed container to carbon dioxide labeled with carbon-14 (^{14}C) or in recent years carbon-11 (^{11}C). Carbon-11 has a short half-life but emits a more powerful ray than ^{14}C, which is often advantageous.* The labeled CO_2 is incorporated into the **assimilates** (products of assimilation and metabolism) by photosynthesis, and in this form it is exported from the leaf in the translocation stream. It is important to know what happens chemically to the tracer in the plant. As a rule, such inorganic ions as phosphate, sulfate, potassium, or rubidium remain chemically unchanged, and most of the ^{14}C in $^{14}CO_2$

---

*^{14}C has a half-life of 5730 years; ^{11}C has a half-life of 20.3 minutes. The positron emitted by ^{11}C has an energy about 6.3 times as strong as the beta particle emitted by ^{14}C; the decay energy of ^{11}C is 12.69 times as strong as that of ^{14}C. The ^{11}C must be prepared just before use, for example, by the bombardment of boric oxide with a beam of deuterons accelerated in a Van de Graaff accelerator (Troughton et al., 1974).

is at first incorporated only into sucrose or some other sugar, but the radioactive carbon is eventually incorporated into every organic compound in the plant.

The radioactive tracer may be detected during transport by bringing a radiation detector into contact with the plant stem or other part. Another method is **autoradiography** (Fig. 7-2). The plant is placed in contact with a sheet of x–ray film, sometimes for several months, and the film is then developed to show the location of the radioactivity. The immediate problem is to immobilize the radioactivity in the plant after harvest. Plants may be placed between blocks of dry ice and then, while still frozen, subjected to a vacuum, allowing the water to sublime (**freeze drying** or **lyophilizing**). Another way of arresting movement is to dismember the plant before making the **autoradiogram.**

The girdling experiments and those utilizing tracers have provided three general conclusions and many more specific ones. In the earliest studies it became apparent that the removal of the bark (containing the phloem) from the stem or trunk (the xylem) had no immediate effects on the growth of the shoot or on transpiration from the leaves. In more recent experiments, water with tritium replacing the hydrogen or with some radioactive solute is observed to move readily through the wood in the transpiration stream even though the bark has been removed. Analysis of the xylem sap shows that it contains mostly dissolved minerals from the soil plus small amounts of various organic compounds, including sugars and amino acids (Läuchli, 1972; Ziegler, 1975). Hence our first conclusion about translocation: *Water with its dissolved minerals moves primarily upward in the plant through xylem tissues.*

In Malphighi's and Hale's experiments, the bark below the girdle dried up and eventually died, while the bark above the girdle swelled somewhat and remained healthy (Fig. 7-3). Clearly, assimilates from leaves, including the products of photosynthesis (**photosynthates**), are necessary for the growth of plant parts that cannot photosynthesize and even for some parts that photosynthesize only at low levels, such as some stems and fruits. These materials move through the bark in the phloem. Indeed, it is possible to kill a tree by girdling the trunk and making the roots dependent upon their stored food, which runs out after a few weeks to a few years, depending on the species.

We shall discuss the anatomy of phloem tissue in a later section. Here it is important to note that detailed studies, especially with radioactive tracers, have shown that assimilates move through the *sieve elements* that form *sieve tubes* in phloem tissue (Fig. 7-4). Hence the second conclusion: *Assimilates, in-*

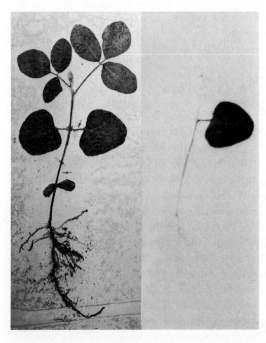

Figure 7-2 Results of an experiment in which autoradiography is used. The purpose of the experiment was to observe the effect of wilting on translocation. The first true leaf on the right of each soybean plant was held for 1 h in an illuminated chamber containing $^{14}CO_2$, which was converted by photosynthesis into radioactively labeled assimilates. After 6 h, the plants were harvested, dried, pressed (left, in each pair), and placed in tight contact with x-ray film. After 2 weeks of exposure, the film was developed (right, in each pair). The dark areas show where the most ^{14}C was located. In both cases, the leaf exposed to the $^{14}CO_2$ had by far the most tracer, but more was moved from the turgid plant (top) than the wilted one (bottom), which was predicted from the Münch hypothesis discussed in this chapter. (Specimens and films courtesy of Herman Wiebe; see Wiebe, 1962.)

Figure 7-3 The effects of girdling the trunk of a tree by removing bark from around the circumference. Note the swelled bark above the girdle compared with that below. The trunk has been cut to reveal the annual growth rings. Note that an entire year's growth was laid down above the girdle but not below. (Specimen courtesy of Herman Wiebe.)

cluding photosynthates, move primarily through sieve tubes in the phloem. This is phloem transport.

Various parts of the plant besides the main stem can be girdled. For example, the bark between a leafy branch and developing fruit can be removed. Again, sugars accumulate in the bark on the side of the leafy branch. Or the girdle can be placed above the leaves on the stem but below a developing shoot tip, in which case sugars still accumulate on the side of the leaves, at the *bottom* of the girdle. Gravity does not govern the movement of materials in the bark; the controlling relationship is the relative positions of the **source** and the **sink**. The leaves, with their photosynthetic capacity, typically constitute the source, but an exporting storage organ such as a beet or carrot root in the spring of its second year is also a source. Cotyledons and endosperm cells of seeds are sources for germinating seedlings. Any growing, storing, or metabolizing tissue might be a sink. Growing fruits, stems, roots, corms, tubers, flowers, or young leaves are examples. Hence the third important point: *Assimilates move from source to sink.*

The first two of the three conclusions, though correct, may obscure the fact that important organic

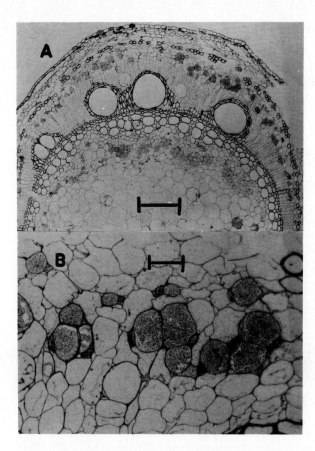

Figure 7-4 Microautoradiographs of 2-μm Epon sections from a morning glory vine, taken from internodal tissue 10 cm below a leaf that had been allowed to photosynthesize for 6 h in the presence of $^{14}CO_2$. The sections were stained with methyl violet. The black spots are silver grains that appear in the developed photographic emulsion previously poured over the sections and allowed to set. Thus, the spots indicate locations of molecules labeled with ^{14}C; they are mostly confined to sieve elements. (**a**) Low magnification; bar indicates 200 μm. (**b**) High magnification; bar indicates 20 μm. (Microautoradiographs courtesy of Donald B. Fisher; see Christy and Fisher, 1978.)

Table 7-1 Comparison of Xylem and Phloem Sap Compositions from White Lupine (*Lupinus albus*).

	Xylem Sap (Tracheal) μg ml^{-1}	Phloem Sap (Fruit Bleeding) μg ml^{-1}
Sucrose	*	154,000
Amino acids	700	13,000
Potassium	90	1,540
Sodium	60	120
Magnesium	27	85
Calcium	17	21
Iron	1.8	9.8
Manganese	0.6	1.4
Zinc	0.4	5.8
Copper	T	0.4
Nitrate	10	*
pH	6.3	7.9

* = not present in detectable amount.
T = present in trace amount.
Source: Pate (1975).

materials are transported in the xylem, and minerals essential to the growth of many sinks are transported in the phloem (Table 7-1). There are sufficient organic and inorganic nutrients in either sap to satisfy the requirements of some sucking insects for their entire life cycles. Neither xylem nor phloem saps provide balanced nutrition for the insects. This is reflected in the honeydew produced by the insects. Xylem feeders (e.g., cicadas, sharpshooters) produce much watery honeydew (several drops per minute) that is high in inorganics. They remove the available organic compounds and secrete the inorganic materials in their honeydew. Phloem feeders (aphids, mealybugs, scale insects, leaf hoppers) produce smaller amounts of honeydew that is high in organics, especially sticky sugars—which accumulate on cars and sidewalks below.

7.2 The Pressure–Flow Mechanism

The model of phloem transport presently favored by most plant physiologists was proposed in its elemental form by E. Münch in Germany in 1926 (see 1927 and 1930 references). Much of what we know about translocation was learned in the process of testing Münch's model. Most tests have been positive, and the negative ones have been reconciled, as we shall see. During the last decade, the weight of informed sentiment has shifted in the direction of this hypothesis. Now there are few holdouts. The history of this model provides a good example of the scientific approach: model building, testing, modification of the model and further testing.

Münch's **pressure flow hypothesis** is simple, straightforward, and based upon a real model that can be built in the laboratory: two osmometers connected to each other with a tube (Fig. 7-5). The osmometers can be immersed in the same solution or in different solutions, which may or may not be connected. The first osmometer contains a solution that is more concentrated than its surrounding solution; the second osmometer contains a solution less concentrated than that in the first osmometer, but either more or less concentrated than its surrounding medium. Water moves into the first osmometer by osmosis, and pressure builds up. Since the osmometers are connected, pressure is transferred from the first osmometer to the second (with the velocity of sound, which is basically a pressure-transference phenomenon). Soon the increasing pressure in the second osmometer causes it to have a more positive water potential than exists in its surrounding medium; as a result, water molecules diffuse out through the membrane, which retains the solute molecules. The total result is osmotic movement of water into the first osmometer from its surrounding solution, bulk flow of solution (water and solutes) through the tube into the second osmometer, and osmotic movement of water out of the second osmometer into its surrounding solution. If the walls of the second osmometer stretch, pressure is relieved even if no water moves out, and if the second osmometer is surrounded by a solution more concentrated than that inside it, water will diffuse into the surrounding solution even without a buildup of pressure.

In Münch's laboratory model, bulk flow (also called mass flow) ceases when enough solute has been moved from the first osmometer to the second to equalize their osmotic potentials and pressure potentials. Münch suggested that the living plant contains a comparable system but with advantages. Sieve elements near source cells (usually photosynthesizing leaf mesophyll cells) are analogous to the first osmometer, but the concentration of assimilates is kept high in these sieve cells by sugars that are produced photosynthetically in the nearby mesophyll cells. The concentration of assimilates in the other end of the phloem system, near the sink cells, is kept low as the assimilates are rendered osmotically ineffective by metabolism, incorporation into protoplasm (growth), or storage as starch or fats. Of course, metabolism, growth, and storage will mostly take place in cells near to the sieve tubes in the sink tissues. The connecting channel between source and sink is the phloem system with its sieve tubes; the surrounding dilute solutions are those of the apoplast, specifically those in cell walls and in the xylem. (Münch proposed the concepts of the symplast and the apoplast in his presentation of the mass flow hypothesis.)

Note that flow through the sieve tubes is *passive*, occurring in response to the pressure gradient caused by osmotic diffusion of water into the sieve tubes at the source end of the system and out of the sieve tubes at the sink end of the system. There is no active pumping of solution by sieve cells along the route, although there is evidence that in bean plants (Giaquinta and Geiger, 1973) metabolism is required to maintain these cells in a condition that will prevent leakage and allow bulk flow of solution through them.

An example of an *active* transport of solutes is that of cytoplasmic streaming, proposed as long ago as 1885 by the Dutch botanist, Hugo de Vries (one of the discoverers of Mendel's paper on genetics). The

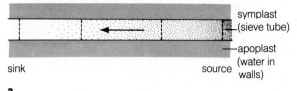

comparable plant structures:

symplast (sieve tube)

apoplast (water in walls)

sink

source

a

dilute solution

concentrated solution

2

1

sink osmometer

source osmometer

b

Figure 7-5 Bottom: A laboratory model consisting of two osmometers and illustrating the pressure-flow theory of solute translocation as proposed by Münch. Note that the concentration of the solute present in the largest amount (represented by black circles) will control the rate and direction of flow, while more dilute solutes (open circles) will move along in the resulting stream. Dashed lines on the left imply that flow will occur if pressure is relieved by expansion of the osmometer as well as by outward movement of water. Top: A schematic suggestion of how the model might apply to the concentrated solutions in the phloem system (symplast) surrounded by the dilute solutions of the surrounding apoplast. Solute concentration is maintained high at the source end of the system as sugars and other solutes are moved into the sieve tubes there; concentrations are low at the sink end as solutes are moved out, which also occurs to some extent along the route from source to sink. Lowered concentration of solutes at the sink end allows water to move out in response to the pressure transmitted from the source end (or in response to even higher concentrations of solutes in the apoplast at the sink.) Sieve tubes do not expand in analogy to the expanding osmometer of the laboratory model, but growth of storage cells at the sink will cause absorption of water from the apoplast, lowering its water potential and thus facilitating exit of water from sieve tubes there.

cytoplasm is known to stream around the periphery of many cells, and any solute that passes from one cell to another will be accelerated in its transport across the cell it has entered by the cytoplasmic streaming in that cell. This must be an important active transport mechanism in many plant tissues (it could be important in moving sugars from leaf mesophyll cells to phloem sieve tubes, for example, or from sieve tubes in storage organs to the storage cells), but since the cytoplasm does not stream in mature sieve elements, cytoplasmic streaming cannot play a role in phloem transport. Streaming is also much slower than transport in sieve tubes.

7.3 Testing the Hypothesis

The Münch hypothesis suggests several ways that it might be tested, as any good model should. We shall consider the following approaches: phloem anatomy, rates of phloem transport, the transported solutes, phloem loading and unloading of assimilates, pressure in the phloem, and some complications.

Phloem Anatomy   A function can often be understood by understanding the structure in which it occurs. An examination of the valves and chambers of the heart makes its function as a pump clear to anyone with some sense of mechanics. Can we discover the mechanism of phloem transport by studying the structure of phloem tissue? Perhaps, if we can understand the structure well enough. Certainly, we cannot hope to understand the mechanism of phloem transport without understanding phloem anatomy.

Phloem tissue   First are the **sieve elements,** which are the elongated living cells, usually without nuclei, in which transport actually takes place (Esau, 1977; Fahn, 1982). In angiosperms, they are connected end to end with pore-filled **sieve plates** between, forming long cellular aggregations called **sieve tubes** (Fig. 7-6). In gymnosperms and lower vascular plants, the existence of sieve plates is not as clear; there are sieve areas with smaller pores on lateral walls and on slanted end walls. Hence the units are called **sieve cells** instead of sieve elements. Second are the **companion cells** (in angiosperms) or **albuminous cells** (in gymnosperms), which are closely associated with the sieve elements or sieve cells and have relatively dense cytoplasm and distinct nuclei. There are usually many plasmodesmata in the walls between sieve elements and their companion cells, with the plasmodesmatal pores frequently being branched on the side of the companion cell. The exact function of the companion cells remains unknown, although they are always present, viable in functioning phloem, and degraded in senescent phloem. In leaves, they apparently absorb sugars and transfer them to sieve elements (phloem loading). Third are the **phloem parenchyma cells,** which are thin-walled cells that are similar to other parenchyma cells throughout the plant except that some are more elongated. They may act in storage as well as in lateral transport of solutes and water. Fourth are the **phloem fibers,** which sometimes are grouped in a bundle. As in other tis-

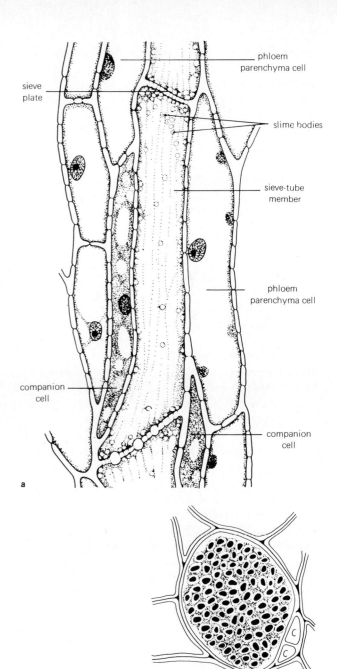

Figure 7-6 (**a**) A longitudinal view of a mature sieve element, companion cell, and phloem parenchyma cells. (**b**) A face view of a sieve plate; the black areas represent holes in the wall.

sues, they are thick-walled cells that provide strength.

The vascular anatomy of the minor veins in leaves is especially important to an understanding of phloem transport. The large veins in a leaf branch into smaller veins and eventually into the minor leaf veins. Each minor vein may contain only one vessel representing xylem and one or two sieve tubes (Fig. 7-7). The vessel is usually above the phloem tissue,

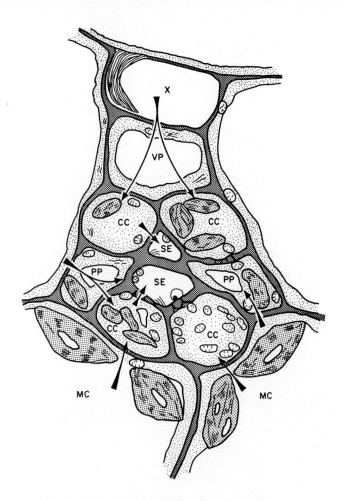

Figure 7-7 Tracing of an electron micrograph of a cross section of minor vein from a tobacco leaf. Arrows illustrate possible assimilate entry routes into the sieve-element–companion-cell complex. X = xylem; VP = vascular parenchyma; CC = companion cell; SE = sieve element; PP = phloem parenchyma; and MC = mesophyll cell. Note that sieve elements are much smaller than companion cells, the reverse situation from that found in stems and roots. (From Robert Giaquinta, 1983.)

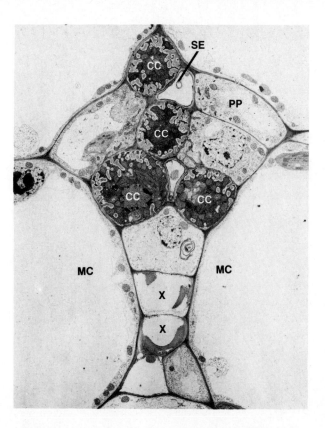

Figure 7-8 Transfer cells in a cross section of a minor leaf vein of groundsel *(Senecio vulgaris)*. In the upper half of the photograph, two sieve elements (small, relatively empty-looking cells) are in contact with four companion cells (dense cytoplasm, wall ingrowths; these are *transfer cells*) and three phloem-parenchyma cells (more vacuolated, less dense cytoplasm, wall ingrowths of faces nearest the sieve elements; also transfer cells). Another parenchyma cell separates the phloem region from the two xylem elements at lower center. Higher magnifications show that the companion cell transfer cells are connected to the sieve elements by plasmodesmata. (Transmission electron micrograph courtesy of Brian E. S. Gunning; see Gunning, 1977.)

and the sieve elements are typically smaller than and surrounded by companion cells. Intermingled with the companion cells are large phloem parenchyma cells, and vascular parenchyma may separate the phloem tissue from the xylem. Companion cells and phloem parenchyma cells sometimes contain chloroplasts, and actively photosynthesizing mesophyll cells (usually consisting of palisade and spongy parenchyma tissues) are often in close contact with the minor vein. Indeed, typically no mesophyll cell in the leaf is separated from a minor vein by more than two or three other mesophyll cells (see Fig. 4-4).

In some species, the companion cells have numerous cell-wall ingrowths, which greatly amplify the membrane surface area of the cell (Fig. 7-8). Cells with such wall ingrowths and expanded membrane surfaces are called **transfer cells.** Although most species do not have transfer cells in their minor veins, they might contribute significantly to the transfer of assimilates from the mesophyll cells to the sieve tubes in those that do, which include many species of the legume and aster families (two of the largest families). Brian E. S. Gunning and co-workers (1974) have calculated that the wall ingrowths in companion cells of *Vicia faba* leaf veins expand the surface area for absorption to over three times the area that would have been available if the companion cells lacked ingrowths.

Transfer cells are not restricted to phloem but occur throughout the plant. They are found in xylem and phloem parenchyma of leaf nodes (where they are common) and in reproductive structures such as

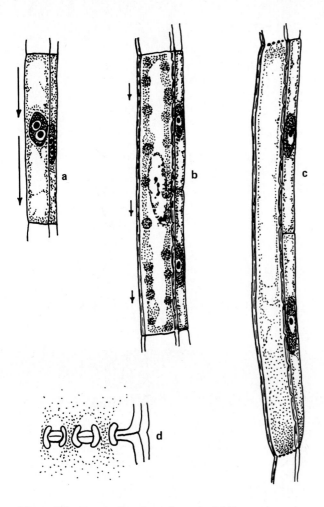

Figure 7-9 Developing sieve elements. (**a**) Young element with companion cells. Long arrows suggest considerable cytoplasmic streaming. (**b**) Sieve element of intermediate maturity. Slime (P-protein) bodies are evident, the nucleus is beginning to disappear, and cytoplasmic streaming has virtually stopped. (**c**) Mature sieve element. The nucleus and tonoplast are no longer evident. (**d**) Longitudinal section of sieve tube through a sieve plate, showing cytoplasmic connections through the pores.

the interface between the gametophyte and the sporophyte of both lower and higher plants. In all vascular plants studied, they are correlated with active transport processes occurring over short distances, such as in salt glands, nectaries, and the connections (haustoria) between a parasite and its host. But do they play a role in phloem loading? Only in certain species, it appears, and that has yet to be demonstrated. In a survey of over 1000 species, it was found that phloem transfer cells associated with sieve elements in minor leaf veins were relatively rare, occurring only in certain herbaceous dicot families.

Phloem development Consider the typical development of a secondary sieve element and its companion cell (Fig. 7-9). A single cambial cell divides twice to produce a sieve element and its accompanying com-

panion cell. The companion cell may divide at least once more. The sieve element expands rapidly and becomes highly vacuolated, with a thin layer of cytoplasm pressed against the cell wall. Minute bodies appear in this cytoplasm. They are generally ovoid in shape, some appearing rather amorphous, while others have a more fibrillar or stranded appearance. Traditionally, they have been called **slime bodies,** but they apparently consist of phloem protein (P-protein, discussed later). As they grow, their boundaries become less defined, and they eventually fuse into a single diffuse mass that is dispersed throughout the cell. At about this time, the nucleus begins to degenerate. Eventually it disappears completely in most sieve elements, although there are a few exceptions in which degenerate or even whole nuclei may remain. The tonoplast (the membrane between the vacuole and the cytoplasm) also disappears at about this stage, but the plasmalemma remains intact. Cytoplasmic streaming has been observed in some developing sieve elements, but when the elements are mature, this activity ceases.

While these events are taking place, the sieve plate is developing. This process begins with small deposits of a special glucose polymer called **callose,** usually around plasmodesmata. (The box essay "A Review of Carbohydrate Chemistry" further describes callose.) The deposits increase in size until they assume the shape of the final pore. The middle lamella first disappears in the center of the deposit, and then the deposits on both sides of the wall fuse as the wall between disappears, so that the callose-lined pore resembles a grommet. The plasmalemma extends through the pore and is thus continuous from cell to cell. The pores are usually 0.1 to 5.0 μm in diameter, much larger than any solute ion or molecule or even virus particle. The sieve elements are usually 20 to 40 μm in diameter, and 100 to 500 μm (0.1 to 0.5 mm) long. Of course, there are numerous exceptions, depending on species, to most of the information just summarized. For instance, in gymnosperms the albuminous cells and the sieve cells do not arise from a single cambial cell.

Some sieve tubes remain functional for several years. This is probably most striking in perennial monocots, such as palm trees, which have little cambial tissue capable of producing secondary phloem. Some sieve elements laid down at the base of a palm tree, for example, apparently must remain functional throughout the life of the tree (which may be well over 100 years), although some new sieve elements are formed by a **primary thickening meristem.**

Phloem ultrastructure Companion cells have unusually dense cytoplasm with small vacuoles. Mitochondria are abundant, as are dictyosomes and endoplasmic reticulum. The nucleus is well-defined until the associated sieve element finally begins to become nonfunctional. Phloem parenchyma cells, in contrast to companion cells, contain large vacuoles and perhaps fewer organelles (although this may be only apparent because of relatively less cytoplasm). Chloroplasts are often conspicuous in these cells. Plastids occur in both types, and indeed, if the vacuole is not well developed in parenchyma, it is not easy to distinguish phloem parenchyma from companion cells.

It is the ultrastructure of the sieve tube that is of immediate interest. Smooth endoplasmic reticulum (ER) occurs as an almost continuous network along the inner surface of and somewhat parallel with the plasmalemma. There are also regions of flattened or convoluted stacks of ER cisternae close to the plasma membrane. In the sieve cells of gymnosperms, the ER exists as a network of smooth tubules close to the plasmalemma but forming massive aggregates at the sieve areas. Mitochondria apparently remain unmodified throughout the differentiation of the sieve element and apparently remain capable of carrying out cellular respiration and other metabolic activities (discussed in Chapter 12). Plastids, sometimes containing starch, protein, or both, occur in young as well as mature sieve elements, but their internal membrane systems remain poorly developed. Microtubules are abundant in the cytoplasm of young sieve elements but disappear at intermediate stages of differentiation. Microfilament bundles are frequently seen in differentiating sieve elements and have been reported in mature sieve elements. Walls of mature elements are nonlignified, rich in cellulose, and often thickened.

In addition to the slime bodies in sieve elements, earlier workers had often seen amorphous material in the **lumen** (the part that was originally the vacuole before disintegration of the tonoplast). The electron microscope has shown that most, if not all, of this slime proved to be a fibrillar proteinaceous material. The diameter of the fibers is on the order of 7 to 24 nm, and the molecular weights vary (in cucurbits, at least) from 14,000 to 158,000. There are clearly several different kinds of this protein, and some of these may be interconvertible. The protein is referred to as **P–protein** (for phloem protein), and slime bodies may be called **P–protein bodies.**

At least since the discovery of P–protein, phloem anatomists have been deeply concerned with the question of whether the pores of the sieve plates are filled with the substance. Clearly, the sieve plates themselves offer a considerable resistance to a passive bulk flow of material as postulated in the Münch model. If the pores are partially or completely blocked with P–protein, the resistance could be greatly increased. Perhaps the evidence that accumulated as electron micrographs became available,

which suggested that the pores are indeed blocked, led to the serious questioning of the pressure–flow hypothesis in the 1960s and to the formulation of alternative theories. Many of these theories suggested that P–protein played some kind of active role in pumping solution through the pores (see, for example, Fensom, 1972; Peel, 1974). Are the sieve plate pores open to the bulk flow of solution or occluded by P–protein?

The earliest studies with the electron microscope were equivocable, because of primitive fixation techniques. Even then, however, some electron micrographs showed occluded pores while others showed open pores. With improving fixation techniques, electron micrographs continued to show that sieve plate pores were predominately occluded. Yet, it was always possible to suggest that the release of pressure caused by cutting the phloem for taking samples might well cause surging and a movement of P–protein from the cell periphery to the pores.

Many attempts have been made to fix material in ways designed to reduce surge artifacts. Such efforts have led to a shift in the observations toward more open pores. The following techniques have been used: rapidly penetrating fixative (acrolein); rapid freezing of whole plants in liquid nitrogen and then transference to chemical fixative; fixation of isolated sieve tubes from tobacco pith cultures grown in the laboratory (presumably they have low internal pressure); and starving or wilting plants to reduce pressure in the sieve tubes before sampling and fixation. Furthermore, some plants such as maize (*Zea*), duckweed (*Lemna*), and some palms always have unplugged pores even without special measures to prevent surging. Carbon-black particles and mycoplasmlike bodies have been shown to pass through sieve pores, and virus particles can replace the P–protein pore plugs in some specimens (not likely if the P–protein were essential for active transport). With these and other evidences, a strong argument can be made that the pores are open in normal growing plants (Fig. 7-10). Nevertheless, in a 1981 review, James Cronshaw concluded that "the question of the nature of the contents of the sieve area pores has not been answered unequivocally," although "most workers agree that there is a mass flow of solution through unplugged sieve plate pores."

The surging that occurs when a sieve element is cut causes P–protein to flow to the sieve plate, blocking the pores. There is also evidence that P–protein coagulates when exposed to air, and in a few cases the sieve cells collapse upon wounding (relatively soft walls with original shape maintained only by turgor pressure). Without one or a combination of these mechanisms, plants might "bleed to death" when injured by grazing or other means. The mechanisms might also cut off nutrients from potential

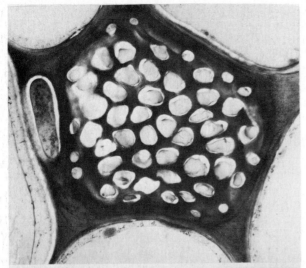

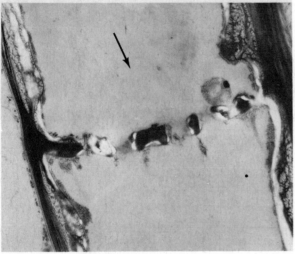

Figure 7-10 Electron micrographs of the sieve plate in soybean *(Glycine max)*, showing rather open pores with relatively small amounts of P-protein fibers. The petiolar tissue containing ^{14}C-sucrose was quick-frozen, freeze-substituted in acetone or propylene oxide, and embedded in Epon. The presence of ^{14}C-sucrose allowed verification of the direction of flow in the sieve tubes. Top, cross section; bottom, longitudinal section (both ×14,000). Arrow indicates probable direction of flow. (Micrographs courtesy of Donald B. Fisher; see Fisher, 1975.)

pathogens. Yet these things make it extremely difficult to study and understand uninjured sieve elements.

The Rates of Phloem Transport   In 1944, Alden S. Crafts and O. Lorenz at the University of California at Davis measured growth of 39 pumpkins from August 5 to September 7, estimating that individual fruits gained an average of 482 g of dry material during their 792 hours of growth. Thus, material moved into each fruit through its peduncle at an average rate of 0.61 g h^{-1}. Sections were taken from the peduncles,

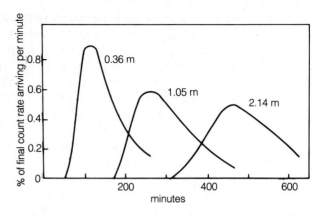

Figure 7-11   Results of a representative experiment by A. Lawrence Christy and Donald B. Fisher (1978). A single source leaf on a morning glory *(Ipomea nil)* vine was labeled by allowing it to photosynthesize with $^{14}CO_2$ for 5 min; then radioactivity was continuously measured with a thin-window Geiger tube pressed against expanding sink leaves on the tips of branches at different distances from the source leaf. All other leaves and the stem above the source leaf were removed. Note how the profiles broaden and the peaks become lower with increasing distance from the source leaf. Times between the peaks suggest translocation velocities of about 300 to 320 mm h^{-1}.

projected with a microprojector onto cardboard, and the cellular outlines traced. After suitable calibration, the outlined phloem tissues were cut from the cardboard and weighed; sieve-tube cross sections were then cut from the phloem sections and also weighed. Crafts and Lorenz estimated that the average cross section of phloem tissue in the peduncles was 18.6 mm^2; of this, about 20 percent consisted of sieve tubes (3.72 mm^2). Thus, material was moving through sieve tubes with a mass transfer rate of 0.61 g h^{-1} ÷ 3.72 mm^2 = 0.164 g mm^{-2}h^{-1}. The **mass transfer rate** is the quantity of material passing through a given cross section of sieve tubes per unit of time.

What about the **velocity** of movement, which is a measure of the linear distance transversed by an assimilate molecule per unit of time? Crafts and Lorenz assumed that the dry material had a specific gravity or density of about 1.5 g cm^{-3} (1500 kg m^{-3}). This figure divided into the rate gives a velocity of about 110 mm h^{-1}. Of course, the material does not move in dry form but as a solute in water. If the solute concentration is 10 percent, its velocity will be about 10 times (100 ÷ 10) that calculated for the dry material, or 1100 mm h^{-1}. Phloem exudates do often consist of 10-percent solutions, but they are usually more dilute than phloem sap. Osmotic potentials of -2 to -3 MPa are common in intact sieve tubes. These values are approximately equivalent to 20- to 30-percent sucrose solutions, so if the assimilates were moving in a 20-percent solution, then they were moving 5 times (100 ÷ 20) faster than the calculated figure for the dry material, or about 550 mm h^{-1}.

Numerous similar measurements have been made, but more meaningful data are now obtained with the short-half-life, energetic ^{11}C. The isotope, incorporated into CO_2, is introduced into the translocation stream via photosynthesis in the leaf, and two or more radiation detectors are placed at measured intervals along the stem. Radiation from the ^{11}C (or ^{14}C) is then measured for brief intervals at different times (Fig. 7-11). The results can be expressed as profiles of radioactivity passing each point on the stem as a function of time, and profiles of radioactivity along the stem can also be determined, especially if several detectors are used. Mathematical models can be developed from such data, and the data and models can be used to test the Münch or other hypotheses (Minchin and Troughton, 1980). Although there is variation among species, studies such as these generally agree with the older work. In most species, the maximum velocity of transport is 500 to 1500 mm h^{-1}.

It is difficult to appreciate these velocities. If a lone sieve element in an angiosperm is 0.5 mm long (gymnosperm sieve cells are longer, about 1.4 mm), at translocation velocities of 900 mm h^{-1} (0.25 mm s^{-1}) an entire element would be emptied and refilled in 2 s. At 300× magnification, an entire phloem element could not be seen at one time in a microscope field; so if one could watch movement of particles of tracer dye through the microscope in sieve elements, velocities would be too rapid to follow easily with the eye. Slow-motion moving pictures would be necessary. Without magnification, movement at 1000 mm h^{-1} is equivalent to the motion of the tip of a 160-mm (6.3-in.) minute hand on a clock. Such movement can easily be detected with the unaided eye.

The Transported Solutes The simplest approach to determining what solutes are contained in phloem sap is simply to cut the phloem and let the sap run out, forming droplets that can then be collected and analyzed. There are problems with this technique, however. We have already seen that bleeding is often rapidly stopped by P-protein and other particulate matter in the phloem that clog the sieve pores. Nevertheless, droplets do often form before bleeding stops. Yet these droplets would be expected to contain some of the P-protein and perhaps other materials from the sieve tubes, and indeed it was shown by Crafts and Lorenz in 1944 that the percentage of nitrogen (found in protein) in phloem exudate from peduncles (flower stems) going to pumpkin fruits was higher than it was in the developed pumpkins. Furthermore, phloem exudate is usually more dilute than intact phloem sap. Releasing the pressure in the phloem by cutting lowers the water potential, and water then moves in by osmosis. These effects of cutting on composition and concentration are more

noticeable in some species than in others (little effect on composition in many trees, for example), but they are often complications to be reckoned with.

If only we had a miniature hypodermic needle that we could insert into a single sieve tube, carefully extracting some of the contents without a sudden pressure release! In 1953, two insect physiologists, J. S. Kennedy and T. E. Mittler, at Cambridge, suggested that, in fact, we have such an instrument. They wondered whether aphids suck to get the phloem sap that nourishes them, using their "hypodermic" mouthpart (**stylet**), or whether the sap is simply forced into their bodies by the pressure in the phloem. When they cut the stylet of a feeding aphid with a sharp razor blade (after an anesthetizing stream of CO_2), leaving the stylet in place, about 1 mm^3 of material exuded from the cut stylet each hour for up to about four days. They said: "This method of obtaining phloem sap is now in routine use in a study of aphid nutrition, and might also be of use to plant physiologists." The method has indeed been in use since then.

In many experiments, it is not even necessary to cut off the insect. Normally the phloem sap passes through the insect and forms droplets called **honeydew** on the aphid's body (Fig. 7-20; honeydew secretion stops when the insect is anesthetized with CO_2). Often radioactive tracers or dyes can be observed in the honeydew, although its composition is no longer quite the same as it was in the plant, as we have already noted.

As we saw in Table 7-1, 90 percent or more of the material translocated in phloem consists of carbohydrates. There are species for which this is not necessarily true, with phloem sap containing as much as 45 percent nitrogen compounds; but sugars make up the great bulk of translocated solutes in the phloem sap of most species. Furthermore, virtually all the sugars transported in the phloem are nonreducing sugars. (See the accompanying box essay for a brief review of carbohydrate chemistry.) Among these, sucrose (common table sugar) is by far the most common. There are a great many plants, perhaps the majority, in which sucrose is nearly the only sugar that is transported. Other sugars, if they occur at all, are present only in trace amounts. (The reducing sugars glucose and fructose, often found along with sucrose in fruits, are sometimes found in phloem exudate, but it has been shown that they are breakdown products of sucrose and are not themselves translocated.)

The major, if not the sole nonreducing carbohydrates that are transported in higher plants, belong to the raffinose series of sugars: sucrose, raffinose, stachyose, and verbascose (Fig. 7-17); or the sugar alcohols: mannitol, sorbitol, galactitol, and myoinositol (Fig. 7-14). Martin H. Zimmermann and

A Review of Carbohydrate Chemistry

To understand modern studies on phloem transport, it is essential to know something about the chemistry of the carbohydrates that are transported. Here is a brief review of carbohydrate chemistry, including even a few compounds that have nothing to do with phloem transport but are mentioned to make the review complete. Eight topics are discussed.

First, the general formula of **carbohydrates** is $(CH_2O)n$; that is, for each C there is an H_2O (water, although it does not exist in this form), thus suggesting the name carbohydrate. The n in the formula means that CH_2O is repeated a certain number of times.

Second, the basic building blocks of carbohydrates are called **monosaccharides** or **simple sugars,** because they are not easily broken down to even simpler sugars. They contain various numbers of carbon atoms (Fig. 7-12) and are named accordingly:

Three-carbon sugars (trioses). These and similar compounds are important intermediates in the metabolic pathways of photosynthesis and cellular respiration.

Four-carbon sugars (tetroses). There are not many of these sugars, although one takes part in photosynthesis and respiration.

Five-carbon sugars (pentoses). These compounds are crucial in photosynthesis and respiration. Two pentoses (ribose and deoxyribose) also form key structural components of the nucleic acids, which are essential for all life. Certain gums peculiar to specific plants (both algae and higher plants) consist largely of pentoses, and the hemicelluloses found in all plant cell walls are rich in pentoses.

Six-carbon sugars (hexoses). These often-discussed sugars take part in many steps of respiration and photosynthesis and constitute the building blocks of many other carbohydrates. Glucose and fructose are key hexoses, but there are several others that occur naturally (Fig. 7-13). For all the mention made of glucose in biology books, in plants the majority of it is bound in polymers and other compounds.

Seven-carbon sugars (heptoses). One of the heptoses is an intermediate in photosynthesis and respiration. Otherwise they are seldom encountered.

Third, the carbohydrates exhibit **stereoisomerism.** If four different atoms or groups of atoms are attached to a single carbon atom, forming a tetrahedral structure, there are two ways to make the attachment, which give mirror-images. Thus, a carbon atom with four *different* things attached can exist as two **stereoisomers,** and molecules with such atoms are said to exhibit stereoisomerism. Two mirror-image isomers of a given compound rotate the plane of plane-polarized light in opposite directions.

Study the hexose sugar (glucose) shown in Fig. 7-13. The top (number 1) carbon has only three things attached to it: a hydrogen, an oxygen (by double bond), and the rest of the molecule. The bottom (number 6) carbon has only three kinds of things attached to it: two hydrogens, an —OH, and the rest of the molecule. Each of the four carbons in between has four different kinds of things attached to it, so each of those four carbons and attached atoms or groups could exist as two stereoisomers (two mirror images). Note in Fig. 7-13 that the names of the sugars are prefaced by the letter D (written as a small capital). This designation indicates their stereoisomeric structure and refers to the position of the —OH on the next-to-the-bottom carbon. If it were on the other side, we would use the letter L. (If *only* this —OH changes from one side to the other, a new sugar is produced with a new name; if the name stays the same, as D-glucose and L-glucose, *all* asymmetric carbons in the two molecules are mirror images of each other.) The chain of carbons in a sugar molecule form a zigzag, but this three-dimensional pattern cannot be represented conveniently on a two-dimensional sheet of paper, so the carbon chain is usually shown as though it were straight.

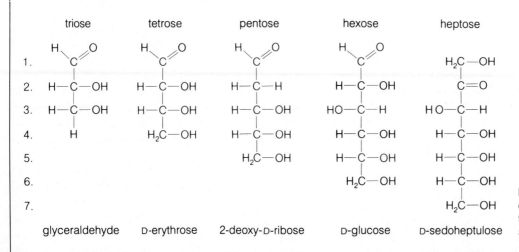

Figure 7-12 Examples of monosaccharides having three to seven carbon atoms.

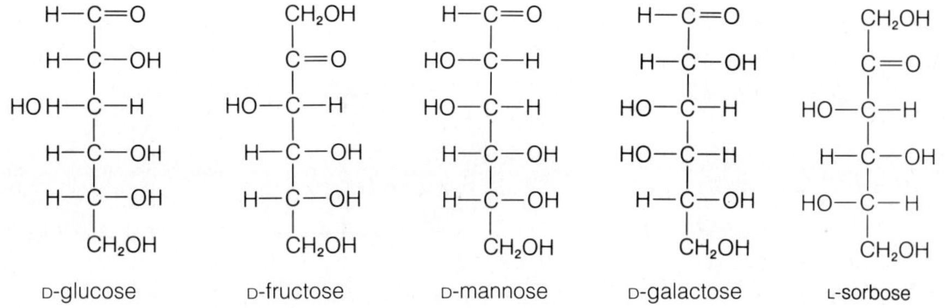

D-glucose D-fructose D-mannose D-galactose L-sorbose

Figure 7-13 (above) Five important hexose sugars. The aldehyde groups (top of all but D-fructose and L-sorbose) are commonly written as —CHO but are expanded here for clarity.

D-mannitol D-sorbitol D-galactitol (dulcitol)

myo-inositol D-galacturonic acid

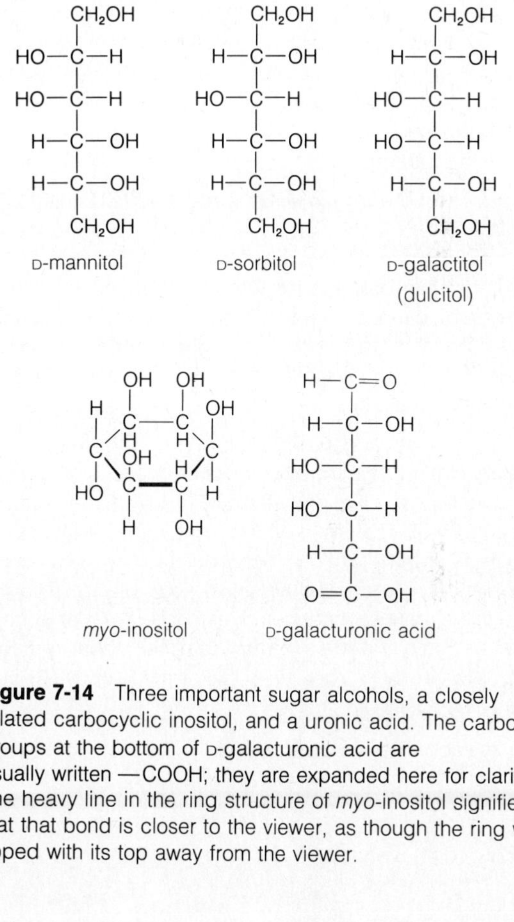

Figure 7-14 Three important sugar alcohols, a closely related carbocyclic inositol, and a uronic acid. The carboxyl groups at the bottom of D-galacturonic acid are usually written —COOH; they are expanded here for clarity. The heavy line in the ring structure of myo-inositol signifies that that bond is closer to the viewer, as though the ring were tipped with its top away from the viewer.

Fourth, the monosaccharides are characterized by the presence of an **aldehyde** —C—H (sugars with the aldehyde are called **aldoses**) or a **ketone** group —C— (ketone sugars are called **ketoses**). The aldehyde or the ketone group is highly reactive in alkaline solution; either one is a reducing agent. In a solution containing one or more oxidizing ions, the aldehyde or ketone group becomes oxidized to an acid group —C—OH (called a **carboxyl**). The oxidizing ion, of course, becomes reduced. This reaction is the basis for several standard reagents that measure so-called **reducing sugars.** In such reagents, an oxidizing cupric ion (Cu^{2+}) is held in solution by some chelating agent (Section 5.5) such as citric or tartaric acid, and the solution is made alkaline with potassium or sodium hydroxide. When a reducing sugar (any of the monosaccharides or their relatives shown in Figs. 7-12 or 7-13) is added to the reagent, the sugars are oxidized to form complex mixtures of sugar acids, and the cupric ions are reduced to cuprous ions (Cu^+), which in turn form cuprous hydroxide (a yellow precipitate) and then become dehydrated to produce cuprous oxide (a brick-red precipitate).

Fifth, the aldehyde or ketone group of the monosaccharides can be reduced as well as oxidized. When it is reduced, it produces another —OH group where the aldehyde or ketone had been located. In that case, all of the carbons then have an —OH group attached. These are called **sugar alcohols** and are important as transported solutes in the phloem of certain species (Fig. 7-14). Note that glucose, fructose, and sorbose can all be reduced to produce the sugar alcohol sorbitol. Note further that when fructose is reduced, the —OH can go on either side of the number 2 carbon; if it attaches to the right side, the product is sorbitol, if to the left side, mannitol.

(*continued on next page*)

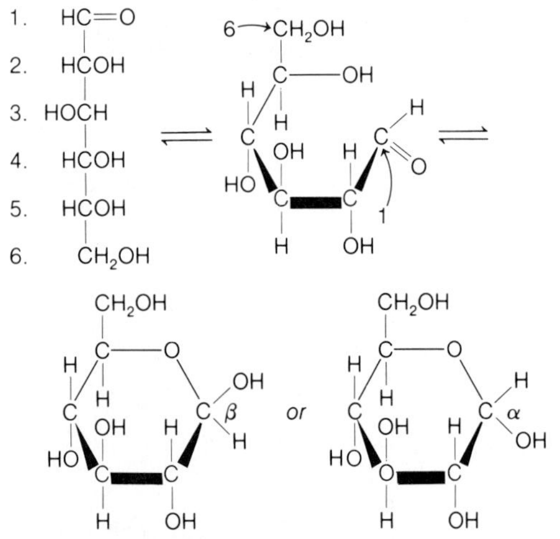

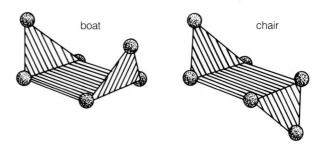

Figure 7-15 Formation of the ring structures of sugars, specifically D-glucose. Note the α and β forms and the heavy line in the rings, indicating that the ring is tipped with its top away from the viewer. The figure also shows that the 6-membered rings naturally assume one of two configurations: a boat or a chair.

(continued from previous page)

Some of the **sugar acids** (produced by oxidation of the aldehyde or ketone group) occur naturally in plants (for example, galacturonic acid in the pectins found in the middle lamella between the walls of adjacent cells). Various groups can also be attached to the sugars in various ways producing **glycosides** (e.g., **glucosides**). For example, methyl groups are attached to some of the galactose residues in pectins.

Sixth, most sugars in solution form rings rather than straight chains. There are two possible ways for the ring to form. As shown in Fig. 7-15, the ═O group on the 1-carbon becomes an —OH group after the ring forms. With the ring drawn as shown, the —OH can point either up or down. The two forms are called alpha (α) and beta (β). The ring form of the monosaccharides does not lie in a plane (i.e., it is not flat); rather, it takes the form of either a boat or a chair (Fig. 7-15).

Seventh, these details suggest how the rings can attach to each other in so many ways. This attachment occurs by removal of a water molecule from two of the sugar molecules (Fig. 7-16). If two monosaccharide molecules (often simply called *units*) hook together, they form a **disaccharide.** Two glucose molecules form the disaccharide maltose—or cellobiose, depending on whether the unit forming the linkage is in the α or the β form. A glucose and a fructose molecule form the disaccharide sucrose (by far the most abundant sugar in the plant kingdom). Three monosaccharide units form a **trisaccharide,** and four a **tetrasaccharide.**

The di-, tri-, and tetrasaccharides are referred to collectively as **oligosaccharides.** In some cases, an aldehyde or a ketone group remains exposed in the oligosaccharide so that the compound has the reducing properties of the monosaccharides. (Maltose and cellobiose are examples.) In other cases, the classical example being sucrose, both reducing groups are utilized to form the linkage between the monosaccharides. In the case of sucrose, the aldehyde of glucose and the ketone of fructose both take part in forming the linkage between the two monosaccharides. Hence, sucrose is a **nonreducing**

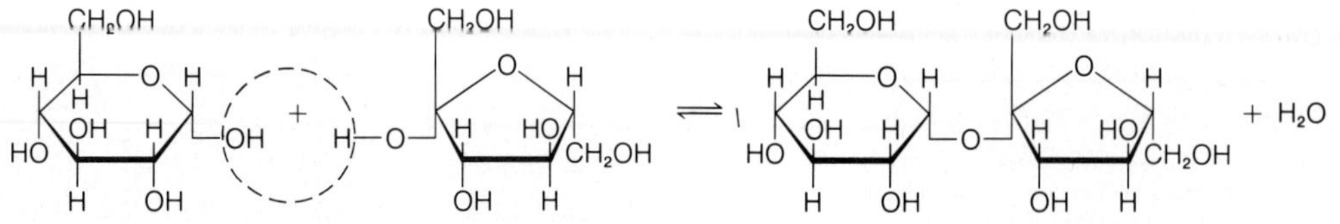

D-glucose   D-fructose   sucrose   + H₂O

Figure 7-16 Formation of the disaccharide sucrose by removal of a water molecule from between two monosaccharides. It is assumed that angles in the rings represent carbon atoms, each with its four bonds. The ring form of D-glucose is called glucopyranose, and the ring form of D-fructose is called fructofuranose.

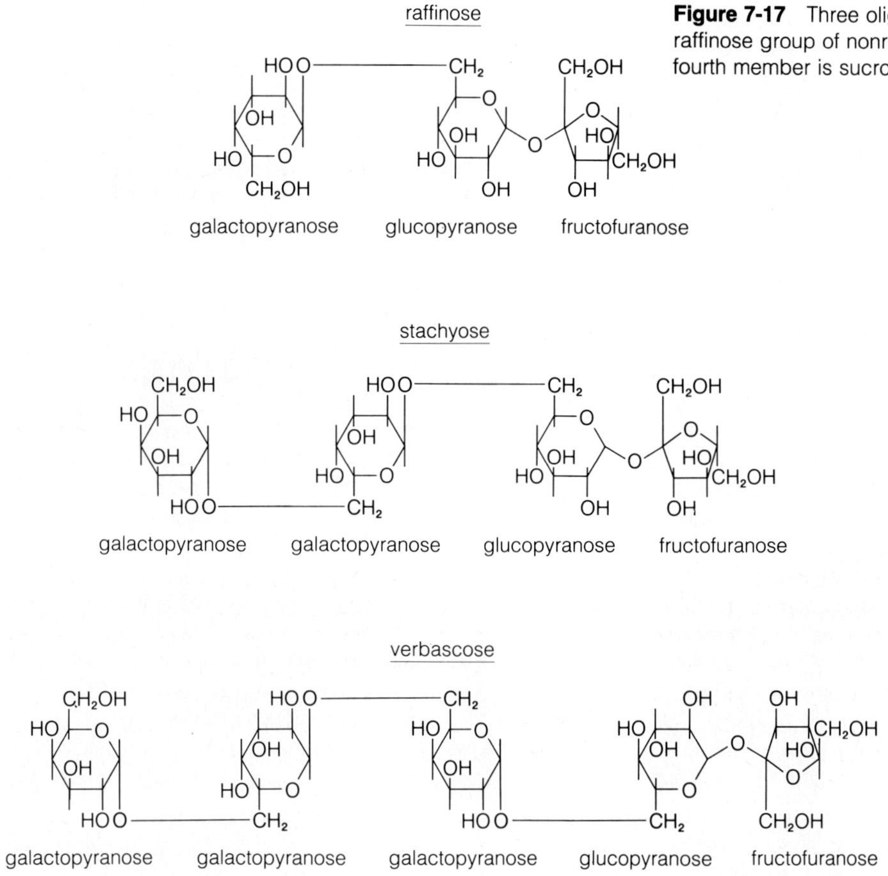

raffinose

galactopyranose · glucopyranose · fructofuranose

Figure 7-17 Three oligosaccharides of the raffinose group of nonreducing sugars. The fourth member is sucrose, shown in Fig. 7-16.

stachyose

galactopyranose · galactopyranose · glucopyranose · fructofuranose

verbascose

galactopyranose · galactopyranose · galactopyranose · glucopyranose · fructofuranose

sugar. Note that the glucose unit forms a six-membered ring (called a **pyranose**), and the fructose forms a five-membered ring (a **furanose**).

A series of nonreducing oligosaccharides that can be thought of as additions of different numbers of galactose molecules to sucrose includes solutes transported in the phloem of several species. This series of oligosaccharides is called the **raffinose group** (Fig. 7-17). Another group of oligosaccharides and higher polymers, collectively called **dextrins,** are breakdown products of starch (Section 12.2). They consist of glucose molecules attached to each other end to end. Since the aldehyde group on one glucose is always free on one end, the dextrins are reducing sugars. Usually dextrins are a mixture of molecules, each with two to several dozen glucose **residues,** as they are sometimes called.

Eighth, when many units hook together, they form such **polysaccharides** as starch, cellulose, callose, hemicellulose, or pectins. When these **complex carbohydrates** are degraded to simple sugars, it is by addition of a water molecule where the original one had (*continued on next page*)

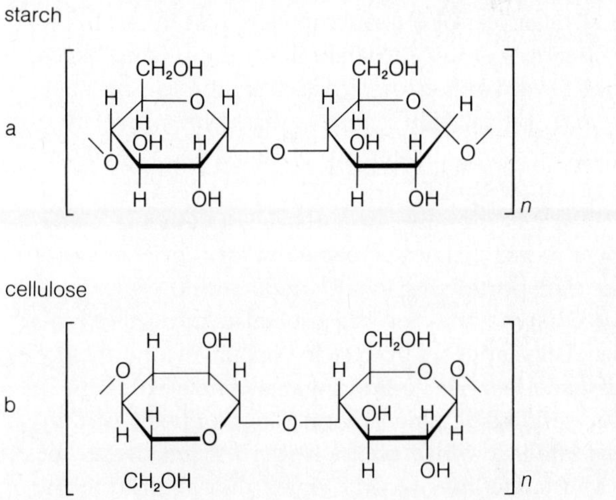

starch

cellulose

Figure 7-18 (**a**) The α linkage between glucose residues, as in starch. (**b**) The β linkage between glucose residues, as in cellulose. Note that the residue on the left in the cellulose segment is upside down relative to the one above it in starch.

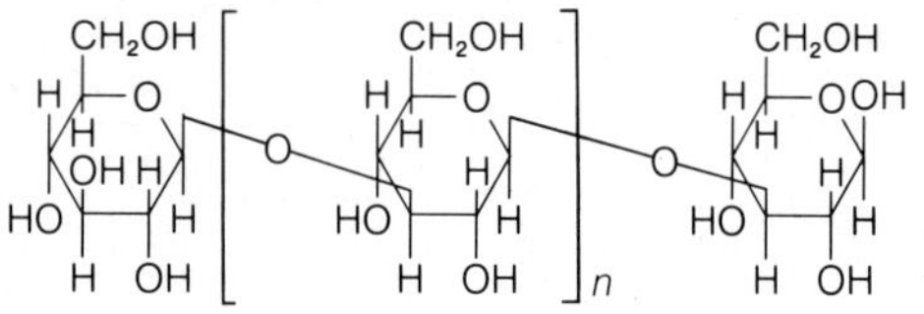

CH₂OH ... callose

Figure 7-19 The molecular structure of callose. Note the interesting 1,3-linkage between β-D-glucopyranose residues, which causes a tight coiling of the molecular callose chain.

(*continued from previous page*)
been removed. This degradation is called **hydrolysis.** It occurs during seed germination, for example, as starch, which is insoluble, is broken down to soluble glucose that can be metabolized and transported within the seedling.

Cellulose molecules consist of glucose units in the β-ring form (Fig. 7-18). A single molecule of cellulose contains 3000 to 10,000 glucose units in an unbranched chain. When β-glucose rings are hooked together to form a long chain, the chain proves to be almost perfectly straight (see Figs. P-4 and P-5).

It is instructive to compare cellulose with starch. Glucose units are the basic constituents of starch, but in this case they have the α linkage (Fig. 7-18), which produces coiled instead of straight chains. How can we account for the **synthesis** (putting together) of two such different molecules as starch and cellulose from the same units, ultimately the open chains of glucose? The answer lies in the highly specific nature and shape of the enzymes responsible for the synthesis of these polysaccharides. The enzyme that synthesizes cellulose is mechanically configured to form only β bonds from substrate molecules, whereas the starch enzyme combines the same molecules to form α bonds.

An important polysaccharide is **callose,** which is composed of β-D-glucopyranose residues linked together by β,1-3 **glycosidic linkages** (linkages between monosaccharides) as illustrated in Fig. 7-19. It is another **glucan,** a compound formed of glucose residues. This interesting linkage produces tightly coiled chains. Callose is important in the formation of the sieve plate, and it also appears almost instantly in various parts of plants that are subjected to mechanical stress (e.g., when shaken). It seems to play a role in the healing of damaged tissue. Its chemical structure is very similar to the storage glucans of several algae.

H. Ziegler (1975) present a list of phloem–sap compositions for more than 500 species belonging to about 100 dicotyledonous families and subfamilies. Usually samples of phloem sap were placed on sheets of filter paper in the field and then later chromatographed and stained to identify the various sugars and to provide some indications of relative amounts (size and intensity of the stained spots on the chromatogram).

The list strongly supports the statements made above; namely, that sucrose is by far the most common transported sugar, although raffinose and stachyose (sometimes verbascose) also appear. Myoinositol also appears in trace to very small amounts in many species. The other sugar alcohols sometimes occur in considerable amounts but only in certain plant families. Since plants in the list are nearly all trees or sometimes woody vines, they might not be representative of flowering plants as a whole. Nevertheless, the list provides some interesting facts about certain families.

The pea family (Fabaceae, subfamilies Papilionoideae, and Mimosoideae) and several other families transport only sucrose with traces of other compounds appearing only rarely. Several families transport mostly sucrose with very small amounts or traces of raffinose, stachyose, and myoinositol in a few (e.g., maple, Aceraceae; cashew, Anacardiaceae; aster, Asteraceae; honeysuckle, Caprifoliaceae; birch, Betulaceae; beech and oak, Fagaceae; fig, Moraceae, and many others). Members of the genus *Buddleia* (butterfly-bush) and of the verbena family (Verbenaceae) transport mostly stachyose with smaller amounts of raffinose and still smaller amounts of sucrose. The staff-tree family (Celastraceae; *Euonymus* is a representative member) transports about equal amounts of sucrose and stachyose with less raffinose, considerable amounts of galactitol in some species, and traces of myoinositol. Some species in the white mangrove family (Combretaceae) transport quite a bit of mannitol, and the olive family (Oleaceae; includes *Fraxinus, Olea, Syringa,* etc.) also transports much mannitol and stachyose, with less sucrose, still less raffinose, and small amounts or traces of verbascose and myoinositol.

There has been considerable interest in the rose family (Rosaceae; one of the largest families), because it includes many fruit trees and other species of

a

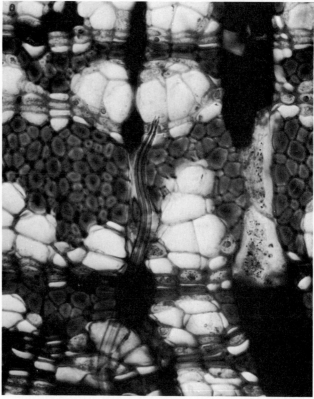

Figure 7-20 Study of translocation in the phloem by the use of aphids. (**a**) An aphid hanging upside down on a branch of a tree. Note the droplet of honeydew being exuded from the insect. (**b**) A cross section of the tree showing an aphid stylet that has penetrated to a sieve element. (Photographs courtesy of Martin H. Zimmermann; see Zimmermann, 1961. For a series of micrographs of aphid stylets penetrating sieve elements, see Botha et al., 1975.)

b

commercial importance and because mostly sorbitol is transported with a little less sucrose and only traces of raffinose, stachyose, and myoinositol (a few with traces of verbascose). The family itself is divided into several subfamilies, and it is interesting that most of these have this pattern of translocated carbohydrates except for the Rosoideae, which includes the genus *Rosa* from which the family takes its name. Members of this subfamily transport no sorbitol but mostly sucrose with traces of raffinose, stachyose, and myoinositol (verbascose in a few species). Some genera from the subfamilies that do transport mostly sorbitol include: *Cotoneaster* (pyracantha), *Crataegus* (hawthorne), *Malus* (apple), *Prunus* (apricot, cherry, etc.), *Pyrus* (pear), *Sorbus* (mountain ash, *Sorbaria* (false spirea), and *Spiraea* (spirea).

It is highly significant that only nonreducing sugars are translocated whereas reducing sugars and their phosphate derivatives are not. Although the reasons for this are uncertain, nonreducing sugars are less reactive and less labile to enzymatic destruction in sieve elements (Arnold, 1968). Indeed, for perhaps the same reason, reducing sugars are seldom abundant in most plant cells. Glucose and fructose, for example, usually occur in cells as their phosphate derivatives, although they do appear as storage sugars in many sweet fruits.

In addition to the carbohydrate composition of

phloem sap, much is also known about the nitrogenous components of both phloem and xylem saps (Pate, 1980). Again, as with carbohydrates, nitrogen components are highly species-specific. In some species much nitrogen is transported in the xylem in the inorganic form of nitrate (NO_3^-), although this is virtually never present in phloem sap. In many other species, nitrogen is transported in the xylem as ureides, amides, or other nitrogen-rich molecules. Furthermore, the same group of organic-nitrogen molecules might carry the bulk of the nitrogen in both the xylem and the phloem channels, but differences in solute composition between xylem and phloem sap have been observed in some species. Alkaloids carry significant amounts of nitrogen in the xylem of certain species, as do certain amino acids that normally do not occur in protein. Amino acids, other organic nitrogen compounds, and the reduction and incorporation of nitrate into organic compounds are discussed in Chapters 8 and 13, but structures of the most important organic compounds involved in nitrogen transport are shown in Fig. 7-21. Note that such compounds often contain more than one nitrogen atom per molecule.

It is important to note the relative nutritional completeness of sieve tube sap (alluded to above in our reference to the nutrition of sap-sucking insects). Many plant parts with no, or minimal, transpiration

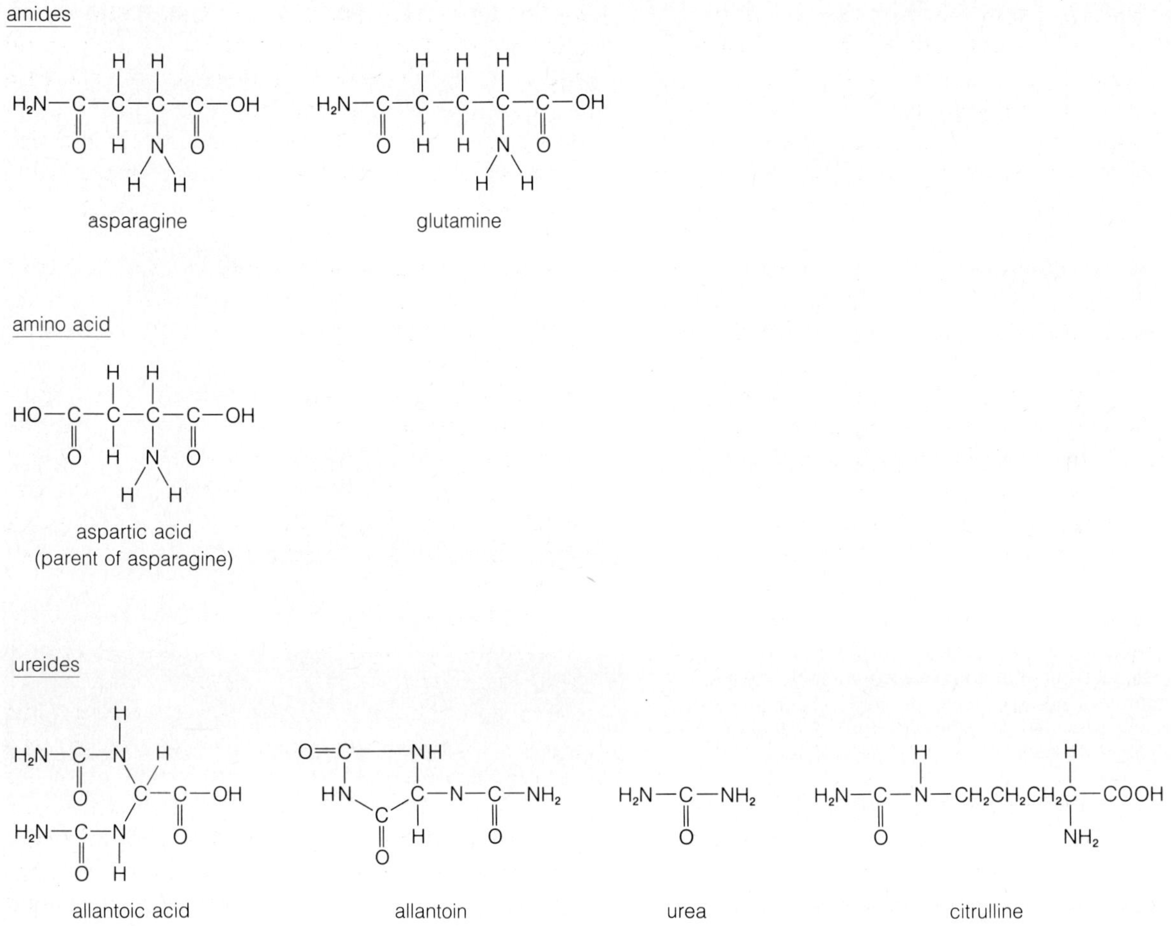

amides

asparagine

glutamine

amino acid

aspartic acid
(parent of asparagine)

ureides

allantoic acid

allantoin

urea

citrulline

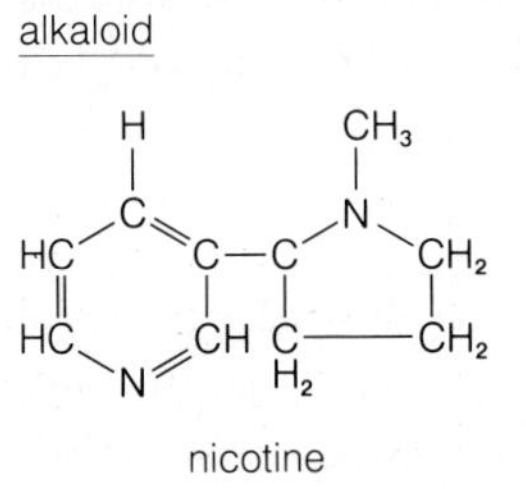

alkaloid

nicotine

Figure 7-21 Examples of organic nitrogen compounds important in transport of nitrogen in both xylem and phloem of many species. Other nitrogen compounds important in transport are discussed in Chapters 8 and 13.

are almost completely dependent on the phloem for nutrients during part or all of their growth (e.g., meristems, fruits, roots). These nutrients include both the organic materials we have been discussing and inorganic nutrients.

Phloem Loading In 1949, Brunhild Roeckl determined the osmotic potentials of photosynthetic cells and sieve sap of *Robinia pseudoacacia* (black locust) using plasmolysis, refractometry, and cryoscopic techniques (Section 2.6). Such measurements have since been repeated by others, and many studies have used radioactive tracers. Typically, the meso-

phyll cells of trees have an osmotic potential of about -1.3 to -1.8 MPa, whereas sieve elements in leaves have an osmotic potential of about -2.0 to -3.0 MPa. Herbaceous plants frequently have somewhat less-negative osmotic potentials in the mesophyll cells. Sugar beets, for example, have an osmotic potential of about -0.8 to -1.3 MPa for mesophyll and phloem parenchyma cells and about -3.0 MPa for the sieve-element–companion-cell complex (Geiger et al., 1973). Since most of the osmotic potential is caused by the presence of sugars in both kinds of cells, it is clear that the sugar concentration is approximately 1.5 to 3 times as high in the sieve elements as in the surrounding mesophyll cells. The process in which sugars are raised to high concentrations in phloem cells close to a source such as photosynthesizing leaf cells is called **phloem loading.**

In recent years, there has been much research on the mechanism and role of phloem loading (Giaquinta, 1983). We shall consider a few of the high points.

The pathway of transport   How does sucrose get from the leaf mesophyll cells where it is synthesized to the sieve tubes in the minor leaf veins? Often it passes through no cells at all or 2 or 3 cells to get to a minor leaf vein. Does it move through the apoplast (cell walls outside of the protoplast) or through the symplast (from cell to cell through plasmodesmata, remaining in the cytoplasm)? Although the evidence is scanty (few species and few studies), it appears clear that movement to the minor veins occurs through the symplast. For example, $^{14}CO_2$ assimilated into carbohydrates in the mesophyll cells does not appear in the cell walls of the mesophyll; there is virtually no leakage from the mesophyll cells. In many plants it has been shown that there are numerous plasmodesmata between mesophyll cells. There are almost always more plasmodesmata where the cells contact each other than where the cell surfaces are exposed to the space within the leaf.

Contrasted to the many plasmodesmatal connections between adjacent mesophyll cells, plasmodesmata are rare or absent between mesophyll cells and adjacent companion cells and sieve elements. This is true for the majority of species that have been examined (e.g., broadbean, maize, sugar beets), but there are a few species in which there is direct symplastic continuity between mesophyll or **bundle-sheath cells** (cells forming a sheath around the vascular bundles of many species) and adjacent companion cells and sieve tubes (e.g., *Cucurbita pepo* and *Fraxinus).* In any case, there is considerable evidence that sugar is transported actively out of mesophyll cells into the apoplast of the minor veins; that is, it is *secreted* into the apoplast (e.g., Geiger et al., 1974; Giaquinta, 1976). From the apoplast, sugar is then absorbed actively, probably into the large companion cells of the minor veins from which it then passes symplastically into the sieve elements (see Fig. 7-7).

There are several lines of evidence for this active loading into companion cells and then sieve elements. One of the strongest involves the use of the sulfhydryl-group modifier, *p*-chloromercuribenzene sulfonic acid (PCMBS). This material apparently does not penetrate the plasmalemma of the cells and thus does not enter the symplast. Nevertheless, it markedly and reversibly inhibited the uptake and phloem loading of sucrose (labeled with ^{14}C) in solutions applied to abraded sugar-beet leaves. There was no effect upon short-term glucose uptake, photosynthesis, or respiration, suggesting that the inhibition occurred at the membrane level and possibly affected the sucrose carrier itself. Furthermore, when sugar-beet leaves were allowed to photosynthesize with $^{14}CO_2$, PCMBS markedly inhibited the translocation of assimilates through the phloem. This strongly suggests that sucrose produced in photosynthesis must enter the apoplast before it is loaded into the phloem.

There is some evidence that the companion cells play the most important role in absorbing sucrose from the apoplast. For one thing, they are much larger and metabolically more active than sieve elements. Furthermore, they sometimes have the cell-wall ingrowths characteristic of transfer cells, as we noted earlier.

Active loading of sucrose into the companion cells produces a very negative osmotic potential in those cells. This leads to an osmotic entrance of water, which then passes in bulk flow across the plasmodesmatal connections between the companion cells and sieve elements, carrying the sucrose along with it. Indeed, it is this high concentration of sucrose produced by loading in companion cells and then sieve elements and the consequent osmotic uptake of water that produces the high pressures and mass flow in sieve tubes. Münch was quite unaware of phloem loading and assumed that the pressures might build up in the mesophyll cells themselves, but phloem loading is a highly appropriate modification of his model.

The selective loading of sugars   Phloem loading has been studied by abrading the leaf surfaces, which destroys the cuticle but ruptures only a few epidermal cells, after which solutions of radioactively labeled sugars are applied. An autoradiograph of leaves made at various times after application of the solutions shows the progress of the loading process. When loading is complete, the minor and major veins are highly radioactive compared with the surrounding interveinal tissue (Fig. 7-22). Using this and other approaches, Donald R. Geiger and his co-workers (1973, 1974) applied various labeled sugars to abraded sugar-beet leaves and studied their absorption into small veins. In these and other studies, it has become apparent that only those sugars that are transported in the phloem are actively loaded into the minor veins. As we have seen, this includes (in various species) the raffinose series of sugars, especially sucrose, as well as sugar alcohols. Such reducing sugars as glucose and fructose are not actively loaded, although small amounts are apparently able

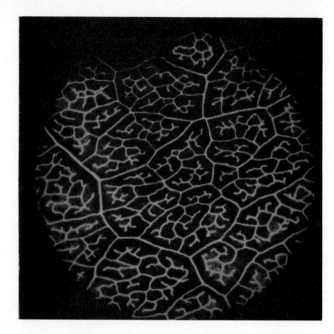

Figure 7-22 An enlarged positive print of an autoradiograph showing phloem loading in sugar beet leaves. Light areas are the minor veins that have accumulated the radioactive sucrose. (Courtesy of Donald Geiger; see Geiger et al., 1974.)

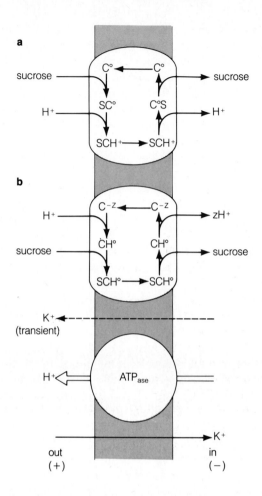

Figure 7-23 A schematic model for sucrose loading across the plasma membrane into sieve elements. The top part of the figure postulates that sucrose is attached to the carrier (C⁰) before the H⁺ is attached; the part just below postulates the opposite, that H⁺ is attached to the carrier (C⁻ᶻ) before the sucrose is attached. The dashed arrow indicates that K⁺ moves out of the cell to maintain electrical balance when sucrose and H⁺ are first transported in. The circle with ATPase inside symbolizes that H⁺ is pumped actively out of the cell, using metabolic energy. This is the driving force, establishing a gradient with more H⁺ outside the cell than inside (lower pH outside). The solid arrow at the bottom shows that K⁺ ions normally enter the cell to maintain electrical neutrality as H⁺ ions are pumped to the outside. (From Giaquinta, 1983.)

to penetrate the companion-cell and sieve-tube membranes passively. This fact in itself is another evidence for an apoplast step between the mesophyll and active loading. Presumably, the selectivity of sugars loaded into the phloem is based on recognition by the sugars of a membrane carrier that transports the sugars across the membrane into the cytoplasm of companion cells.

Comparable studies with amino acids have shown that certain ones are preferentially loaded. Again, these prove to be the compounds that are readily transported in the phloem. This is also true of minerals; those that are readily transported in phloem (phosphorus, potassium, reduced nitrogen) are readily loaded; those that are not readily transported (calcium, iron, boron) are not (Chapter 5). This may be true even for synthetic compounds such as the herbicides 2,4,5-T (relatively immobile in phloem), 2,4-D (intermediate), and maleic hydrazide (most mobile; Field and Peel, 1971; McReady, 1966). But in emphasizing the selectivity of the loading process, we should not lose sight of the fact that many substances can also enter the phloem by passively diffusing in along their own concentration gradients. This is apparently true for several growth regulators, for example.

Sucrose/proton cotransport mechanism In many systems, including bacteria, algae, yeast, fungi, and animal cells, transport of organic molecules such as

sugars and amino acids is linked with transport of hydrogen ions (Chapter 6). Several recent studies (reviewed by Giaquinta, 1983) suggest that sucrose loading into the phloem occurs by such a cotransport system. As noted in Chapter 6, protons are pumped out through the plasma membrane using the energy from ATP and an ATPase carrier enzyme, so the pH outside the cell in the apoplast becomes much lower (more acidic) than inside the cell. Protons then diffuse back into the cell, and their movement across the membrane is coupled to a carrier protein that transports sucrose or other sugar into the cell along with the hydrogen ions. The sugar carrier might first combine with the sugar and then the proton before transporting both inwardly across the membrane, or the carrier might combine with the proton first and then the sugar (Fig. 7-23).

There are several evidences for such a cotransport mechanism. For one thing, the contents of sieve tubes are known to have a relatively low proton concentration (quite alkaline: pH 7.5 to 8.5) and high potassium concentration (100 to 200 mM). The exact pH of the apoplast at the loading sites is not known, but there are indications that it is quite acidic compared with phloem sap (pH 5.5 to 6.0; comparable to xylem sap). If the pH of the apoplast of a leaf is increased, neutralizing its acidity, sucrose uptake at low external sucrose concentrations is markedly inhibited. Of course, the addition of alkaline buffer to the leaf could have other effects as well as the one predicted by the model; but in any case, the results are those that would be expected. When sucrose uptake in sugar-beet leaves is inhibited by PCMBS (last section), active proton secretion by the phloem minor veins is also inhibited.

In another system, addition of sucrose to *Ricinus* cotyledons floating on a solution caused the external medium (representative of the apoplast) to become more alkaline (0.1 to 0.2 pH unit) for about 15 to 30 min, after which it returned to the original pH value (Komer et al., 1980). This is accounted for by entry of protons with the sucrose, raising the pH, followed by an active proton secretion to reestablish the proton gradient across the membrane. Sugars such as methylglucoside, fructose, glucose, and raffinose, which (in this case) do not compete with the sucrose carrier, did not cause the increase in pH, indicating that the proton cotransport system is specific for sucrose in this system. As the sucrose was first absorbed, there was a transient movement of K^+ out of the cells followed by return of the K^+ into the cells as more protons were excreted, reestablishing the electrochemical gradient.

The role of metabolism in transport Münch's model suggests that flow of sap through the sieve tubes is a passive phenomenon generated only by the high pressures in the leaf minor veins (or other sources) and the lower pressures at the sink. Thus the pressure-flow model does not immediately suggest that metabolic energy might be required along the pathway to maintain flow, although some metabolism might be required to maintain the phloem tissues in a condition suitable for transport and to prevent leakage of sugars through the plasma membranes of sieve elements. Early studies seemed to suggest that any inhibition of metabolism (e.g., by low temperatures or respiration inhibitors) along the pathway did inhibit transport. This requirement for metabolism was often cited by those who presented alternative theories to pressure flow. Indeed, it was

suggested that metabolic energy was required along the pathway to move solutes across the sieve plates (e.g., by some pumping or peristaltic contraction of P-protein in sieve pores). Hence these alternative theories were often referred to as *active* theories as contrasted to the passive pressure-flow mechanism.

Studies have shown that the apparent inhibitory effects of low temperature or anoxia (lack of oxygen) in certain species were really only transient effects, and that phloem transport continued after an adjustment period of 60 to 90 min (Geiger and Sovonick, 1975; Watson, 1975; Sij and Swanson, 1973). Thus, maintenance of the phloem transport system for bulk flow of sap apparently requires only a minimum of metabolic energy. Of course, metabolic energy is required for phloem loading, as we have seen.

The development of loading capacity Young leaves normally act as sinks rather than sources. This is true even after they develop some photosynthetic capability. At a certain time, however, they begin to export carbohydrates through the phloem, although import of carbohydrate may continue for a while through different vascular strands. What accounts for the change-over from an import to an export mode of phloem transport? The development of phloem loading capacity by the minor-vein companion cells could account for this switch from import to export (Giaquinta, 1983). Once sucrose begins to be actively loaded into the companion cells and then the sieve elements, water will enter by osmosis and flow will begin out of the minor veins: The leaf will become a source instead of a sink.

Phloem Unloading Removal of sucrose and other solutes from the sieve elements at the sink end of the system has been much less studied than phloem loading. There is, nevertheless, accumulating evidence that an active unloading process plays an important role. Indeed, the degree of unloading could determine the sinks into which most translocation occurs. Unloading of solutes at the sink end of the transport system would be another highly appropriate modification of Münch's original hypothesis. Unloading would insure that turgor pressures at the sink end of the system remained low, since low sugar concentrations would allow water to move osmotically in response to the pressure transmitted from the source (see Fig. 7-5). Furthermore, the solute unloaded at the sink could then be actively absorbed into developing fruit or other cells where concentrations could reach values as high or higher than occurred in the sieve tubes at the source.

Pressure in the Phloem Sieve tubes contain solutions under pressure, as indicated by the feeding habits of aphids. There are a few cases in which exudation after cutting occurs for several hours to many

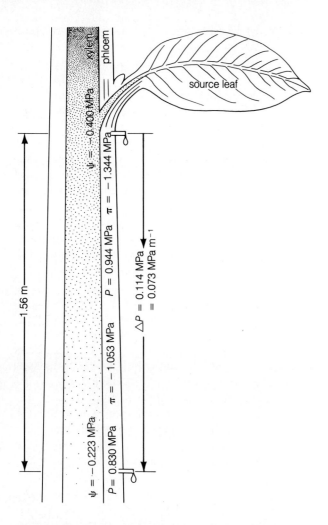

Figure 7-24 The osmotic quantities in phloem sieve tubes and xylem (apoplast) of a young willow *(Salix viminalis)* sapling. The osmotic potentials (π) were determined on phloem sap exuded from aphid stylets; water potentials (ψ) of the apoplast (bark samples) were determined with a vapor psychrometer system (see Figs. 2-5 and 2-9); and pressures (P) in the sieve tubes were calculated by assuming that water potential of the phloem sap was in equilibrium with that of the surrounding tissues ($P = \psi - \pi$) including xylem. Note that there is a positive pressure gradient (ΔP) in the sieve tubes from the apex toward the base, even though there is an opposite gradient in the water potential of the apoplast (caused by tension in the xylem plus matric forces). The pressure gradient of about 0.07 MPa m^{-1} is ample to drive a pressure flow of sap through sieve tubes. (Data are averages from several experiments of S. Rogers and A. J. Peel, 1975.)

days. When the top of a sugar palm trunk or its influorescence is tapped, for example, as much as 10 liters of sugary sap may drip out of the cut sieve tubes in a day, and the Palmyra palm in India produces on the order of 11 liters of sap from cut phloem each day (reviewed by Crafts, 1961). This sap consists of about 10 percent sucrose and 0.25 percent mineral salts and has probably been diluted by water moving osmotically from the xylem (apoplast) into the phloem

after pressure is released. Clearly, pressure exists in sieve tubes, as a pressure flow mechanism requires.

Münch postulated that a pressure *gradient* must occur in the phloem sufficient to account for flow from source to sink. For many years, pressures in the phloem could not be measured directly, so they were calculated by comparing osmotic potentials in the phloem with water potentials of the surrounding apoplast. At equilibrium (probably seldom achieved), water potential in the sieve tubes would be equal to water potential of the apoplast, so pressure in the sieve tubes (P_i) would equal apoplast water potential (ψ_e) minus sieve-tube osmotic potential (π_i). ($P_i = \psi_e - \pi_i$, since $\psi_e = \psi_i$.) Gradients in osmotic potential in the sieve tubes from source to sink have often been measured, with the most negative values at the source (e.g., Housley and Fisher, 1977; Rogers and Peel, 1975). But we saw in Chapter 4 that there is also a gradient in xylem (apoplast) water potential, with the most negative values in the transpiring and photosynthesizing leaves; that is, at the source for much phloem transport. So the existence of a pressure gradient from source to sink in the phloem will depend upon the relative steepness of the osmotic gradient in the sieve tubes compared with the water-potential gradient in the apoplast. Most calculations that take these factors into account suggest that there is a pressure gradient in sieve tubes from source to sink (Fig. 7-24).

Nevertheless, it would be satisfying to measure pressures directly, and several workers have done this. One approach was to attach a pressure gauge to a cut palm shoot, and still another was to apply a pressure cuff similar to those used in measuring blood pressure, increasing pressure in the cuff until wound exudate stopped. Pressures as high as 2.4 MPa were measured.

H. T. Hammel (1968) inserted a specially prepared hollow needle of microdimensions into the bark of red oak *(Quercus rubrum)* trees. A glass capillary partially filled with dyed water and sealed at one end was attached to the needle. When the needle penetrated a phloem tube, sap moved into the needle and capillary, compressing the gas in the glass tube (as in the method illustrated in Fig. 2-10). Hammel measured pressures at two points on the trunk, one 4.8 m higher than the other, with average values 0 to 0.3 MPa higher for the uppermost sampling point, although there was considerable variation. This is what the Münch model predicts.

John P. Wright and Donald B. Fisher (1980) glued glass capillary tubes, sealed at one end, to stylets severed from aphids feeding on phloem sap in weeping willow *(Salix babylonica)*. Pressures were calculated by measuring compression of the gas in the capillary tubes as in Hammel's experiments. Values of up to 1.0 MPa (accuracy ±0.03 MPa) were measured, and pressures remained stable for several hours. They also calculated pressures by measuring sucrose in the phloem exudate (*refractometry*) and leaf water potential by a psychrometer method. The two methods agreed well, when account was taken of the amino acids and K^+ ions in the phloem exudate (not measured by refractometry).

Two Problems with Pressure Flow The most simplistic view of the mass flow hypothesis suggests that substances should move in the phloem not only in the same direction but at the same velocity. Thus, several workers (e.g., Biddulph and Cory, 1957; Fensom, 1972) measured the velocity of flow of different tracer substances (e.g., ^{14}C-sucrose along with ^{32}P and $^{3}H_2O$). When velocities between two points along the transport system were measured, ^{14}C-sugars often moved most rapidly, with ^{32}P-labeled phosphates moving more slowly, and $^{3}H_2O$ moving slowest of all. At first glance, it would seem that if the water moves more slowly than the solutes it is supposed to be carrying, a mass flow hypothesis could not be tenable.

There are two important complications, however. First, water and other substances exchange rapidly along the pathway. That is, much water passes out through sieve-tube membranes into surrounding tissues, while much water from these tissues moves into the sieve tubes. Sucrose and phosphate do not pass as readily through the membranes along the way, so they might appear to move much faster than the carrier water molecules. Second, it is simplistic to imagine that the phloem transport system consists only of inert tubes. Sieve elements are alive and contain cytoplasm with mitochondria, P–protein, and other substances. Thus, solutes may be metabolized or otherwise interact (e.g., by adsorption) to different degrees along their transport route.

Another much-discussed potential obstacle to the pressure-flow hypothesis is that it is incompatible with movement of two different substances in opposite directions in the same sieve tube at the same time. Does such true **bidirectional transport** occur? Beginning in the 1930s and continuing until at least the 1970s, workers attempted to answer this question. Numerous experiments suggested that such transport does occur, but so far it has always been possible to devise alternative explanations (reviewed by Peel, 1974).

In the best and most recent experiments, two different tracers are applied to different points, and their movement is followed. There is no question that bidirectional movement does occur. Tracers applied to young leaves may move **basipetally** (toward the base), while tracers applied to older leaves below may move **acropetally** (toward the tip), so that the two pass each other in the stem, moving in opposite directions. Such bidirectional movement may even occur in a leaf petiole. But do both tracers move in the same vascular bundle or the same sieve tube? Many studies have indicated that they do not. Carol A. Peterson and Herbert B. Currier (1969) abraded the stem surface of several species and applied fluorescein, which was absorbed into intact phloem tissue. After various time intervals, they cut sections from above and below the point of dye application. The tracer moved both up and down, but after fairly short time intervals it was never present in the same bundle both above and below a treated area; each bundle and sieve tube translocated the dye in only one direction. With longer time intervals, the dye could move to a node in one bundle, pass laterally to another bundle, and move back down the stem in an opposite direction.

Studies with aphids might settle the question of whether or not tracers can move in opposite directions in a single phloem cell. For example, fluorescein was applied to one leaf at one node and $^{14}CO_2$ to another at another node with the honeydew from aphids on the stem between being collected on a slowly rotating cellophane disk (Eschrich, 1967; Ito and Peel, 1969). Much of the honeydew contained both tracers.

Are these experiments conclusive? No, because there are alternatives to true bidirectional transport. For one thing, a stylet inserted into a sieve tube might itself become a sink so that substances flow from both directions toward the stylet. It is also likely that the direction of flow changes with time in a given sieve tube, perhaps as source and sink roles change, or perhaps in response to more subtle hormone mechanisms. Thus, if the experiment lasts an hour or more (as is usually the case), then flow was possibly reversed during the experimental period, so that some of the honeydew was produced from one source, while the rest was produced from another. Furthermore, there is considerable lateral transport between sieve tubes and even between the phloem and the xylem, especially at the nodes but in the internodes as well; in some species, sieve elements have lateral sieve pores. So it is possible that two tracers going in opposite directions but in different vascular bundles or sieve tubes, or even in xylem and phloem, become mixed by lateral transport.

Evidence for bidirectional movement contributed strongly to the impetus to find an alternative to the pressure-flow mechanism. At this point, however, other evidences for mass flow are so strong that most workers in the field accept the explanations just presented that reconcile apparent bidirectional trans-

The Specificity of the Phloem Loading System

John E. Hendrix

John E. Hendrix has been a colleague of co-author Cleon Ross at Colorado State University since 1967. His research involves various aspects of carbohydrate translocation in squash and other members of the Cucurbitaceae family. More recently, he has developed a research program concerning carbohydrate translocation and partitioning within winter wheat plants.

As an undergraduate student in agriculture at Cal State Fresno, I was introduced to plant physiology by an excellent teacher, Dr. Ralph McCoy. If it were not for him, I likely would not have become a plant physiologist and may not have attended graduate school.

My first research effort as a graduate student at Ohio State University involved a study of phosphate absorption by bean plants as controlled by *p*H. However, I found myself surrounded by graduate students of Dr. C. A. Swanson, all of whom were working on various aspects of carbohydrate translocation. They were working with such diverse plants as cucumber, bean, sugar beet, apple, ash, and grape. Among them was D. R. Geiger (now of University of Dayton) who demonstrated the error of earlier workers who had concluded that hexoses are translocated through the phloem of sugar beets. Dr. Geiger has continued to contribute to our understanding of phloem function. He worked out methodology to study the amounts of carbohydrate being translocated from individual leaves. He has also contributed to our understanding of phloem

loading. He has educated many excellent graduate students, among them Dr. Robert Giaquinta (now of Dupont) who has markedly enhanced our understanding of phloem loading.

Another graduate student colleague was Terry Weidner (now of Eastern Illinois University), who demonstrated that stachyose is the major carbohydrate translocated by cucumber. About the same time, Gorham and Webb, working in Canada, demonstrated the same for squash plants. Under the·guidance of Dr. Swanson, I decided to study the labeling pattern of the hexose moieties of translocated stachyose (see Fig. 7-17 for structure) when $^{14}CO_2$ is supplied to leaf blades and incorporated photosynthetically. I selected squash over cucumber, because it had longer petioles, so I could study translocation over a longer distance without involving the complex vasculature of a node. Since it has been demonstrated that translocated sucrose becomes equally labeled in each of its hexose moieties (glucose and fructose) when $^{14}CO_2$ is supplied to leaf blades, I hypothesized that the glucose, fructose and galactose moieties of stachyose would become equally labeled in squash plants supplied with $^{14}CO_2$. However, to my surprise, that did not happen. After 15 to 30 min of photosynthetic labeling with $^{14}CO_2$, the glucose and fructose moieties of translocated stachyose were equally labeled, but the galactose moieties contained about three times the radioactivity of the glucose and fructose moieties. Only after 70 min of labeling did the amount of ^{14}C in all hexose moieties of stachyose become the same.

Starting at about the same time that I was doing that work with squash, Otto Kandler and co-workers in Germany demonstrated that stachyose synthesis in bean seeds involves the two step addition of galactose units to preformed sucrose, the first addition yielding raffinose (Fig.

---

port with a pressure-flow mechanism. Furthermore, pulse-labeling experiments with $^{11}CO_2$ clearly show a distinct peak of radioactive carbon from one source moving in only one direction along the path of transport (e.g., Troughton et al., 1974; see also Christy and Fisher, 1978, work with ^{14}C).

Pressure Flow: A Summary Let us return to the laboratory model of the Münch system (Fig. 7-5). It is easy to make it work and to list the requisites for suitable function: (1) an osmotic gradient between the two osmometers, (2) membranes that allow the establishment of a pressure gradient in response to the osmotic gradient, (3) a channel between the two osmometers that allows flow, and (4) a surrounding medium with a higher water potential than that in the osmometer with the most negative osmotic po-

tential. If these requisites are met in the plant, the system must function as it does in the model. Clearly, the osmotic system (the symplast) with surrounding membranes exists in the plant, and pressures are observed in the transport system, although they must be more widely and accurately measured to see if the gradient is sufficient to drive flow over long distances, as in trees. The surrounding medium with high water potential is the hydrated apoplast. The sieve plates have offered the biggest problem all along. Will they allow rapid enough flow to account for observed rates of translocation? Or do they provide too much resistance?

Recently, several workers have calculated the resistance based on assumptions that the pores are open (as the best evidence now suggests) or that they are partially occluded. Comparing the calculated re-

7-17) followed by a second galactose addition yielding stachyose. Galactinol supplies the galactose units. This is one of the few examples of a monomeric unit being added to a growing polymer without the direct involvement of nucleotide or phosphate intermediate. The same system appears to operate in squash leaves. When $^{14}CO_2$ was supplied to those leaves, the galactose moiety of galactinol became labeled more rapidly than the hexose moieties of sucrose, thereby yielding the rapid labeling of the galactose moieties of stachyose.

I have continued my studies of carbohydrate translocation in squash plants at Colorado State University. During these studies of stachyose translocation, ^{14}C-sucrose was consistently isolated from the squash petioles (~15% of the total ^{14}C). It was not known if that ^{14}C-sucrose had been translocated as sucrose, or if it had been derived from translocated ^{14}C-stachyose after that stachyose had been unloaded from sieve tubes. To test the possibility that sucrose could be loaded into sieve tubes in squash leaves, ^{14}C-sucrose was supplied to leaf blades, and the petioles were extracted 30 to 60 min after labeling started. Most of the petiole ^{14}C was in sucrose. This indicated that squash leaf phloem could load sucrose as well as stachyose into sieve tubes. This is especially interesting, because Geiger and co-workers have demonstrated that sugar beet leaf phloem can load no common sugar other than sucrose.

Later, when ^{14}C-sucrose was supplied to squash leaves, most of the translocated ^{14}C appeared in sucrose, not stachyose. However, we found that ^{14}C-stachyose also occurred in petioles of squash leaves supplied with ^{14}C-sucrose. As the experimental time increased, the proportion of the ^{14}C in stachyose increased. I assumed that most of the ^{14}C of stachyose would be in the hexose moieties supplied by sucrose (glucose and fructose). Again my hypothesis was in error. Upon hydrolysis of the stachyose, we found that the hexose moieties of stachyose had the same labeling pattern as when $^{14}CO_2$ was supplied, i.e., the greater amount of ^{14}C in the galactose moieties. These results indicate that the portion of the supplied ^{14}C-sucrose not directly loaded into the sieve tubes enters mesophyll cells. In the mesophyll cells, that sucrose is hydrolyzed and metabolized just as hexoses derived from photosynthesis.

Since it is clear that squash leaves can load at least two sugars into the phloem, work is now in progress to determine if the proportion of the two sugars translocated in the day and night are the same. Another area of interest, which I would like to investigate, involves the possible competition between sucrose and stachyose in phloem loading. Do both sugars use the same loading site or do they each have their own specific site? Still another area in which I have done a little work involves a study of the possible phylogenetic significance of stachyose translocation in the Cucurbitaceae. This is a large family with many genera, tribes, and two subfamilies. Most if not all of the members of one subfamily are succulents. It would be interesting to discover if these succulent Cucurbitaceae translocate stachyose.

Understanding the specificity of the phloem loading system is useful in the development of agents that might be used to influence plant growth or control disease. As an example, if one could design a molecule that would readily control root diseases and is readily translocated through the phloem, the control of such diseases would be easier and require less material than is the case when toxic material is applied to the soil. The approach of applying such an agent to the leaves might also have the benefit of not subjecting beneficial soil organisms to as much of the toxic agent as is common today.

---

sistances with measured or calculated pressure gradients suggests that resistance is not too high; known pressure gradients are great enough to produce flow. J. B. Passioura and A. E. Ashford (1974) showed that rates of translocation (specific mass transfer) ten times greater than the highest previously recorded in the literature could be induced by restriction of the sieve-tube cross section available for transport. This suggests that phloem transport rates are not limited by sieve-plate resistance.

7.4 Partitioning and Control Mechanisms

What controls the amounts and directions of phloem transport? It has long been known, for example, that lower leaves transport relatively more to roots than to young leaves, fruits, or seeds, and that the flag leaves of grasses (e.g., on wheat) and other upper leaves transport preferentially upward to young stems or to developing fruits and seeds. Why? What is in control?

These questions are of high current interest, because agricultural yields depend on the amount of assimilate that is transported to the harvested organ compared with the amount transported to other organs. As a matter of fact, agricultural yields of many species have been improved during the past few decades as the **harvest index** (the ratio of the harvest yield to shoot yield) has been increased, mostly by breeding. This is true for oats, barley, wheat, and peanuts, for example (reviewed by Gifford and Evans, 1981). Attempts to increase photosynthetic efficiency of the leaves have so far met with little if any success, but breeding efforts to increase the partition of assimilates to storage organs of interest to the farmer have been quite successful. The promise of understanding more about what controls partitioning is that yields might be increased further. Most of what is understood so far fits nicely into the context of a pressure-flow mechanism. Consider two current lines of active research and then examine a few complexities.

Photosynthesis and Sink Demand Although control mechanisms could be diverse and much remains to be learned, several cases are now known in which the rate of photosynthesis of the leaves is strongly influenced by the demands of the sink (reviewed by Gifford and Evans, 1981). For example, if potato tubers are removed during their development, photosynthesis of the leaves drops markedly. Short-term responses could be caused by effects on stomatal aperture, but this explanation does not apply to the more lasting effects that are often observed. There are cases in which senescent (aging) leaves can be rejuvenated to full photosynthetic performance when the sink/source ratio is increased substantially. On the other hand, rapid growth of a sink can sometimes compete with leaves for remobilizable nitrogen, leading to senescence of the leaf and a drop in its photosynthetic capacity.

It is not obvious how sink demand might regulate photosynthesis of the leaves. The simplest explanation would be that assimilates pile up in the leaves, causing a product inhibition of photosynthetic reactions. But attempts to test this hypothesis have usually led to negative results. Furthermore, as we saw in our discussion of phloem loading, assimilates are apparently secreted into the apoplast and then actively absorbed by companion cells. Thus, it is not apparent that assimilates would accumulate in the vicinity of the chloroplasts (i.e., within the mesophyll cells). For further description of relations between photosynthesis and sink demands, see Section 11.4.

Phloem Unloading We have already seen that active unloading probably occurs from the sieve tubes in the sink regions. According to the pressure-flow model, this would strongly direct the flow of other assimilates toward those regions. That is, if sucrose or other solutes were pumped out of the sieve tubes into a developing tuber or fruit, the osmotic potential in those sieve cells would become much less negative, and pressure transmitted from the source areas would raise the water potential even further so water would diffuse out into the apoplast (see Fig. 7-5). This would decrease the pressure in the sink end of the sieve tubes, thereby increasing the pressure gradient between source and sink and leading to further

flow toward the sink region. Of course, sucrose or other solutes might then be actively loaded into storage cells at the sink. Clearly, the unloading process could be critical as a control mechanism in carbon partitioning, so it is quite likely that efforts in the near future will be directed toward a broader understanding of the process.

Some Complexities of Controls Certain bits of evidence suggest additional control mechanisms. For example, light might be involved in indirect ways in controlling translocation and partitioning at several steps. Light has been noted to foster active transport of solutes across membranes of living systems, partly by forming the ATP needed for active transport. The mechanism is unknown. There is also evidence that light might stimulate uptake of sugars by sink tissues such as apical meristems (Thaine et al., 1959) or cotton hypocotyls (Hampson et al., 1978). Pressure in sieve elements might also influence loading (Smith and Milburn, 1980).

There is also considerable evidence that growth regulators (discussed in Chapters 16 and 17) help direct translocation (e.g., Patrick, 1979). In most cases, the growth regulators are released from the sinks, although hormones from sources have also been considered possible. Application of cytokinins to a leaf, for example, sometimes causes that leaf— specifically the point of application—to become a sink. Similar observations have been reported for IAA, ethylene, gibberellic acid, and abscisic acid. Combinations of growth regulators can have additive, synergistic, or inhibitory effects (Gifford and Evans, 1981). In most cases, however, the systems that were studied involved very low translocation rates and may not reflect the most important controls for intact plants. There is much to learn about growth-regulator effects on translocation and especially on partitioning.

We are learning a great deal about how fruit and vegetable composition is controlled by phloem transport, xylem interchange with phloem, and import of xylem sap (Pate, 1980). Since many developing seeds, fruits, or other food storage organs transpire at a low rate, if at all, they essentially subsist on a diet made up of phloem sap. A developing potato tuber, for example, must not transpire and may even absorb water directly from the soil, but an ear of wheat or a peapod does transpire, thereby gaining some of its nutrients from the xylem stream. A model for fruit growth based on extensive measurements by J. S. Pate and his co-workers at the University of Western Australia has been worked out for the white lupine (*Lupinus albus*; Pate et al., 1977). Phloem sap supplies about 98 percent of the carbon, 89 percent of the nitrogen, and 40 percent of the water entering the fruit from the parent plant. The remaining nitrogen and water are supplied by xylem sap, and the remaining carbon comes both from xylem sap (as part of organic nitrogen compounds) and from photosynthesis in developing pods. Asparagine and glutamine carry 75 to 85 percent of the nitrogen of xylem and phloem saps, and sucrose carries 90 percent of the carbon of phloem. The nitrogen from the amides (asparagine and glutamine) is used to synthesize a wide variety of seed proteins, as shown by studies with ^{15}N. And much of the nitrogen from the two amides passes through the amino acid arginine, a compound that is almost absent from phloem sap in this plant.

The phloem sap changes significantly in its composition while underway from leaves to sinks such as developing fruits or tubers. In white lupine, Pate and his co-workers (1979) found that the phloem sap entering the developing fruits is more dilute in sucrose and much richer in certain amino acids than is the sap exported from leaves. Apparently, as the sap passes through the stem, sucrose is lost by transfer to adjacent tissue (unloaded somehow), and amino acids are loaded into the phloem. The amino acids must come from pools stored in the stems but originally obtained from xylem sap. These brief examples suggest the vast possibilities of accomplished and future research to determine how composition of fruits, seeds, and other storage organs is determined by partitioning of solutes in phloem and xylem saps. Hopefully, such research will lead to increased agricultural yields as well as to a better understanding of whole-plant physiology.

two

Plant Biochemistry

8

Enzymes, Proteins, and Amino Acids

Living cells are energy-dependent chemical factories that must follow chemical laws. The chemical reactions that make life possible are collectively referred to as **metabolism.** There are thousands of such reactions occurring constantly in each cell, so metabolism is an impressive process. Plant cells are especially impressive as to the kinds of compounds they can synthesize. Hundreds of compounds must be formed simply to produce the organelles and other structures present in living organisms. Plants also produce a whole host of complex substances called secondary metabolites that probably protect them against insects, bacteria, fungi, and other pathogens. Furthermore, they produce vitamins necessary for themselves and, incidentally, for humans, and hormones, by which cells in different parts of the plant communicate to control and coordinate developmental processes.

Some reactions form large molecules such as starch, cellulose, proteins, fats, and nucleic acids. We call formation of large molecules from small molecules **anabolism** (from Greek *ana,* upward). Anabolism requires an input of energy. **Catabolism** (from Greek *kata,* downward) is the breakdown of large molecules to small molecules, and this process often releases energy. Respiration is the major catabolic process that releases energy in all cells; this involves oxidative breakdown of sugars to CO_2 and H_2O.

Both anabolism and catabolism consist of **metabolic pathways,** in which one compound, A, is converted to another, B; then B to C, C to D and so on until a final product is formed. In respiration, glucose represents A, and CO_2 and H_2O are the final products of a metabolic pathway involving some 50 distinct reactions. Other metabolic pathways generally have fewer reactions.

How can a cell control whether most of its metabolic pathways will be anabolic or catabolic, and which specific ones will operate? For example, all cells contain glucose, which can undergo respiration,

or can be converted by anabolism into starch in plastids or into cellulose of cell walls. For a cell to function and develop properly, metabolic pathways must be carefully controlled. *The first control is related to energy.* Only certain reactions in metabolic pathways can occur without energy input, whereas other potential reactions would require so much added energy that they do not occur and are considered essentially impossible. Thus plants can form the cellulose and starch they require from glucose (with the help of energy from the sun), but they cannot convert that glucose into light energy that would allow them to photosynthesize.

The science of thermodynamics helps us understand the control of metabolism through energy input or output (Chapter 1). However, thermodynamic laws say nothing about how fast possible chemical reactions can occur. For example, they state that when H_2 and O_2 combine to form H_2O much energy will be released, but they cannot predict how rapidly that reaction will occur. In fact, if an inorganic catalyst such as finely ground platinum is present, H_2 and O_2 react so fast that an explosion occurs! Therefore, *the second type of control in metabolic pathways has to be one of rates.* A limited amount of heat release (or heat absorption) in any reaction is tolerable and common, but explosions are unacceptable.

Cells control which metabolic pathways operate and how fast by producing the proper catalysts, called **enzymes,** in the proper amounts at the times they are needed. Nearly all chemical reactions of life are too slow without catalysts, and enzymes are much more versatile and powerful catalysts than any metal ions or other inorganic substances that plants could absorb from the soil. Thus enzymes generally speed reaction rates by factors between 10^8 and 10^{20}. Compared with man-made catalysts, enzymes are frequently 10^8 to 10^9 times as effective catalytically (Schulz and Schirmer, 1979). Enzymes are also much more specific than inorganic or even synthetic organic catalysts in the kinds of reactions they catalyze, so thousands of reactions can be controlled without formation of toxic by-products. Finally, enzymes respond to environmental changes so that control is possible for plants living in a variety of climates. These advantages of enzymes are accompanied by a disadvantage: Enzymes are large protein molecules that are formed only at considerable energy cost to an organism. Let us investigate in more detail the nature of enzymes and how they function so effectively.

8.1 Enzymes in Cells

Enzymes are not uniformly mixed throughout cells. The enzymes responsible for photosynthesis are located in chloroplasts; many of those essential for aerobic respiration occur exclusively in mitochondria, while still other respiratory enzymes exist in the cytosol. Similarly, most enzymes essential to synthesis of DNA and RNA and to mitosis occur in nuclei. Enzymes that govern the steps in metabolic pathways are sometimes arranged in membranes so that a kind of assembly-line production process occurs. Thus the product of one reaction is released at a site where it can immediately be converted to a related compound by the next enzyme involved in the pathway, and so on, until the metabolic pathway is completed and a quite different compound is formed (Creasy and Hrazdina, 1982).

The above methods of compartmentalization almost surely increase the efficiency of numerous cellular processes for two reasons: First, they help ensure that the concentrations of reactants are adequate at the sites where the enzymes that act upon them are located. Second, they help ensure that a compound is directed toward the necessary product and not diverted into some other pathway by action of a competing enzyme that can also act upon it elsewhere in the cell. However, this compartmentalization is often not absolute, nor apparently should it be. For example, the membranes surrounding chloroplasts allow outward passage of certain sugar phosphates produced by photosynthesis. These compounds are then acted upon by numerous enzymes outside the plastids that are involved in cell-wall synthesis and respiration essential to growth and maintenance of the plant.

8.2 Properties of Enzymes

Specificity and Nomenclature One of the most important properties of enzymes is their specificity. Each enzyme acts on a single **substrate** (reactant) or a small group of closely related substrates that have virtually identical functional groups capable of undergoing reaction. With some enzymes, specificity appears to be absolute, but with others there is a gradation in their abilities to convert related compounds to products. As explained later, specificity results from combinations between the enzyme and substrate in a lock-and-key process.

More than 4500 different enzymes have been discovered in living organisms, and the number grows as research continues. Each enzyme has been named according to a standardized system and has also been given a simpler common or trivial name. In both systems, the name commonly ends in the suffix **-ase** and characterizes the substrate or substrates acted upon and the type of reaction catalyzed. For example, **cytochrome oxidase,** an important respiratory enzyme, oxidizes (removes an electron from) a cytochrome molecule. **Malic acid dehydrogenase** removes two hydrogen atoms from (dehydrogenates) malic acid. These common names, although conveniently short, do not give sufficient information about the reaction catalyzed. For example, neither tells the acceptor of the removed electron or hydrogen atoms.

The International Union of Biochemistry lists longer but more descriptive and standardized names for all well-characterized enzymes. As an example, cytochrome oxidase is named cytochrome $c:O_2$ oxidoreductase, indicating that the particular cytochrome from which electrons are removed is the c type and that oxygen molecules are the electron acceptors. Malic acid dehydrogenase is called L-malate:NAD oxidoreductase, indicating that the enzyme is specific for the ionized L form of malic acid (malate) and that a molecule abbreviated as NAD is the hydrogen atom acceptor. Table 8-1 lists six major classes of enzymes based on the types of reactions they catalyze, with a few examples.

Reversibility Enzymes increase the rate at which chemical equilibrium is established among products and reactants. At equilibrium, the terms reactants and products are arbitrary and depend upon our point of view. Under normal physiological conditions, an enzyme has no influence on the relative quantities of products and reactants that would eventually be reached in its absence. Thus if the equilibrium state is unfavorable for formation of a compound, an enzyme cannot change this.

The equilibrium constant depends upon the chemical potentials (activities) of all compounds involved in the reaction (Equation 3, Chapter 1). If the chemical potential of reactants is very high compared with that of the products, the reaction might proceed only toward product formation, because of the chemical law of mass action. Most **decarboxylations,** in which carbon dioxide is split out of a molecule, are

Table 8-1 The Major Enzyme Classes and Subclasses.

Class and Subclass	General Reaction Type

Oxidoreductases
  Oxidases
  Reductases
  Dehydrogenases
	Remove and add electrons or electrons and hydrogen. Oxidases transfer electrons or hydrogen only to O_2.

Transferases Kinases	Transfer chemical groups Transfer phosphate groups, especially from ATP
Hydrolases	Break chemical bonds (e.g., amides, esters, glycosides) by adding the elements of water
Proteinases	Hydrolyze proteins (peptide bonds)
Ribonucleases	Hydrolyze RNA (phosphate esters)
Deoxyribonucleases	Hydrolyze DNA (phosphate esters)
Lipases	Hydrolyze fats (esters)
Lyases	Form double bonds by elimination of a chemical group
Isomerases	Rearrange atoms of a molecule to form a structural isomer
Ligases or Synthetases	Join two molecules coupled with hydrolysis of ATP or other nucleoside triphosphate
Polymerases	Link subunits (monomers) into a polymer such as RNA or DNA

Source: Modified from Wolfe, *Biology of the Cell*, 2nd ed., 1981, p. 45.

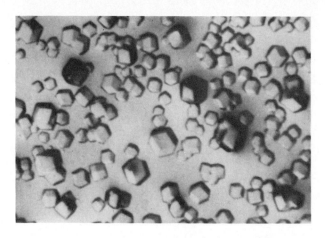

Figure 8-1 Crystals of pure ribulose bisphosphate carboxylase. Note the dodecahedron shape; the largest crystals are more than 0.2 mm in diameter. (Courtesy S. G. Wildman.)

examples of such reactions, because the CO_2 can escape; its concentration and, hence, its chemical potential remain low. **Hydrolytic reactions,** involving cleavage of bonds between two atoms and addition of the elements of H_2O to those atoms, are also irreversible. For example, hydrolysis of starch to glucose by amylases, hydrolysis of phosphate from various molecules by phosphatases, and hydrolysis of proteins to amino acids are essentially irreversible processes. Other enzymes, using different substrates with higher energy levels, carry out synthesis of starch and proteins and the addition of phosphate to various molecules. In fact, large molecules such as fats, proteins, starch, nucleic acids, and even certain sugars are synthesized by one series of enzymes and degraded by another. Synthetic and degradative enzymes are often kept separate from each other by membranes or are formed at different times, so that competition between degradation and synthesis is minimized.

Chemical Composition Every known enzyme has a protein as a major part of its structure, and many contain nothing other than protein. However, some proteins appear to have no catalytic function and are not classified as enzymes. For example, proteins of microtubules and microfilaments and some of the proteins in membranes seem to perform a structural rather than a catalytic function. Other proteins, such as cytochromes that transport electrons during photosynthesis and respiration, are not enzymes, but electron carriers. Furthermore, several storage proteins in seeds also have no known enzymatic functions. The major role of seed storage proteins is to act as a reservoir of amino acids for the seedling after germination, not as enzymes.

Proteins consist of one or more chains (**polypeptide chains**) of tens to hundreds of amino acids. The composition and size of each protein depend upon the kind and number of its amino acid subunits. Commonly, 18 to 20 different kinds of amino acids are present, with most proteins having a full complement of 20. The total number of amino acid subunits varies greatly in different proteins, and so protein molecular weights also vary. Most plant proteins so far characterized have molecular weights of at least 40,000 grams mole^{-1}, yet that of **ferredoxin,** a protein involved in photosynthesis, is only about 11,500 and that of **ribulose bisphosphate carboxylase,** another photosynthetic enzyme, is over 500,000. The latter is composed of eight small, identical polypeptide chains and eight larger, identical polypeptide chains. The chains of such complex enzymes are frequently held together by noncovalent bonds, often ionic and hydrogen bonds, and can be separated *in vitro*. Nevertheless, if care is taken to prevent chain separation during extraction, even complex enzymes such as ribulose bisphosphate carboxylase can be isolated as homogeneous crystals (Fig. 8-1).

Amino acids can be represented by the general formula:

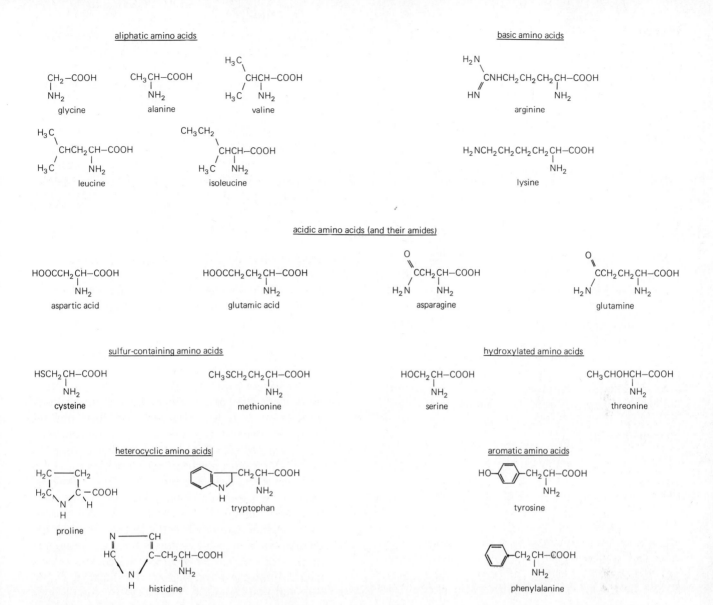

Figure 8-2 Molecular structures of 20 amino acids present in most proteins.

$$R-\underset{\underset{NH_2}{|}}{\overset{\overset{H}{|}}{C}}-\overset{\overset{O}{\parallel}}{C}\diagdown_{OH} \qquad \text{or} \quad RCHNH_2COOH$$

The -NH_2 is the **amino group** and the -COOH is the **carboxyl group.** These two groups are common to all amino acids, with slight modification of the amino group in proline. R denotes the remainder of the molecule, which is different for each amino acid. Figure 8-2 shows the structures of the 20 amino acids commonly found in proteins. The R groups cause amino acids to differ greatly in physical properties, such as water solubility. The aliphatic types in the upper left of the figure and the aromatic types at the

lower right are much less soluble in water (more hydrophobic or water-fearing) than the more hydrophilic (water-loving) basic, acidic, and hydroxylated types.

Structures of two amides, glutamine and asparagine, that occur in most proteins are included in Fig. 8-2. (Technically, amides are amino acids, because they have the general amino acid structure given above. They are amides because the R portion of the amino acid has an amino group connected to a carbonyl carbon.) These amides are formed from the two amino acids that have an additional carboxyl group as part of R, glutamic acid and aspartic acid. The amides are structural parts of most proteins. They also represent especially important forms in which nitrogen is transported from one part of the

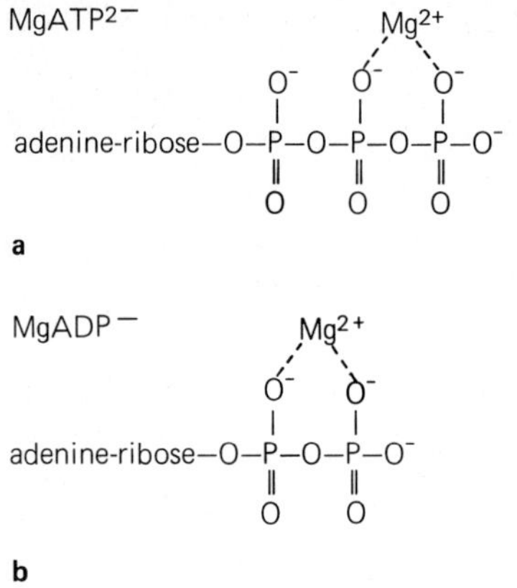

MgATP^{2-} Mg^{2+}

$$\text{adenine-ribose} - O - \overset{\overset{O^-}{|}}{\underset{\underset{O}{\|}}{P}} - O - \overset{\overset{O^-}{|}}{\underset{\underset{O}{\|}}{P}} - O - \overset{\overset{O^-}{|}}{\underset{\underset{O}{\|}}{P}} - O^-$$

a

MgADP$^-$ Mg^{2+}

$$\text{adenine-ribose} - O - \overset{\overset{O^-}{|}}{\underset{\underset{O}{\|}}{P}} - O - \overset{\overset{O^-}{|}}{\underset{\underset{O}{\|}}{P}} - O^-$$

b

Figure 8-3 The Mg^{2+} chelate of (**a**) ATP and (**b**) ADP.

plant to another and in which surplus nitrogen can be stored (Sections 7.3 and 13.4).

The union of amino acids and amides into polypeptide chains of proteins occurs by **peptide bonds** involving the carboxyl group of one amino acid and the amino group of the next, as summarized in an oversimplified form in R8-1. The vertical arrow indicates the peptide bond:

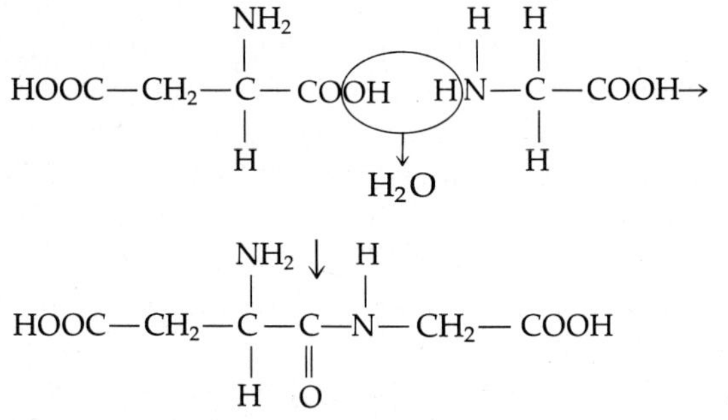

$$\text{HOOC} - \text{CH}_2 - \overset{\overset{NH_2}{|}}{\underset{\underset{H}{|}}{C}} - \text{COOH} \quad \overset{\overset{H}{|}}{H}N - \overset{\overset{H}{|}}{\underset{\underset{H}{|}}{C}} - \text{COOH} \rightarrow$$
$$H_2O$$

$$\text{HOOC} - \text{CH}_2 - \overset{\overset{NH_2}{|}}{\underset{\underset{H}{|}}{C}} - \overset{\overset{H}{|}}{\underset{\underset{O}{\|}}{C}} - N - \text{CH}_2 - \text{COOH}$$

(R8-1)

When aspartic and glutamic acids, each of which has two carboxyl groups, form peptide bonds with other amino acids, only the carboxyl group adjacent to the amino group participates. The other carboxyl group remains free and gives acidic properties to the protein. When lysine and arginine, each of which has two amino groups, form peptide bonds, the amino group farthest from the carboxyl group is always free. The nitrogen atom of each of these groups possesses two electrons that can be shared by H$^+$ in the cells; as a result, H$^+$ are attracted to these basic nitrogen atoms, causing them to become positively charged.

Proteins rich in aspartic and glutamic acids usually have net negative charges in cells, because these amino acids lose a H$^+$ ion during dissociation of the carboxyl group not involved in a peptide bond. Alternatively, those rich in lysine and arginine usually have net positive charges. These charges are important, because whether an enzyme is catalytically ac-

tive or whether it is bound to another cellular component frequently depends upon whether one of its free amino or carboxyl groups is charged or uncharged. For example, chromosomes contain five major kinds of positively charged proteins rich in lysine or arginine called *histones*. These histones are held to negatively charged DNA by ionic bonds, which are important in controlling the structure and genetic activity of the chromosomes. The different net charges on various enzymes allow us to separate them by their chemical and physical properties. Their functions and properties can then be studied without interference by other enzymes.

Prosthetic Groups, Coenzymes, and Vitamins In addition to the protein parts of enzymes, some contain a much smaller, organic nonprotein portion called a **prosthetic group.** Prosthetic groups are usually attached tightly to the protein by covalent bonds and are essential to catalytic activity. An example is found among some of the dehydrogenase enzymes involved in respiration and fatty acid degradation. Here a yellow pigment called a *flavin* is attached to the protein. The flavin is essential to enzyme activity because of its ability to accept and then transfer hydrogen atoms during the course of the catalyzed reaction. Some enzymes contain prosthetic groups to which is attached a metal ion (e.g., iron and copper in cytochrome oxidase, Chapter 12). Other proteins, *glycoproteins* (from the Greek *glykys*, or sweet), contain a group of sugars attached to their protein parts. Such attached carbohydrates can contribute to enzymatic action or to protection of the enzyme against temperature extremes, internal destructive agents such as proteases, and perhaps also against pathogens and herbivores.

Many enzymes that do not have prosthetic groups require, for activity, the participation of another organic compound or a metal ion, or both. These substances are usually called **coenzymes,** although the metal ions are often called **metal activators.** Coenzymes and metal activators are generally not tightly held to the enzymes, but often no sharp distinction between coenzymes and prosthetic groups can be made. *Several vitamins synthesized by plants form parts of coenzymes or prosthetic groups required by enzymes in plants and animals, and this explains their essentiality for life. Several mineral elements also act as enzyme activators.*

The magnesium ion acts as a metal activator for most enzymes that use ATP or other nucleoside di- or tri-phosphate as a substrate. A stable chelate between ATP and Mg^{2+} is formed, probably having the structure shown in Fig. 8-3a. The enzyme-substrate complex is then a Mg-ATP-enzyme complex. Mg^{2+} also combines with ADP, as shown in Fig. 8-3b. Furthermore, Mn^{2+} can combine with ADP or ATP in a similar way, forming a chelate that is often as active as that formed with Mg^{2+}. This combination of cations with a substrate rather than the enzyme may prove important for other divalent cations, but direct combination of certain enzymes with manganese, iron, zinc, copper, calcium, and potassium also occurs, as noted with iron and copper for cytochrome oxidase.

Amino Acid Sequences The number of different ways in which amino acids could theoretically be arranged in proteins is a staggering figure. Consider as a simple example a small enzyme with a molecular weight of 12,500, consisting of 100 amino acids with an average molecular weight of 125. With the usual 20 *different* amino acids, the number of possible arrangements would be nearly 20^{100} (10^{130}). The number of different proteins of all sizes and kinds in nature does not even begin to approach this figure,* although estimates of the existence of as many as 10,000 unique proteins in plants have been made. Considering this, how all the various forms of life can be so different from each other is no longer puzzling. We know that the arrangements of nucleotides in genes that code for these proteins determine their amino acid sequences.

Methods used to determine the amino acid sequence in proteins are tedious and require the availability of a pure protein. As a result, we presently know the complete sequences for relatively few proteins (several dozen). Such studies are important, however, because the results are necessary for us to learn how enzymes catalyze reactions. Furthermore, the comparison of sequences in proteins having the same function in different organisms provides a powerful tool in evolutionary studies. For example, ferredoxins from angiosperms, a green alga, and certain photosynthetic and nonphotosynthetic bacteria have been sequenced. The similarities and differences have been used, with the help of computers, to construct tentative family trees (Cammack et al., 1971). The same kinds of comparisons have been made with cytochrome *c* molecules from many organisms, including primitive and advanced plants and animals. Hemoglobin sequencing has already provided several clues to animal evolution. Comparison of one of the histone molecules (histone IV, also called F2A1) found in nuclei of both calves and peas showed that the same amino acids occur in 100 of the 102 amino acid positions present in each. These results suggest that many mutations in genes that control this protein have been eliminated by natural selection during the last 1.5 billion years or so; there-

fore, this protein and each amino acid in it play an important role in the lives of various organisms (Spiker, 1975).

Three-Dimensional Structures of Enzymes and Other Proteins The simplest proteins consist of only one long polypeptide chain, but results from numerous techniques such as X-ray diffraction, viscosity measurements, and centrifugation studies show that such chains are usually coiled and twisted to form somewhat spherical or globular molecules (Schulz and Schirmer, 1979). An example of a rather simple globular protein with 124 amino acid residues in only one chain is the enzyme *ribonuclease* (which hydrolyzes RNA), shown in Fig. 8-4. *The three-dimensional structure of a polypeptide chain or a protein with several polypeptide chains is determined by the kinds of amino acids present and the sequence in which they are arranged.* Each chain and the whole protein then spontaneously attain the lowest free energy configuration consistent with its amino acid composition and sequence at the existing cellular conditions of *pH*, temperature, ionic strength, and so forth. For each chain, this configuration begins to be attained as it is synthesized on ribosomes; for complex proteins the final configuration is stabilized only after all of the chains in it come together.

In the cytosol, the more hydrophobic amino acids in a protein (such as valine, leucine, isoleucine, methionine, and often tyrosine) become concentrated on the inside of the structure where they are shielded from water, while the more hydrophilic amino acids (such as serine, glutamic acid, glutamine, aspartic acid, asparagine, lysine, histidine, and arginine) are more commonly exposed to the surface where they are in contact with water of the cytosol. The folding or coiling of a polypeptide chain so that some portions of it are in contact with other portions takes place in part because of stabilizing attractive forces between certain *R* groups that are located some distance apart in the chain. Another factor that affects folding is the repulsion of hydrophobic *R* groups by water, which brings these hydrophobic groups close together. Some important kinds of intrachain bonding that result from these forces are shown in Fig. 8-5.

In complex proteins with more than one polypeptide chain, such as ribulose bisphosphate carboxylase, bonds that hold the chains together are probably similar to those that hold separate parts of the same chain together in its three-dimensional structure (Fig. 8-5), although **disulfide bonds** (S-S) seem to be more rare. In some, the chains are identical, in which case **homopolymers** are made; but usually they are different, and **heteropolymers** result. Each polypeptide chain in a heteropolymer is coded for by a different gene. Thus, heteropolymers with two or more different polypeptide chains require the

---

*If each of the 10^{130} possible kinds of proteins with 100 amino acids occupied 10^3 cubic nanometers, all 10^{130} taken together would fill our known universe 10^{27} times!

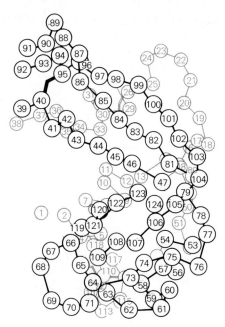

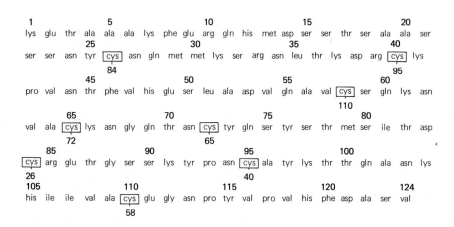

1				5					10					15					20
lys	glu	thr	ala	ala	ala	lys	phe	glu	arg	gln	his	met	asp	ser	ser	thr	ser	ala	ala	ser

				25					30					35					40
ser	ser	asn	tyr	[cys]	asn	gln	met	met	lys	ser	arg	asn	leu	thr	lys	asp	arg	[cys]	lys
				84														95	

				45					50					55					60
pro	val	asn	thr	phe	val	his	glu	ser	leu	ala	asp	val	gln	ala	val	[cys]	ser	gln	lys	asn
																110			

				65					70					75					80
val	ala	[cys]	lys	asn	gly	gln	thr	asn	[cys]	tyr	gln	ser	tyr	ser	thr	met	ser	ile	thr	asp
		72						65											

				85					90					95					100
[cys]	arg	glu	thr	gly	ser	ser	lys	tyr	pro	asn	[cys]	ala	tyr	lys	thr	thr	gln	ala	asn	lys
26									40										

105				110					115					120					124
his	ile	ile	val	ala	[cys]	glu	gly	asn	pro	tyr	val	pro	val	his	phe	asp	ala	ser	val
				58															

Figure 8-4 Three-dimensional configuration of ribonuclease A. The active site of the enzyme is in the cleft at the left center of the molecule. The positions of individual amino acid residues are marked by the numbered circles: these correspond to the sequence list above. Gln and asn represent glutamine and asparagine. Disulfide bridges are indicated by a bent heavy line. Cysteine residues joined by disulfide linkages are enclosed in boxes in the sequence list. The number indicated by the arrow at the bottom of each box shows the amino acid residue to which the disulfide linkage is joined. (From R. E. Dickerson and I. Geis, 1969, The structure and action of proteins, W. A. Benjamin, Inc., Menlo Park, Calif. Copyright © 1969 by Dickerson and Geis.)

presence of two or more different corresponding genes in the cell. In ribulose bisphosphate carboxylase, the gene coding for the smaller polypeptide chain is a nuclear gene, whereas the gene that codes for the larger polypeptide chain is in the chloroplast.

Isozymes During the 1940s a technique called **electrophoresis** was developed to separate proteins, and this technique led to an important new discovery about many enzymes. Electrophoresis is essentially the separation of dissolved proteins or other charged molecules in an electrical field. A mixture of enzymes is placed in a buffered solution or in an inert medium such as a layer of starch gel or a column or slab of polyacrylamide gel that is wetted with a buffer at a controlled pH. The various R groups in each enzyme ionize to an extent controlled by their chemical nature and the pH. For example, if the pH is 7, enzymes that are rich in aspartic and glutamic acid have a net negative charge because of their dissociated carboxyl groups. Alternatively, enzymes that are rich in lysine or arginine are more likely to be positively charged at this pH. Each enzyme in the mixture attains a different charge; and if these differences are great enough, the enzymes can be separated by an electrical current from a negative electrode inserted at one end of the solution or gel to a positive electrode at the other end.

The enzymes migrate in the electrical field, the distance depending upon their net charge and size. After they have migrated, their positions in the gel can be detected, for example, by incubating the gel with the proper substrate, and then chemically staining the reaction product.

When a particular enzyme is investigated by electrophoresis, it is often found that more than one stained zone appears in the gel, indicating the presence of more than one enzyme that can act on the same substrate and convert it to the same product. Such enzymes are referred to as **isozymes** or **isoenzymes**. Differences between isozymes frequently result from the presence of a different gene that codes for each isozyme, or, if it is a heteropolymer, for each of the polypeptide chains in it. If only one polypeptide chain is present, two isozymes could result from genetic coding by each of the two allelic genes derived from a different parent. Other genetic possibilities leading to isozymes exist.

The importance to a plant of having different isozymes that are capable of catalyzing the same reaction apparently is that their isozymes differ somewhat in their responses to various environmental factors (Simon, 1979; McNaughton, 1972). Sometimes one isozyme will exist in one tissue and another in a different tissue. Different isozymes can sometimes

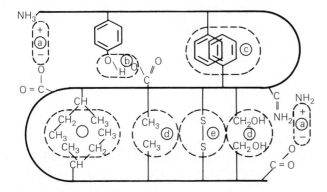

Figure 8-5 Probable types of bonding responsible for holding one polypeptide chain close to another. (**a**) Electrostatic attraction. (**b**) Hydrogen bonding. (**c**) Interaction of nonpolar side chain groups caused by repulsion of each by water. (**d**) Van der Waals attractions. (**e**) Disulfide bonding between former —SH groups of cysteine molecules. (Modified after C. B. Anfinsen, 1959.)

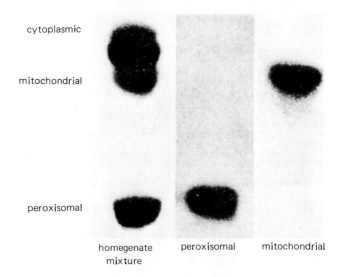

Figure 8-6 Separation of NAD$^+$-malate dehydrogenase isozymes of spinach leaves by starch gel electrophoresis. In the homogenate mixture at left, three isozymes are detectable. Two of these correspond to isozymes extracted from isolated peroxisomes and mitochondria; the third resides in the cytosol. (From I. P. Ting, I. Führ, R. Curry, and W. C. Zschoche, 1975, in Markert, 1975, pp. 369–383.)

even be found within the same cell. Figure 8-6 illustrates the separation of three isozymes of malate dehydrogenase from various organelles in spinach leaf cells, one isozyme from mitochondria, one from peroxisomes, and one from the cytosol. Each isozyme is exposed to a different chemical environment within the cell, and each participates in a different sequence of reactions (metabolic pathway). The environmental factors that influence growth and development of cells that contain some isozymic types can influence other cells with other isozymes differently.

Even if an isozyme in one tissue has the same amino acid sequence as another isozyme in a different tissue, they might differ in charge, catalytic ability, or response to environmental changes. These differences result from small but often critical modifications in structure occurring after isozymes are synthesized (Trewavas, 1976). The differences are independent of the genes coding for the isozymes. Several kinds of modifications can occur, including **phosphorylation** (in which a phosphate is esterified to the hydroxyl group of one or more serine or threonines present), **glycosylation** (in which one or more sugars are attached to the enzyme), and **methylation** or **acetylation** (in which one or more methyl or acetyl groups are added). Such modifications can even occur in the same cell at different stages in its development. They allow another kind of control over the kind of reactions occurring in a given cell or part of a cell at a particular time in its life. As an example, the pyruvate dehydrogenase enzyme complex (a connected group of proteins that dehydrogenates and decarboxylates pyruvic acid during respiration) is made inactive by the simple reversible addition of one or a few phosphate groups (Randall and Rubin, 1977).

8.3 Mechanisms of Enzyme Action

Only the most energetic molecules are usually able to change during chemical reactions. Such molecules temporarily become more energetic than others of the same kind by being subjected to different numbers and types of collisions. If we could analyze the energies in a population of such molecules, statistical predictions indicate that a distribution similar to the hypothetical values of Fig. 1-3 would be found. The curves in that figure show that a temperature rise greatly increases the number of molecules that have relatively high energies. Note that the higher-temperature curve is more skewed to the right than the lower. If we assume that only those molecules represented by the shaded areas in the figure have enough energy to react in the absence of enzymes, the area under the higher-temperature curve is about twice as large as that under the lower-temperature curve. Therefore, twice as many molecules will react in a given period of time at the higher temperature.

But how do enzymes increase reaction rates? Do they cause a shift in the frequency distribution curve similar to that caused by a temperature increase? The answer is no, but to understand this we must consider another aspect of the problem. Figure 8-7 shows that as substrates react to form products, an energy barrier must be overcome. This barrier is called the **energy of activation**. A temperature rise, which also increases reaction rates, increases the number of

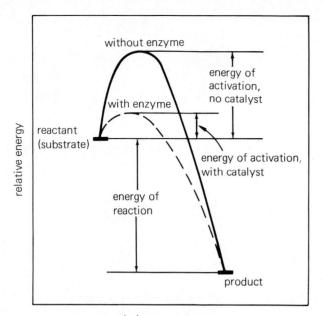

Figure 8-7 Energy diagram for a metabolic reaction occurring in the presence and absence of an enzyme. Reacting substrate molecules must pass over an "energy hump" (accumulate activation energy) to allow formulation of new chemical bonds present in the product, even though the product may be at a lower free energy level than the substrate. A catalyst, such as an enzyme, lowers the activation energy required, thus increasing the fraction of molecules that can react in a given time.

molecules that have energies equal to the average base level energy plus the activation energy. In contrast, enzymes lower the activation energy so that a much greater fraction of the substrate molecules have sufficient energy to react without a temperature increase. In other words, enzymes do not shift the curves of Fig. 1-3, but in their presence, the shaded area of the figure is displaced far to the left.

The enzyme and substrate or substrates combine temporarily during the course of the reaction to form an **enzyme-substrate complex.** The electrical charges on this complex change the shape of the substrate, causing some of the substrate bonds to break; then the bonds rearrange to form products much more rapidly than in the absence of the enzyme.

The enzyme-substrate complex first hypothesized by the great organic chemist Emil Fischer, in about 1884, assumed a rigid lock-and-key union between the two (Fig. 8-8a). The portion of the enzyme to which the substrate (or substrates) combines as it undergoes conversion to a product is called the **active site.** If the active site were rigid and specific for a given substrate, reversibility of the reaction would not occur, because the structure of the product is different from that of the substrate and would not fit well. As contrasted to a rigidly arranged active site, Daniel E. Koshland (1973) found evidence that the active site of enzymes can be induced by close approach of the substrate (or product) to undergo a change in conformation (shape) that allows a better combination between the two. This idea is now widely known as the **induced-fit hypothesis** and is illustrated in Figs. 8-8b and 8-9. Apparently, the structure of the substrate is also changed during many cases of induced fit, thus allowing a more functional enzyme-substrate complex.

The kinds of bonds between enzymes and substrates can be covalent, ionic, hydrogen, and van der Waals. The covalent and ionic bonds are most important with respect to the activation energy for a reaction, but the more numerous hydrogen bonds and van der Waals interactions contribute to the structural orientation of the enzyme-substrate complex. Even when strong covalent bonds are formed, they are usually very rapidly broken to release new product molecules. Both covalent and noncovalent bonds are

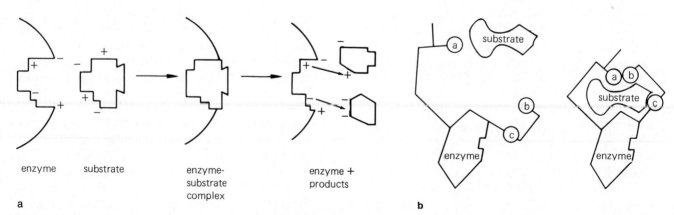

Figure 8-8 (a) The lock-and-key model of the active site, as hypothesized by Emil Fischer. The active site is considered to be a rigid arrangement of charged groups that precisely matches complementary groups of the substrate. (b) A modified conception of the active site, as advanced by D. E. Koshland. Here the catalytic groups A and B must be aligned, but the orientation of these groups is altered by the approaching substrate, resulting in a better fit. (From Wolfe, 1972.)

usually formed between parts of the *R* portions of amino acid residues, not between the atoms involved in the peptide bonds. Thus enzymes must have the proper amino acids (composition) and have them in the right places (sequence).

8.4 Denaturation

The previous discussion of enzyme substrate complexes and three-dimensional structures of proteins suggests that if the structure of an enzyme is altered so that the substrate can no longer bind with it, catalytic activity will be eliminated. Numerous factors cause such alterations and are said to cause enzyme **denaturation.** In many cases, denaturation is irreversible. High temperatures easily break hydrogen bonds and often cause irreversible denaturation. You can't unboil an egg! Extreme heating causes the formation of new covalent bonds between different polypeptide chains or between parts of the same chain, and these bonds are so stable that they break at negligible rates.

Cold temperatures are nearly always maintained during extraction and purification of enzymes to prevent heat denaturation. This is true even though enzymes normally exist undenatured in cells at higher temperatures. We do not fully understand why purification of enzymes at temperatures identical to those at which cells normally exist causes denaturation, but we suspect that extraction and purification procedures remove or dilute substances that normally protect the enzymes. Alternatively, homogenization of cells often releases and allows exposure of enzymes to denaturing substances from subcellular compartments (e.g., vacuoles) that *in vivo* are prevented by membranes from contacting such enzymes. A few enzymes are known to be inactivated by *low* temperatures during purification. Again, a change in structure is the cause.

Oxygen or other oxidizing agents also denature numerous enzymes, often by causing disulfide bridges to be formed in chains in which the SH groups of cysteine are normally present. Reducing agents can denature in the opposite manner, by breaking disulfide bridges to form two SH groups. Heavy metal cations such as Ag^+, Hg^{2+}, Hg^+, or Pb^{2+} can denature enzymes, and it is for this reason that great concern has developed about the presence of these ions in the environment. Many organic solvents also denature enzymes.

When enzymes are dry, they are much less susceptible to heat denaturation than when they are hydrated. This is mainly why dry seeds and dry fungal or bacterial spores can resist high temperatures and why the presence of steam in autoclaves, which are used for sterilization, increases the effectiveness of the treatment above that obtained in a dry oven at the

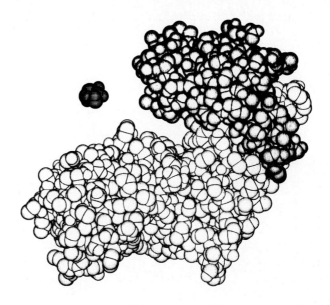

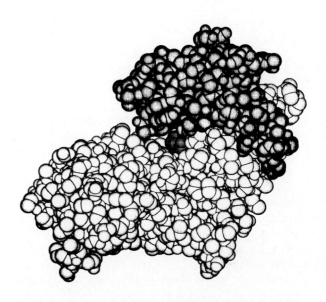

Figure 8-9 Induced fit of yeast hexokinase and glucose, one of its substrates. Without glucose (top), hexokinase has an open cleft that partially closes when glucose is bound (bottom). (From W. S. Bennett, Jr. and T. A. Steitz. 1980. *J. Molecular Biol.* 140: 211–230. With permission from the *Journal of Molecular Biology*, copyright 1980, by Academic Press Inc. (London) Limited.)

same temperature. The dry state also prevents enzyme denaturation by cold temperatures during winter in seeds, buds, and other parts of perennial shrubs and trees.

8.5 Factors Influencing Rates of Enzymatic Reactions

Enzyme and Substrate Concentrations: Either Can Be Limiting Catalysis occurs only if enzyme and substrate form a transient complex. The reaction rate is thus dependent upon the number of successful

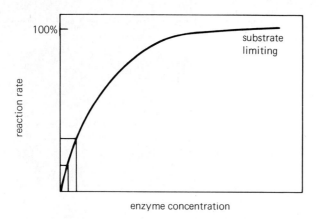

a

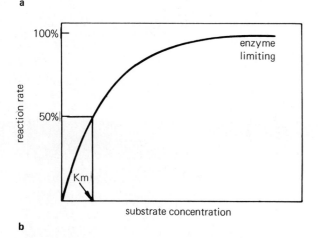

b

Figure 8-10 (**a**) Effects of enzyme concentration on rate of reaction when substrate concentration is held constant. (**b**) Effect of substrate concentration on reaction rate when enzyme concentration is held constant.

collisions between them, which in turn depends upon their concentrations. If enough substrate is present, doubling the enzyme concentration usually causes a two-fold increase in rate (Fig. 8-10a). With the addition of still more enzyme, the rate begins to become constant because substrate becomes limiting.

Figure 8-10b shows the effect of substrate concentration on the reaction rate when the enzyme concentration is held constant. There is usually an apparent direct proportionality between rate and substrate concentration until the enzyme concentration becomes limiting. At this substrate concentration, the addition of more substrate causes no further rise in the reaction rate, because nearly all enzyme molecules are then combined with substrate. When this occurs, there are no more free enzyme active sites to cause catalysis. To increase the speed of reaction then requires the addition of more enzyme.

Figure 8-10b also illustrates another useful fact about enzymes that can be obtained from such graphs. This is the substrate concentration required to cause half the maximal reaction rate, a value

named the **Michaelis-Menten constant (K_m).** K_m values are more or less constants independent of the amount of enzyme present, at least within reasonable limits. The values vary somewhat with pH, temperature, and ionic strength, and also with the kinds or amounts of coenzymes present when these are required. Most enzymes so far studied have K_m values between 10^{-3} and 10^{-7} molar, although exceptions exist. If an enzyme catalyzes a reaction between two or more different substrates, it will have a different K_m value for each.

Certain advantages result from knowing the K_m value for an enzyme of interest. First, if we can measure the concentration of the substrate in that part of the cell in which the enzyme resides, we can predict whether the cell needs more enzyme or more substrate to speed up the reaction. Some studies indicate that respiration enzymes are usually not saturated by substrates. The same is probably true with many photosynthetic enzymes in leaves exposed to bright light, because CO_2 fixation can be increased by increasing the concentration of this gas around such leaves. Second, K_m values represent approximate inverse measures of the affinity of the enzyme for a given substrate; thus, the lower the K_m the more stable is the enzyme-substrate complex. It has frequently been determined that for an enzyme that can catalyze reactions with two similar substrates (e.g., glucose and fructose), the substrate for which the enzyme has the lower K_m is the one most frequently acted upon in the cell. And third, the K_m gives an approximate measure of the concentration of the enzyme's substrate in the part of the cell in which reaction occurs. For example, enzymes that catalyze reactions with relatively concentrated substrates such as sucrose usually have relatively high K_m values for these substrates, and enzymes that react with hormones or other substrates present in very low concentrations have much lower K_m values for their substrates.

pH The pH of the medium influences enzyme activity in various ways. Usually there is an optimum pH at which enzymes function, with decreased activity at higher or lower pH values. Sometimes a plot of activity versus pH, as in Fig. 8-11, gives an almost bell-shaped curve, whereas with another enzyme the curve is almost flat. The optimum pH is often between 6 and 8, but is higher or lower for some enzymes. Extremes of the pH scale usually cause denaturation.

Apart from denaturation effects, the pH can influence reaction rates in at least two ways. First, enzyme activity often depends upon the presence of free amino or carboxyl groups. These can be either charged or uncharged, depending on the enzyme, but only one form is presumed effective in a given case. If an uncharged amino group is essential, the

pH optimum will be relatively high, whereas a neutral carboxyl group requires a low pH. Second, the pH controls the ionization of many substrates, some of which must be ionized for the reaction to proceed.

Reaction Products The rate of an enzymatic reaction can be determined by measuring the rate of disappearance of substrate or the rate of product appearance, or both. By either method the reaction is usually observed to proceed more slowly as time passes. This rate decrease is sometimes caused by denaturation of the enzyme while the reaction is being measured, but other factors are also involved. One of the most important factors is the continuous decrease in concentration of substrate or substrates and accumulation of products. As products accumulate, their concentrations sometimes become high enough to cause appreciable reversibility of the reaction, provided that the relative chemical potentials of products and reactants allow reversibility. In some cases, products can inhibit the forward reaction by combining with the enzyme in such a way that further formation of the enzyme-substrate complex is inhibited.

Inhibitors Many "foreign" substances can block the catalytic effects of enzymes. Some are inorganic, such as several metal cations, and some are organic. Both types are usually classified according to whether their effect is competitive or noncompetitive with the substrate. **Competitive inhibitors** usually have structures sufficiently similar to the substrate that they are able to compete for the active site of the enzyme. When such a combination of enzyme and inhibitor is formed, the concentration of effective enzyme molecules is lowered, decreasing the reaction rate. The inhibitor itself sometimes undergoes a change caused by the enzyme, but that change is not essential for inhibition. Addition of more of the natural substrate overcomes the effect of a competitive inhibitor. A classic example of competitive inhibition is caused by *malonate* (^-OOC—CH_2—COO^-), the doubly-charged anion of *malonic acid*, on the action of *succinate dehydrogenase*. This enzyme functions in mitochondria to carry out an essential reaction of the Krebs cycle (Chapter 12). It removes two H atoms from succinate and adds these to its covalently-bound prosthetic group *flavin adenine dinucleotide, (FAD)*, forming fumarate and enzyme-bound $FADH_2$:

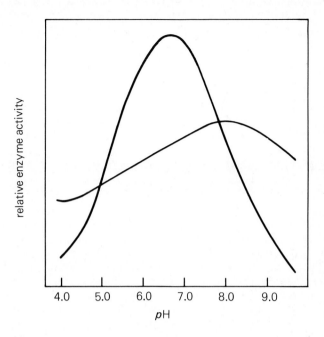

Figure 8-11 Influence of pH upon activity of two different enzymes. The pH optimum and shape of the curve vary greatly among enzymes and depend upon reaction conditions.

Malonate combines reversibly with the enzyme in place of succinate, but because hydrogen removal cannot occur, no reaction takes place. In this way, succinate dehydrogenase molecules bound to malonate are unable to catalyze the normal dehydrogenation of succinate, and respiration is poisoned. Interestingly, beans and certain other legumes contain unusually high concentrations of malonate, probably in the central vacuole where it cannot affect respiration. Another much more powerful inhibitor of succinate dehydrogenase is *oxaloacetate* (^-OOC—CH_2—CO—COO^-), a normal intermediate in the Krebs cycle (Burke et al., 1982). Oxaloacetate poisoning of succinate dehydrogenase is prevented partly because unusually low concentrations of oxaloacetate exist in mitochondria.

Noncompetitive inhibitors also combine with enzymes but at locations other than the active site. This effect is not overcome by simply raising the substrate concentration. Noncompetitive inhibitors generally show less structural resemblance to the substrate than do competitive inhibitors. Toxic metal ions and compounds that combine with or destroy essential sulfhydryl groups often are noncompetitive inhibitors. For example, excess O_2 can oxidize –SH groups that are close to each other, removing the H atom from each and forming new disulfide bridges, thus changing the structure of the enzyme so that its active site no longer can combine well with the substrate or substrates. Heavy Hg^{2+} ions can replace the

$$
\begin{array}{l}
\text{COO}^- \\
| \\
\text{CH}_2 \\
| \\
\text{CH}_2 \\
| \\
\text{COO}^- \\
\text{succinate}
\end{array}
\;+\; \text{enzyme—FAD} \rightleftharpoons
\begin{array}{l}
\text{COO}^- \\
| \\
\text{CH} \\
\| \\
\text{HC} \\
| \\
\text{COO}^- \\
\text{fumarate}
\end{array}
\;+\; \text{enzyme—FADH}_2
\qquad \text{(R8-2)}
$$

Plant Proteins and Human Nutrition

Humans and other animals depend on plants for many of their amino acids, so the composition of seed, leaf, and stem proteins is important in diet. We and other animals use these amino acids to build our own proteins and as a food (energy) source. Although adult humans can synthesize most of the amino acids they require from carbohydrates and various organic nitrogen compounds, eight amino acids must be provided in the diet. These are leucine, isoleucine, valine, lysine, methionine, tryptophan, phenylalanine, and threonine. Furthermore, adequate amounts of the sulfur-containing amino acid cysteine can apparently be formed only when sufficient methionine (another S-amino acid) is provided, and we use phenylalanine to synthesize tyrosine, which otherwise would be essential.

Most of the proteins in the human diet come from seed proteins, especially those of the cereal grains rice, wheat, and corn (maize). Approximately two thirds of the world's population depends on wheat or rice as the principal source of calories and protein. Maize is important in many tropical and subtropical parts of Central and South America. A smaller but still important contribution is made by legume seeds such as beans, peas, and soybeans. Soybeans are an unusually rich, fairly well-balanced protein source; about 40 percent of their dry weight is protein compared to about 10 percent for most cereal grains (Table 8-2).

Compared with most animal proteins, cereal grain proteins are low in lysine, while legume seeds are low in methionine. For example, the lysine content of total protein from seeds of 12,561 wheat cultivars averaged 3.14 percent on a weight basis compared with 6.4 percent in whole egg protein (Table 8-2). Bean seed proteins averaged only 1.0 percent methionine compared with 3.1 percent in egg proteins. Plant breeders are making some progress in introducing new cultivars or hybrid species with increased protein contents and increased percentages of essential amino acids. Examples include the opaque-2 and fluory-2 cultivars of maize, both of which are considerably richer in both lysine and tryptophan than are commonly grown cultivars (Harpstead, 1971; Larkins, 1981).

Table 8-2   Protein Content and Amino Acid Composition of Selected Food Legumes and Cereals.

Food	Protein	Amino Acid Composition (Percent of Total Protein)								
		Lysine	Methionine	Threonine	Tryptophan	Isoleucine	Leucine	Tryosine	Phenylalanine	Valine
Soybean	40.5	6.9	1.5	4.3	1.5	5.9	8.4	3.5	5.4	5.7
Peas	23.8	7.3	1.2	3.9	1.1	5.6	8.3	4.0	5.0	5.6
Beans	21.4	7.4	1.0	4.3	0.9	5.7	8.6	3.9	5.5	6.1
Oats	14.2	3.7	1.5	3.3	1.3	5.2	7.5	3.7	5.3	6.0
Barley	12.8	3.4	1.4	3.4	1.3	4.3	6.9	3.6	5.2	5.0
Wheat	12.3	3.1	1.5	2.9	1.2	4.3	6.7	3.7	4.9	4.6
Rye	12.1	4.1	1.6	3.7	1.1	4.3	6.7	3.2	4.7	5.2
Sorghum	11.0	2.7	1.7	3.6	1.1	5.4	16.1	2.8	5.0	5.7
Maize	10.0	2.9	1.9	4.0	0.6	4.6	13.0	6.1	4.5	5.1
Rice	7.5	4.0	1.8	3.9	1.1	4.7	8.6	4.6	5.0	7.0
Whole egg	12.8	6.4	3.1	5.0	1.7	6.6	8.8	4.3	5.8	7.4

Data from Orr and Watt, 1957 and Johnson and Lay, 1974.

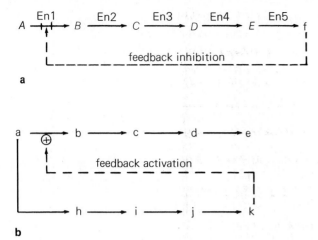

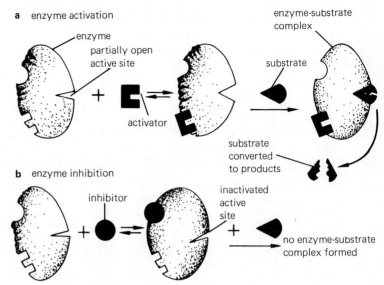

Figure 8-12 (**a**) Feedback inhibition; (**b**) feedback activation.

Figure 8-13 A hypothetical model illustrating how the presence of activators and repressors might influence allosteric enzymes, thus affecting reaction rates. In (**a**), attachment of the activator to the enzyme opens the active site, allowing it to combine with the substrate. In (**b**), attachment of the inhibitor closes the active site, preventing attachment of the substrate.

H atom on a sulfhydryl group, forming heavy mercaptides that are often insoluble, and Ag^+ ions act similarly.

Most poisons affect plants and animals by inhibiting enzymes. Some of these will be discussed later in relation to the specific processes that are affected. Enzymes are also inhibited noncompetitively by any protein denaturant, such as strong acids or bases, or by high concentrations of urea or detergents, which break hydrogen bonds. A relatively simple treatment of these and other aspects of enzyme kinetics and enzyme inhibition is given by Engel (1977).

8.6 Allosteric Enzymes and Feedback Control

We have mentioned that numerous *foreign* ions or molecules can inhibit enzymatic action, in most cases by altering the configuration of the enzyme so that it cannot effectively form a complex with the substrate. However, several enzymes can also be altered by *normal* cellular constituents, with resulting decreases or increases in their functions. Such effects are important mechanisms for homeostatic control at the metabolic level. The more common case is inhibition of a particular reaction by a metabolite that is chemically unrelated to the substrate with which the enzyme reacts.

To understand this, consider an example in which a compound A is converted by a series of enzymatic reactions via intermediates $B, C, D,$ and E to an essential product F (Fig. 8-12a). After this number of reactions, compound F no longer bears much structural resemblance to A. Nevertheless, F can sometimes reversibly combine with the first enzyme

to inhibit its combination with A. This is an example of **feedback inhibition** or **end-product inhibition.** Its advantage is that it provides a rapid and sensitive mechanism to prevent oversynthesis of compound F, because the feedback inhibition occurs only after F has built up to a level sufficient for cellular needs. Later, when the amount of F in the cell has been reduced (say by incorporation into a structural component of the cell), F molecules dissociate from enzyme number 1 and allow it to become active again. Known cases of feedback inhibition nearly always involve action of a product of a metabolic pathway acting upon the first enzyme of that pathway. A well-studied example of feedback inhibition in plants occurs in formation of the nucleotide *uridine monophosphate (UMP),* beginning with aspartic acid and carbamyl phosphate. The pathway requires five enzyme-catalyzed steps, but only the first enzyme, *aspartic transcarbamylase,* is susceptible to feedback control by UMP; no other reaction is blocked in a similar way by other reactants and products in the pathway (Ross, 1981).

To understand cases in which a metabolic process increases activity of an enzyme, consider the situation in which another compound, *a,* is converted by an enzyme to compound *b* in a series of reactions that lead to *e;* yet *a* is acted upon by a competing enzyme to initiate a second reaction leading to product *k* (Fig. 8-12b). Here the cellular levels of *e* and *k* depend on the relative activities of the first enzymes unique to their pathways of formation. Over-synthesis of k is prevented by its activation of the competing enzyme converting a to b. Other more complicated kinds of feedback loops are also known, but primarily in bacteria (Ricard, 1980).

Enzymes that combine with and respond, either negatively or positively, to small molecules such as F or k are called **allosteric enzymes.** The sites at which combination with the smaller molecules occurs are called **allosteric sites** (*allo* = other, i.e. different from the active site). Sometimes an allosteric site is on a polypeptide chain that is different from the chain containing the active site. The small molecules undergoing reversible binding to allosteric sites are called **allosteric effectors.** One result of allosteric binding is to lock the enzyme into a different configuration such that its K_m for the normal substrate is either increased or decreased. A second result is a change in the maximum reaction rate without a change in the K_m. Figure 8-13 shows a diagram of how a single enzyme with two different allosteric sites could be activated by one allosteric effector and inhibited by another. The activation case is another example of induced fit, but here it is an allosteric effector instead of the substrate that alters the shape of the enzyme so that it now combines more readily with the substrate.

9

Photosynthesis: Chloroplasts and Light

Photosynthesis is essentially the only mechanism of energy input into the living world. Organisms that can obtain energy by oxidizing compounds not synthesized by photosynthesis are known, but they are of little quantitative significance in the overall energy budget. They include chemosynthetic bacteria that obtain energy by oxidizing inorganic substrates such as ferrous ions and sulfur dissolved from the earth's crust, or H_2S released from volcanic action. Because of its importance to life, we shall devote three chapters to photosynthesis.

Like energy-yielding oxidation reactions upon which all life depends, photosynthesis involves oxidation and reduction. The overall process is an oxidation of water (removal of electrons with release of O_2 as a by-product) and a reduction of CO_2 to form organic compounds such as carbohydrates. You realize that the reverse of this process, the combustion or oxidation of gasoline or carbohydrates in wood to form CO_2 and H_2O, is a spontaneous process that releases energy. It is the similar yet effectively controlled process of respiration that keeps all organisms

alive (Chapter 12). During both combustion and respiration, electrons are removed from carbon compounds and passed downhill, energetically speaking, where they and H^+ combine with a strong electron acceptor, O_2, to make stable H_2O. Considered in this way, photosynthesis uses light energy to drive electrons uphill away from H_2O to a weaker electron acceptor, CO_2. These relations are summarized in Fig. 9-1. In this chapter, we emphasize the light-harvesting apparatus of plants, the chloroplast, and the way in which it accomplishes this uphill transport of electrons.

9.1 Historical Summary of Photosynthesis Research

Before the early eighteenth century, scientists believed that plants obtained all of their elements from the soil. In 1727, Stephen Hales suggested that part of their nourishment came from the atmosphere and that light participated somehow in this process. It was not known then that air contains different gaseous elements. In 1771, Joseph Priestley, an English clergyman and chemist, implicated O_2 (although this dephlogisticated air, as he called it, was not known to be a molecule) when he found that green plants could renew air made bad by the breathing of animals. Then a Dutch physician, Jan Ingenhousz, demonstrated that light was necessary for this purification of air. He found that plants, too, made bad air in darkness. This caused him to recommend (needlessly) that plants be removed from houses during the night to avoid the possibility of poisoning the occupants.

In 1782, Jean Senebier showed that the presence of the noxious gas produced by animals and plants in darkness (CO_2) stimulated production of purified air (O_2) by plants in the light. So by this time the participation of two gases in photosynthesis had been

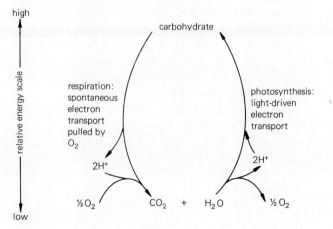

Figure 9-1 Contrasting energy relations of photosynthesis and respiration.

demonstrated. Work of Lavoisier and others made it apparent that these gases were CO_2 and O_2. Water was implicated by N. T. de Saussure when, in 1804, he made the first quantitative measurements of photosynthesis. He found that plants gained more dry weight during photosynthesis than could be accounted for by the amount by which the weight of CO_2 absorbed exceeded the weight of O_2 released. He correctly attributed the difference to an uptake of H_2O. He also noted that approximately equal volumes of CO_2 and O_2 were exchanged during photosynthesis.

The nature of the other product of photosynthesis, organic matter, was demonstrated by Julius Sachs in 1864 when he observed the growth of starch grains in illuminated chloroplasts. The starch is detected only in areas of the leaf exposed to the light. Thus, the overall reaction of photosynthesis was demonstrated to be as follows:

$$nCO_2 + H_2O + \text{light} \rightarrow (CH_2O)_n + nO_2 \qquad \text{(R9-1)}$$

In this reaction, (CH_2O) is simply an abbreviation for starch or other carbohydrates with an empirical formula very close to this.

A further important discovery was that of C. B. van Niel, who in the early 1930s pointed out the similarity between the overall photosynthetic process in green plants and that in certain bacteria. Various bacteria were known to reduce CO_2 using light energy and an electron source different from water. Some of these use organic acids such as acetic or succinic acid as electron sources, while those to which van Niel gave primary attention use H_2S and deposit sulfur as a by-product. The overall photosynthetic equation for these bacteria was believed to be as follows:

$$nCO_2 + 2nH_2S + \text{light} \rightarrow (CH_2O)_n + H_2O + 2S \qquad \text{(R9-2)}$$

When R9-2 is compared with R9-1 above for green plants, an analogy can be seen between the role of H_2S and H_2O, and of O_2 and sulfur. This suggested to van Niel that the O_2 released by plants is derived from water, not from CO_2. This idea was supported in the late 1930s by work of Robin Hill and R. Scarisbrick, in England, that showed that isolated chloroplasts and chloroplast fragments could release O_2 in the light if they were given a suitable acceptor for the electrons being taken from water. Certain ferric (Fe^{3+}) salts were the earliest electron acceptors provided, and they became reduced to the ferrous (Fe^{2+}) form. This light-driven split of water in the absence of CO_2 fixation became known as the **Hill reaction.** It showed that whole cells were not necessary for at least some of the reactions of photosynthesis and that the light-driven O_2 release is not mandatorily tied to reduction of CO_2.

More convincing evidence that the O_2 released is derived from H_2O came in 1941 from results of Samuel Ruben and his associates. They supplied the green alga *Chlorella* with H_2O containing ^{18}O, a heavy, nonradioactive oxygen isotope that was detected with a mass spectrometer. The O_2 released in photosynthesis became labeled with ^{18}O, thus supporting van Niel's hypothesis. For technical reasons, Ruben's experiments could not prove that O_2 came entirely from H_2O, but later work of Alan Stemler and Richard Radner (1975) seems to provide such proof. We must, therefore, modify the summary equation for photosynthesis given in R9-1 to include two H_2O molecules as reactants:

$$nCO_2 + 2nH_2O + \text{light} \xrightarrow{\text{chloroplasts}} (CH_2O)_n + nO_2 + nH_2O \qquad \text{(R9-3)}$$

In 1951, it was found that a natural plant constituent, the vitamin B (niacin or nicotinamide)-containing coenzyme called **nicotinamide adenine dinucleotide phosphate** (commonly abbreviated **$NADP^+$**), could also act as a Hill reagent by accepting electrons from water in reactions occurring in isolated chloroplasts. This discovery again stimulated photosynthesis research, because it was already known that the reduced form of $NADP^+$, **NADPH,** could transfer electrons to a number of plant compounds, and it was immediately suspected that its normal role in chloroplasts was reduction of CO_2. This suspicion proved correct. Thus, one of two essential functions of light in photosynthesis is to drive electrons from H_2O to reduce $NADP^+$ to NADPH. The other function is to provide energy to form ATP from ADP and $H_2PO_4^-$, as described below. (Hereafter $H_2PO_4^-$ is abbreviated **Pi,** as in Chapter 6.)

This conversion of ADP and Pi to ATP in chloroplasts was discovered in the laboratory of Daniel Arnon in 1954. Prior to that, the only important mechanism known to form ATP was respiration, especially those reactions occurring in the mitochondria called oxidative phosphorylation (Chapter 12). Arnon found that ATP was synthesized in isolated chloroplasts only during light, and the process became known as **photosynthetic phosphorylation,** or simply **photophosphorylation.** This process of ATP formation by photophosphorylation can be summarized by reaction R9-4:

$$\text{ADP} + Pi + \text{light} \xrightarrow{\text{chloroplasts}} \text{ATP} + H_2O \qquad \text{(R9-4)}$$

Photophosphorylation in chloroplasts accounts for much more ATP formation in leaves during the light than does oxidative phosphorylation in the mitochondria of those leaves, and so it is clearly of great quantitative significance. Notice, however, that our summary equation for photosynthesis (R9-3) says

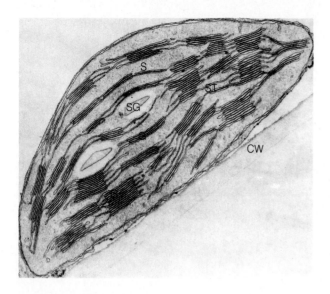

Figure 9-2 Oat leaf chloroplast. S, stroma; ST, stroma thylakoid; G, granum; SG, starch grain; CW, cell wall. (Courtesy P. Hanchey.)

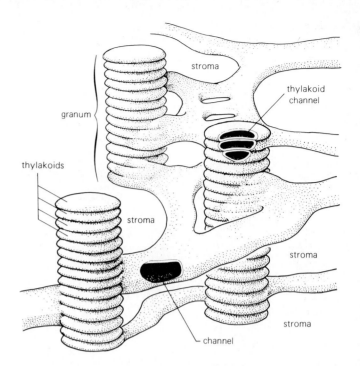

Figure 9-3 A three-dimensional interpretation of the arrangement of the internal membranes of a chloroplast, emphasizing the relation between stroma thylakoids and grana. Note the channels in both kinds of thylakoids. (Redrawn from an original by T. E. Weir.)

nothing about ATP, NADPH, or $NADP^+$. The reason for this is that once ATP and NADPH are formed, their energy is used in the process of CO_2 reduction and carbohydrate synthesis, and ADP, Pi, and $NADP^+$ are again released. So ADP and Pi are rapidly converted to ATP by light energy, and the ATP is just as rapidly broken down when photosynthesis is occurring at a constant rate. How ATP and NADPH are used to help fix CO_2 is a subject of Chapter 10, but in the remainder of this chapter we shall be concerned with chloroplasts and the intricate processes by which they trap light energy and use it to form ATP and NADPH.

9.2 Chloroplasts: Structures and Photosynthetic Pigments

Chloroplasts of many shapes and sizes are found in various kinds of plants. They arise from tiny **proplastids** (immature, small, and nearly colorless plastids with few or no internal membranes). Proplastids are derived most commonly only from the unfertilized egg cell. They divide as the embryo develops, and when leaves and stems are formed they develop into chloroplasts (Possingham, 1980; Virgin and Egneus, 1983). Young chloroplasts also actively divide, especially when the organ containing them is exposed to light. Most chloroplasts are easily seen with the light microscope, but their fine structure can be discovered only by electron microscopy. Each chloroplast is surrounded by a double membrane system or envelope that controls molecular traffic into and out of them. Within the chloroplast is another series of membranes that contains photosyn-

thetic pigments. Assuming that organisms can logically be segregated into five kingdoms, chloroplasts occur in almost all members of the Kingdom Plantae and in certain algae or algal-like members of the Kingdom Protista. Only colorless, parasitic angiosperms are known exceptions. Cyanobacteria and other photosynthetic bacterial members of the Kingdom Monera have no chloroplasts, but they still have photosynthetic pigments embedded in specialized membranes as in chloroplasts. No chloroplasts occur in the Fungi or Animalia. We shall discuss only the chloroplasts typical of vascular plants, bryophytes, and many green algae in the Kingdom Plantae.

The structure of a chloroplast from an oat leaf is shown in Fig. 9-2. Each of the internal membranes containing photosynthetic pigments seems to be the external surface of a flattened tube or sac, called a **thylakoid** (Greek *thylakos,* sac or pouch). In certain regions the thylakoids are stacked to form **grana** (single stack, **granum**). The longer thylakoids that connect one granum to another extend through the chloroplast matrix called the **stroma,** so these members are usually referred to as **stroma thylakoids.** Stroma thylakoids often extend into and make up part of one or more grana, and in those locations there is no apparent distinction between them and the grana thylakoids. Figure 9-3 illustrates a three-dimensional interpretation of the relation between

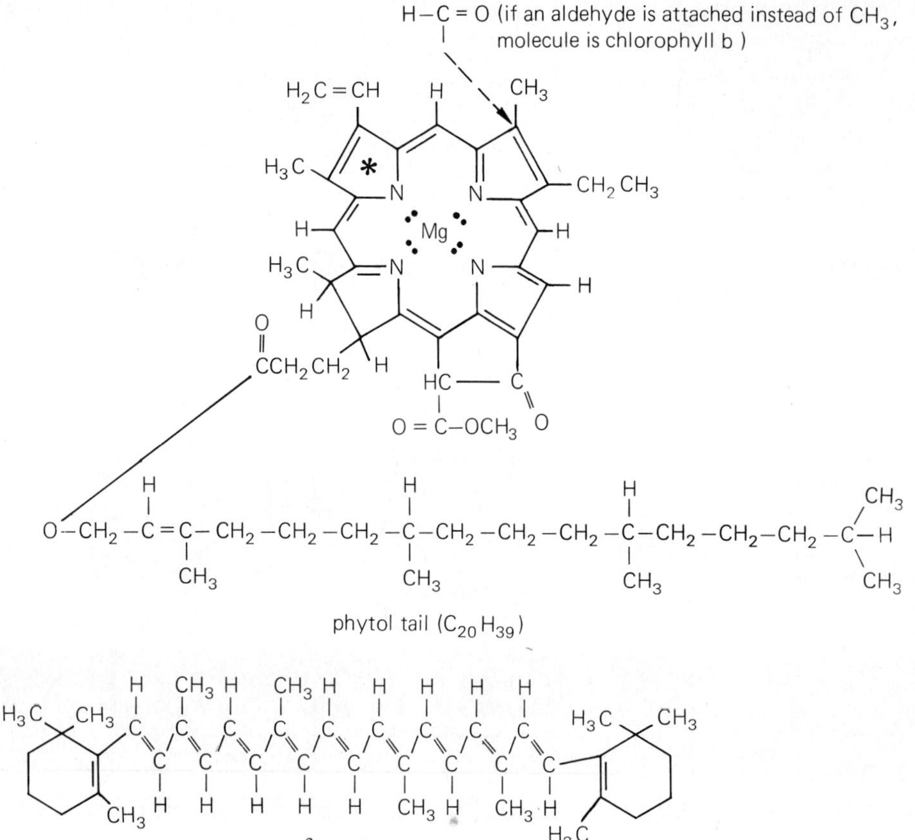

Figure 9-4 Structures of some chlorophyll and carotenoid pigments. (**a**) Structure of chlorophyll *a* and its relation to chlorophyll *b*. The tetrapyrrole ring structure on top is made of four pyrrole rings (see asterisk) and gives the green color, while the hydrophobic $C_{20}H_{39}$ phytol tail common to both chlorophylls probably extends into the interior of the membrane. Chlorophyll *a* is blue-green; chlorophyll *b*, yellow-green. (**b**) *β*-carotene, a yellow carotenoid with the empirical formula $C_{40}H_{56}$. (**c**) Lutein, a yellow xanthophyll with the empirical formula $C_{40}H_{56}O_2$. (**d**) Lycopene, a reddish carotene with the empirical formula $C_{40}H_{56}$. Lycopene is not found in chloroplasts but gives the red color to tomato fruits.

thylakoids of grana and stroma. Note that there is a cavity, which we call a **channel,** between the two membranes of each thylakoid. This channel is filled with water and dissolved salts, but it plays a special role in photosynthesis.

The pigments present in thylakoid membranes consist largely of two kinds of green **chlorophylls, chlorophyll *a*** and **chlorophyll *b*.** Also present are yellow to orange pigments classified as **carotenoids.** There are two kinds of carotenoids, the pure hydro-

carbon **carotenes** and the oxygen-containing **xanthophylls.** Certain carotenoids (especially *violaxanthin*, a xanthophyll) also exist in the chloroplast envelope, giving it a yellowish color, whereas chlorophylls do not occur in the envelopes. The structures of chlorophylls *a* and *b* and some carotenoids are shown in Fig. 9-4. In most plants, including green algae, *β*-carotene and *lutein* are the most abundant carotenoids in thylakoids.

Electron microscopy gives us no information

about how chlorophylls and carotenoids are arranged in thylakoids, but other techniques show that all chlorophylls and most or all carotenoids are embedded within and attached by noncovalent bonds to protein molecules. Altogether, chloroplast pigments represent about one-half of the total lipid content of thylakoid membranes, while the other half is composed largely of galactolipids with small amounts of phospholipids (for structures, see Fig. 6-12). Few, if any, sterols are present, as contrasted to other plant membranes. The galactolipids and phospholipids make a bilayer typical of membranes described in Chapter 6. The fatty acid portions of thylakoid lipids are rich in linolenic acid (with three points of unsaturation) and linoleic acid (with two points of unsaturation). These fatty acids with two or three double bonds cause thylakoid membranes to be unusually fluid, and compounds within them are relatively mobile, including the embedded proteins and their attached pigments.

9.3 Some Principles of Light Absorption by Plants

To find out how light causes photosynthesis, we must learn something about its properties. As mentioned in Appendix B, light has a *wave nature* and a *particle nature*. Light represents the part of radiant energy that has wavelengths visible to the human eye (approximately 390 to 760 nanometers, nm). This is a very narrow region of the electromagnetic spectrum.

The particulate nature of light is usually expressed in statements that light comes in **quanta** or **photons,** discrete packets of energy, each having a specific associated wavelength. The energy in each photon is inversely proportional to the wavelength, so the violet and blue wavelengths have more energetic photons than the longer orange and red ones. One mole (6.02×10^{23}) of photons has been called an **Einstein,** although the term Einstein is now being discouraged because a mole is an SI unit and an Einstein is not. Quantitative and other aspects of light relations are in Appendix B.

A fundamental principle of light absorption, often called the **Stark Einstein Law,** is that any molecule can absorb only one photon at a time and this photon causes the excitation of only one electron. Specific valence (bonding) electrons in stable ground state orbitals are those usually excited, and each can be driven away from its ground state in the positively charged nucleus a distance corresponding to an energy exactly equal to the energy of the photon absorbed (Fig. 9-5). The pigment molecule is then in an *excited state*, and it is this excitation energy that is used in photosynthesis.

Chlorophylls and other pigments can remain in an excited state only for short periods, usually a bil-

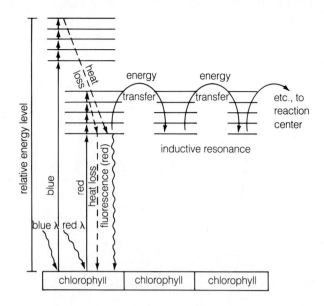

Figure 9-5 Simplified model to explain how light energy striking a chlorophyll molecule is given up. Note that excitation by blue or red light leads to the same final energy level (often called the first excited singlet). From here, the energy can be lost by decay back to the ground state (heat loss or fluorescence of red light) or can be transferred to an adjacent pigment by inductive resonance. Each time a pigment transfers its excitation energy to an adjacent pigment, the excited electron in the first pigment returns to the ground state.

lionth (10^{-9}) of a second or even much less. As shown in Fig. 9-5, the excitation energy can be totally lost by heat release as the electron moves back to ground state. This is what is now happening to electrons in the ink of the words you are reading. A second way that some pigments, including chlorophyll, can lose excitation energy is by a combination of heat loss and fluorescence. (**Fluorescence** is light production accompanying rapid decay of excited electrons.) Chlorophyll fluorescence produces only deep red light, and these long wavelengths are easily seen when a concentrated solution of either chlorophyll *a* or *b* or a mixture of chloroplast pigments is illuminated, especially with ultraviolet or blue radiation. In the leaf, fluorescence is greatly minimized because the excitation energy is used in photosynthesis.

Figure 9-5 helps explain why blue light is always less efficient in photosynthesis than red. After excitation with a blue photon, the electron always decays extremely rapidly by heat release to a lower energy level, a level that red light produces without heat loss when it is absorbed. From this lower level, either additional heat loss, fluorescence, or photosynthesis can occur.

Photosynthesis requires that energy in excited electrons of various pigments be transferred to an energy-collecting pigment, a **reaction center.** We

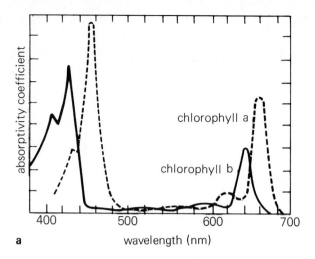

a

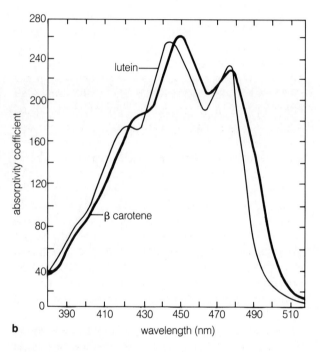

b

Figure 9-6 (**a**) Absorption spectra of chlorophylls *a* and *b* dissolved in diethyl ether. The absorptivity coefficient used here is equal to the absorbance (optical density) given by a solution at a concentration of 1 g/l with a thickness (light path length) of 1 cm. (From F. Zscheile and C. Comar, 1941, *Botanical Gazette* 102:463.) (**b**) Absorption spectra of β-carotene in hexane and of lutein (a xanthophyll) in ethanol. The absorptivity coefficient used is the same as that described in Fig. 9-6a. (Data from F. Zscheile et al., 1942, *Plant Physiology* 17:331.)

shall explain later that there are two kinds of reaction centers in thylakoids, both of which consist of chlorophyll *a* molecules that are made special by their association with particular proteins and other membrane components. Figure 9-5 illustrates that the energy in an excited pigment can be transferred to an

adjacent pigment, and from it to another pigment, and so on until the energy finally arrives at the reaction center. There are various theories to explain energy migration within a group of neighboring pigments, one of which is inductive resonance. We won't discuss this theory, but we emphasize that excitation of any one of numerous pigment molecules in a thylakoid allows momentary collection of the light energy in a chlorophyll *a* reaction center.

Leaves of most species absorb more than 90 percent of the violet and blue wavelengths that strike them and almost as high a percentage of the orange and red wavelengths (see Fig. 3-12). Almost all of this absorption is by the chloroplast pigments. In the thylakoids, each photon can excite an electron in a carotenoid or chlorophyll. Chlorophylls are green because they absorb green wavelengths ineffectively and instead reflect or transmit them. We can measure the relative absorbance of various wavelengths by a purified pigment with a spectrophotometer. A graph of this absorption as a function of wavelength is called an **absorption spectrum.** The absorption spectra of chlorophylls *a* and *b* are given in Fig. 9-6a. They show that very little of the green and yellow-green light between 500 and 600 nm is absorbed *in vitro* and that both chlorophylls absorb strongly the violet and blue and the orange and red wavelengths.

Some of the carotenoids in thylakoids also transfer their excitation energy to the same reaction centers as do chlorophylls. The absorption spectra of β-carotene and lutein are given in Fig. 9-6b. These yellow pigments absorb only the blue and violet wavelengths *in vitro*. They reflect and transmit green, yellow, orange, and red wavelengths, and this combination appears yellow. Evidence indicates that β-carotene is much more effective than lutein or other xanthophylls in transferring energy to either of the two reaction centers of green plants (Goedheer, 1979). However, in brown algae a xanthophyll called *fucoxanthin* is highly effective in energy transfer, and in dinoflagellates a related xanthophyll, *peridinin*, performs the same function. Besides the function of carotenoids as light-harvesting pigments that contribute to photosynthesis, they have a second function in thylakoids; this is to protect chlorophylls against oxidative destruction by O_2 when irradiance levels are high (see solarization, Section 11.3).

When we compare the effects of different wavelengths on the rate of photosynthesis, always making sure not to add so much energy of any wavelength that the process becomes saturated, we obtain an **action spectrum.** Action spectra for photosynthesis and other photobiological processes help identify the pigment involved, because these spectra often closely match the absorption spectrum of any pigment that participates. In Fig. 9-7 are plotted the relative rates of photosynthesis for several species as a function of

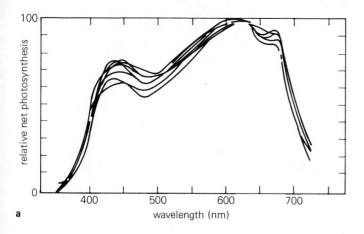

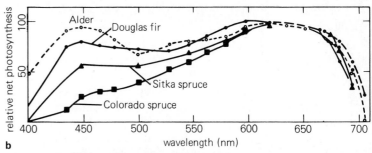

Figure 9-7 Action spectra of various plants. (**a**) 22 species of crop plants. (From K. J. McCree, 1972, *Agricultural Meteorology* 9:191–216.) (**b**) Some trees. (Drawn from data of J. B. Clark and G. R. Lister, 1975, *Plant Physiology* 55:401–406.)

the number of photons of each wavelength striking a unit area of the leaf.*

Figure 9-7a plots action spectra for 22 herbaceous dicot and monocot crop plants, and Fig. 9-7b shows data for four trees. All the plants show a major peak in the red region and a distinct lower peak or at least a shoulder (the conifers) in the blue, both of which result mainly from light absorption by chlorophylls. The conifers show less response in the blue because their waxy blue-green needles reflect more of the blue light and also because they contain high amounts of unidentified yellow pigments that are not photosynthetically active but that still absorb some of the unreflected blue light in competition with chlorophylls. Such pigments are said to screen the blue light, and evidence indicates that in both conifer needles and bean leaves these yellow pigments are water soluble and are not in chloroplasts (Balegh and Biddulph, 1970).

Compared with the absorption spectra of purified chlorophylls and carotenoids, the action of green and yellow light in causing photosynthesis of seed plants and the absorption of these wavelengths by leaves is surprisingly high. Nevertheless, carotenoids and chlorophylls are apparently the only pigments that absorb this light. The main reason that the action spectra are higher than the absorption spectra for

yellow and green wavelengths is that, although the chance of any such wavelength being absorbed is small, those wavelengths not absorbed are repeatedly reflected from chloroplast to chloroplast in the complex network of photosynthetic cells. With each reflection a small additional percentage of these wavelengths is absorbed, until finally half or more are absorbed by most leaves and cause photosynthesis. This internal reflection does not occur in a spectrophotometer cuvet containing dissolved chlorophyll, and so the absorbance of green wavelengths is very low (Fig. 9-6a). Furthermore, the *in vitro* absorption by these pigments in an organic solvent occurs at shorter wavelengths than when they are present in chloroplast thylakoids. When *in vivo*, the absorption of the carotenoids shifts from the blue into the green, and much photosynthesis in the green at about 500 nm results from absorption by active carotenoids.

Both chlorophylls show only small *in vivo* shifts in the blue region, but chlorophyll *a* shows several shifts in the red. The association of chlorophyll *a* with itself to form dimers and trimers and with thylakoid proteins causes additional peaks to occur in the red region (Brown, 1972). We are interested in two of these minor peaks, those at about 680 and 700 nm, because they result from special chlorophyll *a* molecules acting as reaction center pigments. These pigments are abbreviated **P680** and **P700**. Their functions will be discussed in more detail.

---

*Slightly different action spectra with lower blue peaks are obtained when we plot the data as a function of the *energy* in each applied wavelength. This is because although a blue photon absorbed by a photosynthetically active pigment is as effective as any other photon, it contains more energy than others and more of its energy is wasted as heat. Therefore, blue light can be as *effective* in photosynthesis as red but never as *efficient*. In many algae, carotenoids and phycobilin pigments (the red phycoerythrins or the blue phycocyanins) absorb light causing photosynthesis (see personal essay of Elisabeth Gantt). The action spectra of these algae are quite different from those of seed plants.

9.4 The Emerson Enhancement Effect: Cooperating Photosystems

In the 1950s, Robert Emerson at the University of Illinois was interested in why red light of wavelengths longer than 690 nm is so ineffective in causing photosynthesis (Fig. 9-7), even though much of it is absorbed by chlorophyll *a in vitro*. His research group found that if light of shorter wavelengths was provided at the same time as the longer red wavelengths, photosynthesis was even faster than could be expected from adding the rates found when either color was provided alone. This synergism or enhancement became known as the **Emerson enhancement effect.**

We can think of enhancement as the long red wavelengths helping out the shorter wavelengths, or as the short group helping out the long red. We now realize two separate groups of pigments cooperate in photosynthesis and that such long red wavelengths are absorbed only by one photosystem, called **photosystem I (PS I).** The second photosystem, **photosystem II (PS II),** absorbs wavelengths shorter than 690 nm, and for maximum photosynthesis at longer wavelengths both systems must function together. In fact, they normally cooperate to cause photosynthesis in all wavelengths shorter than 690 nm, including the red, orange, yellow, green, blue, and violet, because both photosystems absorb those wavelengths. The importance of Emerson's work is that it indicated the presence of distinct photosystems, which have now been isolated and their functions clarified. Reaction 9-5 summarizes how PS I and PS II use light energy to oxidize H_2O and cooperatively transfer the two available electrons in it to $NADP^+$, thus forming NADPH:

$$PS\ II \xrightarrow[H_2O\quad 1/2\ O_2\ +\ 2\ H^+]{+\ light} PS\ I \xrightarrow[NADP^+]{+\ light} NADPH + H^+ \qquad (R9\text{-}5)$$

9.5 Photosystems I and II: Composition, Functions, and Locations in Thylakoids

When isolated chloroplasts are treated with the proper mild detergent, the two photosystems are removed from thylakoids and can be separated by polyacrylamide gel electrophoresis (Section 8.2). Analysis of the separated green bands shows that PS I contains chlorophyll *a*, small amounts of chlorophyll *b* and some β-carotene attached by noncovalent bonds to several proteins (Malkin, 1982; Anderson, 1982; Barber, 1983). One of the chlorophyll *a* molecules is somehow made special by its chemical environment such that it absorbs light near 700 nm as well as at shorter wavelengths and so is called P700. It is P700 that is the reaction center for PS I and to which all surrounding chlorophyll *a* and β-carotene molecules in that photosystem transfer their energy, as shown in Fig. 9-5. Also present are at least two iron-containing proteins similar to ferredoxin (Section 8.2)

in which each of four iron atoms in each protein is bound to two sulfur atoms; these are called **Fe-S proteins,** as opposed to cytochrome and related proteins that contain one iron bound in a heme group (**hemeproteins,** e.g., hemoglobin). The Fe-S proteins are primary electron acceptors for PS I, meaning that electrons are first transferred from the reaction center of PS I (P700) to one of those proteins (which one is not known). Only one of the four iron atoms present can accept an electron; such acceptance reduces it from the Fe^{3+} to the Fe^{2+} valence. Subsequently, this Fe^{2+} is reoxidized to Fe^{3+} during the overall electron transport pathway, as described later.

Photosystem II also contains chlorophyll *a* and β-carotene (connected to two major proteins), and again little chlorophyll *b* is present. The reaction center is **P680,** a chlorophyll *a* molecule in a chemical environment different from that of P700 or other chlorophylls *a*. Also present in PS II is its primary electron acceptor, now believed to be a colorless chlorophyll *a* that lacks Mg^{2+}. This molecule is called **pheophytin,** abbreviated **Pheo** (Klimov and Krasnovskii, 1981; Cogdell, 1983). Closely associated with Pheo, P680, and protein that binds P680 is a quinone historically called **Q** because of its ability to quench fluorescence of P680 by accepting its excited electron. Finally, PS II contains one or more proteins containing bound manganese, called the **manganese protein** (Amesz, 1983). It is thought that four Mn^{2+} ions are bound to one or more proteins in PS II and that a chloride ion bridges two Mn^{2+} together. If so, this explains the essential functions of both Mn^{2+} and Cl^- to photosynthesis mentioned in Chapter 5. The Mn^{2+} is likely oxidized to Mn^{3+} and then reduced again as electron transport occurs. The manganese protein is part of the inner side of the thylakoid membrane close to the channel and is probably involved directly in the first step of H_2O oxidation (Barber, 1984).

Besides these photosystems, two other major green bands can be separated from chloroplasts by electrophoresis. Each contains both chlorophylls *a* and *b* and xanthophyll carotenoids but very little β-carotene; all these pigments are bound to proteins. These green bands represent **light-harvesting complexes** of pigments and protein, one of which functions with PS I and the other mainly with PS II. Their function is to harvest light energy by absorbing it and transferring it to the proper photosystem, where it eventually reaches P700 or P680.

It was estimated that each granum (a stack of thylakoids) contains about 200 units each of PS I and PS II (Junge, 1982). The amounts in stroma thylakoids are more variable, because these thylakoids differ much more in length and surface area than do thylakoids in grana. Although it was first thought that equal amounts of PS I and PS II occur in chloroplasts, it is now clear that the ratio varies considerably depending on species and growth conditions (Melis and Brown, 1980). Some species (C-4 plants, Section 10.3) have few grana and correspondingly less PS II and more PS I. In contrast, other species that normally grow in shade have a large number of grana with numerous stacked thylakoids, and their chloroplasts are enriched in PS II with less PS I. These and other results show that stroma thylakoids contain primarily PS I, whereas grana contain mainly PS II (Anderson, 1982; Anderson and Melis, 1983; Barber, 1983).

The distinct physical locations of PS I and PS II raise problems of cooperation between them, because they jointly function to transfer electrons from H_2O to $NADP^+$. They cannot be too far apart or such transfer would be ineffective or impossible. Two mobile electron carriers are thought to carry electrons from PS II to PS I, thus providing the necessary connection. One carrier is a small copper-containing protein named **plastocyanin**. Plastocyanin is bound loosely to the inside of thylakoid membranes next to the channel (Fig. 9-3). When its copper becomes reduced from Cu^{2+} to Cu^{1+} by PS II, it can move along the membrane carrying an electron to PS I where it is reoxidized to the Cu^{2+} form. It then shuttles back to PS II and picks up still another electron. The other carrier system is actually a group of quinones, mainly **plastoquinones** (PQ), that moves laterally and vertically *within* the fluid membrane (Trebst, 1978). Plastoquinones carry two electrons and two H^+ from PS II to PS I; as we shall see, both are important for photosynthesis. Reduction and oxidation of PQ is shown in R9-6.

(R9-6)

Action and cooperation of the photosystems require even more electron transport components. Another complex of proteins (without chlorophylls and carotenoids) isolatable from thylakoids contains two cytochromes, one called *cytochrome b_6* and the other *cytochrome f*, along with another iron-sulfur protein. The iron in each of these proteins can pick up or give up one electron as reduction and oxidation occur. This complex lies physically between PS I and PS II and is important for electron transport between them. Finally, one more cytochrome, *cytochrome b_3*,

and another Fe-S protein, *ferredoxin*, participate in photosynthetic transport. The function of cytochrome b_3 is still uncertain, so we have left it out of our models. However, the function of ferredoxin is clear. Ferredoxin transfers electrons from other Fe-S proteins of PS I directly to $NADP^+$, completing the overall light-driven electron transport process.

A final component of thylakoids necessary for photophosphorylation is a complex of proteins called the *ATPase* or *coupling factor (CF)* complex. This complex can, under different conditions, either hydrolyze ATP to ADP and Pi or synthesize ATP from Pi and ADP by photophosphorylation. ATP synthesis is strongly favored by light-driven electron transport, and electron transport is favored by photophosphorylation, so CF couples the two processes together. The CF complex is composed of a spherical headpiece that is attached to the stroma side of the thylakoid membrane and a stalkpiece that extends across the lipid bilayer to the channel side of the membrane.

9.6 Transport of Electrons from H_2O to $NADP^+$

Figure 9-8 shows a tentative and incomplete model that summarizes known relationships among most thylakoid components responsible for light-driven electron transport from H_2O (lower left) to $NADP^+$ and for photophosphorylation (upper right). In this model PS II, the cytochrome b_6, Fe-S protein, cytochrome f complex, and PS I are shown as circles, representing globular proteins. The light-harvesting complexes surround both photosystems and transfer light energy to them. The Fe-S, cytochrome complex lies between PS I and PS II, which can be separated by fairly wide distances (e.g., from grana to stroma thylakoids), but mobile plastoquinone (PQ) and plastocyanin (PC) molecules carry electrons from PS II to PS I. Further details are described below.

The electron transport components of the model can be compared to a bucket brigade. Just as people in a bucket brigade rapidly move buckets of water toward a fire, so these electron transport components move electrons rapidly from H_2O to $NADP^+$. First, follow only the heavy arrows starting with H_2O (lower left). These arrows represent a pathway called **noncyclic electron transport,** because the electrons driven from H_2O to $NADP^+$ never cycle back (as opposed to a cyclic pathway represented in part by light arrows in Fig. 9-8 and described later). The formation of ATP by this electron transport is called **noncyclic photophosphorylation.** The overall process of noncyclic electron transport requires light energy, because H_2O is thermodynamically difficult to oxidize and $NADP^+$ is difficult to reduce.

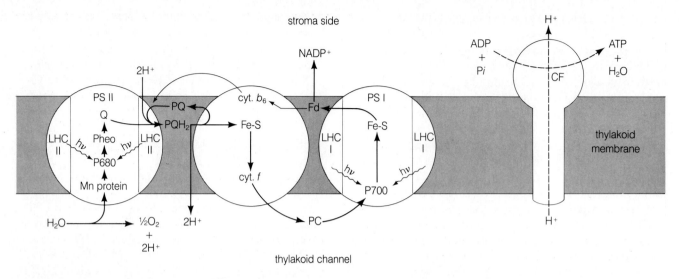

Figure 9-8 A summary of light-driven electron transport in a chloroplast thylakoid. Heavy arrows represent the pathway of noncyclic electron transport from H_2O to $NADP^+$. Lighter arrows represent the cyclic electron-transport pathway. Wavy lines indicate transfer of light energy to P680 or P700. Dashed lines through CF represent transport of H^+ from thylakoid channel to stroma with accompanying photophosphorylation. Abbreviations and explanations are in text. (Modified from models of J. M. Anderson, M. A. Selak and J. Whitmarsh, 1982, R. J. Cogdell, 1983, and N. Carillo and R. H. Vallejos, 1983.) For simplicity, galactolipids and phospholipids of the membrane bilayer are omitted.

When a photon is absorbed by a pigment in the light-harvesting complex associated with photosystem II (LHCII) (Fig. 9-8, left), its energy is transferred by inductive resonance (Fig. 9-5) to P680. This excites an electron in P680 so much that it can be removed by Pheo. Loss of the electron causes P680 to become positively charged, and it then attracts an electron from an adjacent Mn-protein. As the Mn-protein becomes oxidized (loses an electron), it in turn strongly attracts an electron from H_2O. Each H_2O can give up only two electrons; each such electron arises from one of the two electrons in both covalent bonds holding H to O in H_2O. As H_2O is oxidized by losing two electrons, it also releases 1/2 O_2 and 2 H^+. This oxidation requires absorption of two photons by PS II and two consecutive excitations of P680 with accompanying consecutive loss of two electrons, one at a time. Each electron moves from Q to plastoquinone (PQ). Reduction of PQ to PQH_2 requires two electrons and two H^+ (R9-6). An important fact regarding photophosphorylation is that it depends on the absorption of these H^+ from the stroma side of the membrane (upper left) as PQ reduction occurs. Equally important is that these same H^+ are transported into the thylakoid channel when PQH_2 is reoxidized to PQ; here these H^+ join those released from H_2O oxidation, and jointly they cause the pH to decrease in the channel.

From PQH_2, electrons move, one at a time, to cytochrome b_6 or to the Fe-S protein and then to cytochrome f in the complex between PS II and PS I.

From here they go to plastocyanin (PC, lower center), and PC is thought to move along the edge of the membrane to PS I where P700 accepts the electron. However, P700 cannot accept an electron unless it previously lost one. Such loss can only occur by light excitation, and the model shows transfer of light energy from LHCI to P700 to excite that reaction center. Excited P700 gives its electron to Fe^{3+} in one of the Fe-S proteins associated with it, and the electron is then passed to a ferredoxin (Fd) molecule, reducing its Fe^{3+} to Fe^{2+}. Ferredoxin then reduces $NADP^+$ by providing two electrons, one at a time, to $NADP^+$. This reduction is catalyzed in the stroma by an enzyme called **ferredoxin:NADP$^+$ reductase** (not shown in the model), as in R9-7:

$$NADP^+ + 2Fd(Fe^{2+}) + H^+ \rightarrow NADPH + 2Fd(Fe^{3+})$$

(R9-7)

The structure of $NADP^+$ shown in Fig. 9-9 indicates how its reduction requires two electrons and one H^+ (equivalent to one electron and one H atom.)

In summary, our model (Fig. 9-8) shows that transfer of one electron from H_2O to $NADP^+$ requires two photons, because excitation of both photosystems is essential. This apparently explains the Emerson enhancement effect, which first indicated cooperation of two photosystems. Functions of several other electron transport components and their physical relationships in thylakoids are also indicated approximately.

Herbicides and Photosynthetic Electron Transport

You might expect that one way herbicides kill weeds is by inhibiting photosynthesis, and for some this is indeed true. Those that do so generally interfere with some reaction of electron transport. Several urea derivatives, most notably monuron or CMU (3-p-Chlorophenyl)-1,1-dimethylurea) and diuron or DCMU (3-(3,4-Dichlorophenyl)-1,-1-dimethylurea) are applied to soils, move in the xylem to leaves, and block electron transport between Q and PQ. Certain triazine herbicides such as simazine and atrazine and certain substituted uracil herbicides, including bromacil and isocil, seem to block at the same step. Maize is tolerant to triazines (but not to the urea or uracil derivatives), because it contains enzymes that detoxify such compounds. During the past 15 years many weedy species have become resistant to the triazines, and this is because a chloroplast mutation has caused them to form a modified thylakoid protein. In the nonmutants this protein binds both Q and atrazine; the modified protein still binds Q so that electron transport can proceed, but it no longer binds atrazine (Hirschberg and McIntosh, 1983).

Other photosynthetic inhibitors include the bipyridylium (also called bipyridynium) herbicides diquat and paraquat (Black, 1977). These compounds, often called viologen dyes, accept electrons from photosystem I before ferredoxin, transfer them to O_2 to produce toxic activated free radical forms of oxygen such as **superoxide** (O_2^-) and hydroxy ($OH^•$). These free radicals destroy unsaturated fatty acids in thylakoids and other cell membranes, causing plant death. Furthermore, the reduction of these herbicides by PS I followed by their rapid oxidation by O_2 allows them to act catalytically; that is, the interception of electrons doesn't permanently reduce the herbicide and prevent its further action. As a result, such herbicides are effective in low doses. They must be used with caution, however, because they are also toxic to animals, including humans.

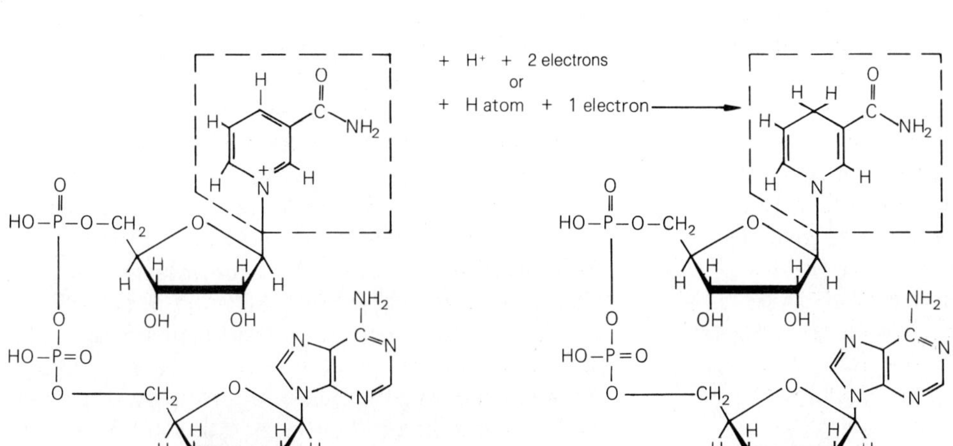

Figure 9-9 Structures of NADP$^+$ (left) and NADPH (right). The part of the NADP$^+$ molecule undergoing reduction, the nicotinamide ring, is enclosed by a dashed line. One electron is added to the nitrogen atom of the nicotinamide, neutralizing its positive charge, and the second electron is added as part of an H atom to its uppermost carbon atom. The relatively complex NADP$^+$ is a combination of two nucleotides, adenosine monophosphate (AMP) (lower half of structure) and nicotinamide mononucleotide (upper half of structure). All **nucleotides** are made of three major parts: (1) a heterocyclic ring, in this case nicotinamide but in other nucleotides a purine or pyrimidine base, (2) the pentose sugar ribose, and (3) phosphate. Phosphate is esterified to the C-5 position of the ribose unit. The two nucleotides in NADP$^+$ are connected in an anhydride linkage between the C-5 phosphate group of each ribose moiety. Notice also that NADP$^+$ contains another phosphate group esterified to the OH group at the C-2 position (see asterisk) of that ribose moiety belonging to AMP. The presence of this additional phosphate is the only way in which NADP$^+$ and NADPH differ from another important electron-carrying coenzyme called **NAD$^+$** (**nicotinamide adenine dinucleotide**). NAD$^+$ and its reduced form, **NADH,** are much less abundant than NADP$^+$ and NADPH in chloroplasts, but they are involved in electron transport during several reactions of respiration (Chapter 12), nitrogen metabolism (Chapter 13), fat breakdown (Chapter 14) and even of a few photosynthesis reactions (Chapter 10).

Let us now relate this model to the summary equation for photosynthesis (R9-3). That equation shows that for each molecule of CO_2 fixed, one O_2 is released and two H_2O molecules are used. The number of photons of light is not specified, but that number is important to calculate photosynthetic efficiencies. Our model requires two photons for each electron transported. Each H_2O provides two electrons, and two H_2O molecules are required, so our model predicts that eight photons would be required to oxidize two H_2O molecules, release one O_2, and provide four electrons. These four electrons could reduce two $NADP^+$, and two NADPH are indeed essential to reduce one CO_2 (Section 10.1). Thus based only on NADPH formation, our model shows an eight-photon requirement to accomplish reduction of one CO_2 molecule. For leaves of many kinds of plants, 15 to 20 photons per CO_2 are usually required (Ehleringer and Pearcy, 1983; Osborne and Garrett, 1983), but usually under the most ideal conditions only 12 photons are probably required (Ehleringer and Bjorkman, 1977).

To understand the apparent discrepancy between our model and the data, we must ask how much ATP is required to cause photosynthesis and how much our model indicates is provided. As will be shown in Section 10.1, three ATPs are needed to reduce one CO_2 to a simple carbohydrate, but slightly more ATP is needed to drive solute accumulation (Section 6.8), cytoplasmic streaming, and to form complex polysaccharides, proteins, and nucleic acids from each reduced CO_2. Therefore, our model ought to account for an excess of three ATP produced per two H_2O molecules split and per two NADPH formed. How much ATP is actually produced during photophosphorylation, and how does that process occur? The following section addresses these questions.

9.7 Photophosphorylation

Formation of ATP from ADP and Pi is highly unfavorable from a thermodynamic basis. Photophosphorylation can only occur because light energy somehow drives it. To understand how that occurs, note that the coupling factor (CF) of Fig. 9-8 causes ATP formation in the stroma and transport of H^+ from the thylakoid channel to the stroma. Each process is favored by the other, but ATP formation absolutely requires H^+ transport. The H^+ ions in thylakoid channels arise from oxidation of H_2O and PQH_2. These oxidations cause the H^+ concentration in the channel (pH 5) to become about 1000 times as great as in the stroma (pH 8) when photosynthesis is occurring. There is thus a strong H^+ diffusion gradient toward the stroma, but thylakoids are quite impermeable to H^+ and other ions except when

transported by CF. *This pH gradient across the membrane provides a powerful form of chemical potential energy largely responsible for driving photophosphorylation.*

The idea that pH gradients could provide energy for ATP formation in chloroplasts, mitochondria, and bacteria was first proposed by Peter Mitchell in England in 1961, but his ideas were not accepted by most biochemists for many years. (Mitchell finally received a Nobel prize for chemistry in 1978.) His theory is called the **chemiosmotic theory** (although it bears no clear relation to osmosis, which we described in early chapters). Direct evidence for the chemiosmotic theory was first obtained by photosynthesis researchers G. Hind and Andre Jagendorf at Cornell University in about 1963. Their work and Mitchell's perseverance sparked thousands of other studies, until now the theory is widely accepted. The theory even explains how **uncouplers** of photophosphorylation work. Uncouplers were named because they remove the interdependence (coupling) of electron transport and phosphorylation. Many are known, including the relatively simple NH_3 and dinitrophenol. Most of them act as "ferryboats," moving into thylakoid channels, picking up a proton, and carrying it back to the stroma where the pH is higher and where the proton is released and reacts with OH^- to form H_2O. Repeated action of such uncouplers destroys the pH gradient across the thylakoid membrane and prevents ATP formation, but electron transport is often sped because it is then thermodynamically easier for electron transport to cause a separation of H^+ across the membrane.

A continuing problem is that we don't understand the mechanism by which CF uses the energy released from the downhill movement of H^+ from channel to stroma to convert ADP and Pi to ATP. (A fairly simple treatment of this problem with historical aspects is given by Hinckle and McCarty (1978), and more recent and advanced reviews are by Hammes, 1983, and Amzel and Pedersen, 1983.) Nevertheless, the number of H^+ that must be transported to form one ATP can be measured without knowing the mechanism, and that number appears to be three. Now, consider that the split of two H_2O molecules will directly release four H^+ and that four more H^+ arise from two H_2O molecules during noncyclic electron transport at the PQH_2 oxidation step. Therefore, because eight photons are required by our model to split two H_2O molecules, these eight photons also provide eight H^+, nearly enough to form three ATP but not enough to form the more than three that are required to convert one CO_2 into complex compounds and maintain other cellular processes.

Formation of additional ATP arises from a pathway of electron and H^+ transport partly separate from the noncyclic pathway described previously. This pathway involves PS I, ferredoxin, the cytochrome b_6 and f complex, plastoquinones, and plastocyanin, but not PS II. Figure 9-8 (see light arrows at cytochrome b_6) shows how one photon of light absorbed only by LHCI (the light-harvesting pigment complex of PS I) could activate P700 and drive an electron all the way back to P700 via those electron carriers also used in the noncyclic pathway. Because the electron cycles from P700 back to P700, we call this **cyclic electron transport.** Absorption of two such photons causes two electrons to cycle and deposits two H^+ in the thylakoid channel when PQH_2 is oxidized. No H_2O is split, because PS II is not involved, so no NADPH is formed. But ATP is produced by CF in response to the decreased pH in the channel; the formation of ATP by this cyclic-electron-transport pathway is, therefore, called **cyclic photophosphorylation.**

Quantitatively, if eight photons involving both photosystems produce eight H^+ by noncyclic electron transport, and if four additional photons absorbed only by PS I produce four more H^+, the 12 photon total will yield 12 H^+ in the thylakoid channel. If exactly three H^+ must be transported by CF to form each ATP, these 12 photons would lead to four ATP molecules. We stated earlier that more than three ATP were needed to convert one CO_2 to complex compounds and that measurements with leaves showed that a minimum of twelve photons were required for each CO_2 used. With both cyclic and noncyclic electron transport, our model is consistent with experimental results; so we can rewrite R9-3 showing approximately how many photons of light are required to convert one CO_2 into a product we shall refer to as (CH_2O), as an abbreviation for carbohydrates such as sucrose and starch. In reality, this (CH_2O) represents all organic forms of carbon in the plant. Proteins and nucleic acids that require (on a weight basis) more ATP to be produced than polysaccharides are more abundant in actively growing cells than in mature cells where polysaccharides predominate. Our new equation (R9-8) therefore shows a *minimum* requirement of 12 photons, depending on the physiological status of the cells and allowing for some uncertainty in photophosphorylation.

$$CO_2 + 2H_2O + 12 \text{ photons} \rightarrow (CH_2O) + O_2 + H_2O \quad \text{(R9-8)}$$

No ATP nor NADPH appear in this summary, because their production is balanced by their use in CO_2 reduction.

9.8 Distribution of Light Energy between PS I and PS II

To maintain maximum efficiency of photosynthesis requires that each photosystem receives input of the same number of photons per unit time so that activation of P680 and P700 can cooperatively drive elec-

trons toward $NADP^+$ (Fig. 9-8). Admittedly, cyclic photophosphorylation occurs without involvement of PS II, but maximum efficiency of noncyclic electron transport and noncyclic photophosphorylation requires simultaneous activation of both photosystems. How can simultaneous activation occur in plastids containing more of one system than the other? The answer is that the photosystems adapt so that **energy spillover** occurs from PS II to PS I or even in the reverse direction. For energy spillover from PS II to PS I, actual movement of some of the LHCII pigments and proteins closer to PS I in stroma thylakoids occurs (Kyle et al., 1983). Here the LHCII pigments transfer more light energy to PS I and less to PS II. Movement is thought to occur because the proteins become more negatively charged by becoming phosphorylated through the action of ATP and a specific *protein kinase* that transfers phosphate from ATP to those proteins. Phosphate groups ionize (lose H^+) to create the additional negative charges on such proteins. These excess negative charges force LHCII proteins with associated pigments apart, and they somehow are attracted toward less negatively charged parts of stroma thylakoids.

The role of light in LHCII movement seems to be as follows: Preferential absorption of light by PS II causes reduction of numerous PQ molecules to PQH_2 (see Fig. 9-8); then these PQH_2 somehow activate the protein kinase (Allen et al., 1981; Allen, 1983; Barber, 1982; Kyle et al., 1983). For energy spillover from PS I to PS II, light absorbed preferentially by PS I causes oxidation of PQH_2 molecules, and this oxidation somehow activates a phosphatase that hydrolyzes the phosphate groups away from LHCII proteins. Loss of phosphate reduces their negative charge so that they then move back to grana thylakoids and donate light energy to PS II. It is still unclear how plastoquinones control activity of the protein kinase and phosphatase and how the resulting changes in charge of LHCII proteins cause them to fit better in one thylakoid than another. Nevertheless, this mechanism certainly increases the cooperativity of the two photosystems and represents an important mechanism to maximize photosynthetic efficiency.

9.9 Reduction Potentials and the Z-Scheme

Our explanation of electron transport above implies that the pathway was solved by learning locations of the two photosystems and the compounds that function with them. Even if we fully understood these locations, the implication would be only partly true. Many approaches were used. Laws of thermodynamics were also important, because these laws determine whether or not one compound is likely to donate an electron to another. Put simply, some com-

Phycobilisomes in Red and Blue-Green Algae

Elisabeth Gantt

Smithsonian Institution

Beth Gantt was born in a German-speaking border town in Yugoslavia, moved with her family to Czechoslovakia, then (illegally!) to English-occupied Germany after World War II. Finally, as a teenager in 1950, her family came to the United States, where she took up U. S. citizenship, made possible because her mother was a U. S. citizen. Some of the rest of her story appears in this essay. She is presently with the Smithsonian Institution.

As so often happens in science, phycobilisomes were discovered by a chance observation. While I was a postdoctoral fellow with Dr. S. F. Conti in the Microbiology Department of Dartmouth Medical School, my assignment was to determine if the primitive red alga *Porphyridium cruentum* possessed mitochondria. If it lacked mitochondria, it could have been considered as an organism in evolutionary transition from the prokaryotic blue-green algae (now considered by many to be cyanobacteria) to eukaryotic red algae. This was of particular interest because both groups have phycobiliproteins as photosynthetic accessory pigments, and it is appealing to think that chloroplasts might have originated as blue-green algal cells that invaded and then lived in red algal cells. By electron microscopy, we found the expected mitochondria, but we also became intrigued by small granules covering the photosynthetic lamellae (membranes). These granules could have been mistaken for ribosomes, except that they proved twice as large. We then hoped that we had discovered a new biological structure.

Although it is always exciting to announce a new structure to scientists, it is much more meaningful to know its composition or function. After results showed that the granular structure contained proteins, we knew that they contained phycobiliproteins. Others knew that phycobiliproteins were present in large amounts in red and blue-green algae, but nobody knew their structural locations. Furthermore, scientists concerned with the energy transfer between algal pigments had postulated that the phycobiliproteins must be closely associated with photosynthetic membranes, because that is where the energy absorbed by the phycobiliproteins is used in photosynthesis.

If the granules we discovered contained the phycobiliproteins, we should expect to find them only in plants containing red pigments. An ultrastructural survey was made that included numerous algal types, lower, and higher plants. It revealed that indeed only red- and blue-green algal species had such structures attached to their photosynthetic membranes. This finding was encouraging. Within another year we learned that the phycobiliproteins and the granules always existed together. Final proof of the granule identification, however, required that they be separated from other cell components to be structurally and spectrally identifiable. Once this was accomplished, it was time for a proper name, and *phycobilisome* was chosen. *Phyco* is derived from the Greek word for algae; *bilin* reflects the similarity in structure to bile pigments; and *some* is from the Greek word soma, meaning body.

Through evolution, aerobic photosynthetic organisms have adapted to the environment in various ways. Blue-green algae, particularly, have had ample opportunity to experiment with light-harvesting systems by being around for billions of years. Under bright sunlight, the light absorbed by chlorophyll is sufficient to run photosynthesis. When light is limiting, the role of the phycobiliproteins becomes particularly important, because they absorb maximally in the spectral region at which chlorophyll absorbs very little.

Very efficient energy transfer from phycobiliproteins to chlorophyll was known to occur, but this required a specific structural association within the phycobilisome and from the phycobilisome to the membrane. This I set out to explore when I came to the Radiation Biology Laboratory at the Smithsonian Institution. Since this laboratory is largely concerned with the effects of light on plants, it was an ideal place for investigating the phycobilisome structure. Several very able investigators have since worked with me on this problem. The major complication in the isolation and

pounds can easily accept electrons and are easily reduced, but such compounds only give up electrons with difficulty. Other compounds do not readily accept electrons but give them up easily. In thermodynamic terms that we can express quantitatively, reactions involving the former compounds have relatively positive **reduction potentials,** and reactions involving the latter compounds have relatively negative reduction potentials. The symbol used to express the standard reduction potential at pH 7 is $\mathbf{E'_0}$. Only if

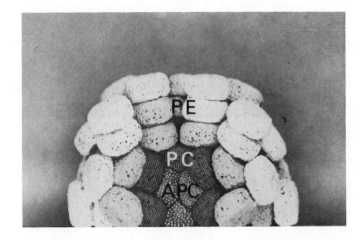

Figure 9-10 This model of a phycobilisome, as it is expected to occur on red algal chloroplasts, *Porphyridium cruentum,* is composed of discs that represent phycoerythrin (PE), phycocyanin (PC), and allophycocyanin (APC). Such an arrangement allows for maximum energy transfer to chlorophyll.

purification of phycobilisomes was their ready solubility in most physiologically compatible media. Had the phycobilisomes been enclosed in a membrane, it would not have required so many experiments, by trial and error, to find a suitable isolation and stabilization medium. Now that an isolation procedure is worked out, it is simple to obtain functionally intact phycobilisomes in a matter of hours. They are most stable in high ionic strength buffers at room temperature.

Phycobilisomes of *Porphyridium cruentum* can be separated into their individual components. The major ones are *phycoerythrin* (a red pigment that absorbs green light), *phycocyanin* (a blue pigment that absorbs orange-red light), and *allophycocyanin* (a blueish-green pigment that absorbs red light). These are linked together by special proteins present in concentrations much lower than those of the pigments. The phycobiliprotein arrangement is shown in the model (Fig. 9-10). In constructing such a model, we had to determine the types of pigments present, the amount of each, and their location. This required many techniques such as kinetic dissociation, protein purification, antisera preparation, immunoelectron microscopy, gel electrophoresis, and fluorescence spectroscopy.

Our greatest excitement came when we discovered, from fluorescence spectra, that a special type of allophycocyanin is the key pigment in the transfer chain. Es-sentially, all the energy absorbed by the other phycobiliproteins is transferred to this terminal pigment, which then funnels it to chlorophyll in the photosynthetic membrane. As long as there is one molecule of this terminal pigment per phycobilisome, the energy transfer will continue because it can transfer the energy much faster than it receives it.

We have begun to understand what it is about phycobilisomes that makes them so efficient, but only when we can control the phycobilisome assembly in the laboratory can we expect to understand them fully. This has now begun in several laboratories. Recent evidence indicates that the terminal phycobilisome pigments in various species have significant similarities, lending support to the hypothesis of a phylogenetic relationship of blue-green algae and chloroplasts of red algae.

For additional information, you may consult:

Gantt, E. 1980. Structure and function of phycobilisomes: light harvesting pigment complexes in red and blue-green algae. Internatl. Rev. Cytol. (eds. G. H. Bourne and J. F. Danielli), 66:45–80. Academic Press, New York.

Gantt, E. 1981. Phycobilisomes. Annual Review of Plant Physiology 32:327–347.

Glazer, A. N. 1982. Phycobilisomes: structure and dynamics. Ann. Rev. Microbiol. 36:173–198.

the E'_0 value for a reaction involving one compound is more negative than that for another compound can the first compound be expected on thermodynamic grounds to readily donate an electron to the second.

For electron transport from H_2O to $NADP^+$, H_2O is difficult to oxidize, because O_2 attracts electrons and protons so strongly (is so easy to reduce). The E'_0 for that reaction is $+0.815$ V, or written in R9-9 as the oxidation reaction occurring in photosynthesis, E'_0 is -0.815 V. Reduction of $NADP^+$ is easier ther-

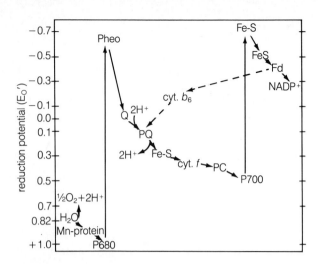

Figure 9-11 The Z-scheme of electron transport. Abbreviations are as in Figure 9-8 and text.

modynamically, and E'_0 is -0.32 V (R9-10). If we add these reactions to obtain R9-11, we define the overall process of electron transfer from H_2O to $NADP^+$ and arrive at a change in E'_0 ($\Delta E'_0$) of -1.14 V.

$$H_2O \rightarrow 1/2\ O_2 +\ 2e +\ 2H^+ \qquad E'_0 = -0.82\ V \qquad \text{(R9-9)}$$

$$NADP^+ + 2e + H^+ \rightarrow NADPH \qquad E'_0 = -0.32\ V \qquad \text{(R9-10)}$$

$$\overline{H_2O + NADP^+ \rightarrow}$$
$$\quad 1/2\ O_2 + H^+ + NADPH \qquad \Delta E = -1.14\ V \qquad \text{(R9-11)}$$

The standard Gibbs free energy change at pH 7 ($\Delta G'_0$) for redox reactions can be calculated from $\Delta E'_0$ values by the equation $\Delta G'_0 = -nF\Delta E'_0$, where n is the number of electrons transferred in the reaction and F is the Faraday constant (96,490 J or 23,062 cal per eV

equivalent). For R9-11, $\Delta G'_0 = (-2$ equiv.$) \times (96,490$ joules/eV equiv.$) \times (-1.14$ electron volts$) =$ approx. 220 kJ or 52.6 kcal. The highly positive value obtained demonstrates the strongly endergonic nature of photosynthesis and indicates how important light energy is for driving the process.

By isolating suspected components of the electron-transport pathway from H_2O to $NADP^+$ and measuring E'_0 values for redox reactions in which they participate, biochemists attempted to place each component next to another with a similar E'_0 value. The goal was to arrange them such that electrons reasonably flowed downhill energetically, from more negative to less negative (or more positive) E'_0 values. The problem became easier after it was recognized that two photosystems were involved and that two photons acted separately to push an electron partway "uphill" on the energy scale. This led to a **Z-scheme** of electron transport first proposed by R. Hill and F. Bendall in 1960. Our Fig. 9-8 is based partly on Z-scheme models described in recent books and reviews, e.g., Kok (1976), Trebst and Avron (1977), Hatch and Boardman (1981), Haliwell (1981), and Cogdell (1983). In Fig. 9-11 we show a Z-scheme in which approximate E'_0 values are listed at the left. These values represent the reduction potential for each redox reaction involving two adjacent compounds in the electron transport pathway. Abbreviations for compounds and specific reactions are as described previously. The advantage of such a model is that it allows you to follow electron transport and to see that each reaction (except the two driven by light) is accompanied by a loss of energy as an electron is transferred downhill toward a compound with a more positive or less negative E'_0 value. Remember, however, that this model, like others, will be modified as research continues.

10

Carbon Dioxide Fixation and Carbohydrate Synthesis

In Chapter 9 we explained how chloroplasts capture light energy to produce NADPH and ATP. Those two molecules are called *reducing power* by some authors, because both help reduce CO_2 after it is fixed into the carboxyl group of a 3-carbon acid, as explained in this chapter. Strictly speaking, however, only NADPH is a reductant; ATP only facilitates reduction. Although plants vary in the mechanisms that fix CO_2 into organic acids, NADPH and ATP are always involved in reduction in such a way that carbohydrates are formed from the CO_2 molecules fixed.

The sequence of reactions involving CO_2 fixation and formation of the complex carbohydrates produced in photosynthesis was solved only after radioactive carbon-14 became available about 1945. Carbon dioxide containing ^{14}C was then prepared, and all molecules produced from it during photosynthetic experiments were tagged with that isotope. Paper chromatography was developed at about the same time, making separation of photosynthetic products possible.

Labeled molecules on paper chromatograms prepared from alcohol extracts of plants that had fixed $^{14}CO_2$ were detected by autoradiography (Section 7.1). In this technique, X-ray film is placed in tight contact with chromatograms in darkness for a few days to a few weeks (depending on the amount of radioactivity present), while radioactivity exposes the film. When the film is developed, dark spots indicate the locations of radioactive compounds. Identities of the compounds and amounts of radioactivity in each can be determined after cutting out the areas of paper that correspond to each dark spot on the film. To measure radioactivity Geiger-Müller tubes were used at first, but now the far more sensitive liquid-scintillation counters are employed. Direct chemical analyses and rechromatography of unknown substances on paper with known compounds (**cochromatography**) were first used to identify radioactive photosynthetic products, but now nuclear

magnetic resonance (NMR) and mass spectrometry provide powerful additional tools for analysis. High-performance liquid chromatography (HPLC) is now a much better and faster separation technique than paper chromatography. These techniques have given us a fairly complete understanding of the pathways of CO_2 fixation and carbohydrate synthesis.

10.1 Products of Carbon Dioxide Fixation

The First Product Paper chromatographic procedures in conjunction with use of $^{14}CO_2$ were applied to the problem of photosynthesis by Melvin Calvin, Andrew A. Benson, James A. Bassham, and others at the University of California in Berkeley, from 1946 to 1953 (see personal essays at the end of this chapter by Drs. Calvin and Bassham). They allowed unicellular green algae such as *Chlorella* to attain a constant rate of photosynthesis and then introduced $^{14}CO_2$ into solutions in which the algae were growing. At various times after the introduction of ^{14}C, algae were dropped into boiling 80 percent ethanol to kill them rapidly and extract any metabolites. Each extract was chromatographed on paper, and autoradiograms were made.

After photosynthesizing 60 seconds in $^{14}CO_2$ the algae had formed many compounds, as shown by the many dark spots on the film in Fig. 10-1a. Amino acids and other organic acids had become radioactive, but the compounds containing most ^{14}C were phosphorylated sugars, shown at the lower right of that autoradiogram. To identify the first product formed from CO_2, time periods were shortened to 7 s and finally even to as little as 2 s (Fig. 10-1c). When this was done, most ^{14}C was found in a phosphorylated three-carbon acid called **3-phosphoglyceric acid (3-PGA).** This acid was the first detectable product of photosynthetic CO_2 fixation in these algae and in leaves of certain plants. (Note that 3-PGA and

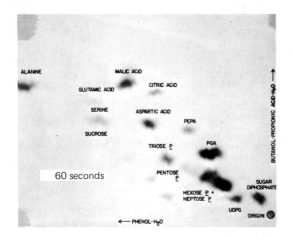

60 seconds

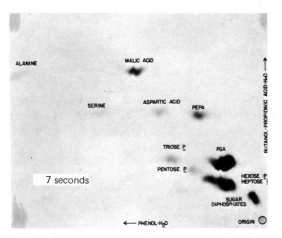

7 seconds

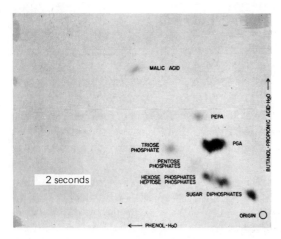

2 seconds

Figure 10-1 Autoradiograms showing the products of photosynthesis in the alga *Chlorella pyrenoidosa* after various times of exposure to $^{14}CO_2$. (Top) 60 seconds, (middle) 7 seconds, (bottom) 2 seconds. Note the increasing importance of 3-PGA and other sugar phosphates as the exposure time is shortened. (From J. A. Bassham, 1965, in J. Bonner and J. E. Varner, eds. Plant Biochemistry, Academic Press, Inc., New York, pp. 883–884.)

most plant acids exist largely in the ionized form, without the H^+ that exist in the carboxyl groups of true acids. Ionized acids exist as salts, with a counter cation, usually K^+; hence, 3-PGA exists as negatively charged 3-phosphoglycerate.)

The Compound Combining with CO_2 The search then began for a two-carbon compound with which CO_2 could react to form 3-PGA. When $^{14}CO_2$ was fed to algae for a short time and the CO_2 supply then suddenly removed, it was expected that the compound with which $^{14}CO_2$ normally combines would accumulate. No two-carbon compound was found. The substance that did accumulate was a five-carbon sugar, phosphorylated at each end, **ribulose-1,5-bisphosphate (RuBP).*** At the same time, there was a rapid drop in the level of labeled 3-PGA. This suggested that ribulose bisphosphate is the normal substrate to which CO_2 is added to form PGA.

The Reaction of CO_2 Fixation Soon (1954) an enzyme was found that catalyzes irreversibly the combination of CO_2 with RuBP to form two molecules of 3-PGA. This is an unusually important reaction. An unstable intermediate product is formed (shown in brackets in R10-1) that splits into two 3-PGAs with addition of H_2O. Thus, if $^{14}CO_2$ is reacting (see * in R10-1), one of the two 3-PGA products becomes labeled with ^{14}C and one remains unlabeled:

$$C^*O_2 + H-C-OH \xrightarrow{Mg^{2+}}$$

with the structure:

$$\begin{array}{c} CH_2OPO_3H^- \\ | \\ C=O \\ | \\ H-C-OH \\ | \\ H-C-OH \\ | \\ CH_2OPO_3H^- \end{array}$$

$$\left[\begin{array}{c} O \quad\quad CH_2OPO_3H^- \\ \| \quad\quad\quad | \\ C^*-C-OH \\ / \quad\quad\quad | \\ OH \quad\quad C=O \\ \quad\quad\quad\quad | \\ \quad\quad H-C-OH \\ \quad\quad\quad\quad | \\ \quad\quad CH_2OPO_3H^- \end{array} \right] \xrightarrow{+ H_2O}$$

$$\begin{array}{c} CH_2OPO_3H^- \\ | \\ H-C-OH \\ | \\ C^*OOH \\ \\ \text{3-PGA} \end{array} \quad + \quad \begin{array}{c} COOH \\ | \\ H-C-OH \\ | \\ CH_2OPO_3H^- \\ \\ \text{3-PGA} \end{array} \quad \text{(R10-1)}$$

*This compound and others with phosphate groups on separate carbon atoms of a molecule were first called diphosphates (thus, ribulose-1,5-diphosphate), but the term *bis*phosphate now distinguishes them from true diphosphates such as ADP in which the two phosphates are linked only to one carbon in a pyrophosphate group.

The enzyme that catalyzes this reaction is commonly called **ribulose bisphosphate carboxylase,** which we shall abbreviate as **rubisco.** Rubisco is functional in all photosynthetic organisms with the exception of a few photosynthetic bacteria. It is important not only because of the essential reaction it catalyzes, but also because it seems to be by far the most abundant protein on earth (Ellis, 1979; Miziorko and Lorimer, 1983). Chloroplasts contain approximately half of the total protein in leaves, and about one-fourth to one-half of their total protein is rubisco, so one-eighth to one-fourth of leaf protein exists as this enzyme, which is therefore important in diets of animals, including ours.

10.2 The Calvin Cycle

Further investigations of radioactive compounds that are formed rapidly from $^{14}CO_2$ showed other sugar phosphates containing four, five, six, and seven carbon atoms. These included the tetrose (4-carbon) phosphate *erythrose-4-phosphate*; the pentose (5-carbon) phosphates *ribose-5-phosphate, xylulose-5-phosphate,* and *ribulose-5-phosphate*; the hexose (6-carbon) phosphates *fructose-6-phosphate, fructose-1,6-bisphosphate,* and *glucose-6-phosphate*; and the heptose (7-carbon) phosphates *sedoheptulose-7-phosphate* and *sedoheptulose-1,7-bisphosphate*. By noting time sequences in which each sugar phosphate became labeled from $^{14}CO_2$ and then degrading each to determine which atoms contained ^{14}C, it was possible to predict a metabolic pathway that related them (Bassham, 1965, 1979).

When ^{14}C-labeled 3-PGA was degraded, most ^{14}C was in the carboxyl carbon, as shown by asterisks in R10-1, but the other two carbons were also labeled. This labeling pattern suggested that the latter two carbons of 3-PGA were not derived from $^{14}CO_2$ directly but were instead formed by transfer of carbon from the carboxyl carbon atom of 3-PGA by some cyclic process. A cyclic pathway that uses 3-PGA to form other sugar phosphates mentioned and that also converts some of its carbons back to RuBP was soon worked out. These reactions have collectively been named the **Calvin cycle,** the **photosynthetic carbon reduction cycle,** or the **C-3 photosynthetic pathway** (because the first product, 3-PGA, contains three carbons). Calvin was awarded a Nobel Prize in 1961 for this work.

The Calvin cycle occurs in the stroma of chloroplasts and consists of three main parts: **carboxylation, reduction,** and **regeneration,** as explained below and summarized in Fig. 10-2. Carboxylation involves addition of CO_2 and H_2O to RuBP to form two molecules of 3-PGA (R10-1). Reduction is of the carboxyl group in 3-PGA to an aldehyde group in 3-phosphoglyceraldehyde (3-PGaldehyde), as in (R10-2):

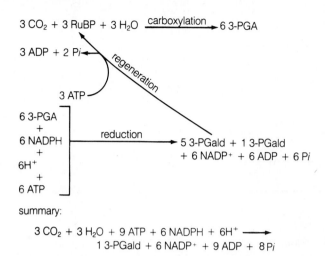

Figure 10-2 A summary of the Calvin cycle, emphasizing carboxylation, reduction, and regeneration phases.

Note that reduction does not occur directly but that the carboxyl group of 3-PGA is first converted to an acid anhydride type of ester in 1,3-bisphosphoglyceric acid (1,3-bisPGA) by addition of the terminal phosphate group from ATP. This ATP arises from photosynthetic phosphorylation (described in Chapter 9), and the ADP released when 1,3-bisPGA is formed is quickly converted back to ATP by additional photosynthetic phosphorylation reactions. The actual reducing agent in R10-2 is NADPH, which donates two electrons to the top carbon atom involved in the anhydride ester group. Simultaneously, Pi is released from that group to be used again to convert ADP to ATP. The $NADP^+$ is reduced back to NADPH in light-driven reactions described in Chapter 9 (Fig. 9-8). R10-2 represents the only reduction step in the entire Calvin cycle; because both of the

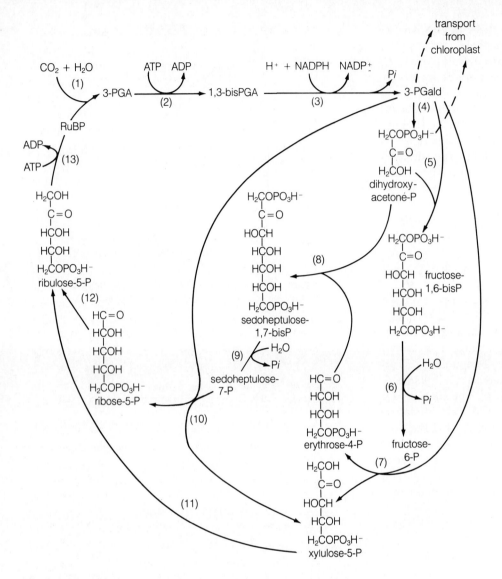

Figure 10-3 Reactions of the Calvin cycle. Detailed reactions include (**1**), upper left, fixation of CO_2 into 3-PGA catalyzed by rubisco, (**2**) phosphorylation of 3-PGA by ATP to form 1,3-*bis*PGA, catalyzed by phosphoglycerokinase, (**3**) reduction of 1,3-*bis*PGA to 3-PGaldehyde, (**4**) isomerization of 3-PGaldehyde to form dihydroxyacetone phosphate, catalyzed by triose phosphate isomerase, (**5**) aldol combination of 3-PGaldehyde and dihydroxyacetone-P to form fructose-1,6-*bis*-P, catalyzed by aldolase, (**6**) hydrolysis of the phosphate from C-1 of fructose-*bis*P to form fructose-6-P, catalyzed by fructose-1,6-bisphosphate phosphatase, (**7**) transfer of upper two carbons of fructose-6-P to 3-PGaldehyde to form the 5-carbon xylulose-5-P, releasing the 4-carbon erythrose-4-P, catalyzed by transketolase, (**8**) aldol combination of erythrose-4-P and dihydroxyacetone-P to form sedoheptulose-1,7-*bis*P, catalyzed again by aldolase, (**9**) hydrolysis of phosphate from C-1 of sedoheptulose-1,7-*bis*P to form sedoheptulose-7-P, (**10**) transfer of the upper two carbons of sedoheptulose-7-P to 3-PGaldehyde to form ribose-5-P and xylulose-5-P, catalyzed again by transketolase. (Note that this is the fourth reaction the 3-PGaldehyde undergoes within the chloroplast, and that it can also be transported out of the chloroplast, upper right.) In reaction (**11**), xylulose-5-P is isomerized by an epimerase to form another pentose phosphate, ribulose-5-P. This ribulose-5-P can also be formed in reaction (**12**) by a different isomerase that uses ribose-5-P as the substrate. In reaction (**13**), ribulose-5-P is converted to RuBP by ribulose-5-P kinase, allowing CO_2 fixation to occur again.

3-PGA molecules produced in R10-1 are reduced in the same way, this step involves use of two of the three ATP required to convert one CO_2 molecule into part of a carbohydrate. Thus, for each CO_2 fixed, two NADP and two ATP are required here. A third ATP is used in the regeneration phase, *making the total requirement 3 ATP and 2 NADPH for each molecule of CO_2 fixed and reduced.*

Regeneration is of RuBP, needed to react with additional CO_2 constantly diffusing into algae or into leaves through stomates. This phase is complex and involves phosphorylated sugars with four, five, six, and seven carbons, as shown in detail in Fig. 10-3. In the final reaction of the Calvin cycle (number 13), the third ATP that is required for each molecule of CO_2 fixed is used to convert ribulose-5-phosphate to RuBP; the cycle then begins again.

We emphasize that three turns of the cycle fix three CO_2 molecules, and there is a net production of one 3-PGaldehyde. Some 3-PGaldehyde molecules are used in chloroplasts to form starch, a major photosynthetic product in most species when photosynthesis is occurring rapidly. Others are transported out of chloroplasts via an antiport carrier system (Section 6.8) in exchange for either Pi or 3-PGA from the cytoplasm (Heldt, 1976; Walker, 1976). Still others are converted to dihydroxyacetone phosphate, a similar 3-carbon triose phosphate that can be transferred out of chloroplasts by the same antiport system. This system helps keep the total amount of phosphate constant in the chloroplast but leads to the net appearance of triose phosphates in the cytosol (Heber and Walker, 1979). These triose phosphates are used in the cytosol to form sucrose, cell-wall polysaccharides, and hundreds of other compounds of which the plant is made. Their transport is especially important, because the numerous other sugar phosphates of the Calvin cycle are largely held within the chloroplast (Walker, 1976).

10.3 The C-4 Dicarboxylic Acid Pathway: Some Species Fix CO_2 Differently

Reactions of the Calvin cycle were thought to have solved CO_2 fixation and reduction in plants, but a new era in photosynthesis research came from a discovery of H. P. Kortschak, C. E. Hartt, and G. O. Burr, made in Hawaii in 1965. They found that sugarcane leaves, in which photosynthesis is unusually rapid and efficient, fix most CO_2 initially into carbon-4 of malic and aspartic acids. After approximately one second of photosynthesis in $^{14}CO_2$, 80 percent of ^{14}C fixed was in those two acids, and only 10 percent was in PGA, indicating that in this plant,

3-PGA is not the first product of photosynthesis. These results were quickly confirmed by M. D. Hatch and C. R. Slack in Australia, who found that some grass species of tropical origin, including maize (corn), displayed similar labeling patterns after fixing $^{14}CO_2$ (Hatch, 1977). Other grasses such as wheat, oat, rice, and bamboo, not closely related taxonomically to tropical grasses, gave 3-PGA as the predominant fixation product.

The new results of Kortschak et al. and Hatch and Slack showed that the primary carboxylation reaction of some species is different from that involving ribulose bisphosphate. Species that produce 4-carbon acids as the primary initial CO_2 fixation products are now commonly referred to as **C-4 species;** those fixing CO_2 initially into 3-PGA are called **C-3 species.** There are, however, a few with intermediate properties, and these are thought to represent evolutionary transitions from C-3 to C-4 species (Rathnam and Chollet, 1980). Most C-4 species are monocots, especially grasses and sedges, although more than 300 are dicots. Among the grasses, sugarcane, maize, and sorghum are important agricultural crops, and numerous range grasses (especially in southern latitudes) are also C-4 plants. The C-4 pathway occurs in certain members of more than 900 known species of angiosperms distributed among at least 19 families. At least 11 genera include both C-4 and C-3 species. Extensive lists of C-4 species are given in references by Krenzer et al. (1975), Downton (1975), Winter and Troughton (1978), and (for North American grasses) by Waller and Lewis (1979). Numerous C-4 sedges are listed by Hesla et al. (1982). All gymnosperms, pteridophytes, bryophytes, and algae that have been studied are C-3 plants, as are nearly all trees and shrubs. A few *Euphorbia* tree or shrub species from Hawaii have the C-4 pathway (Pearcy, 1983).

Considering that there are about 285,000 species of flowering plants (monocots and dicots), the presence of the C-4 pathway in roughly 0.4 percent of those investigated seems hardly worth mentioning. The great attention they have received arose largely because of the economic importance of some of them and because under high light and warm temperatures they can photosynthesize more rapidly and produce substantially more biomass than can C-3 plants. Investigations of the C-4 pathway have also taught us much about limitations to photosynthesis in C-3 plants and have furthered general ecological knowledge about factors that control productivity in various climates (topics covered in Chapters 11 and 23).

The reaction by which CO_2 (actually HCO_3^-) is converted into carbon-4 of malate and aspartate (R10-3) is through its initial combination with **phosphoenolpyruvate (PEP)** to form **oxaloacetate** and Pi.

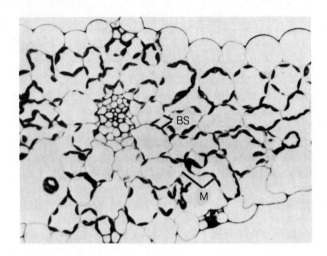

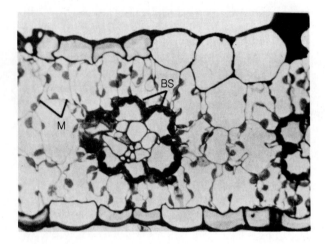

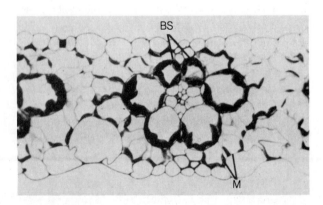

Figure 10-4 Leaf cross sections of a C-3 monocot (oat, top) and C-4 monocots (maize, middle, and Rhodesgrass, bottom). (From S. E. Frederick and E. H. Newcomb, 1971, Planta 96:152–174.)

Oxaloacetate is not usually a detectable product of photosynthesis, but it can be found when precautions are taken to prevent its rapid conversion to malic and aspartic acids and to prevent its unusual susceptibility to destruction in isolation and chromatography procedures.

$$H^*CO_3^- + \begin{array}{c} COOH \\ | \\ C-OPO_3H^- \\ \| \\ CH_2 \end{array} \rightarrow \begin{array}{c} COOH \\ | \\ C=O \\ | \\ CH_2 \\ | \\ {}^*COOH \end{array} + Pi$$

PEP   oxaloacetic acid   (R10-3)

Phosphoenolpyruvate carboxylase (PEP carboxylase), an enzyme that is apparently present in all living plant cells, is the catalyst involved (O'Leary, 1982). The reasons it is of special importance in leaves of C-4 species are that it is unusually abundant there and that a cyclic pathway that maintains a constant and relatively plentiful supply of PEP is also present. In leaves of C-3 species and in root, fruit, and other cells that lack chlorophyll, regardless of the species, other isozymes of PEP carboxylase are present. Here the major function of the enzyme seems to be to help replace Krebs-cycle acids used in synthetic reactions (Chapter 12) and to help form malate needed in charge-balancing functions. In all cases, PEP carboxylase appears to reside in the cytosol outside any organelle such as the chloroplast.

Reactions converting oxaloacetate to malate and aspartate in C-4 plants are shown in R10-4 and R10-5:

$$\begin{array}{c} COOH \\ | \\ C=O \\ | \\ CH_2 \\ | \\ {}^*COOH \end{array} \xrightarrow{NADPH + H^+} \begin{array}{c} COOH \\ | \\ HO-C-H \\ | \\ CH_2 \\ | \\ {}^*COOH \end{array} + NADP^+$$

oxaloacetic acid   L-malic acid   (R10-4)

$$\begin{array}{c} CH_3 \\ | \\ H-C-NH_2 \\ | \\ COOH \end{array} \quad \begin{array}{c} COOH \\ | \\ H-C-NH_2 \\ | \\ CH_2 \\ | \\ {}^*COOH \end{array} + \begin{array}{c} CH_3 \\ | \\ C=O \\ | \\ COOH \end{array}$$

L-alanine   L-aspartic acid   pyruvic acid   (R10-5)

Formation of malate in R10-4 is catalyzed by *malate dehydrogenase,* with the necessary electrons provided

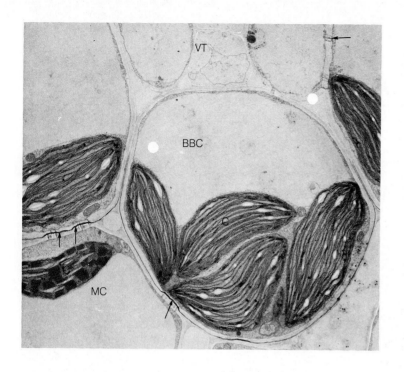

Figure 10-5 Electron micrograph of adjacent mesophyll cell (MC) and bundle sheath (BBC) in the C-4 plant crabgrass (*Digitaria sanguinalis*). Note abundant grana and lack of starch in the mesophyll cell chloroplast, but absence of grana and presence of several small starch granules in bundle sheath chloroplasts. Arrows mark plasmodesmata where passage of organic acids is suspected to occur. Vascular tissue (VT) is shown at top. (From C. C. Black et al., 1973.)

by NADPH. Interestingly, malate dehydrogenase is a chloroplast enzyme, which means that oxaloacetate must move into the chloroplast for reduction to malate. This movement occurs by another chloroplast antiport or shuttle system in which oxaloacetate is transported in on a carrier that also moves malate out. Formation of aspartate from malate occurs in the cytosol and requires another amino acid such as alanine as the source of an amino group. This type of reaction is referred to as **transamination,** because transfer of an amino group is involved (Chapter 13).

It became evident that there is a division of labor between two different kinds of photosynthetic cells in C-4 species, mesophyll cells and bundle sheath cells (Campbell and Black, 1982). Both kinds of cells are required to produce sugars, starch, and other plant products. One, or occasionally two, distinct layers of tightly packed, often thick-walled **bundle sheath cells** almost always surrounds the leaf vascular bundles and separates them from the predominant mesophyll cells. This concentric arrangement of bundle sheath cells is described as **Kranz** (German, halo or wreath) **anatomy** (Laetsch, 1974). As contrasted to C-3 plants, in which a much less distinct bundle sheath is often also present, bundle sheath cells of many C-4 plants contain far more chloroplasts, mitochondria, other organelles, and smaller central vacuoles. A comparison of cross-sectional leaf anatomy in representative C-3 and C-4 grasses is shown in Fig. 10-4. Chloroplasts of bundle sheath cells frequently contain nearly all the leaf

starch, with little present in chloroplasts of more loosely arranged surrounding mesophyll cells. Studies with separated bundle sheath and mesophyll cells confirmed earlier suggestions that malate and aspartate are formed in mesophyll cells, and that 3-PGA, sucrose, and starch are produced mainly in bundle sheath cells. Rubisco exists only in bundle sheath cells, as do most Calvin cycle enzymes, so the complete Calvin cycle occurs only in bundle sheath cells. On the other hand, PEP carboxylase occurs mainly in mesophyll cells. Thus C-4 species really use both kinds of CO_2 fixing mechanisms.

The reason CO_2 first appears in malate and aspartate is largely that it is mesophyll cells into which CO_2 first penetrates after stomatal entry, because the activities of PEP carboxylase are high there, and because rubisco is not present there. Most CO_2 that has been recently fixed into carboxyl groups of malate and aspartate is rapidly transferred, perhaps via abundant plasmodesmata (see arrows in Fig. 10-5), into bundle sheath cells. Here those compounds undergo decarboxylation with release of CO_2 which is then fixed by rubisco into 3-PGA. The principal source of CO_2 for bundle sheath cells is therefore C-4 acids formed in the mesophyll. Sucrose and starch are ultimately formed from 3-PGA in bundle sheath cells using Calvin cycle reactions and other reactions not yet mentioned. The division of labor referred to above involves the trapping of CO_2 into C-4 acids by mesophyll cells and after transfer of these acids to bundle sheath cells, the decarboxylation and re-

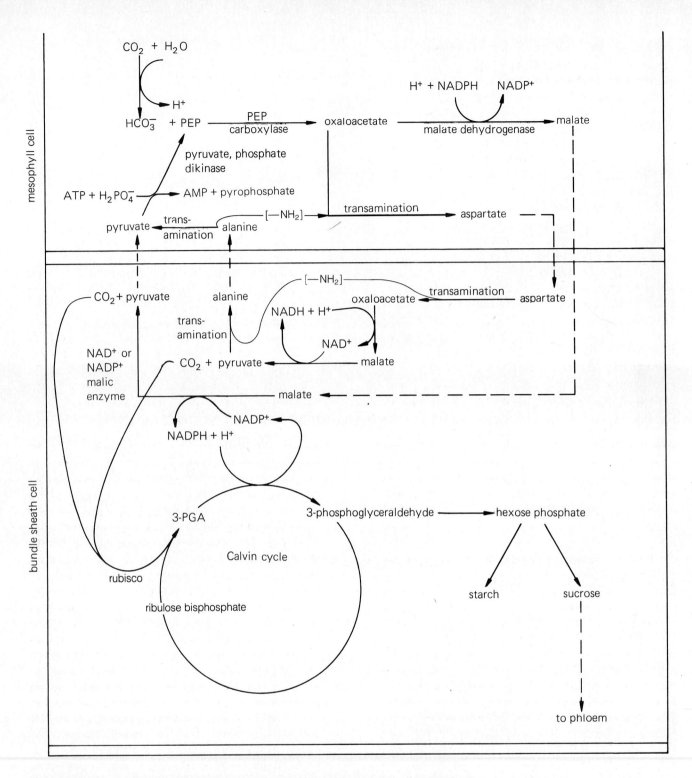

Figure 10-6 A summary of metabolic division of labor in mesophyll and bundle sheath cells of C-4 plants. CO_2 is initially fixed into C-4 acids in the mesophyll, then these acids move into the bundle sheath (probably as K^+ salts) where they are decarboxylated. The CO_2 thus released is fixed via the Calvin cycle in the bundle sheath chloroplasts. Sucrose and starch are common products, as shown. After decarboxylation of the C-4 acids, three-carbon molecules such as pyruvate and alanine move back to the mesophyll cells where they are converted to PEP so that CO_2 fixation can continue there.

fixation of CO_2. The 3-carbon acids (pyruvate and alanine) resulting from decarboxylation of C-4 acids are then returned to mesophyll cells where they are again carboxylated to keep the cycle going. The C-4 acids formed in mesophyll cells seem only to be carriers of CO_2 to bundle sheath cells. This idea is illustrated in our model of the C-4 pathway and its relation to the Calvin cycle in Fig. 10-6.

Two additional aspects of this model require explanation. First, by what mechanisms are aspartate and malate decarboxylated in bundle sheath cells? Surprisingly, three such mechanisms were discovered, which depend on the species, two of which are shown in Fig. 10-6. In the so-called **aspartate formers** (species that form more aspartate than malate), aspartate that moves into bundle sheaths is changed back to oxaloacetate by transamination. Next, that oxaloacetate is reduced to malate by an *NADH-dependent malate dehydrogenase*. (**NADH** is a coenzyme capable of transferring two electrons and is nearly identical in structure to NADPH, as explained in Fig. 9-10. The oxidized form of NADH is NAD^+. These two coenzymes operate mainly in mitochondria, while $NADP^+$ and NADPH are much more important in chloroplasts.) The malate is oxidatively decarboxylated by a *malic enzyme* that uses NAD^+ as the electron acceptor. *Pyruvate*, CO_2, and NADH are the products. The pyruvate is then converted to alanine by another transamination. As alanine moves back to mesophyll cells, the nitrogen in it replaces that lost when aspartate was transported to bundle sheath cells.

Malate formers (species that form mostly malate) transfer malate into bundle sheaths where it is also oxidatively decarboxylated to CO_2 and pyruvate, but with a malic enzyme that uses $NADP^+$ in strong preference to NAD^+ (Fig. 10-6, left center). The NADPH formed by this enzyme helps reduce 3-PGA to 3-PGaldehyde (see R10-2), as indicated in the model. The third decarboxylation system, not shown in Fig. 10-6, operates mainly in aspartate formers and involves oxaloacetate formed in bundle sheaths. In this system, catalyzed by *PEP carboxykinase*, oxaloacetate reacts with ATP to release CO_2, PEP, and ADP. The CO_2 is fixed by rubisco and converted to carbohydrates by the Calvin cycle, and the ADP is converted back to ATP by photosynthetic phosphorylation.

The second aspect of our model that requires explanation is how the 3-carbon acids that are transported back to the mesophyll cells regenerate PEP needed for continued fixation of CO_2 there. Alanine is converted back to pyruvate by another transamination, then an unusual chloroplast enzyme named *pyruvate, phosphate dikinase* uses ATP to convert pyruvate to PEP and PPi:

$$
\begin{array}{l}
\text{COOH} \\
| \\
\text{C}{=}\text{O} + \text{ATP} + \text{P}i \ \rightleftharpoons \\
| \\
\text{CH}_3 \\
\text{pyruvic acid}
\end{array}
$$

$$
\begin{array}{l}
\text{COOH} \qquad\qquad\qquad\quad \text{O}^- \quad\ \text{O}^- \\
| \qquad\qquad\qquad\qquad\qquad\ \ | \qquad\ | \\
\text{C}{-}\text{OPO}_3\text{H}^- + \text{AMP} + \text{HO}{-}\text{P}{-}\text{O}{-}\text{P}{-}\text{OH} \quad \text{(R10-6)} \\
\| \qquad\qquad\qquad\qquad\qquad\quad \| \qquad\quad \| \\
\text{CH}_2 \qquad\qquad\qquad\qquad\qquad \text{O} \qquad\ \ \text{O} \\
\text{PEP} \qquad\qquad\qquad\qquad \text{pyrophosphate}
\end{array}
$$

In leaves, pyruvate, phosphate dikinase has been found only in mesophyll cells of C-4 plants and in CAM plants (Section 10.6), but nowhere in leaves of C-3 plants, so that enzyme presumably evolved for the unique task of converting pyruvate back to PEP.

Calculations of the energy required to operate the C-4 pathway, with the additional Calvin cycle, indicate that for each CO_2 fixed two ATP besides the three needed in the Calvin cycle are required. These two additional ATP are necessary to regenerate ATP from AMP (by reactions not shown), so that PEP synthesis can be maintained for continued CO_2 fixation. There is no additional NADPH requirement, because for each NADPH used to reduce oxaloacetate in the mesophyll cells, one is also regained during malic enzyme action in bundle sheath cells. In spite of the apparent inefficiency with respect to ATP utilization in C-4 species, these plants almost always show more rapid rates of photosynthesis on a leaf area basis than do C-3 species *when both are exposed to high light levels and warm temperatures*. The C-4 species are adapted to and apparently evolved in areas of periodic drought, such as tropical savannas. When temperatures reach 25 to 35°C and light irradiances are high, C-4 plants are about twice as efficient as C-3 plants in converting the sun's energy into production of dry matter. Ecological and environmental aspects of photosynthesis are deferred until the next chapter, but it is useful to mention now that the comparatively low efficiency of most C-3 species results largely from light-enhanced *loss* of part of the CO_2 that they fix by a phenomenon called photorespiration; little or no such loss occurs in C-4 plants.

10.4 Photorespiration

Otto Warburg, a famous German biochemist who devoted much of his attention to photosynthesis, recorded in 1920 that photosynthesis in algae is inhibited by O_2. This inhibition occurs in all C-3 species

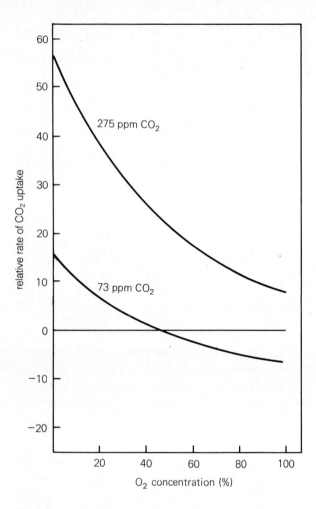

Figure 10-7 The Warburg effect: inhibition of photosynthesis in soybean (C-3) plants by O_2. Normal air contains 20.9 percent O_2 and 0.034 percent CO_2 (340 ppm by volume). The light intensity was equal to about one-sixth maximum sunlight; the temperature was 22.5°C. Negative values represent a net loss of CO_2 by respiration. (From M. L. Forrester, G. Krotkov, and C. D. Nelson, 1966, Plant Physiology 41:422–427.)

studied since and is termed the **Warburg effect.** Figure 10-7 illustrates this effect for C-3 soybean leaves exposed to two different CO_2 concentrations, one a near-normal CO_2 concentration and one a reduced CO_2 level. Note that even the normal O_2 concentration of 21 percent is inhibitory compared with zero O_2 at both CO_2 levels. Furthermore, inhibition by O_2 is greater at the lower than at the higher CO_2 concentration. Thus existing atmospheric levels of O_2 inhibit photosynthesis in C-3 plants. In contrast, photosynthesis in C-4 species is hardly affected by varying O_2 concentrations.

To understand these different effects of O_2 on C-3 and C-4 species, remember that net CO_2 fixation is the amount by which photosynthesis exceeds respiration, because respiration continuously releases CO_2. Respiration of C-3 leaves during darkness is

small compared with photosynthetic rates (roughly one-sixth), yet respiration of C-3 plants in light is much faster than in darkness. Data indicating more rapid respiration in light than darkness for C-3 plants were obtained with an infrared CO_2 analyzer and first published by John P. Decker in the 1950s, but physiologists were slow to accept his conclusions. We now know that respiration in leaves of C-3 species is often two or three times as rapid in light as in darkness and that under field conditions it causes release of about one-fourth of the CO_2 fixed by photosynthesis simultaneously. Respiration in illuminated photosynthetic organs occurs by two processes: the process that occurs in all plant parts even during darkness and a much more rapid process known as **photo-respiration.** The two processes are spatially separated within the cells; normal respiration occurs in the cytosol and in mitochondria, and photo-respiration occurs in chloroplasts, peroxisomes, and mitochondria in a cooperative way (Tolbert, 1981; Huang et al., 1983; Ogren, 1984).

Loss of CO_2 by photorespiration in C-4 species is almost undetectable, which is the principal reason that those species show much higher net photosynthetic rates at high light intensities than do C-3 species. To understand why photorespiration is so much higher in C-3 than C-4 plants, we must first understand the chemical reactions of photorespiration. As we shall see, photorespiration is stimulated by three factors other than high light levels: high O_2 levels, low CO_2 levels, and high temperatures.

In 1971, W. L. Ogren and George Bowes at the University of Illinois theorized from various data that carbons 1 and 2 of ribulose bisphosphate were the precursors to glycolic acid, a 2-carbon acid. They showed experimentally that O_2 could inhibit CO_2 fixation by rubisco, thus apparently explaining the Warburg effect. They also showed that rubisco catalyzes an oxidation of ribulose bisphosphate by O_2. *Thus rubisco is also an oxygenase.*

The two products of rubisco action on ribulose bisphosphate and O_2 are 3-PGA and *phosphoglycolic acid,* a 2-carbon phosphorylated acid. Using heavy oxygen ($^{18}O_2$), it was shown that only one of the O_2 atoms is incorporated into phosphoglycolate, the other apparently being converted to water (OH^- ion), as in R10-7:

$$^{18}O_2 + \begin{array}{c} CH_2OPO_3H^- \\ | \\ C{=}O \\ | \\ H{-}C{-}OH \\ | \\ H{-}C{-}OH \\ | \\ CH_2OPO_3H^- \end{array} \longrightarrow \left[\begin{array}{c} CH_2OPO_3H^- \\ | \\ {}^{18}O{-}C{-}OH \\ | \\ C{=}O \\ | \\ H{-}C{-}OH \\ | \\ CH_2OPO_3H^- \end{array} \right] \longrightarrow$$

ribulose bisphosphate

$$\underset{\substack{\\ \text{phosphoglycolic acid}}}{\overset{\displaystyle \text{CH}_2\text{OPO}_3\text{H}^-}{\underset{\substack{^{18}\text{O}}}{\overset{\displaystyle |}{\underset{\displaystyle \diagdown}{\overset{\displaystyle \text{C}}{\diagup}}}}\text{OH}}} \quad + \quad \underset{\text{3-PGA}}{\overset{\displaystyle \text{COOH}}{\underset{\displaystyle \text{CH}_2\text{OPO}_3\text{H}^-}{\overset{\displaystyle |}{\text{H}-\text{C}-\text{OH}}}}} \quad + \quad {}^{18}\text{OH}^-$$

(R10-7)

Molecular O_2 and CO_2 thus compete for the same rubisco enzyme and for the same ribulose bisphosphate substrate. Oxygen fixation seems to represent about two-thirds of the total O_2 absorbed during photorespiration, the remainder coming from oxidation of phosphoglycolate. This competition between O_2 and CO_2 explains the greater inhibition of photosynthesis in C-3 plants at low than at higher CO_2 levels (Fig. 10-7). The affinity of rubisco for CO_2 is much greater than for O_2, but O_2 fixation in all plants can occur because the O_2 concentration in leaves or cells of algae is much higher than that of CO_2. (Atmospheric concentrations of O_2 average 20.9 percent by volume and CO_2 about 0.034 percent, Chapter 11.) At any given time, rubisco enzymes are fixing about one-third to one-fourth as much O_2 as CO_2 (Miziorko and Lorimer, 1983). When temperatures are warm, the ratio of dissolved chloroplastic O_2 compared to CO_2 is higher than when temperatures are cool, so O_2 fixation by rubisco occurs faster, and photorespiration then indirectly slows growth. Photorespiration is light dependent, because ribulose bisphosphate formation occurs much faster in light than in darkness. This is so because operation of the Calvin cycle needed to form ribulose bisphosphate requires ATP and NADPH, both light-dependent products, as explained earlier. Furthermore, light causes release of O_2 from H_2O directly in chloroplasts, so chloroplastic O_2 is more abundant in light than in darkness when it must diffuse inwardly through leaf surfaces with closed stomates.

Photorespiration is essentially absent in C-4 plants for two simple reasons: Rubisco and other Calvin cycle enzymes are present only in bundle sheath cells, and the CO_2 concentration in those cells is maintained too high for O_2 to compete with CO_2. High CO_2 concentrations in bundle sheath cells are kept high by rapid decarboxylation of malate and aspartate transferred there from mesophyll cells. If bundle sheath cells are separated from mesophyll cells, their source of CO_2 from C-4 acids is removed; they will then photorespire, but not when in an intact leaf.

Phosphoglycolate formed in R10-7 represents the source of CO_2 released in photorespiration. Its phosphate group is first hydrolyzed away by a specific phosphatase found in chloroplasts of C-3 plants, releasing Pi and glycolic acid. The glycolate then moves out of chloroplasts into adjacent peroxisomes. **Peroxisomes** are small organelles that contain several

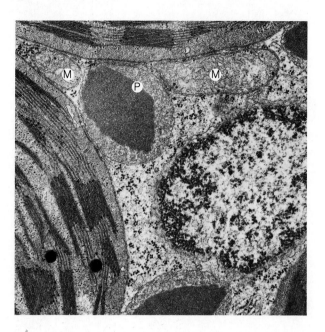

Figure 10-8 Close association of chloroplasts, peroxisomes (P), and mitochondria (M) in a leaf cell. The crystalline-like matrix in these peroxisomes is due to the enzyme catalase, but many peroxisomes containing catalase show no such matrix. (Courtesy Eugene Vigil.)

oxidative enzymes and represent, with glyoxysomes of fat-rich seeds (Chapter 14), the two major types of plant microbodies. Peroxisomes exist almost exclusively in photosynthetic tissue and often appear in electron micrographs in direct contact with chloroplasts (Fig. 10-8). In peroxisomes, glycolate is oxidized to glyoxylic acid by glycolic acid oxidase, an enzyme containing riboflavin as part of an essential prosthetic group:

$$\underset{\substack{\\ \text{glycolic acid}}}{\overset{\displaystyle \text{CH}_2-\text{COOH}}{\underset{\displaystyle \text{OH}}{\overset{\displaystyle |}{|}}}} + O_2 \longrightarrow \underset{\substack{\\ \text{glyoxylic acid}}}{\overset{\displaystyle \text{HC}-\text{COOH}}{\underset{\displaystyle \text{O}}{\overset{\displaystyle \|}{\|}}}} + H_2O_2$$

(R10-8)

Glycolic acid oxidase thus transfers electrons (present in H atoms) from glycolate to O_2, reducing the O_2 to H_2O_2 (hydrogen peroxide). Nearly all of this H_2O_2 is then broken down by catalase (another peroxisomal enzyme) to water and O_2:

$$2H_2O_2 \rightarrow 2H_2O + O_2 \qquad \text{(R10-9)}$$

The ultimate fate of glyoxylate can vary somewhat. Small amounts of it are probably oxidized to CO_2 and formic acid (HCOOH) by H_2O_2 not removed by catalase, but most of it is converted to glycine by a transamination reaction with an amino acid. Then, after transport into mitochondria, two molecules of glycine are converted to one molecule of serine (a

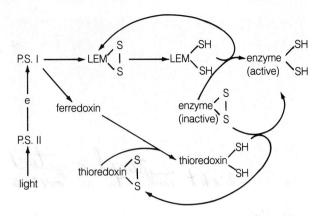

Figure 10-9 Postulated mechanisms for activation of certain photosynthetic enzymes by the LEM system (top) and the thioredoxin system (bottom).

3-carbon amino acid), one molecule of CO_2, and one NH_4^+ ion (Sarojini and Oliver, 1983). *This mitochondrial reaction is the major source of the CO_2 that is released in photorespiration.* It is also important because of the NH_4^+ released; this has to be reincorporated into amino acids so that glycine formation can continue, and this is energy-expensive (Section 13.3).

Serine is then converted to 3-PGA by a series of reactions that involves loss of its amino group and gain of a phosphate group from ATP. Reactions by which glycolate is converted to 3-PGA via glyoxylate, glycine, serine, and other 3-carbon acids are known as the **glycolate pathway** (Tolbert, 1979; 1981). The 3-PGA is then converted to sucrose and starch in chloroplasts. This pathway, therefore, is a way to conserve an average of three-fourths of the carbons split off ribulose bisphosphate (as phosphoglycolate) when O_2 reacts with it (one CO_2 lost for every two 2-carbon acids).

An interesting question is how photorespiration persists in C-3 plants and has not been eliminated through evolutionary selection pressures, because it certainly reduces their net CO_2 fixation and growth rates. The answer is unclear, but some experts suggested that it is a means of removing excess ATP and NADPH produced at high light levels (see Ogren, 1984). Thus both ATP and NADPH are needed to regenerate ribulose bisphosphate from 3-PGA formed during O_2 fixation, so both molecules would be used up in photorespiration without CO_2 fixation. They postulated that this use of excess "reducing power" prevents high light levels from causing damage to chloroplast pigments (see solarization, Chapter 11). Others suggested that photorespiration is a necessary consequence of the structure of the enzyme rubisco, a structure that evolved to fix CO_2 in ancient photosynthetic bacteria when atmospheric

CO_2 concentrations were high and those of O_2 were low. According to this hypothesis, as O_2 accumulated in the atmosphere by the photosynthetic splitting of H_2O by algae and early land plants, rubisco necessarily began to fix O_2 simply because its active site for CO_2 could not discriminate effectively between the two similar gases. Whether natural selection or genetic engineering methods can eventually force modification of rubisco to favor CO_2 even more is an interesting and important question (Wildner, 1981). Some evidence indicates that evolution is indeed favoring CO_2 fixation (Jordan and Ogren, 1981).

10.5 Light Control of Photosynthetic Enzymes in C-3 and C-4 Plants

We emphasized the role of light in providing ATP and NADPH needed for CO_2 fixation and reduction. In addition, light regulates the activities of several chloroplast photosynthetic enzymes. Such enzymes exist in an active form in light and an inactive or much less active form in darkness. Carbohydrate production from CO_2 is, therefore, shut off tightly at night because of enzyme inactivity, closed stomates, and a deficiency of ATP and NADPH.

In C-3 species, five Calvin cycle enzymes are activated in light: rubisco, 3-phosphoglyceraldehyde dehydrogenase, fructose-1,6-bisphosphate phosphatase, sedoheptulose-1,7-bisphosphate phosphatase, and ribulose-5-phosphate kinase. The function of each enzyme is given in the legend to the complete Calvin cycle in Figure 10-3. In C-4 species, PEP carboxylase, $NADP^+$-malate dehydrogenase, and pyruvate, phosphate dikinase of mesophyll cells are also light activated.

Mechanisms of light activation are indirect, and light's energy is not absorbed by the colorless enzymes directly. Part of the difficulty in clarifying these mechanisms is that light changes the levels of many compounds that have an activation effect on specific enzymes within the chloroplast. Nevertheless, most evidence indicates that ribulose-5-phosphate kinase, 3-phosphoglyceraldehyde dehydrogenase, $NADP^+$-malate dehydrogenase, and both phosphatases have *disulfide (S–S) groups* that are reduced to two *sulfhydryl groups (–SH plus –SH)* when they are light activated. Each light activation reduction causes an important modification in enzyme structure so that the enzyme functions much faster. Reduction occurs using electrons that are derived from the light-dependent split of H_2O but that are not used to reduce $NADP^+$ to NADPH. Two mechanisms for use of these electrons have been proposed (Fig. 10-9). In the first, electrons reduce iron-sulfur proteins of the thylakoid membranes. These proteins are similar to ferredoxin and are called *light effect mediators* (Anderson et al., 1982). Collectively, they are frequently referred to as the *LEM system* (Fig. 10-9, top). LEM proteins are thought to contain disulfide bonds that become reduced by light to sulfhydryl groups, which then reduce directly the disulfide groups of the photosynthetic enzymes, thus activating them. A second system is the *ferredoxin-thioredoxin system* (Buchanan, 1980, 1984), which involves reduction of ferredoxin by the usual photosynthetic electron transport system (Fig. 10-9, middle). An enzyme called *ferredoxin-thioredoxin reductase* next reduces one or more *thioredoxins*, small proteins that contain an important disulfide bond (Holmgren, 1981). Again, two sulfhydryl groups are produced when the disulfide bond is reduced. Then reduced thioredoxins reduce and activate the photosynthetic enzymes, just as the LEM system proteins are proposed to do.

There is still disagreement about the relative importance of these two systems, but there is no disagreement that disulfide and sulfhydryl groups are involved and that the electrons necessary for reduction and activation are derived photosynthetically from H_2O. Inactivation in darkness involves oxidation of the enzymes, most likely by oxidized LEM proteins or oxidized thioredoxin.

Activation of rubisco and PEP carboxylase is generally not as great and is even less understood than for the five enzymes mentioned above. Reduction and disulfide-sulfhydryl changes are probably not involved. For rubisco, three factors caused by light seem important (Heldt, 1979; Perchorowicz and Jensen, 1983). First is a rise in pH of the chloroplast stroma from about 7 to 8 caused by light-driven H^+ transport from the stroma to thylakoid channels (Fig. 9-8). Rubisco is much more active at pH 8 than at lower pH. Second is the transport of Mg^{2+} from the thylakoid channels to the stroma accompanying the pH change, important because rubisco requires Mg^{2+} for maximal activity. (High pH and Mg^{2+} levels also activate phosphatases acting on fructose-1,6-bisphosphate and sedoheptulose-1,7-bisphosphate.) Third is formation of a still-unidentified intermediate (or intermediates) of the Calvin cycle that accumulates in light and activates rubisco.

Activation of PEP carboxylase in light has not yet been explained well. Although it, too, requires Mg^{2+} and a pH near 8 for maximal activity, there is no reason to believe that light alone causes either factor to rise in the cytosol, where the enzyme exists, so that it could operate better than at night. Control of activity of the final light activated photosynthetic enzyme, pyruvate, phosphate dikinase of C-4 mesophyll chloroplasts, is quite complex (Chapman and Hatch, 1981; Nakamoto and Edwards, 1983), and it will not be covered here.

Figure 10-10 Miscellaneous succulents. From left to right, *Opuntia, Aloe obscura, Echeveria corderoyi, Crassula argentea, Agave horrida.* (New York Botanical Garden photograph courtesy Arthur Cronquist.)

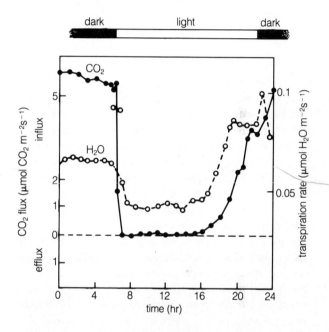

Figure 10-11 CO_2 fixation and transpiration rates of the CAM plant *Agave americana* during alternate light and dark periods. (From T. F. Neales, A. A. Patterson, and V. J. Hartney, 1968, Nature 219:469–472.)

10.6 CO_2 Fixation in Succulent Species (Crassulacean Acid Metabolism)

Numerous species living in arid climates have thick leaves with relatively low surface-to-volume ratios and accompanying low transpiration rates. Their cells have unusually large vacuoles relative to the thin layer of cytoplasm. Such species are frequently referred to as **succulents** (Fig. 10-10). As discussed in Chapter 3, many open their stomates and fix CO_2 into organic acids primarily at night. Figure 10-11 illustrates the daily cycle of transpiration and CO_2 fixation in one such species, *Agave americana,* a century plant.

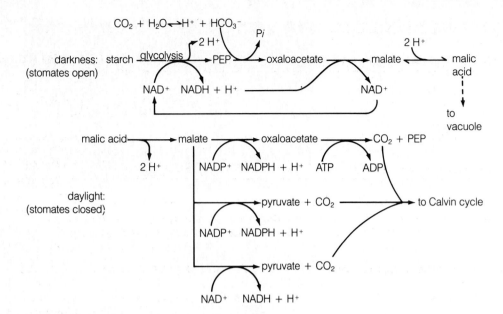

$$CO_2 + H_2O \leftrightarrow H^+ + HCO_3^-$$

darkness: starch —glycolysis→ PEP → oxaloacetate → malate ⇌ malic acid
(stomates open)

to vacuole

malic acid → malate → oxaloacetate → CO_2 + PEP

daylight:
(stomates closed)

→ pyruvate + CO_2 → to Calvin cycle

→ pyruvate + CO_2

Figure 10-12 A summary of CO_2 fixation in CAM plants.

The metabolism of CO_2 in succulents is unusual, and because it was first investigated in members of the Crassulaceae, it is called **crassulacean acid metabolism (CAM).** CAM has been found in hundreds of species in 20 families (nearly all of which are angiosperms), including the familiar Cactaceae, Orchidaceae, Bromeliaceae, Liliaceae, and Euphorbiaceae (Szarek and Ting, 1977; Kluge and Ting, 1978). Not all CAM plants are succulents, and some succulents, such as most halophytes (salt-lovers), do not possess CAM. Species with CAM usually lack a well-developed palisade layer of cells, and most of the leaf or stem cells are spongy mesophyll (Gibson, 1982). Bundle sheath cells are present but, in contrast to those of C-4 plants, are quite similar to the mesophyll cells.

The most striking metabolic feature of CAM plants is formation of malic acid at night and its disappearance during daylight. This formation of acid at night is detectable as a sour taste and is accompanied by net loss of sugars, starch, or glucose polymers similar to starch (Black et al., 1982). The most abundant acid in CAM plants is malic, but citric and isocitric acids which are derived from it accumulate to a lesser extent in some species. However, citric and isocitric acids show little change in concentrations during light and darkness. PEP carboxylase in the cytosol of CAM plants is the enzyme responsible for CO_2 fixation into malate at night (in contrast to its low activity in C-4 plants in darkness), but rubisco becomes active during daylight as in C-3 and C-4 plants. The role of rubisco is identical to its function in bundle sheath cells of C-4 plants, that is, refixation of the CO_2 lost from organic acids such as malic.

A model consistent with what we know about CO_2 fixation in CAM plants is shown in Fig. 10-12. During darkness, starch breaks down by glycolysis

(Chapter 12) as far as PEP. CO_2 (actually HCO_3^-) then reacts with PEP to form oxaloacetate; then oxaloacetate is reduced to malate by an NADH-dependent malate dehydrogenase. Most of the malate is pumped into and stored in the vacuole as malic acid until daybreak. During daylight, malic acid diffuses passively out of the vacuole and is decarboxylated by one of the three mechanisms also present in C-4 plant bundle sheath cells (see Fig. 10-6). The mechanism employed depends on the species. The CO_2 released by each of the three decarboxylation reactions is refixed by rubisco into 3-PGA of the Calvin cycle, which then leads to formation of sucrose, starch, and other photosynthetic products. Pyruvate formed by decarboxylation is converted to PEP by pyruvate, phosphate dikinase, as in C-4 plants, and PEP is then partly respired, partly converted to sugars and starch by reverse glycolysis, and partly converted to amino acids, proteins, nucleic acids, and so forth by reactions to be described in Chapters 12 to 14 (Kluge, 1979).

Thus CAM plants, like C-4 plants, use PEP carboxylase to form malate, later decarboxylate the malate to release CO_2 by one of three mechanisms, and then refix the CO_2 into Calvin cycle products using rubisco. In C-4 plants there is a spatial separation between mesophyll and bundle sheath cells for malate formation and decarboxylation, and both occur in daylight. In CAM plants both processes occur in the same cells, one process at night and the other in daylight, and the large central vacuole stores malic acid that would otherwise cause too low a cytoplasmic pH at night. Low permeability of the tonoplast to H^+ resulting from malic acid ionization in the vacuole must be especially important in CAM plants, because the vacuole pH often becomes as low as 4 at night.

An interesting question is what causes rubisco,

not PEP carboxylase, to fix CO_2 in CAM plants during daylight. Both enzymes are present, both have about equal affinities for dissolved forms of CO_2, and the cytosol location of PEP carboxylase should allow it to encounter incoming CO_2 before that CO_2 reaches rubisco in the chloroplast. Part of the answer is that CAM plants convert their PEP carboxylase from an active form to an inactive form during daylight (O'Leary, 1982). Their inactive form present during daylight has less affinity for PEP and is inhibited strongly by the malate released from the vacuole then, but their active form has a higher affinity for PEP and is much less inhibited by the malate that it forms at night. Changes in activities of still other enzymes that favor CO_2 fixation by PEP carboxylase only at night are described by Holtum and Winter (1982) and in the book by Ting and Gibbs (1982).

Although the ability of a plant to perform CAM is genetically determined, it is also environmentally controlled. In general, CAM is favored by hot days with high light levels, cool nights, and dry soils, which predominate in deserts. High salt concentrations in soils that lead to osmotic drought also favor CAM. Several CAM plants switch to a greater rate of CO_2 fixation in daylight by a C-3 photosynthetic mode after a rainstorm or when night temperatures are high. Stomates then remain open longer during morning daylight hours. These facts suggest that evolution for CAM is favored only by water stress. Nevertheless, CAM also occurs in some underwater fern allies (e.g., some members of the genus *Isoetes*) living in shallow ponds or in oligotrophic lakes in which CO_2 levels are lower during daylight than at night (Keeley and Morton, 1982). The occurrence of CAM in such species suggests that this process has also been selected for by diurnal changes in CO_2 availability. Finally, CAM can be induced by marked day-night changes in temperatures in alpine *Sempervivum* spp. of the central Alps in Austria. Even with adequate moisture, night temperatures near freezing followed by high irradiance levels near noon cause leaf temperature changes so large (up to 45°C) that CAM then occurs (Wagner and Larcher, 1981).

10.7 Formation of Sucrose, Starch, and Fructans

In each of the three major pathways by which CO_2 is fixed, the principal leaf storage products accumulating in light are usually sucrose and starch. Free hexose sugars such as glucose and fructose usually are much less abundant than sucrose in photosynthetic cells, although the opposite is true in many nonphotosynthetic cells. In many grass species (especially those that originated in temperate zones, including the Hordeae, Aveneae, and Festuceae tribes)

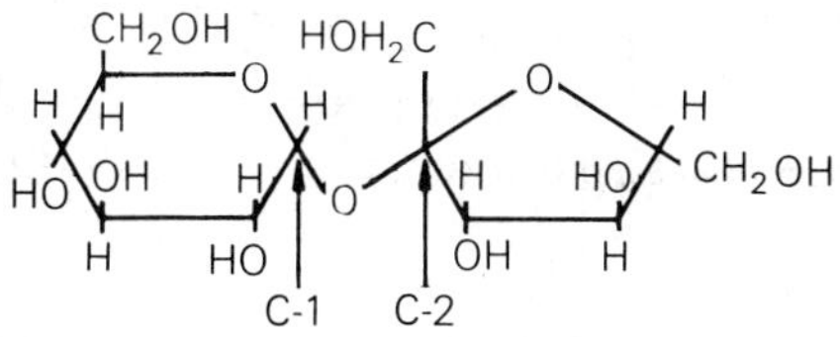

Figure 10-13 The structure of sucrose, a disaccharide made of a glucose unit (left) and a fructose unit (right) connected between carbons 1 and 2, as shown.

and also in a few dicots, starch is not a major product of photosynthesis; in these plants sucrose and fructose polymers called **fructans** (or **fructosans**) predominate in leaves and stems, but starch is the predominant storage polysaccharide in roots and seeds. In this section we summarize sucrose, starch, and fructan formation from photosynthetic products; their utilization as energy sources during respiration is explained in Chapter 12.

Synthesis of Sucrose Sucrose (Fig. 10-13) is especially important because it is so common and abundant in plants and because we consume so much of it as table sugar. It acts as an energy source in photosynthetic cells and is readily translocated through the phloem to growing tissues. It is commercially important in sugar beet roots and sugar cane stems because of its unusual abundance there.

Synthesis of sucrose occurs in the cytosol, not in chloroplasts where the Calvin cycle occurs. Free glucose and fructose are not important precursors of sucrose; instead, phosphorylated forms of those sugars are. As mentioned in Section 10.1, triose phosphates (3-phosphoglyceraldehyde and dihydroxyacetone phosphate) exported from chloroplasts in photosynthetic cells serve as precursors of hexose phosphates and of sucrose. These triose phosphates (3-carbons) become converted into one Pi and one fructose-6-phosphate (6-carbons), part of which is changed to glucose-6-phosphate and then glucose-1-phosphate. (Details of those reactions, which also occur in respiration, are in Fig. 12-4.) Glucose-1-phosphate and fructose-6-phosphate contain the two hexose units needed to yield the disaccharide sucrose, but combination of those units is indirect because energy must be provided to activate the glucose unit. This energy is provided by *uridine triphosphate (UTP)*, which is a nucleoside triphosphate similar to ATP but which contains the pyrimidine base uracil instead of the purine base adenosine. The UTP acts by reacting with glucose-1-phosphate; the two terminal phosphates of UTP are removed together as PPi, and the phosphate of glucose-1-P becomes esterified to the remaining phosphate in UTP to form a molecule called *uridine diphosphate glucose (UDPG)*. The glucose in UDPG can be considered to be activated, because it can now readily be transferred to an acceptor molecule such as fructose-6-phosphate. This reaction and others involved in the major pathway of sucrose formation are summarized by reactions R10-10 through R10-15 below (Whittingham et al., 1979; Avigad, 1982). The enzymes that catalyze each reaction require Mg^{2+} as a cofactor, and this is one of many reasons why magnesium is essential for plants.

$$UTP + glucose\text{-}1\text{-}phosphate \rightleftharpoons UDPG + PPi \quad \text{(R10-10)}$$

$$PPi + H_2O \rightarrow 2\ Pi \quad \text{(R10-11)}$$

$$UDPG + fructose\text{-}6\text{-}phosphate \rightleftharpoons$$
$$sucrose\text{-}6\text{-}phosphate + UDP \quad \text{(R10-12)}$$

$$sucrose\text{-}6\text{-}phosphate + H_2O \rightarrow sucrose + Pi \quad \text{(R10-13)}$$

$$UDP + ATP \rightleftharpoons UTP + ADP \quad \text{(R10-14)}$$

A calculation of the total energy cost to the plant in forming one sucrose molecule can be made by adding the reactions to arrive at R10-15, which shows that only one ATP is needed to form the glycosidic bond that connects glucose to fructose in sucrose:

$$glucose\text{-}1\text{-}phosphate + fructose\text{-}6\text{-}phosphate +$$
$$2H_2O + ATP \rightarrow sucrose + 3\ Pi + ADP \quad \text{(R10-15)}$$

Because three ATP molecules are required in the Calvin cycle for each carbon in each hexose of sucrose (36 total ATP), the one additional ATP needed to form the glycosidic bond in sucrose is a small additional requirement.

Formation of Starch The major storage carbohydrate of plants is starch (Jenner, 1982). In leaves of most species, starch accumulates in chloroplasts where it is formed directly from photosynthesis. In storage organs, it accumulates in **amyloplasts,** where it is formed following translocation of sucrose or other carbohydrate from leaves. Thus starch always exists in a plastid. The amounts of starch in various tissues depend on many genetic and environmental factors, but in leaves the light level and duration are especially important. Starch builds up in daylight when photosynthesis exceeds the combined rates of respiration and translocation; then some of it disappears at night by the last two processes.

You have eaten many nonphotosynthetic starch-rich organs, such as potato tubers, banana fruits, and seeds of cereal grains and legumes so common in our diets. Besides food crops, most native perennial plants store starch before and during the dormant period, and this starch is used as energy for regrowth

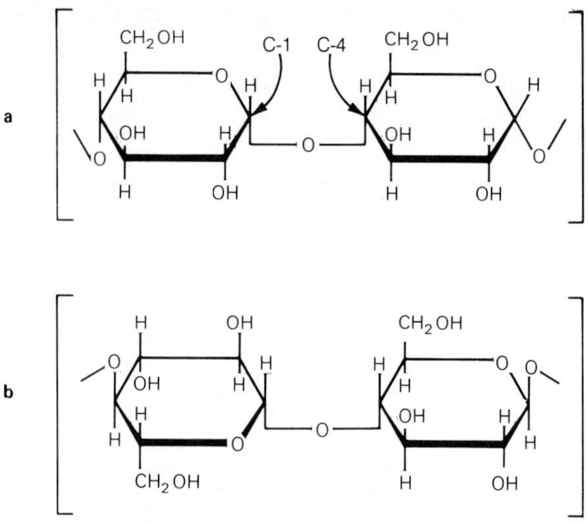

Figure 10-14 (**a**) The alpha (α) linkage between glucose residues, as in starch. (**b**) The beta (β) linkage between glucose residues, as in cellulose.

the next growing season. In deciduous trees and shrubs, starch is stored largely in amyloplasts of young twigs, in the bark (phloem parenchyma cells), in living xylem parenchyma cells, and also in some root parenchyma storage cells. Many herbaceous perennial grasses and dicots store starch in roots, bases of stems (crowns), or in underground bulbs or tubers. In stems, cortex and pith cells are frequent sites of starch storage, both in annuals and perennials. Maple sugar trees store starch in twig and trunk xylem parenchyma during late summer and early fall, then convert this to sucrose in early spring when it can be collected as maple sugar. Collection occurs by tapping the flow from tree trunks, an osmotic flow resulting from turgor pressure in xylem caused by conversion of relatively few starch grains into many more sucrose molecules.

Two types of starch are present in most plastids, *amylose* and *amylopectin,* both of which are composed of D-glucoses connected by α-1,4-bonds (Fig. 10-14a). The α-1,4-bonds cause starch chains to coil into helices (Fig. 10-15). Amylopectin consists of branched molecules, the branches occurring between C-6 of a glucose in the main chain and C-1 of the first glucose in the branch chain. The number of glucose units present in various amylopectins ranges from 2000 to 500,000. Amyloses are almost entirely unbranched and contain a few thousand sugar units, the number depending on the species and environmental conditions. Amylose becomes purple or blue when stained with the iodine-potassium iodide solution, a mixture that produces the reactive I_5^- ion (Banks and Muir, 1980). Amylopectin reacts much less intensely with this reagent, and it exhibits a purple to red color. The

AMYLOSE

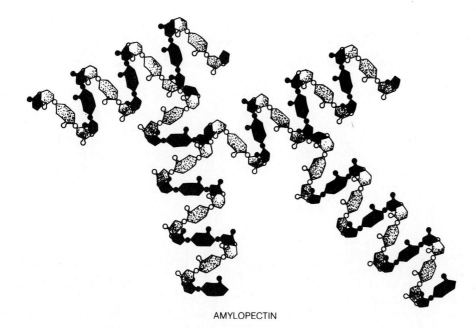

AMYLOPECTIN

Figure 10-15 A schematic representation of a small part of starch molecules. Amylose and amylopectin are similar, except that amylopectin is branched.

iodine test is often used by students and researchers to determine whether starch is present in cells. The percentage of amylopectin in starch from most species varies from about 60 to 100 percent (Akazawa, 1976). Potato tuber starches contain about 78 percent amylopectin and 22 percent amylose. These ratios are similar for starches of banana fruits and the seeds of pea, wheat, rice, and field maize.

Starch formation occurs mainly by one process involving repeated donation of glucose units from a nucleotide sugar similar to UDPG called *adenosine diphosphoglucose, ADPG* (Jenner, 1982; Preiss, 1982a, 1982b). Formation of ADPG occurs in chloroplasts and other plastids using ATP and glucose-1-phosphate. Reaction R10-16 summarizes starch formation from ADPG, in which a growing amylose molecule with a glucose unit having a reactive C-4 group at its end combines with C-1 of the glucose being added from ADPG:

ADPG + small amylose (*n*-glucose units) $\rightarrow$

larger amylose (with $n + 1$ glucose units) + ADP   (R10-16)

Starch synthetase, which catalyzes this reaction, is activated by K^+, which is one reason K^+ is essential for plants and probably why sugars, not starch, accumu-

late in plants deficient in K^+ (Chapter 5). Various isozymes of starch synthetase occur in different plants and parts of plants (Preiss and Levi, 1979).

Branches in amylopectins between C-6 of the main chain and C-1 of the branch chain are formed by various isozymes of enzymes summarily called the *branching* or *Q enzyme*. Surprisingly little is known about how they catalyze branching, but branch bonds do not result from glucose transferred from ADPG. Rather, branching enzymes transfer short units from a growing starch molecule to the same or another starch molecule to form the α-1,6 linkage.

Much remains to be learned about starch formation and its control, especially regarding branching. Nevertheless, a few facts about control are important. High light levels and long days favor photosynthesis and carbohydrate translocation, so the long days of summer cause accumulation of one or more starch granules in chloroplasts and starch storage in amyloplasts of reserve organs of non-photosynthetic cells. Furthermore, starch formation is favored in chloroplasts by bright light, because the enzyme that forms ADPG is activated allosterically by 3-PGA and inhibited allosterically by P*i* (Preiss, 1982a, 1982b, 1984). Levels of 3-PGA increase in light as CO_2 is fixed, but levels of P*i* decrease as it is added to ADP to form ATP during photosynthetic phosphorylation.

Exploring the Path of Carbon in Photosynthesis (I)

James A. Bassham

James A. Bassham, whom most of his friends call "Al," participated as a graduate student in elucidating the pathway of carbon in photosynthesis and is now a professor of chemistry at the University of California, Berkeley. He tells us his story:

When I think back to my days as a graduate student 30 years ago, the first two impressions that enter my mind are those of good fortune and excitement. The good fortune had to do with my becoming involved in that particular research at that time and place.

The chain of events that led me there began with my having to fill out a form of entry to the University of California at Berkeley and deciding to put down chemistry in the appropriate blank space because I had received a chemistry prize in high school. Then, during my freshman chemistry class at Cal one spring morning, our section instructor, Professor Sam Ruben, chose to put aside his usual laboratory lecture notes and tell us a little about his research. He and other scientists had discovered a new radioisotope of carbon, called carbon-14, and were putting it to many uses, of which the most interesting to me was as a tracer to map the pathway of carbon fixation by green plants making sugar and using sunlight. This was certainly more interesting than precipitating various sulfides with hydrogen sulfide on the porch of the freshman chemistry laboratory as part of my training in qualitative analysis. The advent of World War II, however, and three years of service in the U.S. Navy were to put such thoughts from my mind until a later date.

The third link in this chain of circumstances followed my return to the University at Berkeley as a graduate student in chemistry, after the close of the war. Since I had been an undergraduate student at Berkeley, the Chemistry Department was reluctant to admit me as a graduate student, it being their wise policy to encourage former undergraduates to go elsewhere for graduate training. After

much discussion, they allowed me to enter for course work only. By some miracle, after the first semester had passed, the Dean invited me to stay on and work for my doctorate degree.

As is usual in such cases, I was directed to interview several professors of organic chemistry so that, hopefully, I would find one willing to accept me to do research on some project for my thesis. The first person I spoke to was a young professor named Melvin Calvin. By coincidence, the first topic he mentioned to me was the mapping of the path of carbon reduction during photosynthesis using carbon-14 as a label. I soon learned that Dr. Ruben had lost his life in an unfortunate laboratory accident during the war, and that the work on carbon-14 was now being carried forward in the newly formed BioOrganic Chemistry Group of the University of California. Professor Ernest Lawrence, director of the Radiation Laboratory (now the Lawrence Berkeley Laboratory), had invited Melvin Calvin to form a division to explore the uses of carbon-14 in investigations of biochemistry and organic reaction mechanisms. The work was already under way.

Professor Calvin then proceeded to tell me of several projects involving organic synthesis and reaction mechanisms, making use of this radioisotope, but he might have saved himself the trouble because my mind was already made up. If at all possible, and if acceptable to him, I was very eager to get going on the work of using carbon-14 to study photosynthesis. He agreed to this and before long escorted me from the old red brick chemistry building across the court to an even older and shabbier wooden frame building that had been constructed as a "temporary" building about half a century before. As most university colleagues will know, temporary buildings built on university campuses are usually good for at least 50 years.

In this building, called the Old Radiation Laboratory (ORL), I soon became acquainted with a small group of people who were to prove instrumental in the work leading to the mapping of the path of carbon in photosynthesis. Of key importance among these was Andrew A. Benson, a young postdoctoral scientist working with Professor Calvin on the photosynthesis project. Andy was an excellent experimentalist and taught me a great deal about how to

Formation of Fructans Our knowledge of fructans and their synthesis is unusually limited considering their importance in plants, especially in stems and leaves of so-called cool season grasses that dominate pastures in temperate climates. Such C-3 grasses provide most feed for cattle and sheep, so the fructans they store are important to animals, including humans, as well as to plants. Besides grasses, fructans have been found in certain organs in six other families, including underground storage organs of

the Asteraceae (composites such as asters and dandelions) and the Campanulaceae, and in leaves and bulbs of the Liliaceae (lilies), Iridaceae (irises), Agavaraceae, and Amyrillidaceae (Meier and Reid, 1982).

Fructans are fructose polymers that are much smaller than the glucose polymers of starch; they usually have 10 to 100 fructose units, and so they are rather water soluble and probably stored largely in vacuoles. All contain one terminal glucose unit, indicating that they are built by adding fructose units

devise apparatus and techniques for solving new kinds of problems in the laboratory. Of course, as those familiar with the mapping of the path of carbon in photosynthesis know, he played a very important role in the identification of various sugar phosphate intermediates, which were to turn out to be the essential compounds in the photosynthetic pathway. There were others who played important roles, visiting scientists, staff scientists, and students, but space will not permit me to describe them all.

Thinking about these former colleagues and the old laboratory brings to me my second main impression of that time: a sense of great excitement. The Old Radiation Laboratory was an exciting place to work, the research project was fascinating, and the people were all, without exception, enthusiastic about what they were doing. I suppose, to some extent, this is because we had a new and, at that time, almost exclusive technique and an important problem to which to apply it. It went far beyond that, however, and, to a large extent, stemmed directly from the personality of Melvin Calvin. He was in the laboratory every morning, or as soon as he could get free from his teaching duties, asking questions about the latest experiments, and laying out a program of new experiments to pursue. Never mind the fact that just the day before he had outlined an experiment which, in our opinion, would take a month or so, he would be in the next morning to ask the progress we had made with it. As we all know, this is an excellent procedure for stimulating graduate students. I don't mean to imply that he was a taskmaster; he was simply motivated by a tremendous enthusiasm for the project and the results we were getting. When we were fortunate to get a new experimental result, it always took on a new and greater importance after he had examined it.

This sense of achievement was further heightened when we developed techniques for analyzing the radioactive products of photosynthesis by using two-dimensional paper chromatography and radioautography. It was a beautiful analytical tool but suffered from one drawback: It required about two weeks from the original experiment until the films were developed to locate and count the ^{14}C in the radioactive spots. Of course, we had to learn to go on doing other experiments without waiting for the results of the previous one. Thus, there was a considerable sense of suspense when the x-ray films were developing in the dark room and the spots first began to appear before our eyes.

Some other phases of the research were even more painstaking and drawn out. For example, the degradation of molecules following short periods of photosynthesis to locate the radiocarbon within the molecule could take several weeks, or even months, before the final product was obtained.

Driven by our excitement to overcome these difficulties, we tended to work long hours and long weeks. When this reached an intolerable point and we felt the need for some mental rejuvenation, Andy Benson would organize an expedition to the high Sierra, and several of us would dash off to the mountains for a strenuous weekend of climbing 14,000-foot peaks. After such excursions, we returned to the laboratory physically exhausted but mentally refreshed. I suspect that Melvin Calvin was always relieved to see us return to the laboratory after these mountain and rock-climbing expeditions with no more serious infirmities than a few sore muscles or blisters on our feet.

About midway through the time when the path of carbon in photosynthesis was being mapped, I had done enough work on my part of the project to justify writing it up as a doctoral thesis. My main reluctance in doing this was the thought that I might then have to depart from the interesting project and find a job somewhere else doing something which was sure to be far less exciting. Fortunately for me, I was invited to stay on in a postdoctoral status and was able to remain a part of the team while the path of carbon in photosynthesis was fully mapped.

The techniques that we developed during that period proved to be extremely valuable for studies of metabolic regulation in plant cells and even in animal cells, and have shaped my whole scientific career since that time. To a large extent, we have been able to maintain in our laboratory a sense of excitement and cooperation over the years, in spite of two moves and growth from a dozen people to over a hundred. The person who has been most responsible for maintaining that sense of excitement and purpose is Melvin Calvin.

onto the fructose of a sucrose molecule. The *inulin* type occurs in species other than grasses and consists of relatively short, nonbranched straight chains. The *levan* or *phlein* type in grasses consists of larger molecules that have short branches. Each branch contains only one fructose unit. Thus far, only two enzymes that form fructans have been identified, both of which form straight chains without branches. One starter enzyme combines two sucroses to form a glu-fru-fru unit (releasing a glucose from one of the sucroses), while the second enzyme adds additional fructose units from sucrose molecules to the terminal fructose formed by the first enzyme. This second enzyme is responsible for addition of most of the fructoses to fructans, and it is a chain elongating enzyme. Each time it adds a fructose unit from sucrose, one glucose is also released from that sucrose. Almost surely, many of those glucoses are converted back to hexose phosphates so that sucrose can be formed again to maintain fructan formation.

Exploring the Path of Carbon in Photosynthesis (II)

Melvin Calvin

F.B.S. 1977

Melvin Calvin was a busy man (now retired). While working for the AEC (Atomic Energy Commission, now DOE, the Department of Energy) in 1974, I [F.B.S.] visited the laboratory he directed on two or three occasions, since it was one of the national laboratories supported largely by the AEC. We had a formal inspection of the lab that year—and found it to be administered in a unique way based on close trust and cooperation among the directing scientists but designed to frustrate thoroughly a Washington bureaucrat (which I was at that time)!

I told Professors Calvin and Bassham that we were planning to revise our text and asked if they would write essays for us, telling about the work on what is now called the Calvin cycle. They agreed, but Calvin was unable to find the time. However, on January 4, 1977, I recorded the following telephone conversation with him:

Frank B. Salisbury: How did you get into science?

Melvin Calvin: Well, it was a very practical consideration. When I was still in grade school, I was concerned with what my future livelihood would be. I looked around and decided that almost everything that I had contact with (for example, in a food store, the food itself and its processing, the cans and the making of them, the paper and the making of it, the dye stuff on the paper), everything had to do with chemistry. This prompted me to say: "I'll try to understand how the food gets made, how it's processed, and how it eventually reaches the grocery store." I realized that how the food got there was an essential activity for human survival, and that I had a good chance of finding a job if I knew anything about it.

F.B.S.: So it really was a practical consideration. But what aimed you at biology, especially plant physiology?

M.C.: Well, partly it was the food thing, and partly (much later) as I learned more of chemistry, it was the mysteriousness of how a plant could use sunshine to make food.

F.B.S.: So the practical approach became a bit of an intellectual challenge as well?

M.C.: Yes, exactly. You've got to make a living, but you want to do something interesting; if you can do both, you really have it made.

F.B.S.: What got you from the grocery store to Ernest Lawrence?

M.C.: That is a long, long trip, that one! I was an undergraduate at Michigan Tech, a graduate student at Minnesota, and then did postdoctorate work in England with Michael Polanyi. It was there that the real interest in photosynthesis, which came long ago, began to be executed. That's when I started working on chlorophyll and

chlorophyll analogues: How do they work electronically? What kinds of things were they? Thus began the marriage of the practicality of food production and the intellectual challenge of energy conversion. So when I came to Berkeley, I started working on synthetic analogues of chlorophyll and heme. I was living in an environment in which radioactive isotopes were being turned up every day, and the idea of having the radioactive isotope of carbon was fairly rampant around the place. Work on the chemistry of the production of sugars wasn't something that happened because we had the carbon; the carbon was something we wanted, and we knew what we wanted it for long before we had it. Then Ruben and Kamen came along with both the carbons, carbon-11 and carbon-14. I was busy with porphyrin chemistry during the early war years, and my background in this and metal complexes led to my association with the Manhattan District for the development of methods of uranium purification and plutonium isolation. That's how I got to Ernest. And it came from chlorophyll, whether you like it or not! By that time, Sam Ruben had been killed, and Ernest said, "Well, you ought to do something with radiocarbon." I said, "Yes, I ought to," and I knew exactly what I had to do.

F.B.S.: What were your essential insights or acts of creativity?

M.C.: You mean in terms of the carbon cycle? Those came after a good deal of work. We knew the experiments that had to be done, and the mechanics of doing them were obvious. It was just a matter of having the material and the time, and with the end of the war we had both. Carbon-14 had appeared just before the war, so the moment we had the time and the opportunity to generate the radiocarbon (which we did in 1944–1945) the work began in earnest, I mean in a serious way! We made carbon-14 at the Hanford and Oak Ridge reactors. Everybody else did, too; we had lots of it. The experiments were easy to design. We did several, and by 1951 we had already mapped a good many of the early compounds on the way to sugar, although actual delineation of the whole sequence didn't occur until somewhat later.

F.B.S.: Putting it all together must have been the creative part.

M.C.: Yes. The first thing we saw were 3s [three-carbon intermediates], then we saw 5s and 6s and 7s. We didn't see the 4s until toward the end of the line. They were an essential part of the puzzle when we did find them. Putting it all together occurred in the early and middle 1950s. It came in bits and pieces and fits and starts. The first important step was the recognition of phosphoglyceric acid as the first product. We kept looking for a 2-C piece because the first thing we saw was 3, and it was logical that CO_2 should add to a 2. But we never did find a 2. As you know, there wasn't one! The reaction was a 5+1 that gives two 3s.

I saw that. I can remember being at home and reading some papers in JACS about the mechanism of decarboxylation of β-keto acids and of dicarboxylic acids (malonic acid and acetoacetic acid), completely mechanistic studies in organic chemistry. I then recognized how a CO_2 would add to a β-keto sugar. It was the recognition that ribulose diphosphate was the acceptor of the CO_2 that allowed me to finally draw the whole thing out. It was done almost all at once, because the pieces and parts had been accumulating for several years.

F.B.S.: That was the white heat of creation?

M.C.: Yes. The actual drawing of the reaction was on a scrap of paper right beside me where I was reading. The article prompted my thinking of the reverse decarboxylation reaction and what I had to do it with. The ribulose diphosphate was the only thing I had to do it with. I tried it with fructose diphosphate, but you come out with one carbon too many. So it had to be a five-carbon and not a six-carbon keto sugar. We had already found the five-carbon sugar, but I didn't know what it was doing there. I know which chair I was sitting in when I put it together! Once you hit it, you know it's right. Everything fits when you hit the right key. And I know when that happened! It was quite an exciting few minutes. The next step was to come back in the lab and pick up the missing pieces, which we did.

F.B.S.: You were the director of the lab with several graduate students and postdocs working with you. How did that interaction work?

M.C.: Oh, that was wonderful! I enjoyed that immensely. That was the best part of the thing, you know. Every day I would come in and say, "What's new?" We would review and then see what the next experiment had to be. Usually there were anywhere from one to six people involved at any one time.

F.B.S.: Were you involved in the laboratory part of the work?

M.C.: Oh, yes. It was a joint effort, the design of the experiments. Some of them I would carry out myself. Some were easy! For example, the identification of phosphoglyceric acid I did personally. And I did it by the way the stuff behaved on an ion exchange column a year or two before we had paper chromatography. One of the graduate students designed a transient experiment with CO_2. Alec Wilson was a boy from New Zealand and did a beautiful job! The experiment pointed the finger at ribulose, because when you shut off the CO_2, the ribulose diphosphate rises. Shutting the light on or off was an easy experiment. But to shut off or turn on the radioactive CO_2 wasn't so easy. We had to build a very special apparatus, and he did it. It was a beautiful technical accomplishment.

F.B.S.: Two final questions: Just what is the nature of creativity in science? What price do you have to pay, and what are the rewards?

M.C.: I have a phrase in response to that to my students. I tell them that it's no trick to get the right answer when you have all the data. A computer can do that. The real creative trick is to get the right answer when you only have half enough data, half of what you have is wrong, and you don't know which half is wrong! When you get the right answer under those circumstances, you are doing something creative! The students learn after a while that this is not a joke. You must be able to sift out the critical points and put it all together in the right way, ignoring what doesn't seem to fit right. If you ignore the right things, you come out with the right answer! But if you pay attention to the wrong things, you don't.

F.B.S.: It's not a mechanical thing: if you have only half the data, you've got to sense or feel your way to the right answer.

M.C.: One which fits your concept of the physical world. The process involves intuition. You can't do it on a part-time basis, obviously. It is done day and night, winter and summer, under all circumstances, at any given moment. You can't tell where you will be when you are doing it. It does intrude on what some people call their private lives. In fact, I don't see how it can work otherwise. So, you ask for price. The price usually is paid by the person's family in the form of neglect and competition. I don't see any way to beat that.

F.B.S.: So what are the rewards? Are they worth it?

M.C.: The satisfaction of guessing correctly and then showing that it is correct is really great! You can usually tell when you are right, even before you've proved it; that's when the satisfaction is the greatest. When something is born and you *know* it is right; then you set about for the next 10 years trying to show it's right. It sounds as though you're making the discoveries over a period of 10 years; you're really not. You make it all at once, and then you spend a long time showing that this is the way it is. Even today, I have some ideas about how photoelectric charge separation occurs—a conversion of solar energy—and I'm beginning to see physically how to do it. We will simulate the thing in a synthetic system within a decade, I would say, probably less. How soon it will be economic in terms of building useful gadgets is another matter. But it's on its way. That's the great sport.

F.B.S.: Now, lastly, your Nobel prize is the only one that qualifies as plant physiology. How has it affected your life?

M.C.: Oh, my! It's made it easier in some ways and more difficult in others. Easier in that I didn't have to spend so much time proving that what I wanted to do was worth doing. On the other hand, the responsibilities that came with it in terms of the students that came here and their expectations, in terms of students everywhere and their expectations of what you can and can't do, is a burden sometimes. I'm beginning to feel it now, a little bit. It's both a privilege and a responsibility.

F.B.S.: Most things of value end up that way, don't they?

M.C.: Yes. They almost always do. You don't get anything for nothing. Some of our modern-day students don't really understand that. They think someone owes them a living. It's hard to get them to realize that you pay for everything. The price is intellectual, physical, and emotional. It's all three, but it has its compensations.

11

Photosynthesis: Environmental and Agricultural Aspects

The amount of photosynthesis occurring on earth is staggering. Estimates vary greatly, but a value of about 70 billion metric tons of carbon (170 billion tons of dry matter with an empirical formula close to CH_2O) fixed per year was obtained recently. Table 11-1 lists estimates of photosynthetic productivity (called *primary productivity* by ecologists) for specific types of ecosystems. Roughly two-thirds of this productivity occurs on land and only one-third in seas and oceans. This vast productivity occurs in spite of the low atmospheric CO_2 concentration (only about 0.034 percent by volume, or 340 μl/l). Most of the CO_2 used by plants is ultimately converted into cellulose, the major component of wood. At present, most of the earth's carbon is in oceanic carbonate deposits, including sedimentary limestones and dolomites. Oil shale, coal, and petroleum are also important carbon reservoirs. The atmosphere has less carbon than other reservoirs, but it still contains about 700 billion metric tons. Roughly 10 percent of the carbon in the atmosphere is used in photosynthesis each year.

11.1 The Carbon Cycle

The amount of CO_2 in the air has increased slowly since about 1850 from about 280 μl/l to slightly over 340 μl/l at present (Rycroft, 1982). Data from 1880 are in Fig. 11-1. The main reason for this increase is the burning of fossil fuels, but removal of forests and grasslands that use more CO_2 in photosynthesis than they release in respiration has also contributed (Stuiver, 1978). The small increases show, overall, that return of the gas to the air is nearly balanced by its use in photosynthesis or that its concentration is otherwise maintained fairly constant. Important return agencies are respiration of plants, microorganisms, and animals, and activities of volcanoes, factories, and automobiles (Fig. 11-2), but oceans and seas are even more important in preventing rapid

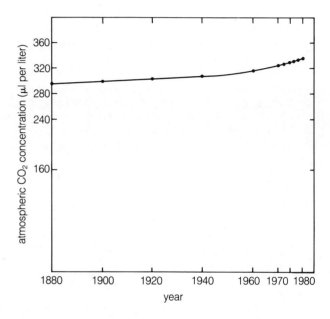

Figure 11-1 Increases in atmospheric CO_2 concentrations from 1880–1980. Data are from several sources and represent values from both the South Pole and Mauna Loa, Hawaii, measurement stations. Mean yearly values were very similar for each station. Data since 1960 are courtesy of C. D. Keeling and NASA. There are significant seasonal variations in values, especially in Hawaii, because of photosynthesis in summer.

rises in atmospheric CO_2 levels. In these, solid and dissolved carbonates equilibrate with CO_2, a change in one eventually affecting the other, so that the concentration of CO_2 on a worldwide scale is buffered by the carbonates in water. Within plant canopies, as in a maize field for example, the CO_2 content can fall to 260 μl/l or less during daylight hours, whereas the content there can reach 400 μl/l during darkness because of respiration by plants and soil microbes. Worldwide increases in atmospheric CO_2 are of concern because they could lead to warming of the earth's surface (Revelle, 1982). Such warming could, within hundreds of years, cause melting of so much ice in polar regions that the ocean would rise and

Table 11-1   Estimates of Net Primary Productivity and Plant Biomass for Various Ecosystems[a]

Ecosystem Type	Area (10⁶ km²)	Net Primary Production, per Unit Area (g/m²/yr) Normal Range	Mean	World Net Primary Production (10⁹ t/yr)	Biomass per Unit Area (kg/m²) Normal Range	Mean	World Biomass (10⁹ t)
Tropical rain forest	17.0	1000–3500	2,200	37.4	6–80	45	765
Tropical seasonal forest	7.5	1000–2500	1,600	12.0	6–60	35	260
Temperate evergreen forest	5.0	600–2500	1,300	6.5	6–200	35	175
Temperate deciduous forest	7.0	600–2500	1,200	8.4	6–60	30	210
Boreal forest	12.0	400–2000	800	9.6	6–40	20	240
Woodland and shrubland	8.5	250–1200	700	6.0	2–20	6	50
Savanna	15.0	200–2000	900	13.5	0.2–15	4	60
Temperate grassland	9.0	200–1500	600	5.4	0.2–5	1.6	14
Tundra and alpine	8.0	10–400	140	1.1	0.1–3	0.6	5
Desert and semidesert scrub	18.0	10–250	90	1.6	0.1–4	0.7	13
Extreme desert, rock, sand, and ice	24.0	0–10	3	0.07	0–0.2	0.02	0.5
Cultivated land	14.0	100–3500	650	9.1	0.4–12	1	14
Swamp and marsh	2.0	800–3500	2,000	4.0	3–50	15	30
Lake and stream	2.0	100–1500	250	0.5	0–0.1	0.02	0.05
Total continental	149		773	115		12.3	1,837
Open ocean	332.0	2–400	125	41.5	0–0.005	0.003	1.0
Upwelling zones	0.4	400–1,000	500	0.2	0.005–0.1	0.02	0.008
Continental shelf	26.6	200–600	360	9.6	0.001–0.04	0.01	0.27
Algal beds and reefs	0.6	500–4,000	2,500	1.6	0.04–4	2	1.2
Estuaries	1.4	200–3,500	1,500	2.1	0.01–6	1	1.4
Total marine	361		152	55.0		0.01	3.9
Full total	510		333	170		3.6	1,841

[a]Units are square kilometers, dry grams or kilograms per square meter, and dry metric tons (t) of organic matter. One metric ton equals 1.1023 English tons.

From Whittaker, 1975.

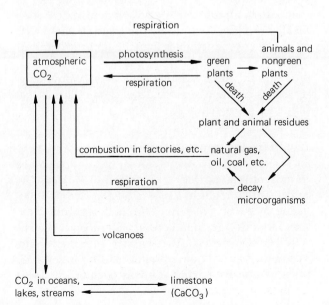

flood many coastal cities. Other accompanying climatic changes, especially in rainfall patterns, would greatly alter both agriculture and natural vegetation. The situation remains uncertain, however, because atmospheric particulate pollutants (the ever-present haze) reflect the sun's rays, and increases in these pollutants could lead to global cooling.

11.2 Photosynthetic Rates in Various Species

Photosynthetic rates of various species living in such diverse conditions as arid deserts, high mountains, and tropical rain forests differ greatly (Table 11-1).

Figure 11-2   The carbon cycle in nature.

Table 11-2 Maximum Photosynthetic Rates of Major Plant Types Under Natural Conditions

Type of Plant	Example	Maximum Photosynthesis (μmol CO_2 m^{-2} s^{-1})[a]
CAM	*Agave americana* (century plant)	0.6–2.4
Tropical, subtropical, and Mediterranean evergreen trees and shrubs; temperate zone evergreen conifers	*Pinus sylvestris* (Scotch pine)	3–9
Temperate zone deciduous trees and shrubs	*Fagus sylvatica* (European beech)	3–12
Temperate zone herbs and C-3 pathway crop plants	*Glycine max* (soybean)	10–20
Tropical grasses, dicots, and sedges with C-4 pathway	*Zea mays* (corn or maize)	20–40

[a]Values are calculated on the basis of one surface of the leaf; for conifers, data are for the optical projection of needles. An extensive list of tree photosynthetic rates was compiled by Larcher, 1969. Data for several C-3 and C-4 crops are listed by Radmer and Kok, 1977. Values for many C-3 native plants are given by Björkman, 1981.

Differences result partly from variation in light, temperature, and the availability of water, but individual species show remarkable differences under specific conditions optimum for each. Species possessing the C-4 pathway of CO_2 fixation generally have the highest photosynthetic rates, while slow-growing desert succulents exhibiting crassulacean acid metabolism (CAM) have among the slowest rates. Table 11-2 summarizes the approximate range of maximum values for a few major groups of plant types representing many different species. Relatively few data are available for CAM plants.

Other interesting groups, for which even fewer results are available, include perennial alpine and arctic species. These usually have short growing seasons: Alpine species have moderate day-lengths and high irradiance levels, and arctic species have long day-lengths and low irradiance levels. Their photosynthesis exceeds respiration so much that they can double in dry weight within a month or less, so carbohydrate accumulation is apparently not a survival problem.

11.3 Factors Affecting Photosynthesis

Many factors influence photosynthesis: H_2O, CO_2, light, nutrients, and temperature, as well as plant age and genetics. Which most limits photosynthesis in natural or agricultural ecosystems? From Table 11-1, we may conclude that higher plants apparently are most limited by availability of water. Deserts are extremely unproductive, while marshes, estuaries, and tropical rain forests are the most productive ecosystems, with the exception of certain irrigated crops. When water potentials become too negative (that is, when water becomes limiting), cellular expansion is first retarded so that growth is reduced. With only a little more water stress, stomates begin to close and CO_2 uptake is restricted. Photosynthesis is then limited by water because of retarded leaf expansion and because of restricted CO_2 absorption. The relationship of water availability to photosynthesis is examined in Chapter 24, and the importance of leaf nitrogen content to rubisco concentrations and photosynthesis is mentioned in Section 13.4.

Table 11-1 suggests that two other factors are important in plant ecosystems: First, alpine and arctic tundras have low productivity, mainly because of low temperatures and short growing seasons. Second, oceans are also low in productivity on a unit area basis, although there are regions that are highly productive. What limits productivity in open oceans? Obviously water is available, and temperatures in many regions are not too cold; sunlight and CO_2 are available, especially in surface waters. Mineral nutrients are the limiting factor. Many organisms settle to the bottom when they die, taking their minerals with them. Thus the surface waters where light and CO_2 are most abundant become impoverished in phosphates, nitrates, and other essential nutrients. When special conditions bring upwellings with nutrients to the surface, a profuse bloom of phytoplankton often results. But most oceans are "nutrient deserts" with much lower productivities than thought several years ago.

With this short summary of planetary productivity, let us examine in more detail certain factors that influence photosynthesis.

Light Effects Figure 11-3 illustrates the rather typical (except for CAM plants) course of CO_2 fixation by photosynthesis during daylight and CO_2 release by respiration at night in a plot of alfalfa. Certain interesting facts can be derived from this graph. First, maximum CO_2 fixation occurs about noon when the irradiance is highest. That light often limits photosynthesis is also shown by reduced CO_2 fixation rates when plants were exposed briefly to cloud shadows.

The figure also shows the relative magnitudes of photosynthesis and dark respiration; in this example the photosynthetic rate reached a maximum of about eight times the nearly constant night respiratory rate. Average ratios for net carbon fixed during daylight compared with night losses by respiration are nearer 6, but vary with plant and environment.

To understand quantitatively how light affects the rate of photosynthesis, we must first examine how much light energy sunlight provides. At the upper boundary of the atmosphere and at the earth's mean distance from the sun, the total irradiance is 1360 J m^{-2} s^{-1} (the **solar constant**), which includes ultraviolet and infrared wavelengths. As this radiation passes through the atmosphere to the earth's surface, much energy is lost by absorption and scattering caused by water vapor, dust, CO_2, and ozone, so that only about 900 J m^{-2} s^{-1} reach plants, depending on elevation, time of day, latitude, and other factors. Of this, about half is in the infrared, roughly 5 percent in the ultraviolet, and the rest (approximately 400 J m^{-2} s^{-1}) has wavelengths between 400 and 700 nm capable of causing photosynthesis, called **photosynthetically active radiation, PAR** (see McCree, 1981 and Appendix B). The actual amount of radiant energy in the PAR range varies with atmospheric conditions, depending mainly on absorption of infrared radiation by atmospheric water vapor on cloudy days.

Direct sunlight at noon in late June in the northern hemisphere corresponds to about 10,000 foot-candles or 108,000 lux. However, foot-candle and lux units are not energy units but are measures of **illuminance,** a subjective description of the ability of the human eye to perceive light (see "Radiation and Light" in Appendix B). We use W m^{-2} (where 1 watt equals 1 J s^{-1}) as a measure of **irradiance** (the flux density of radiant energy per unit time) and joules as a measure of total energy.

In any environment the irradiance varies with season, time of day, and spatially with elevation and within a plant canopy. During the summer in the United States, each square meter of land surface receives about 6–13 million joules of PAR per day, depending on cloud cover, location, and date. About 80 percent of this PAR is absorbed by a representative leaf, although this value varies considerably with leaf structure and age. The remainder (mostly green wavelengths) is transmitted to lower leaves or the ground below or is reflected to the surroundings. Of that absorbed and potentially capable of causing photosynthesis, more than 95 percent is usually lost as heat; thus less than 5 percent is captured during photosynthesis.

Let us now investigate how varying the irradiance affects photosynthetic rates, first when

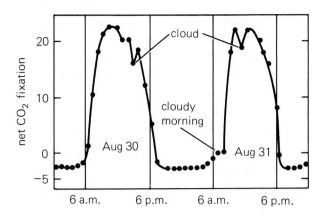

Figure 11-3 Photosynthesis in an alfalfa plot over a two-day period in late summer. The effect of periods of cloud cover can be noted. Negative CO_2 fixation values during hours of darkness indicate the respiration rates. (From data of M. Thomas and G. Hill, 1949.)

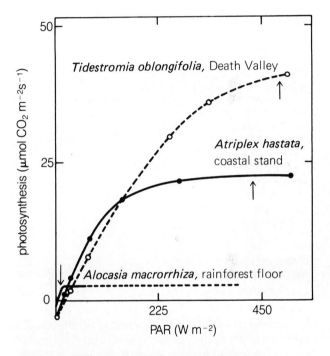

Figure 11-4 Influence of light on photosynthetic rates in single, attached leaves of three species native to different habitats. Maximum irradiances to which the plants are normally exposed are indicated by arrows. The light compensation points are indicated on the graph where the lines cross the abscissa. (Redrawn from J. Berry, 1975.)

single leaves are exposed to normal air with about 320 μl/l CO_2. There is, of course, no net CO_2 fixed in darkness (except in CAM plants), and even in dim light the respiratory loss of CO_2 sometimes exceeds that used in photosynthesis. The irradiance at which photosynthesis just balances respiration (net CO_2 exchange is zero) is called the **light compensation point** (Fig. 11-4). This point varies with the species, with

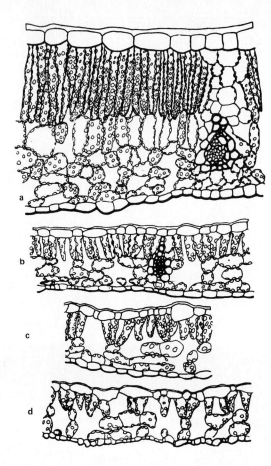

Figure 11-5 Cross sections of leaves of sugar maple (*Acer saccharum*), an unusually shade-tolerant tree, exposed to different light intensities during growth. (**a**) Leaf from south side of isolated tree. Note thick cuticle over the upper epidermis and long palisade parenchyma cells. (**b**) Leaf from center of crown of an isolated tree. (**c, d**) Leaves from base of two forest trees. All trees were growing near Minneapolis, Minnesota. (From H. C. Hanson, 1917.)

the irradiance during growth, and to some extent with the temperature at which measurements are made and the CO_2 concentration, but it is usually less than 2 percent of maximum sunlight (approximately the irradiance in a well-lighted classroom). Only when the irradiance is above the light compensation point can dry weight increases occur. Differences in light compensation points are caused primarily by differences in respiration rates (including photorespiration, which is probably negligible at such low irradiance levels). When respiration is slow, the leaf requires less light to photosynthesize rapidly enough to balance the CO_2 being lost, so the light compensation point is then also low.

Figure 11-4 shows responses to irradiance changes exhibited by single leaves of three dicot species after growth in their native habitats. The upper curve is for a C-4 perennial shrub species, *Tide-*

stromia oblongifolia, that grows under unusually hot and arid summer conditions at high light levels in Death Valley, California; the middle curve is for *Atriplex patula* subspecies *hastata*, a C-3 plant that grows along the Pacific coast of the western United States; and the lowest curve is for *Alocasia macrorrhiza*, which grows on the floor of a rainforest in Queensland, Australia. The energy of PAR received each day by the first and last species during their growth differs by a factor of 300.

The *Alocasia* responses are typical of many species native to shady habitats (**shade plants**), including most house plants. First, these species exhibit much lower photosynthetic rates under bright sunlight than do crop plants or other species grown in open areas. Second, their photosynthetic responses are **saturated** (no longer increase with increasing light levels) at much lower irradiances than are those of other species. Third, they usually photosynthesize at higher rates under very low irradiance levels than do other species. Fourth, their light compensation points are unusually low. These characteristics cause them to grow slowly in their natural shady habitats, yet they survive where species with higher light compensation points could not photosynthesize and would die.

The photosynthetic light responses of single leaves of *Tidestromia oblongifolia* shown in Fig. 11-4 are typical of C-4 species native to sunny habitats and of such C-4 crops as maize, sorghum, and sugar cane. Such leaves show no rate saturation up to and even beyond full sunlight and have maximum rates at least twice those of most C-3 species (at optimum temperatures for each). For such C-4 crops, rates as high as 30 to 40 μmol CO_2 fixed m^{-2} s^{-1} are not uncommon. The *Atriplex hastata* responses in Fig. 11-4 are representative of many C-3 crop species, such as potatoes, sugarbeets, soybeans, alfalfa, tomatoes, and orchard grass. Individual leaves of these species show photosynthetic light saturation at irradiances one-fourth to one-half full sunlight. However, peanut and sunflower are two C-3 species that do not become rate-saturated until nearly full sunlight and that show maximum rates almost as high as those of C-4 crops. Most trees native to temperate climates show maximum rates intermediate between those of typical C-3 crops and shade plants and are often saturated by irradiances as low as one-fourth full sunlight.

The more rapid photosynthesis of C-4 species under high irradiances results in a lower water requirement per gram of dry matter produced, although CAM plants have much lower requirements than either C-4 or C-3 species. Table 11-3 compares this and several other photosynthetic characteristics of C-3, C-4, and CAM plants, most of which are

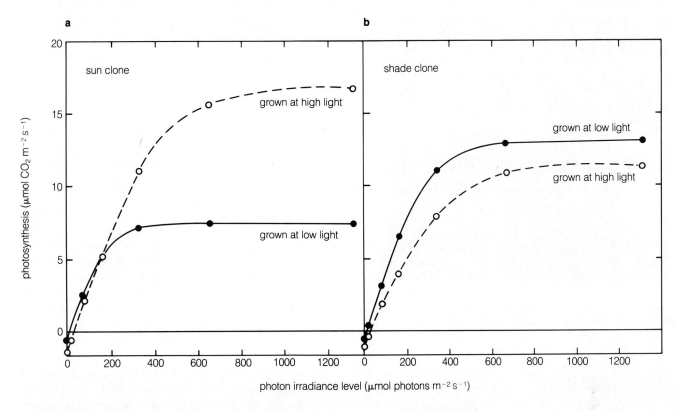

Figure 11-6 Differences in ability of sun clones and shade clones of *Solidago virgaurea* to adapt photosynthetically to high irradiance levels. Dashed lines represent photosynthetic rates of each type after being grown at high light levels; solid lines represent rates after growth at low levels. (**a**) The sun clone adapted to high light during growth; it then required more light to saturate photosynthesis and photosynthesized faster than did plants of the same clone previously grown under low light. (**b**) The shade clone behaved differently; it photosynthesized less rapidly after growth in high than low light. (From O. Björkman and P. Holmgren, 1963. Used by permission.)

described in Chapter 10 or in the following sections. A good review of the ecophysiology of C-3 and C-4 plants is by Pearcy and Ehleringer (1984).

Adaptations to Sun and Shade In trees, shrubs, and to some extent in herbaceous plants, many leaves develop in the shade of others and attain characteristics during development much like those of true shade plants. These are called **shade leaves**, as opposed to **sun leaves** that develop in bright light. In dicots, shade leaves are often larger in area but thinner than sun leaves. Sun leaves become thicker than shade leaves because they form longer palisade cells or an additional layer of such cells (Fig. 11-5). On a weight basis, shade leaves also generally have more chlorophyll, especially chlorophyll *b*, mainly because each chloroplast has more grana than do those of sun leaves. In addition, grana of *Alocasia* and certain other shade plants develop far more thylakoids in the grana, even up to 100 per granum (Björkman, 1981). On the other hand, chloroplasts of shade leaves have less total stroma protein, including rubisco, and probably less thylakoid electron transport protein than do

sun leaves (Boardman, 1977; Björkman, 1981). Thus shade leaves invest more energy in producing light-harvesting pigments that allow use of essentially all the limited amount of light striking them. Furthermore, chloroplasts in leaves exposed to deep shade become arranged by phototaxis within the cells in patterns that maximize light absorption (Section 19.8). The petioles of dicots also respond to the direction and intensity of light by bending (Section 18.3), causing the leaf blades to move into less shaded regions. All of these factors seem to allow net CO_2 fixation under low irradiance levels with minimum energy cost to produce and maintain the photosynthetic apparatus.

To what extent can sun plants (or sun leaves) adapt to shade and shade plants (or shade leaves) adapt to bright sunlight? Mature leaves show much less adaptation to shade or sun than growing leaves, but adaptation of whole plants of some species to either condition during development is considerable, especially adaptation to shade. Of course, there are genetic limits to the extent of adaptation. Some plants seem to be obligate shade plants (e.g., *Alocasia*);

Figure 11-7 One method of shading spruce seedlings to prevent solarization during reforestation planting. (Courtesy Frank Ronco.)

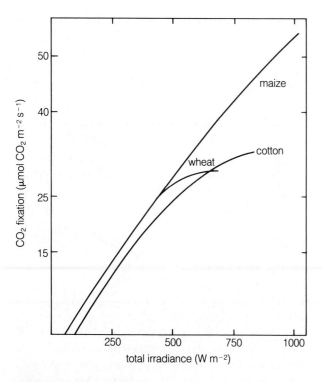

Figure 11-8 Effect of total solar radiation intensity at the top of the canopy on net photosynthetic rates in maize, wheat, and cotton plants. [Drawn from data of (maize) D. N. Baker and R. B. Musgrave, 1964; (wheat) D. W. Pukridge, 1968; (cotton) D. N. Baker, 1965.]

others are obligate sun plants (sunflower, *Helianthus annuus*). But most are facultative shade or sun plants. Facultative C-3 and certain C-4 sun plants adapt somewhat to shade by producing morphological and photosynthetic characteristics similar to those of shade plants (Björkman, 1981). Thus their light compensation points decrease (mainly because they respire much more slowly), they photosynthesize much more slowly, and photosynthesis is saturated at lower irradiance levels. They gradually develop the ability to grow in shade, but this growth is slow.

The reverse adaptation from shade to sun conditions is less common. Shade plants usually cannot be moved to direct sunlight without inhibited photosynthesis and eventual death of the older leaves within several days. Some interesting data were obtained with two different types of *Solidago virgaurea*, one native to open habitats and the other native to shaded forest floors. Clones of each type were made, and these were then grown at high and low irradiance levels. Their photosynthetic responses to various light levels were then measured (Fig. 11-6). Note that the shade clone previously grown under high irradiance levels photosynthesized more slowly than the shade clone grown under low light. Sun clones of the same species, however, photosynthesized much faster after growth under high rather than low irradiances, as expected.

Some conifers are also sensitive to excess light. A dramatic example of this is Englemann spruce *(Picea engelmannii)* in the central and southern Rocky Mountains. When seedlings of this species are transplanted in the open during reforestation work, they usually become chlorotic and die. These symptoms result from a phenomenon known as **solarization,** a light-dependent inhibition of photosynthesis followed by oxygen-dependent bleaching of chloroplast pigments. A major function of certain carotenoid pigments is protection against solarization by absorbing excess light energy that is released as heat instead of being transferred to chlorophylls. In Englemann spruce and some other species, this protection is insufficient. If such seedlings are shaded with logs, stumps, or brush, their survival rate is much higher than when unshaded (Fig. 11-7). Light sensitivity is an important factor in plant succession, because unusually sensitive species never become established except in the shade of others; these are often climax species that can reproduce in their own shade. Other factors also contribute to shade tolerance, one of which in dicots is the ability to form broad, thin leaves at the expense of reduced root systems.

Light Effects in a Plant Canopy The curves of Fig. 11-4 show how photosynthesis in single leaves changes with irradiance level, but we may reasonably ask whether a whole plant, an entire crop, or a forest would exhibit similar response curves. Figure 11-8 shows that stands of C-4 crop plants respond in almost a linear fashion to increases in light level measured on top of the canopy. This curve is much like that for a single leaf of maize or sugar cane (Fig. 11-4), except that it shows even less indication of light saturation. Figure 11-8 also lists typical results for two C-3 crops, cotton and wheat. For them we see a distinct tendency toward light saturation, but only at much higher light levels than in the single-leaf C-3 example of Fig. 11-4 (*Atriplex hastata*).

The principal reason for differences in light responses of single leaves and whole plants or groups of plants is that the upper leaves absorb much of the incident light, leaving less for the lower leaves. In this situation, exposure to a higher irradiance may saturate the upper leaves, but more light is then transmitted and reflected toward the shaded leaves below that are not saturated. As a result, single plants, crops, or forests as a whole probably seldom receive enough light to maximize the photosynthetic rate. This is consistent with the alfalfa data of Fig. 11-3, in which temporary cloudiness decreased photosynthesis.

Availability of CO_2 Photosynthetic rates are enhanced not only by increased irradiance levels but also by higher CO_2 concentrations, unless stomates are closed by drought. Figure 11-9 illustrates how increasing CO_2 levels in the air increase photosynthesis in a C-3 plant at three different irradiance levels. Here the additional CO_2 decreased photorespiration by increasing the ratio of CO_2 to O_2 reacting with rubisco. Photorespiration decreases with increasing CO_2 to O_2 ratios, which leads to faster net photosynthesis. Note that at high CO_2 concentrations, high irradiance levels increase photosynthesis more than at low CO_2 concentrations, and that to saturate photosynthesis a higher CO_2 concentration is required at high than at low irradiance levels. In contrast, photosynthesis of C-4 species is generally saturated by CO_2 levels near 400 $\mu l/l$, just above normal atmospheric concentrations, even at high irradiance levels in which demands for CO_2 are greatest. Some C-4 species are even saturated by normal atmospheric CO_2 concentrations (Edwards et al., 1983).

The difference in CO_2 requirement between C-4 and C-3 species can be observed easily if CO_2 levels are decreased below atmospheric levels. If irradiance levels are above the light compensation points for each, net photosynthesis of C-3 species usually reaches zero at CO_2 concentrations between 35 and 45 $\mu l/l$ (Bauer and Martha, 1981), while C-4 plants continue net CO_2 fixation down to levels between 0 and 5 $\mu l/l$. The CO_2 concentration at which photosynthetic fixation just balances respiratory loss is

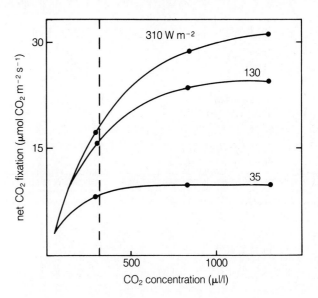

Figure 11-9 Effects of atmospheric CO_2 enrichment on CO_2 fixation in sugar beet leaves. Intact, fully developed leaves from young plants were used. Fixation rates for three different irradiance levels of PAR are shown. The dashed line represents the present atmospheric CO_2 concentration. Higher CO_2 levels increased CO_2 fixation more at increasing irradiance levels. At the highest level, which is only slightly less than that obtained from full sunlight, the highest CO_2 concentration nearly saturated the fixation rate, but at lower intensities, this rate was saturated by lower CO_2 concentrations. Leaf temperatures were between 21 and 24°C. (Redrawn from data of P. Gaastra, 1959.)

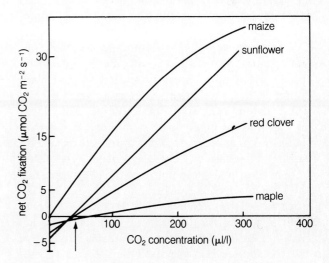

Figure 11-10 Influence of reduced CO_2 concentrations on photosynthetic rate in C-4 (maize) and C-3 plants. Artificial lights providing approximately the same energy as sunlight in the 400–700 nm region were used. (From J. D. Hesketh, 1963.) Arrow indicates compensation points.

Figure 11-11 Left, a C-3 plant (wheat) and right, a C-4 plant (maize) grown in a soilless medium of perlite inside an airtight chamber in which the CO_2 supply became depleted. The maize remained green except at the leaf tips, while the wheat leaves were brown and apparently dead. (Courtesy Dale N. Moss.)

called the **CO_2 compensation point**, a few examples of which are illustrated in Fig. 11-10. Note that the value for maize appears to be zero, whereas that for the C-3 species sunflower and red clover is about 40 $\mu l/l$.

The difference in CO_2 compensation points for C-4 and C-3 species is exhibited dramatically by contrasting responses when a plant of each type is placed in a common sealed chamber in which photosynthesis can occur (Moss and Smith, 1972); the plants must be grown hydroponically in a soilless medium such as sand, perlite, or vermiculite to avoid CO_2 release by soil microorganisms. Both plants fix CO_2 until the CO_2 compensation point of the C-3 plant is reached, but the C-4 plant will photosynthesize at still lower CO_2 concentrations using CO_2 lost by respiration, including photorespiration, from the C-3 plant. As a result, the C-3 plant will usually die within a week or so, but the C-4 plant continues to grow longer. There is a net transfer of CO_2 from one plant to the other (Fig. 11-11).

The lower CO_2 compensation points in C-4 than in C-3 species arise from the much lower photorespiratory release of CO_2 by C-4 plants. The difference in compensation points essentially disappears if the O_2 concentration to which the plants are exposed is decreased from the normal 21 percent down to about 2 percent. In this case the CO_2 compensation points of the C-4 species remain the same but those of the C-3 species also approach zero, because insufficient O_2 is present to compete with CO_2 for rubisco, so photorespiration becomes negligible.

During the summer growing season, insufficient CO_2 is a common cause of suboptimal photosynthesis of C-3 plants, especially for leaves exposed to bright

light. Even slight breezes can enhance photosynthesis by replacing CO_2-depleted air in the boundary layer around a leaf. Students sometimes ask whether the usual limiting factor in photosynthesis is CO_2 or light. The answer is that both can be limiting for C-3 plants and both usually are, but not for the same leaves. The upper, more illuminated leaves will usually respond to increases in CO_2, while the lower leaves may be CO_2 saturated but will respond to additional light. Thus an increase in either factor increases CO_2 fixation of a whole plant or crop. For C-4 plants, light usually limits growth of shaded leaves, unless water or temperature are limiting factors.

Greenhouse crops usually lack enough CO_2 for maximal growth, and this is especially serious in winter when greenhouses are closed. Some growers fertilize the air with CO_2 released from high-pressure tanks or other sources, thereby obtaining increased yields of many ornamental and food crops during the winter months. CO_2 levels are usually not allowed to exceed 1000 $\mu l/l$, because such concentrations are frequently toxic or cause stomatal closure, sometimes even reducing photosynthesis (Hicklenton and Jolliffe, 1980). In summer, greenhouses are usually cooled with evaporative cooling systems in which outside air is drawn across wet pads in greenhouse walls. Increased growth in such greenhouses must result in part from increased CO_2 levels caused by the incoming fresh air.

Temperature The temperature range over which plants can photosynthesize is surprisingly large. Certain bacteria and blue-green algae photosynthesize at temperatures as high as 70°C, whereas conifers can photosynthesize extremely slowly at -6°C or below. In some antarctic lichens, photosynthesis occurs at -18°C with an optimum near 0°C. In many plants exposed to bright sunlight on a hot summer day, leaf temperatures often reach 35°C or higher with photosynthesis continuing.

The effect of temperature on photosynthesis depends on the species, the environmental conditions under which the plant was grown, and the environmental conditions during measurement. Desert species have higher temperature optima than do arctic or alpine species, and desert annuals that grow during the hot summer months (mostly C-4 species) have higher optima than those that grow there only during winter and spring (mostly C-3 species). Crops such as maize, sorghum, cotton, and soybeans that grow well in warm climates usually have higher optima than do crops such as potatoes, peas, wheat, oats, and barley that are cultivated in cooler regions. In general, optimum temperatures for photosynthesis are similar to the daytime temperatures at which the plants normally grow, except that in cold

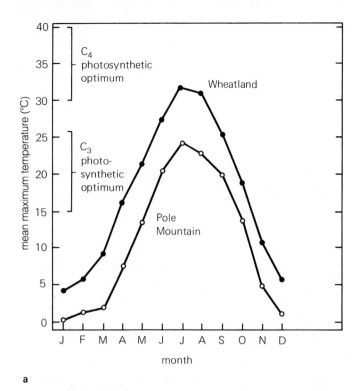

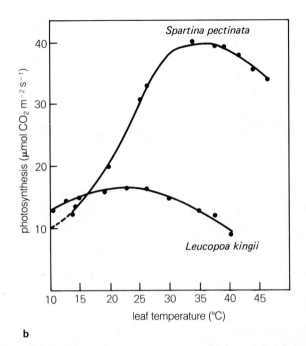

a

b

Figure 11-12 Effect of temperature on photosynthesis in grasses native to the northern great plains. (**a**) Mean maximum temperatures during various months at two Wyoming sites. The Wheatland site on the Laramie River has an elevation of 1470 m.; the Pole Mountain site near Laramie is at 2600 m. Grasses native to the Wheatland site are primarily C-4 species; those at Pole Mountain are mainly C-3 species. The temperature optima for the C-4 species are about 15°C higher than for the C-3 species. (**b**) Temperature-photosynthesis response curves for two species native to the sites in Fig. 11-12a. *Spartina*, a C-4 plant, has a higher temperature optimum and greater photosynthetic rates at most temperatures than does *Leucopoa*, a C-3 plant. All measurements were made on whole plants in the field on cloudless days near noon. Stomates of both species remained open at temperatures up to 40°C, but partially closed at higher temperatures. (Data of A. T. Harrison.)

environments the optima are usually higher than air temperatures. Figure 11-12 illustrates the general relationship for two grasses native to the Great Plains in Wyoming, *Spartina pectinata* (prairie cordgrass), a C-4 plant, and *Leucopoa kingii* (king's fescue), a C-3 plant. *Spartina* grows at a lower elevation (Wheatland site, Fig. 11-12a) than does *Leucopoa* (Pole mountain site), and the mean day temperatures are higher for the Wheatland site. Figure 11-12b shows that the optimum photosynthetic temperature for *Spartina* is near 35°C, compared with about 25°C for *Leucopoa*.

Although there are exceptions, C-4 plants generally have higher temperature optima than do C-3 plants, and this difference is controlled largely by lower rates of photorespiration in C-4 plants. Normal temperature increases have little influence on the light-driven split of H_2O or diffusion of CO_2 into the leaf, but they more markedly influence biochemical reactions of CO_2 fixation and reduction. Thus increases in temperature usually increase photosynthetic rates until enzyme denaturation and photosystem destruction begin. However, respiratory CO_2 loss also increases with temperature, and this is especially pronounced for photorespiration, largely because a temperature rise increases the ratio of dissolved O_2 to CO_2 (Hall and Keys, 1983). As a result of O_2 competition, net CO_2 fixation in C-3 plants is not promoted by increased temperature nearly as much as one might expect. The promoting effect of a temperature rise is nearly balanced by increased respiration and photorespiration over much of the temperature range at which C-3 plants normally grow, so a rather flat and broad temperature response curve between 15 and 30°C often occurs. Since photorespiration is of little significance in C-4 plants, they often exhibit optima in the 30 to 40°C range (Table 11-3). There is also evidence that at high temperatures ATP and NADPH are not produced fast enough in C-3 plants to allow increases in CO_2 fixation, so formation of ribulose bisphosphate becomes limiting.

The most dramatic example of photosynthesis at high temperatures in an angiosperm was found in the shrublike C-4 plant *Tidestromia oblongifolia* by Olle Björkman and co-workers at the Carnegie Institution

Table 11-3 Some Photosynthetic Characteristics of Three Major Plant Groups

Characteristic	C-3	C-4	CAM
Leaf anatomy	No distinct bundle sheath of photosynethetic cells	Well-organized bundle sheath, rich in organelles	Usually no palisade cells, large vacuoles in mesophyll cells
Carboxylating enzyme	Ribulose bisphosphate carboxylase	PEP carboxylase, then ribulose bisphosphate carboxylase	Darkness: PEP carboxylase Light: mainly ribulose bisphosphate carboxylase
Theoretical energy requirement (CO_2:ATP:NADPH)	1:3:2	1:5:2	1:6.5:2
Transpiration ratio (g H_2O/g dry weight increase)	450–950	250–350	18–125
Leaf chlorophyll a to b ratio	2.8 ± 0.4	3.9 ± 0.6	2.5–3.0
Requirement for Na^+ as a micronutrient	No	Yes	Yes
CO_2 compensation point (ppm CO_2)	30–70	0–10	0–5 in dark
Photosynthesis inhibited by 21% O_2?	Yes	No	Yes
Photorespiration detectable?	Yes	Only in bundle sheath	Detectable late afternoon
Optimum temperature for photosynthesis	15–25°C	30–47°C	≈35°C
Dry matter production (tons/hectare/year)	22 ± 0.3	39 ± 17	Low and highly variable

Slightly modified from more extensive table of Black, 1973.

in Stanford, California. As mentioned before, this plant grows in Death Valley, California, and does so under the hottest natural environment in the Western Hemisphere. As contrasted to species that grow in this area only in winter or spring, *Tidestromia* grows in the hot summer months. It has a remarkable photosynthetic optimum at an air temperature of 47°C (117°F) and, as expected, proved to be a C-4 plant. Such high temperatures are tolerated well by *Tidestromia* and other comparable species, but the photosystems of most species are destroyed by excess heat.

Leaves of many species (even when mature) can adapt somewhat to temperature if they are exposed for a few days to different temperatures; this helps plants adjust to seasonal changes (Berry and Björkman, 1980; Osmond et al., 1980; Öquist, 1983). One example of this that involves substantial temperature changes is conifers that photosynthesize in both summer and winter. Their temperature optima are higher in summer. Of course, there are genetic limits and considerable genetic variability in the extent of adaptation to temperature, just as to irradiance levels.

Leaf Age As leaves grow, their ability to photosynthesize increases for a time and then, often even before maturity, begins slowly to decrease. Old, senescent leaves eventually become yellow and are unable to photosynthesize because of chlorophyll breakdown and loss of functional chloroplasts. However, even apparently healthy leaves of conifers that persist several years usually show gradually decreasing photosynthetic rates during successive summers. Many factors control net photosynthesis during leaf development (Sěsták, 1981).

Carbohydrate Translocation An internal control of photosynthesis is the rate at which photosynthetic products such as sucrose can be translocated from leaves to various sink organs. It is often found that removal of developing tubers, seeds, or fruits (strong sinks) inhibits photosynthesis after a few days, especially in adjacent leaves that normally translocate to these organs. Furthermore, species that have high photosynthetic rates also have relatively high translocation rates, consistent with the idea that effective transport of photosynthetic products maintains rapid CO_2 fixation. Severe infection of leaves by pathogens often inhibits photosynthesis so much that these leaves become sugar importers rather than exporters; the adjacent healthy leaves then gradually photosynthesize much faster, suggesting that enhanced translocation from them has removed some limitation to their CO_2 fixation. We do not fully understand the mechanism of these relations (see discussion in Section 7.4), but one factor in some species is buildup of starch grains in chloroplasts when translocation is slow and photosynthesis is fast. Such starch grains press thylakoids unusually close together in chloroplasts and physically prevent light from reaching the thylakoids and causing photosynthesis. Another probable factor is feedback inhibition of photosynthesis by sugars or perhaps other photosynthetic products when translocation is slow (Herold, 1980; Wardlaw, 1980; Gifford and Evans, 1981; Azcón-Bieto, 1983).

11.4 Photosynthetic Rates, Efficiencies, and Crop Production

Many crop physiologists, ecologists, and plant breeders are concerned with how environmental factors and plant genotypes can be altered to increase yields of agronomic and forest crops (Evans, 1980; Gifford and Evans, 1981; Johnson, 1981; Hanover, 1980; Marzola and Bartholomew, 1979; Kommedahl and Williams, 1983). But efficiencies are one thing, and crop yields are another. Maximum efficiencies for all species are obtained only at low irradiance levels, not in bright sunlight (with long days) when yields are highest. Considering supplies of total PAR to land area on growing crops, overall biomass production efficiencies are always much below 18 percent.* Many crops, including forest trees and herbaceous species, convert only 1 to 2 percent of the PAR striking the field during the growing season into stored carbohydrates (Wittwer, 1980; Good and Bell, 1980). Much PAR is wasted by striking the bare ground between young plants before leaves have grown enough to absorb it; this is true for both C-3 and C-4 plants, ignoring evergreens. Furthermore, only 40–45 percent of the sun's energy is in the PAR region, so the theoretical maximum efficiency from all the sun's energy is only about 8 percent (45 percent of 18 percent).

At temperatures from 10 to 25°C and normal atmospheric CO_2 and O_2 levels, efficiencies are about the same for both C-3 and C-4 plants (Ehleringer and Pearcy, 1983; Osborne and Garrett, 1983). Both require about 15 photons of PAR to fix one molecule of CO_2. At lower temperatures or under only 2 percent O_2, photorespiration of C-3 plants is essentially eliminated, and they become more efficient than C-4 plants, requiring only 12 photons per CO_2 fixed. However, under the same conditions C-4 plants still require at least 14 photons, in part because they re-quire three ATP molecules for each CO_2 to operate the Calvin cycle and two more to operate the C-4 pathway (Section 10.3). The number of photons required by C-4 plants depends on the mechanism by which they decarboxylate C-4 acids in the bundle sheath and how effectively CO_2 can leak out of the bundle sheath cells (Pearcy and Ehleringer, 1984). The photon requirement varies from 14 to 20. Photosynthetic efficiencies never exceed the theoretical 18 percent of absorbed PAR, and there is no known way by which this can be increased.

In normal air, increasing temperatures gradually decrease efficiencies of C-3 plants, with efficiencies of C-4 plants remaining constant. As temperatures rise above 30°C, efficiencies of most C-3 plants become lower than those of C-4. This efficiency crossover with increased temperature results from lower net photosynthesis in C-3 plants because of faster CO_2 loss by photorespiration. The absence of detectable photorespiration in C-4 plants, even above 30°C, gives them a substantial efficiency advantage at high but not at low temperatures and especially in non-shaded conditions (Björkman, 1981).

Agriculturists, including foresters, are concerned with productivity of economic plant parts, not with efficiencies or total plant weight, so their goal is to increase both the percentage and the amount of PAR energy that goes to harvestable products. The percentage of weight in the harvestable product compared with aboveground plant biomass is called the **harvest index**. For important crops, including wheat, rice, barley, oats, and peanuts, harvest indices averaging near 50 percent have been reached (Austin et al., 1980; Hargrove and Cabanilla, 1979; Johnson, 1981). For cereal grains, which feed most humans, such high harvest indices have come from breeding programs resulting in varieties that convert less PAR to leaves and stems and more to seeds. A comprehensive analysis of three-fold increases in Minnesota maize yields since the 1930s indicates that introduction of hybrid cultivars accounts for most of that increase (Cardwell, 1982). Photosynthetic efficiencies of very few crops have been improved by breeding (Gifford and Evans, 1981; Zelitch, 1982; Gifford et al., 1984).

Even though overall reduced vegetative growth relative to seeds is advantageous in cereals, cultivars of other crops (e.g., sugar beets and potatoes) that produce extensive leaf cover early in the season are desirable, because they intercept more PAR than cultivars that produce relatively more early stem or root growth (Allen and Scott, 1980). Much research has concerned architectures of plant canopies in relation to productivity. A term called the **leaf area index (LAI)** is widely used to indicate the ratio of leaf area (one surface only) of a crop to the ground area upon which that crop grows. LAI values up to 8 are

---

*The theoretical 18% is calculated as follows. Assume that 12 moles of photons represent the minimum number of photons needed to fix one mole of CO_2 (R9-8) and that an average photon in the PAR region (400–700 nm) has a wavelength of 550 nm. From Planck's equation relating photon energy and wavelength in Appendix B, you can calculate that one mole of such photons has an energy of 217,000 J (51,900 cal). Twelve moles of photons would therefore have an energy of 2.6×10^6 joules. This is the input energy. The output energy, one mole of fixed carbon in carbohydrate, has an energy of about 0.48×10^6 J. Efficiency equals output energy/input energy or 18 percent.

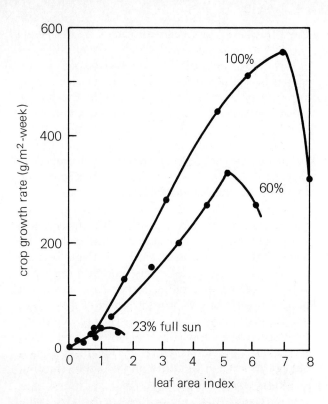

Figure 11-13 Growth of sunflower plant communities (100 plants/m²) at various leaf area indices and light intensities given as percent of full sunlight. At full sunlight, the optimum LAI is 7; the optimum at 60 percent full sunlight is only 5; and at 23 percent full sunlight, it is only 1.5. (From A. C. Leopold and P. E. Kriedemann, 1975.)

common for many mature crops, depending on species and planting density. Forest trees have LAI values of about 12, and many shaded leaves then receive less than 1 percent of full sunlight. Productivity rates increase somewhat with LAI because of more total light interception, but larger LAI values often cause no more increases and then even decreases on a ground-area basis (Fig. 11-13), probably because of respiratory CO_2 loss from shaded leaves and stems.

Increased stem elongation is often an advantage for plants competing for light, but in a uniformly growing cereal crop no such advantage occurs, and increased grain yields are obtained with dwarf or semidwarf varieties that allocate relatively more photosynthate to grain than to stems. Plant breeders have also provided cultivars with other alterations in canopy structure that increase yields. For example, computer results predicted that at LAI values above 2 or 3, depending on the species, cereal grain varieties having erect leaves near the top of the canopy should photosynthesize more rapidly than those having more horizontally oriented leaves in this location. Indeed, yields of the erect leaf types have been significantly greater. It is expected that further cooperation between physiologists and geneticists will bring about increased yields of several other species. Elimination of photorespiration in future agricultural and forest crop plants appears to be a worthwhile goal (Somerville and Ogren, 1982; Zelitch, 1982), even though no success with crops has so far been obtained.

12

Respiration

All active cells respire continuously, often absorbing O_2 and releasing CO_2 in equal volumes. Yet, as you know, respiration is much more than a simple exchange of gases. The overall process is an oxidation-reduction in which compounds are oxidized to CO_2 and the O_2 absorbed is reduced to form H_2O. Starch, fructans, sucrose or other sugars, fats, organic acids, and, under rare conditions, even proteins can serve as respiratory substrates. The common respiration of glucose, for example, can be written as in R12-1.

$$C_6H_{12}O_6 + 6\ O_2 \rightarrow 6\ CO_2 + 6\ H_2O + energy \qquad \text{(R12-1)}$$

Much of the energy (approximately 2,870 kJ or 686 kcal mol^{-1} of glucose) released during respiration is heat. When temperatures are low, this heat might stimulate metabolism and benefit growth of certain species, but usually it is just transferred to the atmosphere or soil with little consequence to the plant. Far more important than heat is the energy trapped in ATP, because it is used later for many essential processes of life, such as growth and ion accumulation.

The above summary equation for respiration is misleading in a way, because respiration, like photosynthesis, is not a single reaction. It is a series of 50 or more component reactions, each catalyzed by a different enzyme. It is an oxidation (with the same products as burning) that occurs in a water medium, near neutral pH, at moderate temperatures, and without smoke! This gradual, stepwise breakdown of large molecules provides a means for converting energy into ATP. Furthermore, as the breakdown proceeds, carbon skeleton intermediates are provided for a large number of other essential plant products. These include amino acids for proteins, nucleotides for nucleic acids, and carbon precursors for porphyrin pigments (such as chlorophyll and cytochromes) and for fats, sterols, carotenoids, anthocyanins, and certain other aromatic compounds. Of course, when these compounds are formed, conversion of the original respiratory substrates to CO_2 and H_2O is not complete. Usually only some of the respiratory substrates are fully oxidized to CO_2 and H_2O (a catabolic process), while the rest are used in synthetic (anabolic) processes, especially in growing cells. The energy trapped during oxidation can be used to synthesize the large molecules required for growth. *When plants are growing, respiration rates increase as a result of growth demands, but some of the disappearing compounds are diverted into synthetic reactions and never appear as CO_2.* Whether carbon atoms in the compounds being respired are converted to CO_2 or to any of the large molecules mentioned above depends on the kind of cell involved, its position in the plant, and whether or not the plant is rapidly growing.

12.1 The Respiratory Quotient

If carbohydrates such as sucrose, fructans, or starch are respiratory substrates, and if they are completely oxidized, the volume of O_2 taken up exactly balances the volume of CO_2 released from the cells. This ratio of CO_2/O_2, called the **respiratory quotient** or **RQ**, is often very near unity. For example, the RQ obtained from leaves of many different species averaged about 1.05. Germinating seeds of the cereal grains and many legumes such as peas and beans, which contain starch as the main reserve food, also exhibit RQ values of approximately 1.0. Seeds from many other species, however, contain much fat or oil. When fats and oils are oxidized during germination, the RQ is often as low as 0.7. Consider the oxidation of a common fatty acid, oleic acid:

$$C_{18}H_{34}O_2 + 25.5\ O_2 \rightarrow 18\ CO_2 + 17\ H_2O \qquad \text{(R12-2)}$$

The RQ for this reaction is $18/25.5 = 0.71$.

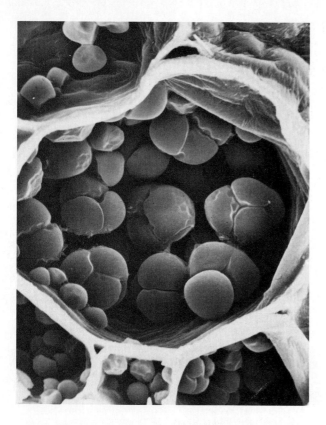

Figure 12-1 Scanning electron micrographs of starch grains in amyloplasts of a rice stem parenchyma cell. Many amyloplasts in these cells contain four starch grains. (Courtesy P. Dayanandan.)

By measuring the RQ for any plant part, information can be obtained about the type of compounds being oxidized. The problem is complicated, because at any time several different types of compounds can be respired, so the measured RQ is an average that depends on the contribution of each substrate and its relative content of carbon, hydrogen, and oxygen. In this chapter, we emphasize respiration of carbohydrates. The utilization of fats is described in Chapter 14. We shall first describe some of the major biochemical aspects of respiration; then these will be used to help explain more physiological and environmental aspects of respiration of various plants and plant parts.

12.2 Formation of Hexose Sugars from Reserve Carbohydrates

Storage and Degradation of Starch As described in Section 10.7, starch is stored as water insoluble granules (grains) that consist of branched amylopectin molecules and unbranched amyloses. Starch accumu-

lated in chloroplasts during photosynthesis is an important carbohydrate reserve in leaves of most species. Starches formed in amyloplasts of storage organs after translocation of sucrose or other nonreducing sugars are also principal respiratory substrates for these organs at certain stages in their development (Fig. 12-1). Parenchyma cells in roots and stems commonly store starch; in perennial species, the starch stored there during the growing season is maintained during winter months and is used in new growth the following spring. Potato tubers are rich in starch-containing amyloplasts, and much of this starch disappears by respiration and translocation of sugars from tuber sections planted to obtain a new crop. The endosperm or cotyledon storage tissues of many seeds contain abundant starch, and most of this also disappears during seedling development. Starch storage in various plant parts is reviewed in detail by Jenner (1982).

Figure 12-2a shows the relation of the starch-storing endosperm to the rest of the seed in maize, while a germinating maize seedling is shown in Fig. 12-2b. These pictures illustrate a situation in which only some of the glucose molecules derived from starch are totally oxidized to CO_2 and H_2O. Other glucoses are converted into sucrose molecules in the scutellum and then moved into the growing root and shoot where some are totally respired and others are diverted into cell-wall materials, proteins, and other substances needed for growth of the seedling.

Most steps in the degradation of starch to glucose can be catalyzed by three different enzymes, although still others are needed to complete the process. The first three include an *alpha amylase* (α-amylase), a *beta amylase* (β-amylase), and *starch phosphorylase*. Some studies show that of these, only alpha amylase can attack intact starch granules, so when β-amylase and starch phosphorylase are involved, they must act on the first products released by α-amylase (Dunn, 1974; Chang, 1982; Steup et al., 1983). Alpha amylase randomly attacks 1,4-bonds throughout both amylose and amylopectin, at first releasing products that are still large and complex. Later, fragments containing about 10 glucose units, the *dextrins*, are released, and eventually α-*maltose*, a disaccharide containing two glucose units (Fig. 12-3a), and glucose are produced from amylose. Alpha amylase cannot, however, attack the 1,6-bonds at the branch points in amylopectin (see Fig. 10-14), so amylopectin digestion stops when branched dextrins of short chain lengths still remain. Many α-amylases are activated by Ca^{2+}, which is one explanation for the essentiality of this element. Beta amylase hydrolyzes starch into β-*maltose* (Fig. 12-3b), starting only from the nonreducing ends. The β-maltose is rapidly changed by mutarotation into

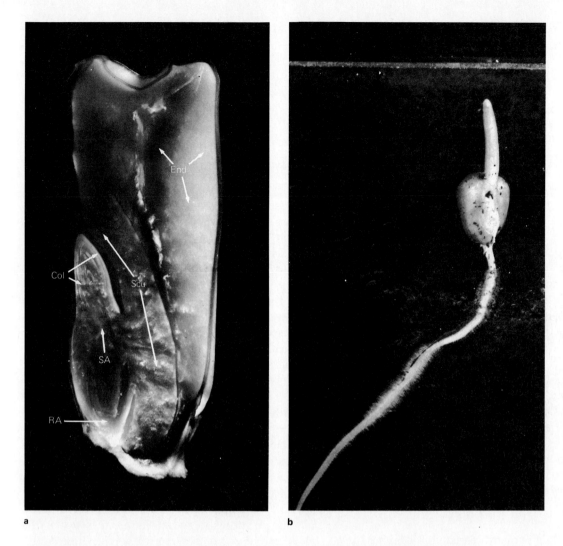

Figure 12-2 (**a**) Longitudinal section of a maize seed, showing the relation of the starch-storing endosperm (End) to the other seed parts. Col, coleoptile; Scu, scutellum or cotyledon; SA, shoot apex; RA, radicle. (From O'Brien and McCully, 1969.) (**b**) Maize seedling being nourished by the endosperm. (From Jensen and Salisbury, 1972.)

the natural mixtures of α- and β-isomers. Hydrolysis of amylose by β-amylase is nearly complete, but amylopectin breakdown is incomplete because the branch linkages are not attacked. Dextrins again remain.

The activity of both amylases involves the uptake of one H_2O for each bond cleaved, so they are *hydrolase enzymes* (Section 8.2). Hydrolytic reactions are not reversible, so no starch synthesis by amylases can be detected. *A general principle is that large molecules are usually synthesized by one series of reactions (pathway) and broken down by another* (Beevers, 1974). For example, we explained in Section 10.7 how polysaccharide synthesis requires an activated form of a sugar such as ADPG, UDPG, or perhaps even glucose-1-phosphate.

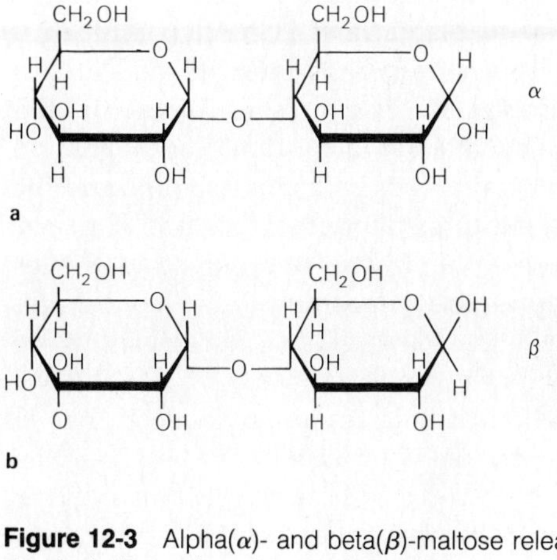

Figure 12-3 Alpha(α)- and beta(β)-maltose released from starch during action of α- and β-amylases.

The amylases are widespread in various tissues but are most active in germinating seeds that are high in starch. In leaves, α-amylase is probably of much greater importance than β-amylase for starch hydrolysis. The α-amylase is located inside chloroplasts, often bound to the starch grains that it will attack. It functions both day and night, although of course during daylight there is a net production of starch from photosynthesis. Furthermore, α-amylase has a pH optimum near 5, so it hydrolyzes starch slowly during daylight when the pH of the chloroplast stroma is near 8 but more rapidly at night when the pH is about 7.

Starch phosphorylase breaks down starch beginning at a nonreducing end. This breakdown is not by incorporating water into the products as amylases do, but by incorporating phosphate. It is therefore a *phosphorolytic enzyme* rather than a hydrolytic one, and the reaction that it catalyzes is reversible *in vitro*:

$$\text{starch} + H_2PO_4^- \rightleftharpoons \text{glucose-1-phosphate} \quad \text{(R12-3)}$$

In spite of the *in vitro* reversibility of this reaction, most experts believe that its only important role is in starch degradation. One reason for believing this is that the Pi concentration within plastids is often 100-times that of glucose-1-phosphate; under these physiological conditions reversibility would be negligible. As will become more apparent later, formation of glucose-1-phosphate avoids the need for an ATP to convert glucose into a glucose phosphate during respiration.

Amylopectin is only partially degraded by starch phosphorylase: The reaction proceeds consecutively from the nonreducing end of each main chain or branch chain to within a few glucose residues of the α-1,6 branch linkages, so dextrins again remain. Amylose, almost entirely free of such branches, is completely hydrolyzed by the repeated removal of glucose units beginning at the nonreducing end of the chain. As are the amylases, starch phosphorylase is widespread in plants, and it is often difficult to decide which enzyme digests most of the starch in the cells concerned. The present theory is that α-amylase is essential for initial attack, as mentioned, and for cereal grain seeds both amylases appear functional, but starch phosphorylase does not. For seeds of other species, for leaves, and for other tissues, starch phosphorylase apparently also contributes, especially after the starch grains are partially hydrolyzed by one of the amylases (Steup et al., 1983).

The 1,6 branch linkages in amylopectin that are not attacked by any of these above enzymes are hydrolyzed by various *debranching enzymes*. Plants contain two types that differ somewhat as to the types of polysaccharides they will attack: a *pullulanase* and an *isoamylase* type (Preiss and Levi, 1979; Ishizaki et al., 1983). Action of these enzymes provides ad-

ditional end groups for attack by amylases or starch phosphorylase and allows complete digestion of amylopectin into glucose, maltose, or glucose-1-phosphate.

Maltose seldom accumulates to any appreciable extent in plants, probably because it is slowly hydrolyzed to glucose by α-amylase and more rapidly by a *maltase* enzyme, as in R12-4:

$$\text{maltose} + H_2O \rightarrow 2 \; \alpha\text{-D-glucose} \quad \text{(R12-4)}$$

The resulting glucose units are now available for conversion into other polysaccharides as described in Section 10.7 or, as stated in the present chapter, for degradation by subsequent respiratory processes.

In summary, the amylases hydrolyze amylose to glucose and maltose, while starch phosphorylase converts amylose to glucose-1-phosphate. The action of all three enzymes on amylopectin leaves a dextrin, the branch linkages of which must be hydrolyzed by debranching enzymes. Maltose is hydrolyzed to glucose largely by maltase.

Hydrolysis of Fructans As mentioned in Section 10.7, the principal carbohydrate food reserve material in some species, most notably stems and leaves of temperate region grasses and parts of members of the Asteraceae and other families, is not starch. Instead, fructans (fructosans) predominate. But even in these species, fructans are seldom, if ever, present in important quantities in seeds. As usual, starch is the main carbohydrate reserve of seeds. Surprisingly little is known about fructan metabolism considering their importance, although it is known that they are hydrolyzed by β-*fructofuranosidase* enzymes having specificity for the particular β-2,1 or β-2,6 links involved. For example, one such enzyme from the Jerusalem artichoke tuber successively cleaves fructose units from inulin until a mixture of fructose and the terminal sucrose unit remain:

$$\text{glucose-fructose-(fructose)}_n + nH_2O \rightarrow$$
$$\text{(fructan)}$$

$$n \text{ fructose} + \text{glucose-fructose} \quad \text{(R12-5)}$$
$$\text{(sucrose)}$$

Hydrolysis of Sucrose The sucrose produced from fructans in reactions such as R12-5 and those sucrose molecules translocated through the phloem from leaves to various receiving (sink) cells must be degraded to glucose and fructose before respiratory breakdown can continue. The major reaction of sucrose degradation is by irreversible hydrolysis to free glucose and fructose by *invertases*:

$$\text{sucrose} + H_2O \rightarrow \text{glucose} + \text{fructose} \quad \text{(R12-6)}$$

12.3 Glycolysis

The group of reactions, collectively called **glycolysis,** converts glucose, glucose-1-P, or fructose (set free by the preparatory reactions described above) to pyruvic acid in the cytosol. (Several reactions of glycolysis also occur in chloroplasts and other plastids, but the complete pathway probably does not.) Glycolysis is the first of three closely related phases of respiration and is followed by the Krebs cycle and electron-transport processes occurring in mitochondria. The individual reactions of glycolysis, now believed to occur in all living organisms, were discovered between 1912 and 1935 by German scientists interested in alcohol production by yeast and by others concerned with breakdown of animal starch (glycogen) to pyruvic acid in muscle cells (Lipmann, 1975; Cori, 1983). The term glycolysis, meaning lysis of sugar, was introduced in 1909 to mean breakdown of sugar to ethyl alcohol (ethanol). However, most cells produce pyruvic acid when normally aerated instead of ethanol. (We shall describe formation of ethyl alcohol by fermentation in Section 12.4.) Furthermore, the common sugars that are broken down are hexoses, so glycolysis has come to mean the degradation of hexoses to pyruvic acid (although many animal biochemists use the term to mean degradation of glycogen, animal starch, to pyruvate).

The individual reactions of glycolysis, the enzymes that catalyze them, and the particular requirements of the enzymes for metal activators are in Fig. 12-4. The overall process, however, (beginning with glucose) can be summarized by R12-7:

$$\text{glucose} + 2\ NAD^+ + 2\ ADP^{2-} + 2\ H_2PO_4^- \rightarrow 2\ \text{pyruvate}$$
$$+ 2\ NADH + 2\ H^+ + 2\ ATP^{3-} + 2\ H_2O \qquad \text{(R12-7)}$$

Glycolysis has several functions. *First, glycolysis converts one hexose molecule into two molecules of pyruvic acid, and oxidation of the hexose occurs.* No O_2 is used, and no CO_2 is released. For each hexose converted, two molecules of NAD^+ are reduced to NADH $(+2H^+)$. These NADH are important, because each can subsequently be oxidized by O_2 in a mitochondrion such that NAD^+ is regenerated and two molecules of ATP are formed. Furthermore, some of these NADH that do not enter mitochondria are used in the cytosol to drive various anabolic, reductive processes. Clearly, these NADH produced during glycolysis are important compounds. Inspection of Fig. 12-4 shows that NADH is formed only at one step in glycolysis, during the oxidation of 3-phosphoglyceraldehyde to 1,3-bisphosphoglyceric acid (see large asterisk in Fig. 12-4).

A second function of glycolysis is production of ATP. The initial utilization of glucose or fructose actually requires input of two ATP molecules in reactions catalyzed by hexokinase, fructokinase, and phos-

phofructokinase (Fig. 12-4 top region). Nevertheless, two ATP are later released for each 3-carbon unit involved (in the reactions catalyzed by phosphoglycerokinase and pyruvate kinase). Thus there is a total production of four ATP per hexose used, and a net production of two ATP per hexose. If glucose-1-P, glucose-6-P, or fructose-6-P is the substrate, one less ATP is required to begin glycolysis, so the net ATP production is three per hexose phosphate. In photosynthetic cells, 3-PGA, 3-phosphoglyceraldehyde, and dihydroxyacetone phosphate produced in the Calvin cycle can become substrates of glycolysis when they are transported into the cytosol from chloroplasts, so you can see that photosynthesis becomes connected to respiration in several ways.

A third function of glycolysis is the formation of molecules that can be removed from the pathway to synthesize several other constituents of which the plant is composed. This function is not apparent in Fig. 12-4 or R12-7, but it will be given special attention in Section 12.11 and Fig. 12-11.

Finally, glycolysis is important because the pyruvate that it produces can be oxidized in mitochondria to yield relatively large amounts of ATP.

12.4 Fermentation

Although glycolysis can function well without O_2, the further oxidation of pyruvate and NADH by mitochondria requires this gas. Thus when O_2 is limiting, NADH and pyruvate begin to accumulate. Under this condition plants carry out **fermentation** (anaerobic respiration), forming either ethanol or lactic acid (usually ethanol), as shown in Fig. 12-5. The two top reactions in Fig. 12-5 consist of a decarboxylation to form acetaldehyde, then rapid reduction of acetaldehyde by NADH to form ethanol. These reactions are catalyzed by *alcohol dehydrogenase.* Some cells contain *lactic acid dehydrogenase,* which uses NADH to reduce pyruvic to lactic acid. Thus either ethanol or lactic acid, or both, are fermentation products, depending on the activities of each dehydrogenase present. In each case, NADH is the reductant, and only under anaerobic conditions is it abundant enough to cause reduction. Furthermore, in some plants NADH is used to cause accumulation of other compounds when O_2 is limiting, especially malate and glycerol (Crawford, 1982; Davies, 1980). The occurrence of fermentation in various plants under oxygen stress will be described in Section 12.13.

12.5 Mitochondria

To understand how pyruvate and NADH produced in glycolysis are oxidized by mitochondria, it is first helpful to understand some of the properties of these

glycolysis

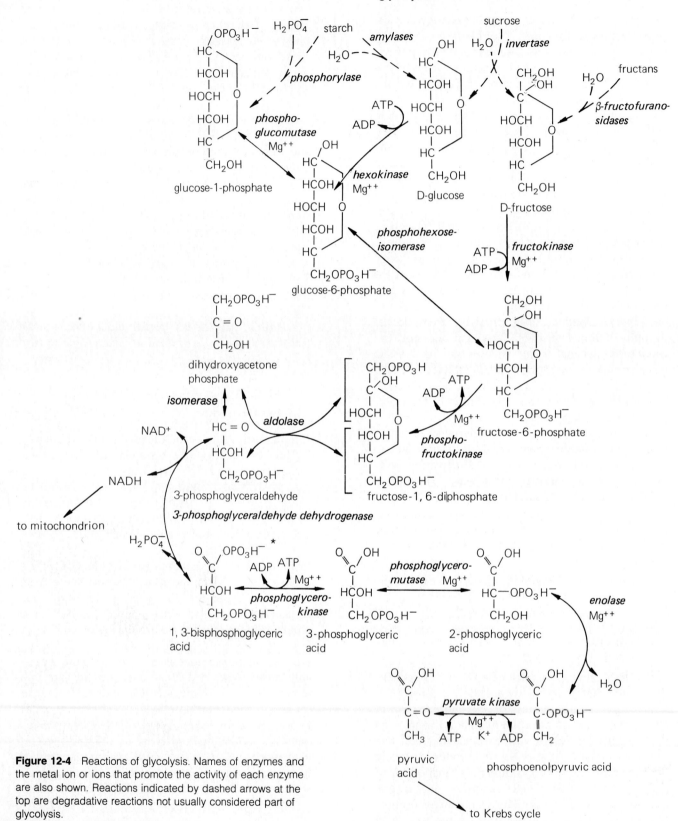

Figure 12-4 Reactions of glycolysis. Names of enzymes and the metal ion or ions that promote the activity of each enzyme are also shown. Reactions indicated by dashed arrows at the top are degradative reactions not usually considered part of glycolysis.

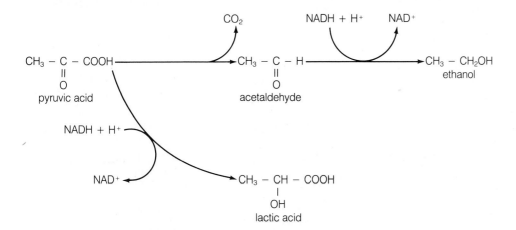

Figure 12-5 Fermentation of pyruvate to form ethanol or lactic acid.

organelles. In some respects, mitochondria are similar to chloroplasts, although functionally they are quite different. Each contains circular DNA that has genetic information used to produce a small percentage of its enzymes, each is formed mainly or entirely by division of the preexisting organelle, and each is surrounded by a double membrane or envelope with an extensive inner membrane system. All living plant cells contain many mitochondria, frequently about 200, and estimates of the maximum number per cell have reached 2000 for immature metaxylem vessel elements from young corn shoots (Malone et al., 1974).

Mitochondria are only a few μm long, not much longer than many bacteria, and although they can be seen with a light microscope, their fine structure is made clear only with the electron microscope. Their general morphology is shown in electron micrographs of earlier chapters (e.g., Fig. 10-8) and is illustrated in a highly idealized form in Fig. 12-6a. However, mitochondria change their shapes rapidly, from long and narrow to more nearly spherical. The inner membrane of the mitochondrial envelope is highly convoluted, protruding into the interior or **matrix** in sheetlike or tubelike patterns in many places. Each such convolution is called a **crista** (plural, **cristae**). In some plant mitochondria, sheetlike cristae are well developed (Bajrachara et al., 1976; Öpik, 1974), but this varies with the type of cell, its age, and its extent

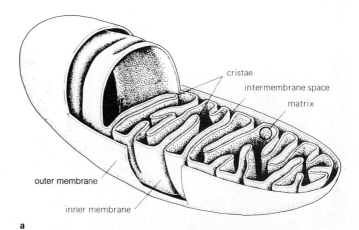

a

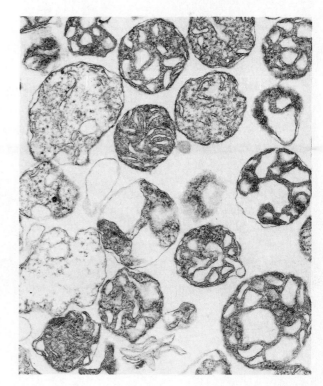

b

Figure 12-6 (a) An idealized drawing of a mitochondrion with cristae. (From Jensen and Salisbury, 1972.) (b) An electron micrograph of mitochondria isolated from maize shoots. The mitochondria have been cut in various planes. The unstained (white) areas correspond to the intermembrane spaces shown in (a). These spaces apparently end adjacent to the outer membrane in regions above and below the plane of sectioning. The matrix of (a) is shown here as the heavily stained regions between the cristae. (From C. Malone et al., 1974.)

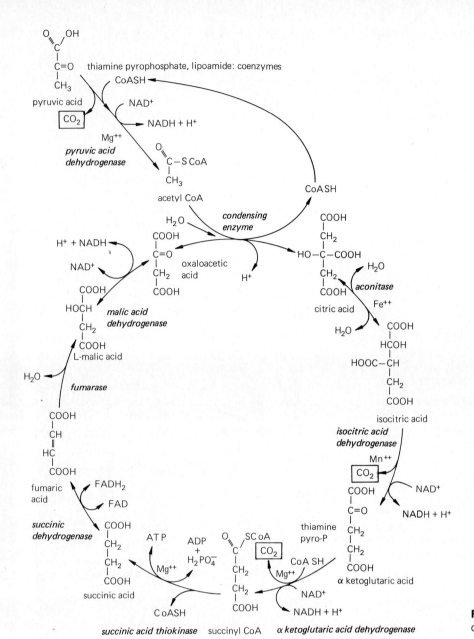

Figure 12-7 Reactions of the Krebs cycle, including enzymes and coenzymes.

of development. In many, one crista is fused to another in the interior of the mitochondria, forming a continuous saclike intermembrane compartment between them (Fig. 12-6b), while other modifications are shown in electron micrographs and drawings by Malone et al. (1974) and Öpik (1974). Regardless of their form, cristae contain most of the enzymes that catalyze steps of the electron-transport system following the Krebs cycle, so the increased surface area they provide is of great importance. The Krebs cycle reactions occur in the protein-rich matrix between the cristae.

12.6 The Krebs Cycle

The **Krebs cycle** was named in honor of the English biochemist Hans A. Krebs, who, in 1937, proposed a cycle of reactions to explain how pyruvate breakdown takes place in the breast muscle of pigeons. He called his proposed pathway the **citric acid cycle,** because citric acid is an important intermediate. Another common name for the same group of reactions is the **tricarboxylic acid (TCA) cycle,** a term used because citric and isocitric acids have three carboxyl groups. It was not until the early 1950s that mitochondria capable of carrying out this cycle were isolated from plant cells.

The initial step leading to the Krebs cycle involves the oxidation and loss of CO_2 from pyruvate and the combination of the remaining 2-carbon acetate unit with a sulfur-containing compound, *coenzyme A (CoA)*, forming *acetyl CoA*

$$(CH_3 - C - SCoA)$$
$$\underset{O}{\parallel}$$

This and another comparable role of CoA in the Krebs cycle are important reasons why sulfur is an essential element.

The reaction of pyruvate decarboxylation also involves a phosphorylated form of **thiamine** (vitamin B_1) as a prosthetic group. Participation of thiamine in this reaction partially explains the essential function of vitamin B_1 in plants and animals. Besides the loss of CO_2, two hydrogen atoms are removed from pyruvic acid during the formation of acetyl CoA. The enzyme catalyzing the complete reaction is called *pyruvic acid dehydrogenase*, but it is actually an organized complex containing numerous copies of three or four different enzymes. The hydrogen atoms removed are finally accepted by NAD^+, yielding NADH. This and other Krebs cycle reactions are described in Fig. 12-7.

The Krebs cycle accomplishes removal of some of the electrons from organic acid intermediates and transfer of these electrons to NAD^+ or FAD. Notice that none of the dehydrogenase enzymes of the cycle uses $NADP^+$ as an electron acceptor. In fact, $NADP^+$ is usually undetectable in plant mitochondria, a situation opposite to that of chloroplasts where $NADP^+$ is abundant but where there is much less NAD^+. Not only are NADH and $FADH_2$ important products of the Krebs cycle, but one molecule of ATP is formed from ADP and P*i* during the conversion of *succinyl coenzyme A* to *succinic acid*. (In mammals, but apparently not plants, formation of ATP at this step requires GDP and GTP, guanosine nucleotides.) Two additional CO_2 molecules (shown boxed in Fig. 12-7) are released in these Krebs cycle reactions, so there is a net loss of both carbon atoms from the incoming acetate of acetyl CoA. The release of CO_2 in the Krebs cycle accounts for the product CO_2 in the summary equation for respiration R12-1, but no O_2 is absorbed during any Krebs cycle reaction.

The primary functions of the Krebs cycle are as follows:

1. Reduction of NAD^+ and FAD to the electron donors NADH and $FADH_2$ that are subsequently oxidized to yield ATP.

2. Direct synthesis of a limited amount of ATP (1 ATP for each pyruvate oxidized).

3. Formation of carbon skeletons that can be used to synthesize certain amino acids that, in turn, are converted into larger molecules (see Section 12.11 and Fig. 12-11 for explanation of the kinds of compounds formed from Krebs cycle intermediates and what prevents the cycle from stopping when intermediates are removed from it). Considering that two pyruvates are produced in glycolysis from each glucose, the overall reaction for the Krebs cycle can be written as follows:

$$2\,\text{pyruvate} + 8\,NAD^+ + 2\,FAD + 2\,ADP^{2-} + 2\,H_2PO_4^- + 4\,H_2O \rightarrow 6\,CO_2 + 2\,ATP^{3-} + 8\,NADH + 8\,H^+ + 2\,FADH_2$$

(R12-8)

12.7 The Electron Transport System and Oxidative Phosphorylation

When NADH and $FADH_2$ produced in the Krebs cycle or in glycolysis are oxidized, ATP is produced. Although this oxidation involves O_2 uptake and H_2O production, neither NADH nor $FADH_2$ can combine directly with O_2 to form H_2O. Rather, their electrons are transferred via several intermediate compounds before H_2O is made. These electron carriers constitute the **electron-transport system** of mitochondria. Electron transport proceeds from carriers that are thermodynamically difficult to reduce (those with negative reduction potentials) to those that have a greater tendency to accept electrons (have higher, even positive reduction potentials). Oxygen has the greatest tendency to accept electrons, and it ultimately does so. Each carrier of the system usually accepts electrons only from the previous carrier close to it. They are arranged in an assembly-line fashion in the inner mitochondrial membrane, and there are several thousand electron transport systems in each mitochondrion.

As in the chloroplast electron-transport system, which is involved in the transfer of electrons from water molecules, the mitochondrial system involves cytochromes (up to four of the *b* type and two of the *c* type) and numerous representatives of quinones, especially *ubiquinone*. Also present are several *flavoproteins* (riboflavin-containing proteins), some iron-sulfur (Fe–S) proteins similar to ferredoxin, an enzyme named *cytochrome oxidase*, and a few other electron carriers not yet identified (Goodwin and Mercer, 1983; Storey, 1980). The cytochromes and cytochrome oxidase both contain iron as part of a heme group. The flavoproteins contain either *flavin adenine dinucleotide (FAD)* or the similar *flavin mononucleotide (FMN)* as bound prosthetic groups. Many of these electron carriers have counterparts in chloroplasts, yet each is unique in structure.

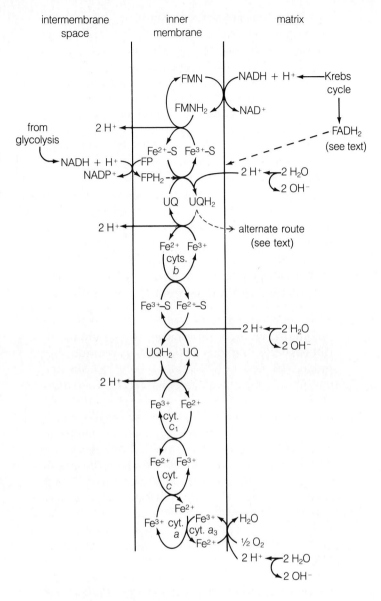

intermembrane space   inner membrane   matrix

Figure 12-8 Reactions of the mitochondrial electron-transport system. Abbreviations are in text, except that FP is an oxidized flavoprotein and FPH_2 a reduced flavoprotein.

phosphorylation is catalyzed by a *coupling factor* or *ATPase*. This ATPase has a stalk and a headpiece much like that of the thylakoid ATPase, and it extends completely across the inner membrane. The headpiece faces and extends into the matrix, while the stalk extends outwardly toward the space between the inner and outer membranes. ATP is formed at or in the headpiece within the matrix and is then transported toward the cytosol in exchange for incoming ADP. This exchange is carried out by an antiport system (Section 6.8) present within the inner membrane, and the ATP then moves readily across the much more permeable outer membrane into the cytosol where it performs most of its functions. Phosphate is also necessary for ATP formation, and it is carried into the matrix by an antiport system that simultaneously moves OH^- out of the matrix to the intermembrane space. A similar antiport system catalyzes exchange of OH^- and pyruvate, and this probably explains how pyruvate from glycolysis gets into the matrix where it is oxidized by pyruvate dehydrogenase (Hanson and Day, 1980).

How are the electron transport components arranged in the inner membrane of mitochondria so that rapid electron flow to O_2 can occur and so that a pH gradient can be established? Unfortunately, far less is known about plant than animal mitochondria, so our answers must be limited and based partly on known and presumed similarities between them (comparative biochemistry). Nevertheless, a few distinct differences between plant and animal mitochondria are known (Palmer, 1979; Storey, 1980; Palmer and Møller, 1982), and these will be mentioned. The plant biochemistry textbook of Goodwin and Mercer (1983) and a review by Elthon and Stewart (1983) describe models of the suspected arrangements of electron-transfer components within plant mitochondria; the interested student may consult those references for details. For our purposes, however, it is important only to emphasize how the orderly flow of electrons from NADH to $FADH_2$ to O_2 occurs so that oxidative phosphorylation results. The important concept is that their arrangement causes H^+ transport from the matrix toward the intermembrane space.

Figure 12-8 indicates the major electron-transport pathway beginning with $NADH + H^+$ formed in the matrix by Krebs cycle enzymes (upper right). The two electrons and two H^+ are passed to a flavoprotein containing FMN, which in turn passes the electrons to a Fe-S protein. The iron in the latter can accept only one electron at a time and accepts no H^+; the two H^+ are somehow transferred into the intermembrane space. This is the first of at least three steps in which a pair of H^+ is moved from the matrix fully across the inner mitochondrial membrane. The reduced Fe-S transfers electrons to ubiquinone (UQ),

The cytochromes and Fe–S proteins can receive or transfer only one electron at a time. Ubiquinone, like the plastoquinone of chloroplasts, receives and transfers two electrons and two H^+ (compare R9-6); the same is true of flavoproteins. This property of ubiquinone and flavoproteins is important in establishing a pH gradient from the matrix (pH about 8.5) to the outside of the inner mitochondrial membrane (pH near 7), because this pH gradient drives formation of ATP from ADP and Pi according to the Mitchell chemiosmotic theory of Section 9.7 (Mitchell, 1979). In mitochondria, ATP formation from ADP and Pi is indirectly driven by the strong thermodynamic tendency of O_2 to become reduced and is called **oxidative phosphorylation.** As in chloroplasts,

which, with 2 H^+ taken from the matrix, becomes reduced to UQH_2. From UQH_2 the electrons move one at a time to various cytochromes b, and the two H^+ from UQH_2 are transferred outwardly. Another Fe-S protein then receives and transfers electrons to the Fe^{3+} in cytochrome c_1 with a third outward transport of a pair of H^+. From cytochrome c_1, electrons are received by cytochrome c, and then their transfer to O_2 to form H_2O is catalyzed by *cytochrome oxidase*. This oxidase contains inseparable a and a_3 components (Fig. 12-8, bottom) and also some other polypeptides that contain, in total, two copper ions that undergo oxidation-reduction between Cu^+ and Cu^{2+} forms. The two coppers are involved in electron transport between the iron components of cytochromes a and a_3 (Brunori and Wilson, 1982).

Although reduction potentials are not shown in Fig. 12-8, the overall $\Delta E'_0$ is from NADH at an E'_0 of -0.32 V to O_2 at an E'_0 of $+0.82$ V, a total change of $+1.14$ V. This is the same change that occurs during photosynthetic electron transport from H_2O to $NADP^+$ (Section 9.9), but in mitochondria the sign preceding $\Delta E'_0$ is positive and 220 kJ of energy are released for each mole of Krebs cycle NADH oxidized. The transport of three pairs of H^+ across the membrane probably causes a sufficient pH gradient to allow formation of three ATP by the ATPase (not shown in Fig. 12-8, but compare Fig. 9-8). Nevertheless, there is considerable evidence that in animal mitochondria cytochrome oxidase also pumps two H^+ across the membrane (Casey and Azzi, 1983). How this pumping occurs has not been explained, and we do not know if the plant cytochrome oxidase also does it. Regardless of the number of H^+ transported, numerous studies with isolated plant mitochondria show that for each Krebs cycle NADH oxidized, three ATP are formed.

For each NADH released in glycolysis (Fig. 12-8, upper left) and for each $FADH_2$ released in the Krebs cycle by oxidation of succinate (Fig. 12-8, upper right), only two ATP are formed. The reason for this is that these NADH and FADH molecules donate electrons to the transport chain only after the first pair of H^+ in the main pathway has been passed into the intermembrane space, so less pH difference across the membrane is created when they are oxidized. For NADH arising from glycolysis in the cytosol, a flavoprotein (FP) containing *NADH dehydrogenase* exists on the outer surface of the inner membrane, as shown in Fig. 12-8. Furthermore, NADPH (as resulting from the pentose phosphate pathway, Section 12.10) can be oxidized by a similar dehydrogenase not shown (Palmer and Møller, 1982; Nash and Wiskich, 1983). This ability of plant mitochondria to oxidize cytosolic NADH and NADPH directly is not shared by animal mitochondria. (Animals have a transhydrogenase enzyme that transfers electrons from NADPH to NAD^+, forming $NADP^+$ and NADH, and they employ special carriers to move electron pairs from NADH into the matrix.) The $FADH_2$ produced in the matrix is oxidized similarly in plant and animal mitochondria; its electrons enter the transport chain at the ubiquinone step, and its H^+ are therefore moved across the membrane. Two ATP are formed from each FADH arising from succinate in the Krebs cycle.

Oxidative phosphorylation from all mitochondrial substrates is uncoupled from electron transport by numerous uncoupling compounds, just as in chloroplasts (Section 9.6). Most uncouplers neutralize the pH gradient by carrying H^+ into the matrix, preventing oxidative phosphorylation but still allowing electron transport to occur. Sometimes electron transport occurs even faster, presumably because there is less "back pressure" from the pH gradient (i.e., H^+ transport accompanying electron flow is easier when the H^+ concentration into which H^+ are transported is lower). Ammonia, primary amines, and dinitrophenol uncouple in mitochondria, as in chloroplasts, and so do numerous other compounds. Still other compounds inhibit either oxidative phosphorylation or electron transport without uncoupling the two processes. For example, two potent phosphorylation inhibitors are *oligomycin,* an antibiotic produced by a *Streptomyces* species, and *bongkrekic acid,* an antibiotic produced by a *Pseudomonas* species that grows on fungal-infected coconuts called "bongkreks" by native Indonesians (Goodwin and Mercer, 1983). Oligomycin inhibits ATP formation by the ATPase, whereas bongkrekic acid prevents ATP formation by blocking the antiport system that carries ADP into the matrix in exchange for ATP. Without this ADP, no oxidative phosphorylation can occur. *Antimycin A*, also from *Streptomyces*, blocks electron transport at or near the cytochrome b to Fe-S protein step (Fig. 12-8). This prevents phosphorylation, but it does not uncouple the two processes.

12.8 Summary of Glycolysis, Krebs Cycle, and the Electron Transport System

When a hexose is completely oxidized to CO_2 and H_2O using these three processes, R12-1 describes the overall reaction. However, R12-1 lists energy as a product, and you now know that much of this energy is trapped in ATP. But how much is in ATP, and how much is lost as heat? To answer this, note that glycolysis yields two ATP and two NADH per hexose used (R12-7). Each such NADH oxidized by the electron transport system yields two ATP, as described above, so glycolysis contributes a total of six ATP per hexose. The Krebs cycle contributes two ATP per

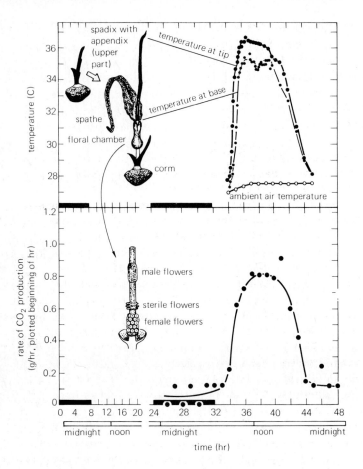

Figure 12-9 Respiration and temperature of a spadix of *Sauromatum guttatum* as a function of time. *Sauromatum* is a Pakistani and Indian genus in the Araceae family. Growth from the corm to a structure about 50 cm tall may occur in about 9 days (drawings at upper left), with a maximum growth rate of 7 to 10 cm/day. If this occurs in constant light, the spathe remains wrapped around the spadix; but after the "normal" time for flowering has passed, a single period of darkness, if it is long enough (bar on the abscissa—two 8-hour dark periods were given in this experiment), will initiate opening of the spathe and a burst in CO_2 production (note extremely large quantities) with a concurrent rise in temperature. The heat apparently serves to volatilize various compounds (especially amines and ammonia), which give an odor of rotting meat. Carrion flies and beetles are attracted and serve in pollination. They enter the floral chamber (lower drawing, somewhat schematic). (Original data. Experiment performed for use in this text by B. J. D. Meeuse, R. C. Buggein, and J. R. Klima of the University of Washington, Seattle.)

hexose or per two pyruvates (R12-8) when succinyl CoA is cleaved to succinate and CoASH (Fig. 12-7). This cycle also produces eight NADH per hexose within the mitochondrial matrix; by oxidative phosphorylation each of these NADH yields three ATP, or 24 per hexose. Each $FADH_2$ from the Krebs cycle yields two ATP by oxidative phosphorylation, or four per hexose (two pyruvates, R12-8). The total contribution of the Krebs cycle is then 30. Adding these 30 to the 6 from glycolysis leads to a total of 36 per hexose completely respired by these processes.

We can also estimate the efficiency of respiration in terms of how much energy in glucose can be trapped in the terminal phosphate bond of ATP. The standard Gibbs free energy change at pH 7 ($\Delta G'_0$) for complete oxidation of one mole of glucose or fructose is -2870 kJ (-686 kcal), so we shall use this as the energy in the reactants of respiration. Among the products, only the energy in the terminal phosphate of ATP is additional useful energy. The $\Delta G'_0$ for hydrolysis of the terminal phosphate in each mole of ATP is about -31.8 kJ (-7.6 kcal), or -1140 kJ in 36 moles of ATP. *Thus the efficiency is about $-1140/-2870$ or 40 percent.*[*] The remaining 60 percent is lost as heat.

12.9 Cyanide-Resistant Respiration

Aerobic respiration of most organisms, including some plants, is strongly inhibited by certain negative ions that combine with the iron in cytochrome oxidase. Two such ions, *cyanide* (CN^-) and *azide* (N_3^-) are particularly effective. *Carbon monoxide (CO)* also forms a strong complex with this iron, preventing electron transport and poisoning respiration. In many plant tissues, however, poisoning cytochrome oxidase by such inhibitors has only a minor effect on respiration. The respiration that continues in this situation is said to be **cyanide resistant respiration** (Solomos, 1977; Day et al., 1980; Laties, 1982; Siedow, 1982). Several fungi and algae and a few bacteria and animals are also resistant to cyanide, azide, and CO, but most animals are not (Henry and Nyns, 1975).

The reason respiration can continue when cytochrome oxidase is blocked is that such mitochondria have an alternative, short branch in the electron-transport pathway at the first step involving ubiquinone (Fig. 12-8). This branch or *alternative route* also allows transport of electrons to oxygen, probably from ubiquinone to a flavoprotein to the oxidase. The terminal oxidase and most other components of this route have not yet been identified, but it is known that the oxidase has a much lower affinity for O_2 than does cytochrome oxidase and that little oxidative phosphorylation is coupled to the pathway; that is, it leads mainly to production of heat, not ATP. This heat production is thought to be beneficial to certain plants,

---

[*]This efficiency of 40 percent based on $\Delta G'_0$ values is probably somewhat unrealistic for conditions occurring naturally in cells where $NAD^+/NADH$ ratios might be as high as 30 in the mitochondrial matrix, and cytosol ATP/ADP ratios might usually be between four and ten. A theoretical review by Ericinska and Wilson (1982) indicates that true efficiencies for electron transport and oxidative phosphorylation are as high as 75 percent in liver and heart mitochondria. Use of $\Delta G'_0$ and $\Delta E'_0$ values to calculate the efficiency of those processes gives only 43 percent. Thus true efficiencies for the entire process of respiration might be considerably higher than 40 percent.

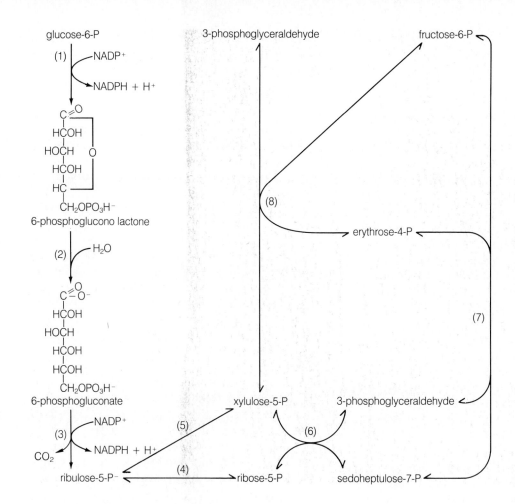

Figure 12-10 Reactions of the pentose phosphate respiratory pathway.

as in the pollination ecology of arum lilies such as *Sauromatum guttatum* and *Symplocarpus foetidus* (skunk cabbage). (See Fig. 12-9 and review by Meeuse, 1975.) On the other hand, the role of the cyanide-resistant pathway in most plants is unclear, because it doesn't usually operate unless cytochrome oxidase of the normal pathway is poisoned by cyanide, azide, or CO (Day et al., 1980; Laties, 1982). An exception is that it does operate when glycolysis and the Krebs cycle occur unusually rapidly, because then the normal electron transport pathway cannot handle all the electrons provided to it. This probably explains its importance in the arum lilies, where rates of glycolysis and the Krebs cycle are somehow decontrolled and heat production occurs partly because normal respiration is so fast. Although there is much more to learn about the cyanide-resistant pathway, the conclusion of Laties (1982) that it often "appears to be lurking in reserve" seems appropriate.

12.10 The Pentose Phosphate Pathway

After 1950, plant physiologists gradually became aware that glycolysis and the Krebs cycle were not the only reactions by which plants obtain energy from the oxidation of sugars into carbon dioxide and water. Much of the research indicating that a different pathway also occurs in plants was performed in the 1950s by Martin Gibbs at Cornell University and by Bernard Axelrod and Harry Beevers at Purdue University. Because five-carbon sugar phosphates are intermediates, this series of reactions is usually called the **pentose phosphate pathway (PPP).** It has also been called the hexose monophosphate shunt and the phosphogluconate pathway.

Several compounds of the PPP are also members of the Calvin cycle, in which sugar phosphates are synthesized in chloroplasts. (Coincidentally, most reactions of the PPP and the Calvin cycle were discovered during the same time period in the early 1950s.) The major difference between the Calvin cycle and the PPP is that, in the latter, sugar phosphates are degraded rather than synthesized. In this respect, the reactions of the PPP are similar to those of glycolysis. In addition, glycolysis and the PPP have certain reactants in common, and both occur mainly in the cytosol; so the two pathways are greatly interwoven. One important difference is that in the PPP, $NADP^+$ is always the electron acceptor, whereas in glycolysis NAD^+ is usually the acceptor.

Reactions of the PPP are outlined in Fig. 12-10.

The first reaction involves glucose-6-phosphate, which can arise from starch breakdown by starch phosphorylase followed by phosphoglucomutase action in glycolysis; from the addition of the terminal phosphate of ATP to glucose; or directly from photosynthetic reactions. It is immediately oxidized (dehydrogenated) irreversibly by *glucose-6-phosphate dehydrogenase* to *6-phosphoglucono-lactone*. This lactone is rapidly hydrolyzed by a *lactonase* into *6-phosphogluconate* (reaction 2), then the latter is irreversibly and oxidatively decarboxylated to ribulose-5-P by *6-phosphogluconate dehydrogenase* (reaction 3). Note that reactions (1) and (3) are catalyzed by dehydrogenases highly specific for $NADP^+$ (not NAD^+). Furthermore, glucose-6-P dehydrogenase is strongly inhibited noncompetitively (allosterically, Section 8.6) by NADPH. In chloroplasts, where an isozyme of this enzyme exists and where the PPP also operates during darkness, light inactivates the enzyme, thereby preventing degradation of glucose-6-P and allowing the Calvin cycle to operate faster. One mechanism of deactivation by light is formation of inhibitory NADPH from $NADP^+$ by the thylakoid electron-transport system (Lendzian, 1980); another is the ferredoxin-thioredoxin system described in Section 10.5 (Buchanan, 1984), while still another is the LEM system described in Section 10.5.

The next reactions of the PPP lead to pentose phosphates and are catalyzed by an *isomerase* (reaction 4) and an *epimerase* (reaction 5), which is a type of isomerase. These and subsequent reactions are similar or identical to some in the Calvin cycle (Fig. 10-3). Important enzymes are *transketolase* (reactions 6 and 8) and *transaldolase* (reaction 7). Note that these last three reactions lead to 3-phosphoglyceraldehyde and fructose-6-P, which are intermediates of glycolysis. As a result, the PPP can be considered an alternate route to compounds subsequently degraded by glycolysis (ap Rees, 1980; Turner and Turner, 1980). Nevertheless, three other functions of the PPP are important. *First is the production of NADPH*, because this nucleotide can be oxidized by plant mitochondria to form ATP. Furthermore, NADPH is used specifically in numerous biosynthetic reactions requiring an electron donor. For these reactions (e.g., formation of fatty acids and of several isoprenoids to be described in Chapter 14), NADH is nonfunctional, although for certain other reduction processes it works well. *Second, erythrose-4-P is produced* in reactions (7) or (8), and this four-carbon compound is an essential starting reactant for production of numerous phenolic compounds such as anthocyanins and lignin (Section 14.4). *Third, ribose-5-P is produced*; this is a required precursor of the ribose and deoxyribose units in nucleotides, including those in RNA and DNA. Clearly, the PPP is just as essential to plants as glycolysis and the Krebs cycle.

12.11 Respiratory Production of Molecules Used for Synthetic Processes

Near the beginning of this chapter, we stated that respiration is important to cells because many compounds are formed that can be diverted into other substances needed for growth. Many of these are large molecules, including lipids, proteins, chlorophyll, and nucleic acids. ATP is needed to form them, and the electrons present in NADH or NADPH are frequently also required. Another process requiring significant quantities of NADH is the reduction of nitrate to nitrite (Section 13.3). We emphasized in the preceding section the importance of the PPP to produce NADPH, ribose-5-P, and erythrose-4-P for anabolic reactions. The role of glycolysis and the Krebs cycle in producing carbon skeletons for synthesis of larger molecules is summarized in Fig. 12-11. You should remember when studying this figure that if carbon skeletons are diverted from the respiratory pathway as shown, not all carbons of the original respiratory substrate (e.g., starch) will be released as CO_2, and not all electrons normally transferred by NADH or NADPH will combine with O_2 to form H_2O. Yet it is essential that some of the substrate molecules be totally oxidized, because use of diverted carbon skeletons to form larger molecules is effective only when oxidative phosphorylation is producing an adequate supply of ATP.

Another important point is that when organic acids of the Krebs cycle are removed by conversion into aspartic acid, glutamic acid, chlorophyll, and cytochromes, for example, regeneration of oxaloacetic acid will be prevented. Thus diversion of organic acids from the cycle would soon cause the cycle to stop if it were not for another mechanism to generate oxaloacetate. In all plants, day and night, there is some fixation of CO_2 into oxaloacetate by the reaction catalyzed by PEP carboxylase (see Chapter 10, R10-3, and Fig. 12-11, left). This reaction is essential for growth processes, because it replenishes organic acids converted into larger molecules and allows the Krebs cycle to continue.

12.12 Biochemical Control of Respiration

To understand environmental effects on respiration of various plants and plant parts described in the next section, it is helpful to learn some of the main biochemical control points and how these controls take place. As a help in understanding control, consider that a photosynthesizing plant must regulate how much carbohydrate is stored in sucrose and starch, for example, compared to how much is respired

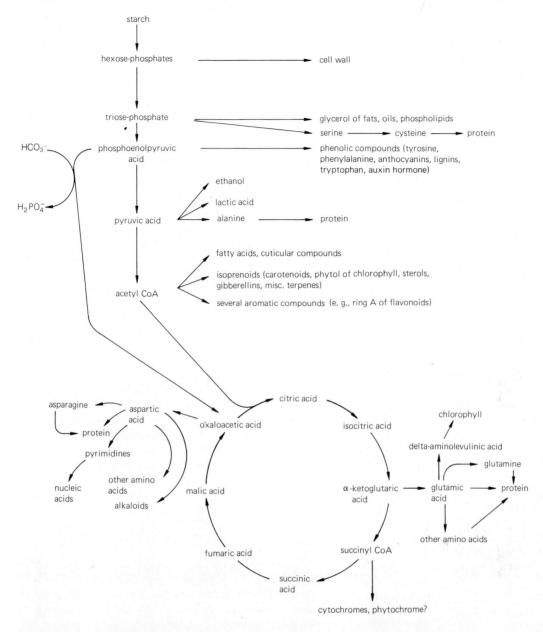

Figure 12-11 Glycolysis and Krebs cycle simplified to show their roles in formation of some other essential compounds. There are many unsolved problems of transport from one organelle to another not shown in this diagram, such as the transport of glutamate from mitochondria to chloroplasts, where it is used in chlorophyll synthesis.

totally and how much is used in growth processes that require formation of membranes, cell walls, and so on. One logical control point should be near the start of glycolysis, because the hexose phosphates used in glycolysis can alternatively be used to form sucrose, starch, and fructans. Indeed, an important control point exists there. Another important control of metabolism might depend on the ATP, ADP, and Pi concentrations; because ATP is the only important product of complete respiration, ADP and Pi levels should help control how fast ATP is formed. ATP formation is surprisingly fast, even though cell ATP

concentrations are only in the millimolar range, or less. Pradet and Raymond (1983) calculated that 1 g of actively metabolizing maize root tips would convert about 5 g of ADP to ATP per day! Such high production rates mean that utilization rates are correspondingly high; otherwise, cells would soon be filled with ATP. Utilization of ATP occurs in many ways, but formation of starch, sucrose, fructans, cell-wall polymers, nucleic acids, and membrane proteins and lipids are all important in considering metabolic control. How important are ATP, ADP, and Pi as control agents?

The View from the Top of Respiratory Peaks

George G. Laties

George Laities (as his family then spelled the name) was born in a military hospital in Sevastopol, Russia, just after the Revolution. Before his first birthday, and under great duress, his family escaped via Constantinople to New York City, where he grew up, deciding to be a county agent—as he tells us in this essay. He is now with the Department of Biology at the University of California at Los Angeles.

When I slipped from an aspiring Agricultural County Agent to a Botanist manqué at Cornell University many years ago, I hadn't the foggiest notion of the richness and satisfaction of plant physiology as a life's preoccupation. I think it was my interest in agriculture that finally led me to the University of California, Berkeley, to become a student of Dennis R. Hoagland in plant nutriton. For whatever reason—perhaps the burgeoning emphasis on mechanism in conjunction with observation in biology—I became involved in studies of barley root respiration, respiration being the undisputed driver of salt uptake. It is hard to imagine the relatively rudimentary understanding of plant respiration in the early 1940s. The Krebs cycle had barely been discovered in the mammals, and plant physiology was not free of the historical dictum that animal respiration was one thing and plant respiration quite another. Accordingly, my self-imposed initiation rite entailed forcing an open door to establish that the tricarboxylic acid cycle indeed ruled supreme in barley roots.

Upon my arrival at Caltech in 1947 to begin what would be an unconscionably prolonged on-and-off postdoctoral stay in the happy ambience of James Bonner, Frits W. Went and Arthur W. Galston, it was just being comprehended in the biochemical world that the gelatinous mass prepared by differential centrifugation of liver homogenates that was capable of carrying out the coordinated tricarboxylic acid cycle in fact comprised mitochondria, organelles until then mainly of interest to cytologists, and only beginning to receive biochemical attention, particularly by Hogeboom and his colleagues at the Rockefel-

ler Institute and Albert Lehninger at the Johns Hopkins University.

While plant mitochondria thereupon occupied my attention, it should come as no surprise that for a time the most trenchant observations to be made were that plant mitochondria behaved much as their mammalian counterparts. This revelation was reenforced during the course of a year that profoundly affected my life, in which I spent time in the laboratories of Hans A. Krebs—then at Sheffield University in the English midlands—and C. S. Hanes, at Cambridge University. I cannot overemphasize the joys and benefits of an extended stay in a cordial foreign land early in the game: expanded horizons, quickened curiosity, and a modicum of tolerance for other ways. (With some luck, one might even learn the language.) It was at Cambridge that C. S. Hanes mentioned in passing an early study carried out with John Barker that showed potato tuber respiration to be *stimulated* by cyanide. This fact disappeared into my memory bank to be recalled many years later.

The Cambridge Botany School was suffused with the aura of F. F. Blackman, whose classic studies in the respiratory physiology of fruit ripening continue to point our direction. Blackman was mainly concerned with the nature and cause of the climacteric—the burst of respiratory activity and associated metabolic changes that attend ripening in many fruits. He also called attention, however, to the remarkable fact that at low external oxygen concentrations (e.g., 3 percent, the concentration often used in modified atmospheres for fruit storage) fruit respiration drops sharply without concomitant glycolysis, that is, without alcohol or lactate production. Blackman recognized this phenomenon for what it was—a subtle feedback system designed to suppress respiration and thus preclude damaging anaerobiosis. Of late we have come back to this problem in the hope that it may yield in the face of our enlarged understanding of regulatory mechanisms. It seems appropriate in retrospect that I should have wound up at the University of California at Los Angeles as a colleague of Jacob Biale, whose pioneering studies of ethylene action in fruit ripening have done so much to extend Blackman's seminal observations.

In the interim, however, my interests encompassed respiratory control in general and, in particular, the relationship between salt uptake and respiratory activity in

Energy Charge In 1968, David E. Atkinson published an article in which he explained many reasons why ATP, ADP, and AMP should be master controllers (also see his 1977 book). He realized that ATP has two "high-energy" phosphate bonds, ADP one, and AMP none, and that a highly active mitochondrial and chloroplast enzyme called *adenylate kinase* catalyzes a freely reversible reaction (ATP + AMP ⇌ 2 ADP) that keeps these nucleotides in equilibrium. He hypothesized that a value he defined as the **energy charge (EC)** that depends on nucleotide concentrations should be important in metabolic control:

$$EC = \frac{(ATP) + 1/2(ADP)}{(ATP) + (ADP) + (AMP)}$$

For nearly all active cells investigated, EC values are between 0.8 and 0.95, but substantially lower values are found in anaerobic or poisoned cells. Atkinson argued theoretically and from some data from studies slices of bulky storage organs. Storage organ slices were long a favored object for the study of salt uptake, largely because of the ready availability of material and its relative uniformity and sturdiness. It was early recognized, however, that whereas so-called aged slices (kept moist in air for several hours) absorbed salt handily, uptake by fresh slices was puny at best. I was convinced that the three- to fourfold increase in the respiration rate on slice ageing could not account for the almost qualitative difference in salt uptake capacity between fresh and aged slices, and accordingly I became engrossed in the respiratory behavior of ageing storage organ slices. To shorten a long story, we grew to understand that the fresh slice—particularly the potato slice—is a metabolic cripple; its respiratory apparatus is in shambles because of the dramatic degradation of membrane phospholipids that results from slicing. When Bruce Jacobson and I measured the carbon mass-isotope ratio of respiratory CO_2 of intact potato tubers and compared it with the $^{13}C/^{12}C$ ratio of the carbon in potato carbohydrate, lipid and protein, we discovered that the respiratory CO_2 of intact potato tubers and aged slices is of carbohydrate origin, whereas the CO_2 from fresh slices is derived from lipid! (We were able to do this because certain steps in biosynthesis lead to a difference in the mass-isotope composition of different products.) When in the course of several hours the lipid debris—primarily free fatty acids—is cleaned up by oxidation, the restored and enhanced conventional carbohydrate-dependent respiration serves to sustain vigorous cell activity, including salt uptake. Thus the respiration of fresh slices largely reflects a scavenger operation and differs in kind from the normal respiration of aged slices. The general metabolic mayhem that attends slicing was found to have yet another result—the impairment or disengagement of the cyanide-resistant electron path. Because with ageing, respiration was also cyanide-resistant, we were led to the view that cyanide resistance is not induced with ageing but rather is restored. In a way it is heartening that none of what we discovered was anticipated; so curiosity is its own reward, and limitless exercise can be derived from jumping at conclusions.

Of late, my interests have veered once again to questions dealing with the nature of the ethylene-induced climacteric in fruits and storage organs. This interest has stemmed in part at least from the seductive, albeit possibly specious analogy between the wound-induced respiration in ageing slices and the ethylene-induced respiration in intact organs. Not least among the points of commonality is the well-developed cyanide-resistant respiration in both objects. With Theo Solomos we noted the correlation between ethylene responsiveness and cyanide resistance in a great number of fruits; but try as we might we have been unable to establish that the cyanide-resistant path normally contributes to the respiration in the absence of cyanide. In stressing the correlation, however, we led others, and perhaps ourselves, down the garden path, and we must now do penance by reminding whomever will listen that the presence of the alternative path does not bespeak its use. While we were off course, the ultimate truth continues to elude us. Knowing that you are drunk is not the same as being sober.

The rush of molecular biology onto the scene—breathtaking, if not devastating in both its conceptual and technical aspects—offers us new ways to pursue old problems, particularly the basis of respiratory release in ageing slices and ethylene-treated organs. We may come to grips with the crusty old question of whether the induced respiration is an on-call response, wherein the respiratory apparatus rises to the occasion to supply energy to an array of biosynthetic events, or whether the eliciting signals cause the synthesis of one or more specific enzymes that, for whatever reason, lead to a quickening of respiration. To this end I have made a reluctant compact with the march of time and, together with a splendid group of graduate students and postdoctoral fellows, have begun an exploration of the influence of slicing and ethylene on gene expression. The exercise has been salubrious, if not comfortable, inasmuch as evidence has been quick in coming that both slicing and ethylene treatment affect gene expression and, in particular, transcription. The extent to which the effects of both are analogous remains to be established, as does the precise nature of the genes affected. We have come to understand with increasing conviction what should have been clear from the outset: The respiratory thermometer is rude and diagnostic, reflecting the behavior of an inordinately complicated machine. We are in no danger of running out of questions regarding how the machine operates.

with animal cells that important enzymes of metabolic pathways that use ATP (e.g., polysaccharide synthesis) should be activated allosterically by high EC values. Important enzymes in pathways that regenerate ATP should be inhibited allosterically by high EC values. He assumed that such enzymes bind two or more such nucleotides on allosteric sites with high affinity, so their response depends on concentration *ratios* of nucleotides rather than on the absolute concentration of only one nucleotide.

Numerous results with enzymes from animal and some microbial cells are consistent with the importance of measured EC values, but some results are inconsistent. Many investigations concerning the importance of EC in plants have now been made (Pradet and Raymond, 1983), but only a few plant enzymes respond to EC as expected. Furthermore, EC values in leaves remain constant when plants are switched from light to darkness or vice versa, yet we know that leaf biosynthesis of starch (ATP utilization) occurs only in light and that respiration (ATP formation) is relatively more important than biosynthesis in darkness. We also know that light activates several photosynthetic enzymes rather quickly, independent of EC (Section 10.5). Light also inactivates chloroplast glucose-6-P dehydrogenase, the rate-limiting enzyme of the PPP, as mentioned above. Our conclusion is that plants seem to have important control mechanisms different from energy charge, but that EC values are likely important in some cases. We now mention a few additional control mechanisms.

Regulation of Glycolysis For glycolysis, **phosphofructokinase** has long appeared to be the enzyme most susceptible to important control. This enzyme catalyzes formation of fructose-1,6-bisP and ADP from ATP and fructose-6-P (Fig. 12-4). This is the first glycolytic reaction that involves a hexose phosphate that cannot also be used to form sucrose or starch, so it represents a control over the entire glycolytic pathway. Activity of phosphofructokinase is inhibited by ATP, PEP, and citric acid, but it is enhanced by Pi (Turner and Turner, 1980). ADP is usually slightly inhibitory or without effect. Inhibition by ATP, PEP, and citric acid, which are formed from or during glycolysis, seems to be a reasonable way to prevent the overproduction of these compounds. Activation by Pi might also be expected, because Pi is used in glycolysis along with fructose-1,6-bisP, but the small inhibition by ADP is unexpected and difficult to explain when it does occur.

Another important regulator of glycolysis is the NAD$^+$/NADH ratio, because NAD$^+$ is essential for glycolysis while NADH is a product. These facts are also true for the Krebs cycle and the electron transport system. Because O$_2$ is so important to oxidize NADH and regenerate NAD$^+$, good aeration favors glycolysis, the Krebs cycle, and electron transport. Little else is known for sure about how the last two processes are regulated, but we shall discuss aeration effects in Section 12.13. The discovery of an enzyme other than phosphofructokinase that synthesizes fructose-1,6-bisP and the discovery of still another enzyme that synthesizes **fructose-2,6-bisP** from ATP and fructose-6-P raise still more questions about control of glycolysis versus formation of sucrose and starch from 3-carbon and 6-carbon sugar phosphates. The review of Preiss (1984) describes some of these controls; they are complex, and we are unsure that they are generally important, so we shall not describe them.

Control of the Pentose Phosphate Pathway We mentioned above that the rate-limiting enzyme of the PPP is the first, glucose-6-P dehydrogenase, and that in chloroplasts the isozyme present is inhibited by NADPH formed in light and is otherwise inactivated in light. Presumably, the isozyme present in the cytosol is also inhibited by NADPH; but regardless, it requires NADP$^+$ as a substrate. Any process that favors conversion of NADPH to NADP$^+$ should therefore speed the PPP. Two such processes are oxidation of NADPH by the electron transport system and oxidation during biosynthesis of fatty acids and isoprenoid compounds such as carotenoids and sterols.

12.13 Factors Affecting Respiration

Many environmental factors influence respiration. The previous description of the individual reactions involved should allow you to understand how such factors affect the overall rate of respiration and its importance to plant maintenance and growth.

Substrate Availability Respiration depends on the presence of an available substrate, and starved plants with low starch, fructan, or sugar reserves respire at low rates. Plants deficient in sugars often respire noticeably faster when sugars are provided. In fact, respiration rates of leaves are often much faster just after sundown when sugar levels are high than just before sunup when levels are lower. We mentioned in the previous chapter that the lower, shaded leaves usually respire slower than do the upper leaves exposed to higher light levels. If this were not true, the lower leaves would probably die sooner than they do. The difference in starch and sugar contents resulting from unequal photosynthetic rates is probably responsible for the lower respiratory rates of shaded than illuminated leaves.

If starvation becomes extensive, even proteins can be oxidized (ap Rees, 1980). These proteins are hydrolyzed into their amino acid subunits, which are then catabolized (degraded) by glycolytic and Krebs cycle reactions. In the case of glutamic and aspartic acids the relation to the Krebs cycle is especially clear, because these amino acids are converted to α-ketoglutaric and oxaloacetic acids, respectively (Fig. 12-11). Similarly, the amino acid alanine is oxidized via pyruvic acid. As leaves become senescent and yellow, most of the protein and other nitrogenous compounds in the chloroplasts are broken down. Ammonium ions released from various amino acids are combined in glutamine and asparagine (as amide groups) during this process, and this prevents ammonium toxicity. These processes are discussed in the next chapter.

Available Oxygen The O$_2$ supply also influences respiration, but the magnitude of its influence differs greatly with plant species and even with different organs of the same plant. Normal variations in O$_2$

content of the air are much too small to influence respiration of most leaves and stems. Furthermore, the rate of O_2 penetration into leaves, stems, and roots is usually sufficient to maintain normal O_2 uptake levels by the mitochondria, mainly because cytochrome oxidase has such a high affinity for oxygen that it can function even when the O_2 concentration around it is only 0.05 percent of that in air (Drew, 1979).

In more bulky tissues with lower surface/volume ratios, diffusion of O_2 from air to cytochrome oxidase in cells near the interior is probably retarded enough to slow respiration rates. One might suspect that in carrot root, potato tubers, and other storage organs, the rate of O_2 penetration would be so low as to cause the respiration inside to be primarily anaerobic. Quantitative data on gas penetration into such organs are meager, but the measurements at hand show that rate of O_2 movement through them is certainly much less than in air. In pure water, O_2 diffusion is about three million times slower than in air (Drew, 1979). However, the French physiologist H. Devaux showed, in 1890, that central regions of bulky plant tissues do respire aerobically, although slowly (for other reasons). He demonstrated the importance of intercellular spaces for gaseous diffusion. We now know that these spaces represent significant amounts of the total tissue volume. For example, in potato tubers approximately 1 percent of the volume is occupied by air spaces, and values for roots from 2 to 45 percent have been observed in various species, the higher values more common among wetland plants (Crawford, 1982). Such intercellular air spaces extend from the stomata of leaves to most cells in the plant, aiding their aerobic respiration. Only tightly-packed xylem parenchyma cells and cells in meristematic regions appear to have no access to such air spaces.

We mentioned in Chapter 6 that diffusion of O_2 through the intercellular space system from the leaves to the roots was probably important in providing O_2 needed for root respiration in waterlogged soils. This emphasizes that O_2 and other gases can move through plant tissues more rapidly than might have been expected for an organism with no lungs or hemoglobin in blood to help transport the gas. In general, the intercellular air system from leaves to roots is more important for grasses and sedges with hollow stems, and indeed those species are generally more tolerant to flooding than most others (Crawford, 1982). (John Dacey's personal essay for Chapter 3 explains how bulk-flow forces air through petioles of water lilies, and this might also be important for waterlogged roots.) Nevertheless, flooding for long periods is toxic to nearly all plants, especially when no detectable O_2 is present around the roots (**anoxic** or entirely anaerobic conditions). Among crop plants, only rice is known to tolerate anoxia for long, although *Echinochloa crus-galli* (barnyardgrass),

a common weed of rice fields, is also unusually tolerant, and so, apparently, are five other species native to wetlands (Barclay and Crawford, 1982). In native plants, tolerance is generally greatest when temperatures are low and respiration is slow (as in winter) and when adequate carbohydrates are stored in fleshy rhizomes or in roots. Substantial differences in tolerance also exist in trees (Gill, 1970; Joly and Crawford, 1982). For certain tropical mangroves (*Rhizophora mangle* and *Avicennia nitida*), roots that grow upward out of the water (*pneumatophores*) transfer oxygen to flooded roots. Thus the flooded roots are really not anoxic, although no doubt **hypoxic** (under reduced oxygen levels). Among conifers, lodgepole pine (*Pinus contorta*) is more tolerant to hypoxia under flooding than is Sitka spruce (*Picea sitchensis*), and part of the difference lies in the greater ability of the pine to transport oxygen to the roots (Philipson and Coutts, 1980). Some species form extensive adventitious root systems when their stems are flooded, and these roots aid absorption of mineral salts and water. Still other species form new roots on the original root system.

Physiologists are interested not only in how some species tolerate hypoxia better than others (in addition to morphological adaptations) but how hypoxia injures plants. Injurious effects are caused by several metabolic imbalances ultimately resulting from insufficient oxygen (Kozlowski, 1984). One effect is retarded transport of cytokinin hormones from young roots to shoots, and as explained in Chapter 17, one important source of these hormones for leaves and stems is probably root tips. Other imbalances include insufficient absorption of mineral salts, especially nitrogen; leaf wilting, accompanied by slower photosynthesis and carbohydrate translocation, because insufficient oxygen reduces the permeability of roots to water; and accumulation of toxins caused by microbes around the roots (Drew, 1979; Bradford and Yang, 1981). As you might predict from the preceding biochemical sections, the supply of ATP is limited because the electron-transport system and the Krebs cycle cannot function without oxygen. Furthermore, products of fermentation accumulate to some extent, especially ethanol in most species, but also lactic acid, malic acid, and rarely glycerol. Both ethanol and lactic acid can become toxic, although the review of Kozlowski (1984) suggests that fermentation products usually do not become toxic because they leak out and move away from roots, especially in turbulent waters.

Some results with rice are interesting. Rice seeds will germinate under hypoxia or anoxia, but they do so by upward protrusion of the coleoptile rather than by protrusion of the radicle. (Radicle protrusion is the common mode of germination in rice and nearly all other species exposed to air.) Roots hardly develop, but the coleoptile continues to grow under hypoxia,

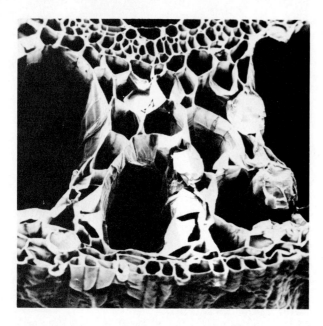

Figure 12-12 Scanning electron micrograph showing aerenchyma formation in a maize root subjected to hypoxia. (From R. Campbell and M. C. Drew, 1983. Used by permission.)

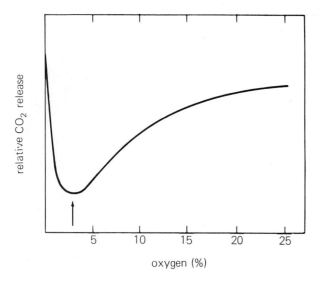

Figure 12-13 The influence of atmospheric oxygen concentration on CO_2 production in apple fruits. On the right side of the arrow increasing O_2 supply increases respiration because of stimulated Krebs cycle activity, yet anaerobic CO_2 release from pyruvate becomes minimal in this region of the curve due to indirect inhibition effects of O_2 on glycolysis. At the left of the arrow, the O_2 concentration is low enough to allow a very rapid breakdown of sugars to ethanol and CO_2. (Redrawn from James, 1963.)

faster than under anoxia and also faster than in air (Atwell et al., 1982; Alpi and Beevers, 1983). Enhanced growth of coleoptiles under hypoxia almost surely results from accumulation of ethylene gas by the young plants, and this plant hormone promotes coleoptile elongation at the expense of root growth (see Section 17.2). Nevertheless, this does not explain why rice plants can tolerate hypoxic conditions unusually well; in spite of much biochemical work, this enigma remains (Alpi and Beevers, 1983).

In both dicots and monocots accumulation of ethylene causes collapse and lysis of mature cortical cells in the root, leading to a tissue with large air spaces called **aerenchyma** (Fig. 12-12). Aerenchyma increases diffusion of oxygen toward the remaining cells when this gas is present in low concentrations (Kawase, 1981).

In addition to roots of flooded trees and plants growing in wetlands, marshes, and estuaries, many seeds, especially large ones, exhibit **fermentation** during imbition of water that leads to germination. Peas, beans, and maize have been studied most, and either ethanol or lactate (or both) is produced. For peas, internal concentrations of ethanol as low as 0.1 percent by volume are toxic, so it is important that most of this ethanol leaches out during imbition. No aerenchyma is formed in seeds.

A surprising result of hypoxic and anoxic conditions was discovered by Louis Pasteur in studies of wine production by yeasts over a century ago. When the O_2 concentration around yeast and most plant cells is decreased gradually below the atmospheric 20.9 percent, CO_2 production from respiration decreases until a minimum is reached, *but lower O_2 concentrations cause a rapid rise in CO_2 production* (Fig. 12-13). What Pasteur found was that yeast cells grew rapidly in air but used little sugar and produced little ethanol and CO_2; under anaerobic conditions they grew slower but used more sugar and produced more CO_2 and ethanol. This phenomenon became known as the **Pasteur effect,** measured by plant physiologists as the oxygen inhibition of carbohydrate breakdown and production of ethanol (or lactate).

The Pasteur effect has been explained biochemically as an allosteric inhibition of phosphofructokinase in the presence of oxygen. In animal cells, energy charge values increase as O_2 is provided in low concentrations, and higher EC values inhibit phosphofructokinase, the activity of which otherwise allows glycolysis and fermentation to occur rapidly. In plants, the role of EC is much less certain, but increased production of ATP and citrate with increased O_2 is known to allosterically inhibit phosphofructokinase and should therefore decrease glycolysis and fermentation (at the left of the arrow in Fig. 12-13).

The Pasteur effect no doubt causes decreased

carbohydrate reserves in flooded soils and probably helps explain why plants with swollen rhizomes or thick roots that store foods can survive anoxia longer, especially at cold temperatures when respiration is slow. The Pasteur effect also has some practical importance in fruit storage, especially with apples. Here the object in storage is to prevent extensive sugar loss and over-ripening. This is done by carefully decreasing O_2 to the concentration at which aerobic respiration is at a minimum but sugar breakdown by anaerobic processes is not stimulated. Additional CO_2 is also added to the air, and the temperature is lowered closer to the freezing point, which further prevents over-ripening. As discussed in Chapter 17, CO_2 inhibits action of a fruit-ripening hormone, ethylene, and this is a probable explanation for its effectiveness in inhibiting over-ripening. Also, low concentrations of O_2 generally slow ethylene production.

Temperature For most plant species and parts, the Q_{10} for respiration between 5 and 25°C is usually near 2.0 to 2.5. (The Q_{10} is the ratio of the rate of a process at one temperature divided by the rate at a temperature 10° lower, see Section 1.3.) With further increases in temperature up to 30 and 35°C, the respiration rate still increases, but less rapidly, so the Q_{10} begins to decrease. A possible explanation for the Q_{10} decrease is that the rate of O_2 penetration into the cells through cuticle or periderm begins to limit respiration at these higher temperatures, at which chemical reactions could otherwise proceed rapidly. Diffusion of O_2 and CO_2 are also sped by increased temperature, but the Q_{10} for these physical processes is only about 1.1; that is, temperature doesn't speed diffusion of solutes through water much.

With even further rises in temperature to 40°C or so, the rate of respiration even decreases, especially if plants are maintained under such conditions for long periods. Apparently, the required enzymes begin to be denatured rapidly at high temperatures, preventing metabolic increases that would otherwise occur. With pea seedlings a temperature increase from 25 to 45°C initially increased respiration greatly, but within about two hours the rate became less than before. A probable explanation is that the two-hour period was long enough to partly denature respiratory enzymes.

Type and Age of Plant Because there are large morphological differences among members of the plant kingdom, it is to be expected that differences in metabolism also exist. In general, bacteria, fungi, and many algae respire considerably more rapidly than do seed-plants. Various organs or tissues of higher plants also exhibit large variations in rates. One reason that bacteria and fungi have so much higher values than plants, based on dry weight, is that they

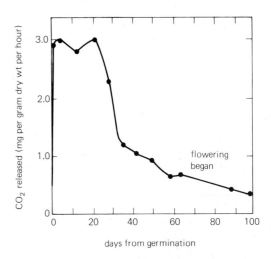

Figure 12-14 Respiration of whole sunflower plants from germination until maturity. The rate gradually declined after the 22nd day, even though the rate for individual parts, such as inflorescences, increased for a time after that. (Drawn from data of F. Kidd et al., 1921, Proceedings of the Royal Society of London, series B 92:368.)

contain only small amounts of stored food reserves and have no nonmetabolic woody cells. Such dead woody cells contribute to the dry weight and strength of vascular plants but not to respiration. Similarly, root tips and other regions containing meristematic cells with large protoplasm contents have high respiratory rates expressed on a dry-weight basis. If comparisons are made on a protein basis, these differences are smaller. In general, there is a fairly good correlation between the rate of growth of a particular cell type and its respiration rate. This results from several factors, such as the use of ATP, NADPH, and NADH for synthesis of proteins, cell-wall materials, membrane components, and nucleic acids; consequently ADP, $NADP^+$, and NAD^+ become available to be used in respiration. Inactive seeds and spores have the lowest (usually undetectable) respiration rates, but here the effect is not entirely due to a lack of growth. Rather, certain changes in the protoplasm, especially desiccation, shut off metabolism. Such seeds and spores generally contain abundant food reserves.

The age of intact plants influences their respiration to a large degree. Figure 12-14 shows how the rate changes in whole sunflowers from germination until after flowering. The rate is expressed as the amount of CO_2 released per amount of preexisting dry weight. The curve is extrapolated to zero time to show the common initial large burst in respiratory activity as the dry seeds absorb water and germinate. Respiration remained high during the period of most rapid vegetative growth but then fell prior to flowering. In this and other examples, much of the respiration in the mature plants is carried out by the young leaves and roots and by the growing flowers.

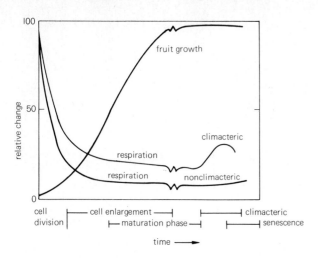

Figure 12-15 Stages in development and maturation of fruits that undergo the climacteric respiration increase and of those that do not. Discontinuities in the lines indicate that the time scale was changed to show differences in development rates of different fruits. The growth pattern may be single or double sigmoid (see Chapter 15). (From J. Biale, 1964, Science 146:880. Copyright 1964 by the American Association for the Advancement of Science.)

Changes in respiration also occur during the development of ripening fruits. In all fruits, the respiration rate is high when they are young, while the cells are still rapidly dividing and growing (Fig. 12-15). The rate then gradually declines, even if the fruits are picked. However, in many species, of which the apple is a good example, the gradual decrease in respiration is reversed by a sharp increase, known as the **climacteric.** The climacteric usually coincides with full ripeness and flavor of fruits, and its appearance is hastened by production in the cells of traces of ethylene, known to stimulate fruit ripening (Biale and Young, 1981). Further storage leads to senescence and to decreases in respiration.

Some fruits do not show the climacteric, including the citrus fruits, grapes, pineapple, and strawberry (lowest curve, Fig. 12-15). Grapefruits, oranges, and lemons ripen on the trees; if removed sooner, their respiration simply continues at a gradually decreasing rate. The advantages and disadvantages to a climacteric are unknown. The biochemical basis for the climacteric respiratory rise is also unclear, but it is currently being actively investigated.

13

Assimilation of Nitrogen and Sulfur

The importance of nitrogen to plants is emphasized by the fact that only carbon, oxygen, and hydrogen are more abundant in them. Although nitrogen occurs in a vast number of plant constituents, most of it is in proteins. Sulfur is only about one-fifteenth as abundant in plants as nitrogen, but it occurs in many molecules, especially proteins. Both elements are usually absorbed from the soil in highly oxidized forms and must be reduced by energy-dependent processes before they are incorporated into proteins and other cellular constituents. Human metabolic systems cannot duplicate this reduction, just as we cannot reduce CO_2. To describe the ways in which nitrate and sulfate are reduced and subsequently combined with carbohydrate skeletons to form amino acids is an important task of this chapter. We shall also discuss fixation of atmospheric N_2 and interconversions of nitrogen compounds during various stages of plant development.

13.1 The Nitrogen Cycle

Nitrogen exists in several forms in our environment. The continuous interconversion of these forms by physical and biological processes constitutes the nitrogen cycle, summarized in Fig. 13-1.

Vast amounts of N_2 occur in the atmosphere (78 percent by volume), yet it is energetically difficult for living organisms to obtain the N atoms of N_2 in a useful form. Although N_2 moves into leaf cells along with CO_2 through stomates, enzymes are available to reduce only the CO_2, so N_2 moves out as fast as in. Most of the N_2 in living organisms arrives there only after fixation (reduction) by prokaryotic microorganisms, some of which exist in the roots of certain plants, or by industrial fixation to form fertilizers (see boxed essay on page 253). Small amounts of nitrogen also move from the atmosphere to the soil as NH_4^+ and NO_3^- in rain and are then absorbed by roots.

This NH_4^+ arises from industrial burning, volcanic activity, and forest fires, while NO_3^- arises from oxidation of N_2 by O_2 or ozone in the presence of lightning or ultraviolet radiation. Another source of NO_3^- is the oceans. Wind-whipped white caps produce minute droplets of water called aerosols, from which the water evaporates, leaving ocean salts suspended in the atmosphere near coastlines. These salts can be brought to the land in rainwater. They are called **cyclic salts,** because they eventually cycle through streams back to the oceans.

Absorption of NO_3^- and NH_4^+ by plants allows them to form numerous nitrogenous compounds, mainly proteins. Manure and dead plants, microorganisms, and animals are important sources of nitrogen returned to the soil, but most of this nitrogen is insoluble and not immediately available for plant use. Nearly all soils contain small amounts of various amino acids, produced largely by microbial decay of organic matter but also by excretion from living roots. Such amino acids can be absorbed and metabolized by plants, yet these and other more complex nitrogen compounds contribute little to the plant's nitrogen nutrition in a direct way. They are, however, of great importance as a nitrogen reservoir from which NH_4^+ and NO_3^- arise. In fact, up to 90 percent of the total nitrogen in soils is estimated to be in organic matter, although in some cases significant amounts exist as NH_4^+ bound to clay colloids (Runge, 1983).

Conversion of organic nitrogen to NH_4^+ by soil microbes is called **ammonification.** In warm, moist soils with near neutral pH, NH_4^+ is further oxidized by bacteria to NO_3^- within a few days of its formation or its addition as a fertilizer. This oxidation, called **nitrification,** provides energy for survival and growth of such microbes, just as does oxidation of more complex foods for other organisms. In many acidic soils, however, nitrifying bacteria are less abundant, so NH_4^+ becomes a more important nitrogen source than NO_3^-. Many forest trees absorb most of their nitrogen as NH_4^+, because of the low

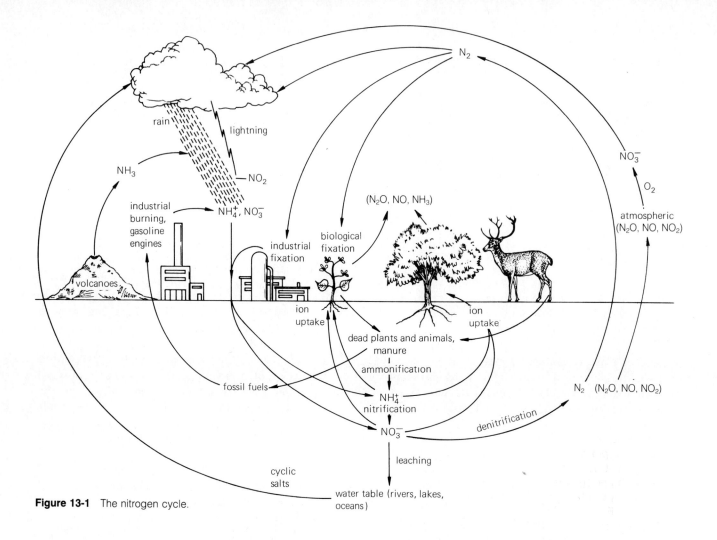

Figure 13-1 The nitrogen cycle.

*p*H common to forest soils and probably because other factors contribute to slow rates of nitrification. Because of its positive charge, NH_4^+ is adsorbed to soil colloids, while NO_3^- is not adsorbed and is much more readily leached. Nitrate is also lost from soils by **denitrification,** the process by which N_2, NO, N_2O, and NO_2 are formed from NO_3^- by anaerobic bacteria. These bacteria use NO_3^- rather than O_2 as an electron acceptor during respiration, thus obtaining energy for survival. Denitrification occurs relatively deep in the soil where O_2 penetration is limited, in waterlogged or compacted soils, and in certain regions near the soil surface where the O_2 concentration is low because of its especially rapid use in oxidation of organic matter. Furthermore, plants lose small amounts of nitrogen to the atmosphere as volatile NH_3, N_2O, NO_2, and NO, especially when well fertilized with nitrogen (Wetselaar and Farquhar, 1980; Duxbury et al., 1982). Oxidized forms of nitrogen in the atmosphere are important ecologically, because when converted to NO_3^- they contribute HNO_3 to acidic rainfall ("acid rain") (Brimblecombe and Stedman, 1982).

13.2 Nitrogen Fixation

The process by which N_2 is reduced to NH_4^+ is called **nitrogen fixation.** It is, so far as we know, only carried out by prokaryotic microorganisms. Principal N_2-fixers include certain free-living soil bacteria, free-living cyanobacteria (blue-green algae) on soil surfaces or in water, cyanobacteria in symbiotic associations with fungi in lichens or with ferns, mosses, and liverworts (Peters, 1978), and by bacteria or other microbes associated symbiotically with roots, especially those of legumes. It is of great importance to the food chain in forests, deserts, fresh-water and marine environments, and even arctic regions.

About 15 percent of the nearly 20,000 species in the Fabaceae (Leguminosae) family have been examined for N_2 fixation, and approximately 90 percent of these have root nodules in which fixation occurs (Allen and Allen, 1981). Important nonlegumes that fix N_2 are primarily trees and shrubs, including members of the genera *Alnus* (alder), *Myrica* (such as *M. gale,* the bog myrtle), *Shepherdia, Coriaria, Hippophae, Ceanothus, Eleagnus, Casaurina,* and seven

Nitrogen Fertilizers and World Energy Problems

For most native species used by humans, economics do not warrant application of nitrogen fertilizers. Growth of such plants depends upon soil nitrogen deposited by rainfall, upon decomposition of previous vegetation, and upon nitrogen fixed by soil bacteria and cyanobacteria, by bacteria or actinomycete-like organisms in root nodules, and by other bacteria in other kinds of associations. On the other hand, most high-yielding crops are fertilized with anhydrous ammonia, liquid ammonia, or $(NH_4)_2SO_4$. Nitrate salts are also used, but they are more expensive. Legume crops are seldom fertilized with nitrogen, because they contain root nodules in which nitrogen fixation occurs, and they usually do not yield more if fertilized.

The major source of nitrogen fertilizers is presently the natural gas and petroleum-dependent industrial formation of ammonia by the Haber-Bosch process. This process involves reaction of N_2 and H_2 at high pressure and temperature. The ammonia can then be converted to $(NH_4)_2SO_4$, NO_3^-, or urea fertilizers. The Haber-Bosch process is expensive, and so are transportation, storage, and application of the fertilizers. Furthermore, crops generally recover only about half of the applied nitrogen, mainly because of losses due to leaching of NO_3^- and to denitrification. Presently, of the total amount of energy required to produce a maize crop in the United States, at least one-third is needed to manufacture, transport and apply the nitrogen fertilizer. Nevertheless, such fertilizers are widely used, and our food supply depends upon them.

R.W.F. Hardy, research director for E. I. duPont de Nemours and Co., compiled data showing a direct relation between worldwide use of nitrogen fertilizers and cereal grain production from 1956 to 1971 (Fig. 13-2). This relation held for both the more-developed and the less-developed countries. Because most of the world's food comes from cereal grains, the continuing importance of such fertilizers is obvious.

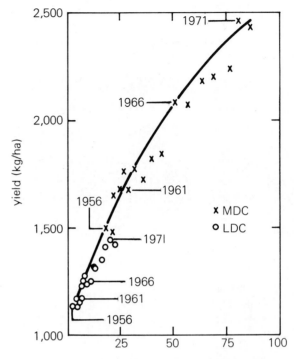

Figure 13-2 The relation between the use of nitrogen fertilizers and the yield of cereal grains from 1956 to 1971 in more developed countries (MDC) and less developed countries (LDC). Although yields leveled off late in this period, they were generally proportional to the amount of fertilizer applied per hectare. (From R.W.F. Hardy, 1975.)

other known genera (Bond, 1976; Torrey, 1978). They are typically pioneer plants on nitrogen-deficient soils; for example, *M. gale* on the bog soils of western Scotland and *Casaurina equisetifolia* on sand dunes of tropical islands. Figure 13-3 demonstrates the important role of root nodules in providing nitrogen to *M. gale*. All plants in this figure were five months old, but only the nodulated group on the left could grow well in the nutrient solution lacking nitrogen salts.

There is considerable interest among foresters to select and breed N-fixing nonlegume trees that can be planted with or before more economically important trees. One goal is to reduce the requirement of nitrogen fertilizers in the lumber industry. Some success

has already been obtained using mixed red alder and Douglas fir populations in the Pacific Northwest of the United States (Gordon and Dawson, 1979). Different alder species have promise for other temperate regions, and species of other genera should be effective in tropical regions (Dawson, 1983).

The microorganisms responsible for N_2 fixation in roots of many species have been identified. In some tropical trees it is cyanobacteria, but in most species actinomycete-like organisms (filamentous bacteria) carry out this process. In the legumes, bacterial species of the genus *Rhizobium* are responsible (Stewart, 1982). A particular *Rhizobium* species is generally effective with only one legume species. The

Figure 13-3 Bog myrtle (*Myrica gale*) plants cultured with (left) and without (right) nodules in a hydroponic solution lacking nitrogen. Plants were grown from seed for 5 months in the solutions. (From Bond, 1963.)

Rhizobia are aerobic bacteria that persist saprophytically in the soil until they infect a root hair or damaged epidermal cell (Fig. 13-4) (Hubbell, 1981; Pierce and Bauer, 1983). Root hairs usually respond to invasion by surrounding the bacterium with a thread-like structure called the infection thread, although such a thread could not be detected in some legumes. The infection thread consists of an infolded and extended plasma membrane of the cell being invaded, along with new cellulose formed on the *inside* of this membrane. The bacteria multiply extensively inside the thread, which extends inwardly and penetrates through and between the cortex cells.

In the inner cortex cells the bacteria are released into the cytoplasm and stimulate some cells (especially tetraploid cells) to divide. These divisions lead to a proliferation of tissues, eventually forming a mature **root nodule** made largely of tetraploid cells containing bacteria and some diploid cells without bacteria (Fig. 13-4). Each enlarged, nonmotile bacterium is referred to as a **bacteroid.** The bacteroids usually occur in the cytoplasm in groups, each group surrounded by a membrane called the **peribacteroid**

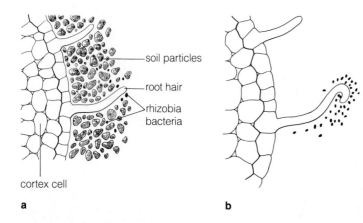

a b c d

Figure 13-4 Development of root nodules in soybean. (**a**) and (**b**) *Rhizobia* bacteria contact a susceptible root hair, divide near it, and upon successful infection of the root hair cause it to curl. (**c**) Infection thread carrying dividing bacteria, now modified and apparent as bacteroids. Bacteroids cause inner cortical and pericycle cells to divide. Division and growth of cortical and pericycle cells lead to (**d**) a mature nodule complete with vascular tissues continuous with those of the root. (Micrograph in (d) from F.J. Bergersen and D.J. Goodchild, 1973.)

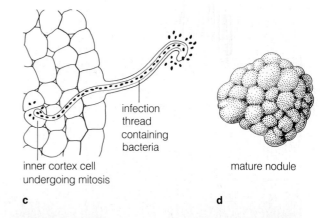

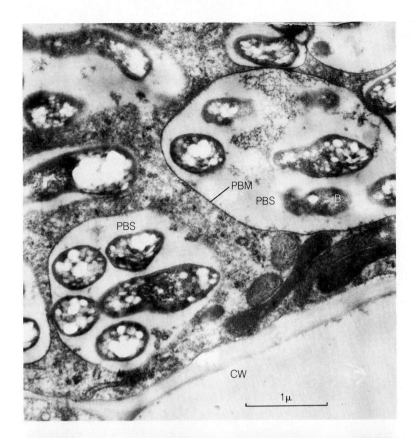

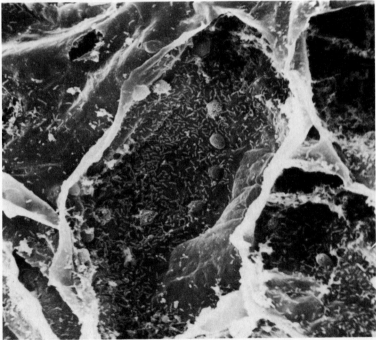

Figure 13-5 (**a**) Transmission electron micrograph of part of a bacteroid-containing cell from a soybean root nodule. Bacteroids (B) are in groups of four to six, each group present in a saclike structure that is surrounded by a peribacteroid membrane (PBM). Cell wall (CW) and a few mitochondria (M) of the nodule cell are visible. Light areas in bacteroids are probably food reserves of poly-β-hydroxybutyric acid. PBS represents peribacteroid space. (From D.J. Goodchild and F.J. Bergersen, 1966.) (**b**) Scanning electron micrograph showing hundreds of bacteroids in a bean (*Phaseolus vulgaris*) root nodule cell. Larger egg-shaped granules probably are starch grains. (Courtesy P. Dayanandan.)

membrane. Between the peribacteroid membrane and the bacteroid group is a region called the **peribacteroid space** (Robertson and Farnden, 1980). Outside the peribacteroid space in the plant cytoplasm is a protein called **leghemoglobin** (Appleby, 1984). This molecule is red because of a heme group attached as a prosthetic group to the globin protein. Leghemoglobin gives legume nodules a pink color, al-though it is much more dilute in nodules of non-legumes. Leghemoglobin is thought to transport O_2 into the bacteroids at carefully controlled rates. Too much O_2 inactivates the enzyme that catalyzes nitrogen fixation, yet some O_2 is essential for bacteroid respiration. Figure 13-5 is an electron micrograph of groups of bacteroids in soybean root nodules.

Nitrogen fixation in root nodules occurs directly

within the bacteroids. The host plant provides bacteroids with carbohydrates, which they oxidize and obtain energy therefrom. These carbohydrates are first formed in leaves during photosynthesis, and then are translocated through phloem to the root nodules. Sucrose is the most common and abundant carbohydrate translocated, at least in legumes. Some of the electrons and ATP obtained during oxidation in the bacteroids are used to reduce N_2 to NH_4^+.

The Biochemistry and Physiology of Nitrogen Fixation The overall chemical reaction for nitrogen fixation (reduction) is summarized in R13-1.

$$N_2 + 6 \text{ electrons} + 8 \text{ H}^+ + \sim 15 \text{ ATP} \rightarrow$$
$$2 \text{ NH}_4^+ + \sim 15 \text{ ADP} + \sim 15 \text{ P}i \qquad \text{(R13-1)}$$

As noted, the process requires a source of electrons and protons, and numerous ATP molecules. Also required is an enzyme complex called **nitrogenase.** The number of ATP molecules needed to fix each N_2 is not yet known; a minimum of two for each of the six electrons involved seems essential, and under some environmental conditions perhaps twice that many are needed. The original source of electrons and protons is carbohydrate translocated from the leaves (and then respired by the bacteria). Respiration of carbohydrate in bacteroids leads to reduction of NAD^+ to $NADH + H^+$ (or $NADP^+$ to $NADPH + H^+$), then NADH or NADPH reduces ferredoxin or similar proteins that are highly effective in reducing N_2 to NH_4^+.

Nitrogenase accepts electrons from reduced ferredoxin or other effective reducing agents as it catalyzes N_2 fixation. Nitrogenase consists of two distinct proteins, often called components I and II. Component I is an Fe–Mo protein, apparently with two molybdenum and 28 iron atoms; component II is an Fe-protein containing four atoms of iron (Yates, 1980). Both molybdenum and iron become reduced and then oxidized as nitrogenase accepts electrons from ferredoxin and transfers them to N_2 to form NH_4^+. ATP is essential to fixation because it binds to component II and causes that protein to act as a stronger reducing agent. Component II transfers electrons to component I, accompanied by hydrolysis of ATP to ADP. Component I then completes the transfer of electrons to N_2. When six such electrons (and 8 H$^+$) are accepted, the two NH_4^+ products are released from the enzyme.

The NH_4^+ is translocated out of the bacteroids before it can be further metabolized and used by the host plant. In the cytosol of bacteroid-containing cells (external to the peribacteroid membrane), NH_4^+ is converted into glutamine, glutamic acid, asparagine, and, in many species, nitrogen-rich compounds called **ureides.** The two principal ureides in legumes are *allantoin* ($C_4N_4H_6O_3$), and *allantoic acid* ($C_4N_4H_8O_4$) (for structures, see Fig. 7-21); like asparagine ($C_4N_2H_7O_4$), they have relatively high C:N ratios. Each of these three compounds represents a major form of nitrogen translocated from nodules to other parts of the plant. Asparagine predominates in legumes of temperate origin, including peas, alfalfa, clovers, and lupines. Ureides predominate in legumes of tropical origin, including soybeans, cowpeas, and various beans (Reynolds et al., 1982a, 1982b). In alder, a nonlegume, another ureide called *citrulline* (see Fig. 7-21) is the major nitrogen compound transported from root nodules.

Asparagine and ureides move from bacteroid-containing cells into pericycle cells adjacent to the vascular bundles that surround the nodule proper. These pericycle cells are modified as transfer cells (Section 7.3), and they seem to actively secrete nitrogen compounds into the conducting xylem cells (Pate, 1980). From here the compounds move into the xylem of the root and shoot to which the vascular bundles of the nodule are connected. Roots of nodulated plants that have not themselves become nodulated seem to receive appreciable nitrogen only after it has moved into the leaves and then back through the phloem along with sucrose. There is, therefore, a cycling process: from nodules upward to shoots in the xylem and then return of excess nitrogen downward to roots via the phloem.

Because of the importance of nitrogen fixation in nature and in agriculture, ecologists, agronomists, and plant physiologists have studied the environmental and genetic factors that control it. The high costs of nitrogen fertilizers have stimulated even more research in the past few years. In general, those factors that enhance photosynthesis, such as adequate moisture, warm temperatures, bright sunlight, and high CO_2 levels, are found to enhance fixation. Consistent with this, the rate of fixation is usually maximal in early afternoon when translocation of sugars from leaves to nodules is occurring rapidly. Early afternoon is also a time when transpiration is especially rapid, and the transpiration stream aids removal of nitrogen compounds from roots and root nodules (Pate, 1980).

Several genetic factors control the rates of nitrogen fixation and yields of legumes. One factor concerns how effectively nodulation occurs, and this depends on the genetically controlled recognition process between the *Rhizobium* species and the legume species or variety. Attempts are being made to increase the efficiency of nodulation by altering genes in various *Rhizobia* and by selecting for more compatible host varieties (Barton and Brill, 1983). Another genetic factor relates to the ability of nitrogenase from all organisms to reduce H$^+$ in competition with N_2. It is estimated that about one-third of the electrons made available for reduction of N_2 in legume nodules are used instead to reduce H^+ to H_2, and the H_2 simply escapes into the soil atmosphere carrying away wasted energy. Nevertheless, certain strains of *Rhizobium* species contain a **hydrogenase** enzyme that oxidizes much of the H_2 to H_2O before it escapes. During this oxidation, ATP is produced from ADP and Pi. There is evidence that soybeans and a few other legumes that contain *Rhizobium* strains with an active hydrogenase yield slightly more than legumes that contain a mutant without hydrogenase, probably because of less energy waste (Eisbrenner and Evans, 1983). Perhaps even more effective *Rhizobia* species or strains with active hydrogenases can be found or developed to increase legume yields through genetic engineering techniques. The goal of incorporating nitrogen-fixing genes into roots of non-legume crops by genetic engineering techniques now seems unrealistic or at least many years in the future, partly because the genes involved in nitrogen fixation and its control are so numerous (Anderson et al., 1980; Postgate and Cannon, 1981).

The stage of growth also influences N_2 fixation. Three important grain legumes, soybeans, pigeon peas, and peanuts, all show maximum fixation rates after flowering when the demand for nitrogen in the developing seeds and fruits increases. These species, as is common to legumes, contain seeds that are especially protein rich. In fact, soybean seeds contain 40 percent protein, the highest percentage of any plant known. Quantitatively, roughly 90 percent of the N_2 fixation in these species occurs during the period of reproductive development and roughly 10 percent during the first two months of vegetative growth. Surprisingly, N_2 fixation provides only about one-fourth to one-half of the total nitrogen in several mature grain legumes grown on soils of normal fertility; the remaining half to three-fourths is absorbed as NO_3^- or NH_4^+ from the soil, mainly during the period of vegetative growth. Even these figures are only rough estimates, and LaRue and Patterson (1981) stated that we have no valid estimates of nitrogen fixation for any legume crop. Nevertheless, yields of grain legumes cannot usually be increased with nitrogen fertilizers, because N_2 fixation is decreased in proportion to the added amount of fertilizer nitrogen absorbed. For nitrate fertilizer, this decrease results from inhibition of attachment of *Rhizobia* to root hairs, abortion of infection threads, slowing of nodule growth, inhibition of fixation within established nodules, and more rapid senescence of the nodules when either NO_3^- or NH_4^+ is added (Robertson and Farnden, 1980; Noel et al., 1982; Carroll and Gresshoff, 1983).

The amount of N_2 fixed by perennial native and legume crop species during various times in the growing season is still being studied but is probably again greatest during reproductive development. The percentage of nitrogen in such species that is derived from fixed N_2 is likely to be greater than that in annual grain legumes such as peas, beans, and soybeans, because nodules are perennial and fixation should begin earlier than in annuals, where nodule development must start anew each year. Furthermore, native N_2-fixing plants often grow in relatively unfertile soils in which the main input of nitrogen is from fixation. For perennial alfalfa fields, from which each crop is removed as soon as it blooms, the main supply of nitrogen is also from N_2 fixation (Vance and Heichel, 1981).

13.3 Assimilation of Nitrate and Ammonium Ions

For plants that cannot fix N_2, the only important nitrogen sources are NO_3^- and NH_4^+. Crop plants and many native species absorb most nitrogen as NO_3^-, because NH_4^+ is so readily oxidized to NO_3^- by nitrifying bacteria (but see boxed essay on page 259). However, climax communities of conifers and of grasses absorb most nitrogen as NH_4^+, because nitrification is inhibited either by low soil pH or by tannins and phenolic compounds (Rice, 1974; Haynes and Goh, 1978). A thorough review of the importance of NO_3^- and NH_4^+ to various species under various environmental conditions is by Runge (1983). We shall treat first the assimilation of nitrate, because of its abundance in most soils and because it must be converted to NH_4^+ in the plant before the nitrogen enters amino acids and other nitrogen compounds.

Sites of Nitrate Assimilation Both roots and shoots require organic nitrogen compounds, but in which of these organs is NO_3^- reduced and incorporated into organic compounds? Roots of some species can synthesize all of the organic nitrogen they need from NO_3^-, whereas roots of others rely on shoots for organic nitrogen. Figure 13-6 shows the kinds of nitrogenous substances present in the xylem transport stream of several herbaceous species during vegetative growth with their roots in sand watered with a nitrate-containing nutrient solution. (The legumes did not contain root nodules.) None of these plants translocated detectable amounts of NH_4^+ to the shoots, but some transported large quantities of organic nitrogen compounds derived from NH_4^+. The cocklebur (*Xanthium strumarium*) and white lupine (*Lupinus albus*) represent extremes among those studied. Cocklebur roots reduce almost no NO_3^-, so they depend upon amino acids translocated in the phloem from the leaves. In the lupine, nearly all NO_3^- is absorbed and converted into amino acids and amides

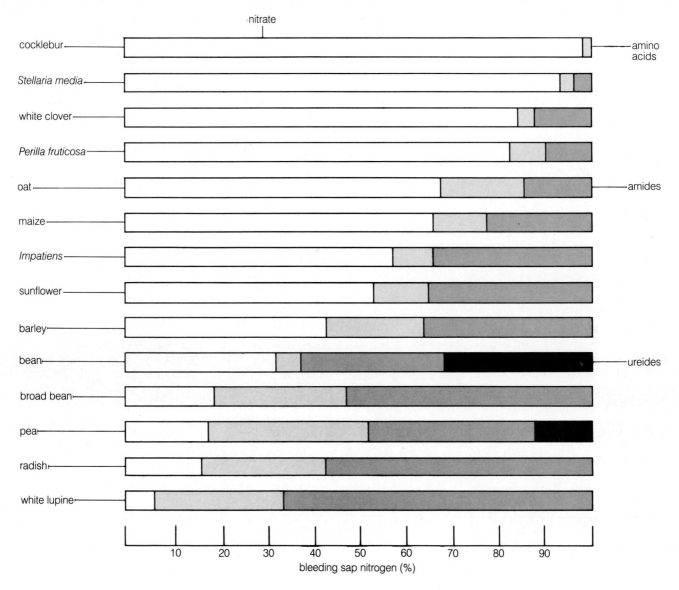

Figure 13-6 Relative amounts of nitrogen compounds in the xylem sap of various species. Plants were grown with their roots in sterile sand watered with a sterile nutrient solution containing 140 mg/liter nitrogen as nitrate (10mM NO_3^-); then the stems were severed to collect xylem sap from the cut stumps. Species at the top of the figure transport primarily nitrate; those at the bottom, mainly amides and amino acids. Two legumes also transport ureides, especially citrulline. (From J.S. Pate, 1973.)

in the roots. Most deciduous trees investigated behave like the lupine and translocate almost no NO_3^- to the shoots; the shoots of these species are provided a diet of organic nitrogen. Much more research is needed with plants of different ages and with conifers, other forest trees, and herbaceous plants grown in nature where the fungal hyphae in mycorrhizae might contribute organic nitrogen compounds to the root cells. Furthermore, the kind of compounds transported by legumes is altered when root nodules are present and nitrogen fixation occurs.

The relative amounts of NO_3^- and organic nitrogen in the xylem depend upon environmental conditions. Even plants that normally do not translocate much NO_3^- do so if provided with excessive

amounts of it in the soil, or if the roots are cold. Under these conditions, reduction of NO_3^- in the roots cannot keep pace with transport to the shoots. Reduction then occurs in leaves and stems, especially during sunny days.

The Process of Nitrate Reduction The overall process of reduction of NO_3^- to NH_4^+ is an energy-dependent one summarized in equation R13-2.

$$NO_3^- + 8 \text{ electrons} + 10 \text{ H}^+ \rightarrow NH_4^+ + 3 \text{ H}_2\text{O} \qquad \text{(R13-2)}$$

The oxidation number of nitrogen changes from +5 to −3.

Sources of the eight electrons (and 10 H$^+$) re-

Many Grasses Also Support Nitrogen Fixation

Although it was long believed that no nitrogen fixation occurs in cereal grains or other grasses, an assay method developed in the 1960s has helped prove that this is not entirely true. The method depends on the ability of all N-fixing organisms to reduce acetylene to ethylene, which then can be rapidly measured by gas chromatography. Sealed soil samples or aqueous samples from lakes, ponds, and streams can be collected in the field and assayed later. Positive results with the acetylene reduction method are usually correlated well with $^{15}N_2$-fixing ability measured by more laborious and less sensitive methods with mass or emission spectrometers.

A few reports of limited fixation by sugar cane and other tropical grasses in the 1960s were verified and widely extended in the 1970s with the new assay. Bacteria living on or near root cells of numerous species show activity (Fig. 13-7). These bacteria reside in a transition zone between soil and root often called the **rhizosphere.** Occasionally, such bacteria even enter the roots, as evidenced in part by the ability of surface sterilized roots to fix nitrogen. The most prevalent bacteria usually identified with active grass roots are members of the genus *Azospirillum*, although reasonably well-defined associations of sugar cane with *Beijerinckia*, *Paspalum notatum* (another tropical grass) with *Azotobacter paspali*, certain wheat varieties with *Bacillus*, and rice with *Achromobacter*-like organisms also exist (Stewart, 1982). Even when such bacteria exist only in the rhizosphere, there is a loose mutualism involved, because some of the nitrogen fixed is absorbed by the roots, and carbohydrates released by the roots nourish the bacteria.

There is still disagreement as to how much nitrogen fixation is supported by grasses, because it is impossible to measure rates for whole crops in the field over an entire growing season. Fixation rates with even the most effective tropical grasses are certainly less than with legumes and other species that have root nodules harboring nitrogen fixers in much more ideal environments (Van Berkum and Bohlool, 1980). Furthermore, fixation rates with cereal grain crops in the United States are much less than those with tropical grasses; unless soils are anaerobic (as when moist or wet) for many hours, rates with these cereal grains are barely detectable, probably because O_2 inactivates nitrogenase. Nevertheless, field inoculations with *Azospirillum*

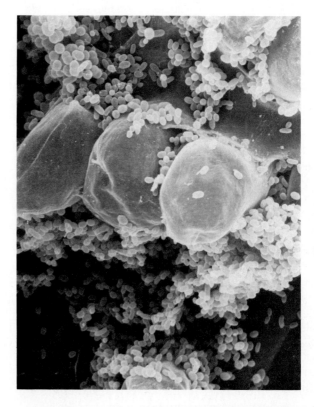

Figure 13-7 Scanning electron micrograph of part of a sorghum (*Sorghum bicolor*) root covered with nitrogen-fixing (*Azospirillum brasilense*) bacterial cells. Three large epidermal cells were being sloughed off into the rhizosphere. (From S. C. Schank et al., 1983; photo by Howard Berg, used by permission.)

species have reportedly increased dry matter yields of various crops in Israel, India, the Bahamas, Australia, and Florida (Schank et al., 1983). Currently, research to increase fixation rates with various grasses is being done. It might be possible to alter soil aeration properties and genetically alter both bacteria and grasses so that certain native grasses and cereal grains can fix greater fractions of the nitrogen they require.

quired for this reduction process will be described shortly, but note that two more H^+ than electrons are used in the reaction. This use of H^+ causes the cell's pH to rise. Continued pH rises would be lethal to plants if they had no way to replace such H^+ ions (Raven and Smith, 1976). About half are neutralized when NH_4^+ is subsequently converted into protein, because that process releases one H^+ for each nitrogen atom involved. Neutralization of the other half occurs by various mechanisms. In algae, aquatic angiosperms, and roots, H^+ ions are absorbed from the surrounding medium (or OH^- ions are excreted into the medium, either case helping to maintain constant cellular pH). In shoots, replacement of H^+ ions occurs by production of malic acid and other acids from sugars or starch. This is part of a biochemical *pH stat* (Davies, 1973), and it occurs because PEP carboxylase much more effectively fixes HCO_3^- and PEP into oxaloacetic acid, the precursor of malic, as the *pH* rises (Section 3.4 and 10.4). It is also believed that some of the malate anions produced during the neutralization process are transported in the phloem back to the roots as K^+ salts. This transport prevents the osmotic potential from becoming too negative as organic acid salts accumulate in the shoot cells. In roots, the transported malate is decarboxylated to pyruvate and CO_2, while the K^+ is recirculated with NO_3^- and other anions back to the shoots through the xylem (Raven and Smith, 1976).

Nitrate reduction occurs in two distinct reactions catalyzed by different enzymes. The first reaction is catalyzed by **nitrate reductase (NR),** an enzyme that transfers two electrons from NADH or, in a few species, NADPH. *Nitrite (NO_2^-),* NAD^+ (or $NADP^+$), and H_2O are the products.

$$NO_3^- + NADH + H^+ \xrightarrow{NR} NO_2^- + NAD^+ + H_2O \quad \text{(R13-3)}$$

This reaction occurs in the cytosol outside any organelle. NR catalyzes a seemingly simple reduction process, yet it is a large and complex enzyme. It contains FAD, a cytochrome, and molybdenum, all of which become consecutively reduced and oxidized as electrons are transported from NADH to the nitrogen atom in NO_3^-. For molybdenum, this may be its only essential function in plants, because when NH_4^+ is used as the nitrogen source no requirement for molybdenum can be demonstrated (Hewitt and Gundry, 1970).

NR has been studied intensively, because its activity often controls the rate of protein synthesis in plants absorbing NO_3^- as the major nitrogen source (Srivistava, 1980; Naik et al., 1982). The activity of the enzyme is affected by several factors. One is its rate of synthesis and another is its rate of degradation by protein-digesting enzymes (proteinases, described shortly). Apparently, NR is continuously synthesized and degraded, so these processes control activity by regulating how much NR a cell has (Somers et al., 1983). Activity is also affected by both inhibitors and activators within the cell. Although it is difficult to separate effects of these factors, abundant levels of NO_3^- in the cytosol clearly increase the activity of NR, largely because of faster synthesis of the enzyme. This is a case of **enzyme induction,** enhanced formation of an enzyme by a particular chemical. Enzyme induction is widespread in microorganisms, but fewer examples in plants or mammals are known. Induction of NR by NO_3^- is an excellent example of substrate induction, because the inducer is also the substrate for the enzyme. Induction of NR is widespread in various parts of various plants. The cells involved apparently conserve energy by not synthesizing NR until NO_3^- is available; then the enzyme begins to appear within a few hours. Opposing this inductive action of NO_3^- is a negative effect of NH_4^+. Thus the extent to which NR synthesis becomes induced can depend in many species on the cytoplasmic concentration of these two ions.

In leaves and stems, light also increases NR activity when NO_3^- is available. How light does this is still unclear and seems to vary with the plant and its stage of development (Abrol et al., 1983). First, in green tissues, light activates one or both chloroplast photosystems. This increases (perhaps by providing ATP) transport of stored NO_3^- from the vacuole to the cytosol where induction of NR then occurs (see Granstedt and Huffaker, 1982). Second, light activates the phytochrome system (Chapter 19), which somehow increases the ability of ribosomes to synthesize various proteins, including NR. Third, light somehow inactivates one or more proteins that are inhibitors of NR. (This inactivation appears similar to the inactivation by light of certain respiratory enzymes by the LEM and thioredoxin systems described in Chapter 10.) Finally, light promotes activity of NR because it increases the carbohydrate supply, and NADH necessary for nitrate reduction is produced from those carbohydrates when they are respired (Aslam and Huffaker, 1984). Physiologically, the overall plant response to these light effects is a large increase in the rate of NO_3^- reduction after sunrise, especially in the shoot.

Reduction of Nitrite to Ammonium Ions The second reaction of the overall process of nitrate reduction involves conversion of nitrite to NH_4^+. Nitrite arising in the cytosol from nitrate reductase action is transported into chloroplasts in leaves or into proplastids in roots where subsequent reduction to NH_4^+ occurs, catalyzed by **nitrite reductase.** In leaves, reduction of NO_2^- to NH_4^+ requires six electrons derived from H_2O by the chloroplast noncyclic electron transport system (R13-4). During this electron transfer, light drives electron transport from H_2O to ferredoxin (denoted by Fd); then reduced ferredoxin provides the six electrons used to reduce NO_2^- to NH_4^+. It is at this step (R13-5) that net use of $2 H^+$ occurs during the overall process of nitrate reduction to NH_4^+ (see R13-6).

$$3 H_2O + 6 Fd (Fe^{3+}) + light \rightarrow 1.5 O_2 + 6 H^+ + 6 Fd (Fe^{2+})$$
$$\text{(R13-4)}$$

$$NO_2^- + 6 Fd (Fe^{2+}) + 8 H^+ \rightarrow NH_4^+ + 6 Fd (Fe^{3+}) + 2 H_2O$$
$$\text{(R13-5)}$$

$$NO_2^- + 3 H_2O + 2 H^+ + light \rightarrow NH_4^+ + 1.5 O_2 + 2 H_2O$$
$$\text{(R13-6)}$$

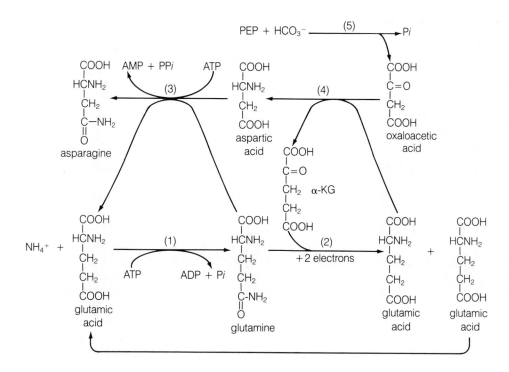

Figure 13-8 Conversion of ammonium (lower left) into important organic compounds.

R13-4 and R13-6 show that three H_2O molecules are required to provide the necessary six electrons in reduced ferredoxin (2 electrons per H_2O split by light energy), even though two H_2O also appear as products of this overall reaction.

Although reduced ferredoxin is the normal donor of electrons to leaf nitrite reductase, the reducing substance of roots is unknown. When nitrite reductase is studied *in vitro*, it only weakly accepts electrons from NADH, NADPH, or naturally occurring flavin compounds such as $FADH_2$. Reduced ferredoxin will provide electrons for isolated root nitrite reductases, but it does not do this *in vivo*, because neither proplastids nor other parts of root cells have detectable amounts of ferredoxin. Although we are still unsure how roots reduce NO_2^- to NH_4^+, it is clear that a carbohydrate supply from the leaves is necessary. Furthermore, there is indirect evidence that NADPH derived from the respiratory pentose phosphate pathway in plastids is the active reducing substance (Dry et al., 1981).

Conversion of NH_4^+ into Organic Compounds Whether NH_4^+ is absorbed directly from the soil or produced by energy-dependent N_2 fixation or NO_3^- reduction, it does not accumulate anywhere in the plant. Ammonium is, in fact, quite toxic (Givan, 1979), perhaps because it inhibits ATP formation in both chloroplasts and mitochondria by acting as an uncoupling agent. Except for traces of NH_4^+ lost to the atmosphere as volatile NH_3 (Fig. 13-1), all NH_4^+ is rapidly converted into the amide group of glutamine. This conversion and other reactions lead-ing to glutamic acid, aspartic acid, and asparagine are summarized in Fig. 13-8 and described briefly below.

Glutamine is formed by addition of an NH_2 group from NH_4^+ to the carboxyl group farthest from the alpha carbon of glutamic acid. An amide bond is thus formed (Fig. 13-8, reaction 1), and glutamine is one of two especially important plant amides (for structures, see Fig. 8-2). The necessary enzyme is *glutamine synthetase*. Hydrolysis of ATP to ADP and P*i* is essential to drive the reaction forward. Because this reaction requires glutamic acid as a reactant, there must be some mechanism to provide it; this is accomplished by reaction 2 catalyzed by *glutamate synthase*. Glutamate synthase transfers the amide group of glutamine to the carbonyl carbon of α-ketoglutaric acid, thereby forming *two* molecules of glutamic acid. This process requires a reducing agent capable of donating two electrons, which is ferredoxin (two molecules) in chloroplasts and either NADH or NADPH in proplastids of nonphotosynthetic cells. One of the two glutamates formed in reaction 2 is essential to maintain reaction 1, but the other can be converted directly into proteins or into other amino acids necessary for synthesis of proteins, chlorophyll, nucleic acids, and so on. Furthermore, some of the glutamate is transported to other tissues where it is used similarly in synthetic processes.

Besides forming glutamate, glutamine can donate its amide group to aspartic acid to form asparagine, the second important plant amide (Fig. 13-8, reaction 3). This reaction requires *asparagine synthetase*, and hydrolysis of ATP to AMP and PP*i* drives it forward. (Interestingly, asparagine synthetase is

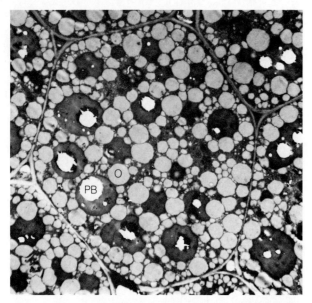

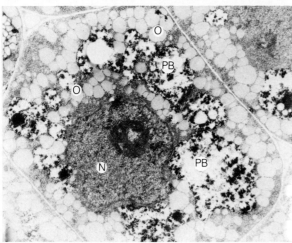

Figure 13-9 (**a**) Electron micrograph of a cortex cell in the radicle of an ungerminated Grand Rapids lettuce (*Lactuca sativa*) seed. The numerous unstained (light grey) structures are oleosomes (O) in which oils and fats are stored, while the larger darkly stained structures are protein bodies (PB). White areas in some of the protein bodies identify sites where most of the reserve phosphorus is stored as phytins. Phytins are calcium, magnesium, and potassium salts of phytic acid, myoinositol hexaphosphoric acid. (Courtesy Nicholas Carpita.) (**b**) Digestion of reserve protein in radicle cortex cells of recently germinated lettuce seeds. Protein bodies (PB) surrounding the nucleus were beginning to fuse to form the large vacuole, and most of the protein in them has disappeared. Numerous oleosomes (O) are still visible. (Courtesy Nicholas Carpita.)

strongly activated by Cl^-, which probably helps explain the uncertain role of chlorine in plants; Rognes, 1980.) A continuous supply of aspartic acid must be present to maintain asparagine synthesis. The nitrogen in aspartate can come from glutamate, but its four carbons probably arise from oxaloacetate (Fig. 13-8, reaction 4). Oxaloacetate, in turn, is formed from PEP and HCO_3^- by action of PEP carboxylase (reaction 5).

Probably because of its high ratio of nitrogen to carbon compared to most other compounds, glutamine is an important storage form of nitrogen in most species. Storage organs such as potato tubers and the roots of beet, carrot, radish, and turnip are especially rich in this amide. In mature leaves, glutamine is often formed from glutamic acid and NH_4^+ produced during protein degradation. It is then transported via the phloem to younger leaves or to roots, flowers, seeds, or fruits where its nitrogen is reused. Glutamine is also produced in roots and root nodules and is then transported through the xylem to shoots. Finally, glutamine can be incorporated directly into proteins in all cells as one of the 20 amino acids (Chapter 8). The other amide, asparagine, performs essentially the same functions as glutamine, especially in legumes of temperate origin in which it is unusually abundant.

Transamination When NH_4^+ containing isotopic ^{15}N is fed to plants or to excised plant parts, glutamic and then aspartic acids are usually the two amino acids that become most rapidly labeled with ^{15}N, as detected with a mass spectograph. (Of course, glutamine is labeled first.) Subsequently, the ^{15}N appears in the other amino acids. The reason for this labeling sequence is that glutamate transfers its amino group directly to a variety of α-keto acids in reversible **transamination** reactions (Givan, 1980). An important example of transamination occurs between glutamate and oxaloacetate, producing α-ketoglutarate and aspartate (see reaction 4 of Fig. 13-8).

The aspartate formed by transamination can transfer its amino group to other α-keto acids to form different amino acids by transamination reactions. Transfer to pyruvate, for example, yields alanine. Alanine and other amino acids can then transfer their amino groups, too, so numerous amino acids are formed by transamination. Pyridoxal phosphate, a vitamin B_6-containing coenzyme, is essential for transaminase enzymes. Reactions by which amino acids other than aspartate are formed are described in biochemical reviews (Miflin, 1980; Lea and Joy, 1983; Bray, 1983). We emphasize, however, that the amino groups of nearly all amino acids in both plants and animals probably passed through glutamine and glutamate on the route from NO_3^-, NH_4^+, or atmospheric N_2. Furthermore, plants synthesize amino acids that animals do not. Besides the 20 amino acids (including the two amides) common in proteins, about 200 nonprotein amino acids have been identified in the plant kingdom. The functions of such amino acids in the plants containing them are mostly unknown, although some are toxic to insects, mammals, and other plants, suggesting ecological defense roles (Bell, 1980; 1981).

13.4 Nitrogen Transformations During Plant Development

Nitrogen Metabolism of Germinating Seeds In storage cells in all kinds of seeds, the reserve proteins are deposited in membrane-bound structures called **protein bodies** (formerly called **aleurone grains**) (Lott, 1980; Pernollet, 1978; Weber and Neumann, 1980). Figure 13-9a shows an electron micrograph of these darkly-stained bodies in cells of a lettuce seed. In this species, protein bodies and oleosomes (oil bodies) are almost the only visible structures. Protein bodies are not pure protein but also contain much of the seed's reserves of phosphate, magnesium, and calcium. The phosphate is esterified to each of the six hydroxyl groups of a 6-carbon sugar alcohol called myoinositol (Fig. 13-10). The product of this esterification is called phytic acid, and ionization of H^+ from the phosphate groups allows Mg^{2+}, Ca^{2+}, Zn^{2+}, and probably K^+ to form salts collectively called **phytin,** or sometimes **phytates** (Maga, 1982; Oberleas, 1983). Phytin is usually bound to proteins in protein bodies.

Imbibition of water by a dry seed sets off a variety of chemical reactions that lead to germination (radicle protrusion through the seed coat) and subsequent seedling development. The proteins in protein bodies are hydrolyzed by *proteinases (proteases)* and *peptidases* to amino acids and amides (note disappearance of most of the protein in the protein bodies of Fig. 13-9b). Membranes surrounding disintegrating protein bodies are not destroyed; rather they fuse to form the tonoplast around the growing central vacuole. Some of the amino acids and amides released during protein hydrolysis in seeds are used to form special new proteins, nucleic acids, and so on in the cells in which hydrolysis occurs, but the majority are translocated through the phloem to growing cells of root and shoot. Release of phosphate and cations from phytin in protein bodies also occurs shortly after germination, and some of these ions are also transported to growing regions via the phloem. Soon, the young root system begins to absorb NO_3^- and NH_4^+, and nitrogen assimilation for another growing plant starts anew.

Traffic of Nitrogen Compounds During Vegetative and Reproductive Stages The general aspects of

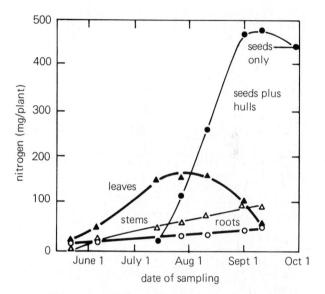

Figure 13-10 Structures of myoinositol (left) and its hexaphosphate, phytic acid.

Figure 13-11 Changes in nitrogen content of various organs of the broad bean (*Vicia faba*) during growth. The extensive accumulation of nitrogen compounds in the fruits (seeds plus hulls) was accompanied by a loss from the leaves and a large uptake from the soil. (Data of A. Emmerling, 1880, Ladw. Versuchsstat 24:113.)

protein transformations in various organs of an herbaceous plant during maturation were demonstrated a hundred years ago. Some of these aspects are illustrated in Fig. 13-11, in which changes in amounts of total nitrogen in roots, stems, leaves, and seeds of a broad bean (*Vicia faba*) plant from the seedling stage until maturity are shown.

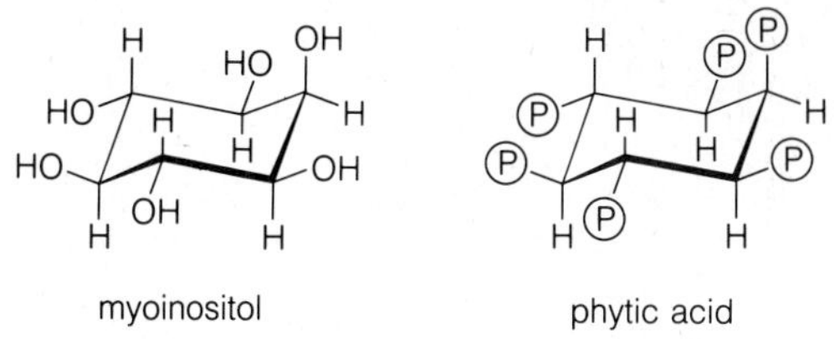

These changes largely reflect degradation and synthesis of proteins, because most of the nitrogen in any plant part is in protein. For leaves, about half of this protein is in chloroplasts. Note that the broad bean leaves actually lost nitrogen during August and September, while the seeds were accumulating it. This transfer of nitrogen compounds from leaves,

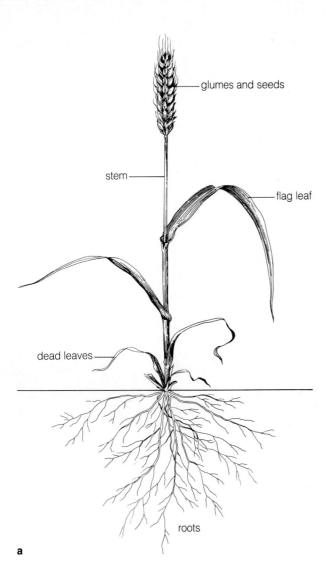

glumes and seeds

stem

flag leaf

dead leaves

roots

a

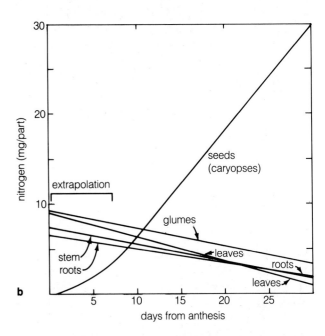

b

Figure 13-12 Changes in nitrogen content of various organs during seed development in a wheat plant. Measurements were made beginning about 6 days after anthesis and continuing until day 30 when seeds were mature. Drawing of plant was made 15 days after anthesis and shows parts analyzed for total nitrogen. On day 15, glumes and flag leaf were green; the leaf just below the flag leaf was yellow at the tip; the second leaf below the flag leaf was half yellow-brown; and other leaves were dead. (From R.J. Simpson et al., 1983. Used by permission.)

especially those that are mature, to developing protein bodies in seeds or to fruits via the phloem is typical of both herbaceous and woody plants. The principal organic compounds translocated are glutamine, asparagine, glutamate, and aspartate, and neither NO_3^- nor NH_4^+ is translocated in significant amounts in the phloem (Section 7.3). The amount of nitrogen transferred from vegetative organs to fruits of the broad bean was much less than that gained by the fruits during the same period (Fig. 13-11). The additional nitrogen demand of seeds in such legumes is usually satisfied by nitrogen from the increased rates of N_2 fixation that occur during seed development. Nevertheless, the nitrogen requirements of legume seeds are so great that loss of nitrogen from leaves, especially those near the seeds, is substantial.

Unfortunately for crop production, one of the major leaf proteins that contains this nitrogen is the abundant photosynthetic enzyme rubisco. As it is hydrolyzed by proteinases, photosynthetic activity decreases considerably during fruit and seed produc-

tion in essentially all crops (Huffaker, 1982). This has been called a "self-destruct" phenomenon, but in soils in which nitrogen is not abundant the hydrolysis of proteins and transport of nitrogen to the seeds is essential for seed production. Chlorophylls also disappear from leaves as proteins are degraded, and the nitrogen in these molecules is apparently also transported to reproductive organs.

In cereal grains and many other annuals that do not fix N_2, transfer of nitrogen from vegetative parts to seeds is sometimes more extensive than in legumes, even though their seeds contain lower percentages of protein than do legume seeds (Sinclair and deWit, 1975). Wheat leaves, for example, can lose up to 85 percent of their nitrogen (and an equal percentage of phosphate) before they die. Figure 13-12 illustrates nitrogen changes in major parts of wheat plants after flowering begins. This extensive conversion of nitrogen from vegetative organs to flowers and seeds is accompanied by a decrease in the rate of uptake of soil nitrogen as reproductive growth begins. Thus wheat and oats can absorb 90 percent of the nitrogen (and phosphate) needed for maturity before they are half-grown. Again, transport of nitrogen from vegetative organs occurs partly at the expense of rubisco degradation. This degradation is probably more of a growth limitation in C-3 than C-4 plants, because C-4 plants contain less than half as much of this enzyme as C-3 (essentially none in mesophyll cells of C-4, Section 10.3).

In perennial herbaceous plants, much of the nitrogen and other elements that are mobile in the phloem move into the crown and roots after seed demands are satisfied. As a result, these elements are available for the next season's growth, and the decaying dead plant parts return less to the soil than otherwise would have occurred. We know less about nitrogen relations in woody perennials, but fruits and seeds again are strong sinks for nitrogen. How much of this nitrogen comes directly from the soil via the xylem and how much comes from mature leaves via the phloem is unknown. However, fruits and seeds always have low rates of transpiration compared with mature leaves, so the xylem must be only a limited supplier of mineral salts to these organs. The total composition of fruits and seeds is much more like that of sieve tubes, suggesting that these organs grow largely on a diet of phloem sap. Much of the nitrogen in this sap comes from the leaves, especially leaves near the fruits. In autumn, but before leaf fall, deciduous woody plants translocate some of the nitrogen in their leaves to ray parenchyma cells of xylem and phloem in both stems and roots; this transport also occurs in the phloem. For apple trees, estimates indicate that up to half of the nitrogen is lost from senescing leaves in this way (Titus and Kang, 1982). The nitrogen is stored mainly in reserve proteins until new growth begins in spring, then amino acids, amides, and ureides appear in the xylem on their way to young leaves or flower buds. Were it not for this conservation process, losses of nitrogen during leaf fall would cause productivity of trees typically growing on nitrogen-deficient soils to be even lower.

RNA molecules are also degraded in mature and senescing leaves and in seed storage tissues. Hydrolytic enzymes called **ribonucleases** are responsible for this degradation. These enzymes generally release purine and pyrimidine nucleotides in which the phosphate group is attached either to carbon 3 or to carbon 5 of the ribose unit. The nitrogen in these nucleotides is probably translocated to other organs only after further degradation and rearrangement of the nitrogen into glutamate, aspartate, and their amides. Less is known about DNA breakdown in plants, except that DNA is much more stable than RNA and that there is much less DNA than RNA present. Even senescent leaves from which all chlorophyll and most of the protein and RNA have disappeared still retain much of their DNA. This DNA remains in the leaf as it is shed, and its nitrogen is cycled back to the soil. No one seems to have studied what happens to the DNA that is broken down when sieve tube elements and conducting xylem cells lose their nuclei.

13.5 Assimilation of Sulfate

Except for small amounts of SO_2 absorbed by the shoots of plants growing near smokestacks, SO_4^{2-} absorbed by roots provides the necessary sulfur for plant growth. Just as the reduction of NO_3^- and CO_2 are energy-dependent reduction processes, so is the reduction of sulfate to sulfide shown in R13-7:

$$SO_4^{2-} + ATP + 8\ electrons + 8\ H^+ \rightarrow$$

$$S^{2-} + 4\ H_2O + AMP + PPi \qquad \text{(R13-7)}$$

Sulfate reduction occurs in both roots and shoots of some species, but most of the sulfur transported in the xylem to the leaves is in nonreduced SO_4^{2-}. Some transport back to roots and to other parts of the plant occurs through phloem, and both free SO_4^{2-} and organic sulfur compounds are transported (Bonas et al., 1982). We know little about SO_4^{2-} reduction in tissues without chlorophyll, but most of the reactions apparently are the same as those occurring in leaves. ATP is essential in each case. In leaves, the entire process occurs in chloroplasts.

The first step of SO_4^{2-} assimilation in all cells is reaction of SO_4^{2-} with ATP, producing *adenosine-5'-phosphosulfate (APS)* and pyrophosphate (PPi). This step is catalyzed by *ATP sulfurylase*. The PPi is rapidly and irreversibly hydrolyzed into two Pi by a *pyrophosphatase* enzyme, then the Pi can be used in mitochondria or chloroplasts to regenerate ATP. These processes are shown in the first two reactions of Fig. 13-13.

The sulfur of APS is reduced in chloroplasts by electrons donated from reduced ferredoxin (Anderson, 1980). For nonphotosynthetic cells, proplastids represent a reasonable but unproven comparable site of reduction, perhaps using NADPH as the electron donor. The oxidation number of sulfur changes from +6 to −2 during APS reduction, which explains why eight electrons are required (R13-7). Reduction in chloroplasts is thought to occur as follows (Fig. 13-13): First, the sulfate group of APS is transferred to the sulfur atom of an acceptor molecule by an enzyme called *APS sulfotransferase* (Schmidt, 1979). The acceptor molecule has not been identified;

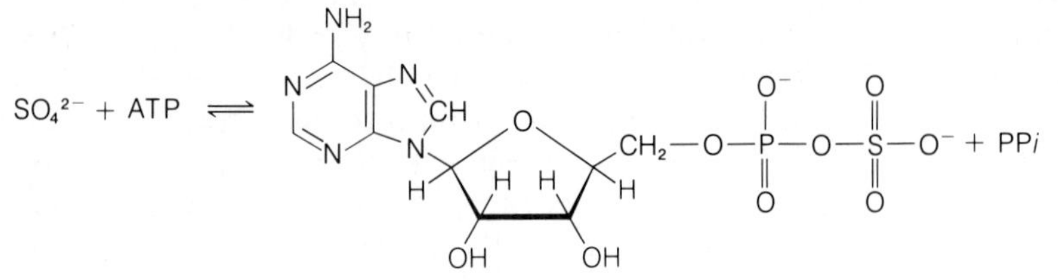

$SO_4^{2-} + ATP \rightleftharpoons$

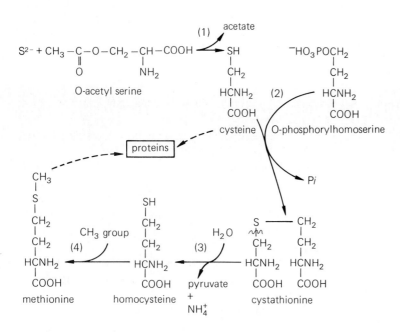

adenosine-5′-phosphosulfate (APS)

$PPi + H_2O \longrightarrow 2\ Pi$

$APS + GSH \longrightarrow AMP + G\!-\!S\!-\!SO_3^-$

$GSSO_3^- + 8\ Fd\ (Fe^{+2}) + 7H^+ \longrightarrow S^{2-} + GSH + 8\ Fd(Fe^{+3}) + 6H_2O$

Figure 13-13 Four major reactions in reduction of sulfate to sulfide.

Figure 13-14 Reactions by which sulfide is converted to cysteine and methionine. Reaction 1, catalyzed by cysteine synthetase, involves replacement of the acetate group in O-acetyl serine by sulfide. Reaction 2, catalyzed by cystathionine synthetase, splits phosphate from O-phosphorylhomoserine and joins the sulfur atom in cysteine to the terminal CH_2 group of the homoserine residue. Reaction 3 is catalyzed by cystathionase, an enzyme that hydrolyzes cystathionine between the S and the carbon shown by the wavy line. Pyruvate and NH_4^+ are released; these products were formerly part of the cysteine molecule. Homocysteine is converted by methionine synthetase to methionine by receipt of a methyl group in reaction 4. N^5-methyltetrahydrofolic acid is the methyl donor for this reaction.

likely candidates are *glutathione* (a tripeptide containing glutamate–cysteine–glycine) and a thioredoxin (described in Chapter 10). This acceptor is noted as GSH in Fig. 13-13, and after it accepts sulfate from APS it is noted as $G\!-\!S\!-\!SO_3^-$. Reduction of the sulfur in the SO_3^- group of $G\!-\!S\!-\!SO_3^-$ by ferredoxin then occurs, producing $G\!-\!S\!-\!SH$. The terminal SH group is then either released as free sulfide or is converted directly into cysteine.

The sulfide (free or bound) resulting from reduction of APS does not accumulate, because it is rapidly converted into organic sulfur compounds, especially cysteine and methionine. Reactions by which these two amino acids are formed are shown in Fig. 13-14 (Giovanelli et al., 1980). Most of the plant's sulfur (90 percent) is in cysteine or methionine of proteins, but small amounts of cysteine are incorporated into co-

enzyme A (see Fig. 12-7), and traces of methionine are used to form *S-adenosylmethionine*. One importance of S-adenosylmethionine is that its methyl group can be transferred to help form lignins and pectins of cell walls, flavonoids such as the brightly colored anthocyanins, and chlorophylls, while another importance is its role as a precursor of the plant hormone ethylene. In some species, especially onion, garlic, and cabbage and its close relatives, odiferous **mercaptans** (R–SH) such as methyl mercaptan and *n*–propyl mercaptan, **sulfides** (R–S–R), or **sulfoxides**

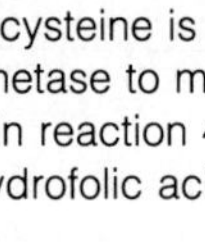

$$R\!-\!\underset{\underset{O}{\|}}{S}\!-\!R$$

accumulate.

Another odiferous compound released in small amounts by leaves of both angiosperms and conifers is H_2S. Production of H_2S seems energy wasteful, because its formation requires both ATP (in APS formation) and reduced ferredoxin. However, its release begins only when the reduced sulfur (cysteine) supply of the leaf is already adequate, that is, in daylight and when the SO_4^{2-} supply is plentiful, so release might represent a mechanism for maintaining a constant cellular level of cysteine (Rennenberg, 1984). Other control mechanisms of sulfate assimilation by various plants involve inhibition of APS sulfotransferase formation by H_2S or cysteine and inhibition of SO_4^{2-} absorption by cysteine.

Although plants, bacteria, and fungi generally reduce and convert sulfur into cysteine, methionine, and other essential sulfur compounds, mammals cannot. Because of this, we and the animals we eat depend on plants for reduced sulfur and particularly for the essential amino acids cysteine and methionine. Because we cannot reduce NO_3^-, plants are equally essential as providers of organic nitrogen.

14

Lipids and
Other Natural Products

 In the preceding six chapters, we emphasized that plants contain an imposing variety of carbohydrates and nitrogen- and sulfur-containing compounds. Many of the reactions by which these substances are formed have been explained, especially in relation to the importance of light in providing energy to drive the reactions in the shoot system. We have seen that light energy is used to drive the reduction of CO_2, NO_3^-, and SO_4^{2-}, processes that humans and other animals cannot accomplish metabolically. In this chapter, we discuss the properties and functions of many other compounds that plants require for growth or survival. Some of these, such as the fats and oils, are important food reserves that are deposited in specialized tissues and cells only at certain times in the life cycle. Others, such as the waxes and the components of cutin and suberin, are protective coats over the plant's exterior. Still others help perpetuate the species by facilitating pollination or by performing a defense role against other competitive organisms.

Besides these compounds, certain plants produce many others, such as rubber, for which no function within them is presently known; tetrahydrocannabinol, the active compound in marijuana, is another such example. Compounds not required for normal growth and development by the metabolic pathways common to all plants are sometimes referred to as **secondary compounds** or **secondary products.** This separates them from primary compounds such as sugar phosphates, amino acids and amides, proteins, nucleotides, nucleic acids, chlorophyll, and organic acids necessary for life of all plants. The separation is not complete because, for example, a compound such as lignin is considered primary and essential for vascular plants (because of its presence in xylem) but not for algae.

In a way, plants can be compared to sophisticated organic-chemical laboratories in which several thousand kinds of molecules can be synthesized,

many of which remain to be discovered. Modern analytical instruments such as gas chromatographs, high performance liquid chromatographs, and mass spectrometers now provide valuable tools to separate and identify these compounds. Structures and biosynthesis of hundreds of secondary compounds are summarized in books by Robinson (1980), Vickery and Vickery (1981), and Conn (1981).

We begin by describing the **lipids.** These are a group of fat and fatlike substances, rich in carbon and hydrogen, that dissolve in organic solvents such as chloroform, acetone, ethers, certain alcohols, and benzene, but that do not dissolve in water. Among these are the fats and oils, phospholipids and glycolipids, waxes, and many of the components of cutin and suberin.

14.1 Fats and Oils

Chemically, fats and oils are very similar compounds, but *fats are solids at room temperatures whereas oils are liquids.* Both are composed of long-chain **fatty acids** esterified by their single carboxyl group to a hydroxyl of the 3-carbon alcohol **glycerol.** All three hydroxyl groups of glycerol are esterified, so fats and oils are often called **triglycerides.** Except where an important distinction is to be made between fat or oil triglycerides, we refer to them as fats. The general formula for a fat is given in Fig. 14-1.

The melting points and other properties of fats are determined by the kinds of fatty acids they contain. A fat usually contains three different fatty acids, although occasionally two are identical. These acids almost always have an even number of carbon atoms, usually 16 or 18, and some are unsaturated (contain double bonds). The melting point rises with the length of the fatty acid and with the extent of its saturation with hydrogen, so solid fats usually have saturated fatty acids. In the oils, one to three double bonds are present in each fatty acid; these cause lower melting points and make oils liquid at room temperature. Examples of commercially important plant oils are those from the seeds of cotton, corn, peanuts, and soybeans. All of these oils principally contain fatty acids with 18 carbon atoms, including **oleic acid,** with one double bond, and **linoleic acid,** with two double bonds. In fact, these two acids, in the order named, are the most abundant fatty acids in nature.

Table 14-1 lists several important fatty acids, with the number of carbon atoms, structure, degree of unsaturation, and the position of double bonds listed for each. The most abundant saturated fatty acids are **palmitic acid,** with 16 carbons, and **stearic acid,** with 18 carbons. Coconut fat is a rich source of **lauric acid,** a saturated acid with only 12 carbons. The seven acids listed in Table 14-1 represent about 90 percent of those occurring in lipids of plant membranes (Chapter 6) and about the same percentage of those in commercial oils from seeds (Harwood, 1980). Seeds of many plants contain a high percentage of fatty acids that are not important in membrane lipids of the same species. Castor beans (*Ricinus communis*), for example, contain **ricinoleic acid** (12-hydroxyoleic acid), which makes up between 80 and 90 percent of the fatty acids in castor oil but which is absent from castor bean membranes and is rare in other species.

Distribution and Importance of Fats Fat storage is rare in leaves, stems, and roots but occurs in many seeds and some fruits (e.g., avocados and olives). In angiosperms, fats are concentrated in the endosperm

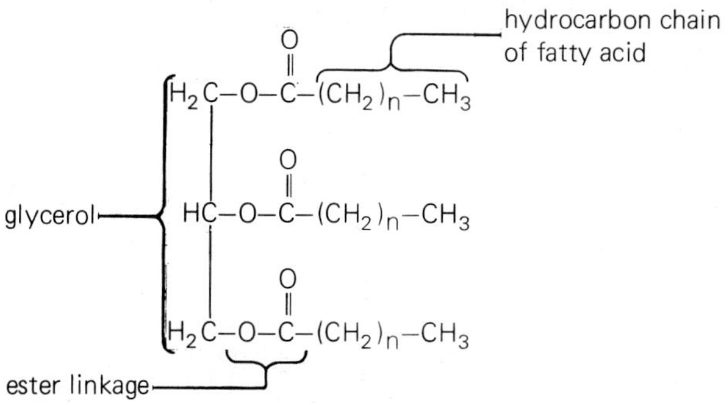

Figure 14-1 The general structure of a fat or oil, both of which are triglycerides.

or cotyledon storage tissues of seeds, but they also occur in the embryonic axis. In gymnosperm seeds, they are stored in the female gametophyte.

Compared with carbohydrates, fats contain larger amounts of carbon and hydrogen and less oxygen; and when they are respired, more O_2 is used per unit weight. As a result, more ATP is formed, demonstrating that greater amounts of energy can be stored in a small volume as fats than as carbohydrates. Perhaps because of this, most small seeds contain fats as the primary storage materials. When these fats are respired, enough energy is released to allow establishment of the seedling, yet the small weight of such seeds often allows them to be scattered effectively by wind. Larger seeds, especially those such as pea, bean, and maize selected by humans for agriculture, often contain much starch and

Table 14-1 Fatty Acids Abundant or Common in Various Plants

Name	Number of Carbons: Number of Double Bonds	Structure
Lauric	12:0	$CH_3(CH_2)_{10}COOH$
Myristic	14:0	$CH_3(CH_2)_{12}COOH$
Palmitic	16:0	$CH_3(CH_2)_{14}COOH$
Stearic	18:0	$CH_3(CH_2)_{16}COOH$
Oleic	18:1 at C-9, 10	
Linoleic	18:2 at C-9, 10; 12, 13	
Linolenic	18:3 at C-9, 10; 12, 13; 15, 16	

Table 14-2 The Chemical Composition of Some Seeds of Economic Importance[a]

Species	Family	Principal Reserve Tissue	Percent Content		
			Carbohydrate	Protein	Lipid
Maize (*Zea mays*)	Poaceae (Gramineae)	Endosperm	51–74	10	5
Wheat (*Triticum vulgare*)	Poaceae	Endosperm	60–75	13	2
Pea (*Pisum sativum*)	Fabaceae (Leguminosae)	Cotyledons	34–46	20	2
Peanut (*Arachis hypogaea*)	Fabaceae	Cotyledons	12–33	20–30	40–50
Soybean (*Glycine sp.*)	Fabaceae	Cotyledons	14	37	17
Brazil nut (*Bertholletia excelsa*)	Lecythidaceae	Hypocotyl	4	14	62
Castor bean (*Ricinus communis*)	Euphorbiaceae	Endosperm	0	18	64
Date palm (*Phoenix dactylifera*)	Arecaceae (Palmae)	Endosperm	57	6	10
Sunflower (*Helianthus annuus*)	Asteraceae (Compositae)	Cotyledons	2	25	45–50
Oak (*Quercus robur*)	Fagaceae	Cotyledons	47	3	3
Douglas fir (*Pseudotsuga menziesii*)	Pinaceae	Gametophyte	2	30	36

[a]The percentages are based on the fresh (air-dry) weights of the seeds, except for date palm, for which the percentages are expressed on a dry weight basis.

From Street and Öpik, 1970. A longer list of analyses of whole seeds is given by T. R. Sinclair and C. T. de Wit, 1975.

only small amounts of fats, but seeds of conifers and those in nuts are usually fat rich (Table 14-2). An encyclopedic list of seed compositions from 113 families is given by Earle and Jones (1962).

Fats are always stored in specialized bodies in the cytoplasm (see Fig. 13-9), and there are often hundreds or thousands of such bodies in each storage cell. These bodies have been called lipid bodies, spherosomes, and **oleosomes** (Latin, *oleo*, oil). For a review of terminology, see Gurr (1980). We prefer the term oleosome suggested by Yatsu et al. (1971), because it correctly indicates that they contain oil and distinguishes them from peroxisomes (Fig. 10-8) and glyoxysomes (see later, this section), which are also spherical bodies. Furthermore, the term spherosomes has been used for years to describe organelles that contain little if any fat (Sorokin, 1967).

Oleosomes can be isolated from seeds in rather pure form, allowing analysis of their composition and structure. Failure of oleosomes to fuse into one large lipid droplet in cells or when isolated suggests that a membrane surrounds each, but that membrane frequently cannot be seen in electron micrographs. When it is visible it appears only about half as thick (approximately 3 nm) as a typical unit membrane (approximately 8 nm). Apparently, the oleosome membrane is indeed a half-membrane whose polar, hydrophilic surface is exposed to the aqueous cytosol and whose nonpolar, hydrophobic surface faces the fats stored inside.

An extensive cytological study of oleosome formation during seed development seems to explain many details, especially how a half membrane arises (Wanner et al., 1981). The main origin of oleosomes is apparently from two sources, the endoplasmic reticulum and plastids. In each case fats apparently accumulate *between* the two layers of phospholipids and glycolipids present in the outer membrane of the plastid envelope or the ER membrane. This accumulation causes separation of the lipid bilayer into two halves with fats forcing them apart until a distinct oleosome swells and pinches off (Fig. 14-2).

Formation of Fats It is now clear that fats stored in seeds and fruits are not transported there from leaves but are synthesized *in situ* from sucrose or other translocated sugar. Although leaves produce various fatty acids present in lipids of their membranes, they seldom synthesize fats. Furthermore, both fatty acids and fats are too insoluble in H_2O to be translocated in phloem or xylem.

Conversion of carbohydrates to fats requires production of the fatty acids and of the glycerol backbone to which the fatty acids become esterified. The glycerol unit (L-α-glycerophosphate) arises by reduction of dihydroxyacetone phosphate produced in glycolysis (Section 12.3). The fatty acids are formed by multiple condensations of acetate units in acetyl CoA. Most of the reactions of fatty-acid synthesis occur only in chloroplasts of leaves and proplastids of

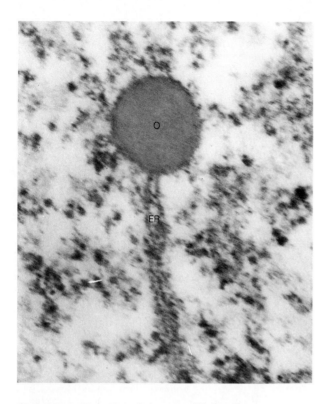

Figure 14-2 Formation of oleosome (O) from the endoplasmic reticulum (ER) in a fat-storing cotyledon of a developing watermelon seed. (Courtesy G. Wanner.)

seeds and roots (Stumpf, 1980b, 1983). The fatty acids synthesized in those organelles are mainly palmitic and oleic. A summary of the many reactions involved in fatty acid synthesis is exemplified for palmitic (as the CoA ester) in R14-1:

$$8 \text{ acetyl CoA} + 7 \text{ ATP}^{3-} + 14 \text{ NADPH} + 14 \text{ H}^+ \rightarrow$$

$$\text{palmityl CoA} + 7 \text{ CoA} + 7 \text{ ADP}^{2-} + 7 \text{ H}_2\text{PO}_4^-$$

$$+ 14 \text{ NADP}^+ + 7 \text{ H}_2\text{O} \qquad (\text{R14-1})$$

Subsequently, CoA is hydrolyzed away when palmitic or other fatty acid is combined with glycerol during formation of fats or membrane lipids.

This summary emphasizes that conversion of acetate units into fatty acids is energy expensive, because almost two pairs of electrons (2 NADPH) and one ATP are needed for each acetyl group present. In illumininated leaves, photosynthesis likely provides most of the NADPH and ATP, and fatty-acid formation occurs much faster in light than in darkness. In darkness and in proplastids of seeds and roots, the pentose phosphate respiratory pathway (Section 12.10) likely provides the NADPH, and glycolysis provides the ATP and the pyruvate from which acetyl CoA is formed.

Although palmitic and oleic are formed in plastids, most of the other fatty acids are formed by mod-

ifying palmitic and oleic in the ER. In seeds, all fatty acids they produce can be esterified with glycerol to produce fats that develop into oleosomes directly in the ER (Fig. 14-2). Alternatively, fatty acids can be transported back to proplastids for oleosome formation. Furthermore, the ER of all cells can convert fatty acids into phospholipids needed for growth of the ER itself or of other cellular membranes (Moore, 1982; Mudd, 1980). In leaves, it is thought that linoleic and linolenic acids are synthesized from oleic and (by elongation) from palmitic in the ER. Then linoleic and linolenic are transported from the ER back to the chloroplasts where they accumulate as lipids in thylakoid membranes.

Conversion of Fats to Sugars: β-oxidation and the Glyoxylate Pathway Breakdown of fats stored in oleosomes of seeds and fruits releases relatively large amounts of energy. For seeds, this energy is necessary to drive early seedling development before photosynthesis begins. Because fats cannot be translocated to the growing roots and shoot, they must be converted to more-mobile molecules, usually sucrose. Conversion of fats to sugars is an especially interesting process, because it occurs largely in fat-rich seeds and fungal spores and in some bacteria, not in humans or other animals.

Most of the reactions necessary to convert fats to sugars occur in microbodies called **glyoxysomes** (Fig. 14-3). Structurally, glyoxysomes are almost identical to peroxisomes of photosynthetic cells (Chapter 10), but many of the enzymes they contain are different (Tolbert, 1981; Huang et al., 1983). In some species they are formed as proglyoxysomes (small glyoxysome precursors) in the cotyledons of developing seeds, then during germination and early seedling development they mature into fully functional glyoxysomes (Trelease, 1984). They persist only until the fats are digested, then they disappear; in cotyledons that emerge above ground and become photosynthetic, they are replaced by peroxisomes (Beevers, 1979).

The breakdown of fats begins with action of **lipases,** which hydrolyze the ester bonds and release the three fatty acids and glycerol (R14-2).

$$
\begin{array}{ccc}
 & \text{H}_2\text{C}-\text{O}-\overset{\displaystyle \text{O}}{\overset{\displaystyle \|}{\text{C}}}-(\text{CH}_2)_n\text{CH}_3 & \rightarrow \text{HOC}(\text{CH}_2)_n\text{CH}_3 \\
\text{H}_2\text{COH} & & \\
| & \text{HC}-\text{O}-\overset{\displaystyle \text{O}}{\overset{\displaystyle \|}{\text{C}}}-(\text{CH}_2)_n\text{CH}_3 & \rightarrow \text{HOC}(\text{CH}_2)_n\text{CH}_3 \\
\text{HCOH} \leftarrow & & \\
| & & \\
\text{H}_2\text{COH} & \text{H}_2\text{C}-\text{O}-\overset{\displaystyle \text{O}}{\overset{\displaystyle \|}{\text{C}}}-(\text{CH}_2)_n\text{CH}_3 & \rightarrow \text{HOC}(\text{CH}_2)_n\text{CH}_3 \\
\text{glycerol} & \text{fat molecule} & \text{free fatty acids}
\end{array}
$$

$$(\text{R14-2})$$

$$palmitate + ATP^{3-} + 7\,NAD^+ + 7\,FAD + 7\,H_2O + 8\,CoASH$$
$$\rightarrow 8\,acetyl\,CoA + AMP^{2-} + pyrophosphate^{2-}$$
$$+ 7\,NADH + 7\,H^+ + 7\,FADH_2 \quad (R14\text{-}3)$$

R14-3 is a generalized summary of β-oxidation that applies wherever the process occurs. In cells other than storage tissues of fat-rich seeds, fatty acids released from membrane lipids have long been thought to undergo β-oxidation in mitochondria, but in all plant cells glyoxysomes or peroxisomes are probably instead responsible (Gerhardt, 1983; Macey and Stumpf, 1983). This now raises the question of how oxidation of the electron-rich products of R14-3 occurs, because glyoxysomes and peroxisomes have no Krebs cycle enzymes and no electron transport system comparable to that of mitochondria. Thus it is unclear how the products of R14-3 can be used as energy sources by the plant. This uncertainty is especially true for reaction products formed in peroxisomes, but remember that peroxisomes are found only in cells of vegetative tissues that store very little fat. In fat-rich seeds the potential energy in $FADH_2$ appears to be totally wasted. This waste occurs because of the absence of an electron-transport system in the glyoxysomes and because glyoxysomes contain an enzyme that transfers H atoms of $FADH_2$ directly to O_2, forming H_2O_2. Each H_2O_2 is then degraded to $1/2\ O_2$ and H_2O by catalase, just as occurs in peroxisomes during the glycolate pathway of photorespiration (Section 10.4). Glyoxysomes can process both NADH and acetyl CoA released in β-oxidation, yet they require help of mitochondria and the cytosol to form sugars. Pertinent reactions that occur in the glyoxysomes are called the **glyoxylate pathway,** namely, conversion of acetate units of acetyl CoA to succinic and malic acids. Details of the glyoxylate pathway and additional mitochondrial and cytoplasmic reactions necessary to convert acetate units to sugars are shown in Fig. 14-4 and described briefly below.

Acetyl CoA reacts with oxaloacetic acid to form citric acid, just as in the Krebs cycle (reaction 1). After isocitric acid (6 carbons) is formed (reaction 2), it undergoes cleavage by an enzyme unique to this cycle, called **isocitrate lyase.** Succinate (4 carbons) and glyoxylate (2 carbons) are produced (reaction 3). The glyoxylate reacts with another acetyl CoA to form malate and free coenzyme A (CoASH, reaction 4). This reaction is catalyzed by another enzyme restricted to the glyoxylate pathway called **malate synthetase.** This malate is transported to the cytosol where it is converted to sugars, as we shall explain.

Succinate produced in reaction 3 moves to the mitochondria for further processing. Here it is oxidized by Krebs cycle reactions 5, 6, and 7 to ox-

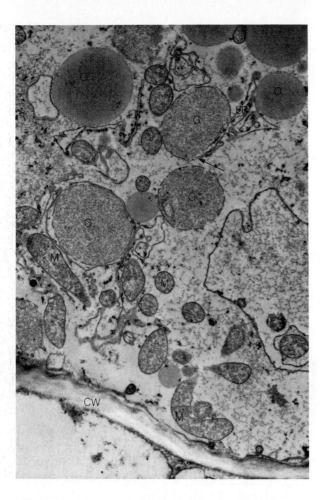

Figure 14-3 Part of a fat-storing cell in the megagametophtye of a ponderosa pine seed germinated 7 days, showing glyoxysomes (G), fat-storing oleosomes (O), endoplasmic reticulum (ER), and mitochondria (M). Most of the fat has been converted to sugars. Some of the glyoxysomes appear to be connected to the ER, from which they arise; note arrow. (From T. M. Ching, 1970, Plant Physiology 46:475-482.)

Apparently, most of the lipase activity is not present in oleosomes but instead takes place in or near the membrane surrounding each glyoxysome (Lin et al., 1982). This means that direct contact between oleosomes and glyoxysomes must occur, and this is frequently observed (Fig. 14-3).

The glycerol resulting from lipase action is converted with ATP to α-glycerolphosphate; this molecule is then oxidized by NAD^+ to dihydroxyacetone phosphate, most of which is converted to sugars by reversal of glycolysis. The fatty acids are taken into the matrix of the glyoxysome where they are first oxidized to acetyl CoA units and NADH by a metabolic pathway called β-**oxidation**, because the beta-carbon is oxidized. Details of β-oxidation will not be presented here but are summarized for palmitic acid by R14-3.

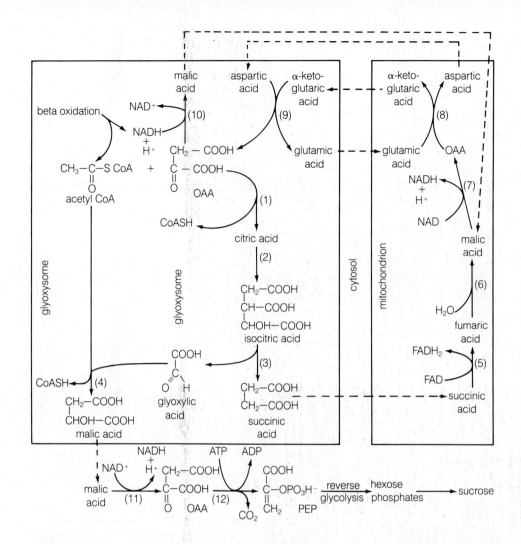

Figure 14-4 Cooperation of glyoxysomes, cytosol, and mitochondria in converting fatty acids of reserve fats to sucrose by the glyoxylate pathway.

aloacetate (OAA), releasing NADH and FADH$_2$. Both NADH and FADH$_2$ are oxidized by the mitochondrial electron-transport system with O$_2$ to form H$_2$O and ATP. Reactions 8 and 9 are transaminations between alpha-keto acids and amino acids that require exchange transport of such molecules between mitochondria and glyoxysomes. Their main function seems to be regeneration of the OAA needed to maintain reaction 1 of the glyoxylate pathway (Mettler and Beevers, 1980). Without these reactions, OAA is converted mainly to malate (reaction 10), because of the large amounts of NADH produced in β-oxidation.

The malate produced by malate synthetase (reaction 4) is enough to account for all of the fatty-acid carbons converted to sucrose carbons (in fact, even more). This malate is first oxidized to OAA by a cytoplasmic NAD$^+$-malate dehydrogenase (reaction 11), then the OAA is decarboxylated and phosphorylated with ATP to yield CO$_2$ and phosphoenolpyruvate, PEP (reaction 12). This reaction is catalyzed by an enzyme we have not mentioned so far called **PEP carboxykinase.** It is likely that ATP produced in the

mitochondria during oxidation of NADH and FADH$_2$ somehow becomes available to drive reaction 12. Once PEP is formed it can readily undergo reverse glycolysis to form hexose phosphates. Sucrose derived from these hexose phosphates is then transported via the phloem to the growing roots and shoot, where it provides much of the carbon needed for growth of those organs.

An overall summary of the conversion of a fatty acid (palmitic) to a sugar (sucrose) is given by R14-4:

$$C_{16}H_{32}O_2 + 11\ O_2 \longrightarrow C_{12}H_{22}O_{11} + 4\ CO_2 + 5\ H_2O \tag{R14-4}$$

This is a respiration process, because O$_2$ is absorbed (during oxidation of FADH$_2$ produced by β-oxidation) and CO$_2$ is released (during conversion of oxaloacetate to PEP). The respiratory quotient (moles of CO$_2$/moles of O$_2$) is 0.36, entirely consistent with measurements of RQ values in numerous fat-rich seeds or seedlings. Although one-fourth of the carbon atoms are lost from fatty acids as CO$_2$, the saving of three-fourths is enough for the ecological requirements of species with fat-rich seeds.

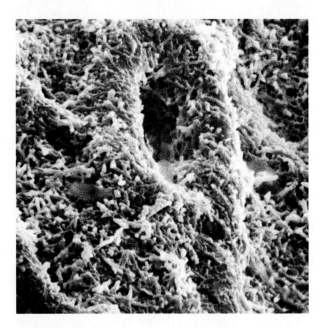

Figure 14-5 Wax on the leaf surface of a carnation. Carnation (*Dianthus* sp.) is a common plant with a prolific layer of wax on the cuticle. The structure of wax on plant surfaces can be as thin flakes, plates, rodlets, or rods. When the wax is in the form of rodlets or rods, it is visible to the naked eye as a bluish "bloom," which can be easily rubbed off the leaf. (From Troughton and Donaldson, 1972.)

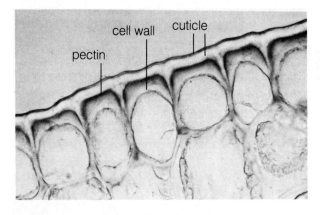

Figure 14-6 Fine structure of the cuticle on the upper surface of a *Clivia miniata* (scarlet kafarlily) leaf. The cuticle (external dark and light regions) covers the pectin portion fused to the outer part of the cell wall. (Courtesy P. J. Holloway.)

14.2 Waxes, Cutin, and Suberin: Plant Protective Coats

The entire shoot system of an herbaceous plant is covered by a **cuticle** that slows water loss from all of its parts, including leaves, stems, flowers, fruits, and seeds (Cutler et al., 1980; Juniper and Jeffree, 1982). A scanning electron micrograph of the cuticle on a car-

nation leaf illustrates the waxy surface structure of the cuticle (Fig. 14-5). Without this protective cover, transpiration of most land plants would be so rapid that they would die. The cuticle also provides protection against some plant pathogens and against minor mechanical damage. It is of additional importance in agriculture, because it repels water used in various sprays containing fungicides, herbicides, insecticides, or growth regulators (Price, 1980). Because of the hydrophobic nature of the cuticle, most spray formulations contain a detergent to reduce the surface tension of water and allow it to spread on the foliage.

Most of the cuticle is composed of a heterogeneous mixture of components collectively called **cutin,** while the remainder consists of overlaying waxes and of pectin polysaccharides attached to the cell wall (Fig. 14-6). Cutin is a heterogeneous polymer consisting largely of various combinations of members in two groups of fatty acids, a group having 16 carbons and one having 18 carbons (Kolattukudy, 1980a, 1980b; Holloway, 1980b). Most of these fatty acids have two or more hydroxyl groups, similar to ricinoleic acid mentioned in Section 14.1. The polymeric nature of cutin arises from ester bonds uniting the various fatty acids. Small amounts of phenolic compounds are also present in cutin, and these are thought to bind by ester linkages the fatty acids to pectins of the epidermal cell walls.

The cuticular waxes include a variety of long-chain hydrocarbons that also have little oxygen. Many waxes contain long-chain fatty acids esterified with long-chain monohydric alcohols, but they also contain free long-chain alcohols, aldehydes, and ketones ranging from 22 to 32 carbon atoms and even true hydrocarbons containing up to 37 carbons. One long-chain primary alcohol from the cuticle with 30 carbons, **triacontanol,** is a plant growth stimulant (Chapter 17). Cutins and waxes are synthesized by the epidermis and are then somehow secreted onto the surface. Waxes accumulate in various patterns, one of which is the rodlike pattern shown in Fig. 14-5.

A frequently less distinct protective coating over underground plant parts is generally called **suberin.** Suberin also covers the cork cells formed in tree bark by the crushing action of secondary growth, and it is formed by many cells as scar tissue after wounding, e.g., after leaf abscission and on potato tubers cut for planting. Suberin further occurs in the walls of non-injured root cells as a Casparian strip and in bundle sheath cells of grasses. A lipid portion (up to half of the total suberin) is a complex mixture of long-chain fatty acids, hydroxylated fatty acids, dicarboxylic acids, and long-chain alcohols. Nearly all members of these groups have more than 16 carbon atoms. The remainder of suberin contains phenolic compounds, of which ferulic acid (see Fig. 14-12) is a major component. As in cutin, these phenolics are thought to bind the lipid fraction of suberin to the cell wall. Thus suberin is similar to cutin in having an important lipid polyester fraction but differs in having a much more abundant phenolic fraction and in the kinds of fatty acids present (Kolattukudy, 1980a, 1980b).

14.3 The Isoprenoid Compounds

Numerous plant products having some of the general properties of lipids form a diverse group of compounds with a common structural unit. They are classified as **isoprenoids** or as **terpenoids.** Some people call them terpenes, but this term should be reserved for those isoprenoids that are pure hydrocarbons (Robinson, 1980). Included are hormones such as the gibberellins and abscisic acid (see Chapters 16 and 17), farnesol (a probable stomatal regulator in Sorghum, Section 3.4), xanthoxin (a growth inhibitor produced by light, Section 17.5), sterols, carotenoids, turpentine, rubber, and the phytol tail of chlorophyll. Hundreds of isoprenoids have been found, and the actual number existing in the plant kingdom is probably in the thousands. Many of these are of interest because of their commercial uses and because they illustrate the ability of plants to synthesize a vast complex of compounds not formed by animals. For most isoprenoids, especially the smaller ones, no function in the plant is presently known. Nevertheless, many influence other plant or animal species with resulting benefit to the species containing them. Such chemicals (exclusive of foods) that influence another species are sometimes called **allelochemics** (Barbour et al., 1980; Whittaker and Feeny, 1971). **Allelopathy** (Greek, *allelon*, of one other; *pathos*, disease) is a special case of allelochemy involving a negative chemical interaction between different plant species. For isoprenoids and other compounds produced by plants, allelochemy against insects and other animal herbivores is much more evident than allelopathy, but several cases of allelopathy are known (Putnam and Heisey, 1983; Lovett, 1982).

Except for isoprene (C_5H_8) itself, the isoprenoids are dimers, trimers, or polymers of isoprene units, in which these units are joined in a head-to-tail fashion:

$$\text{(head)}-CH_2-\overset{\overset{\displaystyle CH_3}{|}}{C}=CH-CH_2-\text{(tail)}$$

isoprene unit

The isoprene unit is synthesized entirely from acetate of acetyl CoA by what is usually called the **mevalonic acid pathway,** because mevalonate is an important intermediate. Three acetyl CoA molecules provide

the five carbons for one isoprene unit, while the sixth carbon is lost from mevalonate pyrophosphate as CO_2. These reactions are described in most biochemistry books, but will not be explained here. A description of a few of the isoprenoids is given below.

Sterols All **sterols** (steroid alcohols) are **triterpenoids** built from six isoprene units. The most abundant in green algae and higher plants are **sitosterol** (29 C), **stigmasterol** (29 C), and **campesterol** (28 C), usually in the order given. Figure 14-7 illustrates the structures of these and a few other sterols, including **cholesterol** (widespread in trace amounts in plants), **ergosterol** (rare in plants but common in some fungi; converted by UV radiation of sunshine to vitamin D_2), and **antheridiol,** a sex attractant secreted by female strains of the aquatic fungus *Achlya bisexualis* (McMorris, 1978). Altogether, more than 150 sterols are known to exist in nature. The reactions by which they are synthesized by plants (especially members of the Solanaceae family) was reviewed thoroughly by Heftmann (1983).

Sterols exist not only in the forms shown but also as glycosides, in which a sugar (usually glucose or mannose) is attached to the hydroxyl group of the sterol, and as esters, in which the hydroxyl group is attached to a fatty acid. Free sterols probably exist in all membranes of all organisms except bacteria, and there is little doubt that their contribution to membrane stability is one of their most important functions. None of the sterol glycosides and esters seems to exist in membranes, and their functions remain largely unknown (Grunwald, 1980; Axelos and Peaud-Lenoel, 1982).

Besides a membrane function, certain sterols have allelochemical activity (Roeske et al., 1976; Harborne, 1982). Relatively few examples have been well documented, even though hundreds probably exist in nature. One example concerns **cardiac glycosides** (sterol derivatives that cause heart attacks in vertebrates but are used medicinally to strengthen and slow the heart beat during heart failure) and is related to coevolution of certain milkweeds (*Aesclepias* spp.), monarch butterflies, and bluejays. The milkweeds produce several bitter-tasting cardiac glycosides that protect them against herbivory by most insects and even cattle. However, monarch butterflies have adapted to these glycosides, and the glycosides their larvae ingest cause vomiting in bluejays that eat the larvae. The birds react to the vomiting experience by rejecting other monarchs on sight alone, so considerable immunity occurs among these butterflies from only one emetic experience. Similar cardiac glycosides present in species of the genus *Digitalis* and in other unrelated species include various **digilanides** that have been used since prehistoric

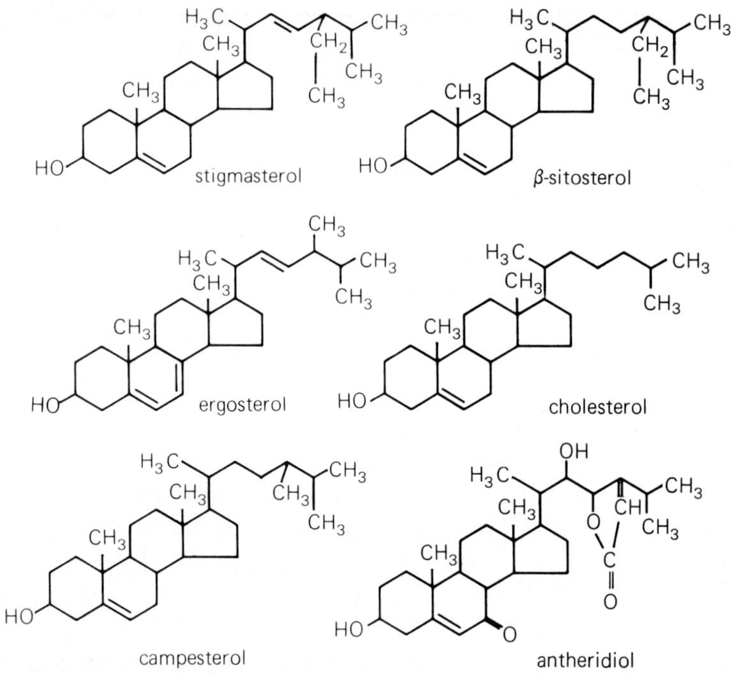

Figure 14-7 Some plant and fungal sterols. Ergosterol probably occurs in trace amounts in some plants, but antheridiol has been found only in certain fungi.

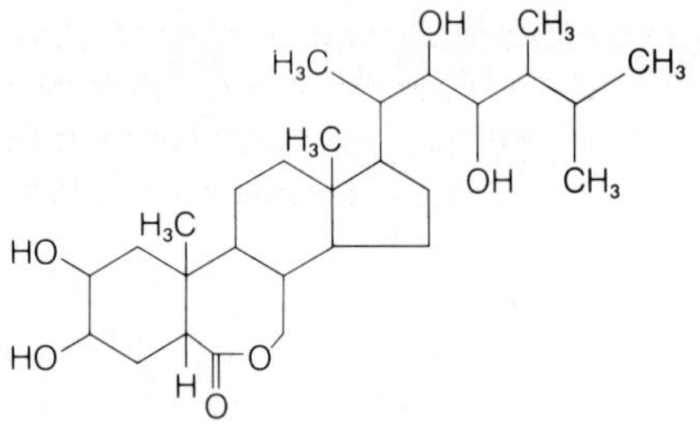

Figure 14-8 Structure of brassinolide.

times as sources of arrow poisons (Robinson, 1980). These are toxic because they inhibit Na-K ATPases of heart muscle membranes. Nevertheless, if heart failure occurs because of hypertension or atherosclerosis, *Digitalis* therapy gives a slower and stronger heart beat. Several million people in the United States with heart disease routinely use digitoxin, digoxin, or some other digilanide from *Digitalis* (foxglove) species (Lewis and Elvin-Lewis, 1977).

Sterols are further important to humans because of their use as precursors for synthesis of certain synthetic animal hormones, including the female ovarian hormone progesterone. (Also see boxed essay.) Several insect-molting hormones (**ecdysones**) exist in plants, and insects rely upon these and other sterols to form the hormones they contain (Heftmann, 1975; Slama, 1979, 1980). Plant sterols are thus vitamins for many insects. Still other sterol derivatives, the **saponins** (sterols attached to a short chain of one or more sugars), cause foaming in the intestinal tract leading to bloat in cattle that eat young alfalfa plants, for example. The cattle are not repelled by these saponins, however; indeed, the young plants are eaten more vigorously than are older plants with much lower saponin contents.

Recently, good evidence was obtained for the existence in plants of certain mammalian steroid **estrogens,** including estrone, estriol, and estradiol (Hewitt et al., 1980). Whether these and related steroids normally function within the plant as sex or growth hormones was long suspected and frequently claimed, but with little evidence. This subject is now being investigated more actively; some positive and many negative results have been obtained (Hewitt et al., 1980; Geuns, 1978, 1982). A recently discovered group of steroid derivatives called **brassins** or **brassinosteroids** have distinct growth-promoting activity in some plants, especially of stems. These compounds were first isolated from bee-collected pollen grains of rape *(Brassica napus),* a mustard (Grove et al., 1979). The structure of one brassin, **brassinolide,** is shown in Fig. 14-8. Interestingly, brassinolide is chemically similar to the ecdysone insect-molting hormones. The importance of brassins to plant physiology and their mechanism of action remains to be demonstrated, but progress is being made (Mandava et al., 1981; Takeno and Pharis, 1982; Cohen and Meudt, 1983).

The Carotenoids Carotenoids are a group of isoprenoids, discussed in Chapter 9 in relation to their functions in photosynthesis. They are yellow, orange, or red pigments that exist in various kinds of colored plastids (**chromoplasts**) in roots, stems, leaves, flowers, and fruits of various plants. Two types of carotenoids exist. The carotenes are pure hydrocarbons, whereas the xanthophylls also contain oxygen, often two or four atoms per molecule. Both types generally contain 40 carbon atoms made from eight isoprene units. Neither type is water soluble, but both dissolve readily in petroleum ether, acetone, and many other organic solvents.

More than 400 different carotenoids have been found in nature, although only a few are found in any given species (Spurgeon and Porter, 1980). β-carotene is the most abundant carotenoid found in higher plants and is the compound imparting the orange color to carrot roots. Lycopene, another carotene, is the compound giving the red color to tomato fruits. Lutein, a xanthophyll, is apparently present in all plants and is the predominant xanthophyll of most leaves. The structures of β-carotene, lutein, and lycopene shown in Fig. 9-4 are typical of several carotenoids.

Two functions of carotenoids in leaves seem well established. As mentioned in Chapters 9 and 11, some of those in chloroplasts participate in photosynthesis, and others prevent photooxidation of chlorophylls. Most yellow flowers contain carotenoids, especially the xanthophyll type, and it is believed that they benefit certain plants by attracting pollinating insects. The abundance of orange β-carotene in carrot roots is probably because carrots have been cultivated and selected by man, and the carotene they contain is both attractive and useful to us because our livers convert it to vitamin A.

Miscellaneous Isoprenoids and Essential Oils

Numerous miscellaneous isoprenoid compounds are present in various amounts among certain members of the plant kingdom (Loomis and Croteau, 1980). In these, isoprene units are condensed into ring compounds commonly containing carbon atom numbers of 10 (the **monoterpenoids**), 15 (the **sesquiterpenoids**), 20 (the **diterpenoids**), or 30 (the **triterpenoids**).

The sesterpenoids with 25 carbons are rarely found. Many of the terpenoids containing 10 or 15 carbons are called **essential oils,** because they are volatile and contribute to the essence (odor) of certain species. These are widely used in perfumes. Some of the volatile hydrocarbons released from plants, including isoprene itself, also contribute to smog and other forms of air pollution. Frits Went (1974) estimated that as much as 1.4 billion tons of volatile plant products, mostly terpenes, are released by plants each year, especially over tropical forests. The Blue Mountains in Australia and the Blue Ridge and Smoky Mountains of Virginia and North Carolina in the United States were probably named because of atmospheric scattering of blue light by tiny particles derived from terpenes.

One of the best known essential oils is turpentine, present in certain specialized cells of members of the genus *Pinus*. The turpentine of some species consists largely of *n*-heptane, although monoterpenoids such as α-**pinene**, β-**pinene**, and **camphene** (Fig. 14-9) are also present. These compounds and the related **myrcene** and **limonene** (Fig. 14-9) represent important terpenoids toxic to tree-killing bark beetles. Such beetles are highly destructive to coniferous forests of North America, causing millions of dollars of damage annually (Mabry and Gill, 1979). Were it not for monoterpenoids present in resin canals, damage would be even greater.

The essential oils sometimes contain hydroxyl groups or are chemically modified in other ways. The structures of two modified monoterpenoids, **menthol** and **menthone,** both components of mint oils, and of **1:8 cineole,** the major constituent of eucalyptus oil, are also shown in Fig. 14-9. Cineole apparently performs an important function in pollination of orchids by male euglossine bees. Such bees are attracted by the fragrance of the orchid flowers, an important constituent of which is 1:8 cineole (Dodson, 1975; Dressler, 1982). This molecule, along with **camphor** (Fig. 14-9), was reported by Muller (see Muller and

α-pinene   β-pinene   menthol   menthone   myrcene   limonene

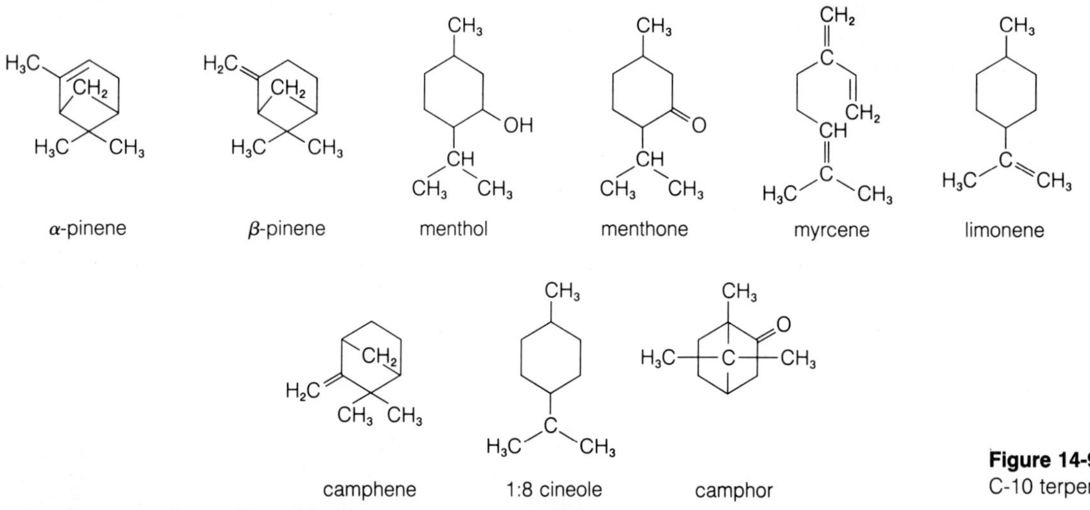

camphene   1:8 cineole   camphor

Figure 14-9 Structures of some C-10 terpenes.

Glaucolide A from various species in the genus *Veronia* repels various lepidopterous insects, white-tail deer, and cottontail rabbits (Mabry and Gill, 1979).

Complex mixtures of terpenes containing 10 to 30 carbon atoms make up the **resins,** which are common in coniferous trees and in several angiosperm trees of the tropics. Resins and related materials are formed in leaves by specialized epithelial cells, which line the resin ducts, and are then secreted inside the ducts where they accumulate. It is speculated that resins protect such trees against many kinds of insects.

Rubber   Rubber is also an isoprenoid compound, the largest of all. It contains some 3000 to 6000 isoprene units linked together in very long, unbranched chains. Most natural rubber is commercially obtained from the latex (milky protoplasm) of the tropical plant *Hevea brasiliensis,* a member of the Euphorbiaceae family. About one-third of this latex is pure rubber. It has been reported, however, that over 2000 plant species form rubber in varying amounts. Various *Taraxacum* (dandelion) species are among the most well-known North American species possessing this ability. However, guayule (*Parthenium argentatum*), a plant common to the southwestern United States, also produces rubber, and it was selected in 1978 by the U. S. Congress for development as a natural rubber crop (Buchanan et al., 1980).

Figure 14-10 Allelopathy in a chapparal-grassland region of coastal Southern California. *Salvia leucophylla* shrubs 1 to 2 meters tall inhibit growth of annual grasses and other species around them, forming a ring nearly devoid of plants.* (From C. H. Muller, 1965. Used by permission.)

Chou, 1972) to be released from *Salvia leucophylla,* a coastal sage, into the California grassland soil in which it grows. Both compounds inhibit root growth of various annual grassland herbs around the Salvia shrubs, so a ring forms around the shrubs that is devoid of the herbs (Fig. 14-10).*

A more complex terpenoid derivative called **glaucolide A** made from three isoprene units is representative of so-called bitter principles largely restricted to the Asteraceae (formerly, Compositae) family. Bitter principles apparently repel, largely by their taste, numerous chewing insects and mammals.

14.4 Phenolic Compounds and Their Relatives

Flowering plants, ferns, mosses, liverworts, and many microorganisms contain various kinds and amounts of **phenolic compounds.** With certain important exceptions, the functions of most phenolics are obscure. Many presently appear to be simply by-

---

*The authors recently learned that Bartholomew (Science 170:1210–1212) showed in 1970 that feeding habits of mice and rabbits, which probably hide in the chaparral by day, can better account for the bare zone.

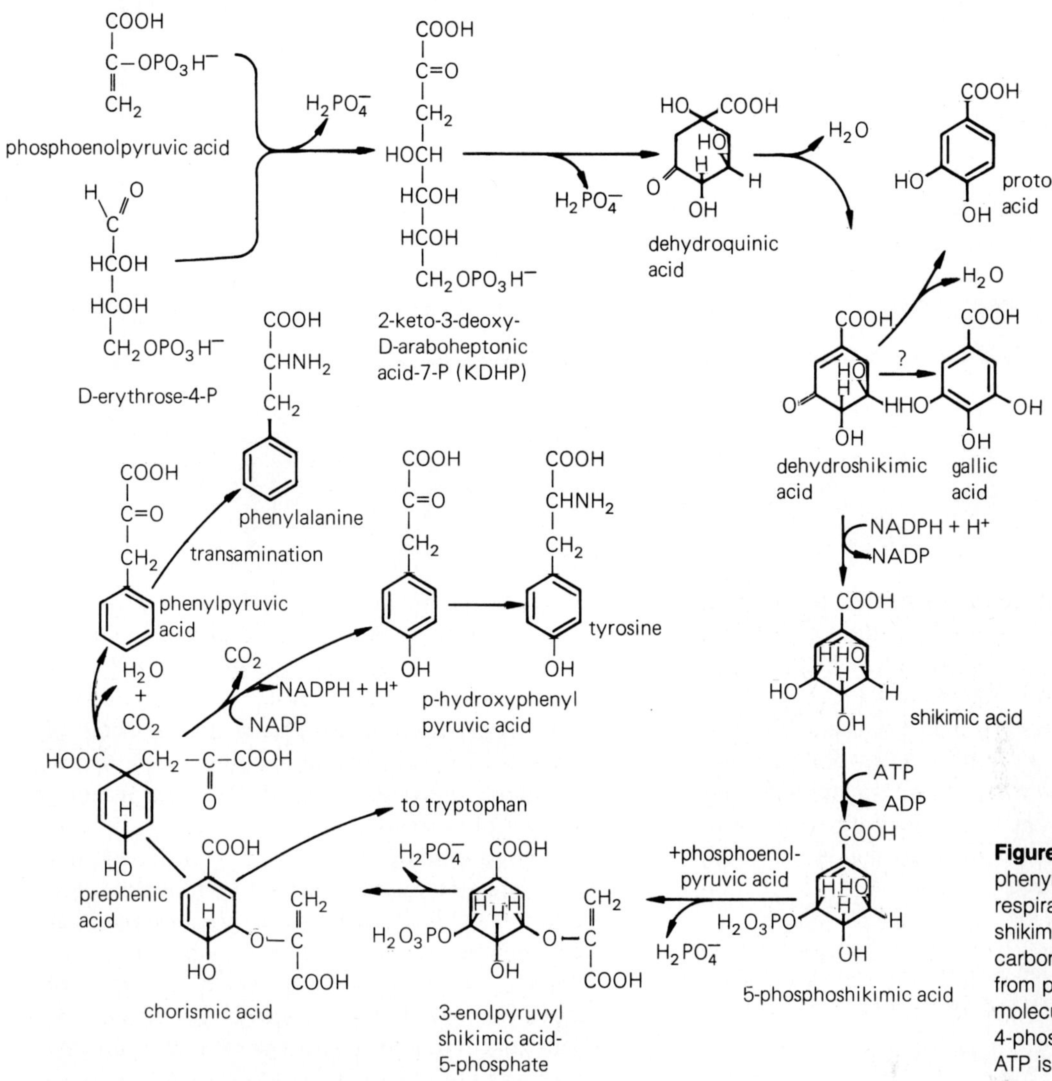

Figure 14-11 Biosynthesis of phenylalanine and tyrosine from respiratory intermediates in the shikimic acid pathway. All of the carbon atoms seem to arise from phosphoenolpyruvate (two molecules) and from erythrose-4-phosphate (one molecule). ATP is also required.

products of metabolism, but this probably reflects in part our ignorance of allelochemical interactions.

All phenolic compounds have an aromatic ring that contains various attached substituent groups, such as hydroxyl, carboxyl, and methoxyl (-O-CH₃) groups, and often other nonaromatic ring structures. The phenolics differ from the lipids in being more soluble in water and less soluble in nonpolar organic solvents. Some, however, are rather soluble in ether, especially when the pH is low enough to prevent ionization of any carboxyl and hydroxyl groups present. These properties greatly aid separation of phenolics from one another and from other compounds.

The Aromatic Amino Acids Phenylalanine, tyrosine, and tryptophan are aromatic amino acids that are formed by a route common to many of the phenolic compounds. Two small phosphorylated compounds are precursors of these amino acids and of many other phenolic compounds. These two are PEP, from the glycolytic pathway of respiration

(Chapter 12), and erythrose-4-phosphate from the pentose phosphate respiratory pathway and from the Calvin cycle (Chapter 10). These two molecules combine, producing a 7-carbon compound that forms a ring structure, which is then converted by several reactions into a rather stable compound called **shikimic acid.** These steps, outlined in Fig. 14-11, make up what has come to be known as the **shikimic acid pathway.** The shikimic acid pathway also exists in fungi and bacteria but not in animals; we require phenylalanine, tyrosine, and tryptophan in our diets because we lack this pathway.

Miscellaneous Simple Phenols and Related Compounds Many other phenolics also arise from the shikimic acid pathway and subsequent reactions. Among these are the acids **cinnamic, p-coumaric, caffeic, ferulic, chlorogenic** (Fig. 14-12), and **protocatechuic** and **gallic** (Fig. 14-11, upper right). The first four are derived entirely from phenylalanine and tyrosine. They are important not because they are

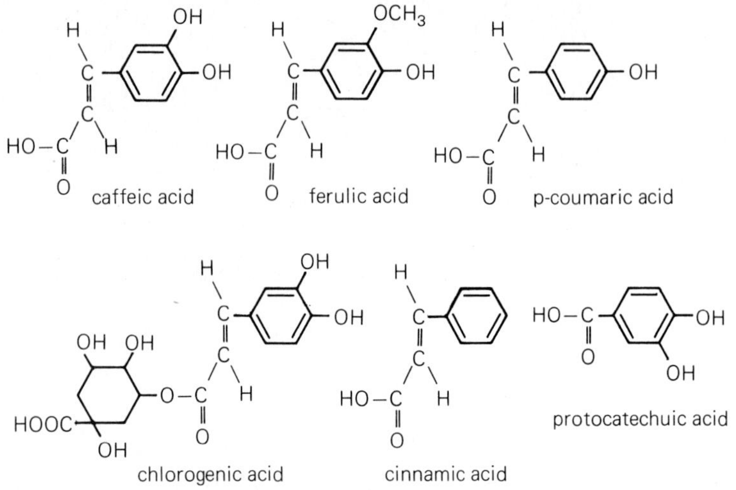

Figure 14-12 Structures of phenolic acids often found in plants. All are shown in the *trans* form. Chlorogenic acid is an ester formed from caffeic and quinic acids.

abundant in uncombined (free) form, but because they are converted into several derivatives besides proteins. These derivatives include phytoalexins, coumarins, lignin, and various flavonoids such as the anthocyanins, all of which will be described shortly.

An important reaction in the formation of these derivatives is the conversion of phenylalanine to cinnamic acid (R14-5). This is a deamination in which ammonia is split out of phenylalanine; it is catalyzed by **phenylalanine ammonia lyase.**

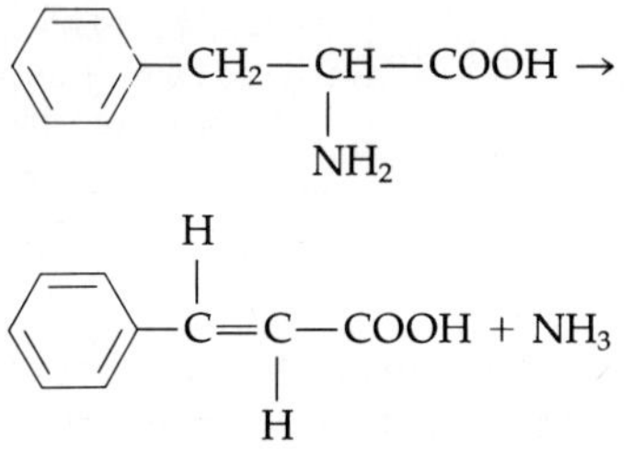

(R14-5)

An analogous reaction with tyrosine can occur in some species, in which **tyrosine ammonia lyase** converts tyrosine to ammonia and p-coumaric acid. This acid can also be formed readily in a wide variety of species by addition of one atom of oxygen from O_2 and an H atom from NADPH directly to the *para* position of cinnamic acid. Subsequent addition of another hydroxyl group adjacent to the OH group of p-coumarate by a similar reaction forms caffeic acid. Addition of a methyl group from S-adenosyl methionine to an OH group of caffeic acid yields ferulic acid. Caffeic acid forms an ester with an alcohol group in still another acid formed in the shikimic acid pathway, quinic acid, thus producing chlorogenic acid.

Protocatechuic and chlorogenic acids probably have special functions in disease resistance of certain plants. Protocatechuic acid is one of the compounds that prevent smudge in certain colored varieties of onions, a disease caused by the fungus *Colletotrichum circinans*. This acid occurs in the scales of the neck of colored onions that are resistant to the pathogen, but it is absent from susceptible white varieties. When extracted from colored onions, it prevents spore germination and growth of the smudge fungus and growth of other fungi.

High amounts of chlorogenic acid might similarly prevent certain diseases in resistant varieties, but the evidence for this is weak (Friend, 1981). Chlorogenic acid is widely distributed in various parts of many plants and usually occurs in easily detectable quantities. In coffee beans the chlorogenic acid concentration is particularly high, and the soluble content of dry coffee reportedly can reach 13 percent by weight (Vickery and Vickery, 1981). It is formed in relatively high amounts in many potato tubers; its oxidation followed by a free-radical polymerization causes formation of large, uncharacterized quinones responsible for the darkening of freshly cut tubers well-known to cooks. **Polyphenol oxidase** enzymes catalyze this reaction, using O_2 as the electron acceptor. It is thought by some that chlorogenic acid and certain other related compounds can be readily formed and oxidized into potent fungistatic quinones by certain disease-resistant varieties but less readily by susceptible ones. In this way, the infection might be well-localized in the resistant plants. Ferulic acid and its derivatives certainly play a role in plant protection, because they form part of the phenolic fraction of suberin.

Gallic acid is important because of its conversion to **gallotannins,** which are heterogenous polymers containing numerous gallic acid molecules connected in various ways to one another and to sugars that are also present. Many gallotannins greatly inhibit plant growth, and the tolerance of plants that contain them probably involves transferring them to vacuoles where they cannot denature cytoplasmic enzymes. Gallotannins and especially other tannins are used commercially to tan leather, because they cross link proteins, denaturing them and preventing their digestion by bacteria. Gallotannins act as allelopathic agents, inhibiting growth of other species around those plants that form and release them (Rice, 1979). Other tannins are even more abundant and widespread in plants than are gallotannins, and their major function seems to be protection against attack by bacteria and fungi (Swain, 1979; Haslam, 1981). Nevertheless, tannins almost surely also act as feeding deterrents against various herbivores, partly because of their astringency (ability to pucker the mouth).

A group of compounds closely related to the phenolic acids and also derived from the shikimic acid pathway are the **coumarins.** More than 500 coumarins exist in nature, although only a few are usually found in any particular plant family. The structures of two coumarins, **scopoletin** and **coumarin** itself, are given in Fig. 14-13. They are formed via the shikimic acid pathway from phenylalanine and cinnamic acid (Brown, 1981).

Coumarin is a volatile compound that is formed mainly from a nonvolatile glucose derivative upon senescence or injury. This is especially significant in alfalfa and sweet clover, where coumarin causes the characteristic odor of recently mown hay. Certain sweet clover strains have been developed that contain low amounts of coumarin and others that contain it in a bound form. These are of economic importance, because free coumarin can be converted to a toxic product, **dicumarol,** if the clover becomes spoiled during storage. Dicumarol is a hemorrhagic and anticoagulant agent responsible for sweet-clover disease (a bleeding disease) in ruminant animals that are fed plants that contain it.

Scopoletin is a toxic coumarin widespread in plants and often found in seed coats. It is one of several compounds suspected of preventing germination of certain seeds, causing a dormancy that exists until the chemical is leached out (for example, by a rainstorm heavy enough to provide sufficient moisture for seedling establishment). It might thus function as a natural inhibitor of seed germination. Numerous other physiological effects of coumarins are known (Brown, 1981), but clear functions for these compounds generally remain to be found.

Also given in Fig. 14-13 are the structures of two coumarin-like compounds named **preocenes** isolated in 1976 from the plant *Ageratum houstonianum.* They cause premature metamorphosis in several insect species by decreasing the level of insect juvenile hormone, thereby causing formation of sterile adults. Decreased hormone levels also lead to reduced pheromone production by male medflies, so their sexual

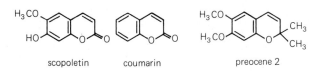

scopoletin   coumarin   preocene 2

Figure 14-13 Structures of two coumarins and of preocene 2, a plant compound that acts as an insect antijuvenile hormone substance. Preocene 1, another antijuvenile hormone substance, also occurs in plants; it lacks the upper methoxyl group on the benzene ring (see W. S. Bowers et al., 1976, Science 193:542-547).

attractance to females is decreased (Chang and Hsu, 1981). Such compounds appear promising as insecticides that have no influence on species other than the target ones.

14.5 Phytoalexins

Since about 1960, various other antifungal compounds that are synthesized by plants when they are infected by certain fungi have been discovered (Deverall, 1977; Stoessl, 1980). These compounds were at first hypothesized to act in a way comparable to the antibodies of animal cells, yet they proved to have little specificity against any given fungus. They are collectively referred to as **phytoalexins** (from the Greek, *phyton,* plant, and *alexin,* a warding-off substance). In general, phytoalexins are much more toxic to fungi than to bacteria, although exceptions might appear. Compounds that act as phytoalexins include **pisatin,** in pea pods, and **phaseolin,** in bean pods. Others that apparently are produced when the plant is invaded and that thus seem to be phytoalexins include **orchinol,** from orchid tubers, **trifolirhizin,** from red clover root, and an **isocoumarin,** from carrot roots. Most phytoalexins are phenolic and products of the shikimic acid pathway, although others are isoprenoid compounds. It appears that nonpathogenic fungi often induce such high, toxic levels of phytoalexins in the host that their establishment is prevented, while pathogenic fungi are successful parasites because they induce only nontoxic phytoalexin levels.

Surprisingly, several kinds of compounds (**elicitors**), even viruses, can induce phytoalexin production, and so can certain kinds of physical injury. A confusing thing about induction by various elicitors is that phytoalexins produced from the shikimic acid pathway as well as isoprenoids from the mevalonic acid pathway can be formed after adding a single elicitor. It is difficult to understand how such different metabolic pathways could be activated by a single

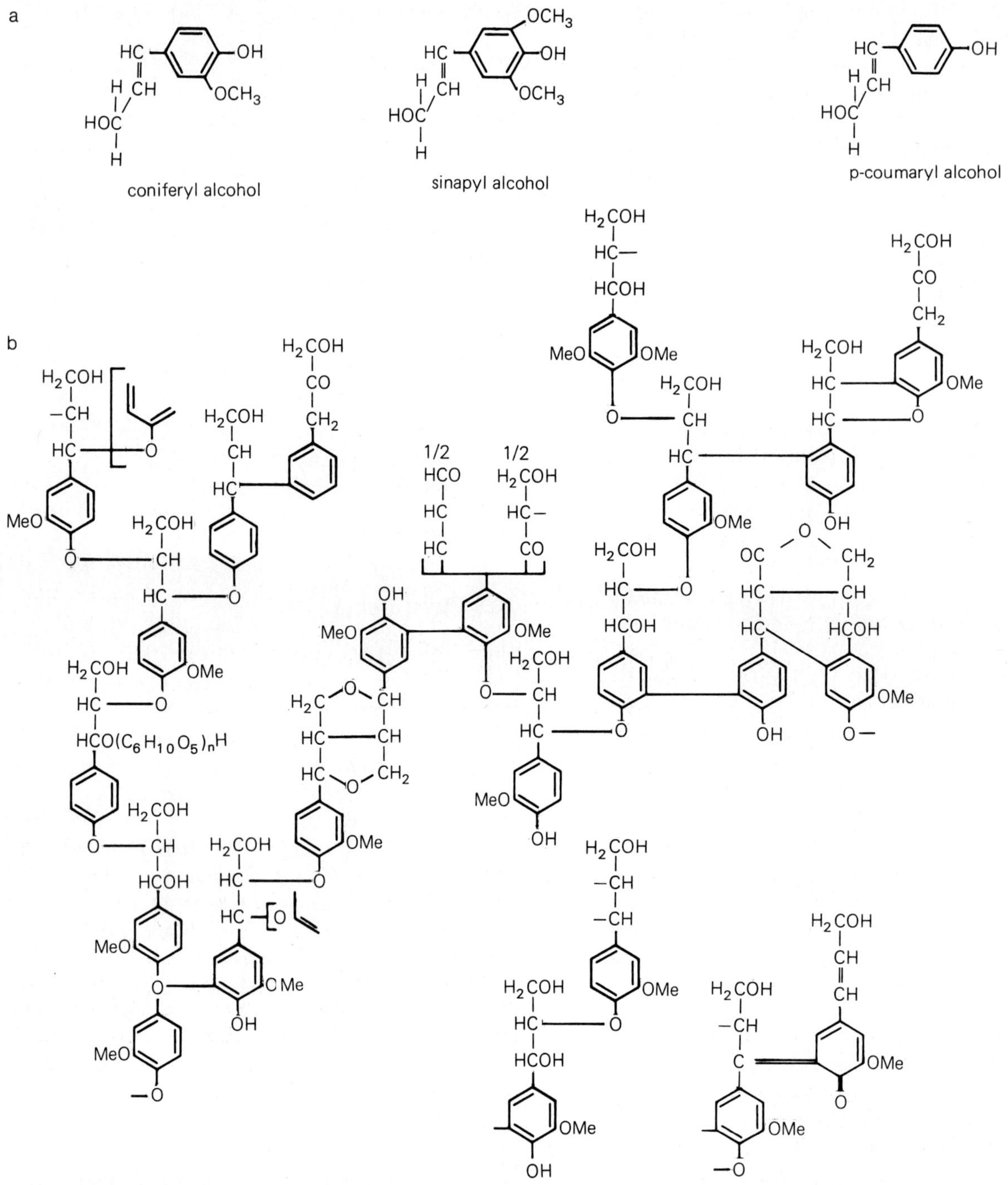

Figure 14-14 (a) Common phenolic subunits found in lignins. (b) Constitutional model of spruce lignin, showing possible mechanisms by which alcohol subunits may be connected. The ratio of alcohols in this lignin is about 14 coumaryl:80 coniferyl:6 sinapyl. Beechwood, a typical hardwood lignin, may have ratios about 5:49:46. Methyl groups are represented as Me. (From K. Freudenberg, 1965, Science 148:595. Copyright 1965 by the American Association for the Advancement of Science.)

elicitor and how different elicitors could activate even a single pathway. Apparently, such exogenous elicitors cause the plant to produce one or a few similar endogenous elicitors that then cause phytoalexin production. The best candidate for such an endogenous elicitor appears to be small polysaccharides formed by breakdown of the plant cell wall (Darvill and Albersheim, 1984). The overall importance of phytoalexins in disease resistance is still controversial, but in a few cases there seems little doubt that resistance is provided.

14.6 Lignin

Lignin is a strengthening material that occurs with cellulose and other polysaccharides in the cell walls of all higher plants. It occurs in largest amounts in wood where it accumulates in the middle lamella, primary walls, and secondary walls of the xylem elements. It usually occurs between the cellulose microfibrils, where it serves to resist compression forces. Resistance to tension (stretching) is primarily a function of the cellulose. The formation of lignin is considered by evolutionists to have been crucial in the adaptation of plants to a terrestrial environment, because only with lignin could rigid cell walls of xylem be built to conduct sap (water and mineral salts) under tension over long distances. Lignin is considered to be the second most abundant organic compound on earth, only cellulose being more abundant. It comprises 15 to 25 percent of the dry weight of many woody species (Gould, 1983). Besides the strengthening function of lignin, it also gives protection against attack by pathogens and consumption by herbivores, both insect and mammalian (Swain, 1979).

Lignin is difficult to study, because it is not readily soluble in most solvents. This insolubility occurs primarily because it has a high molecular weight (probably more than 10,000) and because in the native state it is chemically united to cellulose and other polysaccharides at various points by ether linkages to the hydroxyl groups of polysaccharides. Much of what we know about lignin structure has been determined by analyzing several large molecules that are intermediates in its synthesis. This is contrasted with our knowledge of polysaccharides, proteins, and nucleic acids, the structures of which were largely determined by analyzing degradation products. In general, lignins contain three aromatic alcohols, **coniferyl alcohol,** which predominates in softwoods of conifers, **sinapyl alcohol,** and **p-coumaryl alcohol.** Lignins from hardwood trees and from herbaceous dicots and grasses contain less coniferyl alcohol and more of the others. The structures of these alcohols are given in Fig. 14-14, along with some proposed

models indicating the ways they might be connected in lignin.

The aromatic alcohols in lignin all arise from the shikimic acid pathway. Phenylalanine is converted to aromatic acids such as coumaric and ferulic; then these are converted to CoA esters (Grisebach, 1981). The esters are reduced to the aromatic alcohols by NADPH, and these alcohols are then polymerized into lignin. An iron-containing enzyme called **peroxidase** catalyzes two separate reactions that lead to polymerization (Mader and Amberg-Fisher, 1982). Peroxidase exists in several isozyme forms, a few of which exist in the cell walls. These isozymes apparently function first by forming H_2O_2 from NADH and O_2. Next they remove an H atom from each of two aromatic alcohols and combine the two H atoms with one H_2O_2 to release two H_2O molecules as by-products. The remaining part of each aromatic alcohol is now a free radical, and several kinds of electronic shifts allow migration of the unpaired electron to other parts of the molecule. Many such free radicals combine spontaneously in various ways to form bonds between the alcohols such as those proposed in Fig. 14-14, so lignins presumably always have variable structures.

14.7 Flavonoids

Flavonoids are 15-carbon compounds generally distributed throughout the plant kingdom (Harborne, 1976; Hahlbrock, 1981). More than 2000 from plants have been identified. The basic flavonoid skeleton, shown below, is usually modified in such a way that even more double bonds are present, causing the compounds to absorb visible light and thus giving them color. The two carbon rings at the left and right ends of the molecule are designated the A and B rings, respectively.

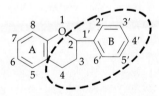

The dotted lines around the B ring and the three carbons of the central ring indicate the part of flavonoids that is derived from the shikimic acid pathway. This part may be compared with cinnamic acid (Fig. 14-12), which is a precursor to it. Ring A and the oxygen of the central ring are derived entirely from acetate units provided by acetyl CoA. Hydroxyl groups are nearly always present in the flavonoids, especially attached to ring B in the 3' and 4' positions

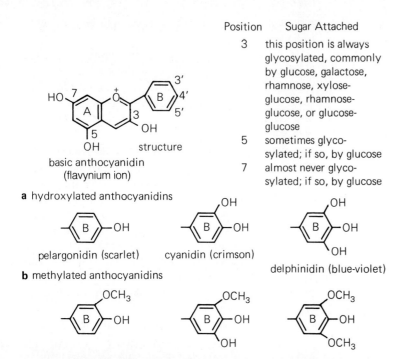

Position	Sugar Attached
3	this position is always glycosylated, commonly by glucose, galactose, rhamnose, xylose-glucose, rhamnose-glucose, or glucose-glucose
5	sometimes glycosylated; if so, by glucose
7	almost never glycosylated; if so, by glucose

basic anthocyanidin
(flavynium ion)

a hydroxylated anthocyanidins

pelargonidin (scarlet)   cyanidin (crimson)

delphinidin (blue-violet)

b methylated anthocyanidins

peonidin (rosy red)   petunidin (purple)   malvidin (mauve)

Figure 14-15 The basic anthocyanidin ring, showing variations of B ring by hydroxylation and methylation to produce various anthocyanins. Anthocyanins are produced by attachment of sugars (glycosylation) to the 3-hydroxyl position of the anthocyanidin, and sometimes also to the 5 or 7 position.

(compare p-coumaric and caffeic acids of Fig. 14-12), or to the 5 and 7 positions of ring A, or to the 3 position of the central ring. These hydroxyl groups serve as points of attachment for various sugars that increase the water solubility of flavonoids. Many flavonoids accumulate in the central vacuole, although their site of synthesis in the cell is uncertain (McClure, 1979; Hrazdina et al., 1980).

Three groups of flavonoids are of particular interest in plant physiology. These are the **anthocyanins,** the **flavonols,** and the **flavones.** The anthocyanins (from the Greek *anthos,* a flower, and *kyanos,* dark-blue) are colored pigments that commonly occur in the red, purple, and blue flowers. They are also present in various other plant parts, such as certain fruits, stems, leaves, and even roots. Frequently, flavonoids are confined to the epidermal cells. Most fruits and many flowers owe their colors to anthocyanins, although some, such as tomato fruits and several yellow flowers, are colored by carotenoids. The bright colors of autumn leaves are caused largely by anthocyanin accumulation on bright, cool days, although yellow or orange carotenoids are the predominant pigments in autumn leaves of some species.

Anthocyanins seem generally absent in the liverworts, algae, and other lower plants, although some anthocyanins and other flavonoids occur in certain mosses. They have only rarely been demonstrated in gymnosperms, although gymnosperms contain other kinds of flavonoids. Several different anthocyanins exist in higher plants, and often more than one is present in a particular flower or other organ. They are present as glycosides, usually containing one or two glucose or galactose units attached to the hydroxyl group in the central ring or to that on the 5-position of ring A, as described in Fig. 14-15. When the sugars are removed, the remaining parts of the molecules, which are still colored, are called **anthocyanidins.**

Anthocyanidins are usually named after the particular plant from which they were first obtained. The most common anthocyanidin is *cyanidin,* which was first isolated from the blue cornflower, *Centaurea cyanus.* Another, *pelargonidin,* was named after a bright red geranium of the genus *Pelargonium.* A third, *delphinidin,* obtained its name from the genus *Delphinium* (blue larkspur). These anthocyanidins differ only in the number of hydroxyl groups attached to the B ring of the basic flavonoid structure. Other important anthocyanidins include the reddish *peonidin* (present in peonies), the purple *petunidin* (in petunias), and the mauve-colored (purplish) pigment *malvidin,* first found in a member of the Malvaceae, the mallow family.

The color of anthocyanins depends first on the substituent groups present on the B ring. When methyl groups are present, as in peonidin, they cause a reddening effect. Second, the anthocyanins are often associated with flavones or flavonols, which cause them to become more blue. Third, they associate with each other, especially at high concentrations, and this can cause either a reddening or a blueing effect, depending on the anthocyanin and the pH of the vacuoles in which they accumulate (Hoshino et al., 1981). Most anthocyanins are reddish in acid solution but become purple and blue as the pH is raised. In larkspur flowers the pH values of epidermal cells containing delphinidin increase from 5.5 to 6.6 during aging, and the color changes from reddish purple to purplish blue (Asen et al., 1975). Because of these properties and the common presence of more than one anthocyanin, there is wide variation in the hues of flower colors in higher plants.

Possible functions of anthocyanins have probably been considered ever since their discovery. One of their useful functions in flowers is apparently the attraction of birds and bees that carry pollen from one plant to another, thus facilitating pollination (Harborne, 1976). Charles Darwin long ago suggested that a fruit's beauty serves as a guide to birds and beasts such that the fruit may be eaten and seeds widely disseminated in the manure. Presumably, anthocyanins contribute to this beauty. Anthocyanins might also play a role in disease resistance, although the evidence for this is weak. Their abundance certainly suggests some functions that have favored their evolution.

Anthocyanins and other flavonoids are of particular interest to many plant geneticists, because it is possible to correlate many morphological differences among closely related species in a particular genus, for example, with changes in the type of flavonoids they contain. A knowledge of the flavonoids present in related species of the same genus gives information that can be used by taxonomists to classify and determine the lines of evolution of these plants (Seigler, 1981).

The flavonols and flavones are closely related to the anthocyanins, except that they differ in the central oxygen-containing ring structure, as follows:

flavonols   flavones

Most of the flavones and flavonols are yellowish or ivory-colored pigments, and like the anthocyanins they often contribute to the color of flowers. Even those flavones and flavonols that are not colored absorb ultraviolet wavelengths and therefore affect the spectrum of radiation visible to bees or other insects that are attracted to flowers containing them. These molecules are also widely distributed in leaves. They apparently function there as feeding deterrents and,

since they absorb UV radiation, as a protection against long-wave UV rays.

Light, especially blue wavelengths, promotes formation of flavonoids (Chapter 19 and Wong, 1976), and these flavonoids apparently increase the plant's resistance to long-wave UV radiation. The anthocyanins have been studied more than other flavonoids regarding effects of light upon their biosynthesis. It has probably been known for centuries that the reddest apples are found on the sunny side of the tree. This is because anthocyanins accumulate in these fruits, and this is increased by light. (The pigments that absorb light that causes formation of anthocyanins and other flavonoids are described in Chapter 19.) The nutritional status of a plant also affects its production of anthocyanins. A deficiency of nitrogen, phosphorus, or sulfur leads to accumulations of anthocyanins in certain plants, as mentioned in Chapter 5. Low temperatures also increase anthocyanin formation in some species, as in the coloration of certain autumn leaves.

Certain species, especially members of the Papilionoideae subfamily of legumes, also accumulate one or more **isoflavonoids,** which differ from flavonoids in that ring B is attached to the carbon atom of the central ring adjacent to the point of attachment in flavonoids. The functions of isoflavonoids are mostly unknown, but some act as allelochemics. For example, *rotenone*, an isoflavonoid from the root of derris (*Derris elliptica*), is a widely used insecticide. Furthermore, isoflavonoid structures resemble those of animal estrogens such as estradiol, and they cause infertility in female livestock, especially sheep (Shutt, 1976). Subterranean clover, in particular, accumulates especially high levels of isoflavones. These compounds cause the serious "clover disease" of sheep, first noted in western Australia in the 1960s as a decline in fertility. They are also suspected to be a factor controlling rodent populations in certain regions. Their infertility effects do not seem to deter grazing animals.

14.8 Betalains

The red pigment of beets is a *betacyanin*, one of a group of red and yellow *betalain* pigments that were long thought to be related to the anthocyanins, even though they contain nitrogen. Neither the red betacyanins nor the other kind of betalain pigments, the yellow **betaxanthins,** are at all structurally related to the anthocyanins, and anthocyanins and betalains do not occur together in the same plant. Betalains seem restricted to 10 plant families, all of which are members of the order Caryophyllales that lack anthocyanins. They cause color in both flowers and fruits from yellow and orange to red and violet, but they

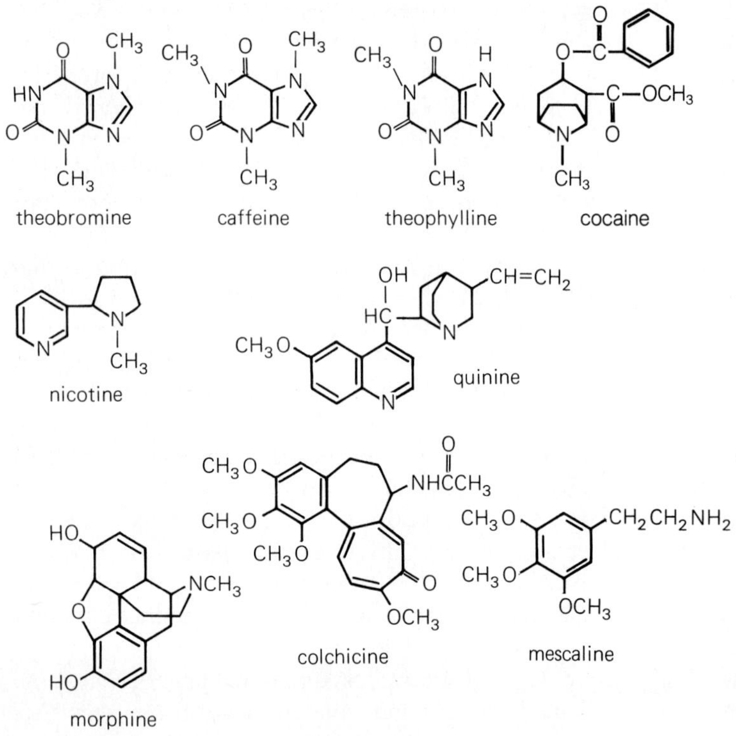

Figure 14-16 Structures of some representative alkaloids.

also give color to vegetative organs in some cases. Like anthocyanins, their synthesis is promoted by light. Betalains also contain a sugar and a remaining colored portion. The most extensively studied member of this group is *betanin* from red beet roots, which can be hydrolyzed into glucose and *betanidin*, a reddish pigment with the following structure:

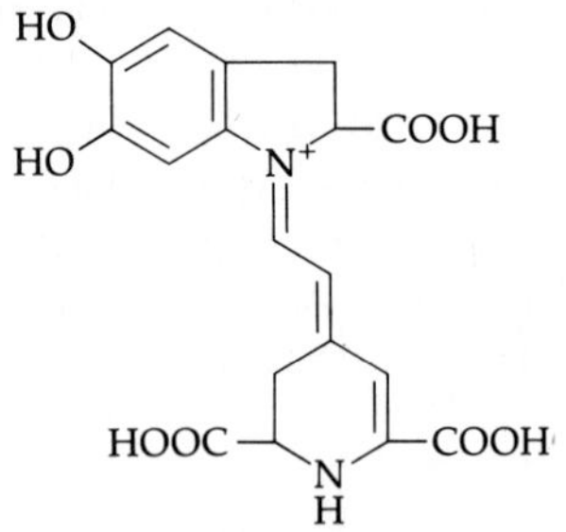

betanidin

Little is yet known of the metabolism or functions of betalains, but a role in pollination comparable to that of anthocyanins in other species seems likely (Piatelli, 1981). Protection against pathogens is another possible function (Mabry, 1980).

14.9 Alkaloids

Many plants contain aromatic nitrogenous compounds called **alkaloids.** Chemically, the alkaloids usually contain nitrogen in a heterocyclic ring of vari-

able structure. This nitrogen frequently acts as a base (accepts hydrogen ions), so many alkaloids are slightly basic, as their name indicates. Most are white crystalline compounds and are only slightly water soluble. They are of special interest because of their dramatic physiological or psychological activity in humans and other animals and the belief that many will prove to have important functions in plants, too.

More than 3000 alkaloids have been found in some 4000 species of plants, most frequently herbaceous dicots, although any given species typically contains only a few such compounds. Relatively few monocots and gymnosperms possess alkaloids. The first alkaloid to be isolated and crystallized was the drug **morphine,** isolated in 1805 from the opium poppy, *Papaver somniferum.* Other well-known alkaloids include *nicotine,* present in cultivated varieties of tobacco; *cocaine,* from leaves of *Erythroxylon coca; quinine,* from cuprea bark; *caffeine,* from coffee beans and tea leaves; *strychnine,* from the seeds of *Strychnos nuxvomica; theobromine,* from cocoa beans; *atropine,* from the poisonous black nightshade (*Atropa belladonna*); *colchicine,* from *Colchicum byzantinum; mescaline,* a hallucinogenic and euphoric drug from flowering heads of the cactus *Lophophora williamsii;* and *lycoctonine,* a toxic alkaloid in *Delphinium barbeyi* (larkspur). The structures of several alkaloids are shown in Fig. 14-16.

Most alkaloids are probably synthesized only in plant shoots, but nicotine is produced only in the roots of tobacco. Many chemical reactions involving formation of some alkaloids are now known, although far more remains to be learned (Waller and Dermer, 1981; Bell and Charlwood, 1980). Synthesis of nicotine has received greatest attention, initially because of its commercial importance, and now because of concern about harmful effects of smoking. *Nicotinic acid (niacin)*, present in NAD and NADP molecules, is a precursor of nicotine. The nitrogen and carbon atoms of nicotinic acid, in turn, arise from a product obtained when aspartic acid and 3-phosphoglyceraldehyde are combined. Other amino acids are precursors of other alkaloids. Very few of the thousands of enzymes that must be necessary for production of the various alkaloids have been demonstrated in the plant kingdom. Furthermore, some experts have speculated that there are many thousands of alkaloids yet to be discovered in plants.

The physiological roles of alkaloids in the plants that form them are unknown, and it has been suggested that they perform no important metabolic function, being merely by-products of other more important pathways. Nevertheless, several examples are known in which they are of ecological importance, providing some survival value to the plant (Robinson, 1979). Plants containing certain alkaloids are avoided by grazing animals and leaf-feeding insects, for example. Others are used by danaid butterflies as substrates for synthesis of their courtship pheromones. Interestingly, larkspur is not avoided by cattle, even when other forage is available, and the lycoctonine in it accounts for more cattle deaths in the United States than any toxin in any other poisonous plant (Keeler, 1975).

three

Plant Development

15

Growth and Development

Think of a favorite plant. Visualize water moving into and across various root cells, up through the xylem into living leaf cells, where hydrogen bonds are broken and the water molecules evaporate into the air spaces of the leaf and finally diffuse out through the open stomates into the atmosphere. Imagine CO_2 molecules diffusing through those stomates into chloroplasts of photosynthetic cells, being fixed into carbohydrates by chemical reactions that are energized by ATP and NADPH arising from light-dependent reactions. Think, too, of photosynthetic products being loaded into phloem sieve tubes and moved to specific sinks. Ions are being selectively and actively absorbed, some of which are assimilated into organic compounds and some of which act as coenzymes. Though no one can explain all that is going on in cells, at least your favorite plant should not seem like a static object. It is a well-organized living thing capable of processing matter and energy in its environment and maintaining a relatively low entropy.

Yet in all this, we have not yet looked at one of the most fascinating things about living organisms: their ability to grow and develop. The continuous synthesis of large, complex molecules from the smaller ions and molecules that are the raw materials for growth leads not only to larger cells but often to more complex ones. Furthermore, not all cells grow and develop in the same way, so a mature plant consists of numerous cell types. The process by which cells become specialized is called **differentiation,** and the process of growth and differentiation of individual cells into tissues, organs, and organisms is often called **development.** Another useful term for this process is **morphogenesis** (Greek *morpho,* form, and *genesis,* origin). These continuing processes take a plant from a fertilized egg to a mighty oak—or a sunflower.

We know that genes govern the synthesis of enzymes, which in turn control the chemistry of cells, and that all this must somehow account for growth and development. We do not know, however, exactly what it is that determines which genes should be transcribed in which cells at a given time. To gain understanding of this is one of the most challenging problems for modern biologists. We do know a great deal about what happens during the growth and development of a plant. We know that chemicals called growth substances or growth hormones often play critical roles in many growth processes. Study of such substances has been an important thrust in plant physiology since early in this century. The two chapters following this introductory one are devoted to a summary of what has been learned, but much of the discussion in subsequent chapters also concerns growth substances. In Chapter 18 we shall examine some specific studies of differentiation, knowledge of which is an important key to the eventual understanding of development.

As the science of plant physiology has evolved, it has become apparent that development can be strongly modified by environment. Light, which plays an important role apart from photosynthesis, is considered in Chapters 19 and 22. Often, light effects are exhibited within the constraints placed by an internal timing mechanism that exists in plants and animals: the biological clock, which is the subject of Chapter 20. Plants also respond strongly to temperature changes; interesting responses, especially to low temperature, are the subjects of Chapter 21.

One of the most striking examples of morphogenesis in plants is the conversion from the vegetative to the reproductive stage. Cells must differentiate in radically new ways, and hormones seem to be involved. Light, particularly the relative lengths of day and night, often modifies or controls flower initiation, and so do temperature changes. We shall examine these effects in Chapter 22. Meanwhile, let us examine some principles of growth and development.

15.1 What Is Meant by Growth?

To most people, growth probably means an increase in size. As most organisms grow from the zygote, they increase not only in volume but in weight, cell number, amount of protoplasm, and complexity. In many studies, we must measure growth. In theory, we could measure any one of the growth features just mentioned, but the two principal methods of measurement determine increases in either volume or weight. Volume (size) increases are often approximated by measuring expansion in only one or two directions, such as length, height, width, diameter, or area. Volume measurements, as by displacement of water, have the advantage that they can be nondestructive; hence, the same plant can be measured at different times. Weight increases are often determined by harvesting the entire plant or the part of interest and weighing it rapidly before too much water evaporates from it. This gives us the **fresh weight,** which is a somewhat variable quantity dependent upon the plant's water status. A leaf, for example, often has a greater fresh weight in the morning than it does at midafternoon simply because of transpiration. Because of problems arising from variable water contents, many people, particularly those interested in crop productivity, prefer to use the increase in **dry weight** of a plant or plant part as a measure of its growth. The dry weight is commonly obtained by drying the freshly harvested plant material for 24 to 48 hours at 70 to 80°C. The leaf that has a lower fresh weight at midafternoon will probably have a greater dry weight, because it photosynthesized and absorbed mineral salts from the soil during the morning. Here dry weight may be a more valid estimate of what we mean by growth than fresh weight. Of course fresh and dry weight measurements are usually destructive and require many samples for statistical significance, although a plant grown hydroponically can be weighed at intervals with little effect on growth.

Sometimes dry weights do not give an adequate indication of growth. For example, when a seed, provided only with water, germinates and develops into a seedling in total darkness, the size and fresh weight increase greatly, but the dry weight decreases because of respiratory loss of CO_2 (Fig. 15-1). Although the total dry weight of such dark-grown seedlings is less than that of the original seed, the growing parts of the stem and root do increase in dry weight as assimilates are translocated from nongrowing parts to the growing regions.

Normally, early stages in seedling development involve the production of new cells by **mitosis** (nuclear division) and subsequent **cytokinesis** (cell division), but normal-appearing seedlings can be produced from the seeds of some species in the absence of mitosis or cell division. When seeds of let-

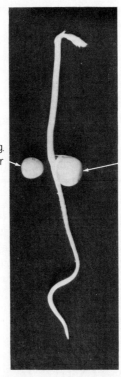

seed
fresh (air dry) weight = 230 mg. dry weight (after 48 hr at 70°C) = 227 mg.

seedling grown 6 days at 20°C fresh weight = 750 mg. dry weight = 205 mg. approximate fresh weight gain = 520 mg. approximate dry weight loss = 22 mg.

Figure 15-1 Changes in fresh and dry weight of a pea seed as it develops into a seedling in darkness. The fresh weight increases greatly because of water uptake, but the dry weight decreases slightly because of respiration. (Photo by C. W. Ross.)

tuce and wheat are irradiated with gamma rays from a cobalt-60 source at levels high enough to stop DNA synthesis, mitosis, and cell division, germination still occurs. Growth continues until seedlings with giant cells are produced. These seedlings, called **gamma plantlets,** can survive up to three weeks but then die, presumably because new cells are eventually necessary. Such gamma plantlets illustrate that even if we could conveniently measure the increase in cell number, this number might be a poor measure of growth. Many other examples of growth without cell division are known, such as growth of certain leaves, stems, and fruits after a certain stage of development. There are also a few examples of cell division without increase in overall size, as in the maturation of the embryo sac. Nevertheless, we use an increase in size as the fundamental criterion of growth, even though there are problems in measuring it.

15.2 Patterns of Growth and Development

Some Features of Plant Growth Growth in plants is not uniformly distributed but is restricted to certain zones containing cells recently produced by cell division in a **meristem.** It is easy to confuse growth (as defined above as an increase in size) with cell division in meristems. Cell division alone does not cause increased size, *but the cellular products of division do grow and cause growth.* Two principal meristematic zones are found near the root and the shoot tips **(apices),** another in the vascular cambium, and another just above the nodes of monocots. The root and shoot apical meristems are formed during embryo development as the seed forms and are called **primary meristems.** The vascular cambium and the meristematic zones of grass leaves are not distinguishable until after germination; they are **secondary meristems.**

Some plant structures are determinate; others are indeterminate. A **determinate** structure grows to a certain size and then stops, eventually undergoing senescence and death. Leaves, flowers, and fruits are excellent examples of determinate structures, and the great majority of animals also grow in a determinate way. On the other hand, the vegetative stem and the root are **indeterminate** structures. They grow by meristems that continually replenish themselves, remaining youthful. A bristlecone pine that has been growing for 4000 years probably could yield a cutting that would form roots at its base, producing another tree that might live for another 4000 years. At the end of that time, another cutting might be taken, and so on, potentially forever. That is, plants can be **cloned** from individual parts. Some fruit trees have been propagated from stem sections for centuries. Although a meristem can be killed, it is potentially immortal since it is indeterminate. A determinate structure is subject to senescence and death, and death is its ultimate fate.

Although there are borderline cases, entire plants are in a sense either determinate or indeterminate. We use different terms, however: **Monocarpic species** (Greek *mono,* single; *carp,* fruit) flower only once and then die; **polycarpic species** (*poly,* many) flower, return to a vegetative mode of growth, and flower at least once more before dying. Most monocarpic species are **annuals** (live only one year), but there are variations on the theme. Many annuals germinate from seed in the spring, grow during the summer and autumn, and die before winter, perpetuating themselves only as seeds. Spring wheats and ryes are commercial annuals that are planted in the spring, but seeds of *winter* wheat or rye germinate in the fall, overwinter as seedlings beneath the snow, and flower the next spring.

Typical **biennials,** such as beet *(Beta vulgaris),* carrot *(Daucus carota),* and henbane *(Hyoscyamus niger)* germinate in the spring and spend the first season as a vegetative rosette of leaves that dies back in late fall. Such a plant overwinters as a root with its shoot reduced to a compressed apical meristem surrounded by some remaining protective dead leaves

(meristem plus leaves called a **perennating bud**). During the second summer, the apical meristem forms stem cells that elongate (**bolt**) into a flowering stalk.

The century plant *(Agave americana)* may exist for a decade or more before flowering once and dying. Though a monocarpic species, it would be called a **perennial,** because it lives for more than two growing seasons. It and many bamboos *(Bambusa* and other genera), which may live more than half a century before flowering once and dying, are excellent examples of the extreme monocarpic growth habit.

Polycarpic plants do not convert all of their vegetative meristems to determinate reproductive ones. All are perennials. Woody perennials (shrubs and trees) utilize only some of their axillary buds for the formation of flowers, keeping the terminal buds vegetative; alternatively, terminal buds may flower while axillary buds remain vegetative. Sometimes a single meristem forms only one flower, as in tulip, while single grass or Asteraceae meristems form an inflorescence or head of flowers (e.g., a sunflower). The bottle brush *(Callistemon* sp.) seems to form a terminal spike of flowers, but the apical meristem remains vegetative and continues to grow the next season, producing leaves and a woody stem. Woody perennials often become reproductive only after they are several years old. Herbaceous perennial dicots such as field bindweed *(Convolvulus arvenis)* or Canada thistle *(Cirsium arvense)* and perennial grasses die back each year except for one or more perennating buds close to the soil. Some form bulbs, corms, tubers, rhizomes, or other underground structures.

The seed contains a miniature plant telescoped into a tiny package: The embryonic root, shoot, and some of the primordial leaves form the **embryo.** The gamma plantlets discussed in Section 15.1 can develop as far as they do because of embryo differentiation during seed formation. Normally, meristematic cells of the root and shoot apices give rise to other cells that divide to form branch roots, still more leaves, axillary buds, and stem and root tissues, including the vascular cambium. Many apical and axillary meristems eventually form flowers. In woody perennials, **lateral meristems (cambium)** produce **secondary xylem** and **phloem** each year, resulting in a growth in diameter of stems and roots.

Steps in Cell Growth and Development Although the patterns we have just discussed seem extremely complex if one thinks of the approximately 285,000 different species of flowering plants, all are accounted for by three rather simple (in appearance, at least) events at the cellular level. The first is **cell division,** in which the volume of one mature cell is divided into two smaller volumes, by no means always

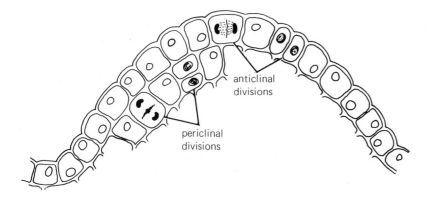

Figure 15-2 The relation of anticlinal and periclinal divisions at the shoot apex.

equal to each other. The second event is **cell enlargement,** in which one or both of the products of cell division increase in volume. The third event is **cellular differentiation,** in which a cell, perhaps having achieved its final volume, becomes specialized in one of the various possible ways. The variety of ways that cells can divide, enlarge, and specialize accounts for the different tissues and organs in an individual plant as well as in the different kinds of plants.

To begin with, cells can divide in different planes. When the new wall between daughter cells is in a plane approximately parallel to the closest surface of the plant, the division is said to be **periclinal.** Alternatively, if the new wall is formed perpendicular to the closest surface, the division is **anticlinal** (Fig. 15-2). Division of a cell (cytokinesis) begins by production of a **cell plate,** which arises by the fusion of hundreds of tiny vesicles, most of which are pinched off from the ends of Golgi vesicles that contain noncellulosic polysaccharides such as pectins (see box essay on carbohydrate chemistry in Chapter 7). As these vesicles fuse, the pectin-rich **middle lamella** is formed. It is bounded by membranes that were formerly part of the vesicles but now become the plasma membranes of the dividing daughter cells (Fig. 15-3). Subsequent formation of the new primary wall of each daughter cell also occurs, in part, by fusion of Golgi vesicles containing other noncellulosic polysaccharides.

What guides the movement of Golgi vesicles to the equator of the cell where the new primary dividing wall is formed during cytokinesis? One theory is that the vesicles migrate along tiny rodlike **microtubules** that extend toward opposite poles of the dividing cell (Section P.3; Hepler and Palevitz, 1974; Hepler, 1976). Figure 15-3 shows numerous microtubules oriented with their long axes perpendicular to the equator. If formation of these microtubules is prevented by such antimitotic drugs as colchicine, Golgi vesicles do not move to the equator of the cell at anaphase. If the drugs are added after anaphase is

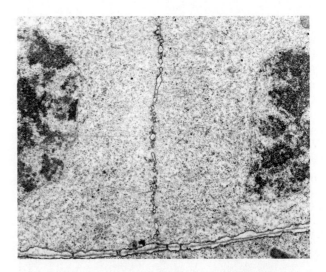

Figure 15-3 Formation of the cell plate during cytokinesis in a cotton root tip. Pectin-rich vesicles pinched off Golgi bodies fuse at the equator to form the new middle lamella and the two plasma membranes in contact with it. Subsequent formation of a primary wall involves noncellulosic polysaccharides secreted from each cell in additional Golgi vesicles into and onto the middle lamella, while cellulose appears to be formed in each plasma membrane without Golgi vesicle involvement. Narrow rodlike microtubules oriented perpendicular to the cell plate might function in guiding Golgi vesicles to this plate. Formation of the nuclear envelope, probably from the endoplasmic reticulum, around each daughter nucleus is nearly complete. Numerous ribosomes (tiny dots) are also visible. (Micrograph courtesy of Dan Hess.)

nearly complete, the cell plate cannot form, cytokinesis cannot occur, and so a binucleate cell is produced.

Not only does the direction of cell division have a lot to do with the formation of various structures, but the direction (or directions) of cell enlargement is also critical. Cell enlargement is largely a matter of absorption of water into the enlarging vacuole as we shall soon see. In such elongate plant organs as stems and roots, the enlargement occurs mostly in one dimension; that is, it is really **elongation.** Of course,

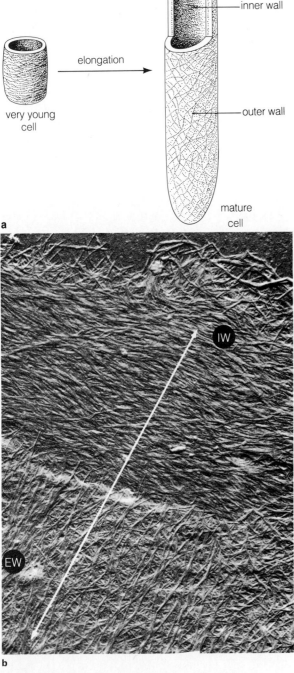

Figure 15-4 (**a**) Changes in orientation of cellulose microfibrils during cell elongation. In the young cell, the microfibrils are oriented almost randomly, but expansion occurs longitudinally because the newly deposited microfibrils on the inner surface of the wall are oriented perpendicular to the cell's long axis. The older microfibrils on the outside of the wall become oriented in the direction of elongation during growth. (**b**) Orientation of cellulose microfibrils in the inner (younger) and outer (older) part of the primary wall. Shown is a leaf hair cell of *Juncus effusus*, a sedge. Note that the microfibrils on the interior of the wall (IW) are perpendicular to the long axis of the cell, those on the exterior of the wall (EW) are parallel to the long axis of the cell, while those in between are intermediate in orientation. The direction of cell elongation is given by the long arrow. (From Jensen and Park, 1967.)

newly formed meristematic cells often enlarge in all three dimensions, but in stems and roots the enlargement soon becomes an elongation.

Primary Wall Changes During Growth What causes cells to elongate mostly in one dimension rather than expanding equally in all directions? In Section P.3, we explained that the primary walls of growing cells consist largely of an amorphous matrix of noncellulosic polysaccharides and some protein, through which run microfibrils of cellulose. It is these cellulose microfibrils, each behaving as a multistranded cable, that minimize extension in the direction of *their* long axes. They do not stretch easily. Wall growth can, however, effectively occur in a direction that allows many of the microfibrils to slide past each other. Growth is favored in a direction at right angles to the microfibril axes. Furthermore, when new cellulose molecules are formed during growth, existent microfibrils are apparently lengthened, allowing some extension parallel to their axes.

Actually, microfibril orientation is not random in some young cells, so growth begins faster along one axis (Fig. 15-4). As growth continues, new microfibrils are deposited into the wall adjacent to the plasma membrane, so the wall retains a near uniform thickness during growth. If the orientation of these new microfibrils is random, growth tends to be equal in all directions (as in fleshy fruits or leaf-spongy-mesophyll cells); but if they are deposited mostly perpendicular to one axis, growth is favored in the direction of that axis (as in elongating roots, stems, and petioles). Figure 15-4 illustrates this. Recent evidence from onion root hairs suggests that the orientation of microtubules and thus possibly microfibrils is much like a spring (a helix or spiral; Lloyd, 1983).

If the pattern of cellulose microfibril deposition is so important in controlling final cell shape, what controls that orientation? No answer is yet available, but there are consistent observations that in growing cells microtubules close to the wall become oriented in the same way as the microfibrils. When colchicine, a drug that prevents microtubule formation, is added to cells, new cellulose microfibrils are randomly rather than transversely oriented. Removal of colchicine allows renewed production of transversely oriented microtubules and microfibrils. If microtubules control microfibril arrangement, we must learn what controls microtubule patterns.

The Cell Cycle Cell biologists, including plant physiologists, are concerned with the series of repeating events called the **cell cycle,** which is reviewed in Fig. 15-5. As you can see, this cycle is concerned largely with the time of DNA replication in relation to nuclear division. There is a period of cell growth before DNA replication (G_1), DNA replication (S),

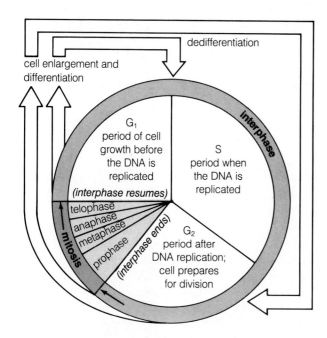

Figure 15-5 A generalized diagram of the cell cycle. There is great variation among different cells in the length of time a cell stays in any phase. In plant cells (less in animal cells), wounding or some other treatment will often cause differentiated cells to again become meristematic, to dedifferentiate. Where they reenter the cell cycle will depend upon where they left it. (Adapted from Starr and Taggart, *Biology*, 2nd ed., 1981, p.131.)

growth after replication (**G₂**), and mitosis. In terms of the events that we have just been discussing, one of the daughter cells produced by mitosis may not continue in the cell cycle but may enlarge and differentiate. If this occurs before DNA replication, the differentiated cell will have the normal diploid number of chromosomes and amount of **chromatin** (genetic material); but in plants it is not unusual for differentiation to occur after DNA replication, so that the differentiated cell has more than the diploid quantity of chromatin. Sometimes chromosomes are further duplicated without cell division, so that the differentiated cell is polyploid. Often these polyploid cells are larger than their diploid counterparts. As the diagram illustrates, differentiated plant cells may sometimes reenter the cell cycle by a process called **dedifferentiation,** after which they again have the ability to divide; that is, they are again meristematic (see Section 18.8).

The Physics of Growth: Water Potentials and Yield Points How do plant cells grow in volume? This growth is primarily caused by an uptake of water, but it also involves synthesis of new cell-wall and membrane materials as well as other substances in the cell.

What causes a cell to absorb water and enlarge? An old hypothesis was that the wall and plasma membrane were extended in small stepwise increments by metabolic activities of the cell; then water entered at each step to fill the void. The opposite and probably correct interpretation is that water (turgor) pressure drives growth by forcing the wall and membranes to expand. The rate of water movement into a cell is governed by two factors: the water potential gradient and the permeability of the membrane to water. Hence, the rate of cell enlargement is also proportional to these factors, to a first approximation.

Equation 2.1 in Section 2.2 shows the basic osmotic relations:

$$\psi = \pi + P$$

Where ψ = water potential, π = osmotic or solute potential, and P = pressure potential or turgor pressure. The difference in water potential ($\Delta\psi$) inside and outside a cell is

$$\Delta\psi = (\pi_e + P_e) - (\pi_i + P_i) \tag{15.1}$$

where e = external and i = internal.

Ignoring outside pressure and including a factor for the permeability of the membrane to water (K), we can write the equation for the rate of cell enlargement as follows:

$$\frac{d \ln L}{d \text{ time}} = K \Delta\psi = K(\pi_e - \pi_i - P_i) \tag{15.2}$$

In equation 15.2, L is cell size; so $(d \ln L)/dt$ is the relative growth rate of the cell ($d \ln L$ = an incremental change in the natural logarithm of cell size; d time = an infinitesimal increment of time).

Equations 15.1 and 15.2 show that the water potential inside a cell could be more negative than water potential outside the cell, making water uptake and growth possible, if *solutes* were to *increase* inside the cell, making the osmotic potential inside more negative or if *pressure* inside the cell were to *decrease*. Measurements have shown that solute concentrations inside certain growing cells remain virtually constant. In those cases, then, the driving force for growth must be a decreasing pressure. This is achieved as the cell wall is "loosened" so that the cellulose microfibrils can slide past each other more easily. As the loosened wall begins to yield (expand), pressure in the cell drops. The wall becomes more **plastic,** stretching irreversibly in response to the incoming water. In plastic stretching (as in bubble gum), the wall does not return to its original dimensions if the pressure that caused the stretching is relieved. In **elastic** stretching (as in a rubber balloon or a stomatal guard cell), the wall returns to its original dimensions when pressure is relieved.

Experiments have shown that growth is proportional to the extent that turgor pressure (P) exceeds a minimum **yield point** or threshold value (Y). This fact can be stated in a second growth equation (Lockhart, 1965):

$$\frac{d \ln L}{d \text{ time}} = m(P - Y) \qquad (15.3)$$

where m represents the proportionality factor known as **wall extensibility.** The yield point (Y) is, of course, expressed in pressure units. The yield point has been difficult to determine experimentally, but it is a valuable concept.

Together, equations 15.2 and 15.3 say that the rate of cell enlargement depends upon five interrelated factors: permeability of walls and membranes to the water that bathes the cells, the difference in osmotic (solute) potential between the inside and outside of the cell, cell turgor pressure, and two wall-yielding properties, extensibility and yield threshold.

In an enlarging root cell or shoot cell, growth can be produced by lowering cell water potential via uptake or biosynthesis of solutes (making π_i of equation 15.2 more negative). Water entering the cell in response to $\Delta\psi$ immediately increases cell turgor pressure (P_i) and will enlarge the cell if the yield threshold (Y of equation 15.3) has been exceeded. Raising (P_i) makes the ψ_i in the cell less negative, and the entrance of water into the cell also makes π_i less negative by diluting the solutes that are present. Both effects lower $\Delta\psi$, which is the driving force for growth. This must be countered if growth is to be maintained, and the logical way to maintain a sufficient $\Delta\psi$ to continue growth is to absorb or otherwise produce solutes in the cell, making π_i more negative. Such solute accumulation often does accompany growth, as is illustrated in Fig. 15-6 with data published in 1941. The sunflower hypocotyl cells in the figure expanded in length 15-fold, yet their solute concentrations remained essentially constant as the increased solutes closely matched cell-size increases. In a more recent study also with sunflower seedlings, the accumulated solutes proved to be mostly glucose, fructose, and K^+ translocated from the cotyledon cells (McNeil, 1976).

What would happen in a tissue growing in water with no access to a solute supply such as mineral salts from a soil solution or sugars derived from photosynthesis? Water uptake would dilute the existing solutes, and π_i would rise toward zero as in the Höfler diagram of Fig. 2-3. Since $\Delta\psi$ remains immeasurably different across the plasma membrane (although ψ_i must be slightly more negative than ψ_e), P_i must decrease; but growth will eventually stop when the yield threshold is reached, unless the threshold decreases. Usually, growth stops in the absence of a solute supply, apparently because the wall either retains its rigidity or becomes even less plastic. Therefore, a plant requires water as the driving force for

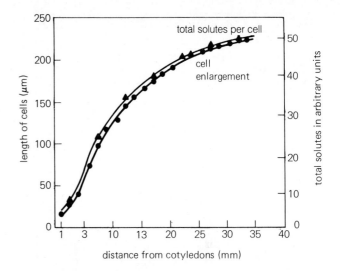

Figure 15-6 Relation between cell size and solute content in epidermal cells of sunflower hypocotyls. The cells are at different stages of development, so the abscissa could logically be labeled *time* or *age* instead of *distance from cotyledons*. Etiolated seedlings 90 h old and 45 to 50 mm tall were used. Lengths and widths of the growing epidermal cells were measured microscopically. Solute concentrations were measured from incipient plasmolysis values and then converted to arbitrary units by multiplying by the cell length. (Cell diameters remained almost constant during growth.) Plants were grown in a peat-sand mixture and were watered only with tap water. (Data from Beck, 1941.)

growth, but continued water uptake almost always requires mineral salt absorption or sugars and other organic solutes provided by translocation or photosynthesis. This fact (along with the essential functions of mineral elements, sugars, and other organic solutes in metabolic processes) is essential to understand how the environment influences growth.

The extreme sensitivity of expansive growth to water stress (Hsiao, 1973) is because the yield threshold (Y) is often very close in value to cell turgor (P_i). When soil dries or solutes are placed in the solution bathing the root, growth stops when P_i equals Y. This is well before P_i reaches zero and before the tissues are wilted. Furthermore, equation 15.3 shows that the growth rate is very sensitive to the extent by which P_i is raised above the turgor threshold (but see the discussion of the mechanics of gravitropic bending in Section 18.4).

Equations 15.2 and 15.3 imply that growth can be modified by changes in permeability, extensibility, or yield threshold. In the alga *Nitella,* the cell apparently modifies its yield threshold to maintain a constant growth rate over a range of π_e values (Green et al., 1971). In higher plants, hormones produce changes in the growth parameters. Auxins, cytokinins, and gibberellins all increase growth of sensitive cells by increasing extensibility and perhaps also by lowering yield thresholds. They do this by somehow loosening the wall, allowing it to be stretched more readily in an irreversible manner.

15.3 Growth Kinetics: Growth through Time

Whole Organs: the S-shaped Growth Curve Many investigators have plotted size or weight of an organism against time to produce a growth curve. Often the growth curve can be fitted with a simple mathematical function such as a straight line or a simple S-shaped curve. The metabolic and physical plant processes that produce the growth curves are too complex to be explained by simple models, but the simple curves are often useful in interpolating or extrapolating from measured data. In addition, the coefficients that must be added so that the equations will fit the curves can be used to categorize the effects of an experimental treatment (such as an irrigation regime or a growth-regulator application) on the growth of observed plants or plant organs.

An idealized **S-shaped (sigmoid) growth curve** exhibited by numerous annual plants and individual parts of both annual and perennial plants is illustrated in Fig. 15-7a. Three primary phases can usually be detected: a logarithmic phase, a linear phase, and a senescence phase (Sinnott, 1960; Richards, 1969). In the **logarithmic phase,** the size increases exponentially with time. This means that the **growth rate** (the increase in size per unit of time) is slow at first (Fig. 15-7b) but continuously increases. The rate is proportional to the size of the organism, meaning that the larger the organism, the faster it grows. A logarithmic growth phase is exhibited by single cells, such as the giant cell of the alga *Nitella*, and by populations of single-celled organisms, such as bacteria or yeasts, in which each product of division is capable of growth and division. Mathematicians have pointed out the analogy between the logarithmic phase and the growth of money invested to draw compound interest. The accumulated interest also draws interest, so the total principal grows exponentially.

In the **linear phase,** increase in size continues at a constant, usually maximum, rate for some time (Fig. 15-7b). We do not understand exactly why the growth rate should be constant in this phase. The **senescence phase** is characterized by a decreasing growth rate (note drop in rate curve in Fig. 15-7b) as the plant reaches maturity and begins to senesce.

Although the curves of Fig. 15-7 are representative of many species, growth curves of other species and organs are often different. In Fig. 15-7 the linear phase is hardly detectable, so the logarithmic and senescence phases are almost continuous. More commonly, the linear phase is extended, as shown for the Swartbekkie pea in Fig. 15-8. The growth rate was constant at just over 2 cm height increase per day for nearly two months. (Senescence phase is not shown, although it occurred later.) The graphs also illustrate results with the Alaska pea, another tall variety, which showed a more sigmoid growth curve

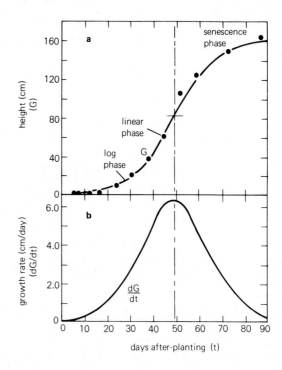

Figure 15-7 A nearly ideal sigmoid growth curve and bell-shaped growth-rate curve. The rate curve in (**b**) is the first derivative (the slope) of the total-growth curve in (**a**). (Drawn from data of W. G. Whaley, 1961.)

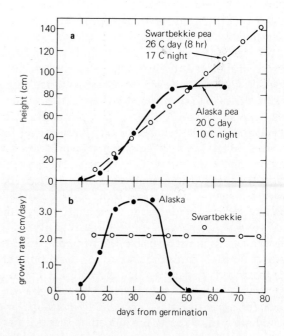

Figure 15-8 (**a**) Growth curves for two tall pea varieties, indicating departure from the typical sigmoid curve, especially in Swartbekkie (note extended linear phase shown by open circles). (**b**) Growth rate curves derived from data in (**a**) as in Fig. 15-7. The bell-shaped curve for Alaska pea differs only in detail from that of Fig. 15-7, but the bell-shape does not even appear for Swartbekkie with its extended constant growth rate. (Data from Went, 1957.)

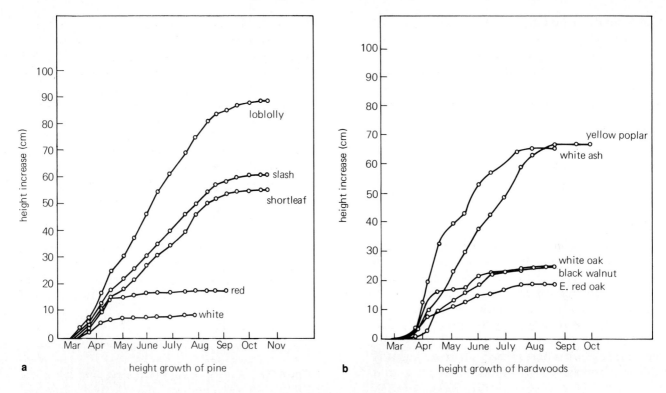

a height growth of pine

b height growth of hardwoods

Figure 15-9 Shoot elongation in pines (**a**) and deciduous trees (**b**) in North Carolina during one growing season (1938). All trees had been planted a few years earlier at the same location, but not all are native to the southeastern United States. Note differences in growth rates and in lengths of the periods of active growth. (Data from Kramer, 1943.)

and a bell-shaped rate curve flattened on top because of the extended linear phase.

Growth of commercial fruits has been studied considerably, no doubt in part because of their economic importance; but there is little information about smaller, less fleshy fruits, even though these predominate (Coombe, 1976). Growth curves of apple, pear, tomato, banana, strawberry, date, cucumber, orange, avocado, melon, and pineapple fruits are sigmoid, while raspberry, grape, blueberry, fig, currant, olive, and all stonefruits (peach, apricot, cherry, and plum) show interesting double-sigmoid growth curves where a first "senescence" phase (flat part of the curve) is followed by another logarithmic phase leading to the second sigmoid part of the curve.

Fewer data are available for perennial species, especially trees, but sigmoid curves would probably be produced, usually with important flat portions caused by winter or dry periods. For specific seasons, data for shoots of trees are readily available, and modified sigmoid curves are indeed observed. Figure 15-9 shows such curves for pines and deciduous hardwood trees. Note that there were important differences in actual height growth and in lengths of the growing period among the various species. Among the hardwoods, all except the poplar essentially stopped elongating before August. Among the

pines, the nonnative (transplanted) red and white species stopped elongating in late spring, while the native species grew taller during a longer time period. Typically, elongation is more rapid under the long days of late spring and early summer (Section 22.2), but there are exceptions.

It is common for trees to cease height growth temporarily in late summer, when temperatures are still warm and days are relatively long (Kramer and Kozlowski, 1979; Zimmermann and Brown, 1971; Perry, 1971). Sometimes growth resumes again before winter dormancy, a deeper dormancy that results in part from the increasing night lengths and decreasing day lengths and in part from the low temperatures of autumn (Chapter 21). Growth in stem diameter (caused by expansion of cells produced by the vascular cambium) continues at a decreasing rate until well after height growth stops. The fact that xylem cells with smaller diameters are produced in summer than in spring is responsible for the contrasting summer and spring wood of the annual rings, from which estimates of age can be made for many tree species. In deciduous trees, photosynthesis continues until the leaves become senescent and yellow; in evergreens, until temperatures become too cold. Because of this, increases in dry weight and radial growth can continue several weeks after stem elongation ceases. Root growth can continue as long as water and nutrients are available and the soil temperature remains high enough; that is, dormancy such as that found in shoots does not occur in the roots so far examined. Because of this, orchard trees often benefit by continued root growth from late summer fertilization and water treatments.

The Flow Analogy of Plant Growth   Because plants grow as meristems produce new cells that then enlarge and differentiate, they leave a record of their growth history—and provide a prophesy of potential future growth. We all know that much of the history of a log can be inferred by examining the growth rings of its cross section. A narrow ring means difficult growing conditions, and a wide ring with large cells means ideal conditions during that year. The same is true at the cellular level at the stem or root tip. We have discussed the stages through which a given cell goes from the time it is produced as a daughter cell by cell division until it differentiates to a specialized cell in a particular plant or organ. Some aspect of this process is occurring at every moment among the cells in the stem or root. Dividing cells are found in the apical meristems, elongating cells a bit farther from the tip, and differentiating cells farther still from the tip. The history of a differentiated cell can be inferred from the younger cells closer to the tip; conversely, the future of a young cell can be inferred by examining the more mature cells farther from the tip. The same is true on a larger scale for the leaves along a stem. The history of leaf production can be inferred by examining the pattern formed by the leaves along the stem, and the future stages of leaf development of a young leaf primordium near the stem tip becomes quite clear from an examination of older leaves farther down the stem.

These features of stems reveal that indeterminate growth in a plant is a *flow process*. An analogy can be drawn to a waterfall. The shape of the waterfall remains constant as long as the flow rate is constant, but at any given moment the molecules of water that combine to produce this form are not the ones that produce the form at any other moment. In a similar way, an upper stem of a Swartbekkie pea plant (Fig. 15-8) appears constant from day to day, but the individual cells making up the tip and youngest leaves are continually changing, flowing, as it were, from the meristematic region of cell division toward the more mature parts of the stem. (Other examples of flowing structures with constant form include the wake of a ship and the flame of a candle.) The realization that plant growth is analogous to fluid dynamics has provided some powerful mathematical tools for the analysis of plant development (Silk and Erickson, 1979; Silk, 1984).

A traditional approach to the study of growth goes back to the extensive studies of Julius von Sachs (whom we met in our discussion of mineral nutrition and other topics), who made marks with India ink at equal intervals along a growing root tip and then examined the distances between the marks at some time (often 24 hours) after marking (reviewed by Erickson and Silk, 1980). Of course, the marks in the elongation zone were farther apart when remeasured, and those in the part of the root where differentiation had occurred were the same distances apart as when they were marked. This approach (which actually goes back even farther to Henri Louis du Hamel du Monceau, who inserted fine silver wires into the roots of walnut seedlings in 1758) is limited, because the rates of elongation of the cells between the marks do not remain constant throughout the interval between marking and measurement some hours later; they are constantly changing, as we have just seen in our discussion of growth as a flow process.

What is needed is to measure the change in length of the cells in very short intervals of time—so short that the growth rates remain constant during those intervals for all practical purposes. The intervals should be on the order of seconds. Such a technique has been developed with **streak photographs,** in which the image of a growing root tip or other organ is focused on a slit in the film plane of a camera, and then the film moves slowly by the slit while the lens remains open. To identify points on the growing organ, one may brush it with a suspension of lamp-black particles. Results of such a technique are shown in Fig. 15-10. The steepness of the streaks (each one representing a particle of carbon black) represents the rate of movement of that point on the root. Since the top of the plant is fastened in place in relation to the camera, the tip of the root is moving fastest and produces the steepest streaks. Points in the differentiated zone are not moving at all, so the streaks appear horizontal.

Such figures can be analyzed in various ways. If we imagine that the tip is standing still, we can plot the velocity of displacement of any point (any streak) as a function of distance from the tip. This velocity can be determined by measuring the distance between the tip and the streak at one time and then measuring the distance at a later time (along a vertical axis to the right of the first measurement). Velocities of displacement can also be determined by measuring the slopes of the streaks at a single point in time and performing appropriate mathematical adjustments. Results of such measurements give data such as those shown in Fig. 15-11a. The familiar S-shaped curve is again evident.

Again, it is possible to measure the slope of that curve to obtain the rate of growth of any individual cell at any given distance from the tip (Fig. 15-11b). This is the familiar bell-shaped curve discussed in

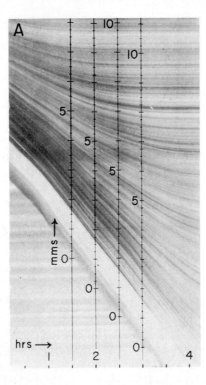

Figure 15-10 Growth of a maize root as recorded in a streak photograph. The root was brushed with a suspension of lampblack and put in a moist chamber mounted in front of a camera. The camera lens was left open, and the film was moved at a constant, slow rate past a narrow, vertical slit mounted directly in front of the film. The black spots on the growing root appear as streaks in the resulting photograph. Near the top of the photograph, the root has stopped growing, so the streaks are horizontal. Since the root is attached above the top of the photograph, the tip is moving most rapidly; streaks made by the tip are steepest. Note curving streaks that represent portions of the root that were growing rapidly when the photograph was started (top left of photo) and then slowed in their growth as their cells began to differentiate (lower right of photograph). The superimposed scales make it possible to measure the distance of any streak from the tip at different time intervals. Such data were used to produce the curves of Fig. 15-11. (Streak photo courtesy of Ralph O. Erickson; from Erickson and Goddard, 1951. See also Erickson and Sax, 1956; Erickson, 1976; and Erickson and Silk, 1980.)

relation to Fig. 15-7. Both curves of Fig. 15-11 show that cells 10 mm from the tip of the maize root are still growing, and Fig. 15-11b shows that the maximum rate of growth occurs at about 4 mm from the tip. Note that if growth of the maize root is a steady flow process, growth rates measured for individual cells along the root provide information about the growth rate of any single cell as a function of time. Hence, the abscissa in Fig. 15-11b could be changed from distance to time if growth is truly a steady flow process.

One of the most fascinating results of the study of stem growth as a flow process comes from an examination of growth of the *epicotyl hook* observed on many dicot seedlings (see discussion of the etiolation syndrome in the introduction to Chapter 19).

The upper part of the stem of many dicot seedlings is bent into a sharp hook (Fig. 15-12). Apparently, the hook protects the seedling as it thrusts up through the soil. Light causes the hook to straighten, but the growth of the hook can be observed over intervals of many hours under dim safe lights of suitable wavelengths. Photographs such as those in Fig. 15-12 allow an analysis of growth, and no lamp black is needed because the surface hairs can be used as markers. Is the hook simply raised in elevation by elongation of cells below, or are new cells continually being produced in the meristem at the tip, enlarging and elongating and eventually differentiating below the hook? Examination of the surface hairs proves that the latter is the case. Cells flow through the hook just as water flows over a waterfall. Since cell divisions stop before the cells reach the hook, the hook must form as cells on the outside elongate more than those on the inside; the stem becomes straight below the hook (or the hook straightens) as cells elongate more rapidly on the inside than on the outside. That is, the form of the hook is determined by internal, closely coordinated factors that control the rate of cell elongation on opposite sides of the stem! The morphogenetic program in control of this phenomenon is completely unknown and in that respect quite representative of our understanding of the ultimate control factors of morphogenesis in general. Whatever these control factors are, they are themselves switched by light of suitable irradiance and wavelength, so that the growth rates that control the formation of the hook become altered in such a way that the hook straightens.

The analogy with the waterfall breaks down when we discuss the control mechanisms. Clearly, the shape of the waterfall is determined by both gravity and the flow channel, the rocks or cliffs that direct where the water will flow. The flow of cells through an epicotyl hook, on the other hand, is somehow internally determined by whatever morphogenetic program is in charge of the organism's growth.

15.4 Plant Organs: How They Grow

Having examined several of the general principles of plant growth, it is now time to consider a few special features of the various plant organs.

Roots

Organization of the young root In the great majority of species, seed germination begins with **radicle** (embryonic root) rather than **epicotyl** (shoot) protrusion through the seed coat (Bewley and Black, 1978). In some species, including sugar pine (*Pinus lambertiana*,) cytokinesis occurs in the radicle before germination is complete. In others (maize, barley, broad

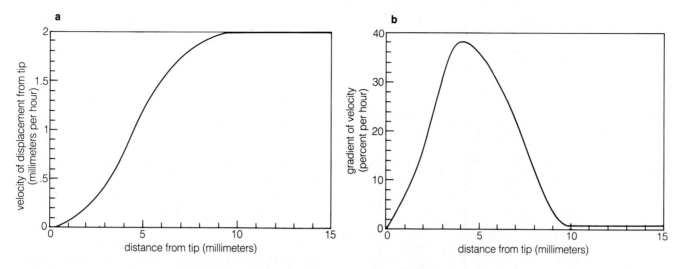

Figure 15-11 Distribution of growth in a maize root. (**a**) If the velocity of displacement from the tip of any point along the root is plotted as a function of its distance from the tip, a sigmoid curve similar to that of Fig. 15-7a is obtained. (**b**) If the growth rate of any point along the root is plotted as a function of distance from the tip, a bell-shaped curve similar to that of Fig. 15-7b is obtained. (Data from Erickson and Sax, 1956.)

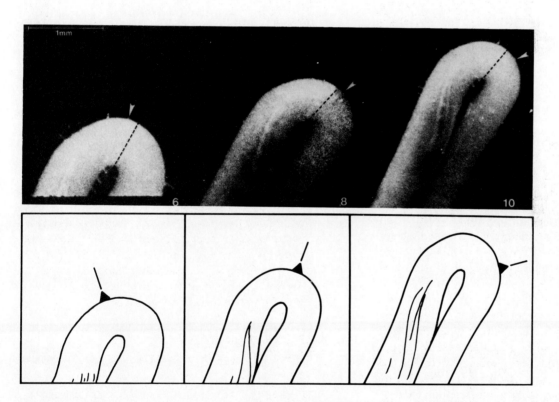

Figure 15-12 Growth of the epicotyl hook of a lettuce seedling photographed at 6, 8, and 10 hours after an initial observation. Bottom is a schematic drawing of the lettuce seedling during growth. The arrow (top) or black triangle (bottom) shows how a point on the hook grows through the hook with time, as discussed in the text. (Figure courtesy of Wendy Kuhn Silk; it is a composite based upon Silk and Erickson, 1978; and Silk, 1980 and 1984.)

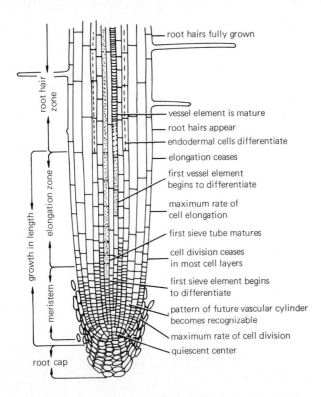

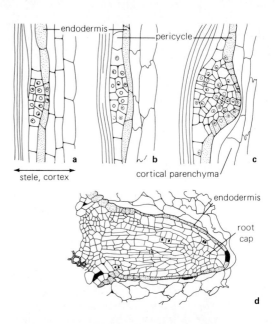

Figure 15-13 Simplified diagram of the growing zone of a root, in longitudinal section. The number of cells in a living root is normally much greater than is shown in this diagram (compare with Fig. 4-9, also somewhat simplified). (From *The Living Plant,* Second Edition, by Peter Martin Ray. Copyright © 1972, by Holt, Rinehart and Winston, Inc. Reprinted by permission of Holt, Rinehart, and Winston.)

bean, lima bean, and lettuce), few if any mitoses occur before radicle protrusion; so elongation is caused by growth of the cells that formed when the embryo was developing on the mother plant. Continued growth of the primary root in the seedling and of branch roots derived from it requires activity of apical meristems. A typical root tip is illustrated in Fig. 15-13.

The oldest cells of the root cap are in the **distal** part (that part farthest from the point of attachment to the rest of the plant, i.e., the tip). In a more **proximal** position (closer to the meristem) are the young cells being formed from the apical meristem. The root cap protects the meristem as it is pushed through the soil and acts as the site of gravity perception for roots (see gravitropism in Section 18.4). Furthermore, it secretes a polysaccharide-rich slime or **mucigel** over its outer surface that may serve the purpose of lubricating the root so that it can slide through the soil. This requires the activity of Golgi vesicles, as shown in Fig. P-12. As the root grows, the mucigel continues to cover its surface as it matures. This mucigel harbors microorganisms and probably influences formation of mycorrhizae, root nodules, and ion uptake in unknown ways (Barlow, 1975; Foster, 1982).

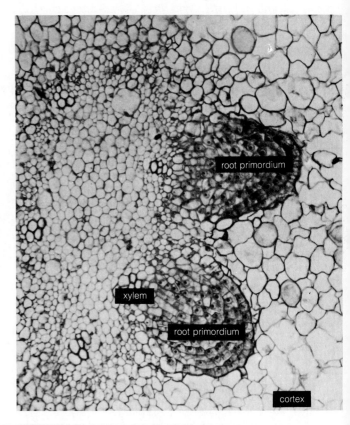

Figure 15-14 The origin of secondary roots. Growth begins divisions in the pericycle that result in the establishment of a small mass of cells. These become the root primordium, which grows outwardly through the cortex. Frequently, the endodermis divides in pace with growth of the branch root, covering it, as in (**d**), until it breaks out of the main root. (From Jensen and Salisbury, 1972.)

Cells produced by divisions in the apical meristem develop into the epidermis, cortex, endodermis, pericycle, phloem, and xylem. Microscopists can observe where cell division is occurring (i.e., where the meristem is) by observation of cells in any of the mitotic stages. Another clever method is to determine where DNA synthesis is occurring, because a doubling of the DNA content usually means mitosis and cytokinesis will follow (see Fig. 15-5). The technique of providing cells with radioactive thymidine that is incorporated into DNA, followed by autoradiography to detect DNA synthesis, is often used. Just proximal to the root cap, there is typically a small zone called the **quiescent center** where division seldom occurs (Clowes, 1975). If the meristem or the root cap is damaged, the quiescent center becomes active and can regenerate either of these parts.

Formation of lateral roots The frequency and distribution of lateral root formation partly controls the overall shape of the root system and hence the zones of the soil that are explored. Lateral or branch roots generally begin development several millimeters to a few centimeters proximal to the root tip. They originate in the pericycle, usually opposite the protoxylem points, growing outwardly through the cortex and epidermis, as illustrated in Fig. 15-14. This growth probably involves secretion by the branch root of hydrolytic enzymes that digest the walls of the cortex and epidermis, although apparently none of the postulated hydrolytic enzymes have been identified. Little is known about why lateral roots form where they do.

Radial growth of roots The roots of gymnosperms and most dicots develop a vascular cambium from procambial cells located between the primary phloem and primary xylem near or in the root-hair zone. This cambium is indirectly responsible for most of the increased width of these roots, because it forms expanding new xylem cells (toward the inside) and phloem cells (toward the outside). Most monocots do not form a vascular cambium, and the small radial enlargement they undergo is caused mainly by increases in diameter of nonmeristematic cells.

After the vascular cambium initiates secondary growth, a **cork cambium (phellogen)** arises in the pericycle. This becomes a complete cylinder that forms **cork (phellem)** toward the outside and, later, some secondary cortex (**phelloderm**) inside (Esau, 1977; Faun, 1982). The epidermis, the original cortex, and the endodermis are sloughed off; so the mature root consists of xylem at the center, vascular cambium, phloem, secondary cortex, cork cambium, and finally cork cells. Water-repellent suberin (Section 14.2) is deposited in the cork cell walls.

As the root system grows, more of it becomes suberized. For example, in mature loblolly pine (*Pinus taeda*) and yellow poplar or tuliptree (*Liriodendron tulipifera*), the surface area of unsuberized roots during the growing season is almost always less than 5 percent of the total. Apparently, suberized roots absorb water and mineral salts through lenticels, through tiny crevices formed by penetration of branch roots, and through holes left when branch roots die (Section 6.4).

Stems The apical meristem of the shoot forms in the embryo and is the place where new leaves, branches, and floral parts originate. The basic shoot tip structure is similar in most higher plants, both angiosperms and gymnosperms. Figure 15-15 shows photomicrographs of the terminal shoots of a representative dicot and a monocot.

In growing stems, the region of cell division is much farther from the tip than it is in roots (Sachs, 1965). In many gymnosperms and dicots, some cells divide and elongate several centimeters below the tip. In grasses, growth also occurs far down the stem but is restricted to specific, periodic regions. Near the shoot tip of young monocots, the leaf primordia are very close together, and the internodes are formed later by division and growth of cells between these primordia. At first, these divisions occur throughout the length of the young internode, but later the meristematic activity becomes restricted to the region at the base of each internode and just above the node itself. These periodic meristematic regions are called **intercalary meristems,** because they are intercalated (inserted) between regions of older, nondividing cells. Each internode consists of older cells at the top and younger cells derived at the base from the intercalary meristem.

Leaves The earliest sign of leaf development in both gymnosperms and angiosperms usually consists of divisions in one of the three outermost layers of cells near the surface of the shoot apex (see Fig. 15-2). Periclinal divisions followed by growth of the daughter cells cause a protuberance that is the **leaf primordium,** while anticlinal divisions increase the surface area of the primordium. Both kinds of divisions are important for further development of leaves and for growth in other parts of the plant.

Leaf primordia do not develop randomly around the circumference of the shoot apex. Rather, each species usually has a characteristic arrangement, or **phyllotaxis,** causing opposite or alternate leaves (Richards and Schwabe, 1969). Alternate leaves are arranged in several species-specific ways much studied by mathematicians as well as plant physiologists. No one knows why a given leaf primordium develops where it does. One theory holds that an already developing primordium sends out an in-

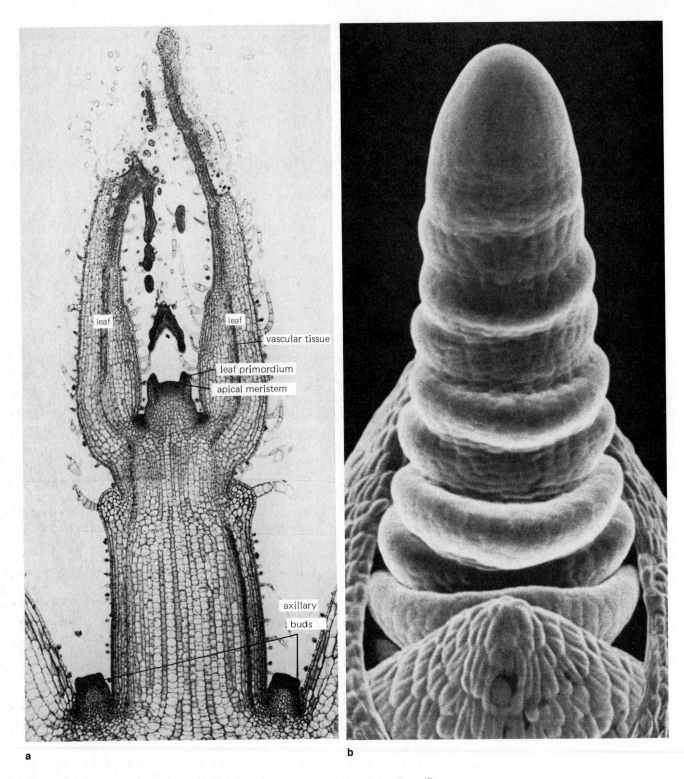

a

b

Figure 15-15 (**a**) Longitudinal section through the apical part of the shoot of a dicot. (From Jensen and Salisbury, 1972.) (**b**) A scanning electron micrograph of the apical meristem of wheat at a late vegetative stage. Leaf primordia in grasses are formed as ridges around the shoot axis. (Micrograph courtesy of John Troughton; see Troughton and Donaldson, 1972.)

hibitor, preventing a new primordium from arising within the inhibitor's field of influence. Another theory suggests that primordia compete for space, so a new primordium can be initiated only when adequate space becomes available.

The shape of the leaf primordium is produced by the magnitude and direction of its cellular divisions and expansions. The direction of expansion is controlled by the yielding properties of the cell wall, so the cell division planes, which are the planes in which the new walls are deposited, affect the primordium shape. Since cell division is accompanied by a coordinated amount of cell expansion, when most of the early divisions are periclinal the primordium appears long and narrow. When more of the divisions are anticlinal, the young organ is shorter and wider.

Subsequent leaf development is highly variable, as shown by the wide variety of leaf shapes. Continued extension outwardly occurs by both periclinal and anticlinal divisions at the primordium tip (apex). Later, often when the leaf is only a millimeter or so long, meristematic activity begins throughout its length. In grass leaves and conifer needles, this activity ceases first at the distal end (tip) and finally resides at the leaf base. An increase in width of the leaf blade in angiosperms results from meristems along each margin of the leaf axis, but these cease activity well before the leaf matures. In grasses, the basal meristem is an intercalary one that remains potentially active for long periods, even after leaf maturity. It can be stimulated by defoliation caused, for example, by a grazing animal or a lawnmower.

Figure 15-16 shows a crabgrass (*Digitaria sanguinalis*) leaf with its base encircling the stem, an encirclement that results from periclinal divisions in the primordium all the way around the shoot apex (see also Fig. 15-15b). The basal meristem in a grass leaf sheath often lies immediately external to an intercalary meristem of the stem.

In dicot leaves, most cell divisions stop well before the leaf is fully grown, frequently when it is half or less of its final size (Dale, 1983). In a bean primary leaf, cell division is complete when it has reached slightly less than one fifth its final area, so the last 80 percent of leaf expansion is caused solely by the growth of preformed cells. This growth occurs over the entire leaf area, although not uniformly. The same is true for many dicots. Cells in the young leaf are relatively compact. As leaf expansion occurs, the mesophyll cells stop growing before the epidermal cells do; so the expanding epidermis then pulls the mesophyll cells apart and causes development of an extensive intercellular space system in the mesophyll, as shown in Fig. 3-4.

A few leaf primordia and sometimes even floral primordia are usually detectable near the shoot apex of the embryo in the seed, but most primordia (espe-

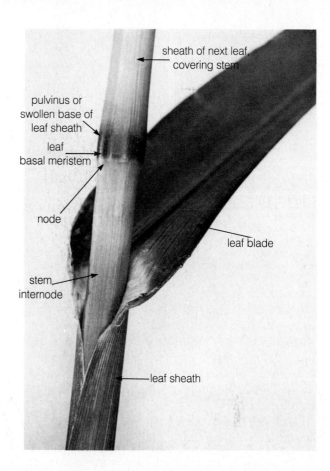

Figure 15-16 The relation of the leaf blade and sheath to the stem in a grass (crabgrass). The cylindrical leaf sheath partly replaces the stem in providing support. (From W. W. Robbins, T. E. Weier, and C. R. Stocking, 1974, Botany, An Introduction to Plant Biology, John Wiley and Sons, Inc., New York. Copyright© 1974 by John Wiley and Sons, Inc. Reprinted by permission of John Wiley & Sons, Inc.)

cially in perennial species) are formed after germination. In conifers and deciduous trees, the early rapid growth of spring usually involves expansion of leaf primordia formed during the previous season and extension of the internodes between these primordia; only in late summer are new primordia formed. These new primordia form part of the bud that is usually dormant during the winter or during a long dry period.

Flowers After establishment of roots, stems, and leaves, flowers and then fruits and seeds form, perpetuating the species and completing the life cycle. Most angiosperm species produce bisexual **(perfect)** flowers containing functional female and male parts, while others such as spinach, cottonwoods, willows, maples, and date palms are **dioecious,** containing **imperfect** staminate (male) and pistillate (female) flowers on different individual plants. **Monoecious**

Figure 15-17 Scale of flower fading in *Ipomea tricolor*. Stage 0 represents the fully open corolla; stages 1 to 4, progressing phases of fading. Flower opening (stage 0) begins at about 06:00 in the morning, while fading and curling in stage 1 begins at about 13:00 of the same day. Curling is caused by turgor changes in the rib cells. The cells in the inner side of the rib lose solutes and water, while the outer rib cells expand, causing curling. (Photographs courtesy of Hans Kende; from Kende and Baumgartner, 1974.)

species such as maize, cocklebur, squash, pumpkins, and many hardwood trees form staminate and pistillate flowers at different positions along a single stem. The reproductive structures of conifers develop in unisexual **cones (strobili).** Most conifers are monoecious, although junipers and certain others are dioecious.

Anthesis, the opening of flowers with parts available for pollination, is sometimes a spectacular phenomenon, usually associated with full development of color and scent. While many flowers remain open from anthesis until **abscission** (falling off), others such as tulips open and close at certain times of the day for several days. Opening is usually caused by faster growth of the inner compared to the outer parts of the petals, but continued opening and closing is probably a response to temporary changes in turgor pressure across the two sides. Opening and closing are influenced by temperature (see Fig. 21-2) and atmospheric vapor pressure, but the major factor is often an internal clock set by the daily dawn and/or dusk signals (see Chapter 20). For example, evening primrose flowers (*Oenothera* species) usually open in the evening about 12 hours before dawn, but they can be rephased to open in the morning by artificially reversing the light–dark cycles. The light influencing this response is absorbed by the flowers themselves.

After anthesis and pollination, the petals eventually wither, die, and abscise. In some species, withering follows anthesis rapidly. For example, in *Portulaca grandifolia* and many morning glories, including *Ipomea tricolor* and *Pharbitis nil*, opening of the flower occurs in the morning, and the corolla withers in late afternoon (Fig. 15-17). Such withering is commonly associated with extensive transport of solutes from the flowers to other plant parts, often to the ovary, with rapid water loss. There is an accelerated breakdown of protein and RNA from petals and sepals during withering, and hydrolytic enzymes such as proteases and ribonucleases are apparently activated by hormonal changes to cause such breakdown. Nitrogenous products such as amino acids and amides are then transported to seeds and other tissues where growth is occurring, so nutrients are conserved. Although withering and color fading are common, certain rose and *Dahlia* species lose petals that are still turgid and that contain most of their original protein.

Seeds and Fruits

Chemical changes in growing seeds and fruits The zygote, embryo sac, and ovule develop into the seed, while the surrounding ovary develops into the fruit. Numerous anatomical and chemical changes occur; we shall emphasize the chemical changes. Frequently, sucrose, glucose, and fructose accumulate in the ovules until the endosperm nuclei become surrounded by cell walls; then the concentrations of these sugars decrease as they are used in cell-wall formation and starch or fat synthesis. These sugars arise largely from sucrose and other sugars transported through the phloem into the young seeds and fruits (Chapter 7). Most of the nitrogen of immature seeds and fruits is present in proteins, amino acids, and the amides glutamine and asparagine. The amino acids and amides decrease in concentration as storage proteins are formed in the protein bodies.

The roles of enzymes and nucleic acids in developing seeds are important to seed longevity. For a mature seed to germinate after remaining alive for long periods, it must either possess all the enzymes necessary for germination and seedling establishment or have the genetic information available to synthesize them. Some of the enzymes essential to germination are produced in a stable form during seed development; others are translated from stable messenger RNA, transfer RNA, and ribosomal RNA molecules synthesized during seed maturation; and still others are formed from newly transcribed RNA molecules only after the seed is planted (Spencer and Higgins, 1982; Higgins, 1984). Thus different seeds control enzyme production in various ways, and there are different mechanisms even in the same seed for control of specific enzymes. Loss of water during seed maturation is critical, leading to important but poorly understood changes in the physical and chemical properties of the cytoplasm. As a result, dry seeds respire extremely slowly and remain alive through extended drought or cold periods.

The chemical composition of edible fruits and the transformation of carbohydrates during ripening have been studied, but little is known about development of nonfleshy, less economically important fruits (Hulme, 1970; Coombe, 1976; Rhodes, 1980). In apples, the concentration of starch increases to a maximum and then decreases somewhat until harvest as it is converted to sugars. In apples and pears, fructose is often the most abundant sugar, but lesser amounts of sucrose, glucose, and sugar alcohols are also present. Grapes and cherries contain about equal amounts of glucose and fructose, but sucrose is often undetectable. The hexose concentration in grapes can reach unusually high values. Concentrations of glucose and of fructose in some varieties reach $0.6\ M$ for each sugar, giving the mature fruits an unusually negative osmotic potential and a sweet taste. During ripening of oranges, grapes, grapefruits, pineapples, and various berries, the organic acids (principally malic, citric, and isocitric) decrease and the sugars increase, so the fruits become sweeter. In lemons, however, the acids continue to increase during ripening, so that the pH decreases and the fruits remain sour. Lemon fruits contain virtually no starch at any time during development, although other fruits (e.g., bananas, apples, peaches) contain much starch when they are immature.

Numerous other changes in fruit composition have been studied, including transformation of chloroplasts to carotenoid-rich chromoplasts, accumulation of anthocyanin pigments, and accumulation of flavoring components. The use of gas chromatography has allowed identification of hundreds of volatile substances such as aliphatic or aromatic esters, aldehydes, ketones, and alcohols contributing to the flavor and aroma of strawberries and other fruits (Nurnsten, 1970). This provides a basis for improvement of fruit flavors by plant breeding and for development of artificial flavoring substances.

Importance of seeds for fruit growth Development of fruits is usually dependent upon germination of pollen grains on the stigma or on this plus subsequent fertilization. Furthermore, extracts of pollen grains added to certain flowers will simulate natural pollination and fertilization by causing ovary growth and wilting and abscission of the petals. Developing

Figure 15-18 Delay of senescence in soybean plants caused by daily removal of flower buds. (Photograph courtesy of A. Carl Leopold; from Leopold and Kriedemann, 1975.)

seeds are also usually essential for normal fruit growth. If seeds are present only in one side of a young apple fruit, only that side of the fruit will develop well, and seeds are essential for normal strawberries.

Normal production of fruits lacking seeds is called **parthenocarpic fruit development.** It is especially common among fruits that produce many immature ovules, such as bananas, melons, figs, and pineapple. Parthenocarpy can result from ovary development without pollination (citrus, banana, and pineapple), from fruit growth stimulated by pollination but without fertilization (certain orchids), or from fertilization followed by abortion of the embryos (grapes, peaches, and cherries.)

Relations between vegetative and reproductive growth Gardeners have long practiced the technique of removing flower buds from certain plants to maintain vegetative growth. A commercial example is the *topping* (removal of flowers and fruits) of tobacco plants, which encourages leaf production. Such an effect on soybeans is shown in Fig. 15-18.

There is a competition for nutrients among vegetative and reproductive organs. Developing flowers and fruits, especially young fruits, possess a large but unexplained "drawing power" for mineral salts, sugars, and amino acids. During the accumulation of these substances by the reproductive organs, there is often an approximately corresponding decrease in the amounts present in the leaves. Studies with radioactive tracers show that nutrient accumulation in developing flowers, fruits, or tubers occurs largely at the expense of materials in nearby leaves. Nevertheless, the situation is more complex than a simple competition for nutrients. In the cocklebur (*Xanthium strumarium*), induction of flowering by long nights causes leaf senescence just as rapidly when flower buds are removed as when they are allowed to develop normally. In other cases, perhaps some inhibitor is transported into the vegetative organs and causes their death prematurely.

There is usually a competition among individual fruits of the same plant for nutrients. For example, fruit size decreases with increasing number of fruits allowed to form on tomato plants or apple trees. The mechanism by which fruits can divert nutrients out of leaves and into their own tissues, sometimes against apparent concentration gradients, is not understood but may be controlled by phloem unloading, as discussed in Section 7.4. Various hormones, especially cytokinins (see Chapter 17), might also be involved.

In general, factors that stimulate shoot growth retard flower, tuber, and fruit development. High-nitrogen fertilization causes luxuriant stem and leaf growth of tomato plants but somewhat reduced fruit

development, while lower nitrogen levels lead to less stem and leaf growth but more fruit development. Similarly, excess nitrogen stimulates leaf growth but inhibits growth of potato tubers or apples and reduces sugar content in sugar-beet roots.

Do processes interfering with vegetative growth also stimulate flower development? Sometimes they do. Heavy pruning, tying branches to the ground, or various other mutilation procedures often stimulate flowering. Furthermore, such commercial growth retardants as Phosphon D, CCC, Amo-1618 (see Chapter 16), and B995 (N-dimethylamino succinamic acid) inhibit growth of stems. This stunting is sometimes accompanied by a more rapid initiation of flower buds or a greater number of flowers per plant. These chemicals are used in commercial chrysanthemum production, for example, but they inhibit flowering in some other species. (These matters are further discussed in Chapter 22.)

It should be evident by now that we cannot proceed much further without a thorough discussion of plant hormones and growth regulators. That is the special purpose of the next two chapters and will often be the theme in the rest of the book.

16

Hormones and Growth Regulators: Auxins and Gibberellins

In the previous chapter, we reviewed some of the growth regions of plants and introduced a few of the numerous effects of certain plant hormones on growth and development. In this and the next chapter, we summarize knowledge about such hormones and related growth regulators. Subsequent chapters present more details about the roles of these compounds in specific developmental processes. There are still only five groups of well-accepted hormones, even though more will almost surely be discovered. The five include three *auxins,* many *gibberellins,* several *cytokinins, abscisic acid,* and *ethylene.*

What is a plant hormone? Most plant physiologists accept a definition that is similar to that developed for animal hormones, even though there is no evidence that the fundamental biochemical actions of plant and animal hormones are the same: *A plant hormone is an organic compound synthesized in one part of a plant and translocated to another part where, in very low concentrations, it causes a physiological response.* The response in the target organ need not be promotive, because processes such as growth or differentiation are sometimes inhibited by hormones, especially abscisic acid. Because the hormone must be synthesized by the plant, such inorganic ions as K^+ or Ca^{2+} that cause important responses are not hormones. Neither are organic growth regulators synthesized by organic chemists (e.g., 2,4-D, an auxin) or synthesized in organisms other than plants. The definition also states that a hormone must be translocated in the plant, but nothing is said about how or how far, nor does this mean that the hormone will not cause a response in the cell in which it is synthesized. Sucrose is not considered a hormone, even though it is synthesized and translocated by plants, because it causes growth only at relatively high concentrations. Hormones are usually effective at internal concentrations of 1 μM or less, whereas sugars, amino acids, organic acids, and other metabolites necessary for growth and development (excluding enzymes and most coenzymes) are usually present at concentrations of 1 to 50 mM.

The principle that plant development is influenced by special chemicals in plants is not new. About 100 years ago, the famous German botanist Julius von Sachs suggested that specific organ-forming substances occur in plants; he supposed that one substance caused stem growth and others leaf, root, flower, or fruit growth. No such organ-specific chemicals were ever identified. Because of their very low concentrations in plants, the first hormone to be identified (indoleacetic acid) was not identified until the 1930s. Because it could evoke so many different plant responses, it was considered by many to be the only plant hormone until numerous effects of gibberellins were discovered in the 1950s. As more hormones were identified and their effects and endogenous concentrations studied, it became apparent that not only does each hormone affect the responses of many plant parts, those responses depend on the concentrations and interactions of the various known hormones and possibly unknown ones. But von Sachs' concept that different tissues can respond differently to different chemicals is certainly valid. Recently, Anthony J. Trewavas has emphasized that differential sensitivity is more important in determining the effects of a hormone than the concentration of that hormone within plant tissues. His personal essay at the end of this chapter summarizes this viewpoint and more, and you might wish to read that essay now and again after reading this chapter and Chapter 17.

16.1 The Auxins

The term **auxin** was first used by Frits Went, who, as a graduate student in Holland in 1926, discovered that some unidentified compound probably caused

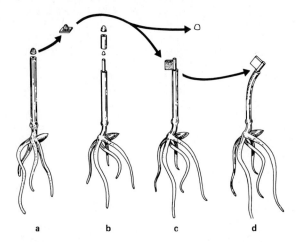

Figure 16-1 The demonstration by Went of auxin in the *Avena* coleoptile tip. Auxin is indicated by stippling. (**a**) The tip was removed and placed on a block of gelatin. (**b**) Another seedling was prepared by removing the tip, waiting a period of time, and removing the tip again (a new "physiological tip" sometimes forms). (**c**) The leaf inside the coleoptile was pulled out, and the gelatin block containing the auxin was placed against it. (**d**) Auxin moved into the coleoptile on one side, causing it to bend. (From Salisbury and Parke, 1964.)

curvature of oat coleoptiles toward light (see his personal essay at the end of this chapter). This curvature phenomenon, called phototropism, is described in Section 18.2. The compound Went found is relatively abundant in coleoptile tips compared to other organs or tissues, and Fig. 16-1 indicates how he showed its existence. The critical demonstration was that a substance present in the tips could diffuse from them into a tiny block of agar. The activity of this auxin was detected by curvature of the coleoptile caused by enhanced elongation on the side to which the agar block was applied.

Went's auxin is now known to be **indole-3-acetic acid (IAA,** Fig. 16-2), and some experts believe that IAA represents the only true auxin hormone, i.e., IAA = auxin. Nevertheless, plants contain two other compounds that cause many of the same responses as IAA and that should be considered to be auxins (Wightman and Lighty, 1982). One of these is **4-chloro-indoleacetic acid (4-chloroIAA),** found in young seeds of various legumes (Fig. 16-2). Another, **phenylacetic acid (PAA**—see Fig. 16-2) is widespread among plants and is frequently more abundant than

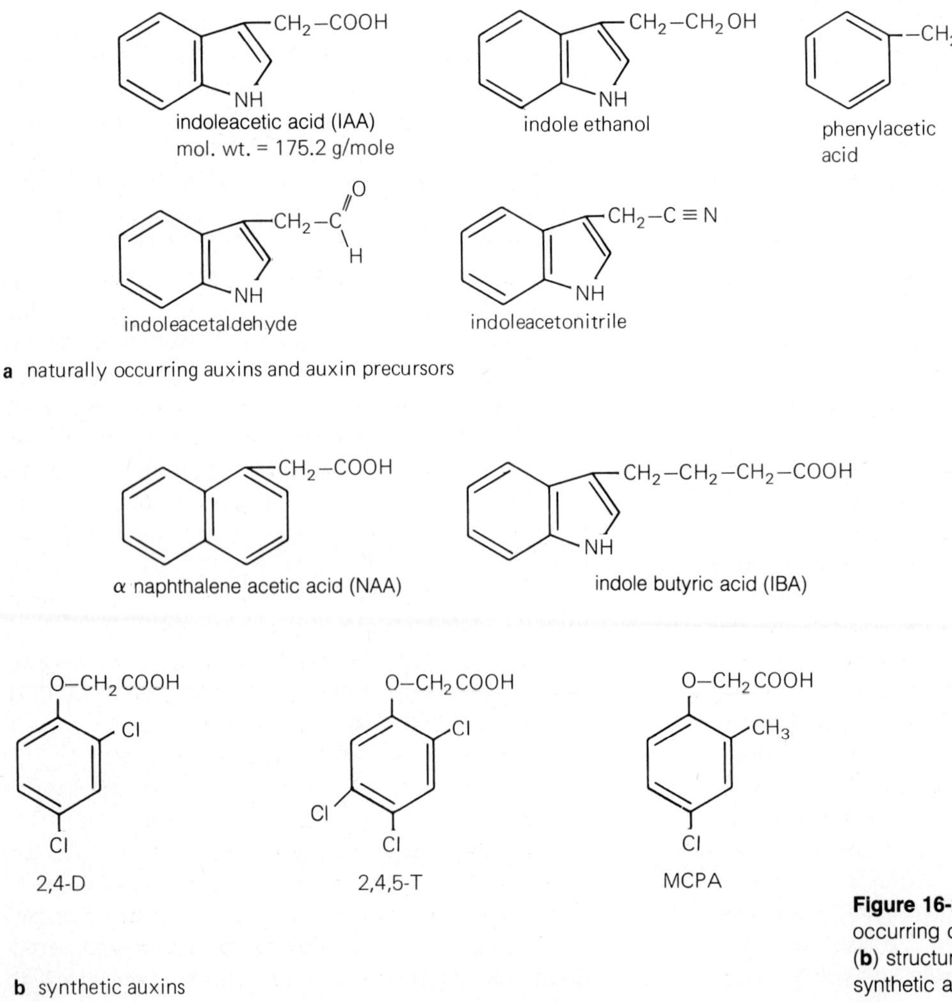

CH_2-COOH

indoleacetic acid (IAA)
mol. wt. = 175.2 g/mole

CH_2-CH_2OH

indole ethanol

$-CH_2-COOH$

phenylacetic acid

$CH_2-C\overset{O}{\underset{H}{\big/}}$

indoleacetaldehyde

$CH_2-C \equiv N$

indoleacetonitrile

a naturally occurring auxins and auxin precursors

CH_2-COOH

α naphthalene acetic acid (NAA)

$CH_2-CH_2-CH_2-COOH$

indole butyric acid (IBA)

$O-CH_2COOH$

Cl

Cl

2,4-D

$O-CH_2COOH$

Cl

Cl

2,4,5-T

$O-CH_2COOH$

CH_3

Cl

MCPA

Figure 16-2 (**a**) Structures of some naturally occurring compounds having auxin activity, and (**b**) structures of other compounds that are only synthetic auxins.

b synthetic auxins

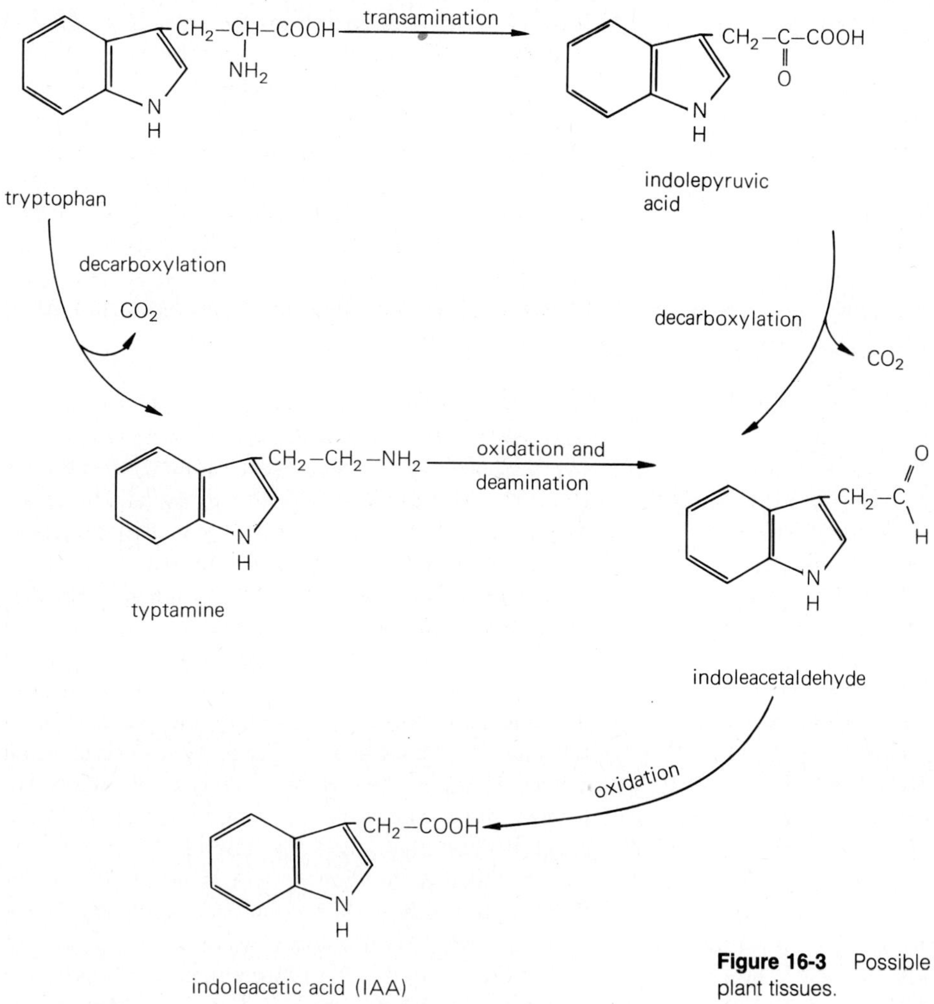

Figure 16-3 Possible mechanisms of formation of IAA in plant tissues.

The diagram shows the following labeled structures and reactions:

tryptophan →(transamination)→ indolepyruvic acid

tryptophan →(decarboxylation, CO_2)→ typtamine

typtamine →(oxidation and deamination)→ indoleacetaldehyde

indolepyruvic acid →(decarboxylation, CO_2)→ indoleacetaldehyde

indoleacetaldehyde →(oxidation)→ indoleacetic acid (IAA)

IAA, although it is less active in causing the typical responses of IAA to be described. Little is known about the transport characteristics ot 4-chloroIAA or PAA and whether they normally function as auxin hormones, although this seems likely. Three additional compounds found in many plants have considerable auxin activity. They are readily oxidized to IAA *in vivo,* and so they are probably active only after conversion. We shall not yet consider them to be auxins, only auxin-precursors. These are *indoleethanol, indoleacetaldehyde,* and *indoleacetonitrile.* Each has a structure similar to IAA, but each lacks the carboxyl group (Fig. 16-2). They are easily oxidized to IAA by various enzymes present in plants.

Similar compounds synthesized by chemists also cause many of the physiological responses common to IAA and are generally considered to be auxins. Of these, *naphthaleneacetic acid (NAA), indolebutyric acid (IBA), 2,4-dichlorophenoxyacetic acid (2,4-D), 2,4,5-trichlorophenoxyacetic acid (2,4,5-T),* and *2-methyl-4-chlorophenoxyacetic acid (MCPA)* (Fig. 16-2) are the best known. Because they are not synthesized by plants, they are not hormones. They are classified as **plant growth regulators,** and many other kinds of compounds also fit into this category. The term auxin has become much more encompassing since Went's discovery of IAA, because so many compounds are structurally similar to IAA and cause similar responses in plants. Without defining an auxin, we emphasize that each auxin-like compound is similar to IAA in having a carboxyl group attached to an aromatic ring.

Synthesis and Degradation of IAA IAA is chemically similar to the amino acid tryptophan and is synthesized from it. Two mechanisms for synthesis are known (Fig. 16-3), both of which involve removal of the amino group and the terminal carboxyl group from the side-chain of tryptophan (Sembdner et al., 1980). The preferred pathway for most species involves donation of the amino group to another α-keto acid by a transamination reaction to form *indolepyruvic acid,* and then decarboxylation of indolepyruvate to form *indoleacetaldehyde.* Finally, indoleacetaldehyde is oxidized to IAA. The enzymes necessary for tryptophan conversion to IAA are most active in young tissues, such as shoot meristems and growing leaves and fruit. In these tissues the auxin contents are also highest, suggesting that IAA is synthesized there.

It seems logical that plants should have mechanisms to control the amounts of hormones as potent as IAA; the rate of synthesis is one mechanism. Control by temporary inactivation is another. In this case, the carboxyl group of IAA is combined covalently with other molecules to form derivatives called **bound auxins** (Cohen and Bandurski, 1982), but perhaps a better term is **conjugated auxins,** because conjugates are bound covalently and compounds can bind temporarily to auxins by hydrogen bonding or other weak forces. Numerous IAA conjugates are known, including the peptide *indoleacetyl aspartic acid* and the esters *IAA-inositol* and *IAA-glucoside*. In general, IAA can be released from these bound forms by hydrolase enzymes, indicating that they are important storage forms of IAA. In cereal grain seedlings, these IAA conjugates are important forms in which IAA can be transported, especially from the endosperm of the seed through the xylem toward coleoptile tips and young leaves. Another process for IAA removal is degradative and involves oxidation by O_2 and loss of the carboxyl group as CO_2. The products are variable, but *3-methyleneoxindole* is usually a principal one. The enzyme that catalyzes this reaction is *IAA oxidase.* Several IAA oxidase isozymes exist, and all or nearly all are identical to the peroxidases involved in early steps of lignin formation (Section 14.6). In beech and horseradish, for example, there were 20 peroxidase isozymes and all had IAA oxidase activity (Gove and Hoyle, 1975). Synthetic auxins are not destroyed by these oxidases and, therefore, persist in plants much longer than does IAA. Conjugated auxins are also resistant to IAA oxidases.

Auxin Transport A surprising thing about the ability of IAA to act as a hormone is the way in which it is transported from one organ or tissue to another. In contrast to movement of sugars, ions, and certain other solutes, IAA usually is not translocated through the phloem sieve tubes nor through xylem (Jacobs, 1979). It will move through sieve tubes if applied to the surface of a leaf mature enough to export sugars, but normal transport in stems and petioles is from young leaves and through other living cells, including phloem parenchyma and parenchyma cells that surround vascular bundles. Even synthetic auxins applied to plants move as IAA does. This transport has features different from those of phloem transport. First, *auxin movement is slow,* only about 1 cm/h in both roots and stems, but still 10 times faster than diffusion would predict. Second, *its transport is polar,* always occurring in stems preferentially in a basipetal (base-seeking) direction regardless of whether the base is normally down or whether it is turned upside down with the apex down. Transport in roots is also polar but preferentially in an acropetal (apex-seeking) direction. Third, *its movement requires metabolic energy,* as evidenced by the ability of ATP synthesis inhibitors or lack of oxygen to block it. Other strong inhibitors of polar auxin transport are *2,3,5-triiodobenzoic acid (TIBA)* and *naphthylthalamic acid (NPA),* although TIBA and NPA interfere specifically with auxin transport and not with energy metabolism. These two compounds are called **antiauxins.**

How can the polar transport of auxins occur? The most popular hypothesis is a chemiosmotic one (Goldsmith, 1977). According to this, cells use energy obtained from ATP hydrolysis in plasma membranes to pump H^+ from cytosol into cell walls. The lower pH of cell walls (about 5.5) keeps the carboxyl group of an auxin less-dissociated than in the cytosol, where the pH is higher (near 7). Because membranes are more permeable to undissociated (noncharged) solutes (Section 6.6), noncharged auxins move from the wall into the cytosol, where the higher pH causes their carboxyl group to dissociate and attain a negative charge. (An alternative suggestion with the same overall result is that negatively charged auxins are moved into the cytosol by symport with H^+.) As the concentration of charged auxin builds up in the cytosol, its outward movement would be favored more thermodynamically, but polar transport would require that the auxin could only move out from the end of the cell opposite to that which it entered. This preferential exit at the basal ends of cells assumes that some carrier in that region of the membrane transports charged auxins out toward the cell wall, where the low pH again causes more of them to become noncharged. This chemiosmotic hypothesis is summarized for IAA in Fig. 16-4.

The crucial problem with this transport hypothesis has been to learn how charged auxins are transported out at the basal ends of cells, because this transport requires the cells to be polarized such that they can absorb at one end and secrete at the opposite end. (We encountered a similar problem for ion transport into dead xylem cells of roots in Section 6.4; we mentioned that living pericycle or xylem parenchyma cells in roots seem to secrete into dead xylem cells only those ions that they absorbed from cells closer to the root surface.) Recently, direct evidence for the polarized location of an auxin transporter at the basal end of pea stem cells was obtained (Jacobs and Gilbert, 1983). This transporter is blocked by NPA, which probably explains how that antiauxin prevents basipetal auxin transport. TIBA apparently blocks at the same site (Goldsmith, 1982). The mechanism of polar transport of auxins still requires more study, but polar transport down the stem from young leaves or meristematic cells of the shoot tip is probably important for control of such processes as renewed vascular cambium activity in woody plants during spring, for growth of stem cells, and probably for inhibition of lateral bud development. It is this transport to coleoptile cells directly below the agar block in Fig. 16-1 that causes the elongation that results in curvature.

Extraction and Measurement of Auxins and Other Hormones A major question confronting us repeatedly is whether a given hormone helps control some particular physiological process. A minimum requirement to determine the answer is to extract the hormone and relate its tissue concentration to the magnitude of the response. The problem is not easy, mainly because hormone concentrations are so low, frequently only 0.01 to 1 μM. We must have a method for analysis that not only is very sensitive but also is highly specific, so that many other cellular components do not interfere with our analyses. This general problem represents one of many in which physiologists with important agricultural, horticultural, and forestry problems have had to collaborate with chemists and biochemists. The first step is to selectively extract the hormone with an organic solvent that will not extract numerous contaminating compounds and will not destroy the hormone we seek. Second, by partitioning the hormone in other immiscible solvents or by using various chromatography procedures, the hormone can be purified much further (Hillman, 1978; Yokota et al., 1980; Brenner, 1981; Morgan and Durham, 1983).

At this stage, the usual technique historically has been to measure the amount of the partially purified hormone by a biological assay, a **bioassay.** Bioassays take advantage of the extreme sensitivity and specificity of certain plant parts to a particular hormone. As an example, for years physiologists tediously analyzed auxins using Went's coleoptile curvature test (Fig. 16-1), measuring the extent of curvature caused by the auxin diffusing from an agar block. Another easier but less sensitive and less specific bioassay for auxins is to cut elongating sections from coleoptiles or from dicot stems, then grow the sections in a petri dish with sucrose, dilute phosphate buffer, and various amounts of the partially purified auxin sample (Fig. 16-5a). Elongation of the sections is quantitatively related to how much auxin is added (Fig. 16-5b). This straight-growth test is subject to interference from inhibitory compounds such as abscisic acid and many phenolics. Cytokinins also inhibit elongation of stem sections (Vanderhoef et al., 1973), although this is seldom a bioassay problem, because cytokinins are chemically different from auxins and do not contaminate reasonably purified extracts of auxins (Yokota et al., 1980). Gibberellins have no influence on elongation of coleoptile sections, and the

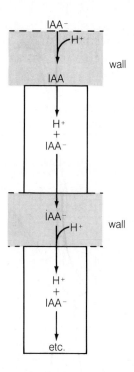

Figure 16-4 Chemiosmotic model to explain basipetal transport of auxin in living cells. ATP-driven proton pumps in the plasma membrane (not shown) keep the wall pH lower than that of the cytosol.

growth promotion of stem sections by gibberellins is usually so weak that they cause no auxin bioassay problems. Nowadays, bioassays for auxins have been replaced with modern instruments (Reeve and Crozier, 1980; Brenner, 1981). However, the principle of bioassays is important, and much of our knowledge about auxin contents of various plant parts was obtained by bioassays. Furthermore, bioassays for cytokinins and gibberellins are still used, even though they are rapidly being replaced with instrumental analyses.

You might ask why we could not just treat an intact stem or coleoptile and measure its growth response as a bioassay. Apparently, enough endogenous auxin is usually supplied to stems and intact plants by basipetal transport from young leaves above, so that additional, exogenous auxin does not enhance growth. Most intact plants will not grow faster regardless of how the auxin is applied or which auxin is used, although cucumber stems and certain flower stalks (scapes) do respond positively to added auxins. Auxin bioassays such as the coleoptile curvature test and the straight-growth test depend on removal of the responding part from the normal auxin supply. In general, a deficiency of a hormone must be created experimentally in a plant to show that adding a hormone has an effect. This is done by removal of the

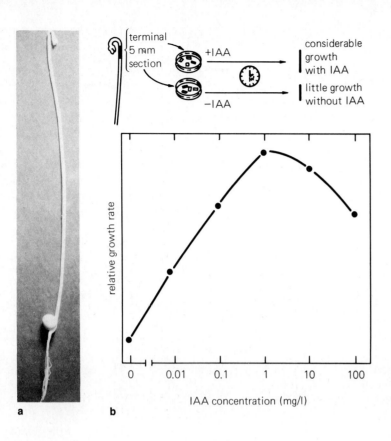

Figure 16-5 (**a**) Week-old pea seedling grown in darkness. The third (upper) internode is used in auxin bioassays, as shown in (b). (Photo by C. W. Ross and Nicholas Carpita.) (**b**) Top, technique used in auxin bioassay using apical sections from etiolated pea stems (epicotyls). Sections are placed in Petri dishes containing sucrose and certain mineral salts. Growth is often measured 12 to 24 hr later. Bottom, influence of the IAA concentration upon growth rate of pea stem sections. Note that auxin concentrations are plotted logarithmically and that an optimum concentration is reached that, when exceeded, results in less growth. (After Galston, 1964.)

source organ: young leaves, for auxins or, for gibberellins, by using plants that are genetically altered so that they respond to exogenous gibberellins.

The role of IAA in elongation of dicot stems and coleoptiles seems well established from studies such as these. Pine hypocotyl sections also grow faster when provided with auxins (Terry et al., 1982; Carpita and Tarmann, 1982). IAA is probably also necessary for growth of fruits (Section 15.4). It is commonly assumed that IAA is essential for growth of leaves, flowers, and grass stems, but even with modern methods of analysis there is still little evidence. Roots have been studied more intensively and require special attention.

Effects of Auxins on Roots and Root Formation
IAA exists in roots at concentrations similar to those in many other plant parts. As first shown in the 1930s, auxins will promote elongation of sections excised from roots or even intact roots of many species (Batra et al., 1975; Scott, 1972), but only at extremely low concentrations (10^{-7} to 10^{-13} M, depending on the species and age of roots). At higher concentrations, elongation is almost always inhibited. The assumption is that root cells usually contain enough or almost enough auxin for normal elongation. Indeed, many excised roots grow for days or weeks *in vitro* without added auxin, indicating that any requirement they have for this hormone is satisfied by their ability to synthesize it. An exogenous auxin supply often causes inhibition of root growth. Part of this inhibition is caused by ethylene, because *auxins of all types stimulate many kinds of plant cells to produce ethylene*, especially when relatively large amounts of auxin are added. Ethylene retards elongation of both roots and stems (Section 17.2).

The question of whether growth rates of roots correlate with their content of IAA has been approached with bioassay techniques and, more recently, with an immunological procedure, a radioimmunoassay (Pengelly and Torrey, 1982; Weiler, 1984). In the latter case, pea roots were grown in sand at different rates by varying their supply of nutrient salts. In general, a fairly good correlation was found between growth rate and IAA content, although this correlation did not always hold for individual parts of a single root. Insofar as correlations are useful, these results suggest that roots do depend upon IAA for elongation and that when well fertilized they do not require exogenous IAA. However, the question as to whether sections from nonfertile roots would grow faster with exogenous auxin was not addressed in that work. Other research shows that intact roots will indeed grow faster when provided with exogenous IAA at concentrations as high as 10^{-8} M, provided that those roots have been pretreated with inhibitors of ethylene biosynthesis (Mulkey et al., 1982). Higher concentrations of IAA (1 μM) were still strongly inhibitory. Furthermore, antiauxins promote growth of maize root sections, supposedly by decreasing the effective tissue concentrations of auxin from a super-optimal to an optimal level (Moloney et al., 1981).

Collectively, these results are not easy to interpret in relation to auxin requirements for root growth. Certainly, the normal presence of IAA in roots suggests a role for that compound in growth, as does the aforementioned correlation between relatively high growth rates and relatively high IAA levels in well fertilized pea roots. Growth promotion by extremely low concentrations of IAA when IAA cannot cause production of inhibitory ethylene confirms that ethylene production nullifies what otherwise would be a more general growth enhancement by *exogenous* auxin. Those results suggest that environmental factors that would increase *endogenous* auxin contents could increase growth only if they also prevented the expected rise in ethylene production, yet pea roots both grew faster and produced more IAA when well fertilized than when slightly nutrient deficient. More data are needed relating endogenous auxin levels and growth rates, but a reasonable conclusion is that roots do depend upon the auxin they contain to elongate. If one could measure the auxin content of specific cells that control elongation of the root, much more could be said about these relations.

There is ample evidence that auxins from stems strongly influence root initiation. In many cases, removal of young leaves and buds, both of which are rich in auxins, inhibits the number of lateral roots formed. Substitution of those organs by auxins often restores the root-forming capacity. Thus there is an important difference in exogenous auxin effects on root elongation, in which an inhibition is observed, and in root initiation and early development, in which promotion is observed (Wightman et al., 1980).

Auxins also promote adventitious root development on stems. Many woody species (e.g., apples, most willows, and Lombardi poplar) have preformed adventitious root primordia in their stems that remain dormant for some time unless stimulated by an auxin (Haissig, 1974). These primordia are often at nodes or on the lower sides of branches between nodes. Apple burr-knots on stems contain up to 100 root primordia each. Even plants without preformed root primordia in stems will form adventitious roots. These result from division of an outer layer of phloem.

Adventitious root formation on stem cuttings is the basis for the common practice of asexual reproduction of many species, especially ornamentals in which it is essential to maintain genetic purity. Von Sachs obtained evidence in the 1880s that young leaves and active buds promote root initiation, and he suggested that a transmissible substance (hormone) was involved. In 1935, Went and Kenneth V. Thimann showed that IAA stimulates root initiation from stem cuttings, and from this demonstration developed the first practical use of auxins. Synthetic

Figure 16-6 Promoting root growth from cuttings by treatment with an auxin. (From E. J. Kormondy et al., 1977.)

auxins such as NAA and IBA (Fig. 16-2) are usually more effective than IAA, apparently because they are not destroyed by IAA oxidase or other enzymes and therefore persist longer. Commercial powders in which cut ends of stems are dipped to facilitate root production usually contain IBA or NAA mixed with inert talcum powder (Fig. 16-6). Propagation from cut leaves is also often promoted by auxins, but the internal hormone balance must also be capable of initiating adventitious buds as well, as is the case in root cuttings. With some species, exogenous auxins inhibit rooting of cuttings, and with other species (apples, pears, and most gymnosperms) rooting is limited with or without added auxin.

The site of adventitious root formation on stems in most species is the physiological basal position away (distal) from the stem apex. Even if cut shoots are inverted in a humid atmosphere, the roots will form near the top, away from the original stem tips and where auxin is presumably accumulated by polar movement. In many species, adventitious roots form near the bases of stems of intact plants, sometimes only as primordia; but sometimes they emerge as the prop-roots do from the nodes on maize stalks. Added auxin often causes emergence of many adventitious roots in the lower internodal stem region, as in tomato plants. Adventitious roots are not restricted to the base of stems but can form on the lower surface of stems that are placed in a horizontal position and kept moist. Higher auxin levels develop in the zone of root emergence prior to root development. In nature, this often allows stems that are weak to develop additional root systems.

Auxin Effects on Lateral Bud Development In stems of most species, the apical bud exerts an inhibitory influence **(apical dominance)** upon the lateral (axillary) buds, preventing or slowing their development. This extra production of undeveloped buds has definite survival value, for if the apical bud is damaged or removed by a grazing animal or a wind storm, a lateral bud will then grow and become the leader shoot. Apical dominance is widespread and has been reviewed carefully by J. R. Hillman from Scotland (1984). It also occurs in bryophytes and pteridophytes and in roots. Another dominant effect of the shoot apex is to cause branches below to grow somewhat horizontally; this horizontal growth often prevents shading of the lower branch and increases photosynthetic productivity of the whole plant.

Gardeners have long practiced the technique of removing the apical bud and young leaves to increase branching. This technique also allows the branches to grow more vertically, especially the uppermost one. In many species, continual removal of the youngest visible leaves is as effective as removal of the entire shoot apex, suggesting that a dominance factor, an inhibitor, arises in young leaves. If an auxin is added to the cut stump after the shoot apex is discarded, lateral bud development and vertical orientation of existing branches are again retarded in many species. This replacement of the bud or young leaves with an auxin suggests that the inhibitory compound they produce is IAA or another auxin, but in some species (e.g., cocklebur, *Xanthium strumarium*) application of an auxin will not replace a bud or young leaves.

Even in species in which auxins will replace the apical bud, there are good reasons to believe that an endogenous auxin is not an inhibitor that normally prevents lateral bud growth. The amount of IAA that must be added to a cut stump (from which the apex was removed) to prevent development of the lateral buds is often 1000-times as great as the IAA content of the apical bud itself. Such high doses cause cell division and elongation of the cut stump, which makes it a nutrient sink that could divert nutrients from the lateral buds and prevent their growth indirectly. Studies with ^{14}C-labeled IAA show that this hormone moves down the stem from the cut surface but not detectably into lateral buds. Furthermore, direct application of IAA to the lateral buds does not inhibit their growth and sometimes even promotes it.

The review of Hillman (1984) emphasized the difficulty of analyzing hormone levels in buds to learn if these levels correlate with the degree of growth inhibition, because it is technically so difficult to analyze many small buds in which hormones are so dilute. Hillman's review also explained that measurements of hormone levels in entire tissues or organs are of little significance without knowing the levels within crucial cells or cell organelles. Nevertheless, he and his colleagues had previously analyzed IAA concentrations in 56,000 whole buds of *Phaseolus* (bean) plants in treatments involving decapitated plants (growing buds) and nondecapitated plants (nongrowing buds). They found more IAA in the growing than the nongrowing buds 24 h after decapitation, which could be interpreted to support the idea that IAA is not the inhibitor that prevents outgrowth of lateral buds. However, growing buds would be expected to synthesize IAA, and it is uncertain in which cells this hormone was present, so this exceptionally laborious study probably tells us little about the natural inhibitor. It also emphasizes the difficulty in performing experiments to verify whether a given hormone causes a particular physiological response. In Chapter 17 reasons are given which suggest that repressed lateral buds are deficient in cytokinins, but even this hypothesis has not been supported by careful analyses of cytokinin levels. Abscisic acid, ethylene, and gibberellins have also been investigated in relation to apical dominance without much evidence that they are important. There are other growth regulators in plants (Section 17.6); one of these might be important, but there is little supporting evidence yet. Other unsatisfying hypotheses to explain apical dominance were reviewed by Phillips (1975), Rubinstein and Nagao (1976), and Hillman (1984).

Possible Mechanisms of Auxin Action Numerous experts have emphasized repeatedly that we still do not understand biochemically how any plant hormone acts. Although this statement is true, it is relative. We do understand many biochemical and physiological processes that are controlled by hormones, although the hormonal effects initiating those processes are unknown. Two facts seem helpful to understand process initiation. *First, hormone effects are amplified greatly.* This seems apparent when we consider that extremely low concentrations of such chemicals cause visible effects that require substantial metabolic redirection of molecules a thousand or more times more concentrated. *Second, the hormone response depends on poorly understood anatomical and physiological conditions of the affected cells, for not all cells are target cells.*

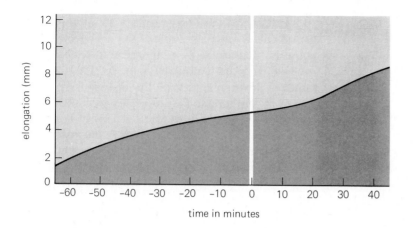

Figure 16-7 Representation of shadowgraph record of growth of excised oat (*Avena sativa* cv. Victory) coleoptile sections. The incubation medium was changed from water to 3 μg/ml IAA at the time corresponding to the vertical white line. The rapid elongation at the beginning of the record is the result of tactile stimulation of the coleoptiles during experimental manipulations. (Reproduced from The Journal of General Physiology, 1969, 53:1–20, by copyright permission of the Rockefeller University Press, courtesy M. L. Evans and P. M. Ray.)

Two assumptions that guide research have followed from these facts and from the fact that different hormones generally have different effects on any cell type. One is that there is a **primary effect** of each hormone on each cell type, from which numerous responses can follow in different kinds of cells. The primary effect varies from one hormone to another. Another assumption is that there is usually only one kind of protein, a **receptor protein,** that first binds a given hormone in any cell. Such receptor proteins could explain specificity (e.g., in terms of enzyme-substrate complexes described in Chapter 8), and they would also explain how only cells that have the proper receptor would be target cells. Cells with the proper differentiation status could produce receptor proteins for two or more hormones if indeed they respond to more than one hormone. And some cells that respond little to any hormone would presumably produce little receptor protein. Amplification of hormonal effects in receptive cells would presumably result from some unknown action of the receptor protein with its bound hormone.

A relevant question about auxin action is whether auxin receptor-proteins can be detected. Indeed they can, and the most interesting ones bind only auxins and do so very strongly (Dodds and Hall, 1980; Rubery, 1981). Nevertheless, we still have no idea about what biochemical responses occur as a result of that binding. (The same applies to proteins receptive of other hormones.) Research concerning hormone receptors has therefore temporarily slowed, but early, even though secondary, effects of auxin action are still being studied actively. Most such research concerns growth promotion by auxins, and we summarize several informative results below.

In the 1950s and 1960s it was learned that growth promotion by auxins requires continued formation of both RNA and proteins (Key, 1969). Although these findings were accepted even then, other work showed that auxins promote growth of excised coleoptile or stem sections within 10 to 15 minutes after application (Fig. 16-7; Evans, 1974). A dilemma was created, be-

cause it was wrongly assumed by many experts that this rapid growth promotion necessarily occurred before auxins could increase production of growth-promoting enzymes and therefore that growth promotion could have nothing to do with effects on transcription or translation. In fact, auxins can increase formation of certain proteins within 10 minutes after they are added to maize coleoptile sections or soybean hypocotyl sections (Zurfluh and Guilfoyle, 1982a, 1982b). Figure 16-8 illustrates changes within 20 minutes in maize coleoptiles.

The Acid-Growth Hypothesis Soon after rapid growth effects were discovered, a hypothesis to explain them was formulated, i.e., the **acid-growth hypothesis** (reviewed by Rayle and Cleland, 1979; Cleland, 1980; Taiz, 1984). This hypothesis states that auxins cause receptive cells to secrete H^+ into their surrounding primary walls and that these H^+ then lower the wall pH so that wall loosening (an increase in plasticity) and growth occur. Presumably, loosening takes place because the low pH activates enzymes that break bonds between growth restrictive polysaccharides within the walls. Regardless of the loosening mechanism, cells would then grow faster under their existing turgor pressure. Much experimental evidence supports the acid-growth theory:

1. When sections are cut from dicot stems or cereal grain coleoptiles and incubated in a dilute buffer to allow them to use up their auxin, they will now elongate faster if provided either an auxin or H^+. (The H^+ ions are usually provided in the form of a buffer at about pH 4, and it is frequently necessary to abrade the cuticle to allow H^+ ions to move into the wall.) Thus H^+ can replace the growth requirement for an auxin.

2. Addition of auxin alone causes the cells to secrete or otherwise release H^+ into the growth medium, especially if the cuticle is abraded. These H^+ first move from the cytoplasm into the wall,

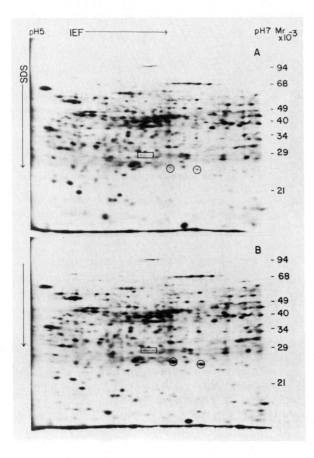

Figure 16-8 IAA-induced increases in certain polypeptides synthesized by messenger RNAs in maize coleoptiles (A, controls). Coleoptile sections in B were exposed to 50 μM IAA for 20 min to promote elongation, then messenger RNAs from sections without auxin or with auxin were isolated. These mRNAs were translated into proteins by a cell-free extract of wheat germ containing ³⁵S-methionine to label all proteins formed. Proteins were boiled with a detergent (sodium docecyl sulfate, SDS), which converts heteropolymers or homopolymers (Chapter 8) into individual polypeptide chains. These polypeptides were then separated by two-dimensional polyacrylamide gel electrophoresis, first from left to right by isoelectric focusing (IEF), which separates on the basis of charge, and then from top to bottom which separates on the basis of molecular weight. The molecular weights in thousands of grams per mole are listed at the right of each photograph. Radioactive polypeptides were visualized as spots or streaks by placing the gel next to a film sensitive to radioactivity (see autoradiography, Chapter 7). (Unpublished data of L. L. Zurfluh and T. J. Guilfoyle). Three boxed or circled polypeptides become more abundant soon after auxin treatment.

so the wall itself must become acidified during this movement. Furthermore, the time required for H⁺ release correlates well with the time required for the cells to begin to grow faster after an auxin is added.

3. Addition of buffers at neutral or slightly alkaline *p*H to cells elongating in the presence of either an auxin or a low *p*H inhibits elongation.

Besides this evidence, results with another compound called *fusicoccin* support the acid-growth hypothesis. Fusicoccin is a diterpene glucoside discovered by plant pathologists in the 1960s as the principal toxin responsible for disease symptoms caused by the fungus *Fusicoccum amygdali* on peach, almonds, and prune trees (for a review of its effects and chemical nature, see Marré, 1979). Fusicoccin has the remarkable ability to cause both H⁺ secretion and cell elongation in many plant parts. Its common mechanism of action is to activate the H⁺/K⁺ pump (Section 6.8); because of this it also causes stomates to open in essentially all species studied. The consistency with which it promotes cell elongation and acid secretion suggests that the acid-growth hypothesis might be correct not only for auxin-induced growth but universally when some chemical or environmental factor causes wall loosening. Consistent with that idea, there are other examples in which the hypothesis does seem to apply (some are discussed in Chapters 18 and 19).

There are several experiments which show that growth can be promoted and walls loosened in the absence of H⁺ efflux. Results with cytokinins are in the next chapter, and results with gibberellins are in Section 16.2 of this chapter. Even auxins can promote elongation when H⁺ ions cannot, as first shown by Larry N. Vanderhoef at the University of Illinois (for a summary review, see Vanderhoef, 1980). He and his colleagues distinguished the elongation effects of an auxin and H⁺ in two ways. They showed, for soybean hypocotyl sections, that elongation in response to a low *p*H only occurs for an hour or two. Others have verified this short response for stem sections from other dicot species and for sections from coleoptiles. After this relatively short period, the sections still respond to auxins for a day or two, but they no longer respond to H⁺. Furthermore, as described in Chapter 17 (see Fig. 17-8), long-term elongation of soybean hypocotyl sections caused by auxins is inhibited by cytokinins, whereas the initial growth period especially sensitive to H⁺ ions is not.

These results indicate that auxins promote elongation in detached stem or coleoptile sections by inducing wall acidification and by another mechanism necessary for sustained growth. Vanderhoef (1980) suggested that sustained growth requires auxin-induced production of cell-wall precursors and that the promotive effects of auxins on transcription described above might be to form enzymes that enhance cell-wall synthesis. These suggestions seem reasonable, yet sustained growth of cells requires much more than wall synthesis. It requires synthesis of the plasma membrane (lipids and proteins), because that membrane grows in contact with the wall. It also requires that the turgor pressure not decrease too much, even though the wall is loosened, because if the turgor drops excessively as the cells expand, insufficient pressure will be exerted against the walls to allow sustained growth. If Vanderhoef's suggestion is expanded somewhat to include enhanced formation of the plasma membrane and maintenance of turgor by enhanced solute uptake from the growth medium, a more encompassing explanation for auxin-induced cell elongation could arise. The importance of a proper concentration of mineral salts in the growth medium for auxin-induced growth (absorption and maintenance of turgor) was demonstrated by Oertli (1975) and Stevenson and Cleland (1981). Whether the effects of auxins on transcription and translation demonstrated by Key (1969) and Zurfluh and Guilfoyle (1982a, 1982b) relate to auxin-induced growth by any mechanism remains to be demonstrated, but this seems likely.

Auxins and Herbicides   In the 1940s, work at the Boyce Thompson Institute in New York established that 2,4-D has auxin activity. Later work there and in England showed that 2,4-D, NAA, and certain other related compounds are effective **herbicides,** or plant killers. Four of the most widely used auxin herbicides have been 2,4-D, 2,4,5-T, MCPA (Fig. 16-2), and derivatives of picolinic acid such as *picloram* (sold under the trade name Tordon). Their popularity derived from their toxicity, their relatively low cost, and their property of affecting dicots much more than monocots (Klingman et al., 1982; Moreland, 1980). Because of this selectivity, they are often used to kill broad-leaf dicot weeds in cereal grains and lawns. For grass pastures and rangelands in which woody perennials such as sagebrush and mesquite are often a problem, 2,4,5-T was particularly effective; but the U. S. Environmental Protection Agency forced its removal from the market, because it contained traces of a powerful toxin (dioxin). Several derivatives of benzoic acid such as *dicamba* also have auxin activity and are more effective than the others against deep-rooted perennial weeds, including field bindweed or wild morning glory (*Convolvulus arvensis),* Canada thistle (*Cirsium arvense),* and dandelion (*Taraxacum officinale).*

These herbicides are formulated as salts of weak bases such as ammonia (amines), as emulsifiable acids, or as esters, and are mixed with an oil or detergent to facilitate spreading and penetration of the waxy cuticle after spraying on the foliage. They are absorbed through the cuticle and are translocated throughout the plant largely in the phloem. Therefore, it is usually best to spray them in the morning of a warm, sunny day so that they are absorbed and translocated along with other photosynthetic products. Windy days are avoided to keep the herbicide from drifting to susceptible neighbor crops.

In spite of much research to determine how they kill only certain weeds, their mechanism of action is generally unknown. Part of their selectivity against broad-leaf weeds results from greater absorption and translocation than in grasses, but more important factors are involved. It is sometimes stated that they cause a plant to "grow itself to death," but this is misleading. Certain parts of some organs do indeed grow faster than other parts, so we see twisted and deformed leaf blades, petioles, and stems because of unequal growth. Much of this results from epinastic effects (Fig. 17-12) that arise from the common property of all auxins to enhance ethylene production, because ethylene is notorious for causing epinasty. But unequal growth that causes twisting could result from inhibition of one part and not promotion of the other. Overall growth of plants is definitely retarded and eventually stopped if enough herbicide is absorbed and translocated. Modern hypotheses suggest that these compounds alter DNA transcription and RNA translation, so that the proper enzymes needed for coordinated growth are not produced properly, but how this is accomplished is unknown.

We shall later consider other auxin functions in relation to phototropism and gravitropism (Chapter 18) and in delaying abscission of leaves, flowers, and fruits (Chapter 17).

16.2 The Gibberellins

Gibberellins were first discovered in Japan in studies with diseased rice plants that grew excessively tall (for a historical review, see Phinney, 1983, and Thimann, 1980). These plants often could not support themselves and eventually died from combined weakness and parasite damage. As early as the 1890s, the Japanese called this the **bakanae** (foolish seedling) **disease.** It is caused by the fungus *Gibberella fujikuroi* (asexual or imperfect stage is *Fusarium moniliforme*). In 1926, plant pathologists found that extracts of the fungus applied to rice caused the same symptoms as the fungus itself, demonstrating that a definite chemical substance is responsible for the disease. In the 1930s, T. Yabuta and T. Hayashi isolated an active compound from the fungus, which they named **gibberellin.** Thus the first gibberellin was discovered as early as IAA; yet because of preoccupation with IAA and synthetic auxins, lack of early contact with the Japanese, and then World War II, the western hemisphere did not become interested in gibberellin effects until the early 1950s.

More than 60 gibberellins have now been discovered in various fungi and plants (51 from seed plants), although no single species contains more than 15 (Phinney, 1979; Jones and MacMillan, 1984) and most species have only a few. The structures of four of these are in Fig. 16-9. All gibberellins have

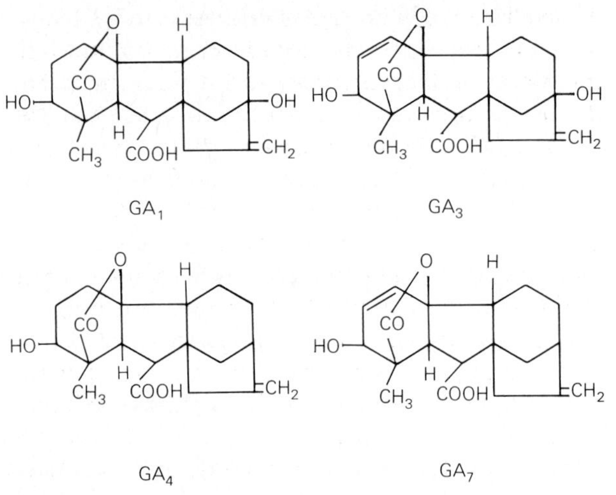

Figure 16-9 Structures of four highly active gibberellins.

either 19 or 20 carbon atoms grouped in a total of either four or five ring systems, and all have one or more carboxyl groups. They are abbreviated GA with a subscript, such as GA_1, GA_2, and GA_3, to distinguish them. All could properly be referred to as gibberellic acids, but GA_3 has been studied much more than the others (because of its availability), so it is commonly referred to as **gibberellic acid.** Gibber-

ellins exist in angiosperms and gymnosperms and probably also in mosses, ferns, algae, and at least two fungi, but apparently not in bacteria. It should also be noted that many of the discovered gibberellins are probably just physiologically inactive precursors of other active ones, and still others are inactive hydroxylated products. Less than half of the gibberellins have reasonably well-documented physiological effects on plants (Hedden, 1979), and further studies indicate that this number is much lower because of metabolic conversions within the cells (Jones and MacMillan, 1984). Furthermore, more than 20 gibberellins have been found in the fungus *G. fujikuroi,* in which they have no known function (Phinney, 1979; Jones and MacMillan, 1984).

Metabolism of Gibberellins As mentioned in Section 14.3, gibberellins are isoprenoid compounds. Specifically, they are diterpenes synthesized from acetate units of acetyl coenzyme A by the mevalonic acid pathway. Geranylgeranyl pyrophosphate (Fig. 16-10), a 20-carbon compound, serves as the donor for all gibberellin carbon atoms. It is converted to copalyl pyrophosphate, which has two ring systems, and then to kaurene. Some of these conversion steps are oxidations occurring in the endoplasmic reticulum. They involve the intermediate compounds kaurenol (an alcohol), kaurenal (an aldehyde), and

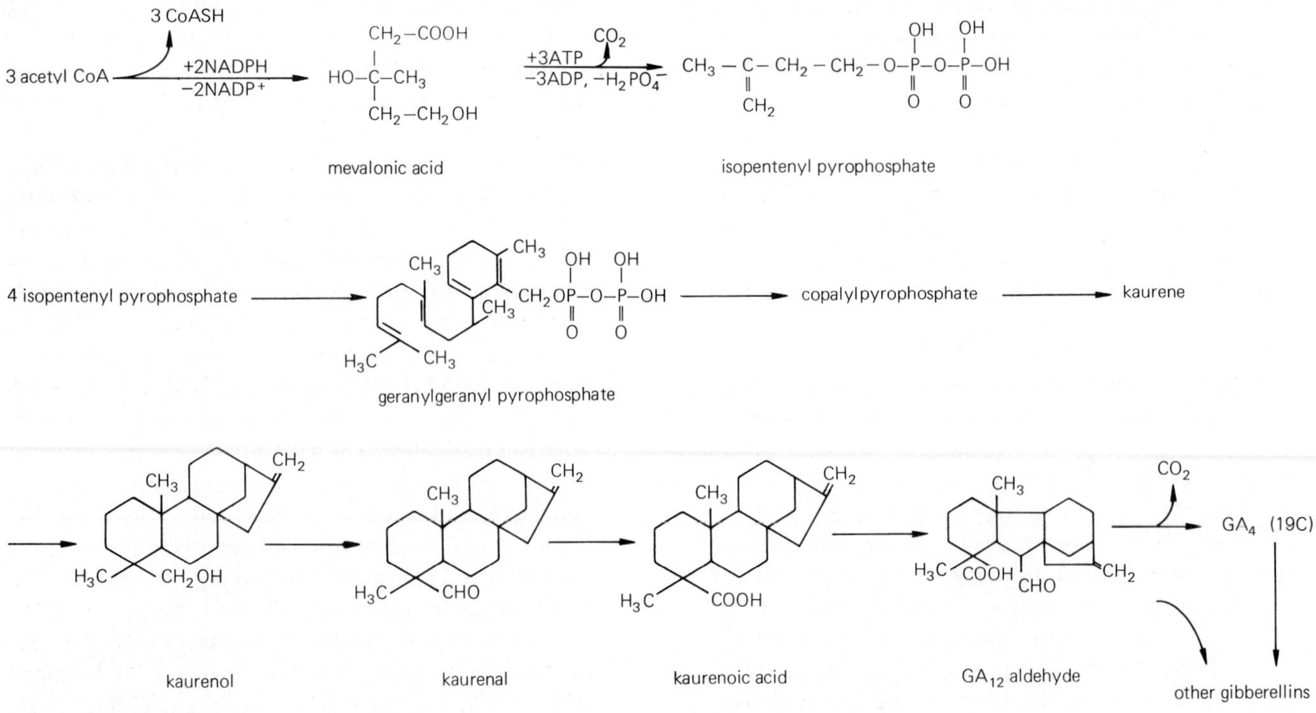

Figure 16-10 Some reactions of gibberellin biosynthesis. Many steps indicated as single arrows actually involve more than one enzyme-catalyzed reaction, especially those before kaurene.

kaurenoic acid. The first compound with a true gibberellane ring system is the aldehyde of GA_{12}, a 20-carbon molecule. From it arise both C-20 and C-19 gibberellins (Phinney, 1979; Crozier, 1981; Railton, 1982; Jones and MacMillan, 1984). In leaves, chloroplasts are a major site of interconversion of gibberellins, although reactions of the pathway up to kaurenoic acid probably occur outside plastids.

Certain commercial growth retardants that inhibit stem elongation and cause overall stunting do so in part because they inhibit gibberellin synthesis. These include *Phosphon D, Amo-1618, CCC* or *Cycocel,* and *ancymidol.* The first two block the conversion of geranylgeranyl pyrophosphate to copalyl pyrophosphate (see Fig. 16-10). Phosphon seems also to inhibit the subsequent formation of kaurene, while ancymidol inhibits a reaction between kaurene and kaurenoic acid. Growth inhibition by each can be completely overcome in many plants by GA_3, which first suggested that their only effects are to inhibit gibberellin synthesis. However, Phosphon D, Amo-1618, and CCC inhibit sterol synthesis in tobacco, in which the stunting effects of Amo-1618 and CCC are overcome not only by GA_3 but also by three sterols (β-sitosterol, stigmasterol, and cholesterol). These results and others strongly indicate that these three inhibitors affect processes other than gibberellin synthesis.

Once formed, gibberellins are usually only slowly degraded, but they can be readily converted to conjugates that are largely inactive. These conjugates might be stored or translocated prior to release at the proper time and place. Known conjugates include **glucosides,** in which glucose is connected in an ether bond to one of the OH groups or in an ester bond to a carboxyl group of the gibberellin. Another important metabolic process is conversion of highly active gibberellins to others less active. For example, Douglas fir shoots, which show little response to most applied gibberellins, can effectively convert GA_4 to the much less active GA_{34} by simply adding an additional hydroxyl group at a key position.

Which parts of the plant synthesize gibberellins? Clearly, if we find these hormones in a plant organ, they might have been either synthesized there or translocated there. Immature seeds contain relatively high amounts of gibberellins compared with other plant parts, and for several species extracts from seeds are fully capable of synthesizing gibberellins. These and other results indicate that much of the high gibberellin content of such seeds results from *in situ* biosynthesis, not from transport.

The ability of other plant parts to synthesize gibberellins is less well established, because no direct biochemical data are available (Crozier, 1981). Nevertheless, the diffusion technique of Jones and Phillips (1964) has helped distinguish between synthesis and transport. An organ was excised, the cut surface was placed in contact with agar, and gibberellins were allowed to diffuse into the agar for several hours. The amounts collected were estimated only by bioassays but could now be analyzed by instrumental methods. These amounts were compared with quantities extractable from similar organs or tissues both before and after diffusion. If the diffusible amounts were significantly greater than the extractable amounts, as they often were, the cells were presumed to have been synthesizing gibberellins or converting them from an inactive form (glucoside?) to an active form.

This diffusion procedure showed that mature leaves have little ability to synthesize gibberellins, just as seems true for auxins. But young leaves are apparently major sites for gibberellin synthesis, again as is true for auxins. This is consistent with the fact that if the shoot tip with young leaves is excised and the cut stump is then treated with either a gibberellin or auxin, stem elongation is promoted relative to cut stems not given hormone. The implication is that young leaves normally promote stem elongation because they transport both hormones to the stem. This is curious, because young leaves are translocation sinks, not sources. For auxins, we explained that their transport is not usually via the phloem but is polarly in other cells, so no problem in explaining their transport arises. But for gibberellins, transport other than by diffusion occurs through both xylem and phloem and is not polar. How gibberellins could be transported effectively from young leaves to cause stem elongation, if indeed this occurs, is unknown.

Roots also synthesize gibberellins in significant quantities, as shown by the diffusion technique; yet gibberellins have little direct effect on root growth, and they inhibit adventitious root formation. These hormones can be detected in the xylem exudates of roots and stems when these organs are excised and root pressure forces the xylem sap out. Inhibitors of gibberellin synthesis decrease amounts of gibberellins in these exudates. Repeated excision of part of the root system causes marked decreases in the concentrations of gibberellins in the shoot, suggesting that much of the shoot's gibberellin supply arises from the roots via the xylem.

Gibberellin-Promoted Growth of Intact Plants
Gibberellins have the unique ability among recognized plant hormones to promote extensive growth of many intact plants. They generally enhance elongation of intact stems much more than they do excised stem sections, so their effects are opposite to those of auxins in this respect. An early demonstration of elongation caused by an ether-soluble substance extracted from bean seeds was made by John W. Mitchell and his col-

Sensitivity to Growth Substances:
A Major Role in Controlling Development

Anthony Trewavas

Anthony Trewavas received his Ph.D. at University College, London, after studies on the metabolic effects of indoleacetic acid. After a further six years at the University of East Anglia he moved to Edinburgh in 1970. His current interests are still plant growth substances but mixed also with calcium regulation and protein phosphorylation.

How do plant growth substances work? What is their role in the development of the intact plant? These persistent questions have puzzled physiologists and biochemists for over 50 years. And the more information we acquire, the more confusing and uncertain their role appears. The theory that has been the conceptual cornerstone of research might loosely be called the hormonal view of growth substances. Originating in the 1930s this theory suggested that growth substances were inductive agents (i.e., hormones) mediating and controlling cell and tissue interactions and cell growth during development; it regarded the concentration of growth substances as of paramount importance.

A very simple test of the validity of these hormonal ideas is that some sort of parallel or preceding change in the concentration of the supposed controlling growth substance should accompany the developmental event itself. In many cases now, major discrepancies in these simple parallels have appeared. Two examples, out of many, illustrate the very real difficulties. Nongrowing buds can be induced to form by exogenous abscisic acid, but, in contrast, the formation of dormant winter buds in autumn involves no changes in endogenous abscisic acid. Again, stems and roots can be made to bend by unequal appli-

cation of auxin, but tropic bending in the same tissues is usually not accompanied by endogenous auxin redistributions that could explain the phenomenon. The feeling is in the air that some reassessment of our basic concepts is necessary. After all, the parallel drawn between plant and animal hormones is only a parallel. Other than economy of concept, there is no reason to suppose that two such different groups of organisms as animals and plants conduct their multicellular relations in the same way.

I believe, like many others, that growth substances are significant factors in plant development. Thus I have sought reasons for these difficulties, and certainly a major one is a strange lack of awareness of the importance of the sensitivity to growth-substances during development. The response of any cell to a growth substance is the combined result of sensitivity and level of growth substance; variations in either could regulate or initiate changes in development itself.

Sensitivity is an operational term; it is measured by assessing the responsiveness of cells and tissues to exogenous growth substances. Although it must in part reflect the possession (or otherwise) of specific receptor proteins, it is more complicated than that; many other internal and external factors are known to modify sensitivity to growth substances. Thus the responsiveness of stomatal cells to abscisic acid or auxin is dramatically modified by their surrounding K^+ content, a parameter which itself changes during stomatal closure. The responsiveness of elongating cells to auxin is greatly modified by sucrose, light, various amino acids, and water availability, as is the sensitivity of abscission zone cells to ethylene and auxin, and so on. Plants live in a constantly changing environment and do not possess a facility for maintaining an unchanging internal composition. Factors that modify sensitivity are therefore likewise subject to continual variation, creating additional difficulties for hormonal type concepts.

In those cases for which information is available, vari-

leagues in 1951 (Fig. 16-11). They were not sure what caused this unusual growth promotion, but they demonstrated that IAA was not responsible. Now we know that seeds of beans and many other dicots are rich sources of gibberellins and that the symptoms Mitchell observed are identical to those caused by several purified gibberellins.

Most dicots and some monocots respond by growing faster when treated with gibberellins, but several species in the Pinaceae family show little or no elongation responses to gibberellins (Pharis and Kuo, 1977), presumably because they already contain sufficient amounts. Cabbages and other species in the rosette form that have short internodes sometimes grow 2 m tall and then flower after GA_3 application, while untreated plants remain short and vegetative. Short bush beans become climbing pole beans, and genetic dwarf mutants of rice, maize, peas, watermelons, squash, and cucumbers exhibit phenotypically tall characteristics of normal varieties when treated with GA_3 or certain other gibberellins. Dwarf meteor peas are sensitive to as little as one billionth of a gram of GA_3, so their growth has long been used for a gibberellin bioassay.

Five different dwarf mutants of maize grow as tall as their normal counterparts after gibberellin ap-

ations in sensitivity frequently parallel, and often precede, changes in development in the intact plant. Thus in growing stems, cells from the meristem embark on a programme of expansion in which they rapidly increase and then decrease their growth rates. Variations in sensitivity to auxin have been found to exactly parallel, but slightly precede, these changing growth rates. In contrast there is great difficulty in seeing any direct relationship between the auxin content of stems and the known elongation characteristics of the constituent cells. Thus the variation in sensitivity more accurately describes and looks, on any simple model, more like the controller of elongation itself. Similar continuous variations in sensitivity to appropriate growth substances seem to accompany fruit ripening, abscission, senescence, cell growth and division in leaves and stems, cambial development, breaking of seed dormancy, amylase production by cereal grain aleurone layers, and growth recovery by excised tissues. Variations in sensitivity could therefore be the controlling factor in all these cases. Faced with this variation there should now be no surprise that growth substance concentrations often fail to correlate with development at all.

Why has sensitivity been so ignored? Dose-response curves, which involve selection of tissue of uniform sensitivity, are probably one major reason. They predispose one to consider only growth substance concentrations, because that is the way a different or graded response is produced. The effect of exogenous growth substances on plant development is probably another reason. But the major obstacle is the general belief that development is specified and controlled in the intact plant by growth substances moving between cells. To suggest an important controlling role for sensitivity variation perhaps implies that cells are making important developmental decisions without external reference. And yet in a sense they do seem to. For example, fruit cells, as well as gradually becoming more sensitive to ethylene in the later stages of fruit growth,

subsequently commence the synthesis of enzymes to make ethylene itself. There is probably far more pre-programming of plant development than we have hitherto thought.

To emphasize sensitivity also questions the validity of what many see as the prevailing dogma in this field and the value of the huge research effort on growth-substance concentrations that supports it. But knowledge and understanding change and any scientific enterprise needs constant questioning.

Variations in sensitivity are not the only problem presently originating for the hormonal theory of growth-substances; other clouds are visible on the horizon. Most phases of development involve all five growth substances, which may interact in complicated ways and probably modify the sensitivity of cells to each other; dose response curves for growth substances are wider than would be anticipated and certainly much wider than known endogenous growth-substance changes. There are problems of growth substance compartmentation and the role of receptor proteins in causing their accumulation in critical cells. The well-known plasticity of plant size and form and the opportunistic character of plant growth raise baffling questions concerning the control of circulating levels of growth substances for which there seems no simple solution. How do we accommodate the fact that the effects of growth substances on development are often mimicked by other environmental or chemical stimuli? In addition, there are the complications of more than 2×10^5 angiosperm species and the likely variation in control systems that this figure implies.

The appreciation of the importance of variations in growth substance sensitivity reinforces the growing realization that plant development is a very complicated process that will require complicated explanations. It is, however, an essential step on what must be a very long but exciting road of exploration.

plication, although at least two others do not respond. Each dwarf that responds contains a mutation on a different gene, and each mutation might control a different enzyme needed in the pathway of gibberellin synthesis. However, only for three mutants has this been reasonably verified (Phinney and Spray, 1982; Jones and MacMillan, 1984). The studies of Phinney and his colleagues indicate that only GA_1 controls stem elongation in maize* and that all dwarf

mutants lack enzymes to convert other gibberellins to GA_1. Growth of normal varieties of maize is not appreciably stimulated by gibberellins, presumably because those plants contain enough gibberellin for growth.

Although numerous dwarfs respond to added gibberellins, we cannot conclude that all dwarfs are deficient in these hormones. Dwarf Japanese morning glory and pea *(Pisum sativum)*, which grow tall when treated with exogenous gibberellins, seem to have as much gibberellin active in bioassays as their normal counterparts, although there is no assurance that the gibberellins found in them are in the growth

---

*Similar new studies by Phinney suggest that only GA_1 controls stem elongation in rice and peas.

Figure 16-11 Growth stimulation of *Phaseolus vulgaris* L. by a gibberellin-containing extract prepared from seeds of the same cultivar (Black Valentine). An ether extract of seeds was evaporated and 125 μg of the residue were mixed with lanolin and applied as a band around the first internode of the plant on the right. Plants were photographed 3 weeks after treatment. Plant on the left was untreated. (From J. W. Mitchell et al., 1951.)

limiting cells or the proper subcellular locations within those cells. The same is true of several semi-dwarf, stiff-strawed wheat cultivars that respond well to fertilizers and are being used as genetic stock for plant breeding experiments. Dwarfing might also be caused by a low auxin content, an inability to respond to endogenous gibberellins or auxins, or high concentrations of ABA or unknown inhibitors, although no examples are known. Promotion of growth of various dwarf plants by gibberellins has been used as one common gibberellin bioassay. The first 38 gibberellins discovered were compared in the dwarf pea, dwarf rice, and three other bioassays (Reeve and Crozier, 1974; Crozier, 1981). In general, those highly active in one bioassay proved active in others, but there were important exceptions that could result from many causes.

Gibberellins Promote Germination of Dormant Seeds and Growth of Dormant Buds The buds of evergreens and deciduous trees and shrubs that grow in temperate zones usually become dormant in late summer or early fall. Dormant buds are relatively hardy during cold winters and drought. Seeds of many noncultivated species are also dormant when first shed and will not sprout even when exposed to adequate moisture, temperature, and oxygen. Dormancy of buds and seeds is often overcome (broken) by extended winter cold periods, allowing growth in the spring when conditions are favorable (Chapter 21). For some species, bud dormancy can also be overcome by the increasing day lengths that occur in late winter, and for seeds of many species dormancy is broken by brief periods of red light when they are moist (Section 19.4).

Gibberellins overcome both kinds of seed dormancy and both kinds of bud dormancy in many species, acting as a substitute for low temperatures, long days, or red light. In seeds, the principal gibberellin effect is to enhance cell elongation so the radicle can push through the endosperm, seed coat, or fruit coat that restricts its growth. Buds have been less carefully investigated, and whether a stimulation of cell division in addition to elongation is necessary is not known.

Flowering As we describe in Chapter 22, the time at which plants form flowers is dependent upon several factors, including their age and certain properties of the environment. For example, the relative durations of light and darkness have key controlling influences on several species. Some species flower only if the light period exceeds a critical length, and others flower only if this period is shorter than some critical length. Gibberellins can substitute for the long-day requirement in some species, again showing an interaction with light. Gibberellins also overcome the need some species have for an inductive cold period to flower (vernalization).

Gibberellins Stimulate the Mobilization of Foods and Mineral Elements in Seed Storage Cells Soon after a seed germinates, the young root and shoot systems begin to use mineral nutrients, fats, starch, and proteins present in storage cells of the seed. The young seedling depends upon these food-reserves before mineral salts can be absorbed from the soil and before extension of the shoot system into the light. The mineral salts are readily translocated via the phloem into and throughout the young roots and shoots, if they are mobile. The seedling has a problem with fats, polysaccharides, and proteins, because these molecules are not translocated. How is the problem solved? We discussed this briefly in Chapters 12, 13, and 14, explaining how storage polymers are converted into sucrose and into mobile amino acids or amides. Gibberellins stimulate these conversions, especially in cereal grains (Akazawa and Miyata, 1982).

The embryo (germ) of seeds of cereal grains and other grasses is surrounded by food reserves present in the metabolically inactive cells of the endosperm; the endosperm in turn is surrounded by a thin, living layer two to four cells thick called the **aleurone layer** (Fig. 16-12). After germination occurs, primarily in response to increased moisture, the aleurone cells provide the hydrolytic enzymes that digest the starch, proteins, phytin, RNA, and certain cell-wall materials present in the endosperm cells. One of the necessary enzymes for these digestion processes is α-amylase, which hydrolyzes starch (Section 12.2). If the embryo is removed from a barley seed, the aleurone cells do not produce and hardly secrete most hydrolytic enzymes, including α-amylase. This suggests that the barley embryo normally provides some hormone to the aleurone layer and that this hormone stimulates aleurone cells to manufacture these hydrolytic enzymes. This hormone, which proved to be a gibberellin, also stimulates secretion of the hydrolytic enzymes into the endosperm where they digest food reserves and cell walls. Reserve mineral elements also become more readily available as a result of gibberellin action. Figure 16-13 illustrates endosperm degradation in barley half-seeds (from which the embryo was removed) in response to as little as 9 trillionths of a gram of GA_3. The increase in α-amylase in aleurone layers of such half-seeds is frequently used as a gibberellin bioassay.

In grass seeds, including barley, gibberellins are probably synthesized in the **scutellum** (cotyledon) and perhaps also in other parts of the embryo. The kind of gibberellin likely depends upon the species, but in barley GA_1 and GA_3 seem most important. Nevertheless, although the aleurone layers of barley, wheat, and wild oats (*Avena fatua*) respond to added GA_3 or certain other gibberellins by synthesizing α-amylase and other hydrolytic enzymes, some cultivated oat and most maize cultivars do not. There is considerable genetic variability among cereal grain seeds relative to gibberellin responses.

Although the aleurone layer is responsible for enzymes that digest some of the reserve foods in the endosperm, there has been evidence for about 100 years that the scutellum also secretes enzymes that cause digestion. The portion of the scutellum that faces the endosperm is composed of a single layer of columnar cells whose internal structure is rich in endoplasmic reticulum and dictyosomes typical of secreting cells (Akazawa and Miyata, 1982). Recent evidence from Akazawa's laboratory in Japan and by

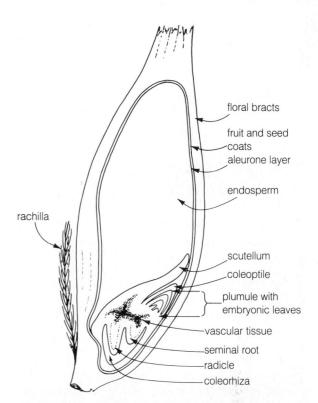

Figure 16-12 Barley seed sectioned to illustrate major tissues. (Original drawing by Arnold Larsen, Colorado State University Seed Laboratory.)

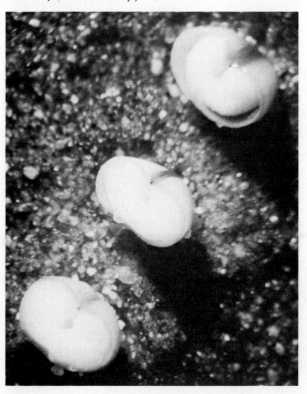

Figure 16-13 Gibberellin-stimulated digestion of the endosperm in barley leaf seeds. The embryo half of each seed was removed before treatment (top to bottom) with 5 μl of 0.1 μM GA_3, 0.001 μM GA_3, or H_2O. (Courtesy J. E. Varner.)

Why a Biologist? Some Reflections

Frits W. Went

With the discovery of auxin, Frits W. Went's fame was assured. In this essay, he tells how that happened while he was a young student working in his father's laboratory at the University of Utrecht in the Netherlands. After completing the doctoral degree, he spent five years in Java, a Dutch possession, and almost 20 years at the California Institute of Technology, where he continued hormone work and developed interests in desert ecology. He moved in 1958 to St. Louis, Missouri, and then in 1964 to the Desert Biology Laboratory at the University of Nevada, where he has continued his desert studies.

Years ago, I tried to discover what made a biologist become a biologist. Soon I found that there are about as many different motivations as biologists, but a few were more prevalent than others. Every human being is born with an enormous amount of intellectual curiosity. If this has not been blunted through unfortunate experiences in youth, then an inspiring teacher may guide this urge for understanding toward biological or other problems. Early acquaintance with plants or animals can direct an inquisitive mind—even without teacher direction—toward the mysteries of life: the problems of growth, form, function, environment, and heredity. The orderly mind may be attracted toward taxonomy or biophysics, whereas the mind intrigued by complexity may select ecology, and the mind trying to understand interrelationships may become a physiologist. The mechanically inclined person might attack biological problems with delicate instruments, whereas the artist might try to solve problems of shape and color in nature.

There is an equal plethora of methodological approaches to the solution of biological problems. Since Francis Bacon in the seventeenth century pointed out the inevitability of cause and effect, the experimental or inductive approach has taken precedence over the old Aristotelian deductive approach, which through pure reasoning forced life into a straitjacket of axioms or preconceived ideas.

Not a few bright minds have been misled by their or others' sophisticated hypotheses. Such complicated but artificial fabrications as the phlogiston theory of the eighteenth century and the ether theory of the nineteenth century have held back real understanding. It is difficult for us to say which present-day ideas may be discarded in the centuries to come, but the Second Law of Thermodynamics might well be one of them.

An important aspect of the motivation to become a biologist may be humanitarian—the urge of the person to take a positive part in the well-being of his fellow man. Most disciplines in biology, particularly agriculture and medicine, have an input to society, and let us not forget teaching, one of the most important motivations of all. This desire to transmit the knowledge gathered by our culture to the generations to come is quite the contrary of the attitude of a professor 100 years ago, who admonished his successor never to tell the students everything he knew!

In my own case, ever since I chose botany over chemistry or engineering as my life's endeavor, I have been intrigued with the form and function of a plant and with its place and role in the environment. When walking in the country, I am forever wondering why a particular plant grows in a particular spot; why it has its particular shape; why it does not grow 100 m farther on; why only some plants grow in the desert or tropics; why a limited number of plants have become weeds; why some plants are closely similar to others that lived 200 million years ago, whereas most are recently evolved; why some trees can be tapped for sugar, whereas most cannot. And I want to look inside the plant to find how it grows and functions, why it branches the way it does, and how it responds to its environment in such a precise way.

Some of these questions have already been answered, at least partially. Yet in many cases the answers don't satisfy me; they are either not general enough, too simple, or blatantly anthropocentric. This means that for me nature is still full of interesting problems.

One of the early problems that attracted me as a student was phototropism. Many of my fellow students thought that our predecessors in my father's laboratory—Blaauw, Arisz, Bremekamp, and Koningsberger, with their doctoral theses on phototropism—had preempted this subject. But such others as Dolk, Dillewyn, and Gorter were fascinated by the unsolved problems of plant responses to environment, and we had almost nightly bull sessions. I had to fulfill my military service obligations, which left only evenings and nights free for more productive projects. We discussed the newest publications of Paal, Seubert,

Nielsen, and Stark—dissecting, interpreting, or repeating them.

This was in early 1926, an exciting time, with the growth substance concept just around the corner. Our discussions were usually based on Paal's theory, that the stem tip normally produces a growth-promoting factor. The hottest point of debate was whether or not in phototropism this factor was destroyed by the light. To wind up the argument, I asserted that I would "prove that the growth regulator from the tip was light stable." Consequently, I had to extract it from seedlings and then expose it to light. For this, I prepared a tiny cube of gelatin, stuck it on a needle, and placed cut stem tips on it all the way around. When I removed the tips after an hour and placed the gelatin cube on one side of the seedling stump, nothing happened at first. But in the course of the night, the stump started to curve away from the gelatin block. It had acquired the capacity of the stem tips to grow! At 3:00 A.M. on April 17, 1926, I ran home to my parents' house nearby, burst into their bedroom, and said excitedly: "Father, come and see, I've got the growth substance."

My father (who was also my major professor) sleepily turned around and said: "Fine. Repeat the experiment tomorrow [which was my day off from military service]; if it is any good, it will work again, and then I can see it."

Then followed an exciting time. I lived for my nights in the laboratory. Every experiment seemed to work, and I learned a lot about the behavior of the growth substance in and outside the stem tip. Of course, I chose this subject for my thesis work. But then, with completion of my military obligation, something unexpected happened. Although I improved my whole experimental procedure, rebuilt the temperature and humidity controls of the lab room, grew much better seedlings, and worked much more cleanly, the growth substance seemed to have disappeared, for none of my test plants responded any more—until I found that bacteria lurking in the gelatin ate all of the growth substance overnight! My procedure had changed, because I prepared the blocks during the day and left them overnight. When I pressed an icebox into service (refrigerators were at that time unavailable), the bacteria ceased their activity and all experiments succeeded again.

To begin with, my approach to scientific problems was rather naïve. I mentioned already that I set out early to *prove* that the growth substance was light stable. I soon learned that experiments cannot prove anything; they can only *test* a hypothesis. Thus, my experiments to test

whether or not auxin is moved along a gradient by decrement unexpectedly showed that its transport was polar. And my tests on the behavior of auxin inside the seedling in unilateral light showed that this deflected the strictly downward auxin stream laterally, thus laying a solid basis for the Cholodny-Went theory of phototropism. Further work suggested that other growth factors were involved in the action of auxin. In an unaccountable way, I was later accused of promulgating theories, whilst I was only presenting experimental facts (which were finally, although many years later, accepted as true).

After receiving my Ph.D., I was faced by a completely new challenge. The working conditions for a tropical botanist in Java—before the blessing of air conditioning—were harder. There was less equipment, which operated less well in the oppressive moist heat, so my main efforts went to ecology or, rather, applied physiology. I found tropical plants admirably suited for work on root initiation, but only after moving to California, where I could again devote full time to physiological problems, could I work out a proper test to study root formation.

Whereas all my auxin experiments were based on more or less logical sequences of deductions leading to the crucial experiments, my ecological work was mostly a set of questions asked of nature, after observations of nature had posed the problems. We constructed a phytotron, in which such environmental factors as temperature, light, humidity, wind, or rain could be controlled. With this new tool I could establish, for example, that the profuse flowering of the desert in certain years depended upon the precise germination response of seeds from desert annuals to temperature and rain and not upon a mystic "survival of the fittest" or a "struggle for existence." In the last 30 years, phytotrons have helped make ecology an experimental rather than a descriptive science; now the extreme complexity of the organism in its total environment can be reduced to experimentally manageable subunits. It is satisfying when laboratory experiments give us a better insight into the mechanism of life, but for me the greatest thrill comes when these new insights help me understand what goes on in nature, when laboratory knowledge is applicable in the field. Thus, nature not only provides the inspiration; it is also the ultimate arbiter. The laboratory is only an interlude between perceiving and understanding.

(See also F.W. Went, 1974, Reflections and speculations, *Annual Review of Plant Physiology* 25:1–26.)

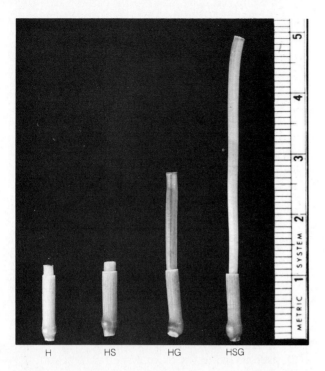

Figure 16-14 Effect of GA and sucrose on the growth of 1 cm oat stem segments. The segments are shown after 60 hr of treatment in Hoagland's nutrient solution (H), Hoagland's + 0.1 M sucrose (HS), Hoagland's + 30 μM GA (HG), and Hoagland's + sucrose + GA (HSG). A centimeter ruler denotes actual size. Elongation of the leaf sheaths did not occur, but growth of cells derived from the intercalary meristem (see Fig. 15-12) accounts for the stem elongation illustrated. (From P. A. Adams et al., 1973.)

Gibbons (1981) in Denmark indicate that the scutellum is probably even more important than the aleurone layer for providing enzymes that digest endosperm reserves in several species. This seems especially true during the first two days when little activity of the aleurone layer can be detected, although the aleurone layer contributes substantially after the seed has germinated. Interestingly, gibberellins have no significant effect on digestion by the scutellum, even though this organ is believed to produce gibberellins that activate the aleurone layer.

Gibberellins have much less dramatic effects on mobilization of food reserves in dicots and gymnosperms than in cereal grains, although the presence of the embryonic axis is still essential in some species for normal degradation of these reserves in the food storage cells (Ilan and Gepstein, 1981). In castor bean, a dicot in which the endosperm remains well-developed in the mature seed, degradation of fats does not require the presence of the embryo, yet fat breakdown is increased by added gibberellins (Marriott and Northcote, 1975). Whether this means that gibberellins are already present in sufficient amounts in the endosperm itself is not yet known. In other dicots and in gymnosperms, degradation of starch and fat is not affected by added gibberellins, but cytokinins sometimes replace the normal role of the embryo in hastening fat breakdown.

Other Gibberellin Effects Gibberellins cause parthenocarpic (seedless) fruit development in some species, suggesting a normal function in fruit growth, and gibberellins formed in young leaves might renew activity of the vascular cambium in woody plants. Other important effects of gibberellins are the delay of aging (senescence) in leaves and citrus fruits and effects on leaf shapes; the latter is a response especially apparent in leaves that show heterophylly or phase changes (Chapter 18).

Possible Mechanisms of Gibberellin Action The many effects of gibberellins suggest that they have more than one important primary site of action. Thus far, research with hormone receptors neither verifies nor denies that idea. Even a single effect such as enhanced stem elongation on whole plants results from at least three contributing events. *First, cell division is stimulated in the shoot apex,* especially in the more basal meristematic cells from which develop the long files of cortex and pith cells (Sachs, 1965). Careful work of Peter B. W. Liu and J. Brent Loy (1976) showed that gibberellins promote cell division because they stimulate cells in the G_1 phase to enter the S phase and because they also shorten the S phase. The mechanism of this has not been elucidated, but the increased number of cells leads to more rapid stem growth, because each of these cells can then grow.

Second, gibberellins sometimes promote cell growth because they increase hydrolysis of starch, fructans, and sucrose into glucose and fructose molecules. These hexoses provide energy via respiration; they contribute to cell-wall formation; and they also make the cell's water potential momentarily more negative. As a result of the decrease in water potential, water enters more rapidly, causing cell expansion but diluting the sugars. In sugar-cane stems gibberellin-promoted growth results in part from increased synthesis of invertase enzymes that hydrolyze incoming sucrose into glucose and fructose (Glasziou, 1969). In dwarf peas, the activities of both invertase and amylase enzymes rise in parallel with growth (Broughton and McComb, 1971). The same is true for amylase in dwarf maize. Less quantitative work with other species indicates that gibberellin-induced stem growth is associated with increases in amylase activity in small water plants and in certain trees, suggesting that the phenomenon is general, but we know of no results for conifers.

Third, gibberellins sometimes increase wall plasticity. An excellent example of this occurs in internodes of oat, in which growth promotion of young cells derived from the intercalary meristem is unusually dramatic. Here no enhancement of cell division occurs. Elongation caused by GA$_3$ is 15-times as great as in the untreated sections (Fig. 16-14), provided that sucrose and mineral salts are present to provide energy and prevent excessive dilution of the cell contents (i.e., prevent a rise in the osmotic potential). A significant increase in wall plasticity occurs (Adams et al., 1975), and a similar phenomenon explains gibberellin-promoted growth in lettuce hypocotyls (Stuart and Jones, 1977; Jones and MacMillan, 1984).

Not only is stem elongation promoted by gibberellins, but so is growth of the whole plant, including leaves and roots. We stated that application of gibberellins directly to leaves does stimulate their growth slightly and does influence their shapes, yet direct application to roots usually has almost no effect on the roots themselves. But if the gibberellin is applied in any manner whereby it can move into the shoot apex, increased cell division and cell growth apparently lead to increased elongation of the stem and in some species to increased development of the young leaves. In species in which faster leaf development occurs, enhanced photosynthetic rates then increase growth of the whole plant, including the roots.

How might gibberellins loosen cell walls and also increase formation of hydrolytic enzymes leading to stem elongation? We have no evidence about wall-loosening mechanisms except that, in contrast to the initial growth burst caused by auxins, H$^+$ ions are not involved (Stuart and Jones, 1978; Jones and MacMillan, 1984). In the oat internodes, there is a lag of nearly 1 h before promotion of elongation can be detected. This should allow ample time for gibberellins to increase synthesis of RNA or enzymes needed to amplify growth responses. For lettuce hypocotyl sections, lags of less than 20 min occur. And intact dwarf peas were reported to elongate faster within 10 min after treatment with GA$_3$ (McComb and Broughton, 1972). In this case, hydrolases that attack cell-wall polysaccharides might be synthesized faster or simply become more active in gibberellin-treated cells.

Fourth, in barley aleurone layers gibberellins increase transcription of the gene that codes for α-amylase (Bernal-Lugo et al., 1981; Mozer, 1980). How this gene activation occurs is unknown, but it certainly suggests that numerous other effects of gibberellins could result from activation of other genes.

Commercial Uses of Gibberellins Considering the numerous effects of gibberellins, it seems logical that

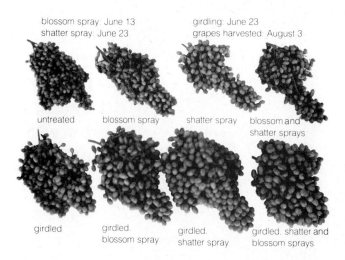

Figure 16-15 Effects of gibberellin and girdling on growth of Thompson seedling grapes. Dates of treatment: blossom spray, June 13; shatter spray, June 23; girdling, June 23. Harvest, August 3. (Courtesy J. LaMar Anderson, Utah State University, Logan.)

they would be used commercially. Major limiting factors have been their cost and their frequent promotion of fresh weights but not of dry weights, especially regarding the possible application in growth of pastures and hay crops. We still must rely on the *Gibberella* fungus to synthesize GA$_3$ at reasonable cost even for physiological experiments. Nevertheless, they are used extensively in the Central and Imperial Valleys of California to increase the size of Thompson seedless grape berries and the distance between them (Fig. 16-15). When applied at the right time and with the proper concentration, gibberellins cause the bunches to elongate so they are less tightly packed and less susceptible to fungal infections. Usually the plants are sprayed twice, once at bloom and again at the fruit-set stage (Nickell, 1978, 1979). Gibberellins are also used by some breweries to increase the rate of malting through their enhancing effects on starch digestion. Celery plants, which are valued for the lengths and crispness of their stalks, respond favorably to gibberellins, but the poor storage qualities of such stalks limit wide use of these hormones in the celery industry. Gibberellins have also been sprayed on fruits and leaves of navel orange trees (when the fruits have lost much of their green color) to prevent several rind disorders that appear during storage. Here the hormones delay senescence and maintain firmer rinds. They are now used commercially in Hawaii to increase sugar-cane growth and sugar yields.

17

Hormones and Growth Regulators: Cytokinins, Ethylene, Abscisic Acid, and Other Compounds

The more we learn about growth and development, the more complex these processes seem to become. In the last chapter, we explained that both processes depend upon IAA and gibberellins, but that these hormones generally influence different parts of the plant in different ways. Thus, whereas IAA promotes cell division that leads to production of lateral roots and adventitious roots, gibberellins enhance cell division in the shoot apical meristem that leads to stem elongation. Both promote elongation of stem cells but apparently by different primary mechanisms. Both increase division and growth of fruit cells causing parthenocarpic fruits, yet this effect varies considerably with the species. In spite of the complexities, we now realize that both hormone types must be considered in understanding growth. In this chapter, we shall discuss the other three kinds of hormones that are presently known (cytokinins, ethylene, and abscisic acid), emphasizing that although each has distinct effects, growth and development normally involve an interplay between all known hormones and probably others yet undiscovered. We shall also mention some additional compounds that are sometimes active as growth substances.

17.1 The Cytokinins

About 1913, Gottlieb Haberlandt discovered in Austria that an unknown compound present in vascular tissues of various plants stimulated cell division that caused cork cambium formation and wound healing in cut potato tubers. This was apparently the first demonstration that plants contain compounds, now called **cytokinins,** that stimulate cytokinesis. In the 1940s, Johannes van Overbeek found that the milky endosperm from immature coconuts is also rich in compounds that promote cytokinesis. In the early 1950s, Folke Skoog and his colleagues, who were then interested in auxin-stimulation of plants grown in tissue cultures, found that cells in pith sections from tobacco stems divided much more rapidly if a piece of vascular tissue was placed on top of the pith, verifying Haberlandt's results. They tried to identify the chemical factor from the vascular tissues, using growth of tobacco pith cells as a bioassay system. These cells were cultured on agar media containing known sugars, mineral salts, vitamins, amino acids, and IAA. IAA itself increased growth for a time by causing enormous cells to be formed, but these cells did not divide. Many of these cells were polyploids with several nuclei. In seeking substances that would promote cell division, they found an adenine-like compound in yeast extracts that was highly active. This led to investigations of the ability of DNA to promote cytokinesis (because DNA contains adenine) and, in 1954, to the discovery by Carlos Miller of a very active compound formed by partial breakdown of aged or autoclaved herring sperm DNA. They named this compound **kinetin.**

Although kinetin itself has not been found in plants and is not the active substance found by Haberlandt in phloem, related cytokinins are present in most plants. F. C. Steward, also using tissue culture techniques in the 1950s, found several cytokinins in coconut milk that enhance cell division in carrot root tissues. The most active of these were later shown by D. S. Letham (1974) to be compounds previously given the common name **zeatin** and **zeatin riboside.** Zeatin had first been identified by Letham in 1964 and almost simultaneously by Carlos Miller, both of whom used the milky endosperm of corn (*Zea mays*) as a source. Since then, other cytokinins with adenine-like structures similar to kinetin and zeatin have been identified in numerous parts of seed plants. None of these cytokinins is present in DNA, nor are they breakdown products of DNA, but some occur in transfer RNA (tRNA) and sometimes in ribosomal RNA molecules of seed plants, yeasts, bacteria, and even primates, and some exist as unbound, free cytokinins. It is the unbound cytokinins that

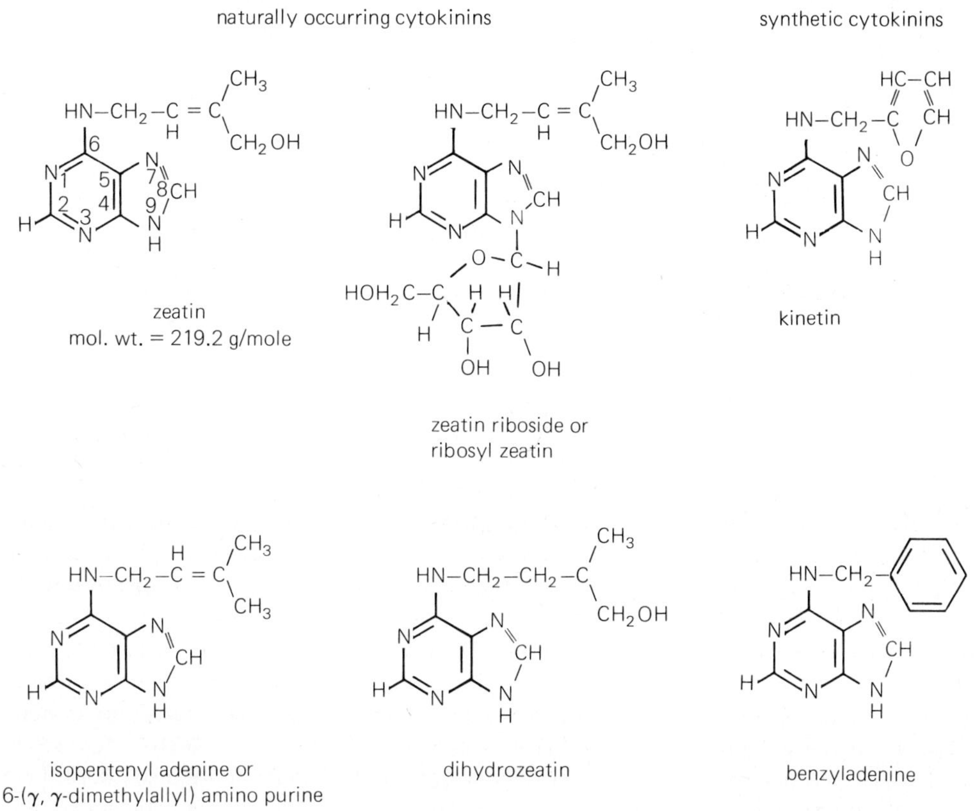

Figure 17-1 Structures of common natural and synthetic cytokinins. All these are adenine derivatives in which the purine ring is numbered as shown for zeatin (upper left). Zeatin and zeatin riboside can exist with groups arranged about the side chain double bond either in the *trans* (as shown) or *cis* (with CH_3 and CH_2OH groups interchanged) configuration. The *cis* form predominates in tRNA-bound cytokinins, but the *trans* form exists in free zeatin and zeatin riboside.

cause known physiological responses described in this chapter, but those in tRNA probably also have unknown functions.

Figure 17-1 (top) shows structures of the free-base form of the three most commonly detected and most physiologically active cytokinins in various plants: zeatin, **dihydrozeatin,** and **isopentenyl adenine (IPA).** Also shown are kinetin and another synthetic cytokinin, **benzyladenine,** which are highly active but which are probably not formed by plants.* A few other cytokinin bases have been found free or in tRNA of various species (reviewed by Letham and Palni, 1983; Horgan, 1984), but we shall not discuss them. Note that all cytokinins have a side-chain rich in carbon and hydrogen attached to the nitrogen protruding from the top of the purine ring. Each cytokinin can exist in the **free-base** form shown or as a **nucleoside,** in which a ribose group is attached to the nitrogen atom of position 9 (see ring-numbering system for zeatin in Fig. 17-1). For example, zeatin riboside is a relatively abundant cytokinin in many plants. Furthermore, the nucleosides can be converted to **nucleotides,** in which phosphate is es-

terified to the 5'-carbon of ribose, just as in adenosine-5'-phosphate (AMP). In a few cases, evidence for formation of nucleoside diphosphates and triphosphates similar to ADP and ATP has also been obtained, but all these nucleotides seem to be less abundant than the free-base or nucleoside forms.

Two questions now arise: How do we define a cytokinin, and should the free bases, nucleosides, and nucleotides all be considered cytokinins? Not every expert would agree on the same definition, but a reasonable one should depend partly on early discoveries that they promote cytokinesis (cell division) in tissues grown *in vitro,* such as cultures from tobacco pith, carrot phloem, or soybean stems. In fact, R. Horgan (1984) defined them as substances which, in the presence of optimal auxin, induce cell division in the tobacco pith or similar assay system grown on an optimally defined medium. Other authors prefer to include in the definition the fact that such compounds are adenine derivatives and that they have common and important effects in addition to promoting cytokinesis. We shall describe these additional effects later, but because all of them promote cytokinesis, *it seems reasonable to define them as substituted adenine compounds that promote cell division in the above-mentioned tissue systems.* The question as to whether the free base, the nucleoside, or a nucleotide form is active has not yet been convincingly answered. Most evidence favors the free base as the

---

*Benzyladenine might be present in some species, because its nucleoside derivative, 6-benzyladenine riboside, was recently found to occur in anise cells (*Pimpinella anisum*) by Ernst et al. (1983).

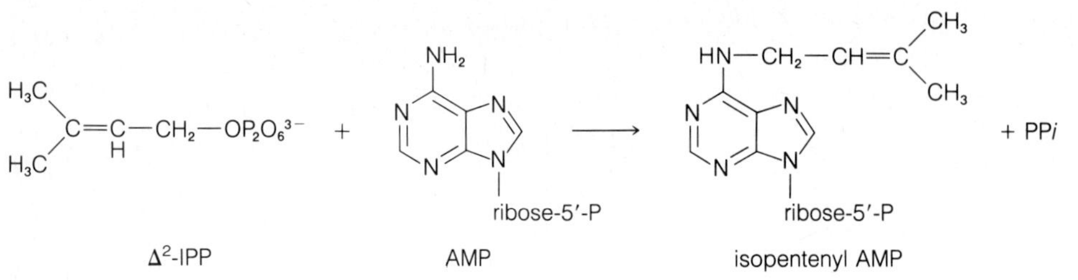

Figure 17-2 Formation of isopentenyl AMP, a precursor of isopentenyl adenine.

active form (Laloue and Pethe, 1982), but the rapidity with which the various forms are interconverted within cells makes this conclusion tentative.

Cytokinins also exist in mosses, brown and red algae, and apparently also in diatoms, and they sometimes promote algal growth. It is likely that they are widespread if not universal in the plant kingdom, but very little is known of their functions except in angiosperms, some conifers, and mosses. Certain pathogenic bacteria and fungi contain cytokinins that are believed to influence disease processes caused by those microbes, and cytokinin production by non-pathogenic fungi and bacteria is thought to influence mutualistic relations with plants, such as formation of mycorrhizae and root nodules (Greene, 1980; Ng et al., 1982).

Cytokinin Metabolism Two important questions about cytokinin metabolism should be asked. How do plants synthesize them, and how do plants regulate the amounts they contain? A breakthrough in our knowledge of biosynthesis came from the demonstration by Chong-maw Chen and D. K. Melitz (1979) that tobacco tissues contain an enzyme (previously discovered in a slime mold) that forms **isopentenyl adenosine-5'-phosphate (isopentenyl AMP)** from AMP and an isomer of isopentenyl pyrophosphate. (The latter compound is a product of the mevalonate pathway and is an important precursor of sterols, gibberellins, carotenoids, and other isoprenoid compounds—see Section 14.3.) The isomer involved is often called dimethylallyl pyrophosphate by those who study the mevalonate pathway, but it can also be properly referred to as Δ^2-*isopentenyl pyrophosphate*, where the prefix Δ^2 means that the molecule has a double bond between carbons 2 and 3. The reaction that occurs in tobacco tissues is shown in Fig. 17-2. Note that pyrophosphate (PPi) is released from the isopentenyl group and that the latter is added to the amino nitrogen attached to C-6 of the purine ring.

The isopentenyl AMP formed in this reaction can then be converted to isopentenyl adenosine by hydrolytic removal of the phosphate group, and iso-

pentenyl adenosine can be further converted to isopentenyl adenine by hydrolytic removal of the ribose group. Furthermore, isopentenyl adenine can be oxidized to zeatin by replacement of one hydrogen by an OH in a methyl group of the isopentenyl side chain (compare structures in Fig. 17-1). Dihydrozeatin is then formed from zeatin by reduction of the double bond in the isopentenyl side chain. These reactions are presently believed to account for formation of the three major cytokinin bases, but other possibilities for biosynthesis should also be considered (Van Staden, 1983).

Cellular levels of cytokinins are also affected by their degradation and by their conversion to presumably inactive derivatives other than nucleosides and nucleotides. Degradation is largely by **cytokinin oxidase,** an enzyme system that removes the five-carbon side-chain and releases free adenine (or, when zeatin riboside is oxidized, free adenosine). Formation of cytokinin derivatives is more complex, because many conjugates can be formed (Letham and Palni, 1983). Briefly, the most common conjugates contain either glucose or ribose. Carbon-1 of either of these sugars can be attached to the hydroxyl group of the side-chain of zeatin, zeatin riboside, dihydrozeatin, or dihydrozeatin riboside. Alternatively, C-1 of glucose can be attached to a nitrogen atom (via a C-N bond) at either position 7 or 9 of the adenine ring system in any of the three major cytokinin bases. No function of any of these conjugates is known, but they might represent storage forms or, in some cases, special transport forms of cytokinins. It is unlikely that such conjugates represent physiologically active forms, but even this has not been proven.

Sites of Cytokinin Synthesis and Transport If we knew how actively the above-mentioned reactions that form isopentenyl AMP, isopentenyl adenine, zeatin, and dihydrozeatin occur in various tissues, we would have good biochemical information about sites of cytokinin biosynthesis. Unfortunately, this information is not yet available, so less direct methods have been used to determine where cytokinins are formed. One method has been to find out where they are most abundant. In general, their levels are highest in young organs (seeds, fruits, and leaves) and in root tips. It seems logical that they are synthesized in these organs, but in most cases we cannot dismiss the possibility of transport from some other site. For root tips, synthesis is almost surely involved, because if roots are severed horizontally, cytokinins are exuded from the xylem of the lower portions remaining (by root pressure) for periods up to four days (Skene, 1975; Torrey, 1976). It is unlikely that these lower portions could store enough cytokinins derived from some other source to act as a rather long-term supplier for the xylem. Evidence such as this has led to the widespread idea that root tips synthesize cytokinins and transport them through the xylem to all parts of the plant. This might explain their accumulation in young leaves, fruits, and seeds into which xylem transport occurs, but the phloem is generally a more effective supply system for such organs with limited transpiration. Although root tips probably do represent an important cytokinin source for various plant parts, small, rootless tobacco plants effectively convert radioactive adenine into various cytokinins (Chen and Petschow, 1978). This observation and other studies with rootless or partially rootless plants indicate that shoots can synthesize some of the cytokinins they require (Carmi and Van Staden, 1983).

Transport of cytokinins (especially zeatin and zeatin riboside) certainly occurs in the xylem, but sieve tubes also contain cytokinins (especially glucosides), as evidenced by their presence in aphid honeydew. Further evidence for transport in the phloem is indicated by experiments with detached dicot leaves. When a mature leaf is cut off from plants of some species and kept moist, cytokinins move to the base of the petiole and accumulate there. This movement is probably through the phloem, not the xylem, because transpiration strongly favors xylem flow from the petiole to the leaf blade. Cytokinin accumulation in the petiole implies that mature leaf blades can supply young leaves and other young tissues with cytokinins via the phloem, provided of course that such leaves can synthesize cytokinins or receive them from roots. Nevertheless, if a radioactive cytokinin is added to the surface of a leaf, very little of that which is absorbed is transported out. These and many other results indicate that cytokinins are not readily distributed in the phloem. Almost surely, young leaves, fruits, and seeds that are transport sinks do not readily transport their cytokinins elsewhere via either xylem or phloem. Our tentative conclusion is that except for delivery from roots via the xylem, transport of cytokinins within the shoot is rather limited.

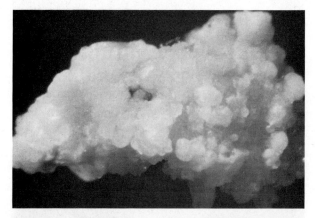

Figure 17-3 (**a**) Callus grown from the scutellum of a rice seed. (**b**) Embryogenic callus that has formed a young shoot (S) and root (R) system. (Courtesy M. Nabors and T. Dykes.)

Cytokinins Promote Cell Division and Organ Formation We explained that a major function of cytokinins is to promote cell division. Skoog and his colleagues found that if the pith from tobacco, soybean, and other dicot stems is cut out and cultured aseptically on an agar medium with an auxin and proper nutrients, a mass of unspecialized, loosely arranged, and typically polyploid cells called a **callus** is formed. Figure 17-3a indicates the general appearance of a callus. If a cytokinin is also provided, cytokinesis is greatly promoted, as already mentioned. The amount of growth from new cells serves as a sensitive and highly specific bioassay for cytokinins and is important in how we define these compounds.

Skoog also found that if the cytokinin-to-auxin ratio is maintained high, certain cells are produced in the callus that divide and give rise to others that develop into buds, stems, and leaves (also see totipotency, Section 18.7). But if the cytokinin-to-auxin ratio is lowered, root formation is favored. By choosing the proper ratio, callus from many species (especially dicots) can be made to develop into an entire

a   b

Figure 17-4 Development of (**a**) tomato and (**b**) petunia plants from a callus, illustrating totipotency. (Photographs courtesy of Murray Nabors and R. S. Sangwan.)

Figure 17-5 Senescence of a trifoliate bean leaf caused by treating the primary leaves of cuttings with the synthetic cytokinin benzyladenine (30 μg/ml) at 4-day intervals. (From A. C. Leopold and M. Kawase, 1964, *American Journal of Botany* 51:294–298.)

new plant. The ability of callus to regenerate whole plants represents a potentially powerful tool to select plants resistant to drought, salt stress, pathogens, and certain herbicides, or having other useful characteristics (see the personal essay of Murray Nabors at the end of Chapter 5 and reviews by Nabors, 1976; 1983; Davies, 1981; Chaleff, 1983; Barton and Brill, 1983).

The way in which a callus forms a new plant is variable. Frequently, with relatively high cytokinin-to-auxin ratios, only a shoot system first develops, then adventitious roots are formed spontaneously from the stems while still in the callus. (Roots can also be promoted to form from stems of young shoots taken out of the callus using common horticultural techniques—see Section 16.1.) This formation of shoots or shoots and adventitious roots by the callus is called **organogenesis.** Sometimes, however, calli become embryogenic (Fig. 17-3b) and form an embryo that develops into a root and shoot; this is called **embryogenesis.** Formation of young plants from calli is shown in Fig. 17-4. Both cytokinins and auxins must usually be added to the medium for embryogenesis to occur, but little information is available indicating how they act as control agents.

Cytokinins Delay Senescence and Increase Nutrient Sink Activities   When a mature but still active leaf is cut off, it begins to lose chlorophyll, RNA, pro-

teins, and lipids from the chloroplast membranes more rapidly than if it were still attached, even if it is provided with mineral salts and water through the cut end (Thimann, 1980). This premature aging or senescence, a yellowing of leaves, occurs especially fast if the leaves are kept in darkness. In dicot leaves, adventitious roots often form at the base of the petiole, and then senescence of the blade is greatly delayed. Something is apparently provided by the roots to the leaf that keeps it physiologically young. This something almost surely consists in part of a cytokinin provided through the xylem. There are two main evidences that a cytokinin is involved: Many cytokinins will partially replace the need for roots in delaying senescence, and the cytokinin content of the leaf blade rises substantially when the adventitious roots form. In sunflowers, the cytokinin content of the xylem sap increases during the period of rapid growth, then decreases greatly when growth stops and flowering begins. This suggests that a reduction in cytokinin transport from roots to shoots might allow senescence to occur faster (Skene, 1975).

How cytokinins retard senescence in detached oat leaves has been investigated extensively by Kenneth V. Thimann, a pioneer in auxin research, and his colleagues at the Thimann Laboratories in Santa Cruz, California (Thimann, 1980; Thimann et al., 1982). When leaves of oats and many other species are cut off and floated on a solution of dilute mineral salts, their senescence is delayed in light or darkness by cytokinins and certain other compounds. Interestingly, all effective compounds maintain the stomates of oat leaves in a more widely open condition in light or darkness. Other compounds that promote senescence cause stomatal closing. This and other evidence led to the hypothesis that cytokinins slow senescence by allowing CO_2 to enter the stomates; this CO_2 inhibits competitively the strongly promotive action of ethylene on senescence. We shall return to ethylene and senescence later in this chapter, but it should be noted that stomatal opening in darkness would be expected to decrease CO_2 concentrations in the leaf by allowing respiratory CO_2 to escape. Furthermore, although CO_2 can antagonize the effects of ethylene, it also increases its production in leaves (Section 17.2).

The delay of senescence by cytokinins appears to be a natural, partially root-controlled phenomenon and is associated with other interesting phenomena. Cytokinins cause transport of many solutes from older parts of the leaf and even from older leaves into the treated zone. A dramatic illustration of this is shown in Fig. 17-5. Here the oldest (primary) leaves of a bean plant were painted at 4-day intervals with the synthetic cytokinin benzyladenine. Normally, these leaves become senescent before the trifoliate leaves above, but in this example the senescence pattern was reversed. The treated primary leaves withdrew nutrients from the adjacent trifoliate, causing it to senesce first. (Note also that the benzyladenine apparently did not move effectively from the treated leaves to the younger trifoliate leaves just above.) Other studies with many dicots and monocots show that if only one part of a leaf is treated, radioactive metabolites added to another part of the same leaf or to an adjacent leaf migrate via the phloem into the treated zone and accumulate there (Patrick, 1976; Gersani and Kende, 1982). The implication is that young leaves can remove nutrients from older ones partly because they are rich in cytokinins and, therefore, that cytokinins enhance the ability of young tissues to act as sinks for phloem transport. Whether these hormones are involved in the normal transport of mobile nutrients into twigs and larger branches of woody plants before leaf fall in autumn is an interesting question. That cytokinins in reproductive structures might have survival value by enhancing the movement of sugars, amino acids, and other solutes from mature leaves into seeds, flowers, and fruits is also an interesting hypothesis.

When certain fungi that cause rust diseases infect leaves, necrotic areas of dead and dying cells are produced. These areas are often immediately surrounded by several green and starch-rich cells, even when the rest of the leaf has become yellow and senescent. These green-islands are rich in cytokinins, probably synthesized by the fungus (Greene, 1980). The cytokinins presumably help maintain food-reserves for the fungus and influence further progress of the disease symptoms.

The ability of cytokinins to retard senescence also applies to certain cut flowers and fresh vegetables. The concentration of cytokinins in rose petals decreases as aging occurs, and applied cytokinins slow this aging process. For most cut flowers, however, exogenous cytokinins cannot overcome the senescence-promoting effects of ethylene produced by the flowers (Halevy and Payak, 1981). The storage lives of brussel sprouts and celery are often increased by relatively inexpensive commercial cytokinins such as benzyladenine, but this is not permitted for foods sold in the United States, even though we are constantly exposed to natural cytokinins in food from plants.

Cytokinins Promote Lateral Bud Development of Dicots If a cytokinin is added to a nongrowing lateral bud dominated by the shoot apex above it, the lateral bud often begins to grow. In early studies of this phenomenon, the synthetic cytokinin kinetin was the main compound used, and growth of the lateral bud only continued for a few days. Continued elongation of the bud could be caused only by adding IAA or a gibberellin to it. Another synthetic cytokinin, benzyladenine, sometimes caused substantially more elongation than kinetin, but its effects have only been studied with a few species. More recently, Indiren Pillay and Ian D. Railton (1983) showed that benzyladenine and zeatin dramatically enhance elongation of pea lateral buds for at least two weeks (Fig. 17-6). Adenine was inactive, while isopentenyl adenine (IPA) and kinetin promoted only short-term growth. The reason that the closely related hormones zeatin and isopentenyl adenine cause such different effects is unknown, but the authors speculated that isopentenyl adenine is only weakly active because it is only slowly hydroxylated to the much-more active zeatin in the buds. Other evidence that quiescent lateral buds cannot synthesize active cytokinins also exists, but we are still uncertain as to how important cytokinins are relative to other hormones and nutritional factors in controlling lateral bud development.

Enhanced lateral branching also occurs in two bacteria-caused diseases in which the pathogen synthesizes a cytokinin. One is a fasciation* disease

---

*In **fasciations,** normally round stems become flattened, and numerous lateral buds develop into branches, often forming a broom-like (witches'-broom) bundle of stems.

Figure 17-6 Effects of various compounds on development of lateral buds in pea stems. The bud at the second node (see arrow) of each plant was treated once with 3.3 μg of the compound indicated when plants were one week old, then growth occurred 16 days before photograph was taken. (From Pillay and Railton, 1983. Reprinted by permission.)

control   adenine   IPA   kinetin   benzyladenine   zeatin

caused by *Corynebacterium fascians* occurring in various dicots such as chrysanthemum, garden peas, and sweet peas. In garden peas, symptoms of this disease can be duplicated by adding a cytokinin to young plants. It has been found that highly pathogenic strains of the bacterium contain a **plasmid** (a small circle of DNA that can occur independently of the main DNA molecule in bacterial cells); nonpathogenic strains do not contain this plasmid (Murai et al., 1980). Furthermore, pathogenic strains synthesize and release into their growth media several cytokinins; the amounts released are roughly proportional to their pathogenicity. These cytokinins are identical to those present in the tRNA of the bacteria, strongly indicating that breakdown of tRNA is at least partly responsible for formation of these cytokinins, which almost surely cause the fasciation disease.

Corynebacterium fascians* also causes certain kinds of witches'-brooms in trees, again accompanied by production of multiple lateral buds that grow into branches. Two other pathogens (*Exobasidium* spp.) that cause witches'-brooms also produce cytokinins. In these cases, too, speculation is that the cytokinins cause the appearance of disease symptoms.

Cytokinins Increase Cell Expansion in Dicot Cotyledons and in Leaves   When seeds of certain dicots are germinated in darkness, the cotyledons remain yellow and relatively small. If they are exposed to light, growth increases greatly, even if the light energy provided is too low to allow photosynthesis. This is a photomorphogenetic effect controlled partly by phytochrome, as described in Chapter 19, but cytokinins are also involved. If the cotyledons are ex-

cised and incubated with a cytokinin, growth is enhanced about two-fold relative to controls without the hormone in either light or darkness. Growth is entirely caused by water uptake that drives cell expansion, because the dry weight of the tissues does not increase. This growth promotion occurs with more than a dozen known species, including radish, sugarbeet, lettuce, sunflower, cocklebur, white mustard, squash, cucumber, pumpkin, muskmelon, and fenugreek. Most of these species contain fats as the major food reserve in the cotyledons. Furthermore, the cotyledons normally emerge aboveground and become photosynthetic in each species. No response has been found in species with cotyledons that remain underground after germination, or in the bean, in which cotyledons emerge but do not become leafy. Figure 17-7 shows the promotive effects of zeatin on radish cotyledon enlargement in both light and darkness and also illustrates that light is effective in the absence of zeatin. Auxins do not promote growth of cotyledons, and gibberellins also have little effect when the cotyledons are cultured in water, so this response provides a useful bioassay for cytokinins (Letham, 1971; Narain and Laloraya, 1974).

Do cytokinins promote cotyledon growth only by increasing expansion of cells already present, or do these hormones promote cell division and expansion and growth of the resulting daughter cells? All results show that they increase both cytokinesis and cell expansion, especially the latter. Remember, however, that cytokinesis does not increase growth of any organ by itself, because it is only a division process. *Therefore, overall growth requires cell expansion, and growth promotion by cytokinins involves faster cell expansion and production of larger cells.*

Because cotyledons in which growth is promoted by cytokinins become photosynthetic organs, we might ask whether true leaves also require cytokinins for growth. Definite promotive effects on intact dicot leaves of some species occur after repeated applications, but the effects are usually small and may arise indirectly through attraction of metabolites from other organs. If discs are cut from dicot leaves with a cork borer and kept moist, cytokinins increase expansion through enhanced cell growth, again suggesting a normal function of cytokinins from some other organ, perhaps roots, in leaf growth. Further evidence that cytokinins from roots promote leaf growth comes from experiments in which some or all of the roots were removed from bean (*Phaseolus vulgaris*) and winter rye (*Secale cereale*). Leaf growth from rootless plants was soon slowed in both species, but application of a cytokinin to the leaves restored much of this growth. So far as we are aware, no studies of effects of cytokinins on growth of conifer needles have been made.

Effects of Cytokinins on Growth of Stems and Roots

Normal growth of stems and roots is thought to require cytokinins, but the endogenous amounts are seldom limiting. Therefore, exogenous cytokinins fail to increase growth of these (or other) organs. Suppose, however, that we stop delivery of cytokinins (and gibberellins) from roots to shoots by removing the roots. Can we now add cytokinins and gibberellins and restore shoot growth, especially stem elongation? In sunflower and pea, growth restoration was unsuccessful, but in soybean success was obtained. These are only three species, all dicots, so conflicting results with so few species justify no general conclusion. Experiments to solve this problem more generally seem easy.

Another important approach to determine the importance of cytokinins to normal growth of stems and roots is to excise sections and grow them *in vitro*, just as in experiments with auxins and gibberellins (Chapter 16). In such experiments the assumption is that excised sections will be depleted of cytokinins when separated from shoot tips or root tips that presumably represent hormone sources. However, nobody has ever demonstrated by actual measurements that excised sections become cytokinin deficient. When root or stem sections are grown *in vitro* with an exogenous cytokinin, elongation is almost always retarded relative to control sections. For example, data showing the sharply antagonistic effects of an auxin and of kinetin on elongation of soybean hypocotyl sections are in Fig. 17-8. Interestingly, although elongation is inhibited, the sections usually become thicker by radial expansion of cells, so the overall fresh weight of the treated sections is not much different from that of control sections.

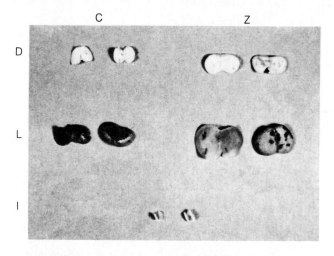

Figure 17-7 Promotion of excised radish cotyledon enlargement by zeatin and light. Cotyledons at the bottom labeled I represent initial cotyledons excised from 2-day-old, dark-grown seedlings before growth studies. Excised cotyledons were incubated for 4 days on filter papers held in Petri dishes containing 2 m*M* potassium phosphate (*p*H 6.4) alone (controls, C) or also with 2.5 μ*M* zeatin (Z). Cotyledons exposed to light (L) received continuous fluorescent radiation at a level near the photosynthetic light compensation point. Cotyledons incubated 4 days in darkness (D). (Unpublished results of A. K. Huff and C. W. Ross.)

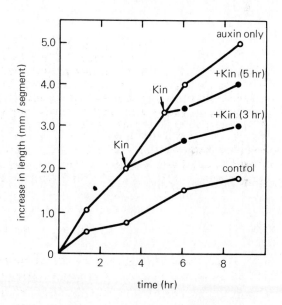

Figure 17-8 Inhibition of auxin-induced elongation in soybean hypocotyl sections by 4 μ*M* kinetin added at different times of incubation (arrows). (From L. N. Vanderhoef et al., 1973.)

What can we conclude from results showing only inhibition of elongation? We might conclude that elongating stems and roots do not require cytokinins. Alternatively, such organs might need the hormones for elongation, but they already contain sufficient amounts. In both cases we could argue that

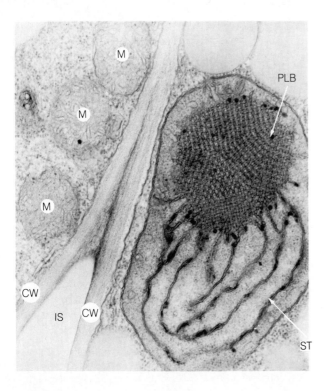

Figure 17-9 Etioplast from a cotyledon of a dark-grown radish seedling, illustrating the prolamellar body (PLB) and stroma thylakoids (ST) radiating from it. Also shown are two adjacent cell walls (CW), an intercellular space (IS) between the walls, and mitochondria (M) in the cell at the left. (Courtesy Nicholas Carpita.)

exogenous cytokinins inhibit *in vitro* growth by causing excess internal concentrations. There seems no easy way to solve this problem without measuring the internal concentrations of cytokinins in excised sections, especially in epidermal cells that probably limit the overall rate of elongation. (Note the relation of this problem to that of assessing the importance of auxins to root elongation described in Section 16.1.) However, there are two cases known in which applied cytokinins do promote elongation: sections from young wheat coleoptiles (Wright, 1966) and intact watermelon hypocotyls, especially those of a dwarf cultivar (Loy, 1980). In the wheat coleoptiles, growth promotion only occurs if the tissues are still young and cell division is still occurring, yet it was found that cytokinins cause growth by promoting cell elongation, not division. In the dwarf watermelons, hypocotyl elongation also occurs, primarily because the rate of cell elongation is increased; this increase results from cytokinins applied either to the shoot tip or to the roots.

In summary, exogenous cytokinins can promote cell elongation in young leaves, cotyledons, wheat coleoptiles, and watermelon hypocotyls, but much remains to be learned about the normal role of these hormones in cell expansion, especially in stems and roots. As usual, even less is known about trees in general and conifers in particular.

Cytokinins Promote Chloroplast Development and Chlorophyll Synthesis If angiosperm seedlings are grown in darkness, we can then remove a young leaf or cotyledon from the seedlings and test whether adding a cytokinin to it has any effect on chloroplast development or chlorophyll synthesis. This is true because in darkness no chlorophyll is formed and because chloroplast development is blocked. Young plastids are then arrested at the stage of proplastids (Section 9.2) or more commonly at the stage of etioplasts. **Etioplasts** (from dark-grown or etiolated seedlings) are yellow because of the carotenoids they contain, and they have an interesting system of internal membranes closely arranged into an internal lattice called the **prolamellar body** (Fig. 17-9). Upon light exposure, the prolamellar body gives rise to much of the thylakoid system found in normal green chloroplasts, and this development is accompanied by formation of special thylakoid proteins that become attached to chlorophylls in the two photosystems and light-harvesting complexes.

Addition of cytokinin to etiolated leaves or cotyledons several hours before they are exposed to light has two main effects: It enhances subsequent (in light) development of etioplasts into chloroplasts, especially by promoting grana formation, and it increases the rate of chlorophyll formation. Much is known about the details of these effects (Parthier, 1979; Guern and Péaud-Lenoël, 1981; Lew and Tsuji, 1982); the principal reason for both is probably that cytokinins enhance formation of one or more proteins to which chlorophylls bind and become stabilized. We suspect that endogenous cytokinins normally increase chloroplast development in leaves in a similar way.

What Are the Mechanisms of Cytokinin Action? The variability of cytokinin effects suggests that they might have different mechanisms of action in different tissues, yet the simpler view is that a common primary effect is followed by numerous secondary effects dependent on the physiological state of the target cells. As with other hormones, amplification of the initial effect must occur, because cytokinins are present in such low concentrations (0.01 to 1 μM). Some promotive effect of cytokinins on RNA and enzyme formation is suspected, partly because their effects are usually blocked by inhibitors of RNA or protein synthesis. No specific effects on DNA synthesis have been observed, although exogenous cytokinins often increase cell division and might normally be required for that process. Many investigators have tried to determine if a special receptor-protein exists in plants that would bind cytokinins and then lead to various physiological effects depending on the cell type, but no such protein with the expected ability to bind tightly only active cytokinins has yet been found. Other approaches must be used to determine how cytokinins act.

One of the most well-investigated cytokinin responses is cytokinesis, studied especially by D. E. Fosket and his colleagues at the University of California, Irvine (Fosket, 1977; Fosket et al., 1981). Fosket (1977) concluded that cytokinins promote cell division by increasing the transition of cells from G_2 to mitosis (see The Cell Cycle, Section 15.2) and that they do this by increasing the rate of protein synthesis. Some of these proteins could be enzymes needed for mitosis. Of course, protein synthesis could be increased by stimulating formation of messenger RNAs that code for proteins, but no increase in messenger RNA production was observed. Instead, the cytokinin effect seems to be specifically on translation. One of several evidences for this is that the ribosomes in treated cells are frequently grouped in large protein-synthesizing polysomes rather than in smaller polysomes or as free monoribosomes characteristic of slowly-dividing untreated cells. There is yet no explanation as to how polysome formation and translation are increased by cytokinins, and no special enzyme or other protein that might lead to mitosis has been discovered in cytokinin-treated cells.

It should be noted that in most other cases of cytokinin action (e.g., growth promotion), their main effect seems to be on translation, as evidenced by increased polysome levels, faster incorporation of radioactive amino acids into proteins, and inhibition of the physiological response by inhibitors of protein synthesis. Formation of special proteins in cytokinin-treated cells is consistent with the idea, described above, that enhanced chloroplast development occurs because of formation of one or more proteins that bind chlorophylls. Important stimulating effects of cytokinins on transcription (leading to synthesis of messenger RNA or other RNAs) is not common, but probably is important in the delay of senescence in detached leaves.

When it is finally discovered how cytokinins increase translation, there will still remain the problem of how newly translated enzymes or other proteins then cause cytokinesis and cell expansion. At Colorado State University, we have attempted to determine how cell expansion of detached cotyledons is increased, without any knowledge of what kind of enzyme is involved. We and others found that when radish cotyledons are grown in weak light, cytokinins increase the internal production of reducing sugars (mainly glucose and fructose), and these sugars act osmotically to cause the water uptake that drives

growth (Huff and Ross, 1975; Bewli and Witham, 1976). We then found for both radish and cucumber cotyledons that cytokinin treatment causes increased plasticity (but not elasticity) of the cell walls; that is, the walls become loosened so they will expand faster irreversibly under the existing turgor pressure (Thomas et al., 1981). We suspect that this wall loosening is much more important for growth promotion than enhanced turgor pressure resulting from reducing sugar production, because cytokinin-treated cotyledons grow with only about 0.15 MPa turgor pressure compared to about 0.9 MPa for untreated cotyledons (Rayle et al., 1982). We also found that whatever the mechanism of wall loosening, it is almost surely not caused by acidification of the wall (Ross and Rayle, 1982), so the acid-growth hypothesis is not applicable. As with auxins and gibberellins, cytokinins cause cells to alter their walls in some way that makes them more plastic, but the nature of this alteration and the enzyme or enzymes that cause it remain to be discovered.

17.2 Ethylene, a Volatile Hormone

The ability of certain gases to stimulate fruit ripening has been known for many years. Even the ancient Chinese knew that their picked fruits would ripen more quickly in a room with burning incense. In 1910, an annual report by H. H. Cousins to the Jamaican Agricultural Department mentioned that oranges should not be stored with bananas on ships, because some emanation from the oranges caused the bananas to ripen prematurely. This was apparently the first suggestion that fruits release a gas that stimulates ripening, but it was not until 1934 that R. Gane proved that ethylene is synthesized by plants and is responsible for faster ripening.

Another historical practice implicating still a different role for ethylene was the building of bonfires by Puerto Rican pineapple growers and Philippine mango growers near their crops. These farmers apparently believed that the smoke helped to initiate and synchronize flowering. Ethylene causes these effects in both species, so it is almost surely the most active smoke component. Stimulation of fruit ripening is a widespread phenomenon, while promoted flowering appears restricted to mangos and most bromeliad species, including the pineapple.

Still another effect of gases was reported as early as 1864. Before the use of electric lights, streets were lighted with illuminating gas. Sometimes the gas pipes leaked, and in certain German cities this caused the leaves to fall off the shade trees. Ethylene causes senescence and abscission of leaves, so again it presumably was responsible.

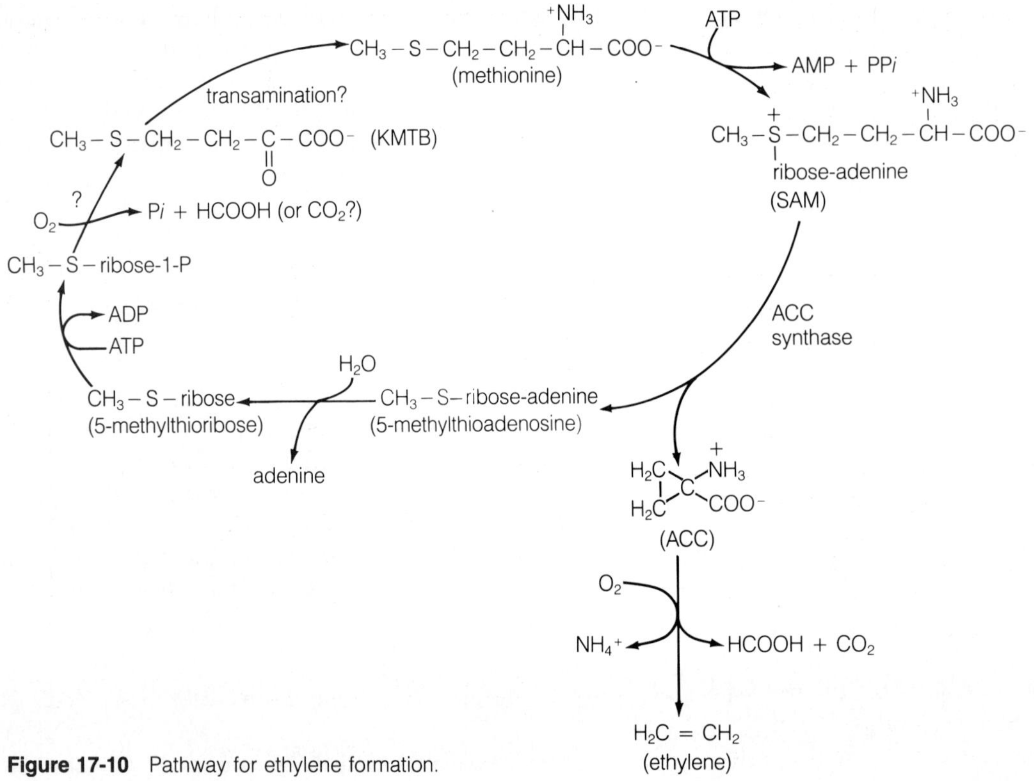

$$^+NH_3$$
$$CH_3-S-CH_2-CH_2-CH-COO^-$$
(methionine)

ATP

AMP + PPi

transamination?

$CH_3-S-CH_2-CH_2-C-COO^-$ (KMTB)
‖
O

$^+NH_3$
$CH_3-S-CH_2-CH_2-CH-COO^-$
ribose-adenine
(SAM)

O$_2$?
Pi + HCOOH (or CO$_2$?)

CH_3-S-ribose-1-P

ACC synthase

ADP
ATP

CH_3-S-ribose ← CH_3-S-ribose-adenine
(5-methylthioribose)   (5-methylthioadenosine)

H$_2$O

adenine

H_2C $^+NH_3$
H_2C C COO^-
(ACC)

O$_2$

NH$_4^+$ HCOOH + CO$_2$

$H_2C = CH_2$
(ethylene)

Figure 17-10 Pathway for ethylene formation.

It was a Russian physiologist named Dimitry N. Neljubow (1876–1926) who first established that ethylene affects plant growth. He identified ethylene in illuminating gas and showed that it causes a **triple response** on pea seedlings: inhibited stem elongation, increased stem thickening, and a horizontal growth habit. Furthermore, leaf expansion is inhibited, and normal opening of the epicotyl hook is retarded. We shall describe these and other effects of ethylene. The book by Abeles (1973) and the reviews by Lieberman (1979) and Beyer et al. (1984) describe other effects in detail.

Ethylene Synthesis Ethylene production by various organisms can often be detected readily by gas chromatography, because the molecule can be withdrawn from tissues under vacuum and because gas chromatography is so sensitive. Only a few bacteria reportedly produce ethylene, and no algae are known to synthesize it; furthermore, it generally has little influence on their growth. However, several fungal species produce it, including some that normally grow in soils. For this reason it might be important in germination of seeds, growth of seedlings, and diseases caused by soil-born organisms. Essentially all parts of all seed plants produce ethylene. In seedlings, the shoot apex is an important site of production. This might result from the high amounts of IAA there, because auxins greatly stimulate ethylene formation. Nodes of dicot seedling stems produce much

more ethylene than do internodes when equal tissue weights are compared. Roots release relatively small amounts, but again auxin treatment causes the rate to rise. Production in leaves generally rises slowly until the leaves become senescent and abscise. Flowers also synthesize ethylene, especially just before they fade and wither, and this gas causes their senescence and abscission. The highest known rate of ethylene release is from fading flowers of *Vanda* orchids: 3.4 μl per g fresh weight per hour (Beyer et al., 1984). In many fruits, little ethylene is produced until just before the respiratory climacteric signalling the onset of ripening, when the content of this gas in the intercellular air spaces rises dramatically from almost undetectable amounts to about 0.1 to 1 μl l^{-1}. These concentrations generally stimulate ripening of the fleshy and nonfleshy fruits that display a climacteric rise in respiration (Section 12.13), if the fruits are sufficiently developed to be susceptible (Coombe, 1976; Biale and Young, 1981). Sections of ripe apple or pear fruits or even apple peelings are often used in laboratory demonstrations as an ethylene source. Nonclimacteric fruits synthesize little ethylene and are not induced to ripen by it.

Interestingly, numerous mechanical and stress effects such as gently rubbing a stem or leaf, increased pressure, pathogenic microorganisms, viruses, insects, water logging, and drought increase ethylene production. This ethylene might contribute to stem shortening and thickening (an example of thigmomorphogenesis—see Chapter 18) in the affected plants. Egyptian civilizations unknowingly took advantage of increased ethylene production resulting from injury by gashing immature sycamore figs to stimulate ripening. When figs only about sixteen days old are gashed, they ripen within as little as four days.

Knowledge of the biochemistry of ethylene synthesis was greatly advanced in 1964 when Morris Lieberman and L. W. Mapson in the Beltsville, Maryland, United States Department of Agriculture laboratory discovered that it is derived from carbons 3 and 4 of the amino acid methionine (reviewed by Lieberman, 1979). A second important advance was made in the laboratory of Shang-Fa Yang at the University of California, Davis, when it was found that an unusual amino acid-like compound **1-amino-cyclopropane-1-carboxylic acid (ACC),** is involved as a close precursor of ethylene (Adams and Yang, 1979). Yang and his colleagues elucidated several other reactions of the pathway for ethylene formation. Figure 17-10 shows this pathway and emphasizes how the sulfur atom of methionine can be conserved by a cyclic, salvage process. Without this salvage, the amount of reduced sulfur limits the amount of methionine and the rate of ethylene synthesis. Other notable features of the pathway are that ATP is essential to convert methionine to S-adenosylmethionine (SAM), and O_2 is needed in the final conversion of ACC to ethylene. (The requirements for ATP and O_2 almost surely explain why ethylene production nearly stops under hypoxic conditions.) Evidence also indicates that four of the carbon atoms of the ribose unit of SAM are salvaged and reappear in methionine. A newly-discovered intermediate, **α-keto-γ-methylthiobutyric acid (KMTB** in Fig. 17-10, upper left), is important in the salvage of these carbons (Kushad et al., 1983). Reactions by which KMTB is formed and converted further to methionine have not been clarified.

Two potent inhibitors of ethylene synthesis have been discovered, both of which are useful tools to investigate the pathway of ethylene formation and to study effects of reduced ethylene production in tissues. These compounds, **aminoethoxyvinylglycine (AVG)** and **aminooxyacetic acid (AOA),** are well-known inhibitors of enzymes that require pyridoxal phosphate as a coenzyme. AVG and AOA block the conversion of SAM to ACC, but they have no other important effect on the pathway. This and other results indicate that **ACC synthase** is a pyridoxal-phosphate-dependent enzyme (Fig. 17-10, right).

The control of ethylene synthesis has received considerable study, especially regarding the promoting effects of auxins and drought stress (Yang et al., 1982; Imaseki et al., 1982; Beyer et al., 1984). In

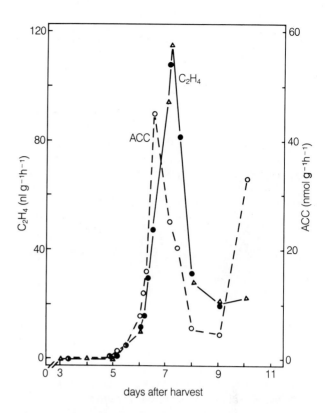

Figure 17-11 Changes in contents of ACC and ethylene in ripening avocado fruits. (From N. E. Hoffman and S. F. Yang, 1980.)

stems of mung bean (*Phaseolus aureus*) and pea seedlings, IAA stimulates ethylene formation several hundred-fold. In these and other tissues auxins induce additional formation of ACC synthase, and the enhanced formation of ACC resulting from action of that enzyme then leads to increased ethylene production. Wounding also increases ethylene production by inducing formation of ACC synthase.

In ripening climacteric fruits, the formation of ACC again limits ethylene formation. In preclimacteric avocado fruits, for example, the concentration of ACC rose from near zero to over 40 nmol g^{-1} of fruit tissue just before a burst of ethylene synthesis occurred (Fig. 17-11). Ripening soon followed. The levels of ACC and ethylene decreased greatly about two days after the peaks occurred, but the level of ACC then rose again without additional ethylene synthesis. Application of ACC to preclimacteric fruits did not cause the rise in ethylene release, indicating that the climacteric is accompanied not only by increased production of ACC but also by an increased ability to convert ACC to ethylene.

A curious autocatalytic (positive-feedback) ability of ethylene to stimulate its own formation occurs in many senescing organs, including leaves, flower

Figure 17-12 Severe epinasty in cocklebur. Plant on the left is an untreated control. Plant on the right was dipped into a solution of 1 m*M* NAA about two days before the photograph was taken. Epinasty in response to ethylene is very similar. (Photo by F. B. Salisbury.)

petals, and over-ripened fruits. This effect results from ethylene induction of ACC synthase, and the increased ACC formed by that enzyme probably explains the ability of one rotten apple in a barrel to spoil the others.

Besides oxygen, other environmental factors that affect ethylene synthesis in leaves are light and carbon dioxide. Light inhibits synthesis, mainly by interfering with the conversion of ACC to ethylene. Carbon dioxide promotes synthesis by enhancing ACC conversion. It seemed logical that the opposing effects of light and CO_2 are explained by the photosynthetic removal of CO_2 in light, and this was found to be true. However, Preger and Gepstein (1984) found that light has an additional inhibiting effect at the ACC-ethylene step which is still not understood.

Ethylene Effects on Plants in Waterlogged Soils and Submerged Plants   The requirement of O_2 to convert ACC to ethylene leads to several symptoms in waterlogged tomato plants (Bradford and Yang, 1981; Kawase, 1981; Jackson, 1982; Bradford, 1983). These symptoms, some of which are also characteristic of other species, include chlorosis of leaves; decreased stem elongation but increased stem thickness; wilting, epinasty, and eventual abscission of leaves; decreased root elongation often accompanied by adventitious root formation; and increased susceptibility to pathogens. In many species, including tomato, aerenchyma forms in the root cortex (see Fig. 12-12), which increases oxygen movement to the roots from the shoots. Furthermore, transport of cytokinins and gibberellins from the roots to the shoot via the xylem is reduced.

Waterlogged soils rapidly become hypoxic (see Section 12.13), because water fills the air spaces and

O_2 replenishment around the roots is reduced by the very slow movement of the gas through water. Ethylene synthesis is then inhibited, because O_2 is required to convert ACC to ethylene, but the ethylene that is synthesized is trapped in the root because its escape through water compared to air is reduced by a factor of about 8,000 (Jackson, 1982). This ethylene then causes some of the cortical cells to synthesize **cellulase,** an enzyme that is partly responsible for degrading their cell walls. These cortex cells also lose their protoplasts and disappear, forming the air-filled aerenchyma tissue. However, even before the aerenchyma develops, ACC accumulates and is transported in the xylem to the shoots. The well-aerated shoots then rapidly convert this ACC to ethylene. The ethylene, in turn, causes leaf epinasty (see Section 18.2). An extreme case of epinasty is shown in Fig. 17-12. Epinasty of the petioles occurs because mature parenchyma cells on the upper side of the petiole elongate in the presence of ethylene, whereas those on the lower side do not. This physiological difference in morphologically similar cells is not understood, but it emphasizes again that only some cells are targets for a given hormone. Ethylene also retards elongation of the stem, increases its radial expansion, causes leaf senescence, and stimulates formation of adventitious roots on the stems (especially in tomato).

Ethylene Effects on Elongation of Stems and Roots   Although ethylene causes epinasty of leaves by promoting elongation of cells on the upper side, it usually inhibits elongation of stems and roots, especially in dicots. (Inhibited stem elongation in peas forms part of the previously-mentioned triple response.) When elongation is inhibited, stems and roots become thicker by enhanced radial expansion of the cells (Fig. 17-13). In dicot stems, altered cell shapes are apparently caused by a more longitudinal orientation of cellulose microfibrils being deposited in the walls, preventing expansion parallel to these microfibrils but allowing expansion perpendicular to them (for review, see Eisinger, 1983). No comparable studies of changes in shapes of cells or cellulose microfibril orientation seem to have been performed with roots, but almost surely the changes are similar.

Root and stem thickening in dicots caused by ethylene are of survival value for seedlings emerging from the soil. In these species, a hook in the epicotyl or the hypocotyl is formed shortly after germination in response to endogenous ethylene, then this hook pushes up through the soil and makes a hole through which the cotyledons or young leaves can be safely drawn. If the soil is excessively compact, the hook and primary root become unusually thick, probably because ethylene is synthesized faster when the compacted cells are subjected to increased mechanical pressure and because the ethylene escapes less rapidly in compacted than in loose soils. This thickness increases the strengths of both stem and root, allowing them to push through the compacted soil. Growth is slow, however, because of retarded elongation. In the cereal grains maize and oat, ethylene has effects on the mesocotyl (first internode, see Fig. 19-7) that are similar to those of dicot stems—inhibited elongation and greater thickening (Camp and Wickliff, 1981). How general this is among members of the grass family is unknown, but the advantage to seedlings in compact soils should be the same as that with dicots. We can also speculate that the loss of gravitropic sensitivity by stems of dicot seedlings is advantageous in compact soils, because a more horizontally-growing stem might more likely encounter a crack in the soil than one that grows upright.

Although retarded stem elongation is common among land plants, certain dicots and ferns that grow with their roots and stems underwater at least part of the time respond to ethylene by enhanced elongation. Among these are *Callitriche platycarpa*, *Ranunculus sceleratus*, *Nymphoides peltata*, and the water fern *Regnellidium diphyllum* (Osborne, 1982). When submerged, the stems of these plants elongate rapidly so that the leaves and upper stem parts are kept at the water surface. Submergence causes ethylene accumulation in the stems, which is responsible for the rapid elongation. A similar phenomenon occurs in stems of deep-water rice (Raskin and Kende, 1984). These examples again emphasize the different responses of somewhat similar cells to a hormone and verify Trewavas' conclusion (see his personal essay in Chapter 16) that variable tissue sensitivity is especially important in hormone effects.

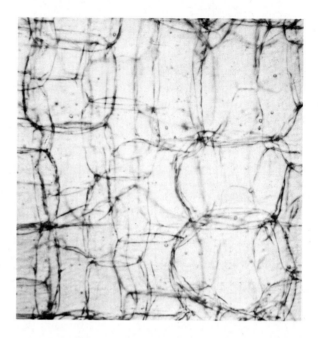

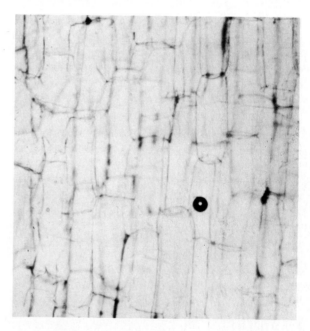

Figure 17-13 Effects of ethylene on cell elongation and radial expansion in the upper internode of pea seedlings. Plants were grown 4 days in darkness, then treated with 0.5 μl/l ethylene (top photo) or left as controls (bottom photo). Cell sections were made 24 hours after ethylene treatments began. (From R. N. Stewart et al., 1974.)

Ethylene Effects on Flowering The induction of flowering in mangos and bromeliads by ethylene mentioned earlier is unusual, because the gas inhibits flowering in most species. Nevertheless, the indirect use of ethylene to promote flowering has been widely used in the pineapple industry in Hawaii. In the 1950s, fields were often sprayed with the auxin NAA, now known to stimulate ethylene synthesis in the plants. Pineapple fields flower faster and, more importantly, mature fruits appear uniformly, the goal being to allow a one-harvest mechanical operation. An ethylene-releasing substance called **Ethrel** (trade name) or **ethephon** (common name) is commercially available. It is 2-chloroethylphosphonic acid (Cl–CH_2–CH_2–PO_3H_2), which rapidly breaks down in water at neutral or alkaline pH values to form ethylene, a Cl^- ion, and $H_2PO_4^-$. Because Ethrel can be translocated throughout the plant, it is now widely used to promote flowering of pineapples (Nickell, 1979) and is used in various aspects of horticulture such as fruit production. It is widely employed in research studies as a source of ethylene.

Relation of Ethylene to Auxin Effects The ability of IAA and all synthetic auxins to increase ethylene production raises the question of whether many auxin effects are really caused by ethylene. Indeed, ethylene appears to be responsible in many cases (see reviews by Abeles, 1973, and Burg, 1973). Included are leaf epinasty, inhibition of stem, root, and leaf elongation, flower induction in bromeliads and mangos, inhibition of epicotyl or hypocotyl hook opening in dicot seedlings, increased percentage of female flowers in dioecious plants, and perhaps apical dominance. Also, release of auxin by germinating pollen grains promotes ethylene production in the stigma, which contributes to senescence of the flower. As described in Section 17.7, abscission of leaves, flowers, and fruits involves interaction between auxins, ethylene, cytokinins, and abscisic acid. Nevertheless, growth promotion, initial stages of adventitious root production, and many other effects of auxins appear to be independent of ethylene production. Only when the auxin concentration becomes relatively high is ethylene production great enough to duplicate certain auxin effects.

Antagonists of Ethylene Action At high concentrations (5–10 percent) CO_2 inhibits many ethylene effects, perhaps by acting as a competitive inhibitor (Abeles, 1973; Dilley, 1978). Because of this inhibition, CO_2 is often used to prevent overripening of picked fruits. Such fruits are stored in an airtight room or container in which the gas composition is controlled. An ideal atmosphere for many fruits contains 5 to 10 percent CO_2, 1 to 3 percent O_2, and no ethylene. Removal of some oxygen is important, because this slows ethylene synthesis. However, if too much O_2 is removed, glycolysis is stimulated by the Pasteur effect (Section 12.13), causing excess sugar breakdown. Another technique useful in fruit storage is to partially evacuate the container, thereby removing O_2 and ethylene from the tissues into the atmosphere.

Inhibition of ethylene action by CO_2, although common, is not universal. One reason for this is that the inhibitory property is lost as the tissue ethylene concentration approaches or exceeds 1 $\mu l\ l^{-1}$, a concentration that gives about half-maximal activity in nearly every response to ethylene studied (Beyer et al., 1984). For this reason and the high concentrations of CO_2 required, it seems unlikely that CO_2 often acts *in vivo* as an antagonist of ethylene action.

A much more effective inhibitor of ethylene action is Ag^+, the silver ion (Beyer, 1976). Among the ethylene effects found by Beyer to be nullified or inhibited by Ag^+ (added as $AgNO_3$) were the triple responses of etiolated pea seedlings, promotion of abscission of leaves, flowers, and fruits of cotton, and induction of senescence in orchid flowers. Since then silver thiosulfate has proven to be even more effective in delaying senescence of cut flowers than silver nitrate (Halevy and Mayak, 1981).

How Does Ethylene Act? Many ethylene effects are accompanied by increased synthesis of enzymes, the kind of enzyme depending on the target tissue (Varner and Ho, 1976). When ethylene stimulates leaf abscission, cellulase and other cell-wall degrading enzymes appear in the abscission layer (Section 17.7). When fruit ripening occurs, necessary enzymes are produced in the fruit cells. When cells are injured, phenylalanine ammonia lyase (Section 14.4) appears, which is an important enzyme in formation of phenolic compounds thought to be involved in wound healing. Nevertheless, we still do not know how ethylene promotes enzyme synthesis.

Two important observations were made that should help elucidate the mode of action of ethylene (see Beyer et al., 1984). Small amounts of ethylene are metabolized in plants, and small amounts are bound. The principal metabolites are oxidation products: CO_2, ethylene oxide, and ethylene glycol. In some tissues there is a good correlation between metabolite production and binding activity, which suggests that binding, oxidation to metabolites, and physiological action occur in a dependent sequence. However, others believe that binding leads to a response without oxidation. The compound that binds ethylene has not been purified and identified. For chemical reasons it is suspected to be a copper-containing protein, and some physiological evidence indicates that it is present in the endoplasmic reticulum or dictyosomes (Golgi bodies).

17.3 Triacontanol and Brassins

Triacontanol is a 30-carbon, saturated primary alcohol first isolated from shoots of alfalfa. It is very insoluble in water (less than $2 \times 10^{-16}\ M$ or 9×10^{-14} g l^{-1}), yet colloidal suspensions of this compound significantly (approximately 20 percent) enhance growth of maize and rice plants when sprayed on the foliage of seedlings at concentrations as low as 0.1 ηg l^{-1} (Ries and Wert, 1982; Laughlin et al., 1983). Both species reportedly respond by increased growth within 10 min. Little is known about the mechanism of action of triacontanol, but it has potential importance for increasing crop yields.

The **brassins** are recently-discovered steroid growth-promoters first isolated from pollen grains of rape plants but now known also to be present in tea, bean, and rice. The nature of these compounds is described in Chapter 14, Section 14.3.

17.4 Polyamines

These are polyvalent cation compounds that contain two or more amino groups, including the amino acids lysine and arginine (Fig. 8-2). Among the most abundant, common, and physiologically active polyamines are **putrescine** (NH$_2$(CH$_2$)$_4$NH$_2$), **cadaverine** (NH$_2$(CH$_2$)$_5$NH$_2$), **spermidine** (NH$_2$(CH$_2$)$_3$NH(CH$_2$)$_4$NH$_2$), and **spermine** (NH$_2$(CH$_2$)$_3$NH(CH$_2$)$_4$NH(CH$_2$)$_3$NH$_2$). These compounds promote growth of some tissues, stabilize membranes, stabilize isolated protoplasts (perhaps by effects on membranes), minimize water stress of various kinds of cells, and delay senescence of detached leaves (for reviews, see Galston, 1983, and chapters in the book by Wareing, 1982). Little is known about their primary mechanism of action, but their positively charged amino groups cause them to combine with negatively charged phosphate groups in DNA and RNA in the nucleus and ribosomes. As a result of this combination, they often increase transcription of DNA and translation of RNA in plant and animal cells (Tabor and Tabor, 1984). Their effects on membrane stability likely result from direct combination with proteins or phospholipids in the membranes, but nothing is known about this yet.

17.5 Abscisic Acid (ABA)

It might seem that plants could regulate all their essential processes and coordinate growth and development by synthesizing and transporting promotive compounds, including hormones, then degrading or altering them when the response is sufficient. Nevertheless, most physiological processes seem also to be influenced by growth inhibitors. The most widespread inhibitor in the plant kingdom appears to be ABA, although phenolic compounds, coumarins, alkaloids, and compounds described later in this chapter are probably also important in some species.

Evidence that bud dormancy might be due to growth inhibitors was first obtained by Torsten Hemberg of Sweden, in 1949. He found that dormant buds of potato tubers and ash trees contained potent growth inhibitors, but the levels of these declined when dormancy was broken. In the 1950s, investigation of growth activity in compounds separated from numerous plant extracts showed a pronounced inhibitory zone on paper chromatograms. The substances in this zone were called the **β-inhibitor**, or *β-inhibitor complex*. B. V. Milborrow (1967) then showed that the principal or sole growth-inhibiting compound in the β-inhibitor complex from many tissues was ABA (although dormant potato buds additionally contain at least one other unidentified phenolic inhibitor). (For a thorough review of the history of inhibitors in various plant parts and their relation to the β-inhibitor complex and the discovery of ABA, see Addicott and Carns, 1983.)

Abscisic acid was first identified and characterized chemically in 1963 by Frederick T. Addicott and his co-workers in California, who were studying compounds responsible for abscission of cotton fruits. They named one active compound *abscisin I*, and the second (much more active) *abscisin II*. Abscisin II proved to be ABA. In the same year it became very likely that two other research groups had discovered ABA. One group was led by Philip F. Wareing in Wales; they were studying compounds that caused dormancy of woody plants, particularly *Acer pseudoplatanus*. They named their most active compound *dormin*. The other group was led by R. F. M. Van Steveninck, first in New Zealand and then in England; they were studying a compound or compounds that accelerated the abscission of flowers and fruits of the yellow lupine (*Lupinus luteus*). After it became evident (in 1964) that dormin and the lupine compound were identical to abscisin II, it was agreed in 1967 to call the compound abscisic acid. ABA appears to be universal among vascular plants. It is also present in some mosses but not in liverworts, algae, bacteria, or most fungi.

Chemistry, Metabolism and Transport of Abscisic Acid ABA (Fig. 17-14) is a 15-carbon sesquiterpenoid synthesized in chloroplasts and other plastids by the mevalonic acid pathway (Section 14.3). Thus early reactions in ABA synthesis are identical to those of isoprenoids such as gibberellins, sterols, and carotenoids. There is some evidence that a small percentage of the ABA in chloroplasts can arise from photochemical or enzymatic destruction of a xanthophyll carotenoid called **violaxanthin** (for review, see Milborrow, 1983). This destruction first forms **xanthoxin** (Fig. 17-14), which can be converted metabolically to ABA. Xanthoxin is a potent growth inhibitor that might participate in phototropism (Section 18.3) and other effects of light, especially blue light (Chapter 19).

Inactivation of ABA can occur in two ways. One is by attachment of glucose to the carboxyl group to form an ABA-glucose ester (Fig. 17-14). Note that a comparable inactivation by attachment of glucose also occurs with IAA, gibberellins, and cytokinins.

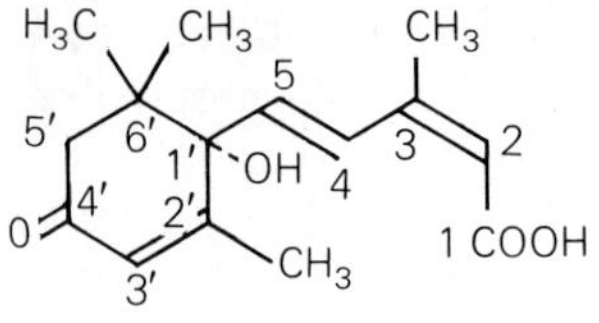

(+) abscisic acid
mol. wt. = 264.3 g/mol

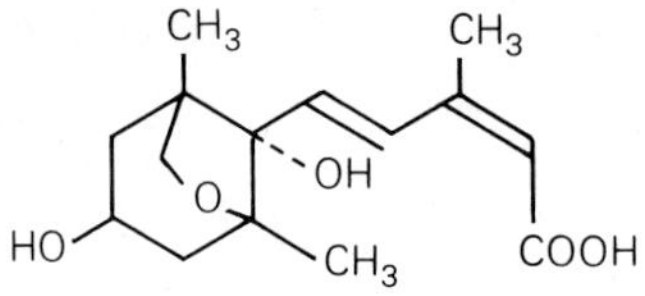

4′–dihydrophaseic acid

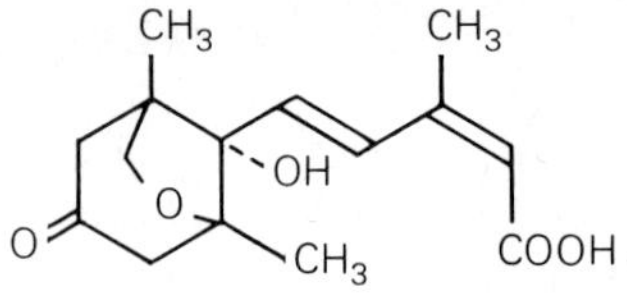

phaseic acid

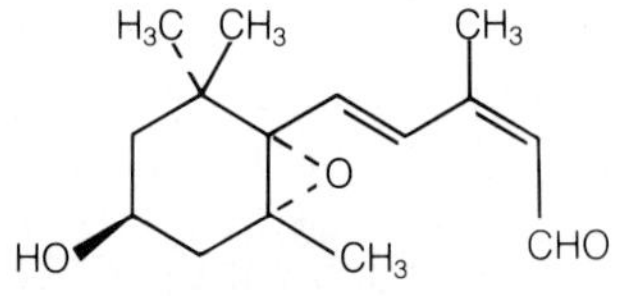

xanthoxin (2-*cis* form)

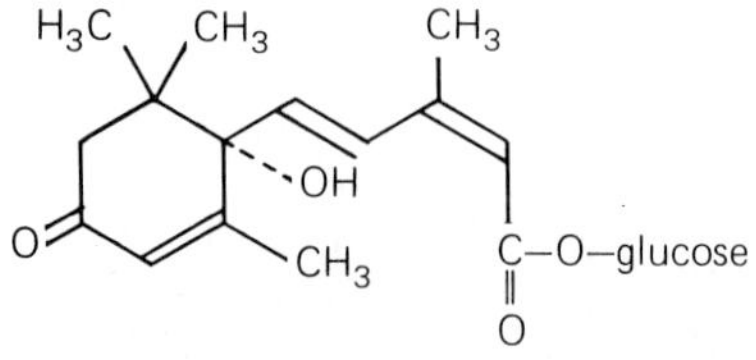

glucose ester of ABA

Figure 17-14 Structure of abscisic acid (ABA) and some related compounds. ABA (upper left) has one asymmetric carbon atom (1′ in the ring). The form synthesized by plants is the dextrorotatory (+) product shown with the S (sinister) configuration about the asymmetric (chiral) carbon atom. But commercial ABA is a racemic (±) mixture. Both forms are biologically active.

Another inactivation process is oxidation with O_2 to form **phaseic acid** and **dihydrophaseic acid** (Fig. 17-14).

Transport of ABA occurs readily in both xylem and phloem and also in parenchyma cells outside vascular bundles. In parenchyma cells there is usually no polarity (as contrasted to auxins), so the movement of ABA within plants is similar to that of gibberellins.

A major function of ABA in plants is probably to cause stomates to close when water stress occurs or when elevated CO_2 levels occur in guard cells (as in the evening, when photosynthesis stops but respiration continues). This function is described in Chapter 3 (for review, see Radin and Ackerson, 1982, and Davies and Mansfield, 1983).

Effects of ABA on Dormancy The most universal response of cells to ABA is growth inhibition. The early results of Wareing and his colleagues that led to the discovery of dormin (ABA) showed that levels of this compound increased considerably in leaves and buds when bud dormancy occurred in the short days of late summer. They also found that direct applications of ABA to nondormant buds caused dormancy. Those results suggested that ABA is a bud-dormancy hormone, synthesized in leaves that detect day length (see Chapter 22) and translocated to buds to induce dormancy. Much additional work with other woody plants argues against such a hormonal role. Perhaps the most convincing result is that direct application of ABA to buds slows or stops growth but does not cause development of bud scales and other characteristics of dormant buds. Other results with ^{14}C-labeled ABA show that extremely little of it moves from leaves to buds when dormancy begins. Furthermore, short-day treatments that induce dormancy in various species cause no rise in ABA levels of buds of several species. However, the latter result seems unconvincing, because analyses of whole buds might not show important changes in ABA levels that occur in relatively few target cells that control whether or not normal dormant buds are produced. Direct localization of ABA by histochemical and cytochemical techniques is needed. It still seems possible that ABA is one hormone that contributes to bud dormancy.

During the last two decades, there have been numerous studies concerning the possible importance of ABA in causing dormancy of seeds (Walton, 1980; Bewley and Black, 1982; Black, 1983; Tillberg, 1983). Exogenous ABA is a potent inhibitor of seed germination (mainly because it slows radicle elongation and delays germination without preventing it). Furthermore, studies with some species show that ABA levels decrease in whole seeds as dormancy is broken by some environmental treatment, for example, exposure to light or cold temperature. Yet other studies show no such decreases. A reasonable conclusion from these results might be that ABA causes seed dormancy in some species but not in others.

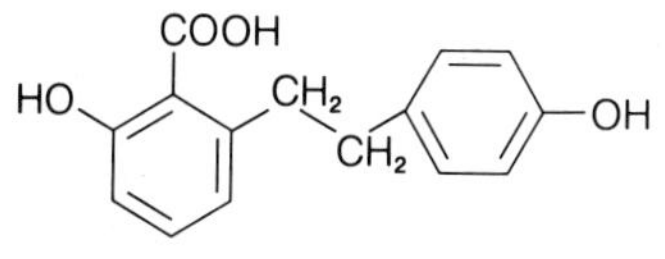

Figure 17-15 Structures of lunularic acid, batasin I, and jasmonic acid.

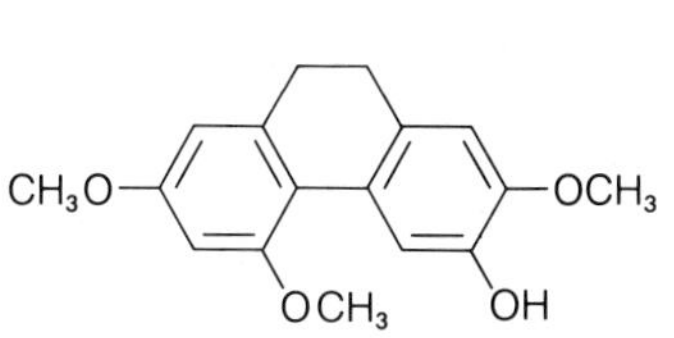

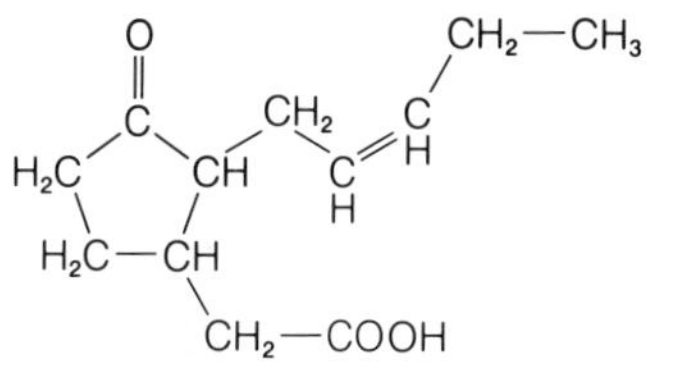

This seems reasonable, because many other compounds are associated with seed dormancy. However, as with buds, it is doubtful if analyses of whole seeds, including storage tissues, could provide the crucial information needed about changes in ABA in the radicle cells that grow to cause germination when dormancy is overcome (also see Section 19.4).

ABA and Abscission The importance of ABA in causing abscission of leaves, flowers, and fruits is controversial. Various experts evaluate published data in different ways. Milborrow (1984) concluded that exogenous ABA causes abscission but much less effectively than exogenous ethylene. Yet Sacher (1983) concluded that endogenous ABA plays a prominent role in senescence of these organs. In an extensive review, Addicott (1983) made a strong case for an important role of endogenous ABA in causing abscission, especially as opposed to the importance of ethylene. It is difficult for us to evaluate published data as well as experts have. But again, we should expect differences among species and the organ that is abscised, and we should require knowledge of ABA concentrations in cells that control abscission (still unmeasured) before we make strong conclusions. These cells are described in Section 17.7.

How Does ABA Act? ABA seems to have three major effects, depending on the tissue involved. One is on the plasma membrane of roots, another the inhibition of RNA synthesis (transcription), and the third an inhibition of protein synthesis (translation). The influence on root membranes is to make them more positively charged, increasing the tendency with which excised root tips stick to negatively charged glass surfaces. This effect is probably involved in the rapid loss of K^+ ions from guard cells (where inhibition of an ATPase is involved), and perhaps in the ability of ABA to rapidly inhibit auxin-induced growth. Interference with synthesis of proteins and other enzymes could explain long-term effects on growth and development, including the

proposed role in dormancy of seeds and buds and the known inhibition of gibberellin-promoted hydrolase activity in cereal grain seeds. The mechanism of both effects is unknown.

17.6 Other Growth Inhibitors

ABA may well be a widespread growth regulator, but other compounds that inhibit growth have been discovered. The structures of these compounds have few similarities (Fig. 17-15). In liverworts, **lunularic acid** is present in **gemmae** (vegetative propagules about 1 mm in diameter formed in gemma cups on the upper surface of thalli of the mother plant). There is evidence, reviewed by Milborrow (1984), that lunularic acid prevents germination of these gemmae until they fall from the mother plant's thallus and the acid is leached out. Furthermore, growth of the entire thallus seems to be partly controlled by lunularic acid in response to daylength conditions. In short days, the content of the inhibitor is low and thalli grow rapidly; in long days the reverse is true. It has been suggested that lunularic acid in liverworts fulfills the role played by ABA in vascular plants, yet except for a well-defined role in closing of stomates (which liverworts lack) we are not certain of the functions of ABA in vascular plants. An extensive survey of nonvascular plants by Gorham (1977) showed that lunularic acid is present in many species of lower plants but not in algae.

Batasins are compounds in yam plants (**Discorea batatus**) that seem to cause dormancy in **bulbils** (vegetative reproductive structures) that arise from swelling of aerial lateral buds. The structure of *batasin I* is in Fig. 17-15. Batasins are concentrated in the skin of the bulbils and are absent from the core. A lengthy exposure to cold temperature (stratification or prechilling—see Section 21.1) that breaks dormancy causes batasins to disappear, whereas their amounts increase during the initial development of dormant bulbils (Hasegawa and Hashimoto, 1975). However, it is not known if batasins are transported into or accumulate within the bud cells whose failure to grow actually causes dormancy.

Jasmonic acid (Fig. 17-15) and its methyl ester (*methyl jasmonate*) occur in several plants and also in the oil of jasmine. These compounds inhibit growth of certain plant parts and strongly promote senescence in detached oat leaves (Milborrow, 1984). Their functions in the plants that contain them remain to be discovered.

17.7 Hormones in Senescence and Abscission

Processes of deterioration that accompany aging and that lead to death of an organ or organism are called **senescence.** Although meristems do not senesce and might be potentially immortal, all differentiated cells produced from meristems have restricted lives. Senescence therefore occurs in all nonmeristematic cells, but at different times. In evergreens, many species retain their leaves for only two or three years before they die and abscise, but bristle cone pine (*Pinus aristata*) retains functional needles for up to thirty years (Sexton and Woolhouse, 1984). In deciduous trees and shrubs, the leaves die each year, but again the stem and root systems remain alive for several years. In perennial grasses and herbs such as alfalfa, the aboveground system dies each year, but the crown and roots remain largely viable. In herbaceous annuals, leaf senescence progresses from old to young leaves, followed by death of stem and roots after flowering. Only the seed survives. For reviews of senescence, see Woolhouse (1978), Thimann (1980), Stoddart and Thomas (1982), and Sexton and Woolhouse (1984).

What causes senescence? Chemical analyses show that leaf senescence is accompanied by early losses in chlorophyll, RNA, and proteins, including many enzymes. Because these and other cellular constituents are constantly being synthesized and degraded, loss could result from slower synthesis or faster breakdown, or both. Slow synthesis is expected when nutrients normally arriving in an organ are diverted elsewhere, as, for example, when flowering and fruit formation occur. One theory for leaf senescence is, therefore, that flower and fruit development cause a competition for nutrients. In Section 15.4, we mentioned competition between vegetative and reproductive organs for nutrients essential for growth and showed in Fig. 15-18 how removal of all flowers postpones leaf senescence in annual soybeans.

Nutrient competition is not the entire explanation of senescence, even for soybeans, because the young flowers could not possibly divert enough nutrients to cause death of the leaves. Numerous ex-

periments of Larry Noodén and his students at the University of Michigan suggest soybean fruits produce a **senescence factor** (a hormone) that moves to the leaves where it causes senescence. In cocklebur plants, short-day and long-night conditions induce flowering and leaf senescence, but even if all flower buds are removed, leaf senescence still occurs. Furthermore, development of male flowers on staminate spinach plants induces senescence as effectively as development of both flowers and fruits on female plants, yet staminate flowers divert much less nutrients than do fruits and seeds (Leopold and Kriedemann, 1975). Another example of senescence already described is the degradation of food reserves and loss of integrity in food storage cells of seeds. We mentioned that endosperm degradation in cereal grains is enhanced by gibberellins moving from the embryo. Taken together, these results indicate that *senescence is also hormonally controlled.* It appears as though organs or tissues acting as nutrient sinks send hormones to other organs, causing them to release some of their nutrients.

If hormones are involved, we should eventually be able to identify them and explain how they act. For gibberellins and cereal grain seeds, we are approaching both answers. We also mentioned that cytokinins can delay leaf senescence in many angiosperms, and that gibberellins are also effective in some. How gibberellins retard leaf senescence is not known. Auxins also delay leaf senescence in some species, but this is less easy to demonstrate than for gibberellins and cytokinins arriving from roots, because leaves normally seem to synthesize enough auxin. Therefore, adding more auxin, even to a detached leaf, usually does not delay senescence.

Contrary to effects of the other hormones, ethylene and ABA promote senescence. In fruits, the effect of ethylene is manifested by rapid ripening followed by abscission; in flowers, the common result is fading, transport of nutrients, withering, and then abscission, while in leaves we observe loss of chlorophyll, RNA and protein, and transport of nutrients, followed by abscission. The effect of ethylene seems to us much more dramatic than that of ABA. To what extent natural rises in the ABA levels contribute to senescence and abscission is still uncertain.

What advantage is there to abscission of senescent leaves, flowers, and fruits? For fruits, the importance in perpetuation of the species is obvious, because fruits contain seeds. For flowers, we suspect that the reasons involve removal of a useless organ that might act as a potential infection source and that, in some species, would shade new leaves the next growth season. We mentioned in Section 15.4 that senescence of flowers is usually accompanied by hydrolysis of RNA and protein. The products are converted into mobile amides and amino acids. Many other large molecules (except for those in cell walls) are also degraded into smaller, more readily trans-locatable forms in which nutrients are conserved by storage in other parts of the plant. For leaves that abscise, a similar extensive salvage of nutrients by mobilization also precedes abscission. This nutrient economy helps forest trees to survive on unfertile soils. Leaves that abscise apparently could not withstand cold winters and would shade new leaves the next spring, so their loss, preceded by nutrient salvage, increases survival and productivity of individual perennial plants.

In most species, abscission of leaves, flowers, or fruits is preceded by formation of an **abscission layer** or **zone** at the base of the organ involved (Kozlowski, 1973; Sexton and Roberts, 1982; Addicott, 1982). In leaves, this zone is formed across the petiole near its junction with the stem (Fig. 17-16). In many compound leaves, each leaflet also forms an abscission zone. The abscission zone consists of one or more layers of thin-walled parenchyma cells resulting from anticlinal divisions across the petiole (except in the vascular bundle). In some species, these cells are formed even before the leaf is mature. Just before abscission, the middle lamella between certain cells in the distal region (that farthest from the stem) of the abscission layer is often digested. This digestion involves synthesis of polysaccharide-hydrolyzing enzymes, most importantly cellulase and pectinases, and their secretion from the cytoplasm into the wall. Formation of these enzymes is accompanied by a rapid rise in respiration in cells of the proximal part of the abscission zone (those cells of the zone close to the stem). This rise is similar to that which occurs in climacteric fruits and also involves increases in polyribosomes characteristic of cells actively synthesizing proteins. Furthermore, one or more layers of these proximal cells increases in size (both length and diameter), while the cells of the abscission zone distal to the breaking point do not. These wall digestion processes accompanied by pressures resulting from unequal growth in expanding proximal and senescent distal cells of the zone apparently cause a break between them. Ethylene causes cell expansion and secretion of the cell-wall degrading hydrolases in some species (Osborne, 1982).

Wound healing in cells proximal to the break point involves formation of a corky layer that protects the plant from pathogen invasion and excess water loss. Compounds synthesized during healing include suberin and lignin.

Several environmental factors promote abscission, especially drought and nitrogen deficiency, but these first hasten senescence. Even though abscission is an active process in cells proximal to the break, the organ being shed nearly always first becomes senescent. What role do hormones play in abscission? Our present ideas about leaves are summarized as follows: An auxin is synthesized in grow-

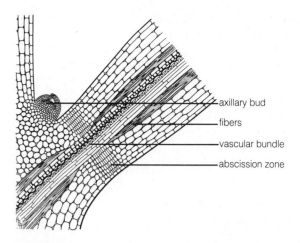

Figure 17-16 The abscission layer. (From F. T. Addicott, 1965, in W. Ruhland, ed., Encyclopedia of plant physiology, vol. 15, part 2, Springer-Verlag, Berlin.)

ing leaf blades and strongly retards senescence and abscission until its level and rate of transport into the petiole decrease too much when the leaf matures. Cytokinins and perhaps also gibberellins arriving from the roots also delay senescence and abscission for a while. Finally, however, degradation processes overcome synthetic processes in the leaf, partly because of competition with growing organs for nutrients, and senescence occurs. Now something, perhaps senescence factor, causes ethylene production in the abscission zone, and ethylene stimulates growth of the proximal cells and formation of cell-wall degrading enzymes in them. Abscission then occurs. No clear-cut role for ABA in this sequence has been established, although increases in ABA might contribute to the senescence preceding abscission and abscission itself.

You should now realize how far we have departed from Julius von Sachs' early idea that plants might have specific hormones responsible for growth or formation of a particular organ. Clearly, the rule is that most physiological processes require interactions among several hormones, and a single hormone has several functions. Furthermore, each process generally depends on the kind of cells involved (target organ) and on the species. Apparently, different species sometimes use different hormones or rely on different interactions among them to accomplish their various functions.

Furthermore, other hormones await discovery. There is evidence for a hormone that moves from leaves to underground stems, where it causes tuber and bulb formation (Chapter 21), for a flower inducing hormone (Chapter 22), for root growth hormones provided by shoots, for a hormone that guides the pollen tube into the embryo sac where fertilization is accomplished, and for a senescence factor. It would obviously be simpler for us to interpret the results if such hormones do not prove to be varying mixtures of known hormones.

18

Differential Growth
and Differentiation

Plant growth and development are complex processes, as we have seen in the last three chapters. In this chapter, we continue this theme by discussing plant movements and certain other aspects of plant morphogenesis.

Plant movements seem to fall into two natural categories: the **nastic movements,** which are triggered by an external environmental stimulus or by an internal timing mechanism, and in which the direction of the stimulus *does not* determine the direction of the movement; and the **tropisms,** in which the direction of the external stimulus *does* determine the direction of the movement. Both nastic and tropistic movements are often the result of differential growth. Nastic movements are stimulated by changes from light to darkness or darkness to light, by touch, by other mechanical stresses, or by the **biological clock** (internal timing mechanism). The tropisms are also often a response to light, and gravity is an important stimulus. Other stimuli such as touch have also received some attention.

We shall discuss two special aspects of differentiation: the **juvenility** that appears in many plant species (e.g., the first leaves on a plant have a different appearance than later leaves) and **totipotency,** the ability of one to a few cells to regenerate an entire plant. Following these discussions, we shall briefly outline some principles of differentiation.

18.1 Nastic Movements

Important nastic movements involve leaves or the leaflets of compound leaves. An upward bending of a leaf (or any organ) is called **hyponasty;** downward bending is **epinasty.** Often, as we shall explain in a moment, these leaf movements occur because of water movement in and out of special cells at the base of the petiole, blade, or leaflet. Such cells are called **motor cells,** and the group of such cells forms an organ called a **pulvinus** (plural: **pulvini**). But hypo-nastic and epinastic leaf movements occur in many plants that have no pulvini. Epinasty, for example, occurs as cells on the top of the petiole or blade (especially the main veins) grow more than those on the bottom. Generally speaking, we think of nastic movements as being reversible—which is typically the case when pulvini are involved—and thus not as growth phenomena. Nevertheless, many hyponastic and epinastic leaf movements do involve growth (irreversible cell elongation), and they also appear to be reversible. For example, epinastic leaf bending changes to hyponastic bending after a reversal of the relative growth rates of cells on the lower side of the stem, compared with rates of cells on the upper side.

Nyctinasty Movement of leaves of many species from nearly horizontal during the day to nearly vertical at night has been recognized for more than 2000 years. As discussed in Chapter 20, such **nyctinastic** (Greek: *nux* = night) movements are rhythmic processes controlled by the biological clock. Here we shall emphasize the anatomical and physiological aspects of such movements.

One of the species most carefully investigated is the silk tree, *Albizzia julibrissin* (Satter and Galston, 1981). It has doubly compound leaves, each leaf bearing several pinnae and each pinna bearing several pairs of opposite pinnules (leaflets) attached to a single rachilla. Such doubly compound leaves often exhibit striking sleep movements, as illustrated in Fig. 18-1. At night, the tips of opposite leaflets press together, rise upward, and become pointed toward the distal end of the rachilla. These movements are caused by pulvini (Fig. 18-2). At night, water moves out of subepidermal cortical cells in the upper (**ventral**) side of the pulvinus and into cortical cells on the lower (**dorsal**) side. Thus, swelling of dorsal cells and compression of ventral cells leads to closing, and a reversal of these changes leads to opening.

What causes water to flow from one side of the pulvinus to another? In 1955, Hideo Toriyama ob-

day

night

Figure 18-1 *Albizzia julibrissin* leaves in normal daytime position and in typical night (sleep) position. (Courtesy of Beatrice M. Sweeney.)

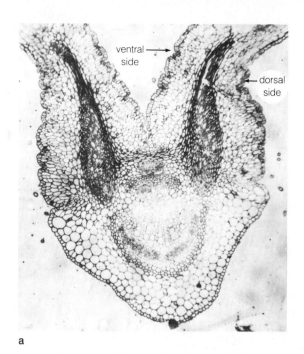

ventral side →

→ dorsal side

a

served, for the comparable process in the sensitive plant (*Mimosa pudica*), that K$^+$ moves out of the cells that lose water. In the early 1970s, Ruth L. Satter, Arthur W. Galston, and others showed that the K$^+$ concentration of *Albizzia* pulvini is unusually high (almost 0.5 *M*) and that leaflet closing is accompanied by loss of K$^+$ from ventral cells and absorption of K$^+$ by dorsal cells, but that the dorsal cells do not receive the K$^+$ which the ventral cells lose. They have also shown that Cl$^-$ shows changes similar to those of K$^+$ (Schrempf et al., 1976). Apparently, other cells or cell walls of the pulvinus act as reservoirs for these ions. Nevertheless, water movement is probably in response to the osmotic effect of ion transport, just as occurs in stomate opening (Section 3.4). The same is true for nyctinastic movements controlled by pulvini in other species investigated. The principal problem remaining is to learn how factors that change from day to night—especially light, temperature, or the biological clock—affect transport of these ions.

S. Watanabe and T. Sibaoka (1983) studied responses of leaflet pairs from detached pinnae of the sensitive plant. During the daytime (from 6:00 to 16:00), leaflet pairs would open in response to blue or

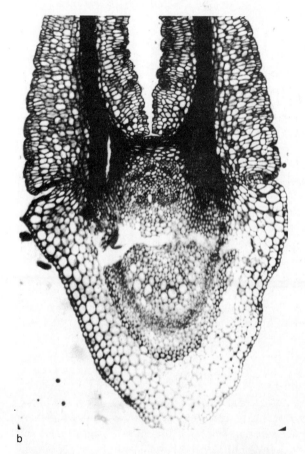

b

Figure 18-2 Longitudinal, dorsiventral sections through *Albizzia julibrissin* pulvinules and adjacent leaflet tissues in the (**a**) open and (**b**) closed positions. The rachilla to which leaflets are attached is shown in transverse section in (**a**) and in oblique section in (**b**). (Courtesy of Ruth Satter; from Satter et al., 1970.)

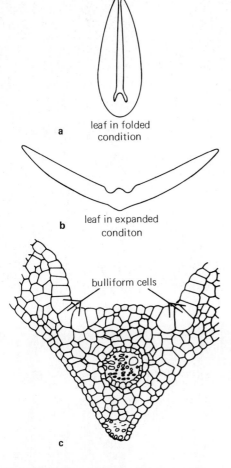

a   leaf in folded
condition

b   leaf in expanded
condition

bulliform cells

c

Figure 18-3 (**a**) Outline drawings of blue grass *(Poa pratensis)* leaf in folded condition and (**b**) in expanded condition. (**c**) A detailed drawing of the midportion of the leaf showing the bulliform cells. Changes in the turgor of these cells control the folding and opening of the leaf. (From Meyer and Anderson, 1952.)

far-red light but not orange or red light (see discussion of the high-irradiance reaction in Chapter 19). At night, they would not respond to the light, but they would open in response to application of auxin (IAA, naphthaleneacetic acid, or 2,4-D).

Olle Björkman and Stephen B. Prowles (1981) have described an interesting response of the redwood sorrel *(Oxalis oregana)*. This shade plant (Section 11.2) photosynthesizes in redwood forests at light levels only 1/200th of full sunlight. Now and again sunflecks penetrate the forest canopy and could damage this delicate species. With only a 10-second lag period, the leaves begin to fold downward, the folding being complete in about 6 minutes. When shade returns, there is a 10-minute lag period, but the leaves revert to their horizontal position in about half an hour. Blue wavelengths of light are sensed by a small pulvinus where each leaflet joins the petiole.

Many grass leaves either fold or roll up when subjected to water stress, and these processes sup-

plement stomatal closure, further minimizing transpiration. Folding and rolling movements are caused by loss of turgor in thin-walled motor cells called **bulliform cells,** as illustrated for Kentucky blue grass *(Poa pratensis)* in Fig. 18-3. The bulliform cells have little or no cuticle and so lose water by transpiration faster than other epidermal cells. As their pressure potential decreases, turgor in cells on the lower side of the leaf causes the folding shown in the figure. This represents but one of several mechanisms by which plants resist drought, as discussed in Chapter 24.

Thigmonasty Nastic movements resulting from touch (**thigmonasty;** Greek: *thigma* = touch) were long thought to occur in only a few species, especially members of the Mimosoideae subfamily in the Fabaceae (Leguminosae) family (Ball, 1969). Now we know that the phenomenon is more widespread. The most notable example is *Mimosa pudica,* the sensitive plant, which has doubly compound leaves with leaflet and pulvinar structures similar to *Albizzia.* Upon being touched, shaken, heated, or treated with an electrical stimulus, its leaflets and leaves rapidly fold up (Fig. 18-4). Only one leaflet need be stimulated; some stimulus then moves throughout the plant. The advantages of this to the plant are uncertain, but one (not very appealing) idea is that the collapse of the leaflets startles away insects before they begin to eat the foliage. Leaflet movement is caused by water transport out of certain motor cells of the pulvinus. Again, an efflux of K^+ precedes water loss, at least in the main pulvinus at the base of each leaf.

How a stimulus can be transported in *Mimosa* has been investigated for many years (Pickard, 1973; Roblin, 1982; Samejima and Takeo, 1980; Sibaoka, 1969; Simons, 1981; and Umrath and Kastberger, 1983). There is evidence for two distinct mechanisms, one *electrical* and the other *chemical.* The electrical response was first studied extensively by J. C. Bose, in India, between 1907 and 1914, and then in better experiments by A. L. Houwink, in Holland, during the early 1930s. The electrical fluctuation is an **action potential,** which is a change in voltage that forms a characteristic peak when plotted as a function of time (Fig. 18-5). Action potentials in *Mimosa* are similar to those occurring in animal nerve cells but much slower. They apparently travel through parenchyma cells of the xylem and phloem at velocities up to about 2 cm s^{-1}, whereas action potentials travel along nerve cells at velocities of tens of meters per second.

The action potential will not pass through a pulvinus from one leaflet to another unless the chemical response is also elicited, in which case several leaflets may fold up. The chemical response was first reported by Ubaldo Ricca, an Italian, in 1916. It is caused by a substance that moves through the xylem

Figure 18-4 Response of the sensitive plant. The tip of one leaf is stimulated in such a manner (in this case, by using a flame) that the entire plant is not jarred. After 14 s, the petiole of the leaf has collapsed, and many of the leaflets have folded up as water leaves pulvinal cells at the bases of petioles and leaflets. As the stimulus travels back along the stem, other leaves collapse and their leaflets fold. This process continues in the last photograph (taken after about 1.5 min). (Photographs by F. B. Salisbury.)

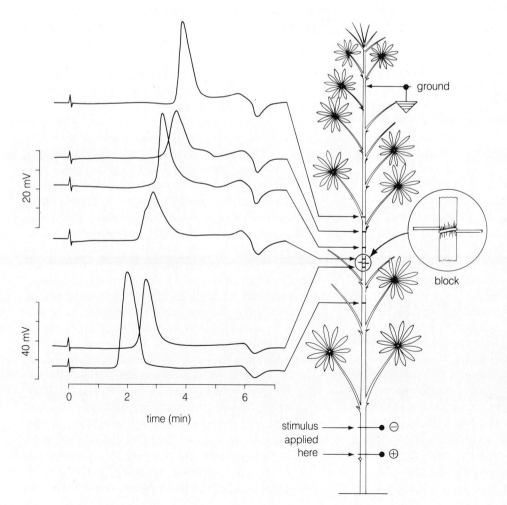

Figure 18-5 An example of action potentials measured in *Lupinus angustifolius*. The purpose of this study was to examine movement of action potentials past a block in the stem produced by tightening a thread around the stem (detail drawing in circle). The drawing of the stem shows the location of the anode (+) and the cathode (−), the block, and the ground, which is necessary to measure the action potentials. Arrows show the locations of electrodes along the stem. The action-potential curves show voltages as a function of time, and the small blip to the left on each curve is an artifact caused by application of the stimulus (the direct electric current between anode and cathode). (From Zawadski and Trebacz, 1982, with permission.)

0 2 5 10 20 30

Figure 18-6 The effect of the number of rubs of the stem (once up and once down, between thumb and forefinger, with moderate pressure) on the growth of young bean plants. From left to right, the number of stimuli were 0, 2, 5, 10, 20, or 30. (Courtesy of Mordecai J. Jaffe; see Jaffe, 1976.)

vessels along with the transpiration stream. Ricca cut through a stem, then connected each cut end with a narrow, water-filled tube. When a leaf on one side of the tube was wounded, a leaf on the other side folded. The active substance can be extracted from wounded cells, applied to a cut branch, and its folding effects then measured. Although still unidentified, it is known as **Ricca's factor.** Its movement elicits electrical responses that travel ahead of it in parenchyma cells from one leaflet to another. Rapidly transported wound signals have also been observed in pea tissues (Davies and Schuster, 1981; Van Sambeck and Pickard, 1976).

One of the few well-studied examples in which an action potential is obviously useful to a plant is the excitation by an insect of one or more sensory hairs of the Venus's-flytrap (*Dionea muscipula*). Action potentials move into the bilobed leaf and cause the lobes to snap shut within a half second or so. This usually traps the insect, which is then digested by enzymes secreted from the leaf, providing nitrogen sources for the plant. About 400 other flowering plants are carnivorous, and the capture mechanisms are diverse and often independent of action potentials (Heslop-Harrison, 1976). In the case of the Venus's-flytrap, Williams and Bennett (1982) have shown that the rapid closing is another example of acid growth (Sec-

tion 16.1). Hydrogen ions are rapidly pumped into the walls of cells on the outside of each leaf in response to the action potentials from the trigger hairs. The protons apparently loosen the cell walls so rapidly that the tissue actually becomes flaccid so that cells quickly absorb water, causing the outside of each leaf to expand and the trap to snap shut.

The tests of Williams and Bennett's hypothesis are a model of the scientific method in action. First, they marked the inside and the outside of the leaf with dots, measuring distances between them before and after closing. The outside expanded about 28 percent, but the inside did not change, confirming that growth on the outside closes the trap. The inside gradually expanded during the 10 hours of reopening, but there was no further change on the outside, so this is a true rapid growth response rather than an action of a pulvinus. Second, their hypothesis suggested that infiltrating the leaves with neutral buffers should neutralize the protons pumped out of the cells and prevent closing. This proved to be the case, but infiltrating the leaves with a sorbitol solution with the same osmotic potential as the neutral buffer did not prevent closing. Third, they predicted that infiltrating the leaves on the outside with acidic buffers (pH 3.0 to 4.5) should cause closing without stimulation of the trigger hairs. This also occurred. Fourth, since hydrogen pumping requires much ATP, they predicted that the amount of ATP should drop rapidly in the tissues during closing. They found a drop of about 29 percent during closing. All their tests were positive, strongly suggesting acid growth in Venus's-flytrap closing!

Thigmo- and Seismomorphogenesis   In the early 1970s, Mordecai J. Jaffe (1973, 1980) began to investigate effects of mechanical stimulation, especially rubbing, on plants that exhibit no rapid movement. Most vascular plants that he investigated responded by elongating more slowly and increasing in diameter, causing short, stocky plants (Fig. 18-6), sometimes only 40 to 60 percent as tall as controls. He called these and similar developmental responses to mechanical stress **thigmomorphogenesis.** Bending the stems also causes these responses, and in nature the bending effects of wind influence plant development in this way. Cary A. Mitchell at Purdue University (1975, 1977; Hamner et al., 1974) has also studied these effects extensively. He has placed plants on a shaker table so that they are mechanically shaken but not physically contacted. The same basic responses appear, and Mitchell terms the phenomenon **seismomorphogenesis** (*seismo* = shaking). In either case, shorter, stronger plants are produced, and these are less easily damaged by the mechanical stresses of nature (especially wind) than are their taller, more slender, counterparts. Actually, the tall slender plants are "unnatural," since they appear only in greenhouses—and then only when conditions are just right for the response (typically, rather low light levels). We should note that in some of Mitchell's experiments, certain shake-table frequencies *promoted* growth, although inhibition was the general rule.

Rubbing by farm machinery and by animals must have inhibitory effects on growth. Spraying tomato plants with water for 10 seconds once a day reduced their growth in a greenhouse to about 60 percent of the growth of controls (Wheeler and Salisbury, 1979). The stockier plants were desirable in greenhouse culture, but unfortunately fruit yields were significantly reduced.

Salisbury (1964) found that simply measuring the lengths of cocklebur leaves with a ruler at daily intervals slowed their growth and caused premature senescence. Others found that young sweetgum (*Liquidamber styraciflua*) trees elongated more slowly and set winter terminal buds when their trunks were vibrated or shaken for only a 30-second period each day. Inhibitory effects of mechanical stress on *flowering* of a few species have also been observed, and it is likely that several additional thigmomorphogenetic responses will be discovered. Such responses may prove to be as common and thus as important to plants as responses to light, temperature, or gravity.

Certainly, they might confuse the outcome of any experiment if controls and treated plants do not receive comparable mechanical stimulation.

What causes these responses? Since the plants at first show no injury symptoms except an altered growth, a change in growth regulator patterns is suspected. The decrease in stem elongation and the increased stem thickening suggest that ethylene production plays an important role, and indeed an increase in ethylene has been observed following mechanical stimulation (see Section 17.2).

How a change in hormone balance might come about is unknown, but Jaffe found that the electrical resistance of bean stems decreases greatly within a few seconds after rubbing, followed by a slower rise back toward the normal level. This finding probably suggests that ions such as K^+ pass rapidly through membranes from the symplast into the apoplast following mechanical stimulation. (The increased ionic concentration would account for the greater flow of current and thus lowered resistance.) A change in membrane permeability might rapidly affect availability of hormones at the normal subcellular sites where they act and might also affect subsequent production of growth regulators by altering the availability of precursor molecules to enzymes that synthesize growth regulators.

18.2 Tropisms: Directional Differential Growth

You are probably familiar with a plant's abilities to respond to the environment by displaying unequal or differential growth. You know that roots grow downward and stems upward in response to gravity (**gravitropism***) and that stems and leaves frequently grow toward the light (**phototropism**). Less apparent is the phenomenon of **thigmotropism,** a response to contact with a solid object exhibited by climbing plants that grow around a pole or the stem of another plant (Jaffe and Galston, 1968). These and other tropisms are responses to environmental stimuli that determine the direction of bending of the plant organ in question, usually by a faster rate of cell elongation on one side of the organ than on the other.

There has been considerable interest in the tropisms for well over two centuries, and they have been studied with modern scientific techniques since at least the time of Sachs during the second half of the last century. It is humbling to realize that, in spite of

---

*Formerly, **geotropism,** which means "tropistic response to the earth." Since the response is actually to the acceleration caused by the gravitational (or other accelerational) force, gravitropism seems like a preferable term—just as several decades ago, heliotropism (response to the sun) was replaced by phototropism (response to light.)

thousands of technical papers reporting studies on the tropisms, we remain baffled by the basic mechanisms. Although we have a vast amount of data, most of it is essentially descriptive or phenomenological, and as you will see, very little is known about basic mechanisms.

Actually, considerable progress has been made in learning about tropisms during recent decades, and the U. S. National Aeronautics and Space Administration is currently supporting considerable research on how plants respond to gravity. This research includes both ground-based studies and experiments under the nominal weightlessness of orbiting satellites. Most studies are presently being conducted in the context of a sequence of postulated events that include the three following steps (Haupt and Feinleib, 1979):

1. **Perception** How does a plant part detect the direction of the environmental stimulus that causes the tropism? Where in the plant is the perception mechanism located? And what is the perception mechanism? It has been difficult to answer these questions for plants, because, in contrast to animals, they do not have specific organs for virtually every function.

2. **Transduction** Whatever the perception mechanism is, how does it convert or transduce its message of stimulus direction to the cells in the organ where tropistic movement occurs? What metabolic or growth-regulator changes occur in response to the environmental stimulus? This has been an especially active area of research for much of this century.

3. **Response** What actually happens during tropistic bending or other responses? Any hypothesis put forth to explain the mechanisms of perception and transduction must account for the observed response. This seems obvious enough, yet the details of each response have been rather neglected for the past few decades—until quite recently. Early researchers in the late 1800s and early 1900s made many careful studies of tropistic responses, discovering that cells on one side of the organ grow more than those on the other side, accounting for bending. Apparently, later researchers assumed that everything possible had already been learned. Yet many of the results of early workers were either overlooked or completely forgotten, and we are just now beginning to appreciate their significance.

Two generalizations have come from work on tropisms during the past two or three decades. Because we must necessarily limit our subject matter in the following pages, these generalizations will not always be apparent. They are: First, similar plant mechanisms often cause different responses. For example, K^+ movement in and out of cells causes such diverse responses as stomatal action and thigmonastic leaf folding in the sensitive plant; K^+ may also be important in some of the tropisms. Second, different mechanisms may produce similar responses in different or even the same organisms. For example, different pigment systems are apparently responsible for phototropic bending in different organisms, although the exceptional ones, which respond to *red* (instead of blue) light, are rare and won't be discussed.

18.3 Phototropism

Coleoptiles and Stems

The basic responses   In 1881, *The Power of Movement in Plants* by Charles Darwin, assisted by his son Francis Darwin, was published (note also an "authorized edition" published in 1896). In this book, the Darwins described many experiments relating to plant tropisms. They observed that reed canary grass coleoptiles would not bend toward dim light if the tips had been cut off. They also studied other monocots and dicots. It was their experiments plus those of others localizing phototropic sensitivity in the tips of coleoptiles that eventually led to the experiments of Frits Went and thus the discovery of auxin (Chapter 16).*

The Darwins found that oat (*Avena*) coleoptiles also bent toward light somewhat when their tips were covered; that is, some phototropic sensitivity occurs below the tips, but the **tip response** is about a thousand times more sensitive than the **base response.** Thus if dim light is used, most of the response is localized in the tip, and this is evident because curvature toward the light begins at the tip and moves gradually down the coleoptile as the stimulus is transmitted from the tip to the tissues below. If higher light levels are used, however, bending begins simultaneously along the entire length of the coleoptile (or, in other experiments, the hypocotyl of such dicot seedlings as sunflowers; see Dennison, 1979, 1984).

What actually happens during phototropic bending of a coleoptile or hypocotyl? The question was asked in studies many decades ago and in more recent studies using photographic techniques (Franssen et al., 1982). Minute glass beads were placed along both sides of coleoptiles or hypocotyls illuminated from one side, and photographs were

---

*The discovery of auxin also depended upon experiments with roots done by the Polish plant physiologist Ciesielski (1873), who was cited by the Darwins.

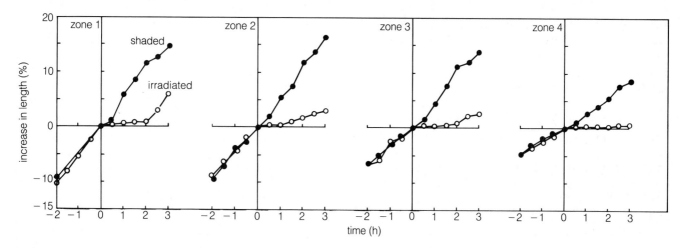

Figure 18-7 Growth rates on the shaded and irradiated sides of oat *(Avena)* coleoptiles that were irradiated from one side with blue light of an intensity high enough to cause second positive curvature. Zone 1 is the most apical and zone 4 the most basal zone. Growth rates were measured by attaching small plastic beads to the coleoptiles and taking photographs with flash units fitted with green filters. Measurements were later made on the projected photographic images. The negative numbers represent hours before the beginning of unilateral irradiation, which began at time zero. Note how the irradiated side stops growing immediately (at least within the 30-minute interval after time 0). In this experiment, growth of the shaded side was at about the same rate after as before irradiation. The difference in growth rates on the two sides accounts for bending toward the light. (From Franssen et al., 1982.)

taken at intervals during phototropic bending. The photographic negatives were then projected onto a wall, and distances between the glass beads were carefully measured on the projected images. This made it possible to plot the rate of extension on the lighted (irradiated) and shaded sides of the coleoptiles or hypocotyls, as shown in Fig. 18-7. When high light levels were used, so that the basal response was elicited, growth on the irradiated side stopped almost instantaneously and completely upon the beginning of irradiation. Growth of the shaded side continued, often at about the same rate as before the beginning of unilateral irradiation. Any explanation of phototropism must account for this strikingly rapid cessation of growth on the irradiated side.

Dose-response relations There are several important quantitative questions that can be asked about the phototropic response. So far, nearly all quantitative work has been done with coleoptiles. Of primary importance is the question of *reciprocity:* Is the response proportional to the duration of the exposure, to its energy level (photon flux density), or to the product of duration times energy level? In the last case, energy and duration bear a *reciprocal* relationship to each other,* so the **law of reciprocity** is said to hold. For practical purposes, it has usually been most

convenient to give a constant low irradiance to coleoptiles or other objects for varying time intervals. Excellent results were obtained by du Buy and Nuernbergk in 1934, and further experiments were carried out by Zimmermann and Briggs (1963)—in the latter case with three different photon flux densities of blue light. Results are plotted as the degrees of phototropic curvature of oat coleoptiles as a function of the product of exposure time multiplied by energy level (Fig. 18-8).

Since the first, ascending part of the curve is the same for all three energy levels, reciprocity holds over this range of bending caused by the relatively low total energy amounts. This part of the curve is called **first-positive curvature**. At higher energy levels (part of the curve labeled C)—meaning shorter exposure times—first-positive curvature is followed by decreasing curvature with increasing energy; this response is called **first-negative curvature**. It is followed by an ascending part of the curve again, called **second-positive curvature**. (Du Buy and Nuernbergk's experiments extended the irradiance times at high energy levels so that they were also able to observe second-negative and third-positive curvatures.) At the intermediate energy levels of Zimmermann and Briggs (labeled B), first-negative curvature is greatly reduced, although second-positive curvature is still quite evident. At the lowest light levels (longest exposure times, labeled A), first-negative curvature has essentially disappeared, although there is a shoulder in the curve suggesting

---

*Energy × Time = Constant, hence: Energy = Constant/Time

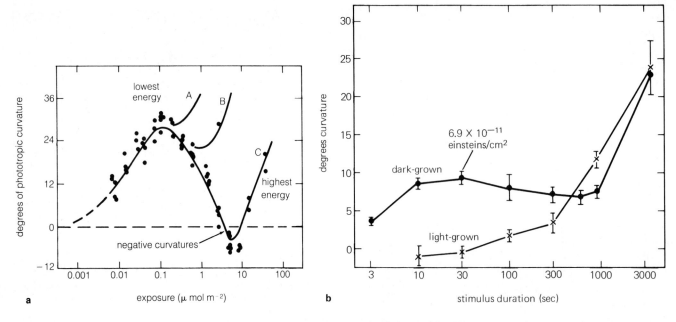

Figure 18-8 (**a**) Phototropic response of oat coleoptiles as caused by increasing energies of blue light at 436 nm. Note first-positive, negative, and second-positive curvatures. Light energy was held constant, and exposure times were varied to give the total exposures shown along the abscissa. Irradiances were 1.40, 0.14, and 0.014 μmol m^{-2} s^{-1} for lines labelled C, B, and A, respectively. Irradiance was with unilateral blue light (435.8 nm) following red light. Points are actual data (to show scatter) but apply only to curve C. (From Zimmermann and Briggs, 1963.) (**b**) Similar dose-response curves for radish seedling phototropism. Stimulus energy was 0.023 μmol m^{-2} s^{-1} with light at 460 nm. Bars represent standard errors. Note that the abscissa is a logarithmic scale in both graphs. (From Everett, 1974.)

second-positive curvature. Clearly, reciprocity does not hold for any part of the dose-response curves except first-positive curvature. Why is that?

Blaauw and Blaauw-Jansen (1970a and 1970b) carried out dose-response experiments with at least 24 different irradiance levels. Their results confirm in detail those of Zimmermann and Briggs, and in their discussion they suggest an explanation. The second-positive curvature occurs at the same *time* at all irradiance levels (about 40 minutes after the beginning of irradiation). Thus, as the irradiance levels decrease, the first-positive curvature is delayed (reciprocity holds), but second-positive curvature comes at the same time regardless of irradiance level, so eventually the two come together, eliminating first-negative curvature.

Thought and a few experiments suggest that light really has two effects in phototropism. First, it acts as a trigger for the bending response; this is what we have been emphasizing. Second, it changes the state of the organ so that its sensitivity to subsequent light is decreased; this effect is nondirectional and is referred to as a **tonic effect.** For example, if a coleoptile is exposed to one second of 0.03 W m^{-2} of unilateral blue light, positive curvature occurs—but only if the coleoptile has previously been in the dark. Much less positive curvature occurs if the coleoptile

has already been exposed to 10 seconds of the same irradiance. In this case, the negative area of the curve is approached. Actually, the tonic effect is easy to observe even when the light comes from above the coleoptile instead of from one side (Meyer, 1969a). When two exposures are given, the tonic effect caused by the first exposure gradually decays away so that after about 20 to 25 minutes the exposure to the second irradiance is the same as if the first exposure had not been given. Recovery has occurred. Hence, Blaauw and Blaauw-Jansen (1970a and 1970b) conclude that the curvatures beyond the first-positive response are not really independent phenomena at all but are the result of the desensitization of the first-positive curvature system. Clearly, the tonic effects of blue light are complex.

In *Avena*, the region of first-positive and first-negative curvatures is shifted to ten-fold higher irradiances by *red* light given just prior to the blue light. The second-positive curvature was shifted to three-fold lower irradiances by the pretreatment with red light (Zimmermann and Briggs, 1963a; discovered by Curry as cited in his 1969 paper). Zimmermann and Briggs found that brief exposures to red light given at intervals over a one to two-hour period had essentially the same effect as continuous exposure. Thus, it was possible to attempt a reversal of the red effect (maximum at 660 nm) with subsequent exposures to far-red light (about 730 nm). Reversal of red-light effects by far-red is a test for the phytochrome system discussed in Chapter 19. This test was successful, suggesting that phytochrome does indeed play a role in determining sensitivity of coleoptiles to the blue light that causes bending—and this is the topic of the next section.

An interesting complication in how plant tissues absorb light has recently been reported (Mandoli and Briggs, 1982, 1983, & 1984). Dark-grown (*etiolated;* see Chapter 19) stems and other tissues conduct light much as the fiber-optic light pipes that are now used in some modern communication systems. A coleoptile tip, for example, that has just penetrated the soil surface, will conduct ("pipe") light down to the primary leaf, mesocotyl, and roots. The relative distribution within the tissue may change so that there is more light on the "shaded" side than on the "light" side. As the light is transmitted through the tissue, some wavelengths are absorbed more than others so that its spectral composition changes. Further spectral changes occur as the etiolated tissue becomes green in response to the light. All of this could be important in phototropism and other light responses discussed in Chapter 19.

Action spectra and the photoreceptor pigment To identify the pigment responsible for any photochemical process, an essential step is to compare the action spectrum for the process (Section 9.3) with the absorption spectra of pigments suspected to be involved. This was done as early as 1909 by A. H. Blaauw* in Holland, who found that blue light was most effective in causing phototropic curvature. Since that time, increasingly detailed action spectra have been measured, and plant physiologists have suggested that one or both of two common yellow pigment systems might be involved in absorbing the light that causes phototropism: carotenoids and riboflavin. (Some yellow pigments are yellow because they absorb blue and/or ultraviolet wavelengths, and the remaining wavelengths combine to produce the sensation of yellow in the human eye.) Figure 18-9 shows the absorption spectra for β-carotene and riboflavin and a representative action spectrum for *Avena* coleoptiles. Although the action spectra differ in detail for the various plant systems known to be phototropically sensitive, blue light is nearly always most

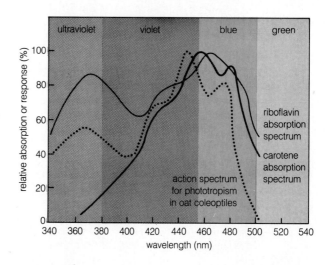

Figure 18-9 The action spectrum for phototropism compared with the absorption spectrum of riboflavin and carotene. (Data assembled from various sources. See especially Thimann and Curry, 1960; Dennison, 1979.)

effective, and the various action spectra resemble that of Fig. 18-9 (Briggs and Iino, 1983). The blue-absorbing pigment (pigments?) has been called **cryptochrome** (Schopfer, 1984). We shall discuss it further in Chapter 19.

As you can see from examining Fig. 18-9, it is not an easy matter to distinguish between the two possible pigment systems by comparing available absorption and action spectra. The strong peak in the ultraviolet (360 to 380 nm) favors riboflavin as the absorbing pigment, but the two peaks in the blue-violet part of the spectrum (about 450 and 480 nm) favor carotene. Nevertheless, evidence accumulating over the past decade or so has favored a flavin pigment as the primary photoreceptor in phototropism. For example, in certain phototropically active fungi where action spectra for various responses are almost identical to that for phototropism in higher plants, a flavin attached to a protein (**flavoprotein**) appears to be the only pigment involved (Munoz and Butler, 1975). Upon light absorption, the flavoprotein becomes oxidized by reducing a *b*-type cytochrome in the plasmalemma (Brain et al., 1977). Certain mutants of higher plants are also exceptionally low in carotene but nonetheless respond phototropically. Furthermore, certain herbicides are known that block the formation of carotenoid pigments but do not eliminate the phototropic response.

In spite of the many decades since the problem of the photoreceptor pigment was first approached, much remains to be done and there are many questions that need to be settled. Very little work has been done with dicots, for example. Dose-response curves have been obtained (Everett, 1974), but virtually no action spectra exist. One interesting complication in

---

*Blaauw performed a number of interesting experiments at the beginning of this century. In one, for example, he reported that *Avena* coleoptiles would curve in response to only about one thousandth of the light from a full moon—but it took 43 hours of exposure to obtain this response! With the highest light levels available to him, he got the same response with an exposure time of only 0.08 seconds.

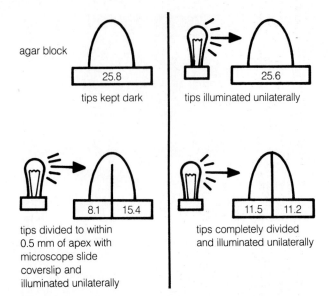

agar block

25.8

tips kept dark

tips illuminated unilaterally

25.6

8.1 | 15.4

tips divided to within
0.5 mm of apex with
microscope slide
coverslip and
illuminated unilaterally

11.5 | 11.2

tips completely divided
and illuminated unilaterally

Figure 18-10 Experiments showing that unilateral illumination of corn-coleoptile tips leads to a transport of auxin from the illuminated to the shaded side of the tips and does not cause a destruction of auxin. Numbers on the agar blocks represent the degrees of curvature caused by application of the blocks to decapitated oat coleoptile stumps. In the partially split tips, part of the auxin was transported laterally above the dividing barrier, but in the completely split tips, this was not possible. (From Briggs, 1963.)

interpretation of action-spectra data is that an inactive screening pigment can cause peaks in the action spectrum (Vierstra and Poff, 1981). That is, although the light absorbed by the pigment may not itself be transduced to a phototropic response, the presence of the pigment accentuates the steepness of the light gradient across the coleoptile. Of course, it is the plant response to the light *gradient* that leads to phototropic bending.

Transduction in phototropism   In 1926, N. Cholodny theorized and Frits Went demonstrated that auxin apparently migrates from the irradiated to the shaded side of a unilaterally irradiated coleoptile (Section 16.1). Figure 18-10 shows results of a similar experiment performed by Briggs and his colleagues in the early 1960s. In the figure, the amount of auxin transported into agar blocks at the base of a coleoptile tip is indicated by the degrees of curvature elicited when the agar blocks were placed upon suitably prepared coleoptiles as in the standard curvature test for auxin (see Fig. 16-1). Tips exposed to light transported as much auxin as those kept in the dark, suggesting that auxin destruction by light did not occur. When the agar and part of the coleoptile were divided with a thin piece of mica and the tip exposed to light from one side, the amount of curvature caused by the blocks was twice as great for the block from the shaded side as for that from the irradiated side. The

division of the entire tip with mica prevented auxin transport in the tip, so auxin amounts were the same on both sides.

For half a century it was assumed that the **Cholodny-Went** theory accounted for the basic transduction mechanism in phototropism: Light from one side caused a transport of auxin toward the shaded side—although no one knew how the gradient in light level across a stem or a coleoptile could lead to auxin transport. This venerable theory is now being questioned. Some investigators point out that the change in auxin distribution during phototropism could be an effect of phototropic bending rather than a cause and that changes in sensitivity of the coleoptile or stem tissue to auxin and perhaps other substances always present might be the basic transduction mechanism—although no real evidence yet exists to support this suggestion. Can the Cholodny-Went theory account for the rapid cessation of growth observed on the bright side of a coleoptile or hypocotyl irradiated at high levels (see Fig. 18-7; Franssen et al., 1982)? Franssen and Bruinsma (1981) found that sunflower hypocotyl tissue responded directly to unilateral light; the tip or young leaves had no particular sensitivity. They further found no unequal distribution of IAA across the hypocotyl, although an inhibitor, xanthoxin, accumulated on the irradiated side.

At the same time that some investigators are questioning the classical Cholodny-Went concept, others are defending the idea vigorously and citing many supporting evidences (such as those of Fig. 18-10). Perhaps the problem will be resolved before we prepare the next edition of this text!

Leaf Mosaics   Leaves as well as coleoptiles and stems respond phototropically to the light. If one half of a cocklebur leaf blade in the light is covered with aluminum foil (simulating natural shading), for example, two responses can be seen. One is an elongation of the corresponding side of the petiole so that it bends and displaces the leaf toward the irradiated side of the blade. The other is an upward bending (hyponasty) of the shaded side of the leaf. Such responses in a number of leaves would presumably move them from the shade into the light whenever this is physically possible. The result is that leaves hardly overlap; instead, they form **leaf mosaics**. Such mosaics are shown by many trees (easily observed by standing under them and looking upward) and by ivy climbing on building walls or especially when growing in a house in which light comes predominantly from a single window throughout the day.

The assumption is that when one part of a leaf blade is shaded by another, the shaded part transports more auxin to its side of the petiole than the brightly illuminated side transports to its corresponding pet-

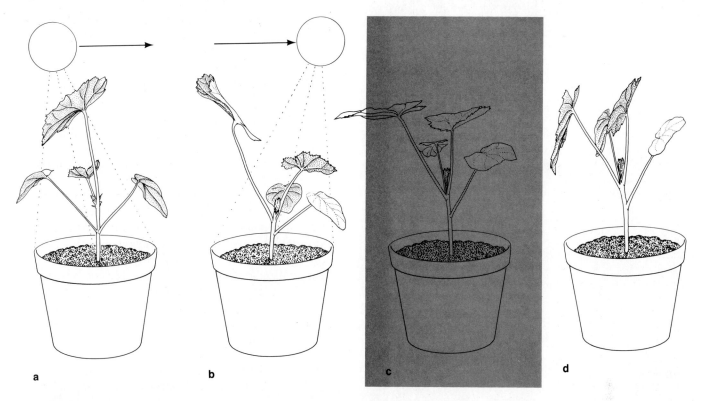

a b c d

Figure 18-11 Solar tracking in a plant with cup-shaped leaves such as some members of the Malvaceae family (e.g., *Malva* or *Lavatera*). The leaves receive directional signals from the sun and swivel around to face it as pulvinus cells at the base of the blade gain or lose water. (**a** and **b**) The leaves track the sun during the day, much as a radio telescope tracks a satellite. (**c**) An hour or two after sunset, the blades are nearly horizontal in the "relaxed" position they maintain during most of the night. (**d**) About an hour before sunrise, the blades face the point on the eastern horizon where the sun came up the day before. (Redrawn from drawings and photographs supplied by Dov Koller.)

iole part. This assumption has not been rigorously tested in the case of leaf mosaics, however, and there is much opportunity for increased understanding of how plants maximize the amount of photosynthetically active radiation to which they are exposed—or sometimes minimize it when irradiances are high enough to be damaging.

Solar Tracking Many plants are capable of **solar tracking** in which the flat blade of the leaf remains at right angles to the sun throughout the day, maximizing the light harvested by the leaf. The phenomenon was studied by the Darwins (1881) and then virtually neglected until H. C. Yin performed some important basic studies, published in 1938. James Ehleringer and I. Forseth (1980) have documented the importance of solar tracking in the desert and in other natural ecosystems, and C. M. Wainwright (1977) has studied a desert lupine (*Lupinus arizonicus*). Most of our understanding of the physiological mechanisms involved in solar tracking, however, depend upon studies of Amnon Schwartz and Dov Koller (1978, 1980) at the Hebrew University of Jerusalem.

To begin with, solar tracking is a true tropism, because the orientation of the leaves is determined completely by the direction of the sun's rays. Nevertheless, the orientation of the leaves is controlled by motor cells in a pulvinus where the blade joins the petiole. Movement of water in and out of these motor cells is completely reversible—and almost certainly controlled by osmotic solutes including K^+. That is, solar tracking is not a growth phenomenon as are the tropisms we have been discussing but depends upon pulvinar action similar to the nyctinastic leaf movements that occur independently of the direction of light.

The pattern of solar tracking is illustrated in Fig. 18-11. The often cup-shaped leaves of such species as cotton, soybean, beans, alfalfa, and various wild members of the Malvaceae including *Malva neglecta* and *Lavatera cretica* follow the sun throughout its daily course across the sky much as a moving satellite is tracked by a radio telescope or radar antenna. At sunset, the leaves are almost vertical, facing the point on the western horizon where the sun is setting. An hour or two later, they have assumed a "resting" horizontal position. An hour or two before sunrise,

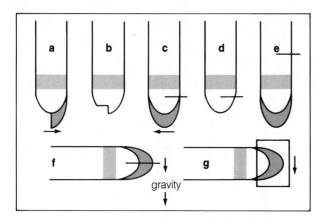

Figure 18-12 Diagramatic representation of various treatments applied to the roots of maize (*Zea mays*) and peas (*Pisum sativum*). The experiments suggest that the lower side of the root cap produces an inhibitor of root growth that is transmitted to the growing cells (shaded). Arrows show the direction of curvature of the root after treatment. (**a**) The vertical tip bends toward the remaining portion of a root cap, but (**b**) removal of cap and a portion of the meristem has no effect. (**c**) Insertion of a horizontal barrier between the cap and the growing zone causes bending away from the side where the barrier was inserted, but (**d**) such a barrier has no effect in the absence of the root cap or (**e**) when the barrier is above the growing zone. Insertion of a horizontal barrier in a horizontal root (**f**) nearly prevents bending (suggested by the short arrow), but a vertical barrier (**g**) has little effect on bending. (From Shaw and Wilkins, 1973. Used by permission.)

the leaves change their orientation so that they are facing the point on the eastern horizon where the sun came up the day before. That is, these leaf movements are controlled not only by the direction of the sun's rays but also by an internal timing system that increases efficiency of light harvesting by prepositioning the leaves to best absorb the rays from the rising sun. If plants are left in total darkness for several days, they will continue to reorientate their leaves each 24 hours to the point where the sun (or artificial laboratory light) first appeared the last time it irradiated the plants. On heavily overcast days, the leaves assume the horizontal "resting" position, but they can change their orientation about four times as rapidly (60° per hour) than the apparent movement of the sun across the sky (15° per hour), so they soon catch up with the sun when it appears from behind the clouds.

Schwartz and Koller (1978, 1980) showed that this response, like other examples of phototropism, is most sensitive to blue light, but detailed action spectra have not yet been obtained. It is the major veins in the flattened, cup-shaped leaves that detect the direction of the sun's rays. They send a message to the motor cells in the pulvinus, leading to reorientation of the leaves. There is some evidence that auxin takes part in transmitting the message. The situation is complex, however, because the apparent message can be either positive or negative. If the rays of light strike a major vein at right angles, the message is neutral. If the rays come from toward the tip of the leaf, the message exported is *negative*, so cells in the pulvinus connecting with the vein lose water, causing the leaf to turn toward the sun. Conversely, if the rays come from toward the base, a *positive* signal is sent from the major vein to the motor cells at its base, causing them to absorb water and again reorient the leaf toward the sun.

An important point remains. Ehleringer and Forseth (1980) and Shackel and Hall (1979) observed "negative solar tracking" in which leaves of some desert plants were maintained parallel to the sun's rays when conditions were especially dry. By this mechanism, they would presumably gain less heat and transpire less.

18.4 Gravitropism

Growth movements toward (positive) or away (negative) from the earth's gravitational pull are examples of gravitropism. Roots are usually positively gravitropic, primary roots generally more so than secondary roots. Tertiary roots and roots of higher orders are hardly gravitropic and often grow almost horizontally. Thus the root system explores the soil more thoroughly than it would if all roots grew straight down, side by side. Stems and flower stalks are generally negatively gravitropic, although the extent to which this occurs is quite variable. The main stem or tree trunk usually grows 180° away from the gravitational stimulus, but branches, petioles, rhizomes, and stolons are usually more horizontal. As with phototropism causing leaf mosaics, these differences allow a plant to fill space and thus absorb CO_2 and light most effectively. The growth of several organs at intermediate angles is given the special term **plagiotropism,** while the vertically oriented stem and root axes exhibit **orthogravitropism.**

Roots Probably the most study of gravitropism has been devoted to roots. Although our summary will be brief, other generalizations about plant gravitropism that also apply to roots are discussed in the following section on stems and coleoptiles.

To begin with, there is excellent evidence that the site of gravity perception is the root cap. The discovery of its sensitivity was foreshadowed by the work of Ciesielski in Poland in 1871 (cited by the Darwins, 1896), who showed that roots of peas (*Pisum*), lentils (*Lens*), and broadbeans (*Vicia*) would not respond to gravity when their tips were cut off until a new root cap and meristem had been formed. More recent experiments have shown that when the root cap of maize or other species is removed by careful microsurgical techniques that do not inhibit growth of the root, there is no gravitropic response until a new cap has been regenerated (Juniper et al., 1966; Volkman and Sievers; 1979).

Furthermore, it is clear that the root cap produces an inhibitor on the lower side of the roots, stopping or slowing growth there so the root bends downward. Figure 18-12 illustrates a series of experiments that demonstrate this. When part of the root cap is removed from a vertical root, the root bends in the direction of the remaining portion, suggesting that an inhibitor is produced by the part that is left. If the entire cap is removed, no bending occurs, even when part of the meristem is removed. Barriers inserted transversely into the root cause bending toward the barrier, providing that the root cap is present and that the barrier is between the cap and the growing zone of the root—again suggesting that an inhibitor comes from the cap. A longitudinal but horizontal barrier in the root greatly reduces bending compared to a root with a vertical barrier, suggesting some downward movement of the inhibitor or its precursor.

Feldman (1981) extracted at least two unidentified growth inhibitors from root caps of maize. In the cultivar he used, light induced gravisensitivity, and light also induced formation of one of the inhibitors.

Near the end of the 1970s, it seemed increasingly clear that the inhibitor from root caps was ABA (Wilkins, 1979), but several conflicting results have become apparent, and by 1984 the identity of the inhibitor was very much in doubt (Wilkins, 1984). The Cholodny-Went theory suggested that auxin was the inhibitor, and some workers cite recent evidence to support this view. There are many conflicting results, however (see Jackson and Barlow, 1981). For example, Wilkins and Wain (1974) failed to find auxin in root caps of maize, while ElAntably (1975) and Pilet (1977) report more IAA than ABA in maize roots. Suzuki et al. (1979) found increased IAA, ABA, and an unidentified inhibitor in the bottom halves of maize roots; they suggest that the unidentified substance might be more important in root gravitropism than IAA or ABA. Some workers report that both IAA and ABA move toward instead of away from the tip in roots.

Both acid growth and calcium ions have been implicated in root gravitropism. Roots embedded in agar containing a pH indicator and allowed to respond to gravity become more acidic on the upper side where most growth is taking place (Mulky and Evans, 1981). We will discuss the Ca^{2+} effects in the next section.

At the turn of the century, Gottlieb Haberlandt (1902) in Austria and B. Nemec (1901) in Czechoslovakia suggested that it was the **amyloplasts,** each containing two or more starch grains, that settle in the large cells of the root cap in response to gravity and thus provide the basic perception mechanism. This hypothesis has been alternately supported and rejected at various times ever since, but recent evidence has supported amyloplasts as the **statoliths** (organelles of gravity perception) of many tissues (Audus, 1979; Juniper, 1976; Volkman and Sievers, 1979). (Cells that contain statoliths are called **statocytes.**) Consider three supporting evidences: First, maize mutants with smaller amyloplasts than the wild type are less sensitive to gravity, and indeed, the quantitative aspects of gravity response proved to be closely correlated with amyloplast size. Second, if roots or coleoptiles are treated with gibberellin and kinetin at elevated temperatures, all starch in the amyloplasts disappears and along with it the response to gravity (Iversen, 1974). Third, gravitropic sensitivity reappears at the same time that starch grains appear in the amyloplasts after destarching (or in the newly regenerated root cap after removal of the cap as noted above). Hillman and Wilkins (1982) add that sensitivity reappears as soon as the new starch grains are able to *settle;* they usually remain suspended in the cytoplasm for a short time after they appear.

It has long been known that the so-called presentation times required to elicit a gravitropic response are closely correlated with the rates of amyloplast settling. In some species (e.g., *Lepidium*), turning the root to its side for as short a time as 20 seconds leads to an observable gravitropic bending that appears some time later, and amyloplasts are known to settle a perceptible distance in this brief time.

How can settling of statoliths lead to the gravitropic response? Although there have been many suggestions (e.g., that contact of amyloplasts with ER close to the plasmalemma some way influences growth-regulator transport into or out of the statocyte), this question remains unanswered. Could other cell organelles settle and cause the gravitropic response? To settle, a statolith must have a density significantly greater than the medium in which it is suspended. Starch has a density of about 1.3, which is greater than that of the cytosol, and as noted, amyloplasts can easily be seen to settle. Densities of other organelles in the cell are often very close to that of the cytosol, so they would not be expected to settle. This is true of the nucleus, for example. There is evidence that some organelles besides amyloplasts do settle, but rates do not correlate with presentation times (Shen-Miller and Hinchman, 1974). Even if an organelle is dense enough, it will not settle if it is too small; it will remain in suspension because of its Brownian movement. This is true of ribosomes and other organelles. Thus amyloplasts remain the most likely candidates for the role of statoliths.

We must mention one other complication in the gravitropic response of roots. In certain cultivars of maize (and also in radish and other species less studied), the root is not gravitropically sensitive until it has been exposed to light. Thus roots grow in all directions in the soil but immediately grow downward if they come close enough to the surface (within a millimeter or two) to be exposed to light. Thus *sensitivity* of the gravitropic mechanism is some way induced by light. Red wavelengths are most effective, and the phytochrome system (Chapter 19) has been implicated.

Stems and Coleoptiles

The response Measurements of growth on the tops and bottoms of coleoptiles, hypocotyls, and stems laid on their sides give the same striking result as observed in phototropic bending: The top of the horizontally placed organ ceases growing almost immediately, while the bottom may continue to grow at about the same rate or may grow somewhat faster or slower (e.g., MacDonald et al., 1983). For several decades, it has been assumed that gravitropic stem bending occurs because growth is more rapid on the bottom than the top of horizontally placed stems, probably because of a movement of auxin from top to bottom (again the Cholodny-Went theory, supported by experiments of Dolk in 1930, experiments quite comparable to those of Fig. 18-10). Careful studies of the mechanics of gravitropic stem bending suggest some important complications.

Consider a simple phenomenon studied in recent years (Wheeler and Salisbury, 1981; Salisbury et al., 1982a) but reported at least as early as 1888 by Ann Bateson and Francis Darwin. If a stem is turned on its side and restrained so it cannot bend upward in the normal gravitropic response, when the restraint is released, it will bend rapidly upward (often to more than 90° and within 1 to 10 seconds). A convenient way to perform this experiment is to place the stem between two stiff wires that are about 2 cm apart and attached to the pot and then wrap the wires and stem with thread. If the degree of bending upon release is plotted as a function of time, it is clear that bending develops slower than it does with unrestrained plants, but after a long enough time (usually days) bending upon release reaches a maximum of perhaps 135° (compared to 90° for a free-bending stem). Actually, bending is quite reversible at any time. The stem can be released from its restraint, the angle of bending measured, and then the stem can be forcibly straightened again and restrained by further wrapping with thread. Upon release some hours later, it will bend to just about the same extent as would a stem that had been restrained for the same total amount of time but that had not been previously released.

Cells on the bottom of such a restrained stem are elongating, although not as much as if the stem were not restrained. Because the stem is held straight and cell divisions are not occurring, cells on the top of the stem are apparently stretched so they also elongate. This seems clear, because bottom cells increase in diameter, while top cells decrease in diameter, and upon release from restraint, cells on the top shorten and thicken, while those on the bottom elongate somewhat and become thinner. This causes the bend. Upon release, cells on the top that were stretched by growth of the cells on the bottom shrink in length while increasing in diameter—apparently conserving their volume—while cells on the bottom do the opposite, elongating and decreasing in diameter to also conserve their volume (Salisbury et al., 1982b; Sliwinski and Salisbury, 1985).

These observations strongly reinforce the point noted earlier that growth not only stops on the top of a stem laid on its side, but it cannot even occur when the uppermost cells are being stretched by elongation of tissues on the bottom of the stem. Note that this situation seems opposite to that described in Section 15.2, where we considered the physics of cell growth. There we concluded that cell growth occurs because wall loosening decreases pressure within the cells, which in turn makes the water potential more negative so that water enters the cells osmotically, driving cell expansion. In a restrained horizontal stem, pressures in the bottom tissues are increasing considerably (preliminary measurements suggest up to 1.8 MPa) at the same time that growth is continuing (Salisbury et al., 1982b; Mueller et al., 1985). True, this refers to pressure in the tissues as a whole including the apoplast, and actual pressures in the protoplasts of individual cells could be different—but probably not less. At the same time, pressures are decreasing in the tissues at the top of the stem as they are being stretched by growth of the bottom tissues, yet no growth occurs in these cells; they return to their original dimensions when the stem is released from restraint and bending is allowed to occur. Clearly, any hypothesis proposed to explain gravitropic stem bending must account for this rapid cessation of growth in the top cells—just as was the case with phototropic bending.

There are further complications that need to be evaluated. If a normal stem is split longitudinally (as children split dandelion flower stalks), the two halves bend outward on the order of 30 to 50° (depending upon species, age of the tissue, etc.). Bending occurs because the outer tissues are under tension with respect to the inner tissues. Attempts so far to locate the tensions and pressures indicate that the tensions occur in the cortical and even vascular tissues, while the pressures are confined mostly to the pith. A series of horizontal cuts are made to measured depths at 2.5-mm intervals into the surface on one side of a vertical stem. These cuts release the tension on that side, and the stem bends away from the cuts, but only if the cuts go well below the epidermis and into the cortical tissues (Salisbury et al., 1984). Thus, you can imagine a stem growing as pith cells take up water and expand against the resistance produced by the rest of the cells in the stem, which are apparently growing with less force. Some change in the balance of tensions and pressures in growing stems or other organs might account for or contribute to the gravitropic bending we have been considering. Such possibilities are being investigated.

Perception   The site of gravity perception in stems is the same as the site of response. That is, much of the stem above the rapidly growing region can be removed, and the rapidly growing region will continue to bend upward in response to gravity if the stem is laid on its side. This is true for coleoptiles, hypocotyls, and true stems. It was well documented by the Darwins (1881), although some textbook writers in recent years have incorrectly emphasized the tip as the site of gravity perception in stems, as the root cap is the site of perception in roots.

As with roots, amyloplasts are thought to be the statoliths of stems. In many species, the amyloplasts are confined to one or two layers of cells that surround the vascular bundles (called the **starch sheath**). This sheath has about the same anatomical relationship to the vascular tissues of stems as the endodermis has in roots, although there are no Casparian strips around starch-sheath cells. The starch sheath generally forms the inner layer of the cortex, which consists of several layers of parenchyma cells and often a layer of collenchyma cells just below the epidermis (Fig. 18-13). Again, the evidence that the amyloplasts in stems play the role of statoliths is strongly suggestive but not conclusive. Amyloplasts can be seen to settle in the starch-sheath cells of stems, and all gravitropically responsive stems so far studied do have amyloplasts. (The distribution is somewhat different in coleoptiles, occurring internal to the vascular tissues instead of outside them.)

Transduction   As in the cases already discussed, the Cholodny-Went hypothesis as it applies to stems and coleoptiles is being seriously questioned by several workers (especially Digby and Firn, 1976; Firn and Digby, 1980). Considering the rapid cessation of growth on the top of a stem laid on its side compared to the often normal growth rates on the bottom, it is difficult to imagine that changes in auxin concentration could occur rapidly enough or reach magnitudes capable of accounting for these differences. Growth rates on the top and the bottom of a horizontal stem can differ by a factor of ten or more. (Parts of the upper surface often do not grow at all; they even

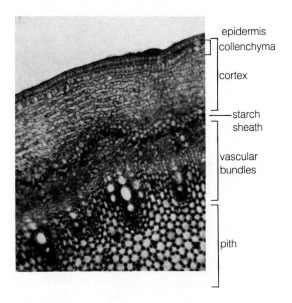

Figure 18-13   Free-hand cross section of a castorbean (*Ricinus*) stem, which has a well-defined starch sheath just outside the vascular tissues. (From Salisbury et al., 1982.)

shrink significantly.) This would mean that the upper cells would have to be almost completely depleted of auxin within a short time after the stem was turned to its side—or that some active inhibitor might move into these cells or cell walls causing the growth cessation, as is thought to be the case in roots. Many experimenters over the past few decades have found only minimal gradients in auxin concentration across gravitropically stimulated stems. Sometimes, with the very best of modern techniques (see Strotman and Bickel-Sandkotter, 1984), no gradients could be detected at all (Mertens and Weiler, 1983). Again, we might validly ask: Are the gradients that have been observed a result of gravitropic bending rather than its cause?

At the same time that the Cholodny-Went hypothesis is being questioned by some, other workers are coming to its defense, reviewing many of the experiments that have supported the hypothesis over the years and providing new experimental data that also seem to support the hypothesis. This topic is currently in such a state of flux that it does not seem warranted even to present tentative conclusions, but consider a few facts and areas of current interest:

Although there are good reasons to wonder whether auxin gradients actually account for gravitropic bending, it still seems quite clear that auxin plays a critical role. It appears to be both necessary and sufficient for gravitropic bending, even if the bending is not produced by auxin gradients. That is, bending will not occur at all when auxin is not present. This was shown by Brauner and Hager (1958),

Studying the Gravitropic Responses of Cereal Grasses

Peter B. Kaufman

Peter B. Kaufman is a Professor of Botany at the University of Michigan in Ann Arbor. He is a dynamic lecturer who has participated in writing general botany textbooks for students in the more applied botanical sciences. As he travels to scientific meetings, he always carries a camera to take pictures for use in his teaching and writing. Here he tells of his current interests.

We embarked on studies of gravitropic responses in grass shoots about 1978 in response to the need for basic research on how cereal grass shoots recover from lodging (falling over due to the action of wind and/or rain)—or in more technical terms, the negative gravitropic curvature (upward bending) response in the shoots. Such studies are of importance for agriculture, because lodging in cereal grasses is responsible for serious losses in grain yield. Any way by which we can understand how the recovery process takes place and can be promoted or how lodging can be prevented is of immense benefit to agriculture. There is another reason why we are interested in this interesting response in cereal grains. We intend to grow such grains as rice and wheat in NASA space vehicles such as the Space Shuttle, Space Lab, and Space Station, where gravity influences are almost absent. We want to grow such plants in space for food (grain production) for occupants of the space vehicles. Further, we want to use grain crops to enrich the cabin atmosphere with oxygen and moisture and to

recycle organic wastes. Finally, we want to grow cereals in space to understand how they grow and how we can control their growth in a near-weightless environment. One of the major questions is: How do cereal grass shoots grow when their gravity-perception mechanisms are not stimulated? And, if they tend to grow in all directions, how can we direct their growth to obtain normal shoot growth and acceptable yields?

Here on earth, we have been vigorously pursuing three basic questions concerned with the mechanisms by which cereal grass shoots grow upwards when they are lodged (gravistimulated). They are (1) how is gravity perceived in the shoots? (2) how are hormones involved in what is called the transduction process? and (3) what is the physiological and metabolic basis of the differential growth response that takes place in the grass pulvini?

It was most exciting for us, with the help of P. Dayanandan, Il Song, Paul Thompson, and C. I. Franklin, when we found that starch grains in amyloplasts are the graviperceptive organelles in the cells near vascular bundles in the leaf-sheath pulvini. They cascade to the bottoms of the cells (statocytes) within 2 minutes, starting within 15 seconds. If the pulvini are treated with alpha amylase, the starch grains disappear, and the pulvini will not curve upward when gravistimulated. However, if after the amylase treatment we feed them with sucrose (a primary product of photosynthesis), the pulvini reform starch and are once again competent to respond to gravity.

Once these starch statoliths fall, how does the gravity stimulus get transduced? We think that the starch grains serve as agents of information transfer. They are associated with esterase activity, easily revealed by treating them with fluorocein diacetate and then examining the cells with

---

who removed the tips of sunflower hypocotyls (presumably a source of auxin) and then gravistimulated the hypocotyls by turning them to the horizontal position. Bending would not occur until auxin was supplied to the tissue by treating the cut stump with IAA dissolved in a lanolin paste. In some experiments the decapitated hypocotyls were gravistimulated and then placed in a refrigerator at temperatures just above freezing where they could be left in an upright position for many hours. When they were removed, allowed to warm to room temperature, and treated with IAA in lanolin, bending of the *vertical* hypocotyls occurred. Brauner and Hager referred to the phenomenon they discovered as the *gravitropic memory* or *Mneme*. Gravitropic bending of decapitated mature stems of sunflower, cocklebur, and tomato is also accentuated by treatment with

auxin in lanolin, even though it seems likely that the rather high concentrations of auxins used and applied in an indiscriminate way to the cut stump of the stem would swamp out the gradients in auxin concentration required by the Cholodny-Went theory (Salisbury et al., 1982a.)

Mulkey and Evans (1982) eliminated gravitropic bending of maize roots by treatment with several inhibitors of auxin transport and with orthovanadate, which inhibits efflux of H^+ in response to auxins. The inhibitors also stopped the previously observed efflux of H^+ on the top of horizontal roots. Wright and Rayle (1983) also used auxin transport inhibitors and other approaches to study the role of auxin in gravitropism of sunflower hypocotyls. They found no asymmetry in acid-growth capacity (sensitivity), but their evidence broadly supports the Cholodny-Went

UV-fluorescence microscopy. Well, this was exciting! Why? Esterase is one of the candidate enzymes for eliciting the release of the auxin IAA from its storage form, inositol ester of IAA. We already know that IAA is not transported downward in gravistimulated pulvini; so here is a mechanism by which it could be released from bound form to an increasing extent from top to bottom of a horizontally placed pulvinus, so that the auxin could move radially from the statocytes and bring about differential cell elongation. The same could hold true for another group of hormones involved in the control of gravitropic curvature in grass leaf-sheath pulvini, namely, the gibberellins (GAs). The GAs also are not transported downward (basipetally) in gravi-stimulated pulvini of cereal grasses; so they must be released differentially from bound form or be synthesized to increasing extents from top to bottom of the pulvinus. We are currently investigating both possibilities for auxin and native GAs in cereal grasses, especially the kinetics of change in amounts of free IAA and its conjugates and of free GAs and their glucosyl ester conjugates during upward bending. We are doing this in collaboration with Dick Pharis at the University of Calgary and have carried out our initial studies on free IAA and its conjugates with Bob Bandurski and Jerry Cohen at Michigan State University and the USDA at Beltsville, respectively. Both are actively helping us with the auxin extraction and purification procedures for identifying the free IAA by GC/MS (gas chromatography/mass spectrometry).

What about the last part of the gravitropic response in cereal grass shoots, namely, the differential growth response? We know that we are dealing with growth that occurs as a result of differential cell elongation. No cell division is involved in graviresponding cereal grass pulvini.

What is the mechanism by which differential cell elongation occurs? We know from recent studies of P. Dayanandan that cellulose synthesis is required; "Daya" also found that both RNA and protein synthesis are necessary. Carrying this a step further, we began to look at the kinds of proteins synthesized in graviresponding pulvini compared with those from upright pulvini. Il Song developed superb methods in our lab for extracting salt- and alkali-soluble proteins and separating them by SDS/PAGE (sodium dodecyl sulfate/polyacrylamide gel electrophoresis) and iso-electric focusing. What we find so far is that early (within 2 hours) after barley shoots are first gravistimulated, at least five new proteins are made in the lower halves of the gravi-responding pulvini where cell elongation is greatest. We think that one of these is cellulase and another is invertase. Top candidates for these newly synthesized proteins are wall-loosening enzymes such as cellulase and pectinase and, of course, enzymes involved in the synthesis of cellulose, hemicellulosic polysaccharides (arabinoxylan in particular), and pectic polysaccharides (polygalacturonans). With the help of Roger O'Neill, we are currently identifying these proteins by immunoelectrophoresis and are examining the activities of key wall-loosening enzymes in the tops and bottoms of gravistimulated pulvini compared with those in upright pulvini ("left" and "right" halves).

Eventually, and we hope soon, we intend to fly our "space oats, barley, wheat, and rice" in the Space Shuttle, Space Lab, and Space Station to learn if possible how we can grow them in space for food, for production of oxygen and water, and to recycle organic wastes. Maybe the next time we write this sketch we can tell you about our space biology experiments with cereal grains grown on Space Lab and the NASA Space Station!

hypothesis. Bandurski et al. (1984) measured an IAA asymmetry in horizontal maize mesocotyls within 15 min after seedlings were placed on their sides. The asymmetry was small, however, with 56 to 57 percent of the IAA in the bottoms of the mesocotyls and 43 to 44 percent in the tops.

There have been studies with growth regulators besides auxin. Gibberellins (GAs), for example, occur in higher concentrations on the bottom of gravi-tropically stimulated stems (reviewed by Wilkins, 1979). But with GAs the problems are even more serious than with auxin, since they move more slowly and since significant gradients are not really observed until long after bending has occurred. In this case, it certainly appears that the GA gradients are the result of gravitropic bending rather than its cause.

In some experiments (Wheeler and Salisbury, 1981, plus references cited therein), ethylene appears to play a positive role in gravitropic stem bending. Four inhibitors of ethylene action or synthesis (Ag^+, CO_2, Co^{2+}, and aminoethoxyvinylglycine, abbreviated (AVG) all reduced the rate of gravitropic bending in cockleburs, tomatoes, and castor beans. Some of the rate of bending could be restored by surrounding the stems with low concentrations of ethylene. At the same time, there is evidence (Clifford et al., 1983) that ethylene is neither necessary nor sufficient for gravitropic bending. It seems to play no role in the gravi-response of dandelion peduncles, for example. What is the ethylene doing when it does promote gravitropic bending? It would be logical to think that the ethylene might in some way contribute to the inhibition of growth on the top of a stem placed hori-

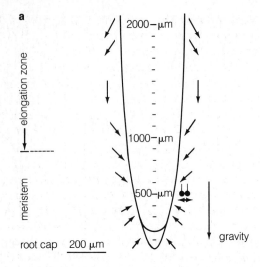

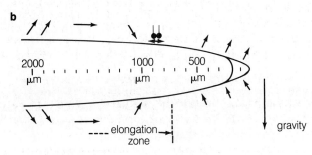

Figure 18-14 (a) The qualitative current pattern measured with a vibrating electrode (symbolized by two heavy dots and vertical lines) around a vertical root. Arrows indicate the direction of current flow, that is, the direction of movement of positive ions. The large arrow indicates the direction of the gravity vector. (b) A hypothetical pattern of current flow of a 24-h-old *Lepidium* root that has been tilted to the horizontal for 20 min. The current pattern is based on a possible interpretation of the measurements of acropetal and basipetal currents. Note the reversal in flow direction on top near the tip. (From Behrens et al., 1982.)

zontally. This would be in keeping with its known inhibitory effect on elongation of vertical stems (Section 17.2). Yet when ethylene is measured in the tissues, it is found to increase at about the same rate as gravitropic bending but in the *bottom* tissues instead of the top tissues.

Clifford et al. (1982) attached a thread to the top of a vertical dandelion peduncle and used the thread to pull the peduncle to one side with a 2-gram weight and pulley system. After the stress was removed, the peduncle bent away from the force, much as it bends when turned to its side and allowed to respond to gravity. This might suggest some participation of ethylene, since ethylene is produced in response to mechanical stress (Section 17.2), but this does not seem very likely in view of the other results of Clifford et al. (1983) mentioned in the previous para-

graph. The stress response might suggest that settling amyloplasts cause their gravieffects by applying mechanical stress within the statocytes.

Evidence is accumulating that calcium ions (Ca^{2+}) play an important role in gravitropism. For one thing, Ca^{2+} concentrations have been observed to be higher on the tops of horizontal stems, and Ca^{2+} is known to inhibit cell elongation (perhaps by overcoming auxin effects). Furthermore, when strong chelators of Ca^{2+} are applied to roots and other tissues, gravitropic bending can be completely inhibited (Lee et al., 1983). Of course, the chelators could be doing something else besides tying up Ca^{2+}, so these experiments are not definitive.

For many decades, there was an interest in electrical phenomena associated with plants and particularly in gravitropism (see discussions of nyctinasty, Section 18.1). There has been a recent revival of interest in the possibility that the gravitropic signal is somehow transduced electrically in plants. Figure 18-14 summarizes some responses observed at the University of Bonn in Germany by the use of minute vibrating electrodes. These techniques allow the measurement of current flow around test roots and clearly indicate that such flow changes during gravitropic stimulation. The meaning of these observations remains unclear, but they could lead to interesting results in the future.

Clinostat experiments Julius von Sachs in 1872 suggested that weightlessness might be simulated for a plant by rotating the plant slowly about a horizontal axis. If the perception time exceeds the rotation time, then the vectors of gravitational stimulation would add algebraically to zero as though the plant being so rotated were truly weightless. Sachs suggested the name **clinostat** for the apparatus that is used to rotate plants as described (usually about their longitudinal stem/root axis, but tumbling end over end should theoretically be the same and does indeed usually produce the same effects). Around the turn of the century, hundreds of clinostat experiments were done in attempts to assess the importance of gravity for plants. Now that space exploration makes possible the use of orbiting laboratories, it is possible to compare the early results with those obtained by allowing plants to grow in the nominally weightless condition of such an orbiting satellite, although little has been done so far.

The most noticeable response of plants on a clinostat is leaf epinasty (downward bending; Section 18.1). In the early 1970s, it was suggested that this and other clinostat responses might be caused not by the simulated weightlessness of clinostating but rather by the mechanical stresses imposed by leaf flopping and other strains (Tibbitts and Herzberg, 1978). Of course, a plant on a clinostat is not really

Figure 18-15 A double exposure showing gravitropism in a wheat stem. The plant was turned on its side for the first exposure (after removing a number of other stems from the pot), and the second exposure was made about 70 h later. Note that only two nodes have responded; the node just inside the rim of the pot is now mature and did not respond at all. After two more days, the stem was completely vertical. (Photograph by Linda Gillespie.)

weightless. As it rotates, its leaves and other organs move in response to gravity. Perhaps these mechanical stresses lead to production of ethylene and leaf epinasty. (Ethylene is known to cause leaf epinasty.) Two kinds of experiments (Salisbury and Wheeler, 1981) suggest that this simple explanation of clinostat-caused epinasty is not correct after all. For one thing, when vertical plants are mechanically stressed in several ways comparable to those produced on a clinostat, leaf epinasty does not appear. Second, when gravity is compensated by the simple procedure of inverting plants every 5 to 30 minutes so that they are upside down half the time, epinasty does appear, just as on a clinostat. The inversions can be done very carefully in an attempt to avoid mechanical stresses, and control plants can be inverted at the same time but immediately returned to the vertical so that they are not upside down half the time; such controls show no epinasty. These experiments suggest that plants grown in an orbiting satellite should display epinasty, and recent experiments in the space shuttle laboratory as well as a few earlier experiments in orbiting laboratories confirm this prediction.

False pulvini of grasses In grass stems, the gravity-sensing organ is located near the nodal region. In the Panicoid grasses, such as corn and sorghum, the base of the internode is swollen and responds to gravity. In addition to this organ, the base of the leaf sheath sometimes has a gravity-sensing region (Gould, 1968). In the Festucoid grasses, such as oats, wheat, and barley, only the leaf-sheath base possesses this specialized organ. The gravity-sensing organ is called a **pulvinus,** although some prefer to call it a **false pulvinus** to distinguish it from the true pulvinus that controls nyctinastic leaf movements in certain dicots by *temporary* changes in cell size (Section 18.1). The grass pulvinus is a highly specialized organ that can both detect and respond to gravity (Dayanandan et al., 1976, 1977). Starch statoliths sediment within 2 to 10 minutes of horizontal placement, and a curvature response is initiated within 15 to 30 minutes of gravity stimulation. Continued differential growth of the pulvinus can bring the shoot to a vertical position after several hours (Fig. 18-15). The pulvinus is made up of parenchyma and collenchyma cells besides small amounts of vascular tissue. Response to gravity consists of differential cell elongation, the lowermost cells elongating most while the uppermost do not grow at all (Fig. 18-16). As with the stems and roots, there are many complications and few positive answers. For example, IAA, GA, and ethylene are all known to accumulate more on the lower half of the pulvinus following gravity stimulation.

Other organs Stamens, flower peduncles, various fruits, leaves, and other organs are also known to be gravitropically sensitive, although relatively little

Figure 18-16 The gravitropic response in a grass *(Muhlenbergia)* pulvinus as shown with the scanning electron microscope. Note the large cells on the bottom (outside of the bend) compared with the much smaller cells on top (inside of the bend). An axillary bud is seen within the leaf sheath on the elongating side. x30 (Micrograph courtesy of Peter B. Kaufman; see Dayanandan et al., 1976.)

study has been devoted to these organs. In experiments in which stems are restrained to the horizontal as described above, leaves can be seen to assume a more or less horizontal position when they are not restrained. Truly, it appears that virtually the entire plant is sensitive to gravity—including those lateral roots, rhizomes, stolons, and branches that grow at some orientation other than the vertical but constant for the given species (plagiotropic growth). Little study has been devoted to such matters.

18.5 Other Tropisms

Several other tropisms have been described over the years, but most have not been studied in any detail. Various workers have reported that plants will grow toward water, certain chemicals (nutrients), electric currents, and so on. These reports should be viewed with skepticism. It is clear, however, that many plant organs, especially tendrils, will respond to touch (thigmotropism). Tendrils bend toward the

point of contact, wrapping around a support. Thigmotropism has been reviewed by Jaffe and Galston (1968).

Circumnutation: Nastic or Tropistic?   Again it was the Darwins (1881) who called our attention to how a stem tip appears to trace a more or less regular ellipse as the stem grows. Darwin called this phenomenon **circumnutation.** Its utility for climbing plants is obvious, because the stem is likely to encounter some kind of support as it circumnutates. If it serves any function in nonclimbing stems, this is not evident.

Darwin suggested that the circumnutation occurred in response to some internal, rhythmical control; that is, that circumnutation was endogenously controlled, a nastic movement. More recently, Johnsson (1971) provided a mathematical model based on the assumption that circumnutation is really a matter of gravitropic overshoot, in which case it would be a tropistic response. As the stem grows to the vertical in response to gravity, it may continue to grow more on what was the bottom side so that it actually passes the vertical—at which time an opposite gravitropic response would begin to bring it back to the vertical—overshooting again and accounting for the oscillation. The mathematical model shows that it is not difficult to convert a back and forth waving to an elliptical or even circular motion.

If the overshoot model is correct, circumnutation should not occur on a clinostat nor in an orbiting space laboratory. In many species, circumnutation does stop on a clinostat (although this could be a secondary effect); in others it does not. In a recent space experiment, sunflower seedlings continued to nutate for several hours of virtual weightlessness (less than about 10^{-3}g; personal communication from Alan Brown and David Chapman). Furthermore, pea stems when laid on their sides continue circumnutations during the upward bending, an observation that is difficult to reconcile with an overshoot theory.

Reaction Wood   As a plagiotropic tree branch elongates, it might be expected to bend downward because of its increased weight and distance from the trunk. This is sometimes observed, but it is resisted by formation of **reaction wood** (Scurfield, 1973; Wilson and Archer, 1977). Reaction wood is the increased xylem produced on either the upper or lower side of a branch by more rapid division of the vascular cambium on that side. In conifer (softwood) limbs, reaction wood forms on the lower side and by expansion *pushes* such limbs more upright, maintaining a more constant angle. Tracheid walls become abnormally thick and contain more lignin and less cellulose than usual. In angiosperm (hardwood) trees, reaction wood forms on top and *pulls* the branch toward the trunk by tension. Angiosperm reaction wood contains more fibers and fewer vessel elements than normal wood. The fibers and fiber tracheids attain secondary walls that are thicker than in normal wood, because they form an additional cellulose-rich layer in the secondary wall. In contrast to conifer reaction wood, reaction wood in hardwoods does not become unusually rich in lignin.

What is reaction wood a reaction to? It could be either a gravitropic response or a response to tensions and pressures (strains) resulting from bending. If a cable is tied to a pine limb, causing the end to bend down (Fig. 18-17), reaction wood forms on the lower side. Either explanation might account for this effect. If the cable causes an upward bend, reaction wood forms on the upper side. In these cases, pressures (compression) on cells of the concave side or tensions on the convex side seem more likely to account for the positions of reaction wood than the direction of the gravitational force. But if a young leader of a pine tree is wrapped into a complete loop, reaction wood forms on the lower side of both horizontal parts of the loop. Since the lower side of the upper part is under pressure and the lower side of the bottom part is under tension, the second hypothesis also proves inadequate. Redistribution of IAA or other hormones might explain these results, but the cause of hormone movement may prove complex.

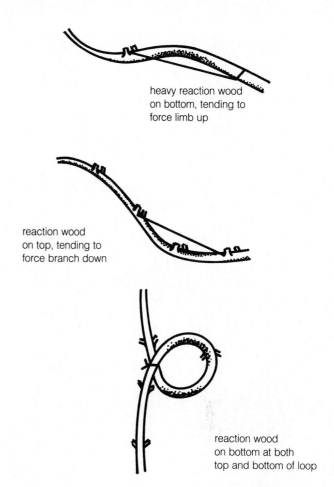

Figure 18-17 Summary of some experiments that cause formation of reaction wood in conifers (indicated by shading).

18.6 Morphogenesis: Juvenility

The life cycles of many perennial species include two phases in which certain morphological and physiological characteristics are rather distinct. After germination, most annual and perennial seedlings enter a rapidly growing phase in which flowering usually cannot be induced. A characteristic morphology, especially evident in leaf shapes, is sometimes produced during this time. Plants having these characteristics are said to be in the **juvenile phase,** as opposed to the **mature** or **adult phase**.

The juvenile phase with respect to flowering varies in perennials from only one year in certain shrubs up to 40 years in beech (*Fagus sylvatica*), with values of 5 to 20 years common in trees. In Chapter 22, this reaching of maturity after which flowering can occur is described as attaining a ripeness to respond. Such long juvenile phases in conifers and other trees pose serious obstacles to genetic programs designed to improve their quality. Another common physiological difference between perennials in the juvenile and adult phases is the ability of stem cuttings to form adventitious roots. In the adult phase, the rooting ability is usually diminished and sometimes lost.

The juvenile and adult morphologies of leaves are examples of **heterophylly.** Heterophylly of an annual dicot is well illustrated by the bean, which always forms simple primary leaves at first and compound trifoliate leaves later (see Fig. 18-5), and by the pea, which has quite reduced, scalelike juvenile leaves. Among perennials, many junipers form needlelike juvenile leaves and scalelike adult leaves. The many species of *Acacia* and *Eucalyptus* often have strikingly different juvenile leaf forms. English ivy (*Hedera helix*), another perennial, has been studied extensively. Its juvenile growth habit is that of a creeping vine, but later it becomes shrublike and forms flowers. Its juvenile leaves are palmate with three or five lobes, while adult leaves are entire and ovate. Although attainment of the adult phase is usually fairly permanent, juvenility in ivy can be induced in shoots that develop from lateral buds of mature stems by treating the leaf just above this lateral bud with GA$_3$. ABA prevents this reversion caused by GA$_3$, suggesting that a balance of gibberellins and ABA might normally be involved in the transition from one state to another. On the other hand, GA$_3$ terminates juvenility and induces flowering in many gymnosperm species in the Cupressaceae and Taxodiaceae families.

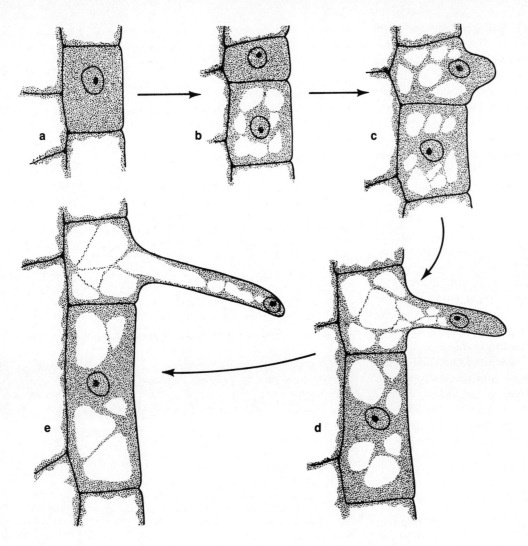

Figure 18-18 (**a**) An unequal cell division of a young epidermal cell precedes formation of a root hair and an ordinary epidermal cell. This division forms (**b**) a trichoblast (smaller upper cell) and an atrichoblast (larger lower cell). (**c, d, & e**) The trichoblast develops into a root hair. (From Jensen and Salisbury, 1972.)

18.7 Morphogenesis: Totipotency

We noted earlier that, in differential growth, cells in a plant become different even though their genes are identical. How do we know they all have the same kinds of genes? First, our understanding of chromosomal duplication and separation during mitosis strongly suggests this. Second, many plant cells are **totipotent.** By this we mean that a nonembryonic cell has the potential to dedifferentiate into an embryonic cell and then to develop into an entire new plant, if the environment is correct. A root parenchyma cell, for example, may begin to divide and produce an adventitious bud and finally a mature flowering shoot. All the genes for production of the whole plant must exist in such differentiated root cells. This could not happen if their genes had been altered during root differentiation. Totipotency is also illustrated by

development of pith callus tissues into new plants, and partial totipotency occurs when adventitious roots develop from stem cells and when xylem and phloem are regenerated from wounded cortex cells. In fact, totipotency might be advantageous to plants mainly because it provides them with a mechanism for healing wounds and reproducing vegetatively by cloning.

In each of these examples of totipotency, several cells cooperate to form primordia from which the whole plant arises. A few examples in which even a single differentiated cell apparently develops into a whole plant are known. Experiments in which plants develop from single cells were pioneered by F. C. Steward and his co-workers at Cornell University in the 1950s, associated with his work on cytokinins (see Section 17.1). He found that cells broke away from pieces of callus derived from carrot-root phloem. When conditions were changed, single cells in the

Table 18-1 Some Principles of Differentiation

Principle	Example or Discussion
1. Once synthesized by enzymes, many large molecules and other structures are arranged into fairly stable three-dimensional structures by spontaneous self-assembly.	After synthesis, polypeptides fold into the most stable structure for their watery medium. Ribosomes, microtubules, and viruses often also assemble from component parts spontaneously. Even certain cell types (e.g., in sponges) can reassemble after being dissociated.
2. Genes control the kinds of enzymes a cell can make, but environment determines whether or not enzymes function effectively.	Nitrate reductase appears in roots in response to nitrate. Temperature often determines enzyme activity.
3. Environment sometimes determines transcription and translation of genetic information into functioning enzymes.	Chromatin (genetic material) from shoot apices of peas would not make globulin *in vitro*, but chromatin from pea cotyledons would do so, as occurs in the intact plant. The genes for globulin were **repressed** in the shoots but not in the cotyledons. Gene repression may be caused by proteins called *histones*.
4. The *position* of a cell in relation to other cells may determine how it responds in differentiation. If the new cells resemble the ones directing their differentiation, the process is called **homeogenetic induction.**	New cambial cells form from cortical cells adjacent to existing procambial cells; the process is called **redifferentiation.** This might be caused by something released from the existing cells. In some cases, plant cells that are already differentiated **dedifferentiate** before becoming differentiated as a new cell type.
5. If the new cells are different from the ones causing the changes, the process is called **heterogenetic induction.**	In some dicots, leaf hairs are located only over vascular bundles, suggesting that the presence of the bundle controls differentiation of the hair cells.
6. Differentiation sometimes seems to be controlled by **field effects,** in which differentiation may occur in "fields" that do not overlap.	A minimum and constant distance is maintained between differentiating stomates on a leaf, so that stomatal patterns are not random. Growth substances could be involved in many of these cases.
7. Tissue differentiation usually requires an initial act of cell division. After *dedifferentiation*, mitosis and cytokinesis occur; then differentiation occurs in the daughter cells. Often, the two daughter cells are not alike.	Cambial cells do not themselves differentiate into xylem or phloem cells. Epidermal cells divide to produce one large and one small cell on a young root surface; the small one is a **trichoblast** and will become a root hair. Similar processes are involved in formation of guard cells, subsidiary cells, and sieve-tube elements and companion cells. This is emphasized by observation of gamma plantlets, in which cell division cannot occur because of radiation treatment; no root hairs, leaf hairs, or stomates are formed (Figure 18-18).

This table is condensed from a much more detailed discussion in the previous edition of this textbook.

cell suspensions would occasionally divide to form multicellular embryoids. From these, new plants capable of producing seeds were formed. Cloning from single cells had been achieved!

Even after Steward's experiments, there was some question about whether single cells were totipotent, because Steward's embryoids had always developed in the presence of many cells in suspension, although each plant apparently came from a single cell. Hildebrandt and Vasil (1965) answered the question by producing entire plants from isolated single cells.

Even haploid pollen grains develop into callus tissues and then whole plants (Sunderland, 1970; Sangwan and Norrell, 1975). Sometimes the plant cells contain predominantly triploid and diploid chromosome numbers, although some of the cells are haploid. Apparently, the diploid and triploid cells result from **endoreduplication** (doubling of chromosomes in mitosis, with lack of subsequent cytokinesis) or nuclear fusion.

18.8 Some Principles of Differentiation

Ultimately, it is the goal of biologists to understand morphogenetic events such as those we have been discussing in this chapter by understanding what is going on at the cellular level and what control mechanisms are involved. We close this chapter with a table (Table 18-1) that reviews some suggestions about such mechanisms. It should be quite evident as you study the table that these suggestions bear little direct relationship to the phenomenological data that we have been discussing. That is, we have a few ideas about how morphogenesis is controlled, and we have vast bodies of information about morphogenetic processes, but it is seldom possible to use the ideas about control mechanisms to understand the observed phenomena. Nevertheless, that remains the goal. Reaching it would be one of the crowning achievements of our time—or any time.

19

Photomorphogenesis

Light is an important environmental factor controlling plant growth and development. A principal reason for this, of course, is that light causes photosynthesis. Furthermore, light influences development by causing phototropism. Numerous other effects of light that are quite independent of photosynthesis also occur; most of these effects control the appearance of the plant, that is, its development or morphogenesis (origin of form). The control of morphogenesis by light is called **photomorphogenesis.** One pigment that absorbs light (red and far-red) effective in causing photomorphogenesis has been identified and named **phytochrome,** but another exists that absorbs violet and blue light; it was named **cryptochrome** (Gressel, 1980), because its chemical nature is unknown.

Some photomorphogenetic effects of light can be noted easily by comparing seedlings grown in light with those grown in darkness (Fig. 19-1). Large seeds with abundant food reserves eliminate the need for photosynthesis for several days. Dark-grown seedlings are **etiolated** (French *etioler:* to grow pale or weak). Several differences caused by light are apparent.

1. Chlorophyll production is promoted by light.

2. Leaf expansion is promoted by light, but less so in the monocot (maize) than the dicot (bean).

3. Stem elongation is inhibited by light in both species. (The maize stem is short and not visible in Fig. 19-1, because it is surrounded by leaf sheaths that extend nearly to ground level in such young plants.)

4. Root development is promoted by light in both species.

All these differences seem related to the need of a seedling to extend its stem through the soil if its leaves are to reach the light. More of the food reserves in the endosperm (maize) or cotyledons (bean)

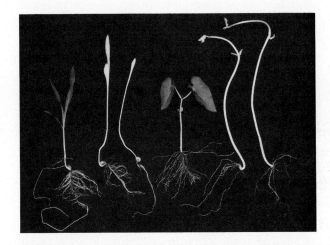

Figure 19-1 Effects of light on seedling development in a monocot (maize) and a dicot (bean). The plant at the left of each group was grown in a greenhouse, while the other representatives of each were grown in continuous darkness for 8 days. (Photograph by Frank B. Salisbury.)

are used to extend the stem upward in darkness than in light, and less food is used to develop leaves and roots and to form chlorophyll, all of which are less important for a dark-grown plant. Still other effects of light on seedlings will be described in Section 19.5.

Besides these light effects, many others are essential to monocots, dicots, gymnosperms, and some lower plants. Such effects sometimes begin with seed or spore germination and often culminate in control over flowering. We shall describe and attempt to explain several of these phenomena in the present chapter, although physiological and biochemical explanations are far from complete. As a start in understanding, it is important to realize a difference between the action of light in causing photomorphogenesis and in causing photosynthesis. In photosynthesis, light provides all the energy for the process. In photomorphogenesis, light in low doses

The Discovery of Phytochrome

F.B.S. 1960

Sterling B. Hendricks

The discovery of a new process in the biological world is always exciting, and when it proves to be an important one, it may be epoch making. This was true for the discovery of phytochrome. In 1970, we asked Sterling B. Hendricks to tell us of the discovery. Sterling Hendricks died on January 4, 1981.

In 1945, Harry A. Borthwick, Marion W. Parker, and I set out to find something about how plants recognize day lengths. Our method was to note changes in flowering induced by breaking long nights with periods of light of various wavelengths and intensities—or, more exactly, to measure action spectra. A low energy of 660 nm (red) light proved most effective in preventing flowering of the short-day soybean and cocklebur and in inducing flowering of the long-day barley and henbane. These oppositely responding plants had the same action spectrum, in all details, for flowering change. The near infrared, 700 to 900 nm, appeared to be ineffective.

Instead of elaborating the finding about flowering, we (with Frits W. Went) next chose to measure action spectra for inhibition of etiolation (the elongation of stems and restriction of leaf expansion that occur in the dark). Suppression of stem lengthening of barley and enhancement of pea leaf size by light gave the flowering action spectrum. It was exciting to know that such diverse displays were related in initial cause.

A need for light in germination of some seeds had been known for about a century. Our first measurements (in 1952 with Eben H. Toole and Vivian K. Toole) were on lettuce seeds, which Lewis H. Flint and E. D. McAllister had studied 17 years earlier. Again, 660 nm light was most effective for promoting germination. Seeds placed in the 700 to 800 nm region of the spectrum, however, germinated less than the 20 percent value of those in darkness (controls). Flint and McAllister had found germination to be suppressed in this region when seeds were initially potentiated by light to 50 percent germination. The differences in experimentation had no immediate import to us.

But one day in 1952, during an action spectrum measurement, the thought came that the 700 to 800 nm region had not been correctly tested in flowering controls. If plants had been first exposed to 660 nm light to potentiate the flowering change, instead of being taken directly from darkness, the 700 to 800 nm region might change them back to their dark condition. The 700 to 800 nm region, in truth, might cause a response similar to darkness rather than being without effect as we had earlier supposed. This region, when tested with lettuce seeds, did indeed suppress potentiated germination from a high to a low value, with maximum effectiveness near 730 nm. The reversal of the potentiated response with short-day plants for flowering also was well enough borne out to support a generalization.

The photoreversibility was the touchstone for deeper understanding of the equivalence of initial action in diverse phenomena. It also led to eventual isolation of the pigment *phytochrome*. The day in 1952, though, held a further nuance that is not so widely appreciated, nor was it quickly grasped by us. We had done quite a different experiment than had Flint and McAllister. The lettuce seeds in our experiments were exposed for short times to low total energy of moderate intensity, while they used high total energy given by continuous exposure at low intensity. The photoreversibility was thus quickly and convincingly evident to us but not to them—as was its implication of determination at the molecular level by quick change in form of a light-absorbing pigment.

In an earlier paragraph, the reversibility of flowering was expressed as ". . . well enough borne out . . . ." The reversal, in truth, was rather poor. The more we tried to enhance it, by increasing the energy through long exposure, the poorer it became. In fact, 730 nm radiation at low intensity for an hour inhibited cocklebur flowering as effectively as did low-energy exposure at 660 nm. This was recognition of what is now called the "high-irradiance reaction." Flint and McAllister were likely dealing with this reaction in lettuce seeds rather than with the simple reversibility of phytochrome. The high-energy reaction, which has many facets in nature, is also thought to arise in part from phytochrome, but in a more complex way than by photoreversibility alone. Finally, our initial objective—how plants recognize day length (or night length)—is somewhat a will-o'-the-wisp. Phytochrome action is involved, but what the action might be and how it ties in with endogenous biological rhythm is still under vigorous debate. [This remains true in 1984.]

sometimes acts like a trigger to initiate a developmental process that depends upon photosynthetic products for its completion. In other examples of photomorphogenesis, light acts not as a trigger but as a modulator, but again the light energy involved is small, and development depends upon photosynthetic products. So photomorphogenetic light initially causes a small change in cells, but this change is somehow amplified greatly so that large morphological events occur. These events vary with the kinds of cells involved, their positions in the plant, and their age (Mohr, 1983). Many genes must eventually become activated and others deactivated during photomorphogenesis, even though we don't know how this occurs. We shall mention some cases in which specific gene products (enzymes) are known to be affected (also see Newbury, 1983).

19.1 Discovery of Phytochrome

The discovery and isolation of phytochrome and the demonstration of its importance as a pigment that controls photomorphogenetic responses represents one of the most brilliant and important physiological accomplishments. Most research leading to phytochrome detection and isolation was accomplished at the United States Department of Agriculture Research Station in Beltsville, Maryland, between 1945 and 1960. The history of phytochrome discovery was summarized by Harry A. Borthwick (1972), who was one of its pioneers, and by Briggs (1976). Sterling B. Hendricks, another pioneer (now deceased), described some aspects of the discovery of phytochrome in a personal essay on page 385.

An important observation had already been made at Beltsville by W. W. Garner and H. A. Allard about 1920. They found that the relative durations of light and dark periods control flowering in certain plants (Chapter 22). Then, in 1938, it was discovered by others that the cocklebur, which requires nights longer than some critical minimum length to flower (a short-day plant), is prevented from flowering by a brief interruption of its dark period with light. Red light proved much more effective than the other wavelengths not only for interrupting the long nights that otherwise induce cockleburs and Biloxi soybeans to flower, but also for promoting expansion of pea leaves. Red light interrupting a dark period was also the most effective color in stimulating flowering of Wintex barley and other long-day plants that require nights shorter than some critical maximum length.

Borthwick and Hendricks then collaborated with experts familiar with seed dormancy in many species. They constructed a large spectrograph that, with a bright light source, could separate various colors of light over distances so broad that potted plants could be lined up and exposed to different wavelengths. They obtained an action spectrum with a peak in the red for promotion of germination of Grand Rapids lettuce seeds, only 5 to 20 percent of which will usually sprout in darkness. It had already been shown in the 1930s that red light promoted germination of such seeds but that blue or far-red light inhibited germination even below that in darkness. *Far-red includes those wavelengths just longer than the red, covering approximately the range 700 to 800 nm.* (Those longer than about 760 nm are invisible to humans and technically are infrared, as shown in Appendix B, Fig. B-2. Visible far-red wavelengths appear dark-red to us.) The

Beltsville group then made a remarkable discovery. When seeds were exposed to far-red just after a promotive red treatment, promotion was nullified; but if red was given after far-red, germination was enhanced. By repeatedly alternating brief red and far-red treatments, they found that the color of light applied last determined whether the seeds germinated or not, red promoting and far-red nullifying that promotion (Fig. 19-2). Furthermore, the inhibition of flowering in short-day plants by red interruption of a long night was largely overcome by immediately following red with far-red.

By then they realized that a blue pigment was present that absorbed red light; but that its concentration was too low to give color to etiolated maize seedlings, in which it was first detected by absorption changes with a spectrophotometer. They also decided that the pigment could be converted by red light to a different form that absorbed far-red (a form that eventually proved olive-green in color) and that the blue pigment could be regenerated with far-red. The green form produced by red light was deduced to be the active form, while the blue form seemed inactive. These ideas were based on physiological and spectrophotometric studies with seeds or etiolated plants, and they needed to be verified by extracting the pigment and studying it *in vitro*. This has been the scientific approach for all biological pigments, including rhodopsin for vision, chlorophylls and carotenoids in photosynthesis, and cytochromes in respiration. In the early 1960s, H. W. Siegelman and other protein chemists purified phytochrome from homogenates (finely-ground suspensions) of cereal grain seedlings by column chromatography and other techniques routinely used to purify proteins. They demonstrated that isolated phytochrome changes color reversibly upon exposure to red and far-red light. Essentially all the early deductions based only upon physiological experiments with whole plants had been verified. Even the absorption spectrum of both forms of phytochrome was measured.

19.2 Physical and Chemical Properties of Phytochrome

Absorption spectra of highly purified phytochrome molecules from angiosperms show maxima in red wavelengths at 666 nm for the red-absorbing bright blue form (P_r) and at 730 nm for the far-red-absorbing green form (P_{fr}) (Vierstra and Quail, 1983a; Smith, 1983). Figure 19-3a shows a modern example of these absorption spectra. The absorption spectrum for P_{fr} shows a shoulder in the red region that is caused by P_r, not P_{fr} (Vierstra and Quail, 1983a), which is present because it is not possible to convert all the P_r to P_{fr} in a phytochrome sample. Figures 19-3b and c illustrate action spectra in the red and far-red for

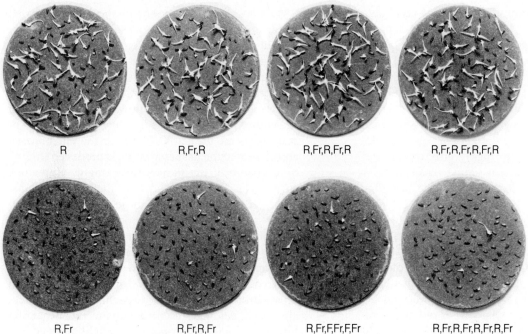

| R | R,Fr,R | R,Fr,R,Fr,R | R,Fr,R,Fr,R,Fr,R |

| R,Fr | R,Fr,R,Fr | R,Fr,F,Fr,F,Fr | R,Fr,R,Fr,R,Fr,R,Fr |

Figure 19-2 Reversal of lettuce-seed germination with red and far-red light. Red exposures were for 1 min and far-red for 4 min. If the last exposure is to red light, seeds germinate; if to far-red, they remain dormant. Temperature during the half hour required to complete the treatments was 7°C; at all other times, it was 19°C. (From Jensen and Salisbury, 1972; courtesy Harry Borthwick.)

various physiological responses. *Similarities of the absorption spectra of phytochrome and action spectra for plant responses give one important evidence that phytochrome is the pigment causing such responses. A second evidence is that responses caused by red are almost always nullified by an immediate subsequent exposure to far-red. A third evidence is that very low irradiance levels of either red or far-red capable of interconverting phytochrome from one form to another also cause these responses.*

In leaves and stems, light absorbed by chlorophyll alters both the action spectra for phytochrome responses caused by P_{fr} and the amount of light (absorbed by P_r) required for the response. Action spectra peaks are shifted toward shorter wavelengths near 630 nm where chlorophyll absorbs less, and much more energy is needed for response (Jose and

Figure 19-3 A comparison of the absorption spectrum of both forms of phytochrome with action spectra for various physiological processes. (Absorption spectra are from R. D. Vierstra and P. H. Quail, 1983a. Action spectra in the middle graph were redrawn from data of M. W. Parker et al., 1949. Action spectra for promotion of bean hypocotyl hook opening in the lower graph were redrawn from R. B. Withrow et al., 1957, and spectra for promotion and subsequent inhibition of Grand Rapids lettuce were redrawn from S. B. Hendricks, 1960.)

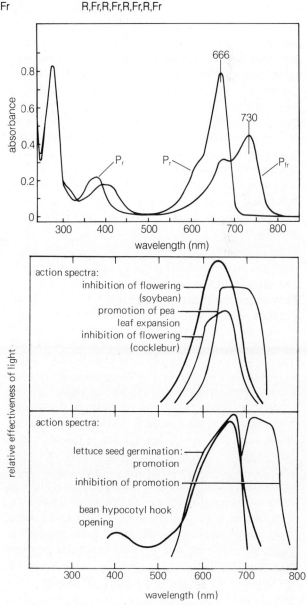

Schafer, 1978). Both P_r and P_{fr} absorb violet and blue light, but low irradiance levels of these wavelengths are much less effective than red or far-red for physiological processes we have described so far. Since neither P_r nor P_{fr} absorbs green light effectively and our eyes are especially sensitive to green, safelights with filters that transmit only low intensity green light are used in physiological experiments in which phytochrome participates. In general, green safelights and green filters are highly effective in phytochrome studies; but they must transmit low irradiances and they must be tested to be sure they cause no response.

Chemically, phytochrome is a protein with a molecular weight varying in different species from 120,000 to 127,000 g mol^{-1} (Vierstra and Quail, 1983b; Vierstra et al., 1984). One prosthetic group called a **chromophore** is attached to the protein. This chromophore is an open chain *tetrapyrrole* similar to the photosynthetic phycobilin pigment of red algae and cyanobacteria. The chromophore absorbs the light that causes phytochrome responses. When P_r is converted to P_{fr} by red light, conjugation of double bonds in the tetrapyrrole chromophore is altered; evidence indicates that the conjugated double-bond system of P_{fr} has one less double bond than that of P_r (Rudiger and Scheer, 1983). Alteration of the chromophore in phytochrome then causes an unidentified subtle change in structure of the phytochrome protein (Pratt, 1982). The change in protein structure is somehow responsible for the physiological activity of P_{fr} and the inactivity of P_r. Keep in mind, though, that active P_{fr} can influence development in numerous ways. Biochemically, we know very little about how P_{fr} acts.

19.3 Distribution of Phytochrome Among Species, Tissues, and Cells

What we have said so far applies to phytochrome of angiosperms. Does phytochrome exist in other kinds of plants? It does so in gymnosperms, liverworts, mosses, ferns, and some green algae, suggesting that it might be present in all photosynthetic organisms except photosynthetic bacteria. Little is known about chemical properties of phytochrome in species other than angiosperms. In a green alga, in several pine species, and in the ancient gymnosperm *Ginko biloba*, *in vivo* absorption peaks for P_r and P_{fr} occur at slightly shorter wavelengths than in angiosperms. However, even angiosperms display some variability in these peaks, because chlorophyll and other neighboring molecules influence the absorption spectra of phytochrome. It is likely that at least the chromophore of phytochrome in all plants is chemically similar.

Phytochrome is present in most organs of all plants investigated, including roots. For many years, quantitative *in vivo* measurements could be performed only with etiolated tissues, in which chlorophyll does not interfere with spectrophotometry. In etiolated plants grown in total darkness, phytochrome is present entirely as P_r, because apparently no P_{fr} can be synthesized in darkness. Phytochrome contents of plant parts are measured *in vivo* by the difference in absorbance of the tissues after P_{fr} is produced from P_r by exposure of the tissues to red light. Numerous such measurements showed that phytochrome occurs in green plants, but these amounts are much lower than in dark-grown plants. When phytochrome was extracted from green and etiolated plants, the differences in concentration measured *in vivo* were confirmed. Etiolated seedlings of dicots and monocots usually contain 30 to 100 times as much phytochrome as light-grown seedlings (Shimazaki et al., 1981). Another recent technique for measuring phytochrome in plants exposed to light takes advantage of the ability of an herbicide, *Norflurazon*, to prevent specifically the formation of chlorophyll in developing leaves and stems (Jabben and Dietzer, 1978; Gorton and Briggs, 1980). Tissues treated with this herbicide contain about the same low amounts of phytochrome as green tissues, showing that light absorption by chlorophyll is not responsible for the phytochrome decrease in light-grown plants. The high concentration of P_r in etiolated seedlings is probably advantageous for them to detect red light and respond to it, but as they respond and become **de-etiolated** in light most of their phytochrome is destroyed.

More recently, Peter H. Quail and his colleagues at the University of Wisconsin showed that the low levels of phytochrome in light-grown plants result from inhibited formation by light, as well as increased destruction (Colbert et al., 1983). Red light causes the inhibition of phytochrome synthesis, and this effect is largely overcome by subsequent far-red. The inhibitory effect is on transcription of the messenger RNA that codes for production of the phytochrome protein. These results are especially interesting, because they show that P_{fr} rapidly prevents additional phytochrome formation and does so by controlling the gene that codes for phytochrome. This represents a remarkable feedback control.

In the early 1970s, Lee H. Pratt and Richard A. Coleman developed an immunological technique for phytochrome determination that allows its identification in specific cells and subcellular organelles. Their original technique is close to a thousand times more sensitive than absorbance methods and is also applicable to green tissues. This basic method has now been improved and made more sensitive and specific (Saunders et al., 1983; Cordonnier et al.,

1983). (See Dr. Pratt's personal essay on pages 386–388 concerning the importance of immunological techniques to research with phytochrome and other aspects of plant physiology.) Both light and electron microscopy can be used in the immunological assays. Their results show that root-cap cells of grass seedlings contain high amounts of phytochrome, consistent with absorbance by the cap of light that stimulates gravitropic sensitivity in certain grasses and other plants. Phytochrome distribution in grass shoots is variable, but oat, rye, rice, and barley seedlings all have high concentrations in the apical regions of the coleoptile, near the shoot apex, and (except for oat) in the growing leaf bases. Subcellularly, phytochrome exists throughout the cytoplasm and is also associated with or is within the plasma membrane and the membranes of chloroplasts.

The ratio of P_{fr} to the total amount of phytochrome in both forms is denoted by ϕ:

$$\phi = \frac{P_{fr}}{P_r + P_{fr}} = \frac{P_{fr}}{P_{total}}$$

Red light of 667 nm converts about 80 percent of phytochrome into P_{fr}, so ϕ is 0.8. Far-red light above 720 nm removes essentially all P_{fr}, so ϕ approaches zero. Even very low irradiances with red and far-red are adequate to establish the photoequilibrium between P_r and P_{fr}, because both forms absorb those wavelengths so effectively. In sunlight or under incandescent light used in growth chambers, the irradiance of far-red photons is only slightly less than that of red photons (see Fig. B-3 or Fig. 19-5). Nevertheless, P_r absorbs red light more effectively than P_{fr} absorbs far-red, so *sunlight acts primarily as a red source that forms more P_{fr} than P_r.* The ϕ value in sunlight is about 0.6.

In most species, some of the P_{fr} gradually disappears even in darkness. Two processes account for this. The first is **destruction,** because after an interval of time in darkness it is no longer possible to regenerate as much P_r in tissues by a far-red exposure; so the total amount of detectable phytochrome is less than before. This destruction probably involves protein denaturation, because it has a high Q_{10} value of 3, typical of protein denaturation processes that occur rapidly as the temperature is increased. The second process is **dark reversion** back to P_r, which usually requires a few hours. (It should be emphasized that destruction and reversion also occur in the light but are not caused directly by light.) Reversion occurs in most dicots and gymnosperms but has not been detected in monocots or in any of the 10 dicot families often classified as part of the order Caryophyllales. Because of destruction and reversion, we must modify our idea that light simply sets up a photostationary reversible state between P_{fr} and P_r.

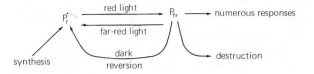

Figure 19-4 A summary of some transformations of phytochrome.

Reversion and destruction (where either is applicable) must be added, as shown in Fig. 19-4.

With a knowledge of these properties of phytochrome and some of the reactions it undergoes, let us consider the physiological processes it controls, beginning with seed germination.

19.4 The Role of Light in Seed Germination

Some Examples of Light-Dependent Germination The importance of light for germination of certain seeds has probably been recognized for hundreds of years, but the first comprehensive study was described by Kinzel in 1907 (Rollin, 1972). Kinzel reported that among 964 species, 672 showed enhanced germination in light. Seeds of most species that respond to light are small and undomesticated, rich in fat but so small that their seedlings might not reach the soil surface and light before their food reserves were used up. Most of our cultivated seeds do not require light, no doubt because of human selection against a light requirement. Seeds of many wild species even show inhibited germination in light, sometimes because of the blue but especially because of the far-red present (Frankland and Taylorson, 1983). Kinzel found 258 of the 964 species he studied to be inhibited by sunlight relative to darkness. The far-red wavelengths of sunlight are nearly always those that are most inhibitory, because they decrease the amount of P_{fr} in the seed to a level below that needed for germination. In some species blue light is also inhibitory; the effects of blue are complex, and we shall not describe them.

Complications in interpretations of light effects on seed germination arise because of the amount of P_{fr} required to promote germination. In some seeds, ϕ values as low as 0.01 are promotive; but in others, values as high as 0.6 are needed for full promotion. Much of this great difference must lie in the variable sensitivity of seeds to P_{fr}; but ϕ values obtained so far are averages for whole seeds and, since they are ratios, do not tell the concentration of P_{fr} in the cells responsible for initiating germination. The responsive cells are normally only those in the radicle or hypocotyl (of dicots). Researchers should use methods to detect P_{fr} concentrations in those cells so that we can properly interpret variable responses of seeds of different species to light. Furthermore, the inhibited germination by sunlight is measured by comparing the germination of those seeds to that of seeds kept in darkness. This comparison is important, because the amount of P_{fr} in dry seeds is determined by environmental conditions as they ripen and mature in fruits on their mother plant. As will be explained shortly, the environment during maturation largely controls the P_{fr} content of mature seeds.

Seeds that require light for germination are said to be **photodormant.** As discussed in Chapter 21, we use *dormancy* as a general description of seeds or buds that fail to grow when exposed to adequate moisture, air, and a favorable temperature for growth. Seeds that normally germinate in darkness but that are inhibited by light also become dormant after the light exposure.

Interactions of Light and Temperature in Photodormant Seeds A further aspect of the light effects of germination is an interaction with temperature. (The temperature during the light treatment can generally be ignored, because the photochemical interconversions of P_r and P_{fr} are relatively independent of temperature.) The chemical reactions controlled by P_{fr} and those influencing destruction of P_{fr} are quite temperature sensitive. An example of crucial temperature control occurs in seeds of Grand Rapids lettuce (*Lactuca sativa*) and peppergrass (*Lepidium virginianum*). Light usually promotes their germination, but extended exposure to 35°C after a single light treatment or exposure to that temperature in continuous light keeps them dormant. Similarly, seeds of the Great Lakes variety of lettuce usually do not require light to germinate; but if they are soaked at 35°C, they become photodormant and then germinate only in light at that or a lower temperature. Still another example is provided by Kentucky bluegrass (*Poa pratensis*), in which temperatures alternating between 15°C and 25°C substitute for light in causing germination.

Evidence shows that temperature responses such as these are often caused by effects on the amounts of P_{fr} in the seeds. High temperature decreases the P_{fr} level mainly by increasing its rate of reversion to P_r, although its rate of formation and the extent to which it combines with other compounds necessary to cause its final effect have not been evaluated. Destruction of P_{fr} in seeds (Fig. 19-4) has been generally ignored, because little evidence for such destruction has been obtained. Both photodormant and nondormant seeds usually imbibe water and swell, unless their seed coats prevent water uptake. Even dead seeds swell! But only nondormant seeds continue to absorb water and grow after imbibition is complete

(after their colloids are fully hydrated). Dormancy is broken by light only when seeds are partially or fully imbibed. The time required for imbibition varies from as little as an hour to almost two weeks, depending on how permeable the seed or fruit coats are to water and how large the seed is (Toole, 1973; Bewley and Black, 1982). Only then is P_r sufficiently hydrated to be transformed to P_{fr}. In seeds that survive many years in the soil, P_r is stable and only awaits the proper combination of moisture, light, and temperature to become P_{fr} and cause germination. When Grand Rapids lettuce seeds are imbibed and exposed to light to form P_{fr}, they can be immediately dehydrated for as long as a year and then will germinate in darkness upon remoistening. This shows that P_{fr} is stable in dry seeds for long periods. That P_r is also stable in dry seeds is shown by the requirement of lettuce seeds for a red-light treatment after they are dried under far-red light. An implication of stable P_r and P_{fr} is that whether a seed requires light to germinate depends on how much P_{fr} was produced in it during ripening on the mother plant.

The amount of chlorophyll that covers the embryo as the seed ripens is especially important in determining whether or not seeds of a given species will be photodormant (Cresswell and Grime, 1981). In general, embryos that are covered during ripening by maternal tissues that contain high amounts of chlorophyll require light to germinate, whereas those that are covered by maternal tissues with little or no chlorophyll do not. The apparent reason for this is that chlorophyll absorbs the red wavelengths and prevents P_{fr} formation in the ripening embryos, so the seeds then require red wavelengths to form P_{fr} that promotes germination. The relative lengths of night and day during ripening also affect photodormancy in many species (e.g., *Chenopodium*); long days usually favor photodormancy whereas short days favor nondormant seeds (Bewley and Black, 1982). The reason for the day-length effect is unknown.

Ecological Aspects of Photodormancy in Seeds What possible ecological benefit is light sensitivity to seeds lying in litter near the soil surface? Answers to such questions frequently involve speculation, and this is no exception. For buried seeds whose germination is promoted by light, germination when they are partly uncovered more nearly assures that the seedlings will be able to photosynthesize, grow, and perpetuate that species. A light requirement for buried seeds might distribute germination over several years and thus help perpetuate the species, since only a fraction of the seeds in a soil might be disturbed and exposed to light in a given season. An unfavorable growing year might otherwise destroy most of the plants. For seeds whose germination is inhibited by light, germination is prevented until they are well-covered by litter, when they would more likely have sufficient water to grow. Koller (1969) described two light-inhibited species that inhabit coarse, sandy soils of the Negev desert, in Israel, in which germination is prevented unless the seeds are well buried, where moisture is more plentiful than at the soil surface.

Another idea is that phytochrome provides seeds with a clue about whether they are covered by a canopy of other plants or exist in an open area. This possibility has been studied extensively by physiological ecologists. The idea developed from two facts. First, far-red light usually inhibits germination of light-requiring seeds and even seeds that germinate moderately well in darkness. Second, leaves in a canopy transmit considerably more far-red than red light. Most of the blue, red, and some of the green wavelengths are removed by leaves through photosynthesis and reflectance, but most of the far-red passes through to seeds below and converts their active P_{fr} to inactive P_r. Figure 19-5 illustrates the spectral distribution of radiation above and under the canopy in a sugarbeet field. The lower curve for radiation filtered by leaves shows a small peak of transmission in the green region (540 nm) and a much larger one in the far-red. Under such a canopy, no more than 10 percent of the total phytochrome would exist as P_{fr}. If seeds ripen on a plant under a canopy transmitting such a high ratio of far-red to red, they are likely to require direct sunlight to form additional P_{fr} before they will germinate. In a forest, many seeds requiring relatively high amounts of P_{fr} might never sprout until a fire, death of old trees, or timber removal eliminates the canopy. Evergreen forests seem especially affected in this respect, but deciduous forests are probably also affected in some climates, depending on when their new leaves develop relative to the temperature and light requirements of seeds below. The light-filtering effect of plant canopies is also important to agriculture, because germination of many weed seeds is promoted by sunlight but inhibited by light filtered from crop plants that have developed above them (Bewley and Black, 1982; Frankland and Taylorson, 1983).

Ecologically, the light sensitivity of seeds from species that inhabit shaded conditions is different from that of pioneer species from more open areas (Grime, 1979, 1981; Bewley and Black, 1982). As you might predict, seeds of species that live in shaded conditions are less inhibited by the far-red light transmitted by plant canopies than are those that invade more open areas.

Is Phytochrome the Only Pigment Active in Germination?   Although both promotive and inhibitory effects of far-red and also blue radiation have often

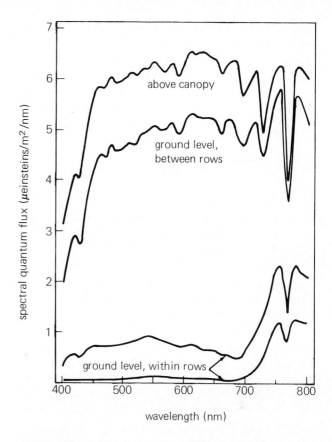

Figure 19-5   Influence of shading upon wavelengths of sunlight present in various regions of a sugarbeet field. Within the rows (bottom two curves) there is much less attenuation of far-red than of the other wavelengths, so shaded plants contain a higher proportion of phytochrome in the P_r form than do unshaded plants. (From M. G. Holmes and H. Smith, 1975.)

been attributed to phytochrome, several results are difficult to interpret this way. In tansy (*Phacelia tanacetifolia*), germination is suppressed not only by extended periods of far-red and blue, but also by red light. If only P_{fr} is important to allow germination, why should wavelengths that cause widely different ϕ values all be inhibitory? Furthermore, the amount of light required for inhibition of this species is much higher than for typical phytochrome responses. The same is true for the far-red and blue inhibition of Grand Rapids lettuce germination and for inhibition in many other species by the far-red transmitted through a leaf canopy. *Whereas most clear-cut phytochrome responses are saturated by energies of red light equal to as little as 100 J m^{-2}, which is less than one-hundredth the energy in all visible wavelengths provided by sunlight during a 1-minute period, these far-red and blue effects frequently require at least 100 times more energy.* These are called **high irradiance reaction(s) (HIR)** (Mancinelli and Rabino, 1978).

We do not yet understand how HIR caused by

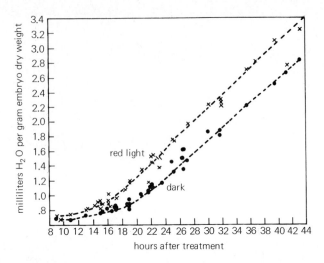

Figure 19-6 Stimulation of growth by red light in naked embryos of lettuce seeds. Seeds were soaked in distilled water for 3.5 h, then some were given a 10-min treatment with a red light source. Endosperm, seed coat, and fruit coat were then removed under a green safelight, and growth (monitored by measuring water uptake) was measured at the time periods indicated. Each point represents the response of one embryo. (From M. W. Nabors and A. Lang, 1971.)

far-red or blue light inhibits germination. Most researchers have ignored blue-light effects, partly because cryptochrome preferentially absorbs blue light and so little is known about that pigment, partly because each form of phytochrome absorbs blue light to some extent (Fig. 19-3a), and partly because blue wavelengths are absorbed much more than red or far-red by seed coats and fruit coats. All these factors make it difficult to evaluate to what extent cryptochrome and phytochrome participate in mediating blue effects on germination of seeds of any species. Nevertheless, there is some evidence that cryptochrome does contribute to blue light promotion of germination at low irradiance levels (Small et al., 1979).

High irradiance reactions caused by far-red have been studied much more than blue effects. All result from absorption by phytochrome, even though the action spectrum peak is usually near 720 nm (Frankland and Taylorson, 1983). There are three important reasons why these effects cannot simply be explained by removal of P_{fr} needed for germination. First, the action spectrum peak is 10 nm shorter than the 730 nm peak of absorption by P_{fr}. Second, high irradiances that give widely different ϕ ratios (and P_{fr} levels) are all inhibitory; for tansy, even red is inhibitory, as mentioned. And third, high irradiances are inhibitory even after a previous treatment with low irradiance red light that promotes germination (by causing P_{fr} formation) has completed its action. That is, formation of P_{fr} by red promotes early in the germination process and high irradiance effects are

noted much later, as late as just before the radicle protrudes through the seed coat. Because of the high irradiances required, numerous experts believe that the inhibition results from a rapid rate of interconversion of P_{fr} and P_r, which they call a fast-cycling rate (Frankland and Taylorson, 1983; Bartley and Frankland, 1982). It is uncertain how a fast-cycling rate is inhibitory, but one of several suggestions is that an activated form of P_{fr} is formed by red light during cycling; this form is presumably very short-lived, because it decays rapidly back to P_{fr}. So it can only exist long enough to act when continually formed by fast-cycling rates. Clearly, there is much more to learn about phytochrome before we speculate about the role of other pigments in germination.

The Nature of Photodormancy We now ignore HIR effects and return to promotion of germination by low irradiance effects. Given that P_{fr} causes photodormant seeds such as lettuce to germinate, why do they not sprout in its absence? To answer this, it is essential to identify that part of the seed in which P_{fr} must be formed. This problem was approached in lettuce seeds by separately exposing the cotyledons and the hypocotyl-radicle tissues with a narrow beam of red light. In *Curcurbita pepo* (pumpkin), a laser beam only 1 mm wide allowed even better resolution of the tissues exposed. In both species, germination resulted only when the hypocotyl-radicle tissues (hereafter referred to in this section as radicle) were exposed. P_{fr} is also formed in the cotyledons when they absorb light, but this does not cause germination. These results were expected, because in nearly all seeds the radicle first breaks through the seed and (or) fruit coats during germination. However, these results also show that formation of P_{fr} in cotyledons does not cause formation of some hormone that is then transferred to the radicle to promote germination.

If we detach the lettuce embryo from the surrounding endosperm, seedcoat, and fruitcoat, the radicle itself now elongates in darkness or after a brief exposure to either red or far-red light, but radicles exposed to red begin to elongate sooner and at a faster rate than those given far-red (Fig. 19-6). (Radicles of embryos kept dark grow, on the average, slightly more than those given far-red light, because far-red decreases their already low level of P_{fr}.) This growth-promoting effect of red is nullified by a short far-red treatment after the red, as expected from germination results with whole seeds (Fig. 19-2). Furthermore, if naked lettuce embryos from imbibed seeds are removed under a green safelight and placed in solutions with negative water potentials, such as polyethylene glycol, those embryos subsequently given red light will absorb water and grow in a solution having a more negative water potential than those kept dark or given far-red (Nabors and Lang, 1971). *Our conclusion is that P_{fr} increases the growth potential of the radicle cells, presumably those in the elongating region, by decreasing their water potential so that they more easily absorb water from soils and germinate.*

These facts suggest that germination of light-requiring seeds fails in darkness because the radicle cannot grow with sufficient force to break through the layers that surround it. Of these layers, the lettuce radicle is restricted almost entirely by the tough endosperm, even though it is only two or three cell layers thick. The endosperm is also the restrictive layer in *Phacelia tanacetifolia* and various *Syringa* (lilac) species. For lettuce, only increased thrust of the radicle seems important, even though the endosperm barrier is weakened greatly after germination is well underway (Bewley and Black, 1982). For other seeds, we might reasonably expect P_{fr} to increase germination either by increasing radicle thrust or by weakening surrounding barriers to its growth, or both. Photodormancy is less a mystery when we consider germination as a struggle between the growth potential of the radicle and the growth-restrictive mechanical effects of surrounding layers. In some cases, the external restriction is great; in others it is of little consequence, and only a small increase in the radicle growth potential caused by P_{fr} is enough to cause germination. Nevertheless, even in these, the removal of P_{fr} by far-red light reduces germination.

Effects of Hormones on Photodormancy In most photodormant seeds, applied gibberellins substitute for the light requirement; and for a few species such as lettuce, cytokinins also substitute for light or partially replace it. Auxins usually do not promote germination of photodormant or nondormant seeds and are instead either innocuous at low concentrations or inhibitory at high concentrations. The role of ethylene is less clear. It cannot break photodormancy, but it can partially overcome other kinds of seed dormancy in cocklebur and in certain peanut and clover varieties. It can also partially overcome dormancy caused by high temperatures in lettuce and certain photodormancy problems in cocklebur (Eshashi et al., 1983). Abscisic acid almost always retards germination, because of its growth inhibitory effects.

Collectively, these results suggest that P_{fr} might break photodormancy by causing synthesis of a gibberellin or a cytokinin or by destroying an inhibitor such as ABA. The evidence about this is presently controversial (Bewley and Black, 1982; DeGreef and Fredericq, 1983), but no one has yet measured hormone changes only in the radicle or hypocotyl cells that are responsible for germination. This seems essential to understand relations among light, growth promoters, and growth inhibitors in photodormancy

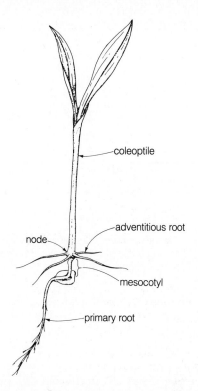

Figure 19-7 Some morphological characteristics of a week-old maize seedling grown in light. The coleoptile has stopped elongating, and two leaves have broken through it and have largely unrolled. The shoot apex is at the node where adventitious (prop) roots originate. The mesocotyl is the first internode formed above the seed storage tissues and the scutellum (cotyledon) in the seed.

and other kinds of seed dormancy discussed in Chapter 21. Analyses of whole seeds for hormone levels seem nearly useless in understanding hormonal aspects of dormancy, because the whole seed is so large relative to the few tissues that control germination.

19.5 The Role of Light in Seedling Establishment and Later Vegetative Growth

Once germination is accomplished, further plant development still remains subject to control by light. We introduced some of these controls in Section 19.1 and Fig. 19-1. We now evaluate these and other effects and ask whether phytochrome is the only pigment involved and how it acts.

Development of Poaceae Seedlings After a grass or cereal grain seed germinates, its coleoptile elongates until the tip breaks through the soil. Between the scutellum (see Fig. 16-12) and the base of the coleoptile is an internode called the **mesocotyl** (first internode, Fig. 19-7) that in some species elongates greatly

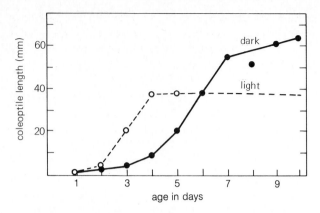

Figure 19-8 Elongation of oat coleoptiles in darkness and in continuous white light. Light at first promotes growth but later is inhibitory. (From B. Thomson, 1954.)

after germination of deeply planted seeds. Elongation of the mesocotyl, coleoptile, and leaves enclosed by the coleoptile is necessary to carry leaves into the light and to establish near the soil surface the adventitious roots produced at the node just above the mesocotyl (Fig. 19-7). Elongation of the mesocotyl has received attention for more than 40 years. All results show that mesocotyl elongation is extremely sensitive to light (Mandoli and Briggs, 1981; Shäfer et al., 1982). In oats, growth of the mesocotyl is slowed over a 54-hour period by continuous red light at photon irradiance levels of only 5×10^{-16} moles of photons per square meter of tissue per second (Schafer et al., 1982). This irradiance level is only about 10^{-7} that in all visible wavelengths provided by moonlight. This effect of light on plants is by far the most sensitive one known, and no green safelights are safe in studies with this response.

Elongation of the coleoptile must equal or exceed that of the leaves it encloses as they grow upward together; otherwise the leaves would grow out of the coleoptile and probably be broken off in the soil. Growth rates of these two organs are coordinated until they reach the soil surface and are exposed to light. After exposure to light, the leaves become green and photosynthetic, and they break through the coleoptile tip. Leaf emergence occurs because light promotes leaf elongation and decreases the extent to which coleoptiles can elongate (although it speeds their early elongation) (Schopfer et al., 1982). Furthermore, breakage of the coleoptile tip by the elongating first leaf stops coleoptile elongation, presumably because this stops auxin transport from the tip to elongating cells below. The light promotion of leaf growth and inhibition of coleoptile growth are phytochrome responses of sunlight. Figure 19-8 shows that the overall effect of continuous light is to reduce coleoptile elongation, even though during the first few days elongation is promoted. The reasonable conclusion from these 1954 results is that light has-

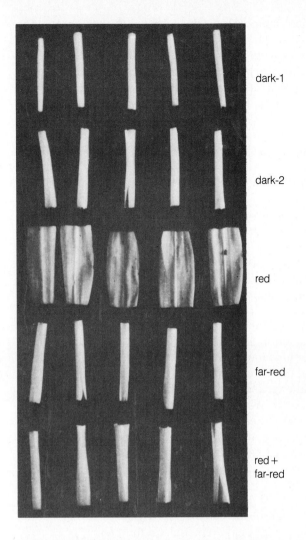

Figure 19-9 Effect of pretreatment with red and far-red light on unrolling of leaf sections from etiolated maize seedlings. Red promotes opening, whereas subsequent far-red treatment nullifies the red effect. (From W. H. Klein, L. Price, and K. Mitrakos, 1963.)

tens growth and maturation of young coleoptile cells, because those cells begin to elongate earlier and mature earlier than in darkness. Effects of light on elongating cells reaching maturity show that elongation is slowed and that maturation occurs earlier than in darkness. More recent results verify those early conclusions and further show that phytochrome is the controlling photoreceptor of light, even though long-term light exposure that results in high irradiance values is necessary to cause the responses (Schopfer et al., 1982).

Figure 19-7 illustrates a maize seedling grown in light from a seed planted near the soil surface. The mesocotyl had elongated very little, and the first two leaves had emerged from the coleoptile. Each of these leaves was rolled up inside the coleoptile, but when exposed to light they began to unroll (flatten out). Rolling was still evident only at the point of departure from the broken coleoptile. Unrolling of grass leaves is controlled by a typical phytochrome response, low irradiance levels of red promoting and subsequent far-red nullifying the red effect (Fig. 19-9). Low energies of far-red are without effect, and low energy blue is only slightly promotive except in rice. Unrolling is caused by more rapid growth of cells on the concave (to be uppermost) than the convex side. Whether this growth is caused by wall loosening, solute production that decreases the cells' osmotic potential, or both, is not known. Nevertheless, exogenous gibberellins and, to a lesser extent, cytokinins replace the need for light and P_{fr} (DeGreef and Fredericq, 1983). These results suggest that P_{fr} causes rolled leaves to form gibberellins or cytokinins which then cause unrolling. This hypothesis might be correct for gibberellins, because P_{fr} promotes gibberellin production and release from young plastids in rolled wheat and barley leaves much sooner than the leaves unroll. No studies showing light effects on cytokinin contents of unrolling leaves are available, so for now it seems safest to conclude that light might induce leaf unrolling by causing production of gibberellins in concave cells. Alternatively, the concave cells might become more sensitive to the hormone levels they already contain when exposed to light (see the personal essay by A. Trewavas in Chapter 16).

Development of Dicot Seedlings In dicots, the cotyledons either remain underground by **hypogeal development,** as in pea, or emerge above ground **epigeally,** as in beans, radish, and lettuce. In either case, a hook is formed near the stem apex that pushes up through the soil and pulls with it the fragile young leaves or cotyledons. (In Fig. 19-1 this hook has moved, as it does in seedlings that develop epigeally, into the epicotyl (stem section above cotyledons) and has opened somewhat, perhaps by slight light exposure during watering.) As mentioned in Section 17.2, this hook forms as a result of unequal growth on the two sides of the hypocotyl or epicotyl in response to ethylene soon after germination. As the hook emerges from the soil, red light acting through P_{fr} promotes opening of the hook. Hook opening apparently results from inhibition by light of ethylene synthesis in the hook. Differential growth that results from faster elongation of cells on the lower (concave) side than on the upper (convex) side causes hook opening (Section 15.3). Accompanying this, light increases leaf blade expansion, petiole elongation, chlorophyll formation, and chloroplast development, as in grass leaves (Fig. 19-1), and P_{fr} also speeds petiole elongation.

Most of the light promotion of leaf growth, at least in dicots, is caused by an HIR (Dale, 1982). A good example is provided by the primary leaves of

bean. Plants grown 10 days under dim red light have slightly larger leaves and substantially more cells than those kept in darkness; but when they are transferred to white light, cell expansion and leaf growth increase greatly. In this case the HIR system causes expansion by enhancing acidification of the cell walls, thus loosening them so that they grow faster under turgor pressure (Van Volkenburgh and Cleland, 1981).

Light effects on chlorophyll formation and chloroplast development result first from a triggering action of P_{fr} that causes production of **delta-aminolevulinic acid (ALA),** probably from glutamic acid (Kasemir, 1983; Castelfranco and Beale, 1983). ALA is the metabolic precursor converted into each of the four pyrrole rings of chlorophyll. Nevertheless, ALA is not converted all the way to chlorophyll without higher irradiance red or blue light. Instead, the metabolic pathway stops when a compound often called **protochlorophyll** is formed. More accurately, protochlorophyll is **protochlorophyllide a,** which differs from chlorophyll *a* (Fig. 9-4) only by the absence of a phytol tail and two H atoms. Protochlorophyllide *a* is rapidly reduced to chlorophyllide *a* in red or blue light, because protochlorophyllide *a,* like chlorophylls, absorbs those photons effectively. Addition of the phytol tail, an isoprenoid formed from the mevalonic acid pathway (Chapter 14), completes formation of chlorophyll *a*; some of the chlorophyll *a* is then converted to chlorophyll *b*. Chloroplast development depends strongly on chlorophyll formation and, therefore, on both of these light effects, although there are other light responses that we shall not discuss (Kasemir, 1983; Virgin and Egnéus, 1983). All of these responses lead, within a few hours, to photosynthesis in grass leaves as they break through the coleoptile and in cotyledons or young leaves of dicots as they break through the soil. Cotyledons of conifers somehow form chlorophyll and become photosynthetic even in darkness, but their needles require light for these processes.

As photosynthesis begins in leaves and cotyledons, stem elongation is inhibited by light. Of course, the seedling cannot elongate after its food supplies are exhausted; but while carbohydrates or fats are still plentiful, light is inhibitory. This inhibition of stem elongation was apparently first recorded by Julius von Sachs in 1852. He observed that stems of many species do not grow as fast during daylight as they do at night. We now realize that blue, red, and far-red all contribute to this phenomenon and that cryptochrome and phytochrome are both responsible. Hans Mohr and others in Germany worked extensively with etiolated seedlings of white mustard (*Sinapis alba*) and measured many responses in these, which are summarized in Table 19-1. They also measured the action spectrum for inhibition of

Antibodies and the Study of Phytochrome

Lee H. Pratt

Dr. Pratt is now a Professor of Botany at the University of Georgia. In 1982, he received the prestigious Charles A. Shull award from the American Society of Plant Physiologists for some of the work he describes in this essay.

Antibodies are proteins synthesized by an animal in response to exposure to a foreign substance called an **antigen.** They are only one component of an animal's immune system, the study of which is referred to as **immunology.** The natural function of antibodies, also known as **immunoglobulins,** is to help the animal destroy such antigens as bacteria and viruses. The goal of this essay is to illustrate that even though antibodies are not made by plants, they nevertheless represent a research tool that can provide invaluable information in every area of plant physiology, including photomorphogenesis. The information thus provided could most often not be gained without their use. I shall summarize here some of the contributions that antibodies have made to knowledge of phytochrome properties and function in the course of our own work with this pigment.

My interest in phytochrome stems from 1963–1964, when my work with this pigment formed the basis of a master's thesis in the laboratory of Winslow Briggs. Following a brief detour into photosynthesis for a PhD in Norman Bishop's laboratory, I returned to phytochrome in 1967 by taking a postdoctorate position with Warren Butler, who had been part of the group responsible for the discovery of this pigment. It was in his laboratory that I was first exposed to the notion that antibodies could be a powerful research tool. At that time, Butler had already prepared antibodies to phytochrome by injecting purified samples of the pigment into rabbits and later harvesting the antibodies by taking blood from them. One of his major goals was to use the antibodies to visualize phytochrome *in situ* by the procedure known as immunocytochemistry. This procedure is nothing more than the use of antibodies to stain an antigen, which in Butler's case was phytochrome. Since antibodies can readily be labeled with such tags as fluorescent dyes or electron-dense metals, and since an antigen can be virtually anything from a plant hormone such as IAA to a protein or even a nucleic acid, immunocytochemistry is an incredibly powerful and generally applicable immunochemical tool. It permits one to see virtually any substance in histologically prepared tissue sections with the convenience of light microscopy or the resolution of electron microscopy.

When I established my own laboratory at Vanderbilt University in 1969, I began independently to explore possible applications of immunological methods. Having been trained as a botanist, my lack of background in immunology prevented me from seeing at the outset the vast potential of antibodies as a research tool. The best that I could do, in my first grant proposal to NSF, was to suggest that I should make antibodies to phytochrome since "cross reactivity between phytochromes isolated from different plant sources and their antibodies should give a relative measure of the extent of homogeneity between phytochromes from different taxonomic groups and from different varieties within the same species." While we eventually did get around to pursuing the immunochemical portion of the work suggested in this initial research proposal, once we began to make antibodies to phytochrome we found even more interesting things to do with them.

One of those more interesting things happened to be the immunocytochemical visualization of phytochrome, as originally proposed by Butler. Although in 1969 we had no intentions of initiating immunocytochemical work with phytochrome, two events occurred to change our plans. Firstly, Ludwig Sternberger, a pioneer in this area, was visiting Vanderbilt University to participate in a short course in immunocytochemistry. Secondly, Richard Coleman, who was then a graduate student working in another laboratory on a cytochemical problem, was working with me part time as a technician. Given that all elements for trying to visualize phytochrome by immunocytochemistry were simultaneously available, including newly developed antibodies to phytochrome, we decided to try the method.

Success was almost immediate, and Richard Coleman became sufficiently interested in the problem to continue his graduate work in my laboratory. For the first time, we could observe phytochrome distribution throughout entire seedlings on a cell-by-cell basis. Not only did we find phytochrome where expected, such as in the tips of grass coleoptiles, but also in somewhat surprising places, such as the cap cells of roots. We also investigated the subcellular distribution of phytochrome by direct observation with both the light and the electron microscope, again obtaining invaluable information that could not have been gained by any other approach, including microspectrophotometry and fractionation of subcellular components. We were able to ask whether phytochrome could be seen in the nucleus, for example, and thereby test visually the hypothesis that phytochrome might be interacting directly with the genome of the cell. (Apart from rare exceptions, we found no phytochrome in the nucleus.) We were also able to test directly the hypothesis that phytochrome upon photoconversion to its active, far-red-absorbing form (P_{fr}) would associate with membranes. John Mackenzie,

Jr., utilizing the methodology developed by Richard Coleman, demonstrated that while the inactive, red-absorbing form of phytochrome (P_r) was distributed throughout the cytosol, P_{fr} did exhibit a discrete distribution, in at least superficial agreement with the hypothesis. As is often the case in science, however, less interesting explanations of his observations remain.

Once we began to appreciate the tremendous potential of immunocytochemistry as a research tool, I decided to sit in on a course in immunology to see what else we might do with antibodies and to determine whether they could help solve a pair of related methodological problems. At that time, about 1974, virtually all biochemical and biophysical work with phytochrome had been done with the pigment as isolated from etiolated plants, even though such plants are clearly atypical. Furthermore, considerable evidence was accumulating to indicate that green plants may contain a pool of phytochrome either not found in etiolated plants or found in only trace amounts. Study of phytochrome in and from green plants was becoming an increasingly important goal. Unfortunately, attempts to work with phytochrome from green plants had almost invariably led to failure for two reasons: (1) Green plants contained much less phytochrome than did etiolated plants, which meant that we needed more rapid and efficient methods for its purification, and (2) the chlorophyll inevitably present in green plants precluded spectral assay of phytochrome, which meant that we needed not only more sensitive but also more selective methods for its detection. We realized that antibodies were the best solution to both problems.

Robert Hunt, then a graduate student in my laboratory, was instrumental in developing the immunochemical methods that we needed. He coupled antibodies against phytochrome to agarose beads in order to prepare an immunoaffinity column to be used for phytochrome purification. With this column, he could immunopurify phytochrome in a single step within a few hours, as opposed to the days that were required for the conventional, column chromatographic procedures then in use. Moreover, the immunopurification procedure was not only more rapid, thereby minimizing time-dependent artifacts (such as proteolysis) that occurred during handling of phytochrome *in vitro,* but was also more efficient, thereby making its purification from green plants more practical. By radiolabeling some of the highly purified phytochrome that he obtained and using antibodies against phytochrome as a probe, Hunt was also able for the first time to quantitate by radioimmunoassay the exceedingly low levels of phytochrome present in crude extracts of green, light-grown plants. From such plants he needed only a few microliters of crude extract for assay, whereas the more conventional spectral assay, if it could have been performed at all, would have required at least 100 times that amount!

Of course, the immunochemical applications mentioned so far have not been the only ones that have been useful for studying phytochrome. With the help of Susan Cundiff, Harry Stone, and Maury Boeshore, all of whom worked in my laboratory as graduate students, and Marie-Michèle Cordonnier, a postdoctoral investigator, we also used antibodies to follow its appearance in developing seedlings, to characterize its fate during the course of the so-called phytochrome destruction reaction, to search for possible changes in its molecular properties *in vivo* as a function of its form, and to initiate comparative studies of phytochrome from different sources, as had been proposed originally to NSF. As was true for the immunocytochemical, immunopurification, and immunoquantitation applications mentioned previously, the resulting information could not have been obtained in most instances without the use of antibodies.

The immunochemical work mentioned so far relied solely on the use of antibodies as found in serum from rabbits injected with phytochrome. Since phytochrome is a large protein, it is a correspondingly complex antigen, possessing perhaps 10 to 20 or so **epitopes,** each of which is that small portion of an antigen against which an antibody is made. Additionally, since a single animal can make against each epitope several thousand chemically distinct antibodies, each the product of a unique cell line, an animal thus has the potential to make many thousands of different antibodies against phytochrome. Nevertheless, each animal makes from this vast repertoire "only" about 5 to 10 antibodies against each epitope, which means that it would be expected to make about 50 to 200 chemically distinct antibodies to an antigen such as phytochrome. Moreover, two animals, even though genetically identical and immunized in exactly the same way, will make a different set of antibodies. The serum of an animal, therefore, contains a large set of heterogeneous antibodies, which can be called **polyclonal antibodies** because they derive from a large number of different cell lines. Furthermore, once an animal dies, it is impossible ever again to produce the same set of antibodies. To make matters even worse, there is always the possibility that the serum of an animal may contain one or more undesirable antibodies, either by chance or because the phytochrome sample injected into the animal contained one or more impurities to which antibodies were also made. To be able to select from such a pool of polyclonal antibodies the one that is best suited to each application, to be able to produce it in limitless quantity, and to be able to produce it free of contamination by undesirable antibodies, would clearly increase by at least an order of magnitude their usefulness as a research tool.

Fortunately, within the past few years procedures have been developed to obtain unlimited quantities of homogeneous antibodies. Marie-Michèle Cordonnier, who had *(continued on next page)*

(continued from previous page)
moved to the Plant Physiology Laboratory at the University of Geneva, learned the technology required to produce these homogeneous antibodies, which are called **monoclonal antibodies** to distinguish them from the heterogeneous polyclonal antibodies present in serum. She then spent a few months in 1981 in my laboratory, which is now in the Botany Department at the University of Georgia, helping us establish and use a facility for producing monoclonal antibodies to phytochrome. Although production is complex in detail, the principle behind making monoclonal antibodies is straightforward. From the spleen of a mouse injected with an antigen such as phytochrome, one isolates cells that synthesize and secrete antibody. These cells are fused with tumor-derived myeloma cells, which can be maintained indefinitely in cell culture, yielding what are called **hybridomas.** The final step is to isolate a single hybridoma cell that secretes an antibody of interest. Since the myeloma cell gives the hybridoma immortality, one can then grow the hybridoma in large quantity in culture to produce a monoclonal cell line (i.e., one that is derived from a single parent cell) that secretes an invariant, chemically defined antibody in whatever quantity one wants over an unlimited period of time.

We now use monoclonal antibodies for all of our immunochemical work. With the help of two postdoctoral investigators, Mary Jane Saunders and David McCurdy, we have been learning which of the 25 or so presently available to us are best for immunocytochemical applications and how to use them for this purpose. Another postdoctoral investigator, Yukio Shimazaki, has developed a simpler and much more sensitive quantitative assay for phytochrome, known as an ELISA (**e**nzyme-**l**inked **i**mmuno**s**orbent **a**ssay). With the ELISA, Shimazaki can now detect less than one femtomol (10^{-15} mol) of phytochrome in a crude plant extract! He has further found that none of our presently available monoclonal antibodies, which were prepared against phytochrome from etiolated oats, will readily detect phytochrome from light-grown oats. It appears that oat plants may contain two pools of phytochrome, consistent with the intriguing possibility that there may be two genes for this chromoprotein, one expressed in the dark and the other in the light. The use of monoclonal antibodies for the study of phytochrome and its function is just beginning. Since we first produced monoclonal antibodies to phytochrome, at least five other research groups have done the same, indicating that by the time this essay appears in print, many other exciting applications of monoclonal antibodies will already have been made.

Not all of our work has required the use of antibodies, nor have antibodies been responsible for all advances in photomorphogenesis. Nevertheless, since all immunochemical methods can at least in principle be applied to the study of any antigen, and since virtually any substance found in a plant can be considered an antigen, it should be evident that antibodies are an irreplaceable and invaluable research tool for any plant physiologist.

---

hypocotyl elongation in etiolated lettuce seedlings. These results are shown in Fig. 19-10. A fairly typical HIR action spectrum was obtained, with peaks in the violet, blue, and far-red at about 720 nm. Note that red light is without effect in lettuce.

Certain other etiolated dicot seedlings show similar responses to far-red and blue and lack of response to red light, but hypocotyls of etiolated cucumber seedlings are inhibited by all three wavelengths (Fig. 19-11), blue being most inhibitory (Black and Shuttleworth, 1976). In cucumber, all these inhibitory effects are caused by light absorbed directly by the hypocotyl. The far-red inhibition deserves special attention, because it is only observed when these and seedlings of other species are young. As seedlings become older, far-red progressively loses its effect. Other species give slightly different results as to the most effective wavelengths and the age at which they are most sensitive, but it is probable that light inhibits hypocotyl growth of etiolated dicots via both cryptochrome and phytochrome

(Schäfer and Haupt, 1983). Low irradiance responses of phytochrome and the HIR system are involved. For the seedlings studied so far (mostly cultivated crops), light inhibits hypocotyl elongation but promotes leaf expansion. Similar results are obtained with pea seedlings, in which epicotyl elongation is inhibited by light and leaf expansion is enhanced.

Photomorphogenetic Effects Later in Vegetative Growth In many well-established but still growing dicots and conifers, other photomorphogenetic processes occur. (Little is known about grasses and other monocots.) If these dicots or conifers grow under a leaf canopy where the light received is primarily far-red, their stems become considerably elongated (Fig. 19-12). This light effect is, therefore, opposite to the retarding effect on elongation of etiolated seedlings mentioned above. Branching of stems is simultaneously retarded in many species under a canopy, so the plant uses more of its energy in raising the stem apex toward the top of the canopy than it does when

Table 19-1   Some Effects of Light on Etiolated White Mustard Seedlings

Inhibition of hypocotyl lengthening

Enlargement of cotyledons

Opening of the hypocotylar (*plumular*) hook

Formation of leaf primordia

Development of primary leaves

Synthesis of anthocyanin

Inhibition of translocation from the cotyledons

Increase of the rate of chlorophyll accumulation (in white light)

Unfolding of the lamina of the cotyledons

Elimination of the lag phase of chlorophyll formation (in white light)

Changes in the rate of cell respiration

Increase in the rate of ascorbic acid synthesis

Changes in the rate of degradation of storage protein

Changes in the rate of degradation of storage fat

Increase of negative geotropic reactivity of the hypocotyl

Increase in the rate of long-term protochlorophyll regeneration

Increase of protein synthesis in the cotyledons

Hair formation along the hypocotyl

Formation of tracheary elements

Decrease of RNA contents in the hypocotyl

Increase of RNA contents in the cotyledons

Differentiation of stomata in the epidermis of the cotyledons

Increase in the rate of carotenoid synthesis

Formation of plastids in the mesophyll of the cotyledons

Differentiation of mitochondria in the cotyledons

For references to these effects, see H. Mohr, 1974.

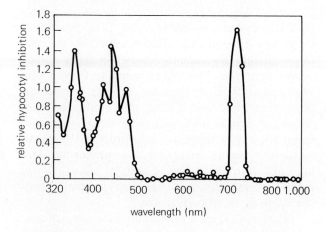

Figure 19-10   Action spectrum for inhibition of hypocotyl elongation in etiolated lettuce seedlings. Data are expressed relative to inhibition by blue light at 447 nm taken as 1.0. Light was applied continuously to the entire seedling for 18 h, starting 54 h after planting the seeds. Hypocotyl elongation was measured at the end of the light treatment. (From K. M. Hartmann, 1967.)

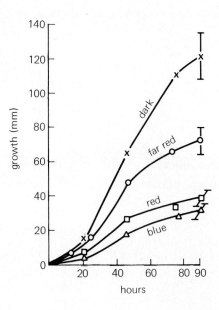

Figure 19-11   Elongation of etiolated cucumber hypocotyls (apical 1-cm section of 5-day-old seedlings) affected by continuous light of various wavelengths. Irradiance levels: blue, 3 W m^{-2}; red, 3.5 W m^{-2}; far-red, 5 W m^{-2}. (From M. Black and J. E. Shuttleworth, 1974.)

Figure 19-12   Growth of *Chenopodium album* after 21 days under two different red/far-red ratios. Both plants were grown to the three-leaf stage under identical conditions; the one on the right was then provided light enriched in far-red. The estimated ratios of ϕ in the two plants were 0.71 (left) and 0.38 (right). Each plant received the same amount of photosynthetically active radiation (400–700 nm). (From D. C. Morgan and H. Smith, 1976.)

Figure 19-13 Growth of Douglas fir *(Pseudotsuga menziesii* [Mirb.] Franco) after 12 months on photoperiods of 12 h (left), 12 h plus a 1-h interruption near the middle of the dark period (middle), and 20 h (right). (From R. J. Downs, 1962.)

unshaded. In agricultural crops planted in rows, plants in the exposed outer rows are often shorter and more highly-branched than are those within the field because of this effect. A similar phenomenon is often seen with plants on greenhouse benches. In thick strands of lodgepole pine, such as are abundant in Yellowstone National Park in Wyoming and many other mountainous areas of the Northwestern United States, the result of retarded branching and death of branches in reduced light is a forest of plants having long, straight trunks that provide excellent timber because they are relatively knot-free. The principle is used in selection of distances between transplanted seedlings in reforestation work. Plants that do not elongate in response to increased far-red radiation (relative to red) are those that normally grow in the shade of others (e.g., on the forest floor). These shade plants seem to be adapted to an environment in which it is impossible to elevate their leaves above the overhead canopy by stem elongation (Morgan and Smith, 1979, 1981; Morgan, 1981).

Branching at the bases of grass stems (called **tillering**) is also controlled in part by the phytochrome system. Tiller (branch) formation is retarded in thick pastures, because the light transmitted to the stem bases is rich in far-red, causing low ϕ values there. Increasing the ϕ value with lights rich in red wavelengths promotes tillering, just as it promotes branching in dicots (Deregibus et al., 1983).

All the effects of far-red light described in this section have been assumed to be caused by decreasing the level of P_{fr} compared with that in unshaded plants, but essentially nothing is known about cycling rates of P_r and P_{fr}. *Results suggest that a major function of phytochrome in nature is to detect mutual shading and to modify growth accordingly.*

19.6 Photoperiodic Effects of Light

In many species, responses to light, especially light absorbed by phytochrome, are influenced by the time of day in which light is received. The effects of light in interrupting the normal dark period or prolonging the normal period of daylight are referred to as **photoperiodic effects** (Vince-Prue, 1975; Vince-Prue and Canham, 1983). These responses concern mainly bud dormancy of perennial plants and production of flowers and seeds by perennials and especially non-perennials. They are described in detail in Chapters 21 and 22. In Fig. 19-13 we illustrate the importance of day length in controlling bud dormancy (and therefore overall growth) of Douglas fir, a conifer. In general, long days promote elongation of stems of most species and short days lead to the changes associated with autumn (e.g., dormancy and frost-hardiness of buds).

19.7 Light-Enhanced Synthesis of Anthocyanins and Other Flavonoid Pigments

Most plants form anthocyanin pigments and other flavonoids in specialized cells of one or more of their organs, and this process is frequently promoted by light. A simple example is the faster development of the red color resulting from an anthocyanin in apple fruits on the south than on the north side of a tree. Production of flavonoids requires sugars as a source of the phosphoenolpyruvate and erythrose-4-P (Sections 10.2 and 12.10) that provide carbon atoms needed for the B ring and as a source of acetate units needed for the A ring (Section 14.8). Sugars, especially sucrose, can arise from degradation of starch or fat in storage organs during seedling development or from photosynthesis in chlorophyll-containing cells. It is no surprise, therefore, that anthocyanin synthesis is increased by light acting photosynthetically in leaves or green apple fruit skins, yet light promotes synthesis of these pigments in organs that photosynthesize little or not at all, including autumn leaves, flower petals, and etiolated seedlings, showing that at least one other pigment participates.

The action spectra for anthocyanin production in several species are shown in Fig. 19-14. In general, maximum responses occur in the red, far-red, and blue regions, whereas green (approximately 550 nm) is almost without effect (Mancinelli, 1983). The peaks in the yellow, orange, red, and far-red regions vary considerably both in wavelength and height in various species. Blue is effective in nearly all, and in sorghum red and far-red are ineffective. A detailed action spectrum in the blue region for synthesis of the aromatic acid precursors of the B ring in gherkin (*Cucumus sativus*) hypocotyls (Smith, 1972) is similar to that of phototropism shown in Fig. 18-9 and to the blue inhibition of hypocotyl elongation in lettuce seedlings shown in Fig. 19-10, suggesting that effective blue wavelengths are absorbed primarily by cryptochrome. Red and far-red wavelengths act independently of photosynthesis in etiolated seedlings, but in green apple skins photosynthesis also contributes. High irradiance levels characteristic of the HIR system are required for these red and far-red effects, and both phytochrome and cryptochrome are probable photoreceptors in most species. Mancinelli (1983) described numerous species and the probable photoreceptors involved.

Numerous attempts to determine the site or sites of light action in biochemical pathways leading to both the A and B rings of flavonoids have been made. The accumulation of flavonoids in many autumn leaves during senescence suggests a relation between protein hydrolysis, phenylalanine appearance, and the use of phenylalanine in ring B formation. Because phenylalanine can be used in various metabolic pathways, control by light of the first step in its conversion to ring B was suspected. This step requires the enzyme phenylalanine ammonia lyase (Chapter 14, R14-5), and light does promote its activity in various organs of many plants (Wong, 1976; Mancinelli, 1983). Nevertheless, several other flavonoid-synthesizing enzymes not mentioned in this book also exhibit increased activity after light treatment (Hahlbrock and Griseback, 1975), indicating that the production of both rings occurs more rapidly in light. As a result, no specific influence of light on a universal rate-controlling reaction of flavonoid synthesis can be identified.

Like the flavonoids, lignins are also formed from the shikimic acid pathway with the participation of phenylalanine ammonia lyase. In seedlings or in immature parts of older plants undergoing xylem differentiation or formation of xylem from the vascular cambium, lignin formation in xylem cell walls is promoted by light. This is partly responsible for the greater stiffness of seedlings grown in light than in darkness.

19.8 Effects of Light on Chloroplast Arrangements

When irradiance levels are high, chloroplasts are usually aligned along radial walls of the cells, becoming shaded by each other against light damage. In weak light and often in darkness, they are separated into two groups distributed along the walls

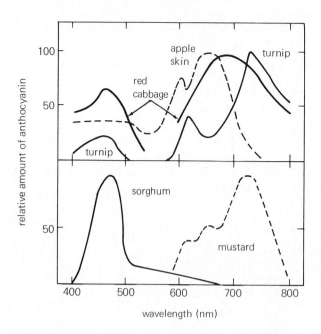

Figure 19-14 Action spectra for anthocyanin formation in various species after prolonged irradiation. The apple fruits contained chlorophyll, but the turnip, cabbage, sorghum, and mustard seedlings were probably chlorophyll free when irradiance began. (Data redrawn from various sources. Apple skin data are from H. W. Siegelman and S. B. Hendricks, 1958; red cabbage and turnip, H. W. Siegelman and S. B. Hendricks, 1957; sorghum, R. J. Downs and H. W. Siegelman, 1963; mustard, H. Mohr, 1957.)

nearest to and farthest from the light source, thereby maximizing light absorption. This movement of plastids depends upon the direction of light as well as its irradiance level and is an example of **phototaxis** (movement of an entire organism or organelle in response to light; Haupt, 1966). In mosses and angiosperms, phototactic responses to both low and high irradiances are maximal under blue wavelengths, and phytochrome does not participate (Inoue and Shibata, 1973). Action spectra suggest that cryptochrome is again involved. In a few algae, however, phytochrome does absorb low irradiance light responsible for movement of the chloroplast or chloroplasts to regions of the cells in which light absorption is increased. In the green alga *Mougeotia*, the effective phytochrome molecules are located in the plasma membrane. In all species, the chloroplast itself does not absorb the light causing its phototaxis. Instead, light absorbed elsewhere in the cell causes chloroplast movements through effects on cytoplasmic streaming, which result from interactions between microfilaments and microtubules. Ecologically, chloroplast movements seem important mainly to increase light absorption at low irradiances and to decrease absorption when irradiances are so high that they might cause solarization (Section 11.3) or other photodestructive effects.

20

The Biological Clock

Change is the only thing that an organism can count on in its environment. Almost nothing is really constant. In a study of the alpine tundra on the north end of Rocky Mountain National Park in Colorado (Salisbury et al., 1968), several kinds of environmental changes appeared: Wind velocities changed significantly in less than a second. Temperatures, light levels, and humidities sometimes changed radically in time intervals from minutes to perhaps 5 or 6 hours. All these and other changes were superimposed on a daily **(diurnal)** cycle. Weather cycles typically lasted several days. In one summer, for example, heavy storms were separated by intervals of from 10 to 22 days, with an average of about 13. Exceptionally clear days were separated by intervals of 5 to 14 days, with an average of about 10. The **annual** seasonal cycle was always evident.

Furthermore, weather trends may be related to the 11-year sun spot cycle, and long-time climatic changes occur over periods of centuries to millenia. **Tidal (lunar)** cycles control the tidal zone environment and could be important in other habitats as well.

It would be to an organism's advantage to anticipate and adjust to these environmental changes. At least three of them—those related to the mechanics of the solar system—are regular enough that this should be possible: the diurnal, lunar (tidal), and annual cycles. For an organism to anticipate and prepare for these regular changes in its environment, it needs a clock and various associated mechanisms. The clock's system should have at least two broad sets of characteristics.

First, it should be accurate. It should not be unduly influenced by capricious elements of an organism's environment, those that cannot accurately be predicted: temperature, light levels during the day (which vary because of clouds and shading), wind velocity, moisture, and so on. Even if a biological

clock were not highly sensitive to these factors, which would be surprising and impressive, it would be even more impressive if such a clock could run with the accuracy achieved by our own mechanical and electronic clock systems. Yet without such accuracy, it might soon get out of phase with the environment and, therefore, be of no benefit to the organism.

There is an alternative to inherent accuracy: The biological clock might frequently and regularly be reset by some dependable feature of the organism's environment. It wouldn't matter much, for example, if the clock were to gain or lose an hour or two a day, if it were reset each day at sunrise or sunset or both. Presumably, the clock might have one status appropriate for daytime and another for night; resetting at dawn and dusk might keep the clock's status in synchronization with the environment. Incidentally, it is conceivable that a "clock" might simply be driven by changing features in the environment. In this case, the organism would only track environmental changes without actually keeping time or anticipating future environmental events.

Second, there must be mechanisms that allow the plant or animal to take advantage of the clock's timekeeping. It is reasonable, for example, that a plant might conserve energy if it could direct and concentrate its available resources to the photosynthetic mechanism during the day and to other metabolic mechanisms during the night. Organisms that live where nights are cooler than days might adjust their temperature optima for critical metabolic processes accordingly. This requires a coupling system so that the clock can control the available resources within the cell, depending upon the time of day. Probably the clock controls metabolism in general and possibly many other features that might function more efficiently when time is measured and events are anticipated.

With these things in mind, we might go to

Biorhythms and Other Pseudoscience

There is talk about biorhythms. The notion is that human behavior is controlled by three cycles, each of which is initiated at the moment of birth: a physical cycle of 23 days, a sensitivity (emotional) cycle of 28 days, and an intellectual cycle of 33 days (Mackenzie, 1973; Thommen, 1973). The first half of each cycle is supposed to be the time when one is most positive in the attribute of that cycle; during the second half, one is supposed to be negative. The days on which there is a crossover from plus to minus or minus to plus are critical days, and if the critical points for two or three cycles fall on the same day (which happens about six times a year for two cycles and once for three), you had better watch out!

The concept was developed from about 1897 to 1932 by certain medical doctors and others in Vienna, Berlin, Innsbruck, Philadelphia, and elsewhere. It is usually presented as a "scientific" doctrine. Some companies such as the Ohmi Railway Company in Japan have calculated the cycles for their employees, warning them of critical days. The accident rate is reported to have dropped by more than 50 percent!

Yet there is little objective evidence to support the hypothesis (Rogers et al., 1974). Most evidence is anecdotal: Such and such movie star is reported to have had a serious accident on a triple critical day! Many adherents to the doctrine swear by it. But, of course, their results might well be examples of the self-fulfilling prophecy: After you have plotted your charts for several months in advance, you will be expecting good and bad days, and subconsciously or otherwise you may adjust your life to meet these expectations. If you keep a careful diary, you might test the theory by plotting your charts for a past interval covered by the diary and looking to see if anything special happened on the good or bad days—but to be objective, you must also note special things that happened on the other days as well.

It should be clear from the discussions in this chapter that the basic premise on which the concept of biorhythms rests has no foundation in scientific observation. Rhythms clearly exist in organisms, including humans, but they have three features at total variance with those of the biorhythm hypothesis: They are typically *circa*, approximating but almost always varying from an exact period length unless they are continually entrained to a cycling environment;

they often vary from individual to individual within a species; and they are relatively easy to shift by various environmental factors. Their periods are plastic and not rigid. The rhythms discussed in this chapter could never maintain exact periods of 23, 28, or 33 days from the time of birth throughout the proverbial four score years and ten of an individual life—the same for everyone.

Speaking of matters pseudoscientific, we find that the botanical sciences have spawned their share. There are those who suggest that talking to your plants, praying over them, or singing to them makes them grow better (and perhaps it does, if you thereby increase the CO_2 concentration around them or take better care of them). There have been several papers purporting that music makes plants grow better. There are even special records on the market, claiming to provide the best music for plants. (This needs much work, but it is possible that sound waves might vibrate the cellular organelles and influence plant growth—but classical music and not rock music?!)

Probably the most notice has been paid to the "experiments" in which a polygraph (lie detector) that was attached to a plant registered wild responses when bad things happened, such as another plant being "murdered" in the same room or brine shrimp being dunked in boiling water. Do plants really have feelings and emotions?

If so, it surely remains to be demonstrated. The "experiments" apparently worked once, but no one has been able to repeat them consistently. And consistent, objective verification is what we must demand in science (Galston, 1974). Truly, progress often depends on startling and unexpected discoveries; but it is as common for such claimed discoveries to be mistaken interpretations or due to poorly designed experiments as for them to be advances in knowledge. We are entitled, even obligated, to test all claims by insisting upon verification by objective observers who thoroughly understand the role of controls in an experiment and who understand all the factors that might influence the outcome. For example, the polygraph responses observed when "murder" was perpetrated near the plant attached to the machine were at about the same level as the "noise" to be expected if the polygraph were attached to an inanimate object. Could the results have been due to coincidence? Of course, and that is likely.

---

nature to observe phenomena that could be manifestations of biological time measurement. Finding such, we could study their features and the roles that they might play in the organism's existence. Indeed, many such phenomena have been observed, and much is known about their manifestations. It appears that virtually all eukaryotic organisms do have biological clocks. Sometimes it is easy to see clock con-

trol of metabolism and activity. There are also examples of highly sophisticated clock responses that we might not have expected.

Confronted with these observations, we immediately ask: What is the mechanism of the clock? How does it work? The observed phenomena imply a few things about the nature of the clock, but at the moment little is known about actual clock

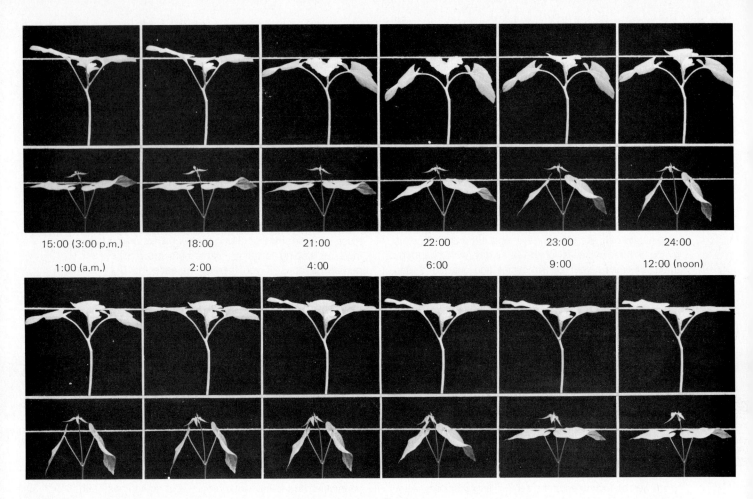

| 15:00 (3:00 p.m.) | 18:00 | 21:00 | 22:00 | 23:00 | 24:00 |
| 1:00 (a.m.) | 2:00 | 4:00 | 6:00 | 9:00 | 12:00 (noon) |

Figure 20-1 Leaf movement in cocklebur (top row) and bean (second row). Plants were photographed at hourly intervals from noon to noon. Twelve photographs were selected for the figure. The bean leaves drop more sharply and somewhat later than the cocklebur leaves. (Photographs by F. B. Salisbury.)

mechanisms. (Several reviews have been written, including Bünning, 1973, 1977; Hillman, 1976; Luce, 1971; Moore-Ede et al., 1982; Satter and Galston, 1981; and Sweeney, 1969, 1983.)

20.1 Endogenous or Exogenous?

As illustrated for two plants in Fig. 20-1, the leaves of many species exhibit one position during the daytime (typically nearly horizontal) and another position in the middle of the night (typically nearly vertical). This observation was made at least as early as 400 B.C.E. by Androsthenes, who was the historian of Alexander the Great.

In 1729, the French astronomer DeMairan was perceptive enough to recognize a fundamental problem in relation to this diurnal cycle of leaf movements in plants. He wondered if the movements were driven by changes in the environment (the daily light-dark cycle), or if they might be controlled by some time-measuring system within the plant. If leaves moved only in response to *external* changes, we could say that timing was **exogenous;** if in response to an *internal* clock, we would say timing was

endogenous. Using the sensitive plant (*Mimosa* sp.), DeMairan observed the movements even after the plants had been placed in deep shade. Since these motions did not require intense sunlight during part of the 24-hour cycle, he suggested that the movements were endogenously controlled.

A few other early workers, including Charles Darwin and Julius von Sachs, were also interested in these rhythms and published preliminary studies relating to them. But the early investigator who probably devoted the most time and effort to this topic was Wilhelm Pfeffer, who from 1875 to 1915 wrote many papers about the leaf movements of the common bean plant (*Phaseolus vulgaris*). (Pfeffer's work on osmosis is mentioned in Chapter 2.) Much of his extensive work is still of interest (Bünning, 1977). When he began, he was skeptical of an endogenous clock, but by the end of his researches he became convinced that such a clock must exist. Ironically, he was unable to provide experimental data sufficiently convincing to convert his contemporaries. During Pfeffer's time, zoologists were also observing and reporting rhythms in animals, especially diurnal rhythms of activity.

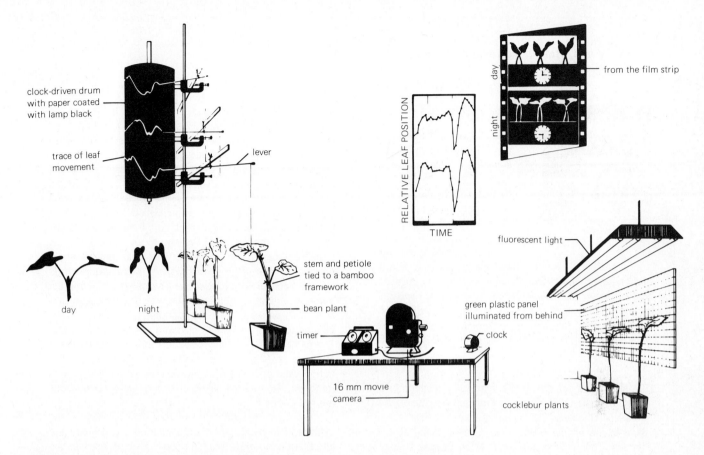

Figure 20-2 Two methods that can be used to record leaf movements. Left, the classical kymograph (clock-driven drum) method of Pfeffer and Bünning; right, a time-lapse photography method. Night photographs in the time-lapse method are time exposures with plants silhouetted against a dim green background, but infrared-sensitive film has been used with infrared light.

The real breakthrough came in the 1920s. Rose Stoppel in Hamburg, Germany had continued the researches of Pfeffer on leaf movements in bean plants. Using the method developed by Pfeffer (Fig. 20-2), she attached the bean-leaf blade (after the stem and the petiole had been fastened to bamboo sticks) with a thread to a lever contacting a moving drum that had been coated with lamp black. The lever traced a record of the leaf movements. (Note that when the leaf is up, the end of the lever tracing the record is down.)

Stoppel observed that when leaf movements were measured in a dark room at constant temperatures, the maximum vertical position was observed at the same time each day. The **period** is the time between the occurrence of a recognizable point on the cycle (e.g., the maximum vertical position) and its occurrence in the next cycle. Since the period was almost exactly 24 hours, Stoppel reasoned (as we have in the introduction above) that a biological clock could probably not be this accurate; some factor in the environment must be driving the clock by resetting it on a daily schedule. Since the plants were in the dark and at constant temperature, this factor

could not be daylight or temperature. Stoppel called it *factor X*.

Two young botanists in Frankfurt, Erwin Bünning and Kurt Stern, were looking for a research problem involving subtle physical factors of the environment, such as the ionic content of the atmosphere. They thought that such a factor might time the leaf movements in Stoppel's experiments. They found that atmospheric ions had no effect on the rhythms, but they identified Stoppel's factor X as the red light she used while watering her plants. When this factor was eliminated, Stoppel's prediction held true: The maximum vertical position of the leaves came about an hour and a half later each day, so the leaf movement cycle was soon out of phase with day and night outside the dark room. Bünning tells this story in his personal essay in this chapter.

The **free-running period** is the period of a rhythm under constant conditions. Since the free-running period was greater than 24 hours, there seemed to be no alternative to an endogenous clock. The rhythms were not simply tracking the normal day and night cycle; they continued in the absence of day and night and at constant temperature, but

Potato Cellars, Trains, and Dreams: Discovering the Biological Clock

F.B.S. 1963

Erwin Bünning

As we've seen in the chapter, it was Erwin Bünning as a post-doctoral fellow, along with several colleagues, who discovered the free-running, circadian nature of the biological clock in plants, a discovery that made endogenous timing seem to be the only acceptable explanation. In 1963–64, one of us [F.B.S.] spent a sabbatical year with Bünning in Tübingen, where he was Director of the Botanical Institute, and had the opportunity to hear him tell of his early work. This is Salisbury's translation of a letter from Bünning, sent in response to a request that he record some of his experiences. The letter is dated June 2, 1970.

The story went something like this. At the Institute for the Physical Basis of Medicine in Frankfurt, the biophysicist Professor Dessauer (an x-ray specialist) became interested in the effects of the ionic content of the air upon humans. Those were the years when people began to be interested in atmospheric electricity, cosmic rays, and so on. Naturally, humans could not be used as experimental objects, and so in 1928 Dessauer searched for botanists to work on plants. One whom he found was Kurt Stern, who lived in Frankfurt; the other was me, who had just finished my doctoral work in Berlin. So we began in

August of 1928 to contemplate the problem. In the process we came upon the work of Rose Stoppel, who had been studying the diurnally periodic movements of *Phaseolus* (common bean) leaves. In the process she had found, as had several other authors, that under "constant" conditions in the darkroom, most leaves reached the maximum extent of their sinking (their maximum night position) at the same time, namely, between 3:00 and 4:00 A.M. Her conclusion: Some unknown factor synchronized the movement. Could this be atmospheric ions? We had Rose Stoppel visit us from Hamburg for two or three weeks so that we could become familiar with her techniques. In our group we always called her *die Stoppelrose* ("stubble rose"), and the name was most appropriate. She was energetic and persistent, so persistent that she just died this January in her 96th year. Her results also appeared in our experiments: The night position usually occurred at the indicated time. Then we investigated the effects of air that had been enriched with ions or air from which all the ions had been removed. The result: Nothing changed—atmospheric ions are not "factor x."

We then decided that our research facilities at the Institute were insufficient. Hence, after *die Stoppelrose* had left, Stern and I moved to his potato cellar, where with the help of a thermostat, we obtained rather constant temperature. Contrary to the practice of Stoppel, who turned on a red "safe" light to water her plants, we went into the cellar just once a day with a very weak flashlight and felt around with our fingers for the pots and recording apparatus so that we could water the plants and so that we could see if

drifted out of phase with outside day and night. Normally, they were reset or entrained to the natural cycle, probably by dawn or dusk or both. The free-running clock betrayed its inaccuracy and thus displayed its endogenous nature.

In the 1950s, Franz Halberg (see 1959 reference) at the University of Minnesota suggested that rhythms with a free-running period of approximately but not exactly 24 hours should be called **circadian.** This term is coined from Latin *circa,* which means "approximately" and from *dies,* or "day."*

---

*Pfeffer, and even A. P. DeCandolla in 1832, had observed rhythms with circadian free-running periods, but they had not attached proper significance to their observations. Yet Antonia Kleinhoonte, working independently in Delft, Holland, arrived at exactly the same conclusions as those of Bünning, Stern, and Stoppel. Results of her and their experiments were published between 1928 and 1932. Stern later emigrated to America and lost interest in the biological clock; Kleinhoonte also lost interest. Bünning, however, continued his studies and is still active despite his retirement.

20.2 Circadian and Other Rhythms

There are several areas of modern knowledge that clearly qualify as *biology* rather than *botany* or *zoology.* Examples are genetics, the pathways of cellular respiration, and the cell theory. Biological clocks and circadian rhythms occur in protistans, fungi, plants, and animals, although they may not occur in monerans (prokaryotes). Rhythms in certain bacteria have been reported but only under cycles of alternating light and dark. Many attempts to discover rhythms in prokaryotes have had negative results.

Several rhythms have been studied in single-celled organisms. Phototaxis in the green alga *Euglena* and a mating reaction in *Paramecium* are good examples. The biological clock has been thoroughly studied in *Gonyaulax polyedra,* a marine dinoflagellate. Beatrice Sweeney (see her personal essay in this chapter) and J. Woodland Hastings, then located at

everything was in order with the recorders. The flashlight was weakened with a dark red filter so that one could see only for a few centimeters' distance. In those days it was the dogma of all botanical textbooks that red light had absolutely no influence upon plant movements or upon photomorphogenesis. We did one other thing differently from Stoppel. Since Kurt Stern's house was a long way from our laboratory, we didn't make our daily control visit in the morning, rather only in the afternoon. The result: Most of the maximal night positions no longer appeared between 3:00 and 4:00 A.M., but rather between 10:00 and 12:00 A.M. Hence we concluded: The dogma is false. Red light must synchronize the movements so that a night position always appears about 16 h after the light's action. That was "factor x." When we eliminated this hardly visible red light, we found that the leaf movement period was no longer exactly 24 h but 25.4 h. [This circadian feature of the clock was the key to understanding its endogenous nature—Ed.]

So that was about how that story went. Naturally, I could also tell you the story about how I came upon the significance of the endogenous rhythms for photoperiodism. That was about in 1934. Of course, I had already often asked myself how such an endogenous rhythm might ever have any selection value (in evolution), and I had already expressed the opinion in 1932 in a publication (Jahrbuch der Wissenschaftlichen Botanik 77: 283-320) that some interaction between the internal plant rhythms and the external environmental rhythms must be of significance for plant development. But there were co-incidences in the story. As a young scientist, one naturally had to allow himself to be seen by the power-wielding people of his field, so that he might receive invitations for promotion. Hence, I traveled in 1934 from Jena to Könisberg in Berlin to introduce myself to the great Professor Kurt Noack. We discussed this and that. He mentioned that discoveries were being made that were so remarkable one simply couldn't believe in them. Such a one, for example, was photoperiodism. He, as a specialist in the field of photosynthesis, must certainly know that it makes no difference whatsoever what program is followed in giving the plant the necessary quantities of light. Then, as I was riding back on the train, the idea came to me— aha, for the plant it does make a difference at which time light is applied, if not exactly in photosynthesis, nevertheless for its development!

I could present a third story, one from very recent years. I have long felt that the daily leaf movement rhythms had no selection value in themselves. As you know, I have recently changed my mind. The movements could indeed be important in avoiding a disturbance of photoperiodic time measurement by moonlight. Before I had begun to test the idea experimentally or even to think about it, it came very simply one night in a dream. The dream (apparently as a memory of one of my visits to the tropics): tropical midnight, the full moon high above on the zenith, in front of me a field of soybeans, the leaves, however, not sunken in the night position, rather broadly horizontal in the day position. My thoughts (in the dream): How shall these plants know that this is not a long day? They had better hide themselves from the moon if they want to flower.

---

the Scripps Institute of Oceanography at La Jolla, California, were the first to work intensively with this organism. They documented three separate rhythms. Most spectacular is a rhythm of **bioluminescence** observable when a suspension of *Gonyaulax* cells is tapped or otherwise jarred, causing them to emit light. The quantity of light they emit follows a circadian rhythm, with the peak of the rhythm normally occurring near midnight. Johnson et al. (1984) have shown that the bioluminescence rhythm matches a rhythm in luciferase, the enzyme that causes emission of the light. There is also a rhythm in cell division, with the maximum occurring near dawn. The third rhythm is in photosynthesis. The quantity of CO_2 fixed under standard test conditions varies according to a circadian rhythm, with the maximum usually occurring near noon; that is, the photosynthetic mechanism is adjusted by the clock to anticipate the environment, just as we suggested in the introduction that it might be. It is interesting to note that we are dealing here with a population of organisms, rather than with individuals as in Bünning's original studies.

Among the fungi, there is a circadian rhythm in formation of conidiospores in *Neurospora crassa* exhibited as a series of dark bands (one per day) in the mycelia growing on agar from one end to the other of a long culture tube. This rhythm is of special interest because a number of *Neurospora* clock mutants are known. Another fungal example is the rhythm of spore discharge in *Pilobolus*.

Many rhythms besides leaf movements have been observed in higher plants. These include petal movements, rates of growth of various organs, concentrations of pigments, stomatal opening and closing, discharge of perfume from flowers, times of cell division, metabolic activity (e.g., photosynthesis and respiration), and even the volume of the nucleus. Of special significance could be the observation that many plant species exhibit a diurnal rhythm in sensitivity of response to certain environmental factors, notably temperature. Many species flower or grow well only when temperatures during the part of the cycle that normally comes at night (subjective night) are lower than temperatures during subjective day. Or light given during subjective night may actually inhibit some plant responses. Thus the clock adjusts a plant's metabolism to coincide with its cycling environment.

Several insects exhibit rhythms that are convenient for study (Saunders, 1976). The time of day at which adult *Drosophila* emerge from the pupae follows a circadian rhythm and provides an intensively studied example, as does the activity of the cockroach. **Activity** or **running cycles** have also been studied in birds and rodents. These rhythms are useful because they often continue under constant environmental conditions for long intervals (months or even years), and they are relatively easy to study. Various automatic devices can be installed in the cage, allowing for continuous recording of an organism's activity. For example, a microswitch attached to the perch can indicate bird activity, and monitored running cages are often used for rodents.

It is important to realize that the rhythms with obvious outward manifestations (leaf and petal movements, activities, and others) may be less important than the internal metabolic changes controlled by the clock. This is true for plants, as mentioned earlier, and many metabolic and physiological cycles (e.g., potassium levels in the blood, urine excretion, and body temperature) have also been documented in animals. These are undoubtedly important in attaining an adjustment between an organism and its environment. A striking observation that may help us realize how subtle these adjustments can be is that an organism may be far more sensitive to toxic chemicals or even to ionizing radiation (particularly x rays) during a part of its circadian cycle. Perhaps animals conserve energy during the inactive part of their cycle by lowering resistance to factors not likely to be encountered. *In any case, anyone doing biological experiments should be aware of the profound effects of the physiological clock on virtually all aspects of an organism's functions.* The time when a treatment is given is often decisive.

Many circadian cycles have also been studied in humans, although in some ways humans are difficult subjects—their cycles may be timed by such factors as a wrist watch! (Social time cues are generally more important in synchronizing the human circadian system than are the environmental factors that time the rhythms in other organisms; see Aschoff et al., 1975; Luce, 1971; Sulzman, 1983.) Nevertheless, detailed studies in special bunkers near Münich, Germany indicate that human circadian rhythms (e.g., sleep versus activity, urinary excretion, temperature, and heart rate) follow the same general principles that apply to the rhythms of other organisms. It was discovered with humans and later confirmed with squirrel monkeys that rhythms with different periods could exist in a single individual and that the different cycles were controlled by at least two separate clocks or pacemakers (Sulzman, 1983). For example, body temperature, urine volume, and urinary potassium excretion all had a free-running period of 25 hours (common also for human activity rhythms), while activity and urinary calcium excretion both had a period of 33 hours in the individual being studied.

Although much study remains to be done, many biologists are intrigued with noncircadian rhythms. Short cycles (minutes to hours) are called **ultradian.** Changes in metabolic components provide examples. Color, activity, and metabolism of fiddler crabs and other organisms have shown cycles in the laboratory that are closely matched to the tides of the bay where the organisms were collected. Lunar rhythms (called **circalunar**) are closely related to tidal rhythms (Palmer, 1975). For example, the grunion (*Leuresthes tenuis*), a small fish living off the coast of Southern California, spawns from late February to early September during three to four nights at the new and the full moon and during the descending tidal series. Rhythms related only to the moon have also been observed. Because of the 28-day period, it has been suggested that the human menstrual cycle might be a circalunar rhythm, but present evidence is against this. The length of the cycle depends on several factors (e.g., it shortens as menopause approaches) that have nothing to do with the lunar cycle.

In certain ground squirrels held under constant conditions, entrance into and termination of hibernation have been shown to follow a rhythm of about a year (circannual rhythm). The amount of daily wheel-running activity of these animals also follows such a cycle. Germination of certain seeds appears to be best at certain times during the year, even though the seeds have been stored under conditions of constant temperature, light, and moisture. Spruyt et al. (1983) have documented an annual rhythm in the sensitivity of dark-grown bean seedlings to irradiation with red light. Opening of the epicotyl hook (see Chapter 19) was maximal between February and June (twice as sensitive as in December), although seeds were germinated under identical conditions. Pigment synthesis alternated in sensitivity with hook opening. Much more work must be done before all the implications of these apparent circannual rhythms are understood.

20.3 Basic Concepts and Terminology

In discussions of biological rhythms, it is helpful to use the terminology applied to physical oscillating systems, although this terminology is sometimes used in a rather special sense in relation to the rhythms. The oscillations may be thought of as having three characteristics (Fig. 20-3): First, the **period** is the time between comparable points on the repeating cycles. Typically, the maxima of the curves are observed and measured, because they show the sharpest changes in slope. Sometimes minima provide a more accurate measurement, or some other point on the cycle might be considered. The term **phase** is used in a specialized sense as any point on a cycle, recognizable by its relationship to the rest of the cycle. The most obvious phase point on a cycle (e.g., a maximum) is called the **acrophase.** Hence, the period is the time between acrophases. In a more general sense, the term *phase* may mean a recognizable portion of a cycle; for example, the part that normally occurs during the light period, the so-called photophil phase.

Second, the **amplitude** is the extent to which the observed response varies from the **mean** (Fig. 20-3). The **range** is the difference between the maximum and the minimum values. Third, one might consider the **pattern** of the cycle. Usually, the common sine wave comes to mind (as in the bioluminescence rhythms of Fig. 20-3), but there are many variations. A sharp maximum might be accompanied by a broad minimum, for example, or the slope of the curve approaching the maximum might be steep while that approaching the minimum might be less so.

When plants or animals are exposed to an environment that fluctuates according to some period, and the rhythms exhibit the same period, they are said to be **entrained** to the environment rather than free-running. As we shall discuss later, this entrainment to the environment can be brought about by several factors, particularly an oscillating light environment, with its **dawn** and **dusk.** Such an entraining environmental cycle is called a **synchronizer** or **Zeitgeber,** a German word meaning *timegiver.* The term *entrainment* is used when the Zeitgeber is a fluctuating environment with several regular cycles. If an environmental stimulus is given only once (e.g., a single flash of light), and the acrophase of the rhythms is shifted in response to it, the rhythm is said to have been **phase-shifted** or **rephased.**

20.4 Rhythm Characteristics: Light

Many investigators have expended much effort in obtaining data relating to the biological clock, especially as it is exhibited by circadian rhythms. Far too much detail has accumulated for discussion here. We shall be able only to consider a few effects of light, temperature, and applied chemicals.

With the discovery that the rhythms had free-running periods not exactly equal to 24 h, it became apparent that they must be entrained by the external

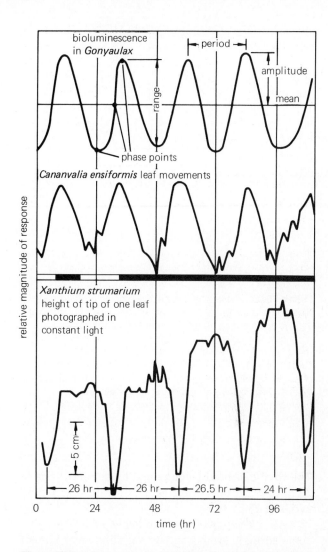

Figure 20-3   Some representative data for various circadian rhythms. Top, bioluminescence intensity in *Gonyaulax* measured for plants kept under constant conditions of dim light. Characteristics of circadian cycles are indicated. Middle, leaf movements of *Cananvalia ensiformis* recorded on a kymograph, so high points on the graph indicate low leaf positions. Light and dark conditions are indicated by the bar. Note gradual shift of the peak during darkness as the cycles progress. Bottom, leaf movement of cocklebur (*Xanthium strumarium*) recorded by the time-lapse photography method illustrated in Fig. 20-2. High points indicate high leaf positions. Period lengths between the troughs are indicated. Note increase in absolute height of the leaves, particularly at the peaks, but also at the troughs. This is largely caused by growth of the stem during the course of the experiment, but the increase in range of leaf movement is also apparent. Light was entirely from fluorescent lamps.

environment to account for the normal 24-h periodicity. The work of Bünning and Stern indicated that entraining factors might be as subtle as a weak red light; hence, light was of obvious interest as a possible Zeitgeber.

One approach was to see if the rhythms could be entrained to some light–dark cycle other than a 24-h one. It was readily apparent that this could be done.

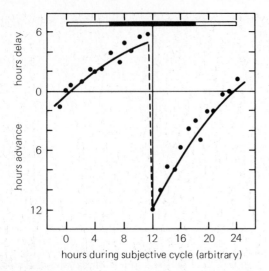

Figure 20-4 Phase shift in petal movements of *Kalanchoe blossfeldiana* following 2-h exposures to orange light given at various times during an extended period of continuous darkness. Bar at the top indicates subjective status of the rhythm; that is, dark part of the bar indicates petal closure or subjective night. As light interruption approaches the middle of subjective night, there is an increasing delay; following the middle of subjective night, there is an advance, which decreases as subjective day is approached. (Data after Zimmer, 1962.)

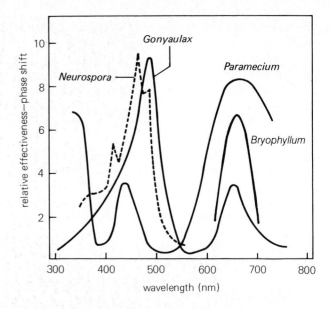

Figure 20-5 Approximate action spectra for phase shifting in the rhythms of *Neurospora, Gonyaulax, Paramecium*, and *Bryophyllum*. (From Ehret, 1960; Munoz and Butler, 1975; and Wilkins and Harris, 1975.)

The rhythms could be entrained by shorter cycles of 20 to 22 h (in rare cases, even 10 to 16 h) or longer cycles of 28 to 38 h.

Another approach was to allow a rhythm to become strongly established by a cycling environment and then to let it run free under constant environmental conditions. In constant darkness, a brief interruption of light was then given at various times during the free-running rhythm. With most organisms, when the flash of light was given during subjective day, there was virtually no effect upon the rhythm. That is, if light comes during the phases typical of day in a natural cycling environment, the following phases of the cycle are not much influenced. When the light interruption comes during early subjective night, however, the rhythm is typically *delayed* (i.e., an acrophase comes later than would have been expected). It is as though the flash of light were acting as *dusk*, but by coming later, it caused a delay. As the light flash is given later and later during subjective night, the extent of the delay increases until a certain point is reached at which the flash of light suddenly results in an *advance* of the rhythm rather than in a delay (i.e., an acrophase comes earlier than expected). The flash of light is acting as *dawn* rather than as dusk (Fig. 20-4).

By carefully studying curves such as those in Fig. 20-4, and knowing the amount of delay or advance caused by a light flash given during otherwise con-

stant conditions at various times during a free-running cycle, it is possible to account for the phenomenon of entrainment. That is, considering the advancing effects of dawn (lights on) and the retarding effects of dusk (lights out) allows us to predict the phases of a rhythm in relation both to the normal 24-h cycle and to cycles other than 24 h in length. Colin S. Pittendrigh, a zoologist, and his co-workers, first at Princeton and then at Stanford University, have pioneered in this approach.

In entrainment by light, some photoreceptor pigment is absorbing the light, and by this absorption it is changed in such a way that it can lead to an advance or a delay of the clock. It would be extremely interesting to understand the photobiochemical mechanism of this response. A first step is identification of the photoreceptor pigment, which is approached by a determination of the action spectrum for the response, as discussed in Chapter 19. Flashes of light of carefully controlled spectral qualities are given at various times to test their effectiveness on phase-shifting or entrainment.

Action spectra have been determined for the various rhythms of *Gonyaulax*, for a mating rhythm of *Paramecium*, for the conidiation rhythm of *Neurospora* (Fig. 20-5), and for *Drosophila* (fruit fly) and *Pectinophora* (a moth). The action spectra are different in the several organisms, but all do have strong responses in the blue part of the spectrum. The response of *Gonyaulax* to red may be due to chlorophyll, but the reason for the even stronger red response of *Paramecium* is not clear. Victor Muñoz and Warren L. Butler (1975) in La Jolla, California isolated a flavoprotein-cytochrome *b* complex from *Neurospora* with an absorption spectrum that closely matches the action spectrum of Fig. 20-5 (see discussion in Chapter 18). They suggest that this or a similar pigment may couple light and the clock in the many organisms, both plant and animal, that respond to blue light.

Lars Lörcher (1957), working in Bünning's laboratory in Tübingen, Germany, found that the rhythms of dark-grown bean plants were most effectively established by red light and that this establishment was reversed by an immediate exposure to far-red light. Furthermore, the leaf-movement rhythms often continue for several days when the plants are maintained at a constant temperature and under continuous light, providing the light is rich in red wavelengths but contains none of the far-red part of the spectrum. When far-red light is present, the rhythms damp out more quickly. These observations implicate the phytochrome system, but Lörcher found that other wavelengths were also effective in entrainment, providing the plants had been grown in the light rather than in the dark.

There are other examples among higher plants in which the phytochrome system has been implicated. For example, Philip J. C. Harris and Malcolm B. Wilkins (1978a, 1978b) in Glasgow, Scotland were able to reset a CO_2-evolution rhythm in *Bryophyllum* (a succulent with CAM) leaves held in the dark only with red light (600 to 700 nm; Fig. 20-5). They were unable to reverse this red resetting with far-red light, but far-red applied simultaneously with or immediately after the red abolished the rhythm completely. E. Simon, Ruth L. Satter, and Arthur Galston (1975), on the other hand, could reset the leaflet-movement rhythm of excised *Samanea* (a semitropical leguminous tree) pulvini with only 5 min. of red light; and the red effect was completely cancelled by subsequent far-red. The situation was not that simple, however. When longer irradiation times were used, blue and far-red could also reset the *Samanea* rhythms, but the resetting curves were qualitatively different, as shown and explained in Fig. 20-6. Similar results have been obtained with other plants (reviewed by Satter and Galston, 1981, and Gorton and Satter, 1983). Clearly, phytochrome can interact with the biological clock in some plants, at least (but not in fungi, animals, or perhaps protistans), but it is equally clear that there are complications that remain to be understood.

For a long time, zoologists simply assumed that the receptor was the eye of the animal with which they were working. But it was shown in the 1950s that gonads of ducks would develop in response to long days, even though the eyes were removed. In the 1960s, Michael Menaker at the University of Texas

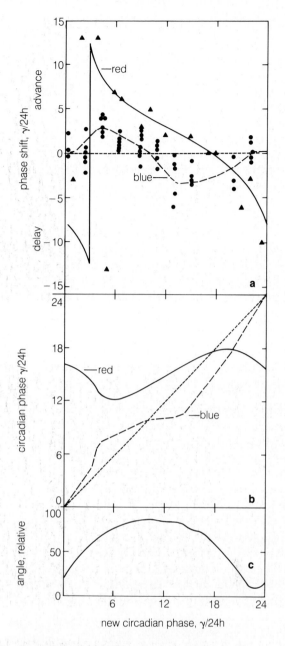

Figure 20-6   (**a**) Phase-response curves that show the effects of a single pulse of red or blue light (given during darkness) on the rhythmic leaflet movements in *Samanea*. Zero phase shift is indicated by the dotted line. Note the sudden shift from delay to advance at about 3 hours. The abscissa shows the time the pulse was given (in *circadian hours*, τ, which sets the natural period of the plants equivalent to 24). Plants were entrained to a 16-h light/8-h dark regime prior to continuous darkness. (**b**) Curves from (a) are replotted with the new phase (shown in circadian hours) plotted as the ordinate. Again, zero phase shift is indicated by the dotted line. (**c**) Rhythmic leaf movement of *Samanea* in continuous darkness. Increasing relative angle represents leaf opening. In (a) and (b), irradiances of red were 5 min at 12 μmol m^{-2} s^{-1}; of blue, 2 h at 5 μmol m^{-2} s^{-1}. (From Gorton and Satter, 1983, with permission. Copyright © 1983 by the American Institute of Biological Sciences.)

in Austin showed that blind sparrows (both eyes removed) could be entrained in their activity rhythm by light signals. He also confirmed with blind sparrows the development of testes in response to long days, as observed in blind ducks. Furthermore, he has shown that a weak green light has an effect upon the activity rhythm of his birds with normal eyes, but that it does not influence the day length response. It is the pineal gland of the brain that actually responds to the light. Enough light, especially red light, penetrates the skull to be effective. This gland has been known since antiquity as the "third eye!" It is especially prominent in birds. Menaker and his co-workers (see Zimmermann and Menaker, 1975) have studied the pineal responses to light and their effects on the clock. The activity and gonadal clocks, at least, seem to be located in the pineal gland; effects are transmitted via hormones rather than nerve impulses.

20.5 Rhythm Characteristics: Temperature

Pittendrigh realized that the clock could be of little value to an organism if the rate at which it ran were strongly dependent upon temperature, as are most metabolic functions. He had heard of the eclosion rhythm of *Drosophila* pupae, discovered by H. Kalmus in Germany. Pittendrigh (1954) studied this rhythm at several temperatures and found the period to be nearly constant over a wide temperature range (shortened slightly at higher temperatures). Thus temperature insensitivity of the biological clock was discovered. Bünning had investigated the question in 1931, but he found that the bean-leaf period was more sensitive to temperature than the animal examples studied later. Bünning did report that the temperature response was unexpectedly low.

Frank Brown and H. Marguerite Webb at Northwestern University in Chicago reported in 1948 that a color change observable in the fiddler crab had a period that was virtually independent of temperature. But Brown and Webb did not deduce from their data a temperature-insensitive clock. Rather, they considered their results to be evidence against an endogenous clock. Indeed, they returned to Stoppel's concept of a factor X in the environment that was responsible for actual time measurement. Brown championed this idea until his recent death. He and his co-workers observed numerous biological responses to such subtle environmental factors as geomagnetic fields. These are of considerable interest in themselves, but no one can see how they might account for biological time measurement. As Pfeffer, nearly a century ago, was unable to convince his colleagues that the clock was endogenous (it took the experiments of Bünning, Stern, Stoppel, and Kleinhoonte to do that), so Brown has been unable to convince most of us that the clock is exogenous. Most workers now believe that the experiments of Brown and Webb with fiddler crabs were just an especially outstanding demonstration of the clock's temperature independence.

Yet we are faced with somewhat of a paradox in our discussion of temperature effects on the biological clock. Changes in temperature of only 2.5°C or less can synchronize the rhythms (act as a Zeitgeber) in *Neurospora* and other organisms, and temperature can also influence the amplitude of the response. Such effects must surely be important in nature. Still, the period of a free-running rhythm is relatively temperature insensitive. So some aspects of the clocks are sensitive and others are insensitive to temperature.

The temperature insensitivity is especially interesting. In *Gonyaulax* the Q_{10} for the effect of temperature on period length is slightly less than 1 (as temperature increases, the free-running period becomes slightly longer), whereas it is equivalent to about 1.3 for the leaf movements of the bean. Various workers have observed Q_{10} values that approach 1.00 very closely. For example, the value is approximately 1.02 for biological time measurement in the flowering of cocklebur (Salisbury, 1963).

20.6 Rhythm Characteristics: Applied Chemicals

If we could find some chemical that would clearly inhibit or promote biological time measurement, then by our speculations about and tests of its mode of action we might gain some insight into the operation of the biological clock. Many chemicals have been studied, but we must be careful in interpreting the results. A given chemical may, and often does, influence the amplitude of the rhythm without influencing its period, the actual indication of time measurement. That is, the physiological manifestation of time measurement (e.g., leaf movements or animal activities) may be easily influenced by some substance such as a powerful respiration inhibitor, although the clock is more resistant (Fig. 20-7). There are some instances in which the amplitude can be completely damped out; yet when the inhibitor is removed, the cycle proves to be in phase with that of control organisms having the same period and to which no inhibitor was applied. This is one aspect of the general problem of **masking.**

In the early 1960s, Bünning and his co-workers reported that several chemical inhibitors applied to bean plants appeared to act directly on the clock itself, rather than on its manifestations (reviewed by

Bünning, 1973). In these cases, the period was lengthened somewhat, and the amplitude was unaffected or only somewhat decreased. These substances include colchicine, ether, ethyl alcohol, urethane (ethyl carbamate), and some others. They are sometimes but not always effective with other organisms. It has been suggested that the feature they have in common is the ability to influence membranes, and that therefore the clock is associated with cellular membranes. Another clear-cut effect of a chemical upon time measurement is that of heavy water, deuterium oxide. Treatment with this substance caused a considerable increase in the phototaxis period of *Euglena* and also of bean-leaf movements, and it slowed the measurement of day length in flowering. Again, heavy water might influence many processes within the plant, but membranes are good candidates. Valinomycin, a compound that alters membrane transport of K^+, also causes phase shifts in beans and in *Gonyaulax*, further implicating membranes (Bünning and Moser, 1972). Lithium ions (Li^+) lengthen the period and otherwise affect the rhythms in several plants and animals (reviewed by Engelmann and Schremf, 1980). Several inhibitors of protein or RNA synthesis do not influence the clock (e.g., actinomycin D), but others clearly do (e.g., cycloheximide and puromycin).

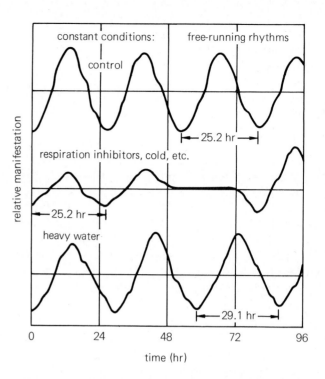

Figure 20-7 Effects of various chemicals and other factors on free-running rhythms. Most compounds that inhibit metabolism reduce the amplitude but do not affect the period (compare center record with control above). Heavy water increases the length of the period (29.1 h compared with 25.2 h for eclosion in *Drosophila*). Curves are schematic and do not represent actual data.

20.7 Clock Mechanisms

What and where is the clock and how does it work? One approach to finding out has been to search for clock mutants in various organisms, especially *Neurospora crassa*, *Drosophila* (the fruit fly), and *Chlamydomonas* (a single-celled, motile green alga). We shall discuss some of the results with *Neurospora*, which has a highly developed genetic system (see reviews by Feldman, 1982, 1983).

Approximately 12 *Neurospora* clock mutants with periods ranging from 16.5 to 29 hours have been found. Seven of the mutants map to a single gene locus called *frq* (frequency), which apparently plays a key role in clock organization. The only phenotypic difference between the mutants and the wild type observed so far is the period of conidial banding. An equal number of mutants are fast and slow, which is surprising, since most mutations are deleterious and would be expected to have phenotypes altered in the same direction. At 25°C, the periods of the *frq* mutants all differ from that of the wild type by some multiple of 2.5 hours. Furthermore, the *frq* cycles are restricted to a 7-hour part of the wild-type cycle, which is shortened in *frq-1* to 2 hours and lengthened in *frq-7* to 14.5 hours. The short-period *frq* mutants retain temperature compensation, but the long-period mutants do not. All these facts (and others we

won't discuss) provide details that must someday find a place in our understanding of how the clocks function. And we must remember that there could be more than a single mechanism, even in a single organism. For example, the leaf-movement rhythm can be separated from photoperiodic timing in several plants (see Chapter 22).

Other particularly significant aspects of clock function that must be explained by a clock model are the long period (compared with the periods of chemical oscillating systems known in the laboratory), the resetting effects of temperature and especially light, temperature insensitivity, the various chemical effects, and the apparent key role of membranes.

Temperature independence of the clock poses an interesting problem, since the Q_{10} for most biochemical reactions is appreciably greater than one. Those involving hydrolysis of ATP, for example, are often about 2.0. How, one might ask, if the living organism is fundamentally a biochemical system, can we account for temperature independence?

We can envision certain feedback systems. The products of one reaction or function might inhibit the velocity of an earlier one. Such feedback inhibition of other processes is well known in living organisms. As temperature increases, the velocity of the first reaction (the time-measuring reaction) might increase,

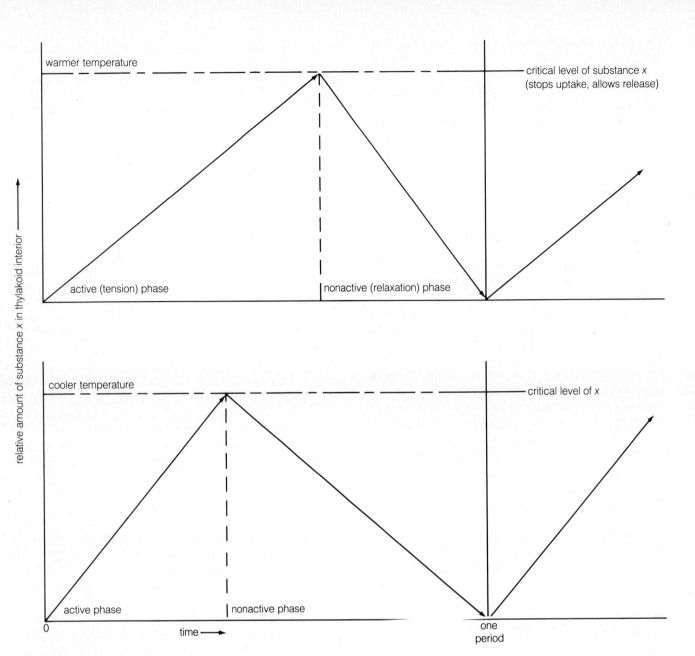

Figure 20-8 Temperature compensation of the circadian period according to Sweeney's model. The model postulates that time keeping depends on the concentration of some substance X (K^+?) in the interior of thylakoid double membranes in chloroplasts of *Gonyaulax* and perhaps other organisms. In the active phase, there is both an uptake and a release of substance X into and from the interior of the thylakoids. At higher temperatures (upper part), substance X is released faster compared with lower temperatures (lower part), thus leading to a reduced net rate of uptake during the active phase. Therefore, the active phase must be longer for the concentration of X to reach the threshold that switches to the passive part, but the passive part is faster at the higher temperatures, accounting for a constant period (temperature compensation). (See Englemann and Schrempf, 1980, for a review of this and other membrane models.)

but the inhibitory product would then increase at a proportional rate, and consequently the reaction would maintain a near-constant rate over a wide range of temperatures. The Q_{10} value of less than 1 observed in *Gonyaulax* could be accounted for by such a scheme, assuming that the inhibitory product were produced somewhat faster with increasing tempera-

ture than the time-measuring reaction is accelerated. It is difficult to imagine any other way to account for this observation. Such a feedback system (possibly overcompensating as in *Gonyaulax*) could be invoked in a **temperature-compensated system** of time measurement based upon biochemical reactions. Furthermore, some chemical reactions catalyzed by enzymes are now known that have temperature coefficients close to 1.

There are several evidences for the possible participation of membranes in the basic clock mechanism (reviewed by Engelmann and Schrempf, 1980). For one thing, slow diffusion across a membrane with a gradual buildup in concentration of some key substance on one side could account for the long period. Indeed, the concentration of some ions, especially K^+ in plants, has been found to oscillate in a circadian manner, and such ions phase-shift certain plant rhythms when applied externally. Several substances known to influence membranes in one way or another also influence the biological clock, as we have seen, and circadian rhythms in permeability and transport properties of membranes have been reported. We can also imagine how light and perhaps temperature might influence membranes and thus the clock. As we have seen, phytochrome apparently plays a role in rephasing plant clocks and is also thought to associate with membranes.

With these and other ideas in mind, several clock mechanisms involving membranes have been proposed (reviewed by Englemann and Schrempf, 1980). Beatrice Sweeney was one of the first with a model in 1972 followed by modifications in 1974 and 1976. Some *substance X*, possibly K^+, is transported into an organelle (thylakoids within chloroplasts in Sweeney's model). When the concentration inside reaches some critical value, active transport is stopped, and substance X leaks out until a critical low concentration is achieved, reinitiating active transport and thus another cycle. The transport phase requires energy, but the leakage phase does not, matching evidences for a tension/relaxation system reported by several authors (e.g., Bünning, 1960). At higher temperatures, leakage is accelerated more than transport, slowing the tension phase but speeding the relaxation phase and thus accounting for temperature compensation (Fig. 20-8). Light shifts the phase by accelerating the leakage rate through the membranes. This causes a delay during tension but an advance during relaxation.

Other membrane models are similar to Sweeney's in their basic concepts, but some include protein synthesis and other features to account for various observations. No one thinks that any one model will prove entirely correct, but at least a start has been made.

20.8 Photoperiodism

Even before the work of Bünning, Stern, Stoppel, and Kleinhoonte, the biological clock had clearly been demonstrated as a component of living plants, although the discovery was not widely recognized by those working on rhythms until perhaps the 1950s.

Wightman Wells Garner and Henry Ardell Allard were two scientists working in the United States Department of Agriculture research laboratories at Beltsville, Maryland. They had made two observations that they could not explain. Maryland Mammoth tobacco (a new hybrid) grew at that latitude to a height of 3 to 5 m during the summer months but never came into flower, although it flowered profusely when only about 1 m tall after being transplanted to the winter greenhouse. They had also noticed that all the individuals of a given variety of soybean would flower at about the same time in the summer regardless of when they had been planted in the spring. That is, the large plants sown in early spring came into flower at the same time as the much smaller plants sown in the summer. Garner and Allard wondered if some factor of the environment might be responsible for flowering of these two species.

They considered various environmental factors that might differ between the summer fields and the winter greenhouses. They tested light levels, temperatures, soil moisture, and soil nutrient conditions, but no combination of these factors resulted in flowering of the tobacco plants. Garner and Allard realized that day length varied throughout the season, as well as being a function of latitude, so they tested what seemed like the remote possibility that this could control flower formation. It did. When the days were shorter than some maximum length, the tobacco plants began to bloom. When the days were shorter than some other maximum length, the soybean plants also began to bloom. Results were published in 1920. Garner and Allard named the discovery of flowering in response to photoperiod **photoperiodism,** after a suggestion from A. O. Cook, a colleague at Beltsville.

Subsequent work seemed to indicate that the length of the dark period was in some ways even more important than the photoperiod (see Chapter 22). The point here is that these plants were capable of measuring the length of the light and/or the dark period. Here was a clear-cut demonstration of biological time measurement. Completely comparable responses were subsequently discovered in animals, including effects on insect life cycles, fur color (e.g., arctic hare), breeding times (gonad size), and migration of many birds and animals.

Some organisms respond when the day length exceeds some minimum. Garner and Allard called such plants (e.g., winter barley and spinach), which flower in response to days longer than some critical length, **long-day plants,** and those such as tobacco and soybean, which flower when the day length is less than some maximum, **short-day plants.** Some of the species they studied showed no response to day length; these they called **day-neutral plants.**

Beatrice M. Sweeney

There is much concern about the role of women in our society. It has often seemed to me [F. B. S.], whether fact or only an illusion, that an unusually high percentage of women has contributed to our understanding about the biological clock. I was discussing this idea with Beatrice M. Sweeney during a symposium on biological clocks several years ago. She agreed to write a few personal thoughts about these matters for use in our textbook. She is located in the Department of Biological Sciences, University of California at Santa Barbara.

Dear Frank:

Until you remarked upon it at the Christmas meeting at Sacramento, I hadn't noticed the unusual number of women in the field of biological clocks. I am not accustomed to thinking about the sex of my scientific colleagues. In this respect, I suppose I am a truly liberated woman. I think I owe the fact that I am not conscious of whether or not scientists are women to my good fortune in receiving my training in Dr. Kenneth Thimann's laboratory at Harvard, where even long ago all graduate students were on an equal footing. We all regarded ourselves as superior, and so our chances of turning out that way were much increased. This brings me back to the subject of women in science, since I often think women have a difficult time believing that they really can do first-class research in a field usurped exclusively for so long by men, at least in the top prestige bracket. In rhythm work, women have, I believe, found relief from their feelings of inferiority, because they recognize in this field a comfortable familiarity. To get up in the middle of the night, perhaps several times, is not strange to them. What matters that it is a series of tubes full of *Gonyaulax* and not a baby that demands their attention?

I notice that the predilection of women for chronobiology does not seem to be declining. The traditions set by such women as Rose Stoppel, Anthonia Kleinhoonte, Marguerite Webb, and Janet Harker are being ably perpetuated by Audrey Barnett, Ruth Halaban, Marlene Karakashian, Laura Murray, Ruth Satter, Therese van den Driessche, and I'm sure you can name others. Of course, I still regard myself as active and expect any day to crack the problem of the basic circadian oscillator.

I would like to say something personal to the young ladies who are perhaps considering a future as scientists, now that the sole production of children is no longer in the interest of world well-being. My work in science has been the very stuff of life to me, endlessly frustrating and rewarding. I should like to relate to these young women what fun it is to work, rather than be a tourist, in strange parts of the world. Imagine the delight of traveling northward inside the Great Barrier Reef on a 60-foot boat usually devoted to the hunting of alligators and to have available a microscope with which to see the details of the strange animals and plants, the familiar yet unfamiliar phytoplankton, the algae growing within the giant clams and coral. I could mention also the jungles of New Guinea at night, ringing with the stridulations of locusts at multiple high frequencies, flashing with fireflies. Or the beaches of Jamaica, white and shining and fringed with the bending fronds of the coconut palms like long eyelashes. The knowledge of flora and fauna acquired in scientific study immeasurably increases the pleasure of viewing an unfamiliar part of the world.

But most of all I'd like to say that the day-to-day research and teaching of an academic profession provides an endlessly varying and interesting way of life, to my tastes infinitely more satisfying than cooking, dusting and shopping, even than bringing up children. Perhaps my four children have benefited rather than suffered from the fact that I have other interests and satisfactions than themselves. At least it is interesting to see that three of the four are in some way pursuing science as a career.

20.9 Photoperiod-Rhythm Interactions

Garner and Allard also did some experiments, now mostly forgotten, in which they subjected plants to cycles of light and darkness that did not add up to 24 hours. The results are complex and not always easy to explain, but it seems clear that when the total cycles deviate much from 24 hours growth is inhibited. It appears that the light-dark cycle must agree at least approximately with the circadian cycles of the plant. When the cycle deviates too much from

24 hours, the rhythmical change in light and darkness can no longer entrain the rhythms, as we have already noted. In addition, plant growth itself is often adversely affected.

Sometimes similar responses can be observed even when plants are subjected to a 24-hour cycle. Tomato plants, for example, are usually said to be day-neutral in as far as their flowering goes. Although short days slightly promote flowering, plants will usually flower over a wide range of day lengths from very short to about 18 h long. There are other effects of day length, however, that are quite easy to observe. Tomato plants held under 16-h days have stems about twice as long as those held under 8-h days (see Fig. 22-2). Furthermore, many tomato cultivars are actually killed when the days account for more than 18 hours of a 24-h cycle. These effects can be observed even when the days are extended by low light levels, so that the plants under long days are not photosynthesizing much more than plants under short days. With sensitive tomato plants, it is even possible to observe a slight but significant inhibition of growth when light is given in the middle of the dark period. Highkin and Hanson (1954) tested this by giving tomato plants either 16 hours of continuous light plus 8 hours of darkness or 14 hours of light with another 2 hours in the middle of the dark period. Tomato plants even seem to be sensitive to the alternation in temperature between day and night; they produce more flowers when night temperatures are lower than day temperatures. This phenomenon was called **thermoperiodism** by Frits Went (discussed in Chapter 21).

20.10 The Biological Clock in Nature

In a cycling environment, it is not difficult to imagine advantages conferred upon organisms by possession of a biological clock. Rhythms of activity in animals, for example, allow a species to occupy a niche not only in space but also in time: A nocturnal animal and a diurnal animal might use the same space but at different times. Plant rhythms in metabolism adjust the plant to the light and temperature environment. Rhythms in flower opening must be closely coupled to the time memory of honeybees. The phenomenon of photoperiodism confers upon the organism possessing it the ability to occupy a particular niche in *seasonal* as contrasted to *diurnal* time. The several species that flower in response to photoperiodism may do so at different times and in sequence throughout the season, providing a rather constant source of nectar for insect pollinators throughout an entire growing season. Given a time during the season with minimal flowers but high availability of pollinators, there would be a selective advantage to any species able to flower during that time.

Leaf Movements   What are the ecological advantages, if any, of the leaf movements? Pfeffer asked this question and settled it in his mind by tying the leaves to a bamboo framework so that the movements could not occur. Since his plants exhibited no apparent ill effects, he concluded that the movements were some by-product of the evolutionary process and of no selective value to the plant.

Charles Darwin suggested that leaf positions might play a role in heat transfer between a plant and its environment (Section 3.6). That is, a horizontal leaf is in a good position for the reception of sunlight during the day, but at night heat might radiate better from a horizontal leaf into space. *Vertical* leaves in a plant community, however, might radiate more to each other. Darwin and his son Francis (1881) performed experiments to test this idea and obtained positive although not very striking results. Similar, more recent experiments (Enright, 1982) confirmed the basic observation, but temperature differences between vertical and horizontal leaves were very small, less than 1°C.

Bünning suggested (see his personal essay) that a horizontal leaf would be in a better position to absorb the light of the full moon (at its zenith at midnight). Since such absorption might upset the photoperiodism response, a vertical leaf position at night could be a protective device to insure successful time measurement in photoperiodism. Even this effect is probably of little consequence, as we discuss in Chapter 22.

Time Memory   While Bünning and Stern were discovering the endogenous nature of the biological clock as it occurs in plants, Ingeborg Behling in Germany (1929) was making an important discovery about the clock in honeybees. She found that it was possible to train honeybees to feed at a certain time during the day. As already noted, this must be an adaptation to plant rhythms of flower opening and nectar production. It is as though the clock in the honeybee can have a "rider" attached to it, indicating the time of day and informing the honeybee 24 hours later that it is time to feed. It remains to be seen whether this is the same clock that controls circadian rhythms.

Humans have a comparable time-measuring system. Our time memory is most frequently manifested by an ability to wake up at a predetermined time. This is particularly impressive, since the person must translate an abstract idea (clock time) into some form that will "adjust the rider" on his or her biological clock. Human time memory can often be most impressively demonstrated under hypnosis, especially with posthypnotic suggestion.

Celestial Navigation   In spite of the extremely intriguing discoveries of the 1920s and 1930s, only a small minority of biologists showed any great interest in the biological clock until the early 1950s. At that time American botanists began to become interested in the work of Bünning, because it seemed to bear a direct relationship to photoperiodism, a topic in which there was considerable interest. Indeed, Bünning had proposed a theory to correlate photoperiodism and the circadian rhythms (see his personal essay). Zoologists, particularly Pittendrigh at Princeton University, also began to take notice of Bünning's discovery and of other work going on in Europe. The discovery of temperature compensation was especially stimulating.

Then Gustav Kramer, K. von Frisch, and others working primarily in Germany found that certain birds and other animals could tell direction on the earth's surface by the position of the sun in the sky. Since this position changes, the organism must be able to correct for the time of day, apparently by the use of some kind of clock. Up until then, some of the manifestations of the clock (e.g., the leaf movements) did not seem to be of much value to the organism. In the case of **celestial navigation,** however, the clock clearly is used, so this spectacular discovery (along with temperature compensation) caught the attention of biologists all over the world and led to a surge of interest in the biological clock.

The Physiological Clock in Humans Celestial navigation is certainly an outstanding and clear-cut application of the biological clock among animals. But human circadian rhythms and other time-measuring abilities are also important in certain obvious ways, and they may clearly increase in importance (in a negative way) as our modern life becomes ever more complex (Thompson and Harsha, 1984). Jet air travel across time zones, for example, has a strong effect upon the internal timing system of the passengers. It may take several days for the clock in a tourist—or a diplomat—to adjust to a new time zone.

All these and other observations suggest that the biological clock offers an exciting and increasingly important field for scientific investigation.

21

Growth Responses to Temperature

Plant growth is notoriously sensitive to temperature. Often a difference of a few degrees leads to a noticeable change in growth rate. Each species or variety has, at any given stage in its life cycle and any given set of study conditions, a **minimum temperature** below which it will not grow, an **optimum temperature** (or range of temperatures) at which it grows at a maximum rate, and a **maximum temperature** above which it will not grow and may even die. Some curves for growth rate as a function of temperature are shown in Fig. 21-1. The growth of various species is typically adapted to their natural temperature environment. Alpine and arctic species have low minima, optima, and maxima; tropical species have much higher **cardinal temperatures**. Plants close to the minimum or maximum temperatures are often under stress—the topic of Chapter 24.

Often different tissues within the same plant have differing cardinal temperatures. A classical and easily demonstrable example of this is the difference in optimum growth temperatures for the upper and lower tepals (collective term for petals and sepals that are similar in appearance, especially in members of the lily family) of the tulip or crocus flower. Studies in Germany, dating from those of Julius von Sachs in 1863, have demonstrated that low temperatures (3 to 7°C) are optimal for growth of the lower tepal tissues, causing tulip or crocus flowers to close, while higher temperatures (10 to 17°C) are optimal for growth of the upper tepal tissues, causing flowers to open (Fig. 21-2). An abrupt change in temperature of only 0.2 to 1°C often results in rapid growth and the opening or closing of tulip or crocus flowers, although the optimal growth temperatures for the two sides of the tepals are about 10°C apart. This temperature-induced movement of the tepals caused by growth is termed **thermonasty.**

Temperatures influence more than tissue growth, however. Often, critical steps in the life cycle are

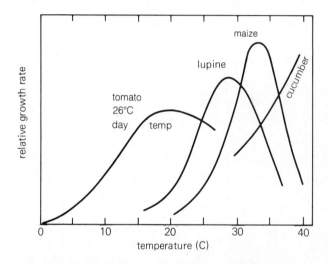

Figure 21-1 Plant growth as a function of temperature for four species. With tomato, day temperature was constant and night temperature varied.

initiated in response to certain temperature treatments: Seeds germinate, flowers are initiated, and perennial plants become dormant or break dormancy. And these developmental responses are often influenced by environmental factors in addition to temperature, including light, photoperiod, and moisture. These interactions are diverse and complex, so the topics of this chapter sometimes stray from its central theme, which is plant response to temperature, especially low temperature.

21.1 The Temperature-Enzyme Dilemma

We usually think of the plant growth response to temperature in terms of enzyme reactions in which two opposing factors seem to operate. With increases in temperature, the increased kinetic energy of the reacting molecules results in an increased rate of reac-

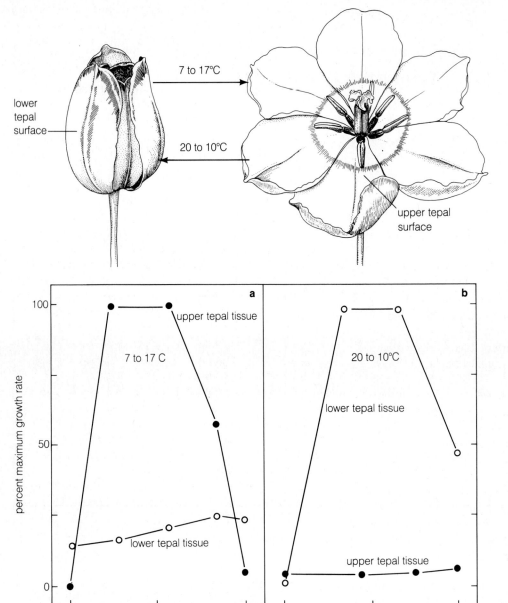

lower tepal surface

7 to 17°C

20 to 10°C

upper tepal surface

Figure 21-2 Effects of changing temperature on opening and closing of tulip flowers. (**a**) As temperature increases, growth of upper tepal tissue speeds up for a short time, while growth of lower tissue remains about constant, causing flower opening. (**b**) When temperature decreases, growth of lower tepal tissue speeds up and growth of upper tissue remains the same, accounting for flower closing. (Figure prepared by Stanley H. Duke based on data of Wood, 1953.)

tion, but increasing temperature also results in an increased rate of enzyme denaturation. Subtracting the destruction curve from the reaction curve produces an asymmetrical curve (Fig. 21-3), with its minimum, optimum, and maximum cardinal temperatures as in Fig. 21-1. The curve applies to respiration, photosynthesis, and many other plant responses besides growth.

Rates of physiological processes or enzymatic reactions are often plotted against the reciprocal of absolute temperature (K), as was first suggested by Svante Arrhenius (Swedish chemist, 1859–1927, 1903 Nobel Prize in chemistry). For a given process occur-

ring within an organism, the **Arrhenius plot** (Fig. 21-4) is normally linear within the range of temperatures in which the organism is capable of living for an extended period. Any curve, break, or inflection in an Arrhenius plot denotes a change in sensitivity to temperature of the process being measured. A sharp drop indicates that protein denaturation exists at the upper temperature limit of the plot, and a discontinuity (break) or inflection (change in the slope) may exist at the lower temperature limit of the plot. Such inflection or discontinuity in an Arrhenius plot indicates that the process being studied is sensitive to low temperature. An increase in the slope of the plot

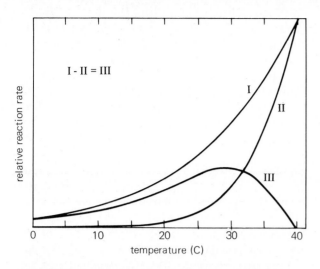

Figure 21-3 Enzyme activity and temperature. I. Rate of reaction with a Q_{10} of 2, which is typical of many chemical reactions including those (at lower temperatures) controlled by enzymes. II. A reaction with a Q_{10} of 6, which is typical of denaturation of protein. III. The expected curve for enzyme denaturation (II) subtracted from the curve for enzymatically controlled reaction rate (I). This curve is typical of the temperature responses of enzymatically controlled reactions over the entire temperature range of such reactions.

below the temperature of inflection indicates that the **energy of activation** (E_a), the minimal energy required for the process being measured to occur, has increased and has become more limiting to the maximal velocity (V_{max}), the maximal rate of the process under given conditions, than at temperatures above the inflection temperature. This situation is commonly observed in physiological processes and enzymatic reactions of plants sensitive to chilling (see Section 24.5). Arrhenius plots of cultivars within the same species can often be quite different for the same process, such as growth (see Henson et al., 1980).

After becoming familiar with positive responses as temperature increases from the minimum to the optimum, it may come as a surprise to learn that certain processes are promoted as temperature decreases toward the freezing point. In **vernalization,** exposing plants to low temperatures for a few weeks results in the formation of flowers, usually after plants are returned to normal temperatures. Low temperatures in the autumn often cause or contribute to the development of dormancy in many seeds, buds, or underground organs, and the low temperatures of winter may be responsible for the breaking of dormancy in these same organs. The interesting paradox of the low-temperature response is that low temperatures first causes dormancy to develop in plants, but then further low temperatures cause a breaking of the dormancy. If we oversimplify dormancy induction by thinking of it as a simple *negative* slowing down of plant processes at decreased temperatures, then we must surely consider the breaking

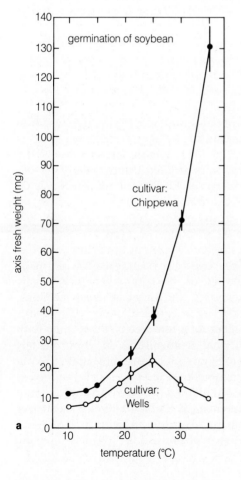

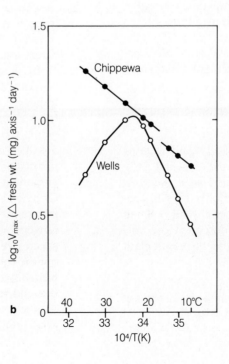

Figure 21-4 The Arrhenius plot as a way to analyze the effects of temperature on germination (or chemical reactions, other plant or animal responses, etc.). (**a**) Effect of germination temperature on fresh weight of the seedling axis of two cultivars of soybeans. (Vertical lines represent standard deviations.) (**b**) Arrhenius plots of the data shown in (a). In such a plot, the ordinate is the logarithm of the response (in this case, fresh-weight increase per axis per day), and the abscissa is the reciprocal of the absolute temperature. (Temperature in degrees Celsius is shown above the line; note that in such a reciprocal plot, low temperatures appear to the right instead of the left.) (From Henson et al., 1980, used by permission.)

of dormancy in an opposite, *positive* sense. This is the fascinating dilemma of the low-temperature response.

We shall consider five different positive responses to low temperatures: The first, vernalization, has been studied extensively, so it provides a good starting place. The second is the breaking of seed dormancy by exposure of moist seeds to low temperatures. This treatment has often been referred to as **stratification**, but the term **prechilling** is now more popular because it is more descriptive.* We shall have more to say about seed dormancy and germination than their responses to temperature. The third process is closely related: the breaking of winter dormancy in the buds of perennial woody plants. The fourth process has been studied less: the low temperature induction of underground storage organs such as tubers, corms, and bulbs. The fifth process has been studied still less: the effects of low temperatures on the vegetative form and growth of certain plants.

In each of these five processes, we are concerned mostly with **delayed effects** (sometimes called **inductive effects**) upon some developmental plant process. Such effects, in which the response appears some time after completion of the stimulus, are also observed in response to other environmental factors, such as day length. In fact, low-temperature and day-length effects are frequently interrelated in the five plant responses.

Ultimately, the low-temperature response could involve activation of specific genes—a switching of the morphogenetic program in response to low temperatures. Could the genes respond directly to low temperature? If not, then what in the cell actually responds, transducing the low-temperature signal to a physiological change? Are the transducers located in the cytoplasm or the nuclei of the cells in which the program is readjusted? Or in other cells? There are few answers to these questions, but they could guide future research.

21.2 Vernalization

Vernalization was described in at least 11 publications in the United States during the mid-19th and early 20th centuries (e.g., in the *New American Farm Book* in 1849), but it was completely overlooked by "establishment" science until 1910 and 1918, when J. Gustav Gassner in Germany described the vernalization of cereals. Much of the early work on plant

development took place in Europe and Great Britain; the United States and Canada were apparently preoccupied with subduing the frontier.

In the 1920s, the term *vernalization* was coined by Trofim Denisovich Lysenko, who, during the reign of Stalin, was allowed to exercise absolute political control over Russian genetic science, decreeing that geneticists there should accept the dogma of the inheritance of acquired characteristics (see Caspari and Marshak, 1965). Vernalization, from Latin, translates into English as "springization," the implication being that winter varieties were converted to the spring or summer varieties by cold treatment. We realize, although apparently Lysenko did not, that the genetic makeup is not changed by the low-temperature treatment. The cold supplied artificially by the experimenter simply substitutes for the natural cold of winter.

The term *vernalization* has been widely misused. Any plant response to cold has sometimes been referred to as vernalization, as has any promotion of flowering by any treatment (even day length). We shall restrict the term **vernalization** to *low temperature promotion of flowering.*

The Response Types There are numerous vernalization responses, depending not only upon species but frequently upon varieties and cultivars within species. In classifying the response types, there are several factors to consider (indicated by italics). To begin with, we may differentiate *delayed* from *nondelayed* responses. Most plants that have been studied respond after a delay, although a few (e.g., Brussels sprouts) form flowers during the cold treatment itself.

An appropriate way to classify the response types is according to the *age* at which the plant is sensitive to cold. The **winter annuals,** especially cereal grasses, were studied during the 1930s and 1940s, particularly in Russia and by Frederick G. Gregory and O. Nora Purvis (see Purvis, 1961) at Imperial College in London. They respond to the low temperatures as seedlings or even as seeds, providing that sufficient oxygen and moisture are present. Petkus rye (*Secale cereale*) seeds are normally planted in the fall of the year, when they usually germinate, spending the winter as small seedlings. Or moist seeds may be exposed to low temperatures in a cold chamber for a few weeks. Plants then form flowers at normal temperatures in approximately seven weeks after growth begins in the spring. Without the cold treatment, 14 to 18 weeks are required to form flowers, but ultimately flowers do appear. Since the cold requirement is a **quantitative** or **facultative** one (low temperatures result in *faster* flowering) but not a **qualitative** or **absolute** one (in which flowering

---

*It is easy to confuse vernalization (an effect on flowering) and stratification or prechilling (an effect on germination), since both processes can occur when moist seeds are exposed to low temperature.

absolutely depends upon cold), we have another basis for classification. Most winter annuals are delayed and quantitative in their response, although some (e.g., Lancer wheat) have an absolute cold requirement.

With Petkus rye there are two interesting complications: Short-day treatment will substitute to a certain extent for low temperature, and flowering of previously vernalized, growing plants is strongly promoted by long days. All winter annuals so far studied are promoted not only by cold but also by the subsequent long days of late spring and early summer.

The **biennials** live two growing seasons, then flower and die (Section 15.2). Examples include several varieties of beets, cabbages, kales, Brussels sprouts, carrots, celery, and foxglove. They germinate in the spring, forming vegetative plants that are typically a rosette (see Fig. 21-6, lower right). The leaves often die back in the autumn, but their dead bases protect the crown with its apical meristem. With the coming of the second spring, new leaves form, and there is a rapid elongation of a flowering shoot, a process called **bolting.** Exposure to the winter cold between the two growing seasons induces flowering. Most biennials must experience several days to several weeks of temperatures slightly above the freezing point to subsequently flower; they have an absolute cold requirement, as contrasted to the facultative winter annuals. Sugar-beet plants may be kept vegetative for several years by not being exposed to cold (Fig. 21-5). Flowering of many biennials is also promoted by long days following the cold, and some may absolutely require this treatment (e.g., the European henbane *Hyoscyamus niger*, Fig. 21-6). Other biennials are day-neutral following vernalization.

There are many species of cold-requiring plants that do not fall readily into the categories of winter annuals or biennials. Flowering of several perennial grasses, for example, is promoted by cold. Some of these have a subsequent short-day requirement for flowering. The chrysanthemum is a short-day perennial that has been studied extensively because of its photoperiodism response. Its cold requirement, which must be met once before it can respond to the short days, was overlooked because plants are propagated vegetatively and cuttings carry the vernalization effect with them. Certain woody perennials have a low-temperature requirement for flowering (Chouard, 1960), and several annual garden vegetables will flower somewhat earlier in the season if they are exposed to a short vernalization treatment (Thompson, 1953).

To summarize: Many different species are promoted to flower by cold periods, in some cases there

Figure 21-5 A 41-month sugar-beet plant kept vegetative by never being exposed to low temperatures. (Photograph courtesy of Albert Ulrich; see Ulrich, 1955. The technician at the Earhart Plant Research Laboratory at the California Institute of Technology is Helene Fox.)

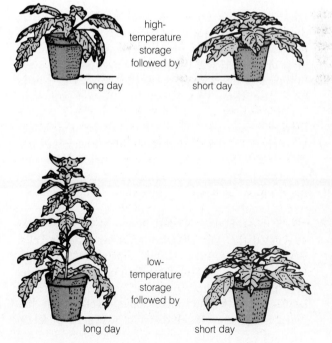

Figure 21-6 Bolting (flowering) response of henbane (*Hyoscyamus niger*), a typical rosette species, to storage at high or low temperature followed by long- or short-day treatment. Only cold followed by long days induces flowering.

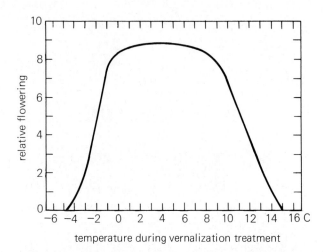

Figure 21-7 Final relative flowering response as a function of temperature during vernalization. The data represent response of moist Petkus rye seeds to a 6-week period of treatment. (From Salisbury, 1963; see Purvis, 1961.)

Physiological Experiments Sometimes in plant physiology we are unable to study biochemical or biophysical events directly but must take a more indirect approach. Whole plants are manipulated in various ways. Results are observed and deductions made, based on our understanding of plant function at the molecular level. These are **physiological experiments;** they typically produce descriptions of **phenomenology.** This chapter includes many examples.

A physiological investigation on vernalization might determine optimum temperatures (Fig. 21-7). Vernalization proceeds at a maximum rate over a fairly wide range of cool temperatures (depending somewhat upon the species), and vernalization occurs even at a few degrees below freezing. Usually the lower limit is set by the formation of ice crystals within the tissues. Another physiological study determines the most effective vernalization times. Minimum lengths for any observable effect vary from 4 days to 8 weeks, depending on species. Saturation times vary from 3 weeks for winter wheat to 3 months for henbane.

If a plant, immediately following vernalization treatment, is exposed to high temperatures, it often does not flower. This reversal is referred to as **devernalization.** To be strikingly effective, devernalizing temperatures must be about 30°C or higher with winter rye, and they must be applied within 4 or 5 days after the low temperature (somewhat longer in other species). Actually, some devernalization can be observed when plants are exposed to any temperature higher than that which will cause vernalization. In winter rye, 15°C is the neutral temperature; any temperature below this speeds flowering, and any higher temperature delays it. Anaerobic conditions given just after vernalization also cause devernalization, even at neutral temperatures. After devernalization, most species can be revernalized with another cold treatment.

is a quantitative and in others a qualitative response, and flowering of many species also requires or is promoted by suitable day length. These responses to environment prepare a plant for the annual cycle of climate. We are not dealing with an endogenous timer as in the previous chapter but with a complex system in which a plant responds to one season by becoming prepared for the next.

Location of the Low-Temperature Response It is the bud, presumably the meristem, that normally responds to cold by becoming vernalized. Only if buds are cooled will the plants subsequently flower. Embryos or even isolated meristems from rye seeds have also been vernalized. In another approach, various parts of a vernalized plant have been grafted onto an unvernalized plant. If a vernalized meristem is so transplanted, it will ultimately flower, but if a meristem from an unvernalized plant is grafted onto a vernalized plant after removal of the vernalized meristem, the growing transplanted meristem remains vegetative (see Lang, 1965b).

S. J. Wellensiek (1964) in the Netherlands has suggested that vernalization requires dividing cells. Several studies support his conclusion, although some seeds respond even at temperatures a few degrees below the freezing point, where cell division seems unlikely and microscopic investigation has failed to reveal it. If cell division or DNA replication in nondividing cells proves to be necessary, it could be significant. We noted in Table 18.1 that DNA must replicate before cellular differentiation. Perhaps only when the DNA is temporarily separated from the chromosomal proteins can gene activation or inactivation occur.

Vernalin and Gibberellins If the apical meristem itself responds to low temperatures, a translocated flowering stimulus or hormone does not seem likely. And in most cases, the vernalization effect is not translocated from one meristem to another, either within the same plant or when a vernalized plant is grafted to a nonvernalized plant. There are exceptions, however, as reported as early as 1937 by Georg Melchers in Germany (see Lang, 1965b). The work has since been extended somewhat in Russia. Melchers grafted a vernalized henbane plant to a vegetative receptor plant that had never experienced low temperature, inducing it to flower. Dissimilar response types will also transmit the flowering stimulus across a graft union; cold-requiring plants can be induced to flower without the cold period by being grafted to a noncold-requiring variety, for example. The reverse, though less clear-cut, will also occur. It must be emphasized, however, that transmission is limited to a few species.

In these experiments, a living graft union must be formed between the two plants, and conditions favoring transport of carbohydrates also favor transport of the stimulus. If the receptor is defoliated or darkened, for example, while photosynthesizing leaves are left on the donor, the receptor must obtain its nutrients from the donor, which favors movement of the vernalization stimulus across the graft union.

Melchers postulated existence of a hypothetical vernalization stimulus, which he called **vernalin.** The logical thing to do would be to isolate and identify it. Many futile attempts have been made, but results with gibberellins show that their properties are similar to those expected for vernalin. Anton Lang found in 1957 (see Lang, 1965b) that gibberellins applied to certain biennials induced them to flower without a low-temperature treatment (Fig. 21-8). Others (e.g., Purvis, 1961) induced winter annuals by treating seeds with gibberellins. It was then shown that natural gibberellins build up within several cold-requiring species during exposure to low temperature. Gibberellins clearly seem to be involved in vernalization.

But is gibberellin equivalent to vernalin? For various reasons, plant physiologists have been reluctant to accept this conclusion. When gibberellins are applied to a cold-requiring rosette plant, for example, the first observable response is the elongation of a vegetative shoot, followed by flower buds developing on this shoot. When plants are induced to flower by exposure to cold, however, flower buds are apparent as soon as the shoot begins to bolt. If the flowering response to applied gibberellins cannot be equated with the natural induction of flowering, several questions come to mind: Could gibberellins induce changes within the plant that in turn lead to flowering or even to production of vernalin? Could several molecules influence the morphogenetic program in the same way?

Mikhail Chailakhyan (1968) in Russia suggested that there are two substances involved in flower formation, one a gibberellin or gibberellinlike material, the other a substance he called **anthesin.** Plants that require low temperatures or long days or both might lack sufficient gibberellins until they have been exposed to the inducing environment, while short-day plants might contain sufficient gibberellins but lack anthesin. Chailakhyan suggested that cold-requiring plants produce vernalin in response to cold, and the vernalin is then converted to gibberellins in response to long days, at least in those plants requiring long days following the cold.

An elegant experiment performed many years ago by Melchers (1937) supports the two-substance

Figure 21-8 Anton Lang's induction of flowering in the carrot by application of GA$_3$ (center) or by vernalization (right). The control plant (left) and the center plant were held at temperatures above 17°C, while the plant at the right was given 8 weeks of cold treatment. The center plant was treated with 10 μg of GA$_3$ each day for the 8 weeks. (Photograph courtesy of Anton Lang; see Lang, 1957.)

point of view. A noninduced short-day plant (Maryland Mammoth tobacco) was grafted to a noninduced cold-requiring plant (henbane), causing the latter to flower. Apparently, each contained one of the essential substances for the flowering process but had to obtain the other from the plant to which it was grafted, which succeeded for the henbane but not for the tobacco. (In Chapter 22 we shall present evidence that gibberellins are less important in flowering than might be implied here.)

Vernalization and the Induced State Flower development typically follows the cold treatment by days or weeks. How permanent is this **induced state,** the vernalized condition of the plant before flowering? One henbane variety requires low temperatures followed by long days for flowering. After vernalization, flowering can be postponed by providing only short days (see Fig. 21-6). No loss of the vernalization stimulus appeared in such plants after 190 days, even though all the original leaves exposed to the cold had died. Only after 300 days was there any loss of the vernalized condition. In many other species, the induced state appears to be highly stable. Certain cereal seeds, for example, can be moistened (40 percent water—too little for germination), vernalized, and then dried out and maintained for months to years without loss of the vernalized condition. Yet there are several other species in which the induced state is far less permanent.

21.3 Dormancy

Only a few plants are able to function actively close to the freezing point. How, then, do plants live where temperatures remain close to freezing for several weeks or months each year? Most commonly, plants become dormant or quiescent, in which condition they remain alive but exhibit little metabolic activity. Leaves and buds of many evergreens show this reduced activity during the winter, and deciduous perennials lose their leaves and form special inactive buds. Seeds of most species in cold regions are dormant or quiescent during the winter. Certain changes occur in the cells of such seeds that allow them to resist subfreezing temperatures (see Chapter 24).

It seems appropriate that temperature itself might play a regulatory role in the survival of plants in cold regions. The dormant or quiescent condition of evergreen leaves and buds often develops in response to low temperatures, the effects often being accentuated by short days. Furthermore, subsequent growth in the spring is often dependent upon prolonged exposure of dormant buds and seeds to low temperatures during the winter. The buds or seeds accumulate or sum the periods of exposure to cold. Thus, they measure the length of the winter and anticipate spring when it is safe to resume growth and lose hardiness.

As we've seen so often in our discussions, the situation becomes complex. Plants commonly respond to multiple environmental cues. Germination of seeds, for example, is influenced not only by temperatures, but also (depending always upon species) by light, breaking of the seed coat to permit penetration of the radicle and perhaps entry of oxygen and/or water, removal of chemical inhibitors, and maturity of the embryo.

Concepts and Terminology Seed physiologists usually define **germination** as the process by which the **radicle** (embryonic root) elongates out through the seed coat. A seed may remain **viable** (alive) but unable to germinate or grow for several reasons. These can be roughly classified into *external* or *internal* conditions. An internal situation easy to understand is an embryo that has not reached a morphological

maturity capable of germination (e.g., in certain members of the Orchidaceae, Orobanchaceae, or the genus *Ranunculus*). Only time will allow this maturity to develop. Germination of seeds of wild plants is often limited in this or some other internal way, but seeds of many domestic plants may be limited only by lack of moisture and/or warm temperature.

To distinguish between these two different situations, seed physiologists have utilized two terms: **Quiescence** is the condition of a seed when it is unable to germinate only because external conditions normally required for growth are not present, and **dormancy** is the condition of a seed when it fails to germinate because of internal conditions, even though external conditions (e.g., temperature, moisture, and atmosphere) are suitable.

There are problems with this terminology. Dormant seeds are frequently induced to germinate by some specific change in the environment such as light or a period of low temperature. Where do we draw the line on conditions "normally required for growth"? Furthermore, in one sense it is *always* the internal conditions that are limiting. (If water is limiting, it is lack of water in the cells of the embryo inside the seed.) In another sense, external conditions always allow germination by influencing the internal ones. We can at least be more precise, stating conditions rather than depending upon the word *normally*: Dormancy* is the condition of a seed when it fails to germinate, even though ample external *moisture* is available, the seed is exposed to *atmospheric conditions* typical of those found in well-aerated soil or at the earth's surface, and *temperature* is within the range usually associated with physiological activity (say, 10 to 30°C). (A seed physiologist will define conditions still more precisely.) Quiescence is the condition of a seed when it fails to germinate *unless* the foregoing conditions are available (Jann and Amen, 1977).

One other term has been widely used in studies in this area. **Afterripening** is used by some authors in reference to any changes that go on within the seed (or the bud) during the breakdown of dormancy. Other authors have used the term in a more restricted sense, limiting it to maturation changes that occur in the embryo during storage (e.g., Leopold and Kriedemann, 1975).

---

*Researchers who study fruit trees (**pomologists**) employ a different terminology (Samish, 1954). The concept of dormancy as defined in this text is called *rest* by them, while the term *dormancy* is used by them in exactly the same sense as *quiescence*. You should be on guard as you study the literature, especially for the term *dormancy*, which may be used in either sense. To avoid confusion, dormancy is never used in the sense of quiescence in this text.

Table 21-1 Some Representative Life Spans for Seeds

	Species	Viability (%) Initial	Viability (%) Final	Age at test	Notes
1.	Sugar maple (*Acer saccharinum*)	—	—	<1 week	
2.	English elm (*Ulmus campestris*)	—	—	ca 6 months	
3.	American elm (*Ulmus americana*)	70	28	10 months	dry storage
4.	*Heavea, Boehea, Thea,* sugar cane, etc.	—	—	<1 y	
5.	Wild oats (*Avena fatua*)	70	9	1 y	buried 8″ in soil
6.	Alfalfa (*Medicago sativa*)	85	1	6 y	buried 8″ in soil
7.	Yellow foxtail (*Setaria lutescens*)	56	4	10 y	buried 8″ in soil
8.	Cocklebur (*Xanthium strumarium*)	50	15	16 y	buried 8″ in soil
9.	Canada thistle (*Cirsium arvense*)	57	1	21 y	buried 8″ in soil
10.	Kentucky bluegrass (*Poa pratensis*)	91	1	30 y	buried 8″ in soil
11.	Red clover (*Trifolium pratense*)	90	1	30 y	buried 8″ in soil
12.	Tobacco (*Nicotiana tabacum*)	89	13	30 y	buried 8″ in soil
13.	Button clover (*Medicago orbicularis*)	—	—	78 y	herbarium
14.	Clover (*Trifolium striatum*)	—	—	90 y	herbarium
15.	Big trifoil (*Lotus uliginosus*)	—	1	100 y	dry storage
16.	Red clover (*Trifolium pratense*)	—	1	100 y	dry storage
17.	Locoweed (*Astragalus massiliensis*)	—	—	100–150 y	herbarium
18.	Sensitive plant (*Mimosa glomerata*)	—	—	221 y	herbarium
19.	Indian lotus (*Nelumbo nucifera*)	—	—	1,040 y	peat bog
20.	Arctic lupine (*Lupinus arcticus*)	—	—	10,000 y	frozen silt, lemming burrows

From various sources. See summaries in Altman and Dittmer, 1962, and Mayer and Poljakoff-Mayber, 1963. Reprinted with permission of Pergamon Press, Inc., New York.

21.4 Seed Longevity and Germination

It is an impressive idea that a living organism can go into a sort of suspended animation, remain alive, but not grow for a long period of time, only to begin active growth when conditions are finally suitable. There have been reports of the successful germination of Emmer wheat from the ancient silos at Fayum (about 6400 years before the present) or the tomb of Tutankhamen at Thebes (4000 to 5000 B.T.P.), for example, but examination of such seeds has shown them to be not only dead but without any high molecular weight nucleic acid components (see Osborne, 1980). Bewley and Black (1982) suggest that at least one of these reports (from 1843; reproduced in their article) must have been a hoax. Nevertheless, the life span of some seeds is indeed great, in some cases exceeding the human life span.

Table 21-1 lists the longevities of several seeds. *Mimosa glomerata* had a probable longevity of 221 years, but a rather typical life span for seeds is from 10 to 50 years. Viable seeds of a lupine (*Lupinus arc-*

ticus) were found in a lemming burrow (along with the remains of a lemming) buried deep in permanently frozen silt of the Pleistocene Age in the central Yukon (Porsild et al., 1967). Material surrounding the seeds was dated by radio-carbon at 14,000 B.T.P., but there is no proof that the seeds themselves were that old (Bewley and Black, 1982). Viable lotus (*Nelumbo nucifera*) seeds found in peat in drained lakes of the Pulantien basin of South Manchuria have been estimated to be anywhere from very young (some results of radio-carbon dating) to one or two thousand years old (archeological dating of the peat). Similar lotus seeds were found in a boat in a lake near Tokyo, and the *boat* was radio-carbon dated at 3000 B.T.P.—which again says nothing about the seeds. But Priestley and Posthumus (1982) have radio-carbon dated portions of a viable lotus seed from Pulantien at about 466 B.T.P. at the time of germination. Of viable seeds dated by circumstantial evidence only, those of *Canna compacta* are probably most credible (Lerman and Cigliano, 1971). Apparently, the seeds had been inserted into young growing fruits of a walnut species, so that at maturity the fruits had healed, producing rattles. The seeds were entirely enclosed within the hardened shells. Shell material was carbon dated at 620 ± 60 years B.T.P.

Storage conditions always influence seed viability. Increased moisture usually results in a more rapid loss of viability, but a few seeds live longest submerged in water. Many domestic seeds, such as those of pea, soybean, and bean, tend to remain viable longer when their moisture content is reduced and they are stored at low temperature. Storage in jars or in the air at moderate to high temperature usually results in dehydration and severe cellular rupture when the seeds are hydrated. Cellular rupture injures the embryo and releases nutrients that are excellent as a substrate for pathogens. Oxygen is generally detrimental to seed life spans. Viability is usually lost most rapidly when seeds are stored in humid air at temperatures of 35°C or warmer. Some seeds remain alive longer when buried in the soil than when stored in jars on a laboratory shelf, perhaps because of differences in light, O_2, CO_2, moisture, and ethylene. A few seeds have an unusually short life span. Seeds of *Acer saccharinum*, *Zizana aquatica*, *Salix japonica*, and *S. pierotti* lose their viability within a week if kept in air. Seeds of several species may remain viable anywhere from a few months to less than a year.

How do long-lived seeds remain viable so long? While a seed remains alive, it retains its stored foodstuffs within its cells; as soon as it dies, some of these begin to leak out. Dormant but viable seeds remain intact on wet filter paper for months; as soon as they die, they are overgrown by bacteria and fungal hyphae, which live on the food that leaks out. What maintains the integrity of the membranes? Frits Went (1974) pointed out that "if the maintenance of the living membranes of 3000-year-old buried *Lotus* seeds required any metabolic processes, then no more than one sugar molecule would have been available each minute per hundred million protein molecules, truly negligible for any maintenance work."

And what happens during germination? Although it is an oversimplification, seed physiologists speak of four stages: (1) hydration or imbibition, during which water penetrates into the embryo and hydrates proteins and other colloids, (2) the formation or activation of enzymes, leading to increased metabolic activity, (3) elongation of radicle cells followed by emergence of the radicle from the seed coat (which is germination proper), and (4) subsequent growth of the seedling. Dormancy can be brought about by interference with any of the first three stages, and covering layers around the embryo—the endosperm, the seed coat, and the fruit coat—play a decisive role in this interference. In some, these layers prevent the entry of water and/or oxygen; in others, they prevent emergence of the radicle by acting as a mechanical barrier (Section 19.4); while in others, they apparently prevent leaching of inhibitors out of the embryos or contain inhibitors themselves. What are the causes of dormancy, what ecological advantages do dormancy mechanisms confer, and how are various forms of dormancy broken to allow germination?

21.5 Seed Dormancy

Impaction and Scarification One of the easiest examples of dormancy to understand is the presence of a hard seed coat that prevents absorption of oxygen or water. Such a hard seed coat is common in members of the Fabaceae (Leguminosae) family, although it does not occur in beans or peas, exemplifying that dormancy is uncommon in domesticated species. In a few species, water and oxygen are unable to penetrate certain seeds because entry is blocked by a corklike filling (the **strophiolar plug**) in a small opening (strophiolar cleft) in the seed coat. Vigorous shaking of the seeds sometimes dislodges this plug, allowing germination. The treatment is called **impaction,** and it has been applied to seeds of *Melilotus alba* (sweet clover), *Trigonella arabica*, and *Crotallaria egyptica*.

Albizzia lophantha is a small, leguminous, understory tree in southwest Western Australia (Dell, 1980). Seeds germinate mostly only in ash beds after fire; less than 5 percent germinate without heat. As it turns out, entry of water into the seed is prevented by a small strophiolar plug until this plug pops out when the seed is heated. Thus the distribution of this plant is controlled by fire and by the presence of a strophiolar plug.

Breaking the seed coat barrier is called **scarification.** Knives, files, and sandpaper have been used. In nature, the abrasion may be by microbial action, passage of the seed through the digestive tract of a bird or other animal, exposure to alternating temperatures, or movement by water across sand or rocks. In the laboratory and in agriculture (when needed), alcohol or other fat solvents (which dissolve away the waxy materials that sometimes block water entry) or concentrated acids may be used. The seeds of cotton and many tropical tree legumes, for example, may be soaked for a few minutes to an hour in concentrated sulfuric acid and then washed to remove the acid, after which germination is greatly improved.

Scarification is of considerable ecological importance. The time required for scarification to be completed by some natural means may protect against premature germination in the autumn or during unseasonal warm periods in winter. Scarification in the digestive tracts of birds or other animals leads to germination after the seeds are more widely dispersed. Dean Vest (1972) demonstrated an interesting symbiotic relationship between a fungus and the seeds of shadscale (*Atriplex confertifolia*) growing in the deserts of the Great Basin. The fungus grew on the seed coats, scarifying them so germination could occur. Fungal growth occurred only when temperatures and moisture conditions were suitable during early spring, the most likely time for survival of the seedlings.

As noted in relation to *Albizzia*, fire is another important natural means of scarification. Several seeds, particularly in situations such as the chaparral vegetation of Mediterranean climates (e.g., Southern California), are effectively scarified by the fires so common there. The result is a relatively rapid recovery of the area following the fires.

Chemical Inhibitors What prevents the seeds in a ripe tomato from germinating inside the fruit? Temperature is usually ideal, and there is ample moisture and oxygen. If the seeds are removed from the fruit, dried, and planted, they germinate rapidly, indicating that they are mature enough for germination. They apparently fail to germinate in the fruit because of chemical germination inhibitors (Mayer, 1974; Mayer and Poljakoff-Mayber, 1975). Tomato juice can be diluted 10 to 20 times and still inhibit germination of many seeds. In many fleshy fruits, the osmotic potential of the fruit juice is simply too negative to permit germination. Other fruits may filter out wavelengths of light that are necessary for germination.

Exogenously supplied ABA blocks the translation of mRNA in seeds, which then inhibits the synthesis of enzymes that are essential for germination. Whether the levels of ABA in dormant seeds are sufficient to cause the same response is unknown. Because of the many observations in which exogenous ABA inhibits germination and dormancy, however, many scientists have speculated that ABA normally induces dormancy (Walton, 1977).

There are also chemical inhibitors present in the seeds, and often these must be leached out before germination can occur. In nature, when enough rain falls to leach inhibitors from the seed, the ground will be adequately wet for survival of the new seedling. This is especially important in the desert where moisture limits more than other factors such as temperature. Vest (1972) found that shadscale seeds contained enough sodium chloride to inhibit them osmotically (see also Koller, 1957). Usually the inhibitor is more complex than table salt (Evenari, 1957; Ketring, 1973), and inhibitors include representatives from a wide variety of organic classes. Some are cyanide-releasing complexes (especially in rosaceous seeds), while others are ammonia-releasing substances. Mustard oils are common in the Brassicaceae (Cruciferae). Other important organic compounds include organic acids, unsaturated lactones (especially coumarin, parasorbic acid, and protoanemonin), al-

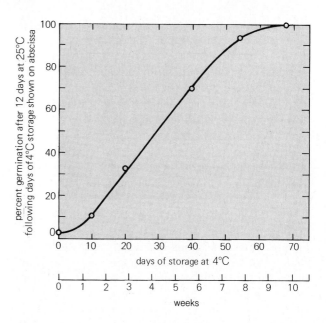

Figure 21-9 Germination of apple seeds as a function of storage time at 4°C. (Data from Villiers, 1972.)

dehydes, essential oils, alkaloids, and phenolic compounds. The most widespread and perhaps the most potent inhibitor is probably ABA.

These substances occur not only in seeds but also in leaves, roots, and other parts of plants. When leached out or released during decay of litter, they may inhibit the germination of seeds or root development in the vicinity of the parent plant (Chapter 14). Substances produced by one plant that harm another are called **allelopathics** (Sections 14.3 and 23.6). (Of course, allelopathics do not produce dormancy in the usual sense.) Actually, some compounds produced by other organisms act as germination promoters. For example, nitrate is a commonly used germination promoter in seed physiology laboratories and is produced by decay of virtually any plant or animal residue.

Before leaving the topic, we should note that many known compounds that are not natural products may strongly influence germination one way or the other. These include many of the growth regulators presently of commercial importance (e.g., Dalapon, others). Thiourea is often used in the laboratory as a germination promoter.

Prechilling Many seeds, particularly those of rosaceous species such as the stone fruits (peach, plum, cherry), many other deciduous trees, several conifers, and several herbaceous *Polygonum* species will not germinate until they have been exposed to low temperatures in the moist condition in the presence of oxygen for weeks to months (Fig. 21-9). Crocker and Barton (1953) listed 62 such species, and numerous others have been found since. Rarely, moist seeds respond to high temperatures, and several seeds respond best when daily temperatures alternate between high and low. The practice of layering the seeds during winter in flats containing moist sand and peat is called **stratification.** Because seeds in the flats are cooled before they will germinate, a more popular and descriptive term than stratification nowadays is **prechilling.** Prechilling in seed laboratories and for physiological experiments is routinely performed in incubators or growth chambers. In nature, the low temperature requirement protects the seed from precocious germination in the fall or during an unseasonal warm period in winter.

What chemical changes go on within the seed during prechilling, allowing it subsequently to germinate when conditions are right? Most seeds, including those that require cold, are rich in fats and proteins but have little starch (Nikolaeva, 1969; Lang, 1965a), and during the cold treatment the embryo of some species grows extensively by transfer of carbon and nitrogen compounds from food storage cells. Sugars accumulate, and these might be required as sources of energy and to attract water osmotically, later causing germination. Even in cold-requiring seeds such as European ash (*Fraxinus excelsior*) where the embryo is already fully developed before stratification, there is a massive degradation of fat in the embryo itself during the cold. The protein content rises, and starch then appears.

Perhaps inhibitors disappear during cold treatment, and/or growth promoters such as gibberellins or cytokinins accumulate (Kahn, 1977). Auxins have little effect on germination, but in many cases gibberellins will substitute for all or part of the cold treatment, just as they often do in vernalization. Perhaps they accumulate during stratification in amounts that overcome dormancy, but most of the data obtained so far argue against this interpretation (Bewley and Black, 1982). Cytokinin effects are usually less dramatic and are much less widespread. In a sense, the far-red-absorbing form of phytochrome (P_{fr}) is a growth regulator that is required for germination of many seeds, as discussed in Chapter 19.

Both inhibitor disappearance and hormone accumulation in whole seeds have been observed, but there are numerous contradictions. In Section 19.3, we discussed similar studies with light-requiring seeds. We concluded that the radicles themselves should be analyzed, because changes in the rest of the seed could mask the important changes in relatively small radicles. I. Arias and co-workers (1976) measured gibberellins in the embryonic axis and in the food storage cotyledon cells of the hazel tree (*Corylus avellana*), a species in which gibberellins fully overcome the stratification requirement. During chilling, there was little accumulation of gibberellin in either part, yet chilling allowed the embryonic axis to synthesize much gibberellin when returned to a germination temperature of 20°C. The much larger cotyledons synthesized relatively little gibberellin, so the GA concentration became 300 times as great in the axis as in the cotyledons. Similar studies are needed with other seeds.

The molecular basis for the breaking of dormancy in seeds is yet to be discovered. A number of substances that limit or alter cellular respiration often promote germination. Compounds that inhibit respiration, such as nitrite, cyanide, azide, malonate, thiourea, and dithiothreitol can often break seed dormancy. On the other hand, Roberts and Smith (1977) and others have shown that elevated levels of oxygen can induce germination of certain dormant seeds.

Where does the dormancy mechanism lie? Does the seed coat contain a chemical that inhibits elongation of the radicle? Does the seed coat or endosperm act as a mechanical barrier to elongation? Or does the radicle itself lack the ability to grow until chilled? Isolated, prechilled embryos from many seeds will subsequently grow when placed at warmer temperatures, but nonprechilled naked embryos will not. Hence, cold temperatures must act directly on the embryo. Consistent with this, growing embryos in prechilled walnut seeds can exert mechanical pressure at least 1.0 MPa greater than nonchilled embryos incapable of breaking the shells. In various species of lilac, including *Syringa vulgaris*, prechilling has no effect on the strength or the inhibitor content of the endosperm, but radicles of chilled embryos will elongate in a solution having a water potential about 0.5 MPa more negative than that in which nonchilled radicles will grow (Junttila, 1973). Although inhibitors are often present in seed coats, direct evidence that they inhibit radicle growth in such seeds seems to be absent. Whether changes in inhibitors and growth promoters in the radicle account for stratification remains to be seen.

Light In Sections 19.1 and 19.4, we saw that light may control germination of many seeds, and we discussed some of the complications in this response. Clearly there are several environmental cues, often interacting in intricate ways, to control the germination process.

21.6 Bud Dormancy

In temperate regions, seed and bud dormancy have much in common, but with buds we are nearly as concerned with the induction of dormancy as we are with breaking dormancy. Bud dormancy almost always develops before fall color and the senescence of leaves. Buds of many trees stop growing in midsummer, sometimes exhibiting a little growth again in late summer before going into deep dormancy in autumn. Flower buds that will grow the next season may form on fruit trees in midsummer. The leaves remain green and photosynthetically active until early autumn, when leaf senescence occurs in response to short, bright, cool days. As chlorophyll is lost, the yellow and orange carotenoid pigments become apparent, and anthocyanins (primarily cyanidin glycoside) are synthesized. Fruits such as apples often mature during this time. Frost hardiness also develops in response to the low temperatures and short days of autumn.

Bud dormancy is induced in many species by low temperatures, but there is also a response to day length, especially if temperatures remain high. With several deciduous trees studied at Beltsville (Downs and Borthwick, 1956), short-day treatment resulted in formation of a dormant terminal bud and cessation of internode elongation and leaf expansion, but often the leaves were retained. Long nights, each interrupted by an interval of light, acted the same as long days. The buds of birch (*Betula pubescens*) detect the day length directly, but leaves usually detect the photoperiod although dormancy occurs in the buds (Wareing, 1956). Perhaps this correlative phenomenon, like others, is caused by a growth regulator, which could be abscisic acid (Section 17.5).

There are always a number of interactions. In the Beltsville study with deciduous trees, the day-length response was observed at temperatures of 21 to 27°C, but at temperatures of 15°C and 21°C there was little stem growth on either long days or short days.

There is considerable genetic variability in dormancy responses within a species. For example, Thomas O. Perry and Henry Hellmers (1973) found that a northern (Massachusetts) race of red maple (*Acer rubrum*) developed winter dormancy in response to short days and cold temperatures in growth chambers, but a southern race from Florida did not. O. M. Heide (1974) studied Norway spruce (*Picea abies*). Austrian trees (47° latitude) stopped elongating at day lengths of 15 hours or less, but trees from northern Norway (64° latitude) stopped when day lengths were 21 hours or less. In either case, it was well before killing frosts. Temperature had little influence, but trees from high elevations stopped elongating at day lengths longer than those required to halt the growth of trees at the same latitude but at low elevations. Heide also found that the roots did not respond to photoperiods applied to the tops. With few exceptions, roots continue to grow as long as nutrients and water are available, until soil temperatures become too cold (Kramer and Kozlowski, 1979). Clearly, such trees are well adapted to the environments where they naturally occur. The Florida maples, for example, because they cannot enter dormancy soon enough in the fall, are restricted to warm, southern climates.

Withholding water frequently accelerates development of dormancy, as does the restriction of mineral nutrients, particularly nitrogen. This is probably important for species that go into dormancy before the high temperatures and drought occurring in dry climates or in the tropics. Situations are also known in which dormancy develops in response to *changing* day length (and even to changing soil temperature).

As dormancy first begins to develop, it can easily be reversed by moderate temperatures and long days (or continuous light). The plant is then only in the initial stages of quiescence and is not truly dormant. Gradually, however, attempts to induce active growth fail, and at this time the plant is said to have reached true dormancy (Vegis, 1964).

Morphology is important in dormancy phenomena. A dormant bud typically has greatly shortened internodes and specially modified leaves called **bud scales.** These scales prevent desiccation, briefly insulate against heat loss, and restrict movement of oxygen to the meristems below. They may also respond to the light environment and/or perform other functions. In a sense, bud scales are analogous to the seed coat.

The hormonal factors involved in dormancy are not known, but in trees abscisic acid has been implicated in the response (Walton, 1980). In the mid-1960s, one group of researchers reported that they were able to induce the formation of resting buds in a number of tree species by feeding ABA through the leaves, but no one has been able to reproduce their results. Phillips et al. (1980) list numerous examples of conflicting data on whether ABA accumulates in dormant tissue. Because of these conflicting data, it is now impossible to conclude that ABA normally causes dormancy.

Overcoming dormancy is also a function of temperature, day length, or both. The temperature effect was studied as early as 1880 to 1914 (see Leopold and Kriedemann, 1975), but the day-length effect has been recognized only since the late 1950s. Since leaves respond to day length in the induction of dormancy and in flowering, it seemed reasonable that leaves are the only organs that respond to day length. But it is now known that dormancy is broken by long days in several leafless trees: beech, birch, larch, yellow poplar, sweetgum, and red oak, for example. Except for beech, these species also respond to cold periods. In other species, cold must be followed by long days. Even in midwinter, certain deciduous species will respond to long-day treatment (particularly continuous light).

A midsummer dormancy occurs in some species (especially evergreens), during which the stems cease to elongate for a period of time. This is typically broken by exposure to more long days.

What organ responds to the long days that overcome dormancy? Apparently the bud scales themselves respond, or enough light penetrates to bring about the response within the primordial leaf tissues inside the bud. Probably both the short-day induction and the long-day breaking of dormancy are phytochrome responses, but the case is not clear-cut. In some studies of the short-day induction, red light is most effective in the night interruption, and its effect is reversed somewhat by a subsequent exposure to far-red light, but this reversal has failed in several other studies.

Dormancy in many buds can be broken by exposure to low temperatures. Days to months may be required at temperatures below 10°C. With fruit trees, 5 to 7°C is more effective than 0°C. Considerable work has been done with fruit trees to determine the minimum cold period required to break dormancy, because this period determines how far south they can be grown in the northern hemisphere. Apples, for example, may require 1000 to 1400 h at about 7°C. Headway has been made in selecting peach varieties with a shorter chilling requirement than normal, for example, which allows them to be cultivated where the winters are warmer. Incidentally, high temperatures following cold will reinduce dormancy in apple trees, a situation closely analogous to devernalization.

The effects of chilling upon the breaking of dormancy are not translocated within the plant but are localized within the individual buds. A dormant lilac bush, for example, may be placed with one branch protruding outside through a small hole in the greenhouse wall. The branch exposed to the low temperatures of winter will leaf out in early spring, but the rest of the bush inside the greenhouse remains dormant.

Several chemical treatments of the bud will break dormancy. For instance, 2-chloroethanol ($ClCH_2CH_2OH$), often called ethylene chlorohydrin, has been used with success for many years. Applied in vapor form, it breaks dormancy of fruit trees. Another simple but often effective treatment is immersion of the plant part in a warm water bath (40 to 55°C). Often a short exposure (15 s) is effective. Applied gibberellins break bud dormancy in many deciduous plants, just as they break dormancy of many cold-requiring seeds and induce flowering of many cold-requiring plants.

21.7 Underground Storage Organs

In many cases, temperature conditions will induce the formation of such underground storage organs as bulbs, corms, and tubers. In some species, dormancy is also broken or subsequent growth influenced by storage temperatures. In other species day length also influences formation of the organs.

The Potato Under usual greenhouse conditions, potato tubers form in response to short days. Such tubers develop from swellings at the tips of underground stems called **stolons**,* which are derived from nodes at the base of the stem in the soil. All the expected features of photoperiodism are present (including a critical night and inhibitory effect of a light interruption during the dark period—see Chapter 22). Although there are differences among cultivars, in one study tuber formation did not require short days but proceeded at any day length (a day-neutral response) when the night temperature was below 20°C. Tuberization was optimal at night temperatures of about 12°C. Such an interaction between photoperiodism and temperature is common, as we explained in relation to vernalization and dormancy.

Since the underground stolons are in darkness, some **tuber-inducing principle** must be produced in the leaves in response to short-day treatment and translocated to the stolons. Low temperatures stimulating tuberization are also detected by the tops of the plants, rather than the underground parts. On long days, no tubers will form at any soil temperatures unless the shoots are exposed to low temperatures. Stolons form over a wide range of soil and air temperatures and day lengths, but only the conditions just described cause these to develop into tubers.

Among the known hormones, auxins are quite inactive in tuber formation, while gibberellins inhibit formation and promote stolon elongation. Ethylene causes increased radial growth and decreased elongation of the stolon tip cells, as it does in other tissues. This stolon swelling is morphologically somewhat similar to that of normal tubers, but these swellings contain little starch. In normal tuber formation, starch formation precedes visual morphological changes. Thus ethylene may not trigger tuber formation, but it could contribute to continued radial expansion. Low concentrations of kinetin stimulate formation of starch-rich tubers, but it is not known whether endogenous cytokinins normally control this process.

Because it is an underground stem, the potato tuber exhibits stem characteristics. Its eyes are the axillary buds, and they remain inactive in response to the presence of the apical bud. When the potato is cut up to produce seed pieces, this apical dominance is lost, and the axillary buds grow if dormancy has been broken. There are practical reasons both to prolong and to break tuber dormancy. The longer tubers can

---

*Stolons are usually defined as aboveground, horizontal stems, as in the strawberry. Underground horizontal stems are **rhizomes**. Potato "stolons" (the term used by physiologists who work with potatoes) are usually underground, but they can be aerial; in darkness, even aboveground potato buds develop into stolons.

be stored during winter and spring in the dormant condition, the higher their price when sold. In potato "seed" certification, however, it is desirable to break dormancy prematurely to test for pathogens in sample tubers. The time normally required to break dormancy is somewhat shorter when the tubers are stored at about 20°C than at lower temperatures, but there is no clear-cut temperature effect. Certainly there is no cold requirement.

It is possible to break dormancy in potato tubers by the chemical treatments that are effective in breaking bud dormancy of aboveground stems (2-chloroethanol, gibberellins, hot water, and so on). Thiourea also causes sprouting but can result in as many as eight sprouts from a single eye rather than the usual single sprout. Dormancy can also be induced or prolonged by spraying such growth regulators as maleic hydrazide or chloropropham on the foliage before harvest or on the tubers after harvest. Storage temperature is also important. Tubers sprout somewhat prematurely at high temperatures, and at low temperatures starch turns to sugar. If a single storage temperature must be used, the ideal compromise seems to be 10°C. Modern potato processing facilities, however, store the tubers at a much lower temperature (about 2°C), and then when they are ready to slice and fry the tubers to make chips, they move them to a higher temperature storage area for several days so that the sugar will be converted back to starch. If this is not done, the sugar caramelizes during frying to produce a dark brown or even black color that is undesirable in the final potato chips.

Bulbs and Corms   There has been little investigation of how bulbs, corms, and rhizomes are induced to form, but much work has been done in Holland, supported largely by the Dutch bulb industry, to determine the optimum storage conditions (primarily storage temperature as a function of time) that will result in the formation of leaves, flowers, and stems at desirable times and with the desirable properties. The approach was to observe the morphology of the bulb carefully in the field during a normal season, and then to repeat these observations with bulbs stored under accurately controlled temperature conditions. The goal was to shorten the time to flowering, a process called **forcing.** This work has been going on since the 1920s (see Hartsema, 1961; Rees, 1972).

Here are a few generalizations: Bulbs must reach a critical size, which often requires 2 or 3 years, before they begin to respond to temperature storage conditions by the formation of flower primordia. In some cases (e.g., tulip), leaf primordia are formed before the flowers, but sometimes leaf and flower formation are nearly simultaneous. Specific temperatures are often required for flower initiation or sub-

sequent stem elongation. The pattern of change and the optimum temperatures usually match the climate at the location where the bulbs are native.

There are several patterns: In some species, flower primordia form before the bulbs can be harvested. This allows little control during storage, so these have not been studied very much. In others, flower primordia form during the storage period after harvest in the summer but before replanting in the fall, making control easier. Figure 21-11 shows a temperature storage regimen designed to cause rapid flowering of tulips in time for Christmas. Note that temperatures inducing flowers are relatively high compared to those effective in vernalization of seeds and whole plants. Nevertheless, there is a parallel response.

In most bulbous irises (Fig. 21-10) the actual flower primordia appear during the low temperatures of winter (9 to 13°C optimum), but a high temperature (20 to 30°C) pretreatment is essential if flower formation is to occur at all. This is a true example of induction similar to vernalization, but the response is to high rather than low temperatures. In each of the examples, the plants are adapted so that their flowering, vegetative growth, and dormancy are nicely synchronized with seasonal changes in temperature.

21.8 Thermoperiodism

Growth rates of the vegetative parts of a plant are always strongly influenced by temperatures, as we have seen (see Fig. 21-1). Sometimes plant form is changed in rather specific ways, and often the responses are delayed. This is often most noticeable in the vegetative structures associated with flowering, as in the bolting stalks of rosette plants formed in response to vernalization and/or long-day treatment. Prechilling of seeds sometimes has a strong delayed effect on growth, in addition to its dormancy-breaking action. If the embryos of peach seedlings are excised from their cotyledons, they germinate without prechilling, but the seedlings are frequently stunted and abnormal. When excised embryos are kept at low temperature, they grow into normal seedlings, so it is prechilling and not the presence of the cotyledons that insures their normality. Since stunted plants often lose their dwarf habit when sprayed with gibberellins, the accumulation of gibberellins or other hormones during prechilling could account for these results, or prechilling could increase the potential to synthesize gibberellins.

Most of the temperature discussion so far has been concerned with the annual temperature cycle, but Frits Went (1957) has described **thermoperiodism,** a phenomenon in which growth and/or de-

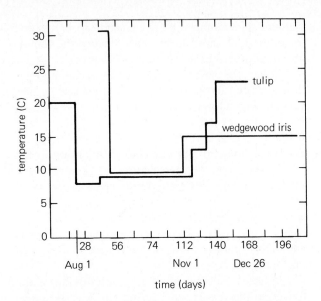

Figure 21-10 Temperature treatment for early flowering of *Tulipa gesneriana* W. Copeland and of *Iris xiphium* Imperator. With the tulip, flower initiation begins and is well under way during the 20°C treatment. Moving to storage rooms at 8 and 9°C provides an acceleration in blooming, so that flowers are produced at Christmas. Continuous 9°C gives equal earliness, but quality is poor unless the 20°C treatment is given first. The bulbs are planted in a controlled-temperature greenhouse about midway during the low-temperature treatment. The temperature is first raised when the leaf tips are visible, then again when they are 3 cm long, and finally when they are 6 cm long. With iris, the short period at high temperature is essential to flowering, although actual initiation of flower primordia does not occur until the bulbs have been moved from low temperature to 15°C, at which time the sprouts are about 6 cm long. Again the 9°C treatment is to insure earliness. At temperatures much above 15°C during the last part of the treatment, abnormal flowers are sometimes produced. Low light levels will also result in "blasted" flowers at this time, especially if the temperatures are not right. If extremely high temperatures (38°C) are used during the first flower-induction period, flower parts are increased or decreased, or tetramerous, pentamerous, or dimerous flowers result. (Data from Annie M. Hartsema, 1961; figure from Salisbury, 1963.)

velopment is promoted by alternating day and night temperatures. We noted before that potato tubers form in response to low night temperatures; fruit set on tomato plants is also promoted by low night temperatures. Stem elongation and flower initiation are also thermoperiodic responses in some species. An original implication of the thermoperiodism concept was that plant productivity was higher under a thermoperiodic environment. For some species, including certain tomato cultivars, this is true, but fluctuating day and night temperatures are not essential for optimum growth of numerous other species. Cocklebur, sugar beet, wheat, oats, bean, and pea grow as well at an *optimum* constant temperature as they do when day and night temperatures vary. An experimenter must be careful to compare various thermoperiodic regimens with the optimum constant temperature rather than some other temperature (Friend and Helson, 1976).

Some plants grow better when the environment fluctuates on a 24-h cycle, presumably to coincide with the phases of their circadian clock. Thus, some species grow poorly when both light and temperature are constant. Varying temperature on a 24-h cycle prevents or reduces the injury caused to tomato plants by continuous light and temperature, if light levels are high enough. Indeed, many thermoperiodic responses interact with the light environment, typically via photoperiodism and probably balances in the phytochrome system.

One of the most spectacular examples of thermoperiodism reported by Went (1957) involved *Laothenia charysostoma* (formerly *Baeria*), a small annual composite commonly seen during spring in California. This plant normally occurs in mountain valleys and foothills and occasionally in the west Mojave Desert. It is extremely sensitive to night temperature. Grown under short-day conditions in Went's experiments, plants survived only two months when the night temperature was 20°C. At lower temperatures, they grew for at least 100 days. They died rapidly at night temperatures of 26°C. Many species do not grow particularly well at night temperatures this high, but how can we account for death at this temperature? The *Laothenia* plants flourish when the day temperature is well above 26°C, providing only that the night temperature is low enough. Other plants native to California acted similarly in Went's experiments.

Perhaps, as pointed out earlier, different tissues within the same plant may have differing cardinal temperatures. For proper growth and development of the entire plant, the temperature range within the day should include near optimal temperatures for the growth of all necessary tissues. Normally, soil temperatures are different from air temperatures, so plants may have differing cardinal temperatures for roots and shoots. By maintaining roots and shoots at the same temperature, optimal growth and development may not be obtained.

21.9 Mechanisms of the Low-Temperature Response

How can we understand positive plant responses to low temperature? We might be dealing with some kind of hormonal or metabolic block. Such a block could be a chemical inhibitor or the lack of some necessary substance within the plant, or both. An inhibitor could disappear, or a growth regulator could arise at low temperatures, influencing flowering, germination, subsequent seedling growth, and so on. Gibberellins and ABA often seem to play roles. Are the mechanisms the same in the several responses we have described? Surely the diversity is great enough that we might not expect a common mechanism, but in many cases there are striking similarities.

Remember the paradox introduced at the beginning of the chapter. If low temperatures reduce the rate of chemical reactions, how can we account for increased production of some growth promoter or increased destruction of an inhibitor at low temperatures compared to high? In the 1940s, Melchers and Lang, and Purvis and Gregory, simultaneously and independently suggested a model (Fig. 21-11), not unlike that of Figure 21-3. There might be two hypothetical interacting reactions, one (I) with a fairly low temperature coefficient or Q_{10}, the other (II) with a higher Q_{10}. Products of reaction I are acted upon by reaction II. If the rate of reaction I exceeds that of II, then the product of reaction I (B) will accumulate; if the reverse is true, the product of reaction II (C) will accumulate. Even if the Q_{10} of reaction I is relatively low, but the reaction progresses at low temperatures more rapidly than reaction II, then we can explain the accumulation of B at low temperatures. With increasing temperature, the rate of reaction II increases much more rapidly than the rate of reaction I, so at some critical temperature, B will be used as fast as it is produced and hence will not accumulate. Reaction II would be devernalization, and the fact that devernalization fails after two or three days at the neutral temperature might indicate a third reaction (III), which converts B to D, a stable end product. Of course, the model is naive speculation, because many other factors could play roles: enzyme synthesis, enzyme activation, membrane permeability changes, phase changes, and so on. In 40 years no one has actually found such a mechanism in organisms, yet the principle has a certain logic and might still turn out to be valid.

The compensated feedback systems discussed in relation to temperature independence in the biological clock (Section 20.7) could provide for an overall reaction with a negative temperature coefficient. The *product* of one reaction might inhibit the *rate* of another. Or, at low temperatures, a substance might

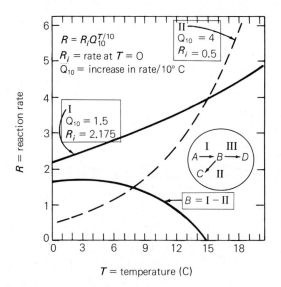

Figure 21-11 Sample curves showing hypothetical reaction rates as a function of temperatures for reactions with Q_{10} values of 1.5 or 4.0. If the reaction with $Q_{10} = 1.5$ is considered to be reaction I in the reactions shown in the circle (and discussed in the text), and the reaction with $Q_{10} = 4.0$ to be reaction II, then the hypothetical product B will be proportional to curve II minus curve I, as shown (curve B). Compare the shape of curve B with the curves in Figures 21-1, 21-3, and 21-7. (From Salisbury, 1963.)

accumulate because another compound inhibiting its production might not. Again, different temperature coefficients would be required. Since gibberellins increase in some seeds and buds as dormancy is broken, they may be equivalent to B or D in Fig. 21-11. Or gibberellins might leak out of a storage compartment when membranes become much more permeable at low temperatures (Arias et al., 1976). In a few species, cytokinins or ethylene could play this role.

What if we are dealing with destruction of an inhibitor at low temperatures rather than the synthesis of a promoter? We have only to reverse the roles of the two hypothetical reactions in the model. The destruction (or conversion) reaction must have a fairly rapid rate at low temperatures and a low Q_{10}. The synthesis of the inhibitor, on the other hand, must be low at low temperatures but have a high Q_{10}.

22

Photoperiodism

The synchronization of organisms with seasonal time is a truly spectacular manifestation. Often, this synchronization is concerned with reproduction: It is appropriate and adaptive for young animals to be born at specific times of year, for all members of a given angiosperm species to flower at the same time (ensuring an opportunity for cross-pollination), and for mosses, ferns, conifers, and even some algae to form reproductive structures in a given season. Many other plant responses, such as stem elongation, leaf growth, dormancy, formation of storage organs, leaf fall, and development of frost resistance, also occur seasonally. Frequently, these seasonal responses are synchronized by photoperiodism. Much of what we see happening in the natural world is happening because plants and animals are able to detect the relative lengths of day and night.

22.1 Detecting Seasonal Time by Measuring Day Length

In a nonmountainous region on the equator, sunrise and sunset occur at the same time each day throughout the year, so the relative lengths of day and night remain constant. Exactly at the poles, the sun remains above the horizon for six months each year and below for the other six months. Again, the day and the night are about equal; each is six months long! As one travels from the equator toward the poles, the days become longer in summer and shorter in winter (Fig. 22-1, top). This is because the equator is tipped 23.5° to the plane of the ecliptic (the earth's orbit around the sun), so that during winter the pole is tipped 23.5° away from the sun and during summer 23.5° toward the sun.

The rate at which day length changes varies during the year (Fig. 22-1, bottom). Near the times of the summer and winter solstices, when days are longest and shortest, there is little change from day to day;

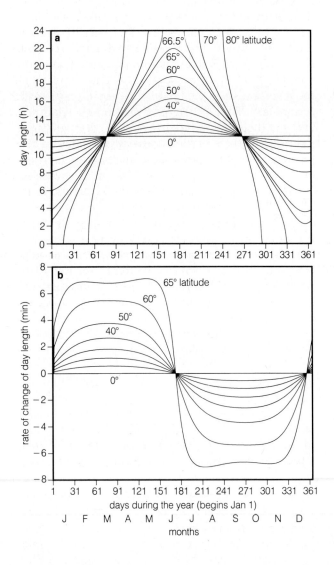

during spring and autumn, the rate of change is much more rapid, as days become longer during spring, shorter during fall. Thus, an organism might detect the season by measuring day and night lengths and how they change, but because the absolute day lengths at any time of year depend so strongly upon

c

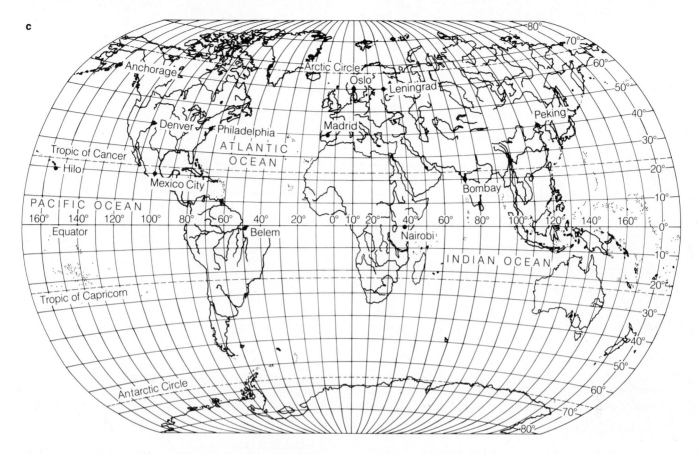

Figure 22-1 (**a**) Day length as a function of time during the year for various northern latitudes shown on the map (**c**). Day lengths are calculated as the time from the moment the upper part of the sun (rather than its center) first touches the eastern astronomical horizon (i.e., the horizon as it would appear on the open ocean) in the morning until the upper part of the sun just touches the western astronomical horizon in the evening. Thus the day length at the equator defined this way is slightly longer than 12 h. The Arctic Circle, 66.5° latitude, is where the center of the sun just touches the astronomical horizon at midnight on the summer solstice, but part of the sun is visible above the horizon at midnight for a few days before and after the summer solstice, as you can see from the day-length curve for 66.5° latitude. (**b**) The rate of change of day lengths as a function of time during the year for several latitudes. Note that the rate of change is quite steady during much of the year (i.e., the curves are relatively flat), especially at far northern latitudes. (The curves were computer-generated and drawn; programs were written by Michael J. Salisbury, based on equations and explanations supplied by Roger L. Mansfield of Astronomical Data Service, 3922 Leisure Lane, Colorado Springs, Colorado 80917.)

latitude, organisms must be "calibrated" to take this into account.

Study of the photoperiodic responses of organisms could contribute important information to our understanding of natural ecosystems, but of the approximately 300,000 species of plants, only a few hundred have been grown with different artificial photoperiods. Not many surprising facts have come to light through this work. As one would expect, plants that grow at latitudes far from the equator respond to longer days than plants growing closer to the equator. It was surprising, however, to learn that many tropical plants respond to day length, detecting the small changes that occur 5 to 20 deg from the

equator (about 1 min day^{-1} at 20° latitude in March or September).

It was also interesting to learn that different ecotypes (see Section 23.4) within a single species often have different responses to day length. In three representative studies, specimens of two short-day plants, lambsquarters (*Chenopodium album*; Cumming, 1969) and cocklebur (*Xanthium strumarium*; McMillan, 1974a), and one long-day plant, alpine sorrel (*Oxyria digyna*; Mooney and Billings, 1961), were collected at various latitudes throughout North America. Alpine sorrel plants were also collected in the arctic and cocklebur plants from all over the world. In each case, the day length that induced

Table 22-1   Representative Species Having Various Flowering Responses to Photoperiodic Treatment[a]

I. Known species that flower in response to a single inductive cycle

SHORT-DAY PLANTS	Approximate Critical Night[b]	LONG-DAY PLANTS	Approximate Critical Day[b]
Chenopodium polyspermum Goosefoot		*Anagallis arvensis* Scarlet pimpernel	12–12.5
Chenopodium rubrum Red goosefoot		*Anethum graveolens* Dill	11
Lemna paucicostata Duckweed		*Anthriscus cerefolium* Salidcherril	
Lemna perpusilla Strain 6746 Duckweed	12	*Brassica campestris* Bird rape	
Oryza sativa cv Zuiho Rice	12	*Lemna gibba* Swollen duckweed	
Pharbitis nil cv Violet Japanese morning glory	9–10	*Lolium temulentum* Darnel ryegrass	14–16
Wolffia microscopia Duckweed		*Sinapis alba* White mustard	ca 14
Xanthium strumarium Cocklebur	8.3	*Spinacia oleracea* Spinach	13

II. Some species that require several cycles for induction[c]

SHORT-DAY PLANTS

1. SDP (qualitative or absolute)

   Cattleya trianae   Orchid
   Chrysanthemum morifolium   Chrysanthemum variety
   Cosmos sulphureus   cv Yellow Cosmos
   Glycine max   Soybean
   Kalanchoe blossfeldiana   Kalanchoe
   Perilla crispa   Purple common perilla
   Zea mays   Maize or corn

   SDP at high temperature; quantitative SDP at low temperature
   Fragaria × *ananassa*   Pine strawberry

   SDP at high temperature; day-neutral at low temperature
   Pharbitis nil   Japanese morning glory
   Nicotiana tabacum   Maryland mammoth tobacco

   SDP at low temperature; day-neutral at high temperature
   Cosmos sulphureus   cv Orange flare   Cosmos

   SDP at high temperature; LDP at low temperature
   Euphorbia pulcherrima   Poinsettia
   Ipomoea purpurea   cv Heavenly Blue   Morning glory

2. Quantitative SDP

   Cannabis sativa   cv Kentucky   Hemp or marijuana
   Chrysanthemum morifolium   Chrysanthemum cultivar
   Datura stramonium   (older plants are day-neutral)
      Jimsonweed datura
   Gossypium hirsutum   Upland cotton
   Helianthus annuus   Sunflower
   Saccharum spontaneum   Sugar cane

   Quantitative SDP; require or are accelerated by low-temperature vernalization
   Allium cepa   Onion
   Chrysanthemum morifolium   Chrysanthemum cultivar

LONG-DAY PLANTS

3. LDP (qualitative or absolute)

   Agropyron smithii   Bluestem wheatgrass
   Arabidopsis thaliana   Mouse ear cress
   Avena sativa, spring strains   Oats
   Chrysanthemum maximum   Pyrenees chrysanthemum
   Dianthus superbus   Lilac pink carnation
   Fuchsia hybrida   cv Lord Byron   Fuchsia
   Hibiscus syriacus   Hibiscus
   Hyoscyamus niger, annual strain   Black henbane
   Nicotiana sylvestris   Tobacco
   Raphanus sativus   Radish
   Rudbeckia hirta   Black-eyed Susan
   Sedum spectabile   Showy stonecrop

   LDP; require or accelerated by low-temperature vernalization
   Arabidopsis thaliana, biennial strains   Mouse ear cress
   Avena sativa, winter strains   Oats
   Beta saccharifera   Sugar beet
   Bromus inermis   Smooth bromegrass
   Hordeum vulgare   Winter barley
   Hyoscyamus niger, biennial strain   Black henbane
   Lolium temulentum   Darnel ryegrass
   Triticum aestivum   Winter wheat

   LDP at low temperature; quantitative LDP at high temperature
   Beta vulgaris   Common beet

   LDP at high temperature; day-neutral at low temperature
   Cichorium intybus   Chicory

   LDP at low temperature; day-neutral at high temperature
   Delphinium cultorum   Florists larkspur
   Rudbeckia bicolor   Pinewoods coneflower

   LDP; low temperature vernalization will substitute (at least partly) for the LD requirement
   Spinacia oleracea   cv Nobel   Spinach
   Silene armeria   Sweetwilliam silene

[a] Mostly from Vince–Prue, 1975 and Salisbury, 1963b.

[b] Critical night or day often depends on conditions (e.g., temperature), age of the plant, number of inductive cycles, and cultivar. Hence, some are not shown, and those that are shown are only representative.

[c] Note that single species often appear in several categories, indicating variabilities of varieties within species. To conserve space, the lists have been greatly abbreviated.

LONG-DAY PLANTS *(continued)*

4. Quantitative LDP

   Hordeum vulgare   Spring barley
   Lolium temulentum   cv Ba 3081   Darnel ryegrass
   Nicotiana tabacum   cv Havana A   Tobacco
   Secale cereale   Winter rye
   Triticum aestivum   Spring wheat

   Quantitative LDP; require or accelerated by low-temperature vernalization
   Digitalis purpurea   Foxglove
   Pisum sativum   Late flowering garden pea
   Secale cereale   Winter rye

   Quantitative LDP at high temperature; day-neutral at low temperature
   Lactuca sativa   Lettuce
   Petunia hybrida   Petunia

DUAL-DAYLENGTH PLANTS

5. Long-short-day plants

   Aloe bulbilifera   Aloe
   Kalanchoe laxiflora   Kalanchoe
   Cestrum nocturnum   (at 23°C, day-neutral at >24°C), night-blooming jasmine

6. Short-long-day plants

   Trifolium repens   White clover

   Short-long-day plants; require or accelerated by low-temperature vernalization
   Dactylis glomerata   Orchard grass
   Poa pratensis   Kentucky bluegrass
   (in these plants, SD is required for induction and LD for development of inflorescence)

   Short-long-day plants; low temperature substitutes for SD effect and, after low temperature, plants respond as LDP
   Campanula medium   Canterbury bells

INTERMEDIATE-DAY PLANTS

7. Plants flower when days are neither too short nor too long

   Chenopodium album   Lambsquarters, goosefoot
   Coleus hybrida   cv Autumn coleus
   Saccharum spontaneum   Sugar cane

AMBIPHOTOPERIODIC PLANTS

8. Plants quantitatively inhibited by intermediate day lengths

   Chenopodium rubrum   ecotype 62° 46′ N at 25°C
   (responds as quantitative intermediate-day plant at 15 to 20°C and as a quantitative LDP at 30°C)   Goosefoot
   Madia elegans   Tarweed
   Setaria verticillata   Hooked bristlegrass

DAY-NEUTRAL PLANTS

9. Day-neutral plants: These are the plants with least response to day length for flowering. They flower at about the same time under all day lengths but can be promoted by high or low temperature or by a temperature alternation.

   Cucumis sativus   Cucumber
   Fragaria-vesca semperflorens   European alpine strawberry
   Gomphrina globosa   Globe amaranth
   Gossypium hirsutum   Upland cotton
   Helianthus annuus   Sunflower
   Helianthus tuberosus   Jerusalem artichoke
   Lunaria annua   Dollar plant
   Nicotiana tabacum   Tobacco
   Oryza sativa   Rice
   Phaseolus vulgaris   Kidneybean
   Pisum sativum   Garden pea
   Zea mays   Maize or corn

   Day-neutral plants; require or accelerated by low-temperature vernalization
   Allium cepa   Onion
   Daucus carota   Wild carrot
   Geum sp.   Avens
   Lunaria annua   Dollar plant

---

flowering was longer for individuals collected farther north. Often, no morphological differences could be detected between individuals having a greatly different flowering response. Charles Olmsted (1944) even found that sideoats grama (*Bouteloua curtipendula*) had short-day strains (*ecotypes*) at the southern end of its range and long-day strains at the northern end. Cultivars and varieties of many other species (e.g., cotton, soybean, rice, wheat, chrysanthemum, and other native grasses) have also been compared, although not always from a wide latitudinal range, and the same diversity has become apparent. Much of this work has been done in Scandinavia (e.g., Bjørnseth, 1981; Hay and Heide, 1983; Juntilla and Heide, 1981). In all these cases the plants flower at some appropriate time. For example, the various cocklebur species in temperate zones flower 6 to 8

weeks before the average killing frost in autumn, allowing time for seed ripening. As you might imagine, there are many implications of photoperiodism for agriculture (Vince-Prue and Cockshull, 1981).

Thus cultivars and varieties are genetically "calibrated" to a habitat. Sometimes one gene will make the difference between a day-length-sensitive variety and one that is not sensitive to day length. In other species, several genes are involved. These results indicate the importance of using a carefully standardized genetic type in a continuing scientific study, and they also indicate that classifications of response types, such as in Table 22-1, are at best only approximations.

Why hasn't the role of photoperiodism in ecology been studied more intensively? Partially, perhaps, because the importance of day length in the life

Some Early History

If day length plays such a decisive role, why was photoperiodism not discovered sooner? To be sure, A. Henfrey had suggested in 1852 that day length might influence plant distribution, but the measurement of time by plants must have seemed unlikely to 19th-century botanists. Even the discoverers seemed to resist their own ideas generated by their own data. Probably the first to realize the role of day length was Julien Tournois, who began studying the flowering and sexuality of hops (*Houblon japonais*) and hemp (*Cannabis sativa*) in Paris in 1910. He noticed the extremely early flowering of his plants in the winter greenhouses, but at first he convinced himself that they were flowering in response to the decreased *quantity* of light rather than its duration. In his third paper, published in 1914, he finally grasped the point: "Precocious flowering in young plants of hemp and hops occurs when, from germination, they are exposed to very short periods of daily illumination." And: "Precocious flowering is not so much caused by shortening of the days as by lengthening of the nights." He had planned further experiments but was killed at the front in World War I.

Across the lines in Heidelberg, Germany, Georg Klebs (1918) had probably also discovered the role of day length. He made plants of house leek (*Sempervivum funkii*) flower by exposing them to several days of continuous illumination. Most scientists of that time thought that nutrition controlled reproduction in plants, but Klebs felt that the additional light, which caused his plants to flower, was acting catalytically and not as a nutritional factor. Since *Sempervivum* was a long-day plant, perhaps flowering required the additional photosynthesis provided by the added light, it could be argued. Tournois's hops and hemp, however, flowered with *less* light; so he tried lower intensities extended over a longer time to see if they would provide the same response as short durations. They did not, so the time factor seemed to be controlling. Garner and Allard (Section 20.8) followed the same line of reasoning, separating in their experiments the effects of light quantity from those of light duration.

Incidentally, nutritional factors often do play at least a quantitative role in plant reproduction. The acceleration of flowering in many species by high carbohydrates and low nitrogen was also being documented during World War I (Fischer, 1916; Klebs, 1918; and Kraus and Kraybill, 1918). The effects of this knowledge on agriculture have been at least as profound as our understanding of photoperiodism. Tomato and apple yields, for example, are increased by withholding nitrogen at appropriate times.

---

of a plant calls no attention to itself until the plant is moved to another day length or to artificial conditions of light and temperature. Plants must be well adapted to the day lengths at the latitudes where they exist, or they could not exist there, so ecologists are not likely to notice the day-length response. Plant physiologists are concerned with such things, but so far they have faced the challenges of understanding the mechanism of photoperiodism rather than its ecological significance.

22.2 Some General Principles of Photoperiodism

Since the days of Tournois, Klebs, and Garner and Allard (see essay in this chapter, "Some Early History"), well over a thousand papers have been published describing studies on photoperiodism. The most striking initial impression to be gained from this vast body of facts might be that there are no broad generalities, no sweeping laws to help us understand the photoperiodism response. Each species and often each cultivar or variety within a species seems to have its own features of response; probably no two respond exactly alike.

Such a situation certainly poses a challenge for a plant physiology student and for the authors of a plant physiology text! But things are not quite as hopeless as they may appear. Though every rule seems to have its exception, some generalizations can nonetheless be made: principles of photoperiodism that apply whether or not the process being controlled is the initiation of a soybean flower or a female pine cone or the development of a potato tuber. In the next seven sections, we shall present seven generalizations. We shall present some experimental data based upon several plant species, and we shall note a few exceptions. In your reading, don't worry about remembering all the details—experts in the field have difficulty doing that—but let those details help you to understand the generalization presented at the end of each section.

We shall document only a few points with specific references to the literature. Several reviews include detailed references (e.g., Bernier et al., 1981; Evans, 1969 and 1975; Salisbury, 1981, 1982; Schwabe, 1971; Vince-Prue, 1975; and Zeevaart, 1976a, 1976b).

22.3 Photoperiod During a Plant's Life Cycle

Photoperiodism is a widespread phenomenon in nature. Garner and Allard in their early papers suggested that bird migrations might be controlled by photoperiod, and soon photoperiodism in birds was demonstrated (Rowan, 1925). Since then, many animal responses to photoperiod have been documented, including several developmental changes in insects, promotion of reproduction in insects, reptiles, birds, and mammals, and fur (pelage) changes in mammals. Virtually every aspect of plant growth and development is influenced by photoperiod (Table 22-1; Hillman, 1979; Vince-Prue, 1975).

Seed Germination To begin with, the germination of certain seeds depends upon the photoperiod applied to the parent plant. Mature seeds of certain species are also influenced in their germination by photoperiod. There are both long-day and short-day seeds for germination, and this has been shown to be a true photoperiodic effect by obtaining the long-day response with an interruption of a long dark period (i.e., interruption of the night for plants on short-day cycles).

Some Features of the Vegetative Shoot Garner and Allard, in a 1923 paper, suggested that stem elongation in response to long days is probably the most widespread photoperiodic phenomenon. Hundreds of papers published since then support this observation. Stem elongation in response to long days has been observed in conifers, where the response is often very strong, and also in both monocots and dicots among the flowering plants (angiosperms). In the few possible exceptions, long days lead to flowering, and flowering may terminate stem elongation. But much more commonly, plants that flower in response to long days do so by the rapid stem elongation we have called bolting. One way to avoid the complication of flowering is to observe effects of day length on stem elongation in plants that are day-neutral for flowering. Again, the elongation in response to long days is often highly apparent (see, for example, the two tomato plants in Fig. 22-2). Related effects are a suppression of branching by long days (promotion by short days) in several species and effects on tillering in grasses. Many temperate-zone grasses, such as barley, tiller more under short days, but rice from the tropics and subtropics tillers more under long days.

Many features of leaves are strongly influenced by day length. Leaf expansion is often promoted by long days, and there are many other responses that have been observed in certain species; decreased stomatal density, increased leaf succulence, increased organic acids, increased chlorophyll (all short-day responses) and increased anthocyanin (some short-day responses, some long-day responses).

Roots and Storage Organs Rooting of cuttings has been promoted by long days, both when applied to the cutting itself and when applied to the mother plant from which the cutting was taken. Considerable study has been devoted to the formation of underground storage organs. Potato tubers are induced by short days as discussed in Chapter 21, where the strong interaction with temperature was emphasized. So-called root tubers of cassava are also promoted by short days, as are tubers and root storage organs such as those of radish and many other species, but bulb formation in onions is a long-day response.

Reproduction A few, not very well-documented examples of the formation of reproductive organs in bryophytes have been reported. In some cases the response was to long days; other species responded to short days. Reproduction in conifers is also influenced by day length, sometimes promoted by short days and sometimes by long days. Most of the rest of this chapter will be devoted to a discussion of flower induction in angiosperms in response to photoperiod. As we shall see, the situation is highly complex with hundreds of examples being known of plants that are promoted in their flowering by short days, by long days, and by a complex assemblage of day length and temperature combinations.

Once the flower has formed in response to day length, its further development is also often strongly influenced by photoperiod. Most typically, the same day lengths that led to the production of flowers also lead to an increased rate of floral development. But plants are known that have a photoperiodic requirement for initiation but are day-neutral for floral development (*Kalanchoe blossfeldiana*, *Fuchsia hybrida*) or the opposite (*Bougainvillea*, *Phaseolus vulgaris*). A strawberry hybrid (*Fragaria* × *ananassa*) is even SD for initiation and LD for development, and *Callistephus chinensis* is LD for initiation and SD for floral development (Vince-Prue, 1975).

Sex expression in many species is often strongly influenced by photoperiod, but no simple relationship exists: Either femaleness or maleness can be promoted by either short days or long days, depending on species. This is true even for different cultivars of cucumber (*Cucumis sativus*).

Many studies have investigated effects of photoperiod and other factors on seed filling, and thus yield, of agricultural plants. In a few cases, photoperiod seemed clearly to influence seed development

SD   LD
Japanese morning glory, ca 35 days

SD   LD
Radish, ca 35 days

LD→SD   LD
Lambsquarters, ca 110 days

SD   LD
Cocklebur, ca 60 days

SD   SD→LD
Henbane, ca 110 days

SD   LD
Tomato, ca 110 days

SD   LD
Muskmelon, ca 46 days

SD   LD
Barley, ca 35 days

SD   SD→LD   LD→SD   LD
Petunia, 54 days (27 SD, 27 LD)

SD   LD

Spinach, ca 35 days

Figure 22-2 *(Facing page and above)* Some representative day-neutral (tomato), short-day (lambsquarters, Japanese morning glory, and cocklebur), and long-day (henbane, radish, muskmelon, petunia, barley, and spinach) plants. Note the strong effects of day length on vegetative form of all plants, but especially tomato, where both short-day and long-day specimens are flowering. In nearly all cases, plants exposed to long days have longer stems. (Photographs by F. B. Salisbury.)

(e.g., soybeans), but the situation is usually complicated by a host of other factors. For one thing, as we have seen, the amount of plant available to produce seeds is often strongly influenced by photoperiod, and this in turn influences yield. Often there is a delicate balance among the various factors. Northern cultivars of soybeans may be so sensitive to photoperiod that they give maximum yields only within a band of latitude about 80 km wide (Hamner, 1969). Cultivars for more southern latitudes are successful over a wider range.

Vegetative reproduction is also strongly influenced by photoperiod, in a few cases at least. For example, strawberry plants typically form runners under long days. Bryopyllum also forms foliar plantlets on its leaf margins under long days (Fig. 22-3).

The Autumn Syndrome As might be expected, several aspects of temperate-zone plants are strongly influenced by the short days of autumn. Typically, the response is also strongly modified by temperature, as discussed in Chapter 21. Features promoted by short days include leaf abscission, reduced stem elongation, reduced chlorophyll production, increased formation of other pigments, dormancy, and development of frost hardiness. Annual plants senesce and die at the end of the growing season, often long before autumn arrives. Frequently this senescence is promoted by the same photoperiod that stimulates flowering, flower development, and seed filling.

Figure 22-3 Foliar plantlets that are produced on the leaf margins of *Bryophyllum* under long days. (Photograph by F. B. Salisbury.)

Spring Rejuvenation in Woody Plants We saw in Chapter 21 that the buds of woody plants often break dormancy in the spring in response to the low temperatures of winter, and often this phenomenon is also promoted by long days. In a few cases, the long days may promote bud break even in the absence of a previous low temperature treatment.

The conclusion of this section: *Many aspects of a plant's life cycle are influenced by photoperiod; long days almost always promote stem elongation, and short days applied to species of temperate regions induce the autumn syndrome; virtually all other plant responses may be promoted by either short days or long days, or plants may be day-neutral in response.*

22.4 The Response Types

In Fig. 22-4, relative flowering is plotted as a function of the hours of light in each 24-h cycle. In a truly day-neutral plant, which is probably rare, flowering is independent of day length, so a horizontal line appears in the figure. Long-day plants* flower sooner on LD, so their curves (solid lines) slope upward to the right. SDPs take longer to flower on LD, so their curves (dashed lines) slope downward to the right. As in vernalization (Section 21.2), there is both a facultative and an absolute response to photoperiod. Curves representing plants with an absolute day-length requirement cross the abscissa at the day length called the **critical day.** It is also possible to speak of the **critical night,** that night length that must be exceeded for flowering of SDPs or inhibition of flowering of LDPs. One other complication is apparent. SDPs will often not flower when only a few

---

*We shall use the following abbreviations: LD(s) = long day(s), SD(s) = short day(s), LDP(s) = long-day plant(s), and SDP(s) = short-day plant(s).

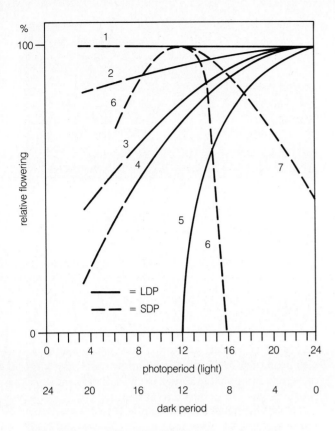

Figure 22-4 Diagram illustrating flowering (and other) responses to various day lengths. Flowering can be measured in various ways such as counting the number of flowers on each plant, classifying the size of the buds according to a series of arbitrary stages (see Fig. 22-8), or taking the inverse of the number of days until the first flower appears. (With many species, there is little or no flowering when days are unusually short.) **1.** A truly day-neutral plant, flowering about the same at all day lengths; such a plant is probably very rare. **2., 3.,** and **4.** Plants quantitatively promoted in their flowering (or other response) by increasing day lengths. The three curves represent three species that are promoted to different degrees. **5.** Qualitative or absolute long-day plant such as henbane; in this example, flowers only when days are longer than 12 h. **6.** Qualitative short-day plant such as cocklebur; in this example, flowers only when days are shorter than 15.7 h, nights longer than 8.3 h. Note that cocklebur also fails to flower if days are shorter than about 5 h but flowers as days get longer (a typical long-day response). **7.** Quantitative short-day plant, flowers on any day length but better under short days. Note that other species, not shown here, have different critical day and night lengths, not just the 12- and 15.7-h days shown for henbane and cocklebur.

hours of light are received each day. This brief day-length requirement is for more than just photosynthesis. Some plants that can be maintained indefinitely in the dark by being fed sucrose nevertheless show the minimum light requirement for the photoperiodism response. Note that this requirement is represented by curves that slope upward to the right. Such curves express the LD response, so SDPs act as LDPs when the days are extremely short!

Note that the SD and LD responses are completely opposite. Line 6 in Fig. 22-2 could represent cocklebur (*Xanthium strumarium*), a classic SDP. Line 5 could represent henbane (*Hyoscyamus niger*), a classic LDP. Both flower when days are about 12.5 to 15.7 h long (nights 11.5 to 8.3 h).

Sometimes the critical day or night can be determined for a population of plants within rather narrow limits (e.g., 5 to 10 min). In other cases, the limits are far less exact and may spread out over one or more hours. With cocklebur, the greater the number of inductive cycles, the more exact is the critical day or night.

Still other ways in which species differ is in their ripeness to respond and in the number of SD or LD cycles required to induce flowering. Species that require only a single inductive cycle for flowering (Table 22-1) have been widely used, because they permit experimental treatments (e.g., application of a chemical) to be performed at various times in relation to the single cycle and without the complications of effects on previous or subsequent cycles.

Representative LD, SD, and day-neutral plants with some critical day lengths are shown in Fig. 22-2 and listed in Table 22-1. The three response types observed by Garner and Allard form the foundation for any classification of photoperiod response types. Since their time, a few other categories have been discovered. To produce flowers, a few species (item 5, Table 22-1) require LDs followed by SDs, as occurs in late summer and fall. When maintained either under continuous LDs or continuous SDs, they remain vegetative. The counterparts to these **long-short-day plants,** the **short-long-day plants** (item 6), require SDs followed by LDs, as in spring. There are a few species, at least (item 7), that flower only on **intermediate day lengths** and remain vegetative when days are either too short or too long. They have counterparts that remain vegetative on intermediate day length, flowering only on longer or shorter days (item 8).

There are several interesting interactions between photoperiod and temperature. We have seen that a vernalization requirement is often followed by a requirement for LDs (Section 21.2), consistent with the LDs of late spring following winter. In other cases, a plant may exhibit a given response type at one temperature but not at another. It may, for example, have a qualitative or quantitative SD response at temperatures above, say, 20°C but be essentially day-neutral at cooler temperatures. Two examples of plants that are absolute SDPs at high temperature and absolute LDPs at low temperature have been reported: poinsettia (*Euphorbia pulcherrima*), and morning glory (*Impomea purpurea*). They are day-neutral only at the intermediate temperature. *Silene armeria,* an LDP, is induced on SDs by either high (i.e., 32°C) or low (5°C) temperatures. Specific day-length requirements only at certain temperatures prove to be quite common.

Vernalization and a given photoperiodic treatment are sometimes interchangeable. For example, in a variety of Canterbury bells (*Campunula medium*), vernalization is fully replaced by exposure to SDs, but LDs are required following either treatment. Still other complications and interactions are known. For example, Japanese morning glory (*Pharbitis nil*) has become a prototype SDP, yet it can be induced under LDs by low temperatures, high-light levels, treatment with growth retardants, removal of roots, and low nutrient levels. Surely such a diversity of response types is of ecological importance.

In this discussion, we have mentioned only flowering responses to photoperiod. There is undoubtedly an equally diverse number of response types with respect to some other plant responses to photoperiod. Most remain to be studied, but examples are known, such as tossa jute (*Corchorus olitorius*), in which stem elongation in response to LDs occurred only above 24°C; below that temperature the plant was day-neutral with respect to stem elongation (Bose, 1974).

The generalization that concludes this section is simply stated: *There is a wide diversity of photoperiodic response types.*

22.5 Ripeness to Respond

Only a few plants respond to photoperiod when they are small seedlings. The Japanese morning glory responds to SDs in the cotyledonary stage, and some species of goosefoot or lambsquarters (*Chenopodium* spp.; Cumming, 1959) respond and flower as minute seedlings. In laboratory studies they can be grown on filter paper in a Petri dish. Most species, such as the cocklebur, must attain a somewhat larger size; cotyledons do not respond. Henbane must be 10 to 30 days old before it will respond to LDs. Certain monocarpic bamboo species and several polycarpic trees will not flower until they are 5 to 40 or more years old, although it is not known whether they then respond to photoperiod. Klebs called the condition a plant must achieve before it will flower in response to the environment *Bluhreife*, which has been translated as **ripeness to flower;** but a more descriptive term is **ripeness to respond.** In many species, the number of required photoperiodic cycles decreases as the plant gets older; that is, the level of ripeness to respond increases with age. Often, the plant finally flowers independently of the photoperiod; it becomes day-neutral.

Individual leaves must also reach a ripeness to respond. In several species, the leaf is maximally sensitive when it is first mature (fully expanded). Cocklebur leaves less than 1 cm long will not respond,

but the half-expanded leaf, the one growing most rapidly, is most sensitive. On the other hand, leaves of the scarlet pimpernel (*Anagallis arvensis*) are most sensitive to LDs when the plant is a seedling; young leaves on older plants are less sensitive.

The concept of ripeness to respond is almost identical to that of juvenility, defined as the condition of a plant before it is mature enough to flower (Section 18.6). The term *juvenility* is often used, however, to emphasize the special morphological features (particularly leaf shape) characteristic of juvenile plants. Another related concept is that of **minimum leaf number,** which is the minimum number of leaves from seedling to earliest flower under the most ideal conditions for flowering.

The conclusion of this section: *Before a plant can flower in response to its environment (particularly day length and temperature), the organs that detect the environmental change, usually leaves or meristems, must reach a condition called ripeness to respond. There is a great diversity among species and plant organs in the age at which they achieve this condition.*

22.6 Phytochrome and the Role of the Dark Period

In 1938, Karl C. Hamner and James Bonner at the University of Chicago wondered about the relative importance of day and night in photoperiodic induction of cocklebur. Which was most important, day or night? They took two experimental approaches. In one series of experiments, days and nights were varied to give cycles that did not equal 24 hours. The critical night, but not the critical day, remained constant, indicating the importance of the dark period. In the other approach, days were interrupted with darkness or nights with light. Dark interruption of the day had little or no effect, but night interruption by light inhibited flowering in SDPs and (in later experiments) promoted flowering in LDPs (Fig. 22-5). This was the discovery of the **night interruption phenomenon.**

Once it was known that a light break during the dark period would nullify the effect of darkness, several possibilities for experimentation immediately became apparent. Researchers could ask: Which is more important, the level of light used (its irradiance) or the total quantity of light energy (as calculated by multiplying its irradiance by the time interval over which it is applied)? Within rough limits, the total quantity of energy proves to be the determining factor (i.e., reciprocity applies; see Section 18.2). Researchers could then ask: How dark is dark? Light applied during the entire dark period is effective (especially in inhibiting flowering of SDPs) at very low irradiances. In some species, for example, it is effective at 3 to 10 times the light from the full moon.

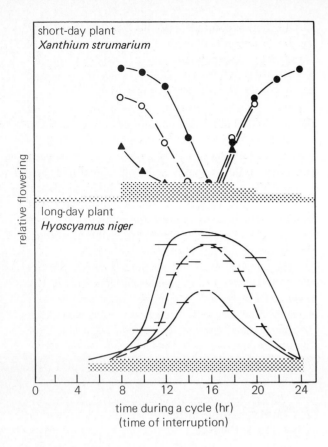

short-day plant
Xanthium strumarium

long-day plant
Hyoscyamus niger

relative flowering

0 4 8 12 16 20 24

time during a cycle (hr)
(time of interruption)

Figure 22-5 Effects of a light interruption given at various times during dark periods of various lengths (shaded bars) on subsequent flowering of an SDP and an LDP. Interruptions with *Xanthium* (cocklebur) were 60 s; with *Hyoscyamus* (henbane), times are indicated by lengths of data lines. (Data for *Xanthium* from Salisbury and Bonner, 1956; for *Hyoscyamus* from Claes and Lang, 1947.)

When is the light most effective? Usually at some constant time after the beginning of an inductive dark period using SDPs or an inhibitory dark period using LDPs (Fig. 22-5). This time of most effectiveness is often equivalent to the critical night.

Which wavelengths of light are most effective? In the early 1940s, it became apparent that red light was considerably more effective than other wavelengths. Action spectra for inhibition in SDPs and promotion in LDPs are typical of those for other phytochrome responses (see Fig. 19-3). Thus in the early 1950s, immediately after far-red reversibility was discovered in lettuce seed germination, cocklebur plants were illuminated in the middle of a long inductive dark period with red light followed by far-red. If the far-red followed the red illumination immediately, plants flowered; if about 30 minutes were allowed to elapse between the red and the far-red exposures, the far-red no longer reversed the effects of the red. Apparently, P_{fr} completes its inhibitory act within 30 minutes in cocklebur.

Let's state a preliminary conclusion: *The dark period plays an important role in the photoperiodic re-*

sponse, since a light break inhibits flowering of short-day plants and promotes flowering of long-day plants. Phytochrome apparently detects the light break, and its effectiveness depends upon the time of irradiation.

Now we shall examine some complications. LDPs are often less sensitive and somewhat more quantitative in their response to a night break than are SDPs. Using four photoflood lamps, for example, flowering is completely inhibited in cocklebur plants by a few seconds of light given about 8 h after the beginning of an inductive dark period. With many LDPs, however, flowering continues to be promoted as the duration of a night break (using comparable high levels of light) increases from seconds to hours. Furthermore, whereas red light is most effective in a night break with SDPs, a mixture of red and far-red wavelengths is almost always more effective with LDPs, whether applied as a night break or during the continuous light that best induces flowering in most LDPs. With SDPs (especially the Japanese morning glory), far-red given at the beginning of a dark period may inhibit if the dark period is exceptionally long but have no effect if it is short!

It would help considerably if we could measure the various forms of phytochrome in leaves at different times during photoperiodic induction. So far, it has not been possible to make these measurements in green tissues. In one study, no phytochrome at all could be found in some of the most sensitive SDPs, including cocklebur, perilla (*Perilla* sp.), and soybean (*Glycine max*), although phytochrome was easily detected in extracts of leaves of the LDPs henbane, spinach (*Spinacia* sp.), beet (*Beta martima*), and barley (*Hordeum vulgare*), and some other SDPs, including Maryland mammoth tobacco (*Nicotiana*) and sorghum (Lane et al., 1963). An alternative approach was to treat plants with an herbicide (called Norflurazon or Zorial) that inhibits chlorophyll and carotenoid formation in white light (Jabben and Deitzer, 1978). The nearly colorless plants were supplied with sucrose through cut leaf ends. Several phytochrome responses were shown to be the same in these plants as in normal dark- and light-grown controls, and flowering was induced in the LD Wintex barley with LDs. Phytochrome was measured *in vivo* (intact shoot tips) in a spectrophotometer and found to drop sharply to about 30 percent of the initial value in 2 to 4 hours after plants were placed in the dark.

An ingenious approach has been developed to provide an indirect measure of the ratio of the two forms of phytochrome at any given time. This is called the **null response technique.** Plants are illuminated with mixtures of red and far-red light. The mixture that provides *no response* must be a measure of P_{fr}/P_{total} at the time of irradiation. If the irradiation has no effect, that ratio must have already been present at the time of irradiation. There is an important limitation to the null-point technique, however: Because of the overlapping absorption spectra of the two forms of phytochrome, no wavelength or mixture of wavelengths can produce more than about 80 percent P_{fr} or less than 1 or 2 percent P_{fr}. Since the determining quantities in the photoperiodic response could exceed these percentages, the null-point technique cannot solve all our problems. It has provided some interesting data, however.

In the SDPs *Chenopodium* and Japanese morning glory, P_{fr} is relatively high during the first 3 to 6 h of an inductive dark period, and then there is an abrupt drop. In LDPs (e.g., perennial ryegrass, *Lolium temulentum*; henbane; and a long-day duckweed, *Lemna gibba*), there is also a drop in P_{fr} during the first 5 h of a dark period. But perhaps the responses to light treatment do not depend only on P_{fr} and P_r. The HIR (Section 19.4) could be operating, phytochrome intermediates could play a role, or some other pigment could be involved. At the moment, we have no satisfactory answer to the problem of phytochrome in photoperiodism, but we can state an operational conclusion anyway: *A night interruption inhibits flowering of short-day plants and promotes flowering of long-day plants. Red light is more effective with short-day plants and a mixture of red and far-red with long-day plants.*

22.7 Detecting Dawn and Dusk

When during evening twilight does a plant begin to measure the dark period? And do plants respond to moonlight? In photoperiodism, plants must detect some photon flux as *light* and some lower flux as *darkness*, thereby discriminating between day and night. The plant must then measure the duration of day, night, or both, and when the duration(s) reach(es) the lengths genetically preprogrammed in the plants, a given process such as flowering or stem elongation must be initiated or controlled.

Figure 22-6 shows irradiance levels measured at 660 nm during twilight at Logan, Utah, in late July 1980 (Salisbury, 1981). The figure also shows red/far-red ratios (660/730 nm). Note the rapid drop in light level during dusk (approximately an order of magnitude each 10 minutes in this case; the rate of drop depends on latitude and time of year) and the shift in spectral quality. Far-red increases relative to red during twilight (see Fig. 23-10), but this increase occurred sooner when there were clouds in the sky than when the sky was clear (July 28 was slightly cloudy). It would also be strongly influenced by atmospheric pollution. The duration of twilight is also strongly a function of latitude and time of year (Fig. 22-7).

A series of experiments was carried out to compare cocklebur sensitivity to light at the beginning of the dark period (i.e., during dusk) with sensitivity after plants had been in the dark for 7 to 9 h

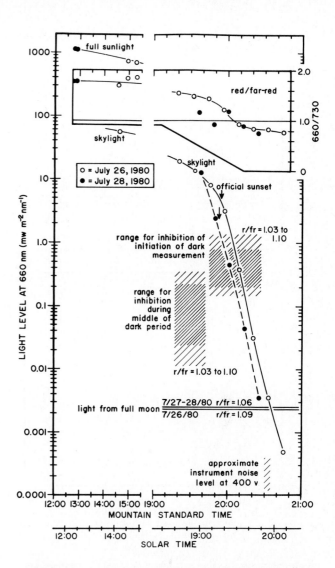

Figure 22-6 Light levels at 660 nm as a function of time on July 26 and 28, 1980, at Logan, Utah. Light from a full moon is also shown. The cross-hatched areas represent ranges of light levels that inhibit initiation of dark measurement in *Xanthium* or inhibit flowering when applied for 2 h during the middle of a 16-h dark period. The inserted graph shows ratios of light levels at 660 nm to levels at 730 nm for all the sunlight and twilight points; other red/far-red ratios are given as numerals. Note the rapid drop in light levels through the range of delay of dark measurement. (From Salisbury, 1981.)

(Salisbury, 1981). Plants were exposed to incandescent light filtered through red Plexiglas, which provided red/far-red ratios similar to those presented in Fig. 22-6. Plants were exposed either during the first two hours of dark periods that varied in length for different groups of test plants, or for two hours beginning 7 h after the plants were placed in the dark. Plants were induced by a single dark period (various lengths for different groups of plants). To delay the initiation of dark measurement, irradiance levels at 660 nm of about 0.2 to 1.0 mW m^{-2} nm^{-1} were required, whereas inhibition during the middle 2 hours of a 16-hour dark period required only 0.01 to

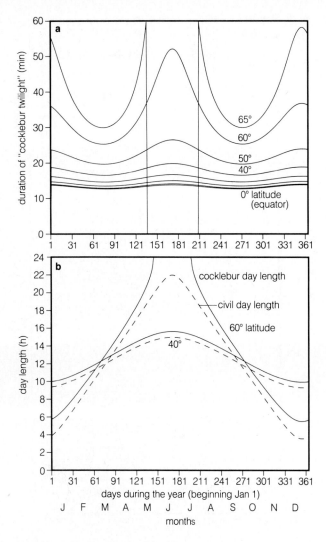

0.3 mW m^{-2} nm^{-1}. These ranges are shown in Fig. 22-6 by shaded areas.

The results suggest that, as far as photoperiodism is concerned, cocklebur plants shift from their daytime mode of metabolism or pigment balance to their dark mode as light levels at 660 nm dropped from about 1.0 to 0.2 mW m^{-2} nm^{-1}. As Fig. 22-6 shows, light levels during twilight passed through this range in only about 5.5 to 11 minutes. The human eye sees objects "after dark," experiences twilight as lasting from 30 to 45 minutes or more, and further detects changes in light level caused by clouds during the day. But cocklebur plants apparently change from day to night almost as if they were controlled by an on/off switch. "Cocklebur twilight" (Fig. 22-7) apparently ended 20 minutes after sunset, at which time the sun was about 4° below the horizon; civil twilight officially ends when the sun is 6° below the horizon. Fig. 22-7 also shows cocklebur twilight added to astronomical day length to produce "cocklebur day length."

Over two decades ago, Takimoto and Ikeda (1961) took a more direct approach to the study of plant sensitivity during twilight. They covered plants at various times during dusk and dawn, compared their flowering with flowering of plants left uncovered, and noted the level of twilight perceived by the plants as darkness. They reported considerable variation among five SD species: *Oryza sativa* (rice) was relatively insensitive to light both in the morning and in the evening; *Glycine max* (soybean), *Perilla frutescens* (perilla), and Japanese morning glory were relatively insensitive at dusk but more sensitive at dawn; and cocklebur was highly sensitive in the evening but less so in the morning.

Considering the moonlight question: As noted, Fig. 22-6 also shows the probable range of light sensitivity of cocklebur plants during the 7th to the 9th hours of a single 16-h inductive dark period (0.01 to 0.3 mW m^{-2} nm^{-1}). Maximum light levels at 660 nm for light from the full moon are also shown. Figure 22-6 suggests that maximum levels of moonlight are not high enough to influence flowering in the middle of a dark period, even though sensitivity to light increases by about an order of magnitude at that time compared to dusk. Nevertheless, a final conclusion about plant response to moonlight will require future research.

22.8 Time Measurement in Photoperiodism

The central characteristic of photoperiodism is measurement of seasonal time by detecting relative lengths of day and night. But how is time measured

Figure 22-7 (**a**) The end of "cocklebur twilight" as a function of time during the year and at different latitudes. Calculations based on the data of Fig. 22-6 suggest that cocklebur begins to respond during evening twilight as though it were night when the sun is 4° below the horizon. Thus the curves show times for the sun to drop 4° below the horizon after official sunset. (Civil twilight ends when the sun is 6° below the horizon; nautical twilight ends when sun is 12° below; and astronomical twilight when sun is 18° below the horizon.) (**b**) Cocklebur day length as a function of time during the year for two latitudes. The dashed lines are the same as those shown in Fig. 22-1 for the two latitudes. The solid lines are derived by taking the cocklebur twilight times of part (a) in this figure, multiplying them by 2 (to account for dawn as well as dusk), and adding them to the day-length curves (something that is not really justified by the data, since *morning* twilight duration was not determined in the experiments discussed in the text and suggested in Fig. 22-6). Note that at 65° latitude, it never gets dark enough during midsummer (i.e., the sun never drops 4° below the horizon) for cockleburs to respond as though it were dark. After sunset, received light is skylight, so mountains have little effect on the end of "cocklebur twilight." (Curves were computer-generated and drawn; programs were written by Michael J. Salisbury, based on equations supplied by Roger L. Mansfield of Astronomical Data Service, 3922 Leisure Lane, Colorado Springs, Colorado 80917.)

in photoperiodism? It seems logical to place photoperiodism in the context of other examples of biological time measurement: circadian rhythms, celestial navigation, and so on (Chapter 20). During the past two decades, plant physiologists have taken this approach to the problem. Why wasn't it taken sooner? Because it seemed equally logical to imagine that time measurement was simply the time required for the completion of some yet-to-be-discovered metabolic reaction(s).

If time is measured by the interval required for some metabolite to be converted to another form, it is analogous to an hourglass that measures the interval of time required for the sand to fall from top to bottom through a narrow opening. In such a system, only one interval of time can be measured, and then some outside influence must restart the system (invert the hourglass). Circadian rhythms that function over long time intervals under constant conditions of light, temperature, and other factors are more like an oscillator such as a pendulum than an hourglass. Is time measurement in photoperiodism analogous to an hourglass or to a pendulum?

Some features certainly seem to have the hourglass mode of operation. Let's consider a simple experiment in which we expose SDPs to dark periods of various lengths and then observe the level of flowering several days later. Such an experiment has been done with many SDPs, but cocklebur is convenient because it will respond to a single inductive night, and the degree of flowering can be observed by examining the apical buds under a dissecting microscope and classifying them according to a series of floral stages (Fig. 22-8). Two or three days after the single long dark period, the buds begin to develop through the stages shown in the figure, but the *rates* at which they develop will depend upon the degree to which they were induced by the long dark period; 9 days after the dark period, plants receiving 10 h of darkness may have only reached stages 1 or 2, whereas plants receiving 16 h of darkness may have reached stages 6 or 7 (Fig. 22-9). It is clear that the flowering stimulus increases from about the 9th hour (the critical night) to the 11th to 16th hour, depending upon conditions. If we are dealing with the synthesis of a flower-inducing hormone during this time, then this might be an hourglass kind of reaction.

During the day, plants are normally exposed to light that converts about 60 percent of the phytochrome pigment to the far-red absorbing form (P_{fr}; Section 19.3). When a night break is given, red light is the most effective, indicating that significant amounts of pigment are in the red-absorbing form (P_r). This suggests that plants detect darkness as their phytochrome shifts from P_{fr} to P_r. When this was discovered in the early 1950s, it was further sugges-

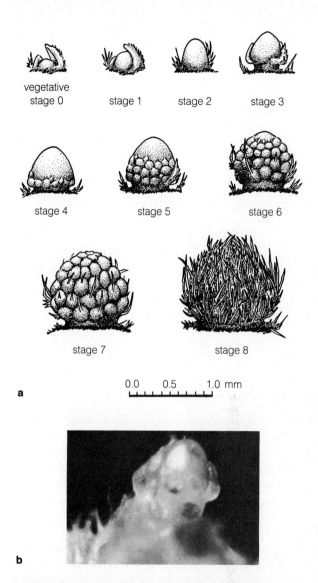

Figure 22-8 (**a**) Drawings of the developing terminal (staminate) inflorescence primordium of cocklebur, illustrating the system of floral stages devised by Salisbury (1955). (**b**) Photograph through a dissecting microscope of a cocklebur inflorescence primordium at stage 3. (Photo by F. B. Salisbury.)

ted that the shift from P_{fr} to P_r might account for time measurement. Perhaps the critical night is the time it takes for P_{fr} to drop below some critical level.

As the data continued to accumulate during the 1950s and 1960s, it became apparent that this explanation was too simple. Phytochrome does shift from P_{fr} to P_r when plants are placed in darkness, and that is how the plant "knows" it is in darkness, but, as we noted in discussing the null technique (Section 22.6), the shift occurs within 3 to 6 hours in many species, far too rapidly to account for time measurement of the critical night. There is evidence that, for functional purposes, the shift may be complete in 1 or 2

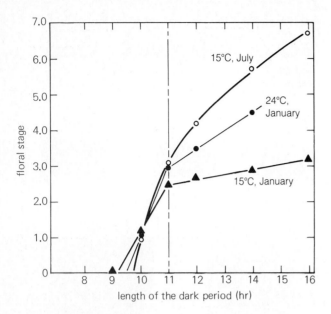

Figure 22-9 Some examples of cocklebur flowering response to different night lengths. (From Salisbury, 1963.)

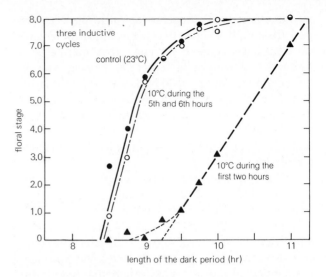

Figure 22-10 Flowering response of cocklebur to various night lengths as influenced by 10°C treatments applied during the first 2 h of the dark period or between the beginning of the fifth and the end of the sixth hours, as compared with controls that received no 10° treatments. Note that low temperature at the beginning of the inductive dark period inhibits flowering (delays time measurement), but low temperatures given later do not. Plants were treated with 3 dark periods, beginning July 18, 1962. Temperature during the dark period other than treatment times was 23°C. (From Salisbury, 1963.)

hours. It seems unlikely, anyway, that simple chemical reactions could account for time measurement in photoperiodism, since such reactions are notoriously sensitive to temperature, whereas photoperiodic time measurement is not. Fig. 22-10 presents data indicating that the initial pigment shift is sensitive to temperature, whereas the subsequent time-measuring mechanism is not.

Let's formulate the first part of our conclusion to this section: *Part of timing in photoperiodism has the characteristics of an hourglass, particularly pigment shift and synthesis of a flowering hormone.*

Bünning suggested in 1937 that plants might use their oscillating-circadian clocks in photoperiodic time measurement. There are two rather simple ways to look for rhythmical phenomena in photoperiodism. Both are illustrated for soybeans in Fig. 22-11. In one illustration, plants were given seven cycles, each including an extended, 64-h dark period, which was interrupted at various times with a 4-h night break. With soybean and many other SDPs and LDPs, there is a rhythmical response to the night breaks: At one time, the light interruption inhibits flowering, and this is repeated about 24 and 48 h later; between the times of inhibition there are periods of promotion. This experiment is certainly suggestive of an oscillating timer. A few species do not respond this way, particularly those such as cocklebur that flower in response to a single inductive cycle. LD barley plants, however, respond in an opposite but analogous way: Supplemental far-red irradiation given during constant light promotes flowering at intervals of about 24 h and fails to pro-

mote when given between these intervals (Deitzer et al., 1979).

In the second experiment of Fig. 22-9, plants were given various combinations of light and dark periods, and their subsequent flowering was plotted as a function of total cycle length. It is evident with soybean (and a few other species) that maximum flowering occurs when the light and the dark periods total 24 h, 48 h, and 72 h. Hence, the dark period is not all-controlling, but the combination of the light and the dark periods plays the decisive role. Again, some species do not show such a response. Cocklebur plants can be kept under continuous light for several weeks, exposed to a single dark period longer than the critical night, and returned to continuous light until flowers have developed.

Yet even with cocklebur, it was possible to observe features similar to those observed in the circadian rhythms. In one set of experiments, cocklebur plants given a 7-h **phasing dark period,** too short to induce flowering. Then they were given an **intervening light period** that varied in duration, after which they were given a 12- to 16-h **test dark period** (depending upon the experiment) that would normally induce flowering. Floral stage after 9 days is shown as a function of length of the intervening light period in Fig. 22-12. With brief intervening light periods there was no flowering, even though the

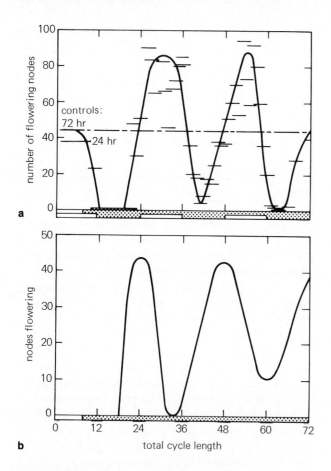

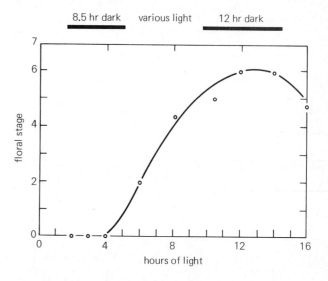

Figure 22-12 Flowering response as a function of length of the intervening light period in an experiment symbolized by the bars above the figure. (Intervening light period is labeled *various light* above the figure.) (From Salisbury, 1965.)

Figure 22-11 Rhythmical responses in flowering of Biloxi soybean (*Glycine max*) (**a**) Flowering response to 4-h interruptions (indicated by scattered horizontal lines) applied at various times during the 64-h dark period. Plants received 7 cycles of 8 h of light followed by 64 h of darkness, as indicated by the top bar at the bottom. Bottom bar shows postulated subjective days and subjective nights.
(**b**) Flowering response of soybeans to 7 cycles, including 8 h of light plus different dark periods to provide the total cycle lengths. When the cycle totalled 24 h, flowering was at a high level; a 34-h cycle produced *no* flowering. (Data from Hamner, 1963, and other publications; figures from Salisbury, 1963.)

subsequent test dark period was several hours longer than the critical night. When the intervening light period was about 5 h, plants began to flower; flowering increased until the light period was 12 h long—a typical long-day behavior (as we noted in our discussion of Fig. 22-4). An optimal dark period for flowering in cocklebur is about 12 h (see Fig. 22-9); combining this with the 12-h optimum light period again gives the magic 24-h value.

An action spectrum was determined for the intervening light period, and red light, even at low levels, was by far the most effective in *promoting* flowering; far-red inhibited. Thus something closely analogous to the night break experiment with soybeans of Fig. 22-9 became apparent with cocklebur:

Red light (P_{fr}) promotes flowering at one time and inhibits it about 12 h later.

In another rather involved set of experiments, cocklebur plants were given a long dark period that was interrupted twice. The first interruption was given 2, 4, or 6 h after beginning of the dark period, and the second interruption was given at various times after the first. Floral stages after 9 days were plotted as a function of the time of the second interruption (Fig. 22-13). Control plants, given only one interruption at various times, were most sensitive about 8 h after the beginning of the dark period, as were plants given their first interruption at 2 or 4 h. Plants given their first interruption 6 h after the beginning of the dark period, however, were most sensitive to the second interruption 18 h after the beginning of the dark period. That is, the time of maximum sensitivity was delayed 10 h when the first interruption was given at 6 h. Such phase shifting is highly reminiscent of circadian rhythms (compare Figs. 20-4 and 20-6). Rather similar results have been obtained with *Pharbitis* (Lumsden et al., 1982).

Do circadian manifestations such as leaf movements compare with time measurement in photoperiodism? Timing is clearly similar in the two situations, but it was possible to separate the leaf movement rhythms from photoperiod timing in cocklebur (Salisbury and Denney, 1974), in *Chenopodium* (King, 1975, 1979), and in Japanese morning glory (Bollig, 1977). Our tentative conclusion for this section: *The measurement of time in photoperiodism has some elements of an hourglass timer, but an oscillating timer is also clearly involved.*

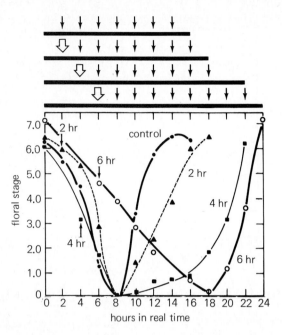

Figure 22-13 Clock resetting in flowering of *Xanthium*. The curve labeled *control* received only one interruption, and flowering nine days later is shown as a function of the time that interruption was given. Plants represented by the other three curves were given a first interruption at 2, 4, or 6 h as labeled; they then received a second interruption at various times, and flowering after 9 days is shown as a function of the time of the *second* interruption. Arrows and bars above the figure indicate how the experiment was done. Large arrows indicate first interruption times when two interruptions were given. (From Papenfuss and Salisbury, 1967.)

22.9 The Florigen Concept: Flowering Hormones and Inhibitors

Not long after photoperiodism was discovered, workers around the world wondered which part of the plant detected day length. It was soon apparent that the leaf responded. Using an SDP, for example, an experimenter could enclose the leaf for 16 h in a black paper envelope, while leaving the rest of the plant under LDs or continuous light. Such a treatment soon induced flowering. A similar experiment with an LDP prevented flowering. Covering the bud but leaving the leaves under LDs did not lead to flowering in SDPs but did in LDPs. (Actually, the green stems of some species will respond to photoperiod if they are given a sufficient number of inductive cycles.)

If the leaf detects the photoperiod, but the bud becomes the flower, there must be some stimulus transmitted from the leaf to the bud. In the 1930s, Michail Chailakhyan in Russia (see his 1968 review) grafted induced plants to noninduced plants held under noninducing day lengths, observing that the flowering stimulus would cross a graft union. He suggested that the stimulus was a chemical substance,

a hormone as opposed to some electrical or nervous stimulus. Chailakhyan coined the term **florigen** (Latin *flora*, flower, and Greek *genno*, to beget).

The grafting studies have provided two important bits of information: Florigen moves only through a living tissue union between the two graft partners and probably only through phloem tissue. Furthermore, florigen frequently seems to move with the assimilate stream. If the receptor partner is defoliated or held under low irradiance levels, movement of assimilates into the receptor is promoted and so is movement of florigen. On the other hand, in some species, hormone is exported from extremely young leaves that would be expected to be importing assimilates, so florigen might move by other mechanisms as well as in the assimilate stream.

Many different varieties or species representing different response types have been grafted to each other (see Lang, 1965; Vince-Prue, 1975). Florigen produced by one response type will often induce flowering in another type; an induced SDP grafted to an LDP in SDs will induce the LDP to flower, for example. In the most extreme example, an induced SDP (*Xanthium strumarium*) caused a vegetative LDP (*Silene armeria*) to flower, although the two species belong to unrelated dicot families. The apparent conclusion is that, although florigen is produced in response to widely different environmental conditions, it is the same compound or physiologically equivalent in virtually all angiosperms.

A third experiment that is difficult to interpret in any way but by the florigen concept was first performed in the early 1950s. Plants from several species that require only a single inductive cycle were defoliated at various times after that cycle, and the level of flowering some time later was plotted as a function of the time of defoliation. Cocklebur results are shown in Fig. 22-14; the SD Japanese morning glory and the LD perennial ryegrass have also been studied. Apparently, when the leaf is cut off immediately following the inductive long dark period (SDPs) or long light period (LDPs), plants remain vegetative, because the hormone has not yet been exported from the leaf. Export is complete some hours later, however, because defoliated plants subsequently flower almost as well as plants from which the leaf was not removed at all.

In spite of the strong evidence for flowering promoters, there is equally strong evidence for inhibitory substances or processes in plant reproduction. Indeed, both promoters and inhibitors must influence flowering. If a plant is induced by exposing only one leaf to suitable photoperiodic conditions, flowering is frequently inhibited by the noninduced leaves. The presence of leaves on a receptor plant in a graft experiment also reduces receptor flowering. These inhibitory effects are often caused by influences on the translocation of assimilates. For example, a long-day leaf growing between an induced short-day leaf and the bud may be exporting assimilates directly to the bud and thereby effectively blocking movement of assimilates—and florigen—from the induced leaf to the bud. Many inhibitor experiments can be understood in this way, and often the explanations are supported by tracer experiments.

Some effects cannot be explained on the basis of translocation, however. If the whole explanation were photosynthesis and transport of assimilates, then light levels should be important. Yet a brief night interruption at low irradiance will often produce a typical long-day inhibitory effect in a single leaf on a short-day plant (Gibby et al., 1971; King and Zeevaart, 1973). In some species, flowering seems to be repressed under noninductive conditions strictly by inhibitors coming from the leaves. Such plants will flower when defoliated. Examples are the LD henbane and various SD varieties of strawberry (*Fragaria* sp.). In some species (e.g., cocklebur and Japanese morning glory) promoters may be dominant although modified by inhibitors; in others there seems to be a true balance (darnel ryegrass), while in still others (henbane, strawberry varieties) inhibitors may be dominant but modified by promoters.

Chailakhyan and I. A. Frolova in Moscow, cooperating with Anton Lang of Michigan State University (Lang et al., 1977), grafted plants of both SD and LD cultivars of tobacco onto a DN cultivar and then grew them under various light conditions. When exposed to SDs, the SD graft partners caused the DNPs to flower earlier, and under LDs, LD partners also promoted earlier flowering of DN partners. These results again confirm the presence of a positive acting flowering hormone. But when the SD graft partners were maintained under LDs, flowering in the DN species was markedly retarded; under SDs, the LD tobacco completely prevented flowering of the DN partner. Here, in one experiment, is clear-cut evidence for both flowering inhibitors produced under unfavorable day lengths and promoters under favorable day lengths.

We need to isolate and identify florigen(s) and flower inhibitor(s). So far, this has not been completely successful, primarily because of the lack of a reliable bioassay. Numerous solvents and application techniques have been tried in attempts to extract induced plants and apply the extracts or fractions to noninduced plants. Sporadic successes have been reported. The best-known method (see Lincoln et al., 1966) extracts quick-frozen, induced cocklebur leaves with organic solvents. Extracts made from vegetative plants have no effect, but extracts from a day-neutral sunflower and even from a fungus (*Calonectria*) have sometimes been effective. Only a low level of flowering is induced in test plants, and GA applied with the

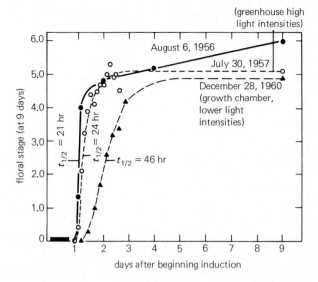

Figure 22-14 Three translocation curves obtained by defoliation of *Xanthium* plants at various times following a 16-h inductive dark period. Times of defoliation are shown, with floral stages of all plants being determined on the ninth day. Numbers on the abscissa represent noon of the indicated day, while the bar on the abscissa suggests the inductive dark period. Approximate times after beginning of the dark period when half the stimulus was out of the leaf are indicated by $t_{1/2}$. Dates refer to the day plants were subjected to the dark treatment. Plants represented by the $t_{1/2} = 46$ h curve were kept under about 400 μmol m^{-2} s^{-1} of fluorescent light in growth chambers (23°C). (From Salisbury, 1963.)

extracts sometimes promotes flower development. Activity in the extracts follows the acid fraction and has been called **florigenic acid.**

The experiment has succeeded in Hamner's laboratory (Hodson and Hamner, 1970) but not always in other laboratories. Ilabanta Mukherjee and Salisbury visited Hamner's laboratory and went through the procedure. Salisbury returned to Utah and, with Alice Denney, was sometimes able to perform the experiment successfully (unpublished). Mukherjee returned to the laboratory of Jan A. D. Zeevaart in Michigan but was never able to obtain an effective extract (a different variety of cocklebur, perhaps?)—although some years previously he had been successful in the laboratory of Dennis J. Carr (1967) in Ireland! Such have been the problems of florigen extraction.

Since florigen apparently moves only in phloem tissue, Charles F. Cleland and A. Ajami (1974) collected honeydew from aphids feeding on induced cocklebur and other species. Using an LD duckweed as a test plant, strong florigenic activity was found. Two active substances proved to be salicylic acid and acetylsalicylic acid—which is common aspirin! Unfortunately, salicylic acid and aspirin are completely ineffective when applied to vegetative cocklebur or other plants. Furthermore, salicylic acid concentrations are equal in the honeydew of both vegetative and induced cocklebur plants. Duckweed flowering is influenced in many complex ways by the nutrient medium, and apparently salicylic acid in the medium is an example.

During recent years, a few investigators have taken another approach. One set of plants is given a cycle that includes a long dark period; another set of plants is given the same cycle, but the dark period is interrupted with a brief interval of light, sufficient to inhibit flowering if the species is an SDP or promote flowering if it is an LDP. Then the plants are analyzed for metabolic components, using the most sophisticated biochemical techniques. So far, only a few differences have been observed, and it has not been possible to ascertain their significance. Nevertheless, the method still seems promising.

Still another approach is that of Sachs (1978), and Sachs and Hackett (1983), who question the basic florigen concept and suggest that photoperiodic induction causes a diversion of nutrients (sucrose, etc.) within the plant, leading to floral initiation. Sucrose might be directed toward the bud that will become induced. Sachs and Hackett are able to present some intriguing experimental evidence in support of their hypothesis, but we shall not discuss it here because of space limitations.

Certainly the lack of a florigen bioassay is a serious problem. When extracts are applied to plants to see if they have florogenic activity, we are entitled to wonder whether the active material is able to penetrate the plants and move to the bud. Perhaps its effects are nullified by the presence of inhibitors produced in the test plants in noninductive conditions. Some workers have tried to overcome these problems by applying these extracts directly to the bud itself. Again, however, successes have been only rather minor.

So far, few attempts have been made to extract and isolate floral inhibitors, although gallic acid appears to be a specific flower inhibiting substance extracted from vegetative plants of the SD *Kalanchoe blossfeldiana* growing in LDs (Pryce, 1972). We must remember that an inhibitory effect might be a process rather than a substance. For example, LD leaves on an SDP might in some way absorb and destroy promoting substances produced in SD leaves.

Because attempts to extract promoters and inhibitors have been so disappointing, other more indirect approaches have been taken. Experimenters have added various antimetabolites to LDPs and SDPs, seeking those that apparently inhibit the flowering process in a specific way. For example, a compound may be effective only when applied during florigen synthesis. Again, results have not been promising. It appears that respiration and protein and nucleic acid synthesis are involved in flowering,

but then these processes are involved in virtually all the life of a plant. Before we are carried further afield, let us state the conclusion for this section: *There is much circumstantial evidence that flower initiation is controlled by hormones: one or more positively acting florigens and one or more negatively acting inhibitors. These substances remain to be identified.*

22.10 Responses to Applied Plant Hormones and Growth Regulators

Since plant hormones and growth regulators influence virtually every aspect of plant growth and development, it is logical to investigate their effects on flowering. Many compounds are now known that will induce or inhibit flowering in one species or another. There is an important potential for practical application of this knowledge, since the induction of flowers often plays such an important role in agriculture. (For example, flowering in sugar cane, sugar beets, and various root crops uses assimilates that might otherwise be stored in the harvested plant part.) Work with hormones and growth regulators could also lead to better understanding of the flowering process. Again, however, there are about as many exceptions as rules. Let's summarize a few tentative observations.

Auxins and Ethylene More often than not, auxins inhibit flowering. In SDPs, this occurs before translocation of florigen from the leaf is complete, after which there may be marginal promotive effects. Promotions have also been observed in LDPs held under days just too short for induction, and auxins clearly promote flowering in some bromeliads, including the pineapple. In pineapple and cocklebur, applied auxins cause production of ethylene, which itself influences flowering in the same way as auxins (inhibition in SDPs, promotion in bromeliads). In the bromeliads, IAA is relatively ineffective, since it is apparently broken down by the plant's enzymes. Thus synthetic auxins such as NAA or 2,4-D must be used. Auxin concentrations required to inhibit flowering usually produce severe epinasties and other responses, and measured plant auxin levels seldom correlate with flowering in any meaningful way. Hence, although endogenous auxins may influence flowering to some extent, they probably do not control it.

Gibberellins Gibberellins (GAs) can substitute for the cold requirement of several species that require vernalization and also for the LD requirement of several LDPs. There are some important exceptions among LDPs (e.g., *Scrofularia hyecium* and *Melandrium* sp.), however, and gibberellins normally fail to replace SDs in SDPs—although again there are a few exceptions (e.g., cosmos and rice). In many species, gibberellins increase in stems and leaves under LDs, so it seems reasonable to assume that gibberellins might account for at least part of the flowering requirement (a florigen complex?) in LDPs (and also for the common stem elongation under LDs). But the situation is not simple. In LDPs, for example, LDs increase both the rate of synthesis and the rate of destruction of GAs, and many studies now show a lack of correlation between extractable GAs and flowering. Jan Zeevaart (1976) states in a review that there is now "conclusive evidence" that flower formation and stem elongation are separate processes, with GA promoting stem elongation but not flower formation in *Silene*, for example. Although GAs were reduced to levels below those that could be detected, flowering occurred. It has also been shown (Wellensiek, 1973) that flower formation and stem elongation are under control of two separate but closely linked genes. Thus flowering seems to be quite independent of GAs in some species—but, of course, the situation could be different in other species.

GAs seem to be particularly important in the formation of cones in conifers. Y. Kato et al. (1958) in Japan, Richard Pharis et al. (1976) in Calgary, Canada, and others have pioneered in this work. Most conifers require several years before they attain ripeness to respond, but Pharis was able to induce the formation of male strobili on Arizona cyprus (*Cupressus arizonica*) when plants were only 55 days old by spraying with GA_3. He and his co-workers were unable to induce cone formation in members of the Pinaceae with GAs, but they now achieve this by applying the less polar GAs (GA_4, GA_5, GA_7, GA_9). Because breeding of conifers is important to the lumber industry, these observations could be significant. Breeding times might be reduced from years to months.

Cytokinins Cytokinins have been observed to promote flower formation in several plants. A combination of a cytokinin (benzyladenine) and GA_5 has induced flowering in one SD variety of chrysanthemum. In another variety, benzyladenine could substitute for the latter part of photoinduction. In most LDPs and SDPs, cytokinins do not affect flowering.

Abscisic Acid When certain SDPs (e.g., Japanese morning glory and *Chenopodium rubrum*) are already slightly induced and then treated with abscisic acid (ABA), flowering is promoted; ABA inhibits flowering in a few LDPs (e.g., *Spinacia*). But no conclusions can be drawn, since ABA inhibits flowering in other SD and LD species and is completely innocuous in still others. ABA is generally, but not always, higher in shoots under LDs.

Sterols A substance called TDEAP (tris-[2-diethylaminoethyl]-phosphate trihydrochloride), which inhibits synthesis of cholesterol in animals, inhibits flowering of cocklebur and Japanese morning glory. Does inhibition by TDEAP imply that florigen is related to cholesterol; that is, that it is a sterol of some kind? Not according to results so far. Application of various sterols does not overcome the effect of TDEAP, nor could any significant changes in sterol fractions be detected during induction. Sterol biosynthesis can be inhibited by compounds other than TDEAP without influencing flowering one way or the other. Thus, we have no idea what TDEAP is doing in the plant, and evidence that florigen is a sterol is at best flimsy. But flowering can be induced in two SDPs by application of sitosterol and lanosterol (in *Chrysanthemum morifolium* ca Princess Anne) and estradiol (in *Callistephus chinesis*; see Vince-Prue, 1975).

γ-Tocopherol Battle et al. (1976, 1977) measured γ-tocopherol (a form of vitamin E) in leaves of cocklebur plants at various times during an inductive dark period and following 20-min light interruptions given at various times during such a dark period. By 10 to 12 h of continuous darkness, the γ-tocopherol had increased in the leaves to levels 5 times those at the beginning of the dark period. Red light interruptions up to and including 8 h after the beginning of darkness maintained the level of γ-tocopherol at about that measured at the beginning, and the effect was completely reversed by far-red light. When the light interruption was given 9 h after the beginning of darkness, however, the substance was almost as high as in dark controls. Just as floral induction reaches a point where it can no longer be reversed by light, so did the level of γ-tocopherol, and at the same time. This is one of the most interesting correlative measurements of a specific substance yet reported.

To state our growth regulator conclusion: *The flowering response is often influenced by applied growth regulators and hormones, but the few patterns that can be tentatively discerned all have many exceptions. Although several compounds will cause flower formation, there is no convincing evidence that florigen is one or more of the well-known plant hormones.*

22.11 The Induced State and Floral Development

To illustrate the uniqueness of studies in flowering physiology, let us briefly consider how the flowering hormone must act in different species. In cocklebur, young leaves that are allowed to grow out on an induced plant can be grafted to vegetative receptors, causing them to flower. The young leaves apparently become induced by the older leaves even under LDs. As a matter of fact, as many as five vegetative cocklebur plants have been grafted in series to an induced plant at the end of the chain, with flowers forming on all plants.

Perilla, on the other hand, acts quite differently. An excised leaf can be induced by SDs and can then be grafted to a series of vegetative receptors for several months, inducing each to flower. But none of the leaves on the receptor plants become induced. Thus the flowering condition in perilla is not as "contagious" as it is in cocklebur.

Upon arrival of florigen at the apical meristems, the course of meristematic development changes from the vegetative to the reproductive mode. In many species, arrival of the stimulus leads to an immediate increase in mitotic activity, and nuclear size often increases, as does the size of the nucleolus. Frequently, there is a buildup in the number of ribosomes, mitochondria, and RNA in the apical cells. The buildup in RNA may occur *before* the stimulus arrives. Does this mean that some other stimulus precedes the main one? Or that florigen at a concentration too low to cause flowering arrives sooner, causing the observed changes in RNA and other factors? (For reviews of cellular changes that accompany conversion of the bud to the floral state, see: Bernier et al., 1974; Jacqumand et al., 1976; Havelange, 1980.)

Our final conclusion: *The induced state has unique properties in different species, and the changes that take place at the flowering apex could help us understand development.*

22.12 Where Do We Go from Here?

Since the first edition of this book was published, much has been written about photoperiodism and the flowering process, yet no major breakthroughs have appeared, and activity in this research field has decreased noticeably; the trend has not been reversed. Most workers apparently feel that we have reached a temporary dead end. There are obvious things to do, but none is really promising. We have realized with increasing vividness that the problems are exceedingly complex. Differences between species are extensive and significant, much more so than we might have imagined one or two decades ago. Studies on reproduction biochemistry have told us virtually nothing. Except within broad limits, there is no way to predict how a given growth regulator will influence flowering of a species not yet tested.

Yet the unanswered questions are central to understanding development, which remains one of the most significant unsolved problems in modern biology. What is time measurement in photoperiodism? What is induction? What is the floral stimulus? How does it act? Hopefully, in the not too distant future, someone will think of new approaches. Then this potentially fascinating and important field will again become a hub of active research.

four

Environmental Physiology

23

Topics in Environmental Physiology

It's high summer in the Rockies. You are on your way to the west coast when you and your traveling companions decide that you will take an extra day to detour through Rocky Mountain National Park over the Continental Divide. You are afraid that the subcompact might begin to boil, but to your great relief, you make it without mishap to the high point on the Trail Ridge Road—the sign says 12,183 feet above sea level! You pull into a parking stall, turn off the engine, and set out for a short tramp on the tundra.

Those little alpine plants are really something, you think to yourself. How can they be so healthy looking when they are covered with snow nine or ten months of the year? They must have to photosynthesize at rapid rates even at low temperatures to succeed in a place like this. How do their genes differ from those of the prairie plants that you saw yesterday?

Just last spring, you completed the undergraduate course in plant physiology. Your copy of Salisbury and Ross (which you decided not to turn in at the student book exchange!) lies on your pile of gear back in the car. Remembering the book's general outline, you begin to formulate questions and observations in your mind: Is there something about the water potential of these plants that allows them to withstand frost? The cells in their leaves must be turgid because of osmosis. The soil seems to have ample moisture; would transpiration cool these plants even below the chilly air temperature? How do they respond to the low temperatures up here? Is there anything special about the way they absorb minerals from this cold soil? Is there anything unusual about translocation of assimilates in these plants? Probably not.

Do their enzymes have special configurations that allow them to function optimally at low temperatures? Photosynthesis must be highly efficient in these plants, and the low temperatures at night probably reduce dark respiration. Do they by any chance utilize C_4 photosynthesis? Do the ones growing almost on the bare rock have some means of nitrogen fixation—mycorrhizae, associated bacteria, or something? The flowers are brightly colored. Are their pigments produced through some special biochemistry? Could the colors be intensified some way by the bright light and intense ultraviolet? Are the plants being "sunburned"?

Being dormant so much of the year could mean that these plants have unusual timing mechanisms. Do they grow slightly under the snow, so that they are ready to go when summer finally comes? What causes them to flower? Do they go dormant in the fall in response to the shorter days? What hormonal mechanisms mediate these responses to environment? Do the leaves fold up at night to resist radiant loss of heat to the cold sky above? Surely auxins, gibberellins, cytokinins, ethylene, and inhibitors play important roles in making these plants what they are, where they are. How do they do it?

With questions like these, you are beginning to think like an environmental physiologist. And, of course, you don't have to be standing on the Trail Ridge to have such thoughts. Perhaps your walk was in the Sonoran Desert near Tucson, Arizona, or the relic grasslands of Nebraska. Maybe you were checking on the yield of a Nebraska wheat field. Were you lucky enough to visit the steaming jungles of Brazil? Or if not, then the near-tropics of the Florida Everglades? Maybe your questions developed as you walked along the Appalachian Trail, or in the Blue Ridge Mountains of Kentucky. Similar questions could develop on a farm in California, in Central Park in New York City, or as you putter in your backyard garden.

In all these situations and thousands more, you can ask related questions about plants in their environments. Indeed, you can develop these questions into a science. Though the questions may be similar, the sciences that develop from them may have different names. If you are a dyed-in-the-wool physiologist, you might be quite happy with the term **environmental physiology.** If your field is agriculture, you may call your science **crop ecology** or even **crop physiology.** Traditionally, if you work with vegetables for kitchen gardens or with ornamentals, you call yourself a **horticulturist,** but you may still be asking the questions of environmental physiology. If your interest is in field crops (cereals, forage crops, root crops, and others), you may call your science **agronomy.** If you are interested in how plants grow in their natural environments, then your field is **physiological ecology,** which also has its applied aspects such as **forestry** and **range management.** Indeed, most students in agriculture and in forestry and range management are required to take a course in plant physiology, mainly for the environmental aspects.

23.1 The Problems of Environmental Physiology

Specific guiding questions for research depend upon the specific field of endeavor. In agriculture, most research is guided by the economics of obtaining the highest maximum yields or the highest possible quality for the lowest cost and energy inputs. Environmental physiology always plays a role in such research.

Studies in physiological ecology apply the methods of physiology to the problems of ecology. Traditionally, these problems have centered on the question of plant and animal distribution. The ever-present assumption is that organisms occur where they do in nature because they are well adapted to their environments. Plants that grow in deserts can withstand drought and high temperatures, for example, while deciduous trees only grow where there is ample moisture and moderate temperature. Hence, studies in physiological ecology may attempt to measure the microenvironments of plants or animals in the field and then duplicate these environments to study the same organisms in the laboratory. But ecology has many interesting problems besides those of distribution.

Near the end of the 1970s, four eminent physiological ecologists were given the assignment of editing four volumes on physiological ecology for the *Encyclopedia of Plant Physiology* (New Series). They thus considered work that was going on all over the world and commissioned various authors to write review articles relating to these areas. Although the emphasis was on natural ecosystems, they did not hesitate to commission many authors who were interested in agricultural and other applied problems. Table 23-1 summarizes the problems of environmental physiology based upon the tables of contents of those four volumes. Often several chapter titles

were concerned with a single topic (suggesting intense interest in that field); in the table such titles have been combined under the single topic. Indeed, many of the titles were modified to fit the format of the table.

To begin with, environmental physiologists must study plant responses to the physical environment. This is reflected by the first large category in the table. Furthermore, much of the emphasis in studies of environmental physiology has been upon water in the soil-plant-atmosphere continuum and in environmental effects on photosynthesis. In addition, a number of other interesting interactions between individual plants and their chemical and biological environments are being investigated. Ecologists, during recent years, have become increasingly interested in the ways individual plants and animals interact with each other. As shown in the table, parasitism, herbivory, and related topics provide examples. We shall consider two of these later in this chapter. The next level of study concerns interactions of plant and animal populations with each other and with the physical environment. Perhaps these topics are being studied more by ecologists than by physiologists. Mineral cycling and energy flow through ecosystems as well as ecosystem productivity have been especially active areas of research, but human influences have also received a great deal of attention—especially in the news media!

In the remainder of this chapter, we shall consider some principles of plant response to environment, including the question of the nature of the environment, and then we shall examine four topics of environmental physiology: the role of genetics, responses to radiation, allelopathy, and herbivory.

23.2 What Is the Environment?

Dictionaries define **environment** as the circumstances, objects, or conditions by which one is *surrounded.** Should environment include everything by which an organism is surrounded? Are a cricket's chirps or low-energy radio waves coming from a distant planet part of a plant's environment? In the broadest sense, yes. But if they have absolutely no effect on the plant, it seems unreasonable to think of them as part of the plant's **operational environment,** which is the complex of climatic, edaphic (soil), and biotic factors that act upon an organism or an ecological community and ultimately determine its form and survival.

---

*If this is the case, is there such a thing as the *internal environment* of a cell? Not if we are careful in our use of the language. It is correct to speak of internal *conditions,* and any cellular organelle such as a chloroplast has its own environment, but we should not speak of *internal environment.*

Table 23-1 Some Topics of Physiological Plant Ecology*

Topic	Some references in this text
I. Responses to the physical environment	
Radiation: Measurement of parameters	Appendix B
Responses to irradiance (quantum flux density)	
Photosynthesis (PAR and PPFD)	Chapters 9–11
Other Responses	Chapter 19
Spectral distribution (radiation quality)	Appendix B
Phytochrome responses	Chapter 19
Other nonphotosynthetic responses (blue light, etc.)	Chapters 18, 19
Photoperiod and biological-clock mechanisms	Chapters 20, 22
Responses to ultraviolet and ionizing radiation	
Temperature: Normal and extreme	Chapter 21 and many others
Wind as an ecological factor	
Energy exchange between a plant and its environment	Chapter 3
Fire as an ecological factor	
The soil environment	
Aquatic environments	Chapter 4
II. Water in the soil-plant-atmosphere continuum	Chapters 1–4
Water in tissues and cells	Chapters 2–4
Water uptake, storage, and transport	Chapters 2–4, 6,7
Water loss through stomates and cuticle	Chapter 3
Plant response to flooding	Chapters 12, 24
Seed and spore germination	Chapters 19, 21
III. Photosynthesis	Chapters 9–11
Ecological significance of different CO_2-fixation pathways	Chapter 11
Modeling of photosynthetic response to environment	
Water use and photosynthesis	Chapter 11
Plant life forms and their carbon, water, and nutrient relations	
IV. Responses to the chemical and biological environment (individual organisms)	
Features of soil chemistry (limestone vs. silicate rocks, soil *p*H, essential or toxic ions)	
The physiology of salt-tolerant plants (halophytes), osmoregulation	Chapters 5, 24
Ecology of nitrogen nutrition (including N_2 fixation)	Chapter 13
Plant-"plant" interactions	Chapter 24
Mutualism (mycorrhiza, lichens)	Chapter 6, 13
Parasitism (viral, bacterial, fungal)	Chapter 6
Competition	
Allelopathy	Chapters 14, 23
Plant-animal interactions	
Pollination, fruit and seed ecology	
Herbivory	Chapter 23
Carnivorous plants	
Mutualism	
Parasitism	
Competition	
V. Ecosystem processes (populations forming a great variety of communities)	
Mineral cycling or transfer	
Energy flow through ecosystems	
Productivity (Photosynthesis again)	Chapter 11
Human influences	
Biocides and growth regulators	
Pollution: atmosphere, water, soil	
Agriculture: controlled or artificial ecosystems	

*This table is based on the tables of contents in the four volumes edited by Lange et al. (1981–1983).

Such a definition helps somewhat, but it may not always be easy to know for sure whether an environmental factor is part of the operational environment. The cricket's chirp is clearly part of the operational environment of another cricket, but we have no reason to believe that it is part of a plant's operational environment. How about the radio waves? We know of no way that such waves could act upon either crickets or plants, but it is possible that such an action remains to be discovered.

George G. Spomer (1973) at the University of Idaho has defined the operational environment in a way that provides further insight. Applying concepts of thermodynamics (Chapter 1), he points out that an environmental factor interacts with an organism only when the factor heats or does work upon the organism or the organism heats or does work on the factor. If the radio waves pass through the plant unchanged, for example, then the plant is also unchanged, and there could be no interaction; the radio waves would not be part of the plant's operational environment. The cricket's chirp does work upon the hearing apparatus of another cricket, so it is part of the second cricket's operational environment. According to Spomer, such an interaction involves a direct transfer of mass (matter) or energy across the boundary between an organism and its environment.

John Muir in his *Sierran Diary* of a century ago said: "When we try to pick out anything by itself, we find it hitched to everything else in the universe." This is a statement of the **holistic concept,** which suggests that everything in the universe interacts with everything else. At some highly theoretical level this might be true (although no one can test the idea), but Spomer's analysis of interactions by energy or mass transfer helps to place the holistic concept in a proper perspective. The transfer-interactions (if they occur at all) might be so infinitesimal that they have no practical consequences. The sound waves from the cricket's chirp act upon the plant as well as the other cricket's ears, but the energy transfer is most probably too small to have any significant effect on the plant's metabolism or other activities—especially compared with energy inputs from other sources. The plant has no way to extract the *information* from the cricket's chirp, although another cricket does.

Factors of the environment that fit Spomer's definition of operational factors include light, heat, water, various gases, mineral elements, and organic substances. These factors may be directly transferred across the boundary between the organism and its environment. Other factors are not operational—temperature, pH, partial pressures of gases, concentrations, water potential—because they are not themselves transferred across boundaries. Instead, they indicate a *potential* for transfer. If the temperature inside an organism is lower than the temperature outside, the difference indicates the potential for transfer of heat across the boundary. Different pH values indicate a potential for transfer of hydrogen ions, different partial pressures of a gas indicate a potential for transfer of the gas, concentrations indicate potentials for transfer of dissolved substances, water potentials indicate potentials for transfer of water, and so on.

The environment—that is, the universe—may be too complex to describe, but if we limit ourselves to factors that are probably part of the operational environment of plants, the task is somewhat simpler. Table 23-2 lists the major energy and mass factors and the potential measurements generally used to describe their levels.

23.3 Some Principles of Plant Response to Environment

Although environmental physiology is a young science, a few general principles have developed. We examine three in the next three sections: the law of the minimum, interactions of factors, and the kinds of plant response to environment.

Saturation, the Cardinal Points, and Limiting Factors Perhaps the most fundamental principle of plant response to environment—and the one most frequently encountered in this text—is that of **saturation.** Organisms respond to virtually any environmental parameter according to a common pattern: As a parameter increases, it reaches a **threshold** above which it begins to have an effect, after which the response increases until the system becomes **saturated** by the parameter. Then, as the parameter level or concentration continues to increase, response remains constant or begins to decrease if the parameter at its high levels becomes toxic or inhibitory. Figure 23-1 shows the expected pattern. Looking back through the text, there are figures that illustrate the phenomenon for temperature (Fig. 21-1), photosynthesis (Figs. 11-4, 11-6, 11-8, and 11-9); mineral nutrition (Fig. 5-4), enzyme action (Fig. 8-10), transport of ions across membranes (Fig. 6-17), and response to auxin (Fig. 16-5).

It is easy to understand these curves between the minimum and the optimum and to understand the concept of saturation. The organism simply utilizes the factor being considered until its capacity for this utilization is used up or saturated. But what about the descending right-hand features present in so many of these examples? Explanations for this part of the curve differ, depending upon the phenomenon considered. When growth is inhibited by high temperature, we have suggested enzyme denaturation as

Table 23-2 Environmental Factors That Are Operational for Many Plants

Factor	Units	Potential Measure	Units
Energy Factors			
radiation (including light)	joules m^{-2} calories* m^{-2}	radiation level (irradiance) compared with absorptivity of pigment	watts meter^{-2} moles of photons m^{-2} s^{-1} calories* m^{-2} min^{-1}*
heat	joules kg^{-1} calories* kg^{-1}	temperature	Kelvin °C*
gravity**			
Mass Factors			
gases	grams or moles	pressure or partial pressure	pascals (N m^{-2}) bars* (0.1 MPa = 1.0 bar)
liquids and solutions water solution (solute in water)	grams or moles grams or moles	densities (concentration) water potential concentration	grams meter^{-3} pascals (N m^{-2}) moles meter^{-3} grams meters^{-3} moles liter*$^{-1}$ (% or ppm*)
hydrogen ions in water solids	grams or moles grams or moles	pH (H$^+$ potential) (concentration) concentration (seldom used)	pH units (as above) grams or moles meter^{-3}

*Not SI unit; should be avoided.

**Gravity does not lend itself to this kind of analysis.

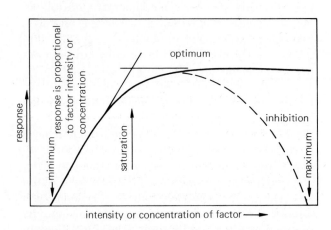

Figure 23-1 A generalized curve showing organism response to an environmental parameter. Minimum, optimum, and maximum are called cardinal points.

the explanation (see Fig. 21-3), but this is not always satisfying. Remember *Laothenia chrysostoma* (Section 21.8) that dies within 30 days when night temperatures are 26°C? Surely enzymes are not being denatured at these temperatures, since day temperatures of 26°C or above are not harmful. The causes are no doubt complex, involving perhaps the production of an inhibitor during warm nights (im-

plying a phytochrome interaction?). Superoptimal concentrations of mineral nutrients might become toxic, because they begin to interact with systems in the organism other than those that were responding on the ascending part of the curves.

In 1840, Justus von Liebig in Germany published a book called *Agricultural Chemistry*. It had an immense impact on thought about plants. It almost became a best seller. In the book, Liebig formulated his **law of the minimum,** which in retrospect can be derived from and understood on the basis of saturation curves. The law states: "The growth of a plant is dependent upon the amount of 'foodstuff' presented to it in minimum quantities." Applied to mineral nutrition or photosynthesis, for example, we might expect the ascending part of the saturation curves to be identical at two levels of one factor when another factor is presented to the organism in relatively small (limiting) amounts. The two curves will have different saturation levels, however, depending upon the different quantities of the second limiting factor. The principle was discussed and the term *limiting factor* proposed in a paper by F. F. Blackman in 1905. It was later illustrated by his experiments on photosynthesis of algae. Such curves are often referred to as **Blackman curves.**

Figure 23-2 illustrates this principle with a simple experiment in mineral nutrition involving two levels

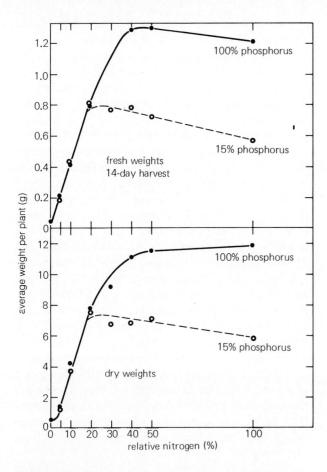

Figure 23-2 Results of an experiment in mineral nutrition. Tomato plants were grown in vermiculite and watered with nutrient solutions containing various concentrations of nitrate (NO_3^-) and one of two concentrations of dihydrogen phosphate ($H_2PO_4^-$), as shown. Curves represent fresh and dry weights of 14-day-old plants. Data are averages of six plants. (From Salisbury, 1975.)

of phosphorus given to plants over a wide range of nitrogen concentrations. The threshold for nitrogen is extremely low, but below that level plants do not grow at all. As nitrogen increases, plants respond the same at both phosphorus levels until phosphorus becomes limiting at the nitrogen saturation level. A higher concentration of phosphorus leads to a higher saturation level for nitrogen.

The practical implications of Liebig's law were and continue to be obvious and important. In agriculture (Liebig was probably the first important agricultural chemist), the challenge is to discover the limiting factor and to remedy it. If plant yield is limited by insufficient nitrogen fertilizer, then more nitrogen is applied. When enough nitrogen has been applied, then perhaps phosphorus becomes limiting and needs to be applied. This approach has had spectacular success since 1840, so that much more food can now be produced on a hectare of land than before. Of course, the limiting factor might be one or

more of many things besides mineral nutrients: water, damage by pests (diseases, insects), competition from weeds, CO_2 concentrations (especially in greenhouses), or the plant's genes (hence the success of breeding programs), to name some important examples.

In physiological ecology studies, a plant's distribution may be limited on one border by the factor presented to it in the "least" amount. This is always a relative matter, since highly disparate quantities of the different elements and environmental parameters are required by plants (see Tables 5-1 and 5-3, for examples).

Interaction of Factors   Unfortunately, things are not as simple as Liebig's law might imply. Under carefully controlled conditions such as those of Figure 23-2, everything may work out as the law predicts. In the real world, things seldom work out so well; the curves are not identical in the ascending parts where only one factor is supposed to be limiting. Figure 23-3 shows a somewhat more typical response, although the experiment is the same as that illustrated in Fig. 23-2.

There are several ways to explain a failure of Liebig's law. Most probably they all boil down to a single idea, however: The extent to which Liebig's law might function (i.e., Blackman curves might be obtained in multiple factor studies) will depend upon the *extent* to which the factors under consideration are able to enter into reactions within the organism. Inability of a factor to enter into such a reaction may depend upon several things, such as restrictions upon diffusion of CO_2 through stomates or movement of ions in the free space, or ultimately the chemical rate constants for the appropriate reactions.

This idea can be understood by thinking of a single reaction that involves two precursors. If the reaction has a large equilibrium constant, so that the precursors are almost completely used up (they enter into the reaction to the greatest extent possible), then one can obtain Blackman curves by plotting the product as a function of precursor concentration (Fig. 23-4a). If, however, the reaction does not go to completion, for whatever reason (the equilibrium constant is small), then Liebig's law does not work as well (Fig. 23-4b).

Can Liebig's law guide our speculations in environmental physiology? It is not a bad starting place. We can begin our studies of agricultural yields or plant distribution by looking for limiting factors. But we will probably have to progress beyond these initial steps to understand the real world, and we may often find more than one (but seldom more than three or four) factors limiting yield or distribution.

There are many complications. In one form of interaction, the minimal or toxic level of one factor (for a given organism) may depend on levels of one or more other factors. Sodium or potassium ions, for example, may be quite toxic to plants at low concentrations when they are supplied to the roots in solution by themselves (along with a suitable anion, such as chloride). Addition of small amounts of calcium ions raises the minimum toxic concentration of the sodium or potassium ions to much higher levels. Another example is the enhancing effect of one drug on the response of an animal to another drug. If two factors interact in such a way that response to them given together is greater than the sum of responses to each given alone, we speak of **synergism.**

The **law of tolerance** is another way to state some of the points we have been discussing. This law says that organisms will not respond to a factor until it is present at some minimum level and, moreover, that they are harmed when the factor is present at a maximum (toxic) level. The range between is the **tolerance** range. Response to temperature, as noted above, provides a good example (see Fig. 21-1e).

Some powerful mathematical tools have been developed to help us understand factor interactions in nature or in controlled experiments. One of these is **regression analysis,** which is a highly valuable tool in such situations as field observations where data must be taken as they come. When an experiment can be carefully designed in advance (e.g., where treatments can be set up according to a table of random numbers), **Fisher's analysis of variance** is used widely and appropriately. (See statistics textbooks for descriptions of these methods.)

Such studies indicate whether or not two environmental factors interact. If they both influence a given response but do not interact, they may be additive in their effects or they may be multiplicative (Fig. 23-5). When they are additive, they act upon different causal sequences that lead to the response.

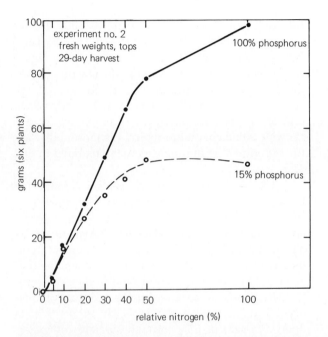

Figure 23-3 Fresh weights and photographs of 29-day-old tomato plants treated as in Fig. 23-2. Curves in the low ("limiting") nitrogen concentrations are no longer exactly superimposed. (From Salisbury, 1975.)

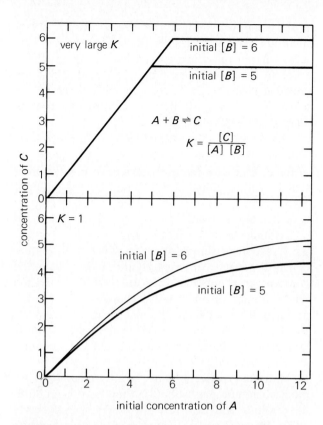

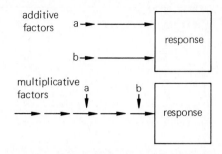

Figure 23-5 An illustration of additive and multiplicative factors according to the analysis of Mohr (1972).

Figure 23-4 An illustration of the principle of limiting factors as it might be observed in a chemical equilibrium reaction. Concentration of a hypothetical product (C) is plotted as a function of initial concentration of one reactant (A) in the presence of two initial concentrations of the other reactant (B). **(a)** With a very large equilibrium constant, virtually all available A enters into the formation of C until B becomes limiting at its two concentrations. This is the ideal limiting factor response. **(b)** If K is only 1, then even at low concentrations, both A and B limit the amount of product formed.

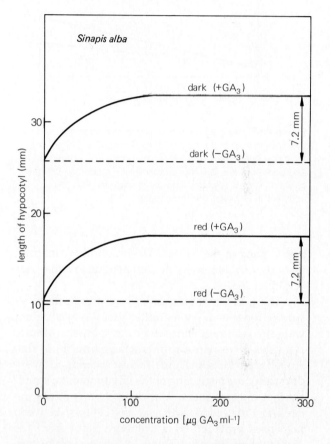

Figure 23-6 An empirical example of "numerically additive behavior" of two factors. Hypocotyl lengthening in the mustard seedling was investigated under the control of light (continuous standard red light) and exogenously applied gibberellic acid (GA_3). It is apparent that the dose response curve for exogenously applied GA_3 is the same with and without light. Note that GA_3 promotes lengthening, whereas red light inhibits lengthening of the hypocotyl. Hypocotyl length was measured 72 h after sowing. (After Mohr, 1972.)

Say that a compound is made in two different compartments in the cell; one factor may influence one compartment and another factor the other compartment. Stem growth in the white mustard plant, for example, can be influenced oppositely by gibberellic acid and by red light (phytochrome: P_{fr}). The two responses are purely additive, as shown in Fig. 23-6. Multiplicative responses are more common. The two factors act on different steps in the same causal sequence (Fig. 23-5), so that the effect of one is always some fraction of the effect of the other. For example, rate of stem growth in the white mustard plant as influenced by red light is determined by the concentration of ions or sucrose in the growth medium (Fig. 23-7). Analysis of variance shows when responses are additive or multiplicative; that is, when they are *not* interactions. Any other result in an analysis of variance indicates an interaction of factors, and there are many kinds of interaction (see Lockhart, 1965, and Mohr, 1972).

The Kinds of Plant Responses to Environment In addition to the quantitative manner in which an organism responds to environment, there are other ways of classifying responses. These are summarized in Table 23-3, the ideas of which have been modified several times since they were originally presented by

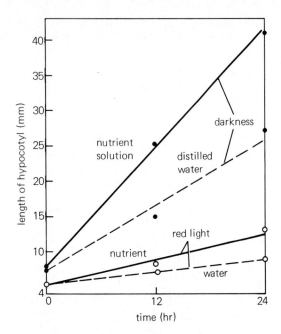

Figure 23-7 An illustration of multiplicative effects. Growth of the hypocotyl of white mustard (*Sinapis alba*) is plotted as a function of time. Red light inhibits compared with darkness, and distilled water inhibits compared with a complete nutrient solution (Knop's solution). The percentage of inhibition is constant at all times. (From Schopfer, 1969.)

Anton Lang at the Annual Meeting of the American Institute of Biological Sciences held at Purdue University in 1961.

The table illustrates three contrasting sets of response types—and some other ideas as well. First, we can contrast immediate responses (photosynthesis in response to irradiance) with those that are delayed for some time (many seconds to many days) after the beginning of the environmental stimulus that initiates them. This distinction is clearly related to the second, which is the contrast between responses provided by the energy input from the environment (photosynthesis, or enzymatically controlled reactions driven by thermal energy) and those responses that are *triggered* by some environmental change but use the energy provided by the plant's metabolism to produce the response, rather than energy derived from the change in the environment that causes the response. In the latter situation, we speak of an **amplification** of the environmental parameter. For example, the energy required to bring about stem bending must be much greater than the energy provided by the relatively few photons that initiate phototropic bending toward a light coming from one side (and often applied for only a brief instant; bending takes much longer). Third, a few plant responses (germination, perhaps) are triggered by some environmental change (that is, they are switched on or off by the change). One good example

is the snapping shut of the Venus flytrap, a carnivorous plant with trigger hairs on the inner surfaces of two opposing special leaves. When these hairs are touched (usually by an insect), acid is secreted into the walls of cells on the outside of the leaves and they grow rapidly, causing the trap to snap shut (Section 18.1).

More commonly, plant responses are not only initiated by an environmental change, but the extent of the change determines the extent of the response; that is, the response is modulated by the environmental change—even though the response may be considerably delayed.

Homeostasis is a particularly interesting organismal response to environment in which some feature of the organism's internal condition is maintained constant or allowed to vary only within restricted limits in spite of much wider changes, often in the same environmental factor, outside the organism. By far the best and most studied examples are among animals (e.g., body temperature), but the phenomenon is also recognized in plants.

Carryover effects could be of tremendous ecological significance, although they have been studied carefully only by a very few workers. In the early 1950s, Harry Highkin, then at the California Institute of Technology, discovered that genetically pure, inbred pea plants grew poorly when the day and night temperatures were equal and held constant (10, 17, or 20°C; see Highkin and Lang, 1966). When the pea plants were grown for several generations under these adverse conditions, each generation (up to about the fifth) grew more poorly than the previous one. When Highkin reversed the situation (germinated seeds from stunted plants under optimum conditions), it required at least three generations to reach the maximum level of growth. Such a carryover of environmental effects from one generation to the next seems quite contrary to most of our concepts of genetics. (The environment does not change the genes.) Nevertheless, the phenomenon is real, and it has since been confirmed by other workers. Highkin and the others were careful to demonstrate that the effects were not caused by genetic selection. Apparently, the developing embryo (or perhaps the stored food material in the cotyledons) is in some way conditioned by the environment and the parent plant so that the effect carries over through a number of succeeding generations.

23.4 Ecotypes: The Role of Genetics

We assume that the distribution of a given species is determined by its genetics, but what if there is genetic diversity within a species? Could the dwarf *Poten-*

Table 23-3 Types of Plant Response to Environment

Type of Response	Characteristics	Examples
1. Direct (nondelayed)	As environment changes, plant response changes immediately (or almost).	Photosynthesis (light level), transpiration (heat load), enzyme-controlled reactions (temperature).
2. Triggered or on-off	Environmental factor crosses a threshold, response begins even if factor returns to original level; often response is delayed. Sometimes amplification occurs.	Germination in response to such factors as low temperature or red light. (Other examples are rare: sensitive plant, carnivorous plants, etc.)
3. Modulated (quantitative), delayed responses	Level of response determined by level (potential) of environmental factor; often both amplification and delay; often interact with biological clock.	Phototropism, gravitropism, many phytochrome responses, vernalization, photoperiodic induction (many responses such as flowering, stem elongation, tuber formation), rhythm setting, and plant growth in response to germination temperature (e.g., peach seedlings). (Many examples).
4. Homeostasis	Maintenance of (nearly) constant internal conditions in spite of changes in the environment; usually (always?) achieved through negative feedback.	Bird and mammal body temperature and blood chemistry, interactions between stomatal aperture and photosynthesis, internal concentrations of growth regulators.
5. Conditioning effects	Gradual changes in the plant in response to continued exposure to some environmental condition.	Development of frost or drought resistance (low water potential).
6. Carryover effects	Effects of growth conditions carried over two or more generations.	Experiments with inbred pea plants (see text); other, less well documented examples.

tilla glandulosa plants of the high Sierra Nevada, for example, have a genetic composition that allows them to get along at relatively low temperatures, while the larger *Potentilla glandulosa* plants found at lower elevations have a genetic composition that allows them to do well only at higher temperatures? When we think about the principles of evolutionary gene-pool change, we might certainly expect such situations. The reproductive processes are relatively slow, which makes the rate of gene flow within the gene pool relatively slow, but climatic pressures on the population differ depending upon location. Hence, we might expect the genetic composition of a population to vary throughout the range of that population.

We can imagine two possible explanations for the dwarfed versus large *Potentillas*: First, their genetic compositions might be alike, but their different appearances might be caused by the different climates to which they are exposed; and second, the differences could be caused by an actual genetic diversity. How do we distinguish between these two possible explanations? Obviously, the thing to do is to bring the different plants together, growing them either in a **uniform garden** or in a controlled environment facility. In the 1920s, Göte Turesson in Sweden developed the uniform-garden approach to a high level of finesse. Jens Clausen, William Hiesey, and David Keck (1940s) of the Carnegie Laboratory at Stanford University in California and others followed suit. As it turns out, both environment and genetics are important.

Environmental effects upon plant morphology (i.e., appearance) and physiology are common. Turesson called plants with similar genetics that exhibit differences caused by varied natural environments **ecophenes.** This is usually not emphasized in discussions of this type, because the genetic differences we are about to discuss are so obviously important. Nevertheless, we should realize that the environment can and does produce many different ecophenes from any uniform genetic stock. Numerous effects of temperature, light, nutrients, and other factors on plant growth and development have been emphasized in this and several other chapters.

Turesson and others also found genetic differences in representatives taken from the different areas of a species's distribution. Turesson called these different genetic representatives of the population **ecotypes.** When the *Potentillas* from the Sierra Nevada, the Coast Range, and other locations were brought together in a uniform garden, they continued to exhibit striking and significant morphological differences (Fig. 23-8). Many species have now been studied, and it seems obvious that different environments will exert different selection pressures, resulting in different genetic compositions that are directly correlated with geography.

As might be expected, selection also works on the physiological responses to environment (Billings, 1970; see also Tieszen et al., 1981). For example, photoperiodic ecotypes have been demonstrated in several species (Section 22.1). Alpine sorrel (*Oxyria dygina*) plants collected from several locations in the

Figure 23-8 Photograph of three *Potentilla glandulosa* specimens grown in a uniform garden at Mather, California and collected on June 5 to 18, 1935. Plants were dug at three locations in the Coastal, Mid-Sierran (Mather), and Alpine stations in California 5 to 13 years previously. (From Clausen, Keck, and Hiesey, 1940.)

arctic flowered only in response to days longer than 20 h, while those collected from the southern Rockies flowered in response to days longer than 15 h, for example. The arctic plants also reached peak photosynthesis rates at lower temperatures than their southern counterparts, but the alpine plants that grow at high elevations and relatively lower CO_2 pressures were more efficient in utilizing CO_2. Any competent taxonomist would classify all the alpine sorrel specimens as the same species, but careful observation revealed a number of morphological differences between the northern and the southern representatives, as well as the physiological differences.

Many other examples could be cited. In one study, Olle Björkman (1968) at the Carnegie Laboratories in Stanford, California studied the enzyme that fixes CO_2 in Calvin-cycle photosynthesis (ribulose-1,5-bisphosphate carboxylase; see Chapter 10). First he examined two ecotypes of goldenrod (*Solidago virgaurea*): one that normally grows in the sun and one that grows in the shade. He found that the sun ecotype had several times as much enzyme as the shade ecotype, even when the shade ecotype was grown in the sun. This finding correlated closely with photosynthesis rates of the two ecotypes. Björkman also studied different species, some collected from the deep shade of a California coastal redwood forest and others collected from several sunny locations, all

close to Stanford. Again, the enzyme proved to be much richer in sun plants than in shade plants.

The carryover effects discussed in the last subsection could be important complications in uniform garden and other studies. Apparent differences in the first generations could be caused by carryover rather than genetics. Regrettably, this possibility has been largely ignored by most workers (but see Clements et al., 1950).

23.5 Plant Adaptations to the Radiation Environment

There are several ways that radiation (the visible portion of which is light) varies in nature. Almost all variations are of potential importance to plants. As an example of what is going on in environmental physiology—or physiological ecology—we shall review the radiation environment and plant responses to it (Smith, 1983).

The Radiation Environment   The basic parameters of the natural radiant-energy environment are controlled by the astronomical characteristics of the earth—a spherical, rotating planet with an atmosphere, its equator tilted 23.5° to the plane of its orbit around the sun. Latitude north and south of the equator, the daily rotation, and the seasons resulting from the tilted equator determine the sun's elevation above the horizon at any point on the earth's surface at any given time of day and on any day of the year. These factors also determine day length (see Fig. 22-1). The sun's elevation above the horizon determines (1) the length of the atmospheric pathway through which the sun's rays must travel to reach a point on earth and (2) the area of horizontal surface that will be irradiated by a given cross-sectional area of the sun's rays. The larger the surface irradiated by a given cross-sectional area (i.e., the farther one is from the equator), the cooler the climate is likely to be. Day length is a function of the sun's elevation when it is at the zenith and the angle the sun's daily path makes with the horizon; the more acute the angle, the shorter the day in winter (down to a 0-h day) and the longer in summer (up to a 24-h day).

These astronomically determined characteristics of the radiant-energy environment can be modified at any time and point on the earth's surface by other factors, including weather, atmospheric composition (e.g., natural or human-caused pollution), shading by topography, shading by vegetation, reflection, and such human-controlled factors as shading by buildings and the presence of glass through which the radiation must pass. We might expect plants to have become adapted to virtually all the natural variations. Consider the following four aspects of the radiation environment and how they can vary.

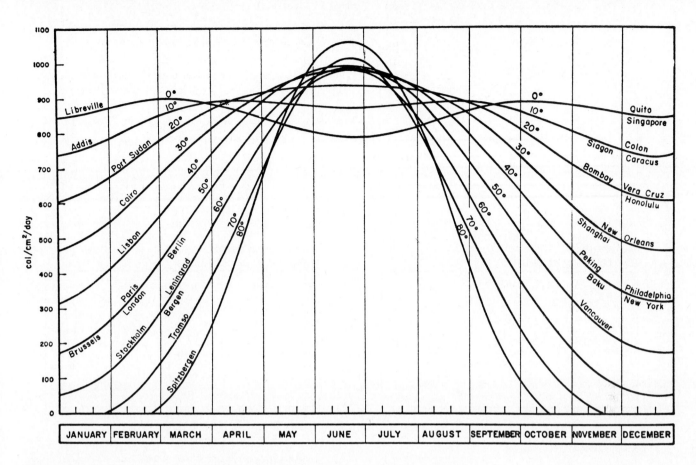

Figure 23-9 Daily totals of the undepleted solar radiation received on a horizontal surface for different geographical latitudes as a function of the time of year, based on the solar constant value of 1.353 kJ m^{-2} s^{-1}. (From Gates, 1962.)

Irradiance or photon flux density In northern latitudes, irradiances at noon and per unit area are greatly reduced compared with those in equatorial regions, because of both the lower solar elevation and the longer atmospheric pathway. Yet total daily photon flux in the north is often greater in summer than that at equatorial latitudes because of the north's long summer days (Fig. 23-9). Of course, weather and other factors often modify both instantaneous and integrated photon fluxes.

Spectral composition: quality At sunrise or sunset at all latitudes (when the sun's elevation is less than 10°), the total light environment (that part of the spectrum to which the human eye is sensitive) becomes enriched in blue (380 to 500 nm) and far-red (700 to 795 nm; Fig. 23-10). Far-red also increases relative to orange-red (595 to 700 nm; see Fig. 22-6). Blue wavelengths are enriched, because much of the irradiation comes from the blue sky (atmospheric scattering); far-red is enriched, because the longer the direct atmospheric path through which the sun's rays pass, the more the rays are enriched in long wavelengths relative to short wavelengths. This is because short wavelengths are scattered, long ones transmitted.

Ultraviolet is greatly reduced by stratospheric ozone. Shade from opaque objects is also enriched in blue wavelengths (skylight), but perhaps the most profound quality change is caused by leaf shading (Section 19.5); chlorophyll absorbs much of the red light, but leaves transmit the far-red, so leaf shade is greatly enriched in far-red. Clouds do not change the spectral composition profoundly, although blue is somewhat enriched, and scattering of all wavelengths is increased. Natural (e.g., volcanic) and human-caused pollution can influence quality in various ways, mostly by reducing blue wavelengths. Light passing through snow can be influenced in various ways, depending on the condition of the snow (Richardsen and Salisbury, 1977).

Moonlight (reflected sunlight) has slightly less blue than sunlight but is otherwise similar. Its irradiance is about 6 orders of magnitude lower than that of sunlight (see Fig. 22-6). Starlight has an irradiance about 2.5 orders below that of moonlight but with a spectrum similar to that of sunlight.

Duration or diurnal cycling of light We have already noted the crucial effects of latitude upon day length, but it is important also to note that the modifying

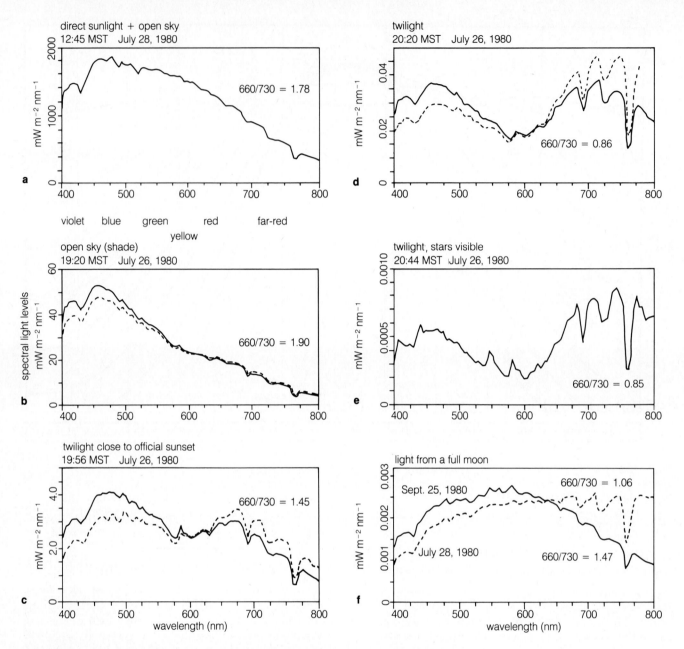

Figure 23-10 Spectral distributions of natural light energies, including sunlight (**a**), skylight measured at four times before and during twilight (**b** to **e**), and light from the full moon (**f**). Note the greatly different ordinate scales of the various curves. The energy levels at the end of the twilight measurement (**e**) are an order of magnitude lower than light from the full moon. Curves in (**b**), (**c**), and (**d**) were made while light levels were changing rapidly, each scan requiring 10 min.; hence, they were "corrected" by lowering the long-wavelength (red) end where the scans began by an amount proportional to the scans taken 12 min. later and by raising the short-wavelength (blue) end in the same way. That is, the curves were "rotated" (by computer) around their center points, down on the red end and up on the blue end. Dashed lines are original data; solid lines are after computer rotation. Dashed line for moonlight (**f**) was obtained with a fiber optics probe aimed directly at the full moon; solid line includes skylight. (From Salisbury, 1982.)

factors of clouds, topographic and leaf shade, reflections, and so on influence day length proportionally much less than the other aspects of the radiation environment. This will depend upon the sensitivity with which a given plant can detect

changes in light level (Section 22.7), but plants are apparently highly sensitive to the slight changes in irradiance that occur rapidly just before sunrise and just after sunset. Thus the length of day detected by a plant may be changed only slightly by clouds and

Table 23-4 Adaptations to the Radiant-Energy Environment

| | Primary Plant Responses | | | |
	1. Irradiance L = low irr. M = medium irr. H = high irr.	2. Spectrum UV = ultraviolet B = blue G = green R = red FR = far-red	3. Duration/cycle P = photoperiodism C = circadian clock	4. Direction important?
1. Photosynthesis, leaf development	H	B–R, G	C	no
2. Seed germination, bud break	L, H	R/FR, B	P	no
3. Etiolation syndrome	L	R/FR		no
4. Stem elongation, apical dominance	H	R/FR	P	no
5. Stem orientation	L	B–UV		yes

6. Leaf orientation:
    circadian
    solar tracking
	M
	H
		R/FR
		B
			C
			C
				no
				yes

7. Reproduction and storage organs	L, H	R/FR	P	no
8. Dormancy	L, H	R/FR	P	no

9. Damage by
    ultraviolet radiation
    photoreversal
	H
	H
		UV
		B
				no
				no

shade (say, 7 to 10 minutes out of 12 hours, which is a little over 1 percent). Some effects on irradiance and spectrum are much greater than this.

Direction The direction of the sun's rays is a function of the sun's elevation and position. Clouds and other shade, by obscuring the sun as a "point" source of radiation, to a greater or lesser degree destroy the directionality of the incoming radiation.

Plant Responses to Radiant Energy In the following paragraphs, we will briefly review the categories of plant response to radiant energy, providing examples of plant adaptations to the varying parameters just outlined (Table 23-4).

Photosynthesis and leaf development There are many photosynthetic adaptations to the light environment. Although terrestrial plants and green algae best use blue and red wavelengths, for example, various brown, red, and blue-green algae (cyanobacteria) also use the green wavelengths that penetrate ocean waters most deeply. The photosynthetic efficiency of certain algae and higher plants is known to be under the control of circadian clocks (Chapter 20), so highest rates are measured (under standard conditions)

during subjective day and lowest rates during subjective night. The photosynthetic characteristics of C_3, C_4, and CAM plants, discussed in Chapter 11, are of profound ecological significance.

Remember the photosynthetic properties of sun and shade plants (see Figs. 11-4 and 11-6, reviewed by Björkman, 1981). Shade plants have an extremely low dark respiration, so the compensating light level is also extremely low, and positive net assimilation is achieved at light levels much below those required by sun plants even to reach compensation. But shade plants are saturated at low levels at which sun plants are just beginning to photosynthesize at moderately high rates. Shade plants never achieve such high rates as sun plants and are actually damaged (**photoinhibited**) by light levels that are not even saturating for sun plants.

When a sun plant is grown at low irradiances (say a twentieth of full sunlight; see Fig. 11-6), it acclimates so that its photosynthesis curve approaches but does not quite reach that of shade plants. Individual leaves on a single plant can also develop either as sun leaves or as shade leaves. The ability to acclimate to low irradiances has been studied mostly with C_3 species, but many C_4 plants also adjust to low irradiances. On the other hand, shade

The Challenge of a New Field: Plant Physiological Ecology

Park S. Nobel

There are many physiological ecologists who began their careers as traditional ecologists, but there are few who began as physicist-engineer-physiologists. Park Nobel is one of those rare ones who came from physics to ecology. Indeed, he is now the recognized leader in this field as the author of a comprehensive text titled Biophysical Plant Physiology and Ecology. *Here he tells the story of his changing and developing research interests.*

A sabbatical year spent in Canberra, Australia in 1973–1974 completely changed my career. I had left for Australia as a laboratory scientist studying the ion and water relations of chloroplasts, especially the use of irreversible thermodynamics to interpret osmotic responses. Following a desire to learn more about environmental matters, I had convinced the Guggenheim Foundation to allow me to make the slight shift to study the chloroplasts of guard cells. However, I soon found that the available techniques required far more chloroplasts than could readily be isolated from guard cells. So, capitalizing on a relatively unused wind tunnel, I spent the year developing equations describing the boundary layers adjacent to cylindrical and spherical objects, an entirely different topic but related to questions students had asked me. This challenged me to call upon my undergraduate training in engineering (Bachelor's of Engineering Physics from Cornell University) and graduate training in physics (Master's in physics from California Institute of Technology), aspects of my back-

ground that had been rather dormant for over a decade. Indeed, after a PhD in biophysics from the University of California, Berkeley, I had steadily moved from the physical sciences that I enjoyed studying to the biological sciences, where I felt that discoveries loomed and that each person could make a significant contribution.

When I returned to Los Angeles, I wanted to continue research on boundary layers of bluff bodies and thought immediately of the magnificent succulent plants that had caught my attention during trips to the desert. Two other professional changes took place at the same time. First, when the department chair asked me to check over the equipment list requested by a faculty member hired when I was away, I noted that it described my chloroplast laboratory almost exactly, and so I said I would exchange the contents of my laboratory for cash, thus burning my cellular bridge but gaining finances to begin in a new field. Secondly, I was asked to join the UCLA Laboratory of Nuclear Medicine and Radiation Biology (now the Laboratory of Biomedical and Environmental Sciences) to study the "biophysics of desert plants." I seized the opportunity and embarked on the study of agaves and cacti, beginning initially with water relations and only many years later returning to my initial interest in boundary layer considerations when that became experimentally logical.

While I was in Australia, an undergraduate (Larry Zaragoza) together with a graduate student (Bill Smith, now at the University of Wyoming) started working in my laboratory on an anatomical problem far afield from chloroplast research but one that had intrigued me for many years based on my teaching a course in plant physiological ecology. Specifically, just as ground area is an appropriate index for discussing productivity by crops and leaf area is appropriate for discussing CO_2 uptake by leaves, mesophyll cell-wall area seemed to me to be the appropriate area for discussing cellular aspects of photosynthesis. Recognition of the importance of mesophyll cell

plants apparently do not become acclimated to high light levels. Grown under such conditions, plants may actually become less photosynthetically capable at all irradiances.

Leaf area and leaf morphology are strongly influenced by light levels during development. Compared to shade leaves, sun leaves have less area per leaf, are thicker (often have more layers of palisade mesophyll consisting of longer cells—see Fig. 11-5), weigh more per unit leaf area, are more densely distributed on the stem with shorter petioles (shade each other more), and are lighter in color (less chlorophyll). Epidermis, spongy mesophyll, and vas-

cular systems are also more developed in sun leaves. Because many treatments have influenced leaf morphology with neutral-density screens that do not change the spectrum appreciably, the enriched far-red of leaf shade must not be essential, but careful studies might show that it contributes to the production of shade leaves.

Seed germination and spring bud break in deciduous plants As discussed in Chapters 19 and 21, seed germination can be promoted or inhibited by light, depending on species and other factors such as temperature, and germination can often be inhibited

surface area per unit leaf area, or A^{mes}/A, allowed a separation of the biochemical effects on photosynthesis occurring within mesophyll cells from effects due to anatomy. For example, sun leaves differed from shade leaves in that the former had a much higher A^{mes}/A, which led to a much greater area within the leaves for CO_2 entry into the mesophyll cells. This in turn led to a more quantitative understanding of the higher photosynthetic capacity per unit leaf area of sun leaves compared to the often four-fold thinner shade leaves on the same plant. Such studies were extended by a postdoctoral fellow in my laboratory, David Longstreth (now at Louisiana State University) and a technician, Terry Hartsock, who helped elucidate the influence of salinity and nutrients on A^{mes}/A. Over the course of a few years we showed that light (more specifically, photosynthetically active radiation, or PAR) primarily influenced A^{mes}/A; salinity, water stress, and temperature influenced both anatomy and cellular photosynthetic properties; while nutrients affected mainly the cellular properties.

Returning now to my studies of desert succulents, the investigations of boundary layers of agaves and cacti had to await an understanding of the environmental responses of these plants. For desert plants, the obvious place to begin was with the water relations. This ranged from analyzing the extension of the growing season caused by the storage of water within the stem of the barrel cactus *Ferocactus acanthodes* to the crucial shifting of water from the succulent leaves to the inflorescence of the common monocarpic century plant, *Agave deserti*. The latter species in its native habitat in the Sonoran Desert was found to have the highest annual water-use efficiency ever reported for any plant: 40 g CO_2 fixed per kg water vapor transpired.

Once the importance of the water status was beginning to be understood, the complicated responses of agaves and cacti to temperature were investigated, often with the use of a computer model. This ranged from seeing what day/night air temperatures led to maximal nocturnal CO_2 uptake by

these CAM plants, to studying the influence of morphology on distributional limits. For the latter a simulation model was developed with Don Lewis that included the conventional energy-budget terms popularized by David Gates, Klaus Raschke, and others, plus additional heat-conduction and storage terms required for the massive organs of succulent plants. This allowed studies on the influence of spines and apical pubescence in protecting the meristem of cacti from freezing, which can extend their northern distributional limits. This was then supplemented with studies on the freezing tolerance of cacti from North and South America. Analysis of the high-temperature tolerance proved equally exciting, since many species of agaves and cacti could tolerate exposure to 60°C with apparently no ill effects.

As another application of modeling, again reflecting my background in physics and engineering, we have been working on the relationships between the morphology of desert succulents and the interception of PAR. Beginning with a field determination that CO_2 uptake by *F. acanthodes* was on the verge of being PAR limited, even during clear days in the high radiation environment of a desert, and incorporating the chance observation that the terminal cladodes (flattened stems) of platyopuntias usually tend to face east–west, we have shown that other morphological features of cacti are adaptations to maximize PAR interception at times of the year most favorable for growth. Thus modeling had again been combined with structural analysis to provide insight into plant physiology.

To close, let me choose a quote from Pasteur, which appears at the end of my most recent book, and which summarizes my own philosophy of plant studies: "Chance favors only the prepared mind." Thus studying the many fields that impinge on plant physiology and ecology, such as calculus, chemistry, physics, and even areas of engineering, may at times seem remote from a direct study of plants, but such fields can provide the tools to enhance overall understanding of the functioning of plants.

strongly by leaf shade or by artificial light enriched in far-red (Morgan and Smith, 1981). This is also true for many species that are light insensitive when tested with white light. Indeed, several hundred species have now been studied, and on the order of 80 percent or more respond to the balance in red/far-red, presumably via the phytochrome system. While it is true that many species normally germinate in spring before canopy leaves have appeared, it also seems clear that many other species germinate only after the canopy has been removed.

Some seeds are known that respond to photoperiod instead of spectral distribution (Vince-Prue,

1975). A few species produce seeds that are promoted in their germination by short days, but promotion by long days is more common (e.g., rice; Bhargava, 1975). It is also known that the photoperiod to which mother plants are exposed while their seeds are developing may strongly influence subsequent seed germination. Frequently, short days applied to the mother plant increase germinability of seeds compared to those that develop under long days (e.g., *Amaranthus retroflexus*, Kigel et al., 1977; *Chenopodium polyspermum*, Pourrat and Jacques, 1975), although a few exceptions have been reported (germinate best when mother plants are exposed to long days: e.g., lettuce, Koller, 1962). (More examples are in Mayer and Poljakoff-Mayber, 1982.)

Spectral effects have not been reported, but a few species are known in which buds become active in response to the lengthening days of spring (e.g., *Betula pubescens, Liquidambar styraciflua;* Downs and Borthwick, 1956), the response often being modified by temperature (e.g., degree of winter chilling).

The etiolation syndrome   Dark-grown seedlings exhibit a syndrome of characteristics that are apparently adaptations to germination at some depth in the soil (Section 19.5; Mohr, 1972). Internodes elongate, leaves do not expand, chlorophyll does not develop, nor do roots grow rapidly. Many dicotyledons have a hook at the tip of the stem that pushes up through the soil, and grasses have a coleoptile that protects the emerging leaf. When the seedling reaches the light, internode elongation slows, leaves expand, chlorophyll develops, roots grow more rapidly, and the hook straightens or the coleoptile is penetrated by the leaf. The entire character of the developing seedling changes (see Fig. 19.5). Phytochrome is in control, since red light is effective and its effects are reversed by far-red light.

Stem elongation and apical dominance   Leaf shade apparently exerts strong control over stem extension and lateral bud growth in many de-etiolated plants (Section 19.5; Morgan and Smith, 1981). This is best shown by providing a constant amount of photosynthetically active light to controls and supplementing this for treated plants with additional far-red. Species that normally grow in open areas will sometimes increase their growth rate as much as 400 percent in response to such irradiation. Axillary buds remain dormant. In certain circumstances this could allow the plants to overtop the canopy and reach full sunlight, while conserving energy by not branching. Species that normally grow under a forest canopy usually do not respond to the increased far-red, which also seems adaptive, since there is seldom any chance for them to overtop their canopy.

Stem elongation is also promoted in many species (probably a large majority) by long days (Section 22.2). Perhaps the effects can best be observed in day-neutral plants, in which flowering is about equal under both LDs and SDs, but stem elongation is much more pronounced under LDs. Leaf number and area are also promoted by LDs in many species. These responses to day length are probably adaptive, since they produce taller, leafier plants during the time of year when growing conditions are likely to be most suitable—if ample moisture is available.

Stem and leaf orientation: phototropism   Phototropic bending of stems toward a light source is a response to low irradiances of blue or ultraviolet radiation,

probably mediated by a flavin pigment (Section 18.3). Such a response might allow a plant to grow laterally around an overhead obstruction. This could be especially important for seedlings growing through the obstructions on a forest floor, for example, or for plants growing inside but near the mouth of caves or under river banks. Most of the time in most ecosystems, however, plants are apparently not responding phototropically. Their stems and branches, rather, bear a constant relation to the direction of the gravitational force.

The leaves of many species, however, exhibit solar tracking, a strong phototropism (Section 18.3). Leaf blades of many plants (e.g., cotton, soybean, beans, and alfalfa) follow the sun during the day, maintaining their blades at right angles to the sun's rays, much as a radio telescope tracks a moving satellite. This maximizes the irradiance on the leaf during the day.

Circadian leaf movements   Study of the circadian sleep movements exhibited by many plants has provided another vast body of literature (Chapter 21). Whether the movements are adaptive has long been argued, and suggested answers are not completely convincing. For example, while horizontal day-time position is rather well suited for photosynthesis, the vertical night position might also effectively lower irradiance from moonlight—if the full moon is almost directly overhead, which is only the case at midnight in the tropics and subtropics and in winter in temperate zones (Salisbury, 1981a, 1981b, 1982). In any case, light is effective in synchronizing the internal plant rhythms with the external cycling environment. Both dawn and dusk are often critical, and phytochrome has been implicated.

Reproduction and storage organs   There are many plant responses to photoperiod, but the most studied is flower induction in angiosperms. Hundreds of species have now been studied, although only a few in any detail (Chapter 22). From an ecological standpoint, the most striking discovery is surely the diversity of response types. There are species, varieties, or cultivars that respond to SDs, LDs, intermediate days, LDs followed by SDs, SDs followed by LDs, and all of these combined with various temperature interactions. Some plants have an absolute requirement for a given day length; others may only be promoted in their flowering by some day length. Different latitudinal ecotypes or cultivars within a species often have different day-length requirements.

A photoperiod requirement for flowering assures that the plant will flower at an appropriate season (e.g., after sufficient vegetative growth has been achieved); this will be nearly the same time each year, within the limits of temperature modification of the photoperiodism response. Furthermore, all the plants in a population will flower at essentially the same time, facilitating cross pollination (and often providing beautiful aspect dominance in an ecosystem).

Responses to twilight, discussed in Chapter 22, have strong ecological implications. In addition, a few species are known that are influenced in their flowering by light levels rather than photoperiod, but study of them has been minimal.

Storage organs are often influenced in their development by photoperiod, and this could also be of ecological significance. Responses are similar to those described for flower induction (Chapter 22).

Dormancy: the Autumn syndrome In many plants, especially deciduous woody perennials, short days lead to leaf senescence and abscission, inhibition of stem elongation, terminal (resting) bud formation, frost hardiness, and other developmental changes of winter dormancy (Section 21.0, Chapter 22). In a sense, this is the ecological opposite of increased stem elongation and leaf growth in response to long days, although that response is as often observed in herbaceous annuals as in woody perennials. Typically, the development of dormancy is also promoted by low temperatures.

Damage Caused by Ultraviolet Irradiation Damage to plants by ultraviolet radiation might be ecologically significant and could become more so if the stratospheric ozone layer is reduced (Caldwell, 1981). However, there is a wide diversity in plant resistance to UV radiation, and UV damage involving DNA can be reversed by high irradiances with blue light (called **photoreactivation**).

Clearly radiation, especially light, is influencing plants in a multitude of ways. The more we learn of these matters especially with reference to individual species, the more we will know about environmental physiology.

23.6 Allelochemics and Allelopathy

Organisms interact in many interesting ways. Chemicals produced by one organism that affect another organism are called **allelochemics** (Barbour et al., 1980; Krebs, 1978; Ricklefs, 1979; Whittaker, 1975). Sometimes a single chemical produced by one organism is harmful to another organism but beneficial to a third organism. Plants of the mustard family secrete mustard oils that irritate many animals and thus prevent them from feeding on the mustard plants. Yet the same oils attract other animals that feed on the mustard plants. One of the oils even stimulates germination of the spores of a fungus that is parasitic on the mustard roots. We are beginning to

appreciate that communities include many complex webs of chemical interactions. Fungi and mycelial bacteria, for example, secrete allelochemics that are lethal to other bacteria. We have learned to use these substances in modern medicine and call them **antibiotics;** they are now part of the ecological interactions of humans.

Allelochemics were suspected in 19th century agriculture because of many observations of "soil sickness" of farmlands. If a piece of ground is continually cropped to one plant, the yields often decrease and cannot be improved by additional fertilizer. Fruit trees, for example, often do poorly in ground where the same species has grown before. Furthermore, it is common for one plant to harm another plant grown in its vicinity (a phenomenon called **allelopathy**). Many experiments of the type illustrated in Fig. 23-11 have been performed. In this experiment, one set of apple seedlings is watered with tap water, another set is watered with water that has percolated through soil with grass growing in it, and the third set of seedlings is watered with water that has percolated through soil with nothing growing in it. Growth of the apple seedlings is apparently inhibited by something produced by the grass plants, since seedlings in the other two treatments grow much better.

In a few cases, the allelochemics have been isolated and identified. It is often difficult to be sure, however, that a particular compound isolated from the roots or leaves of one plant actually plays a toxic role in nature. It appears that juglone (5-hydroxynaphthoquinone) from the roots and hulls of black walnut (*Juglans nigra*) is an allelopathic. It will kill tomato and alfalfa plants (in the field, up to 25 m from a walnut tree), but bluegrass seems to be stimulated close to walnut trees. Closely related English walnut (*J. regia*) and California walnuts (*J. hindsii* and *J. californica*) do not produce juglone.

It is worth noting that many plants produce allelochemics that are called *drugs* by medical and veterinary doctors. In fact, many early physicians (including Linnaeus) were botanists. These substances often cause changes in the physiology or psychology of organisms that consume them. Narcotics, for example, produce drowsiness or sleep or lessen pain. Harmful drugs are called poisons; beneficial ones are medicines. Most exert both effects, depending on concentration and other factors. Humans cultivate tobacco, tea, coffee, and other drug plants. Properly used, some of the drugs are beneficial; we are also aware of their harmful effects. Alcohol is a waste product of yeast metabolism (fermentation) that does not help the yeast (although urine, another waste product, sometimes helps animals establish territories). Alcohol typically builds to levels that harm the very organisms that produce it. Yet it is an allelochemic and a drug.

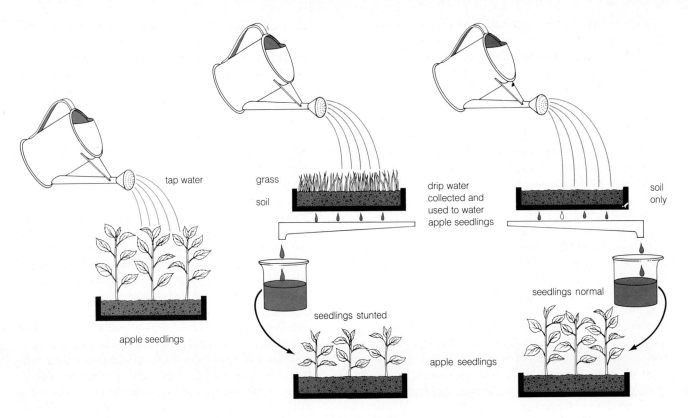

Figure 23-11 An experiment used to demonstrate substances produced by grass plants and harmful to apple seedlings. Plants in the flat on the left are control plants watered with tap water. Apple seedlings (bottom left) are stunted by water that has passed through soil supporting grass plants; water that has passed through soil without grass plants does not harm the apple seedlings (bottom right).

23.7 Herbivory

Many plants seem to have evolved defenses against herbivores. Some of these defense mechanisms are structural—as anyone who picks roses knows—but some plants also contain metabolic products that make them distasteful or toxic to herbivores. Examples include the drugs from plants we have just mentioned as well as others: nicotine, morphine, caffeine (these three are alkaloids; see Section 14.9), peppermint oil, various tannins, phenols, and unsaturated lactones.

Did these substances evolve in response to grazing by herbivores? That is, were genes to produce these compounds preserved by selection? Or are they just waste products from plant metabolism? If so, they would occur in plants even if there were no herbivores, and it is only coincidence that they often reduce grazing. Perhaps the substances originally appeared as by-products but are now critical in plant defense. Certainly it is difficult to account for the high concentrations of noxious substances that occur in some plants if they are merely by-products.

Bruce B. Jarvis and co-workers (1981) of the University of Maryland found that a Brazilian shrub, *Baccharis megapotamicia*, contained relatively large quantities of an antibiotic known to be highly toxic, not only to certain microorganisms but also to most insects and other plants. Since the shrubs grown in the United States contained none of the antibiotic, the researchers suggest that the shrubs absorb the compound from fungi growing around their roots in their native Brazilian marshes. The shrubs not only absorb the antibiotic and are not harmed by it; they also alter it chemically so that it is even more powerful than it is in its original form. The researchers note that few competitive plants grow where the shrubs grow naturally.

Some plant-herbivore associations become quite complex. An example that is often cited is the mutualistic system of defense that has been achieved between certain acacia trees (*Acacia* spp.) and their ant inhabitants. Swollen-thorn acacias have large hollow thorns in which the ants live. The ants feed on special modified leaflets and on enlarged nectaries that supply sugar; hence the ants are herbivores. The ants continually patrol the leaves and branches of the acacia tree and bite or sting both vegetation and potential herbivores, especially other insects, that touch the acacia tree. They also clear away plants from the ground beneath the acacia tree, which then has a volume of space around it that is free from both animal herbivores (other than the ants) and plant competitors. It is revealing that the ant-acacia species lack the protective toxins possessed by other acacias in the same area. The ants and the acacias have evolved such a mutualistic relationship that acacia trees usually die when ants are removed. And, moreover, some ant species survive only on acacia trees.

There are many modern studies of herbivory. In some cases, the herbivores have apparently evolved mechanisms that overcome the plant's defenses. One example is the timing of an insect life cycle so that grazing occurs at the time during the year when the protective compounds are at their lowest levels in the plants that are consumed. Another example is the metabolic ability to detoxify the offending material, as in the insects that thrive on mustard oils.

John P. Bryant (1981), of the Institute of Arctic Biology at the University of Alaska, studied herbivory between snowshoe hares and four species of Alaskan trees: Alaska paper birch, aspen, balsam poplar, and green alder. When these trees are young (e.g., invading a newly disturbed area), and when the hare population is especially high, a condition that occurs at about ten-year intervals, hare browsing, may severely damage these and other trees and shrubs. But the new shoots produced by the damaged plants are highly unpalatable to the hares. Bryant showed that relatively large quantities of terpene and phenolic resins accounted for the bad taste. He dried twigs of feltleaf willow, known to be highly palatable to the hares, and covered them with various amounts of the terpenes and resins extracted from the new shoots. Twigs covered with amounts of resins about half of those present in new shoots were almost completely avoided by the hares—even when there was nothing else to eat. It is as if the Alaskan trees conserve their metabolic energy reserves in their early growth by "taking a chance" that they will not be browsed by hares. In the second flush of growth, they take no chances but expend their energy resources to produce the protective allelochemics.

Jack C. Schultz and Ian T. Baldwin (1983) from Dartmouth College measured the concentration of tannins in sugar maple trees. Such compounds not only tan leather; they also precipitate digestive enzymes in the stomachs of many herbivores. Schultz found much higher concentrations of tannins in leaves that had grown back after earlier leaves had been consumed by insects. He also found that nutrient contents of leaves on sugar maple and oak trees varied considerably from leaf to leaf, apparently making it necessary for the insects to expend energy in foraging for the most nutritious leaves.

On the other hand, herbivores have often developed sophisticated defense mechanisms against the allelochemics. E. A. Bernays and S. Woodhead (1982), of the Centre for Overseas Pest Research in London, report that a tree locust survives better and grows faster when certain phenoles are added to its plant food. These compounds are toxic to most insects. Patricia J. Berger and co-workers (1981) at the University of Utah found that the montane vole became reproductive when fed a chemical found in the spring vegetation of mountain meadows. Adding 6-methoxybenzoxazolinone to rolled oats used to feed the voles in captivity led to increases in testes weights in males and to a high proportion of pregnancies in females. No testes weight increases or pregnancies were observed in voles receiving only rolled oats without the chemical. In this interaction, an animal's life cycle is timed to the life cycle of the plants that will provide food for adult animals and their offspring. This is especially significant in mountain meadows where the coming of spring is tied to the vagaries of weather. We have seen that in more dependable environments, both plants and animals are often timed in their life cycle events by photoperiodism.

A spectacular example of a plant-insect interaction is provided by two Asian moths studied by a group of researchers headed by Jerrold Meinwald (1982) of Cornell University, teamed with another group in Germany led by Dietrich Schneider. They report that the plant diet could well be crucial to the sexual functioning of the moths. The males develop from their abdomens long retractable tubes covered with hair and called *coremata*. These structures release chemical compounds that probably attract females. The researchers showed that the moths can only produce suitable compounds when they ingest certain plant alkaloids. When the male moths were fed diets deficient in these alkaloids, they not only failed to produce the sex-attractants but also failed to produce full-sized coremata, a situation said to be without precedent.

24

Stress
Physiology

An important branch of environmental physiology is concerned with how plants and animals respond to environmental conditions that deviate significantly from those that are optimal for the organism in question—or, in a broader sense, for organisms in general. This sub-science of **stress physiology** can contribute to our understanding of what limits plant distribution in natural environments. In such a context, it would be a part of physiological ecology. Most research in the field, however, is concerned with how adverse environmental conditions limit agricultural yields. One of the first difficulties encountered is how to define the word *stress*.

24.1 What Is Stress?

The term has been difficult to define, partly because the plants found in what might certainly appear to be the most stressful conditions on earth—hot deserts, salty soils, or high mountain tops—often appear healthy and to be species not found anywhere else. If they are flourishing and apparently can't even survive under other "less stressful" conditions, is it valid to think of them as being stressed? Actually, many such plants grow better under less stressful conditions if they are planted and cared for. It has been suggested that they normally do not occur in more moderate situations because they cannot compete with the plants already growing there (Barbour et al., 1980). They may well be stressed in their native habitats.

In 1972, Jacob Levitt (see also 1980) proposed a definition of biological stress derived from physical science. Physical *stress* is any force applied to an object (e.g., a steel bar); *strain* is the change in the object's dimensions caused by the stress. Levitt suggested that **biological stress** is any change in en-

vironmental conditions that might reduce or adversely change a plant's growth or development (its normal functions); **biological strain** is the reduced or changed function.

This definition takes us back to our discussion of limiting factors and the law of tolerance (Section 23.3). When environmental conditions are such that the plant is responding maximally (at or close to the optimum part of the curve in Fig. 23-1), it is not being subjected to stress. Any change in environmental conditions that results in plant response less than the optimum might be considered stressful. Of course, such a concept is sometimes easier to discuss in a theoretical way than it is to apply. Consider a plant suddenly subjected to reduced light levels. Since photosynthesis is immediately reduced, applying Levitt's definition, we say that the reduced light levels are the *stress* and the reduced photosynthesis is the *strain*. Probably stem elongation will be promoted, however, so does that mean that high light levels were a stress so far as stem elongation goes, causing the reduced stem elongation observed before the light levels were decreased? Probably we would conclude that the promoted stem elongation rates were actually the strain, because they led to taller stems with less mechanical strength—but that could clearly be an advantage if the leaves were thereby carried above shading competitors into higher light levels. It's all a question of what is "best" or "normal" for a given plant, and clearly the answer to this question will often be highly subjective, depending upon circumstances and judgements. Most studies in stress physiology have been concerned with conditions that are much more obviously stressful—and that will be true also for most of the discussions in this chapter.

Levitt went on to define **elastic biological strain** as those changes in an organism's function that return to the optimal level when conditions return to those best suited to the organism studied (i.e., when biological stress has been removed). If the functions do not return to normal, the organism is said to exhibit **plastic biological strain.** The analogy with physical objects is clear: A deformation strain in a steel bar, for example, that disappears when the stress is removed was an elastic deformation; if the bar remains deformed (bent) after the stress is removed, the deformation was a plastic one.

Some plant physiologists have resisted use of Levitt's definitions (e.g., Kramer, 1980). One of the main difficulties is that workers have often used the term *stress* in the sense that biological strain was defined by Levitt. Recognizing the difficulties (mostly those of introducing new terms that may sometimes make it more difficult to understand the older literature), we, along with many experts in the field of stress physiology, nevertheless find Levitt's terminology precise and relatively easy to apply and thus shall do so here.

In general, plant physiologists have emphasized in their studies such plastic strains as those caused by the stresses of frost, high temperature, limited water, or high salt concentrations. Elastic strains in plants, such as the photosynthesis example just mentioned (if the low light level does not persist too long, photosynthesis will resume at its original rate when the original light level has been restored), have been less studied by stress physiologists, although they must be extremely common and have been emphasized in studies of animal stress.

Levitt (1972, 1980) listed some other definitions. He suggested that we should distinguish between avoidance and tolerance (hardiness) to any given stress factor. In **avoidance** the organism responds by somehow reducing the impact of the stress factor. For example, a plant in the desert might avoid the dry soil by extending its roots down to the water table. If the plant develops **tolerance,** on the other hand, it simply tolerates or endures the adverse environment. Creosote bush is a good example of a desert plant that is tolerant toward drought. It simply dries out but survives anyway; it tolerates or endures the dryness of its protoplasm.

24.2 Stressful Environments

Recalling our discussion of the operational environment (Section 23.2; Spomer, 1973), we realize that an environmental stress means that some potential in the environment differs from the potential within the organism in such a way that there is a driving force for transfer of energy or matter into or out of the organism, leading to biological strain. Low outside water potentials, for example, provide a driving force for loss of water; low temperatures can lead to loss of

heat. What are the environmental limits for the existence of life on our planet? We might look for answers by examining those environments in which the stresses seem to be greatest compared with environments in which the organisms are most productive. We will begin by reviewing some climatic conditions that produce stressful environments on our planet. The discussion illustrates the variety of stresses to which plants are exposed.

Deserts and Less-Dry Areas A desert is an area of low rainfall—an area of **drought**—with less than about 200 to 400 mm of precipitation per year, depending upon temperatures, potential for evaporation, season of precipitation, and other factors (Fig. 24-1). Deserts often have sparse but fascinating vegetation (Fig. 24-2). The most extensive deserts occur in the so-called **horse latitudes,** which range approximately 20 to 30° north and south of the equator. In these regions, air that has ascended in other latitudes descends and is thereby compressed and warmed. Warm air holds more moisture than cold air, so precipitation does not occur.

North of the northern horse latitudes (and on the southern tip of South America, south of the southern horse latitudes) deserts also occur. Global air movements in the northern and southern temperate zones are predominately from west to east. Storm systems moving this way in the northern hemisphere rotate in a counterclockwise direction (clockwise in the southern hemisphere), so a storm center is preceded by south winds and followed by north or west winds. As these storm systems approach a mountain range, the rising air expands and cools and can then hold less moisture, resulting in precipitation on the western slopes, which are typically covered with lush forests. On the eastern slopes, the air descends, compresses, warms, and can hold more moisture. These areas have low precipitation and are called **rain-shadow deserts,** because they occur in the "rain shadows" of mountains. The descending warm and dry winds on the eastern slopes of the Rockies are called **chinooks** or snow eaters. North and east of the Alps, such winds are called the **Föhn.** The deserts of the Great Basin occur in the rain shadow east of the Sierras, but the plains east of the Rocky Mountains are not rain-shadow deserts, because they receive moisture moving north from the Gulf of Mexico.

Deserts in the horse latitudes are typically hot and dry all year, while most rain-shadow deserts are cold during winter. Because air above deserts is usually dry, it absorbs relatively little incoming sunlight or outgoing long-wave thermal radiation. Thus deserts are hot during the daytime and relatively cold at night. Air warmed in desert valleys rises, while air that cools at higher elevations flows down canyons and gullies, especially during the night. Hence, wind

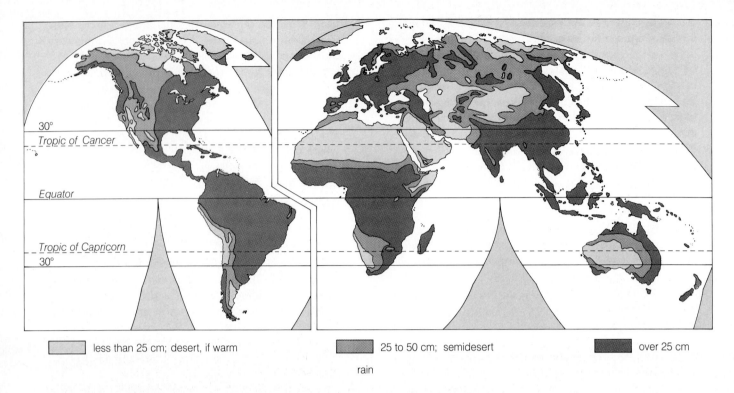

| less than 25 cm; desert, if warm | 25 to 50 cm; semidesert | over 25 cm |

rain

Figure 24-1 Map of the world showing areas of extremely low (less than 250 mm) and low (between 250 and 500 mm) precipitation (usually as rain). Horse-latitude deserts usually occur between north or south latitudes of about 20 to 30°. By far the most striking example is the Sahara Desert of North Africa, extending to the deserts of the Arabian peninsula and the Near East, but the horse-latitude deserts of Mexico, South America, South Africa, and especially Australia are also evident. Rain-shadow deserts are important in North America (the Great Basin northward through Canada to the Arctic), where they are caused by the Sierra Nevada, the Cascades, the Rocky Mountains and other ranges. Rain-shadow deserts are also important in Central Asia (the Gobi Desert and others), where they are caused by the Himalayas and other mountain ranges. Note the low precipitation in the Arctic—in some areas as low as the Sahara and other extreme deserts. Because of the low temperatures in polar regions, evaporation is greatly reduced compared with warm areas such as the subtropics; hence more water is available for plant growth. Yet moisture as well as low temperatures can limit plant growth in the low-rainfall polar regions. (From Jensen and Salisbury, 1984.)

is common and can lead to dune formation, although the great sand dunes associated with deserts, as portrayed by movie makers and others, are not as common as we are led to believe (except in the Sahara Desert). Perhaps so-called **desert pavement** is more common. This consists of a surface layer of small stones, finer material having been eroded away, mostly by wind.

Desert soils are often salty, because the low rainfall does not leach away the salts as they form by weathering of soil particles and rock. The actual status of a desert soil depends considerably upon the time during the year when rain does fall. Mediterranean climate zones, for example, usually occur just to the north of the northern horse latitudes or to the south of the southern horse latitudes. The horse latitudes shift away from the equator during summer, so Mediterranean climates experience summer drought that can last from six to nine months. Yet

they may have considerable rainfall during the winter months when the horse latitudes have shifted toward the equator. Technically, their abundant winter moisture keeps these areas from qualifying as true deserts. Because of the high precipitation during part of the year, these soils are usually less salty than other desert soils—but at the same time, they have some of the characteristics of desert soils (e.g., alkalinity). Summer showers in the Arizona desert (from the Gulf of Mexico, mostly) lead to a unique and relatively lush desert vegetation, but temperatures are high, as is evaporation, and total precipitation is low enough that these areas qualify as true deserts.

Actually, the uncertainties of world climatic patterns mean that some regions in the temperate zone that are normally blessed with ample rainfall for productive agriculture may experience droughts that extend for several weeks to months and/or reduced precipitation that may last for several years—for example, the dust bowls of the southern plains of the United States in the 1930s or the serious drought in Europe during the summer of 1983. Thus **water stress** (water potentials negative enough to damage plants) occurs in many parts of the world besides deserts.

Tundras and Other Cold Areas   **Tundras** are areas on the earth's surface where temperatures are too low to permit the growth of trees (Fig. 24-3). Such areas occur on the tops of mountains (**alpine tundra**) and in the far north (**arctic tundra**). (Antarctic tundras are very limited in extent.) The polar tundras occur because the sun is relatively low in the sky even in summer (and actually below the horizon for days to months in winter poleward of the arctic and antarctic circles), so the slanting solar rays must follow a long pathway through the atmosphere and then strike horizontal surfaces at such acute angles that the energy of a given cross section of solar radiation is spread over a relatively large horizontal area. These combined effects (long atmospheric pathway and low solar angle) become increasingly important as one moves from the equator toward the poles and result in cold temperatures occurring at decreasing elevations. That is, the **upper tree limit (tree line)** occurs at lower and lower elevations when one moves from the equator (where it may be at 6000 to 7000 m) until it reaches sea level just above the arctic circle.

The alpine and the arctic tundras share similar average low temperatures but otherwise differ considerably in such factors as levels of radiation, which is much lower in the arctic but is extended over longer days in summer. Light levels increase in alpine tundras because of a thinner layer of scattering atmosphere. Even when the sky is overcast, the diffused light measurable within clouds shrouding the

Figure 24-2   The desert east of Phoenix, Arizona, on the northern edge of the horse latitudes. A giant Saguaro (pronounced sah-WAH-ro) cactus dominates the picture, with a much smaller cholla (CHAW-yuh) at its base and to the left. Numerous desert shrubs dominate the vegetation. The Superstition Mountains are in the background. The photograph was taken in mid-July. (Saguaro is *Cereus giganteus;* cholla is an *Opuntia* species, a common desert genus with many species. Photograph by F. B. Salisbury.)

Figure 24-3   Alpine tundra high above tree line (note trees at center right) on the northern border of Rocky Mountain National Park, Colorado, U.S.A. in the Mummy Range. The elevation is about 3420 m (11,200 ft) above sea level, and the area is one of several sites for a study in the physiological ecology of tundra plants. Because of different slopes, snow accumulation areas, substrates, and other features, several distinct vegetation types can be recognized in the tundra, but all are characterized by low, rapidly growing plants, often with colorful flowers. (Photograph by F. B. Salisbury.)

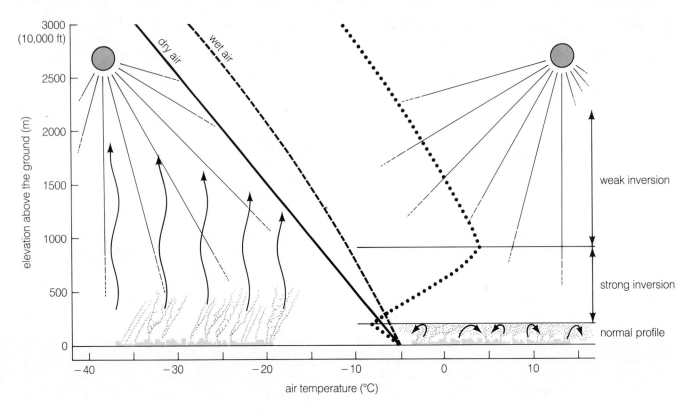

Figure 24-4 The adiabatic lapse rate and atmospheric inversion. The dividing lines between the two parts of the figure show temperature profiles (temperature as a function of elevation) that are examples of the adiabatic lapse rates for dry and wet air. These show the temperature of a volume of air that is allowed to expand adiabatically (without exchanging heat with its surroundings) and thus to cool as it rises through the elevations shown. If the temperature gradient is steeper than the adiabatic lapse rate, as is usually the case, air warmed by contact with the warm ground rises by convection (left side of figure). If the gradient is less steep than the lapse rate, air does not rise by convection (right side). The dotted line shows an actual temperature profile measured with a sounding balloon on a cold December 31, 1963, above Salt Lake City, Utah. Below about 200 m, the temperature gradient is steeper than the lapse rate, and air rises—but only to the bottom of the layer of air in which the profile is actually reversed from its normal condition. The air gets warmer instead of colder with increasing elevation up to about 900 m. This is an unusually strong inversion—and a collector of atmospheric pollutants. But the inversion continues even above 900 m, because the temperature gradient is still not as steep as the lapse rate, even for wet air. (From Jensen and Salisbury, 1984.)

tundra is much brighter than the light below clouds in the valleys. Highest irradiances are observed when the sky is partly cloudy and direct rays of the sun and reflected rays from the clouds both fall on the same area. Irradiances may be as high as 1500 W m^{-2} (solar constant = 1350 W m^{-2}). Ultraviolet radiation can be high and perhaps important for organisms in alpine tundras but is very low in the arctic. The fact that tundra vegetation is so similar in alpine zones and in the arctic (often the same or closely related species) testifies to the importance of low temperature.

Why are temperatures low in alpine tundras? If a volume of gas expands or contracts without exchanging heat with its surroundings, the expansion or contraction is said to be **adiabatic.** As air rises, it expands because pressures are lower at higher elevations. If

the expansion is adiabatic, the temperature of the air will decrease, because there will be less heat per unit volume of air. Dry air in the earth's atmosphere cools about 1°C for each 100 m of vertical rise. Thus, if dry air at 30°C on the valley floor is raised by global winds to the mountain top 1500 m (4921 ft) above, it will cool to about 15°C unless it is warmed or cooled by the mountainside or sunshine on the way up. This rate of cooling is an expression of the **adiabatic gradient,** or **adiabatic lapse rate,** for dry air in the earth's atmosphere (Fig. 24-4). Adiabatic cooling during expansion is the basic reason that higher air is nearly always cooler than lower air.

There is another important reason why alpine tundras are cool. At night the tundra surface radiates more heat into the sky than radiates to it from the thin atmosphere above, so the alpine surface cools faster at night than lower areas. The cooled air contracts, becomes more dense, and often flows down canyons or off slopes, replacing warmer air that is rising in the valleys.

During the day, high gusty winds are characteristic of alpine tundra—partly because of rapid temperature fluctuations but also because the mountain peaks deflect the global air movements that always occur high in the atmosphere. These winds increase evaporation from plants and soil, but humidity is usually high, which reduces evaporation. Virtually everything fluctuates rapidly in the alpine environment: wind velocity, temperature, light level (under partially cloudy skies, which are common), and humidity (see introduction to Chapter 20). The winds in winter mean that much of the alpine tundra is blown free of snow, while deep snow drifts form in more protected areas. Some plants must thus endure the frigid temperatures of exposed areas while other plants avoid the winter extremes under an insulating thick blanket of snow.

We use the tundras as examples of regions on earth where low temperatures are especially important, but freezing temperatures are also extremely important every year in temperate zones, and even in Mediterranean and subtropical climates frost can be damaging to native and especially crop plants. Even in the tropics, unusually low temperatures (although they may be as much as 10°C above freezing) can cause chilling injury to certain sensitive plants such as bananas.

Other Stressful Environments There is a great variety of stressful environments on the earth's surface in addition to those we have described. Some are of limited scale; others may be nearly worldwide in distribution. Flooding, for example, can produce a stressful condition quite opposite to the extremely negative water potentials of deserts, but damage results from the exclusion of oxygen rather than from the high water potentials. High temperatures, often close to or even above the boiling point, occur in hot springs, near volcanoes, in piles of decaying organic matter, and in extremely hot deserts, such as Death Valley, California. Living organisms often occur in these situations as we shall see. We have already discussed (Chapters 11 and 23) stresses caused by low light levels in forests and also at depths in bodies of water. There are many spots on earth where soils are either highly acidic or highly alkaline, or where they are deficient in several nutrients (as are sand and some volcanic material) or specific nutrients (especially nitrogen). The open oceans are deficient in many nutrients, because these are carried to the ocean depths in the bodies of dead **plankton** (single-celled plants and other organisms) and other organisms as they die and settle to the bottom. Thus low nutrient concentrations (potentials) provide special problems over the nearly two-thirds of the earth's surface covered with water. The extreme pressures and total darkness that exist at great depths in the ocean, where organisms nonetheless live, surely seem stressful. Yet the organisms are alive and seem to be doing well (often existing on the "rain" of organic matter from above), and the concept of stress comes readily to our minds because of the environments to which we are ourselves adapted.

24.3 Water Stress: Drought, Cold, and Salt

Although there are many complications, a creosote bush growing in the desert, a white mangrove growing in a coastal forest with its roots in salt water, and a white spruce living in the north woods are all stressed, at least at certain times during the year, by a common factor: negative water potentials (water stress). We will begin our survey with the plants of the desert.

The Xerophytes Ecologists sometimes classify plants according to their response to water: **Hydrophytes** grow where water is always available as in a pond or marsh; **mesophytes** grow where water availability is intermediate; and **xerophytes** grow where water is scarce. Solutes strongly influence water potential and can have specific toxicities, so ecologists further classify plants that are sensitive to relatively high salt concentrations as **glycophytes** and those that are able to grow in the presence of high salts as **halophytes** (**halophiles,** if not a plant).

All the plants of the deserts are called xerophytes, but different species survive the drought in different ways. Fig. 24-5 expands the concepts of avoidance and tolerance introduced above. As you can see, there are several forms of avoidance, but tolerance is always a simple matter of endurance. H. L. Shantz (1927) used four terms in classifying xerophytes. The terms are descriptive and are shown in Fig. 24-5: *escape, resist, avoid,* and *endure.* As you can see from the figure, xerophytes in the desert are actually exposed to a wide range of water potentials. Plants such as the palms that grow at an oasis where their roots reach the water table, or other plants such as mesquite (*Prosopis glandulosa*) and alfalfa (*Medicago sativa*) that have roots that extend as much as 50 m down to the water table, never experience extremely negative water potentials. They are *water spenders.* They certainly avoid the drought. Of course, such plants must be able to use the available soil water while they are extending their roots to the water table. (In the Judean Hills of Israel certain trees can

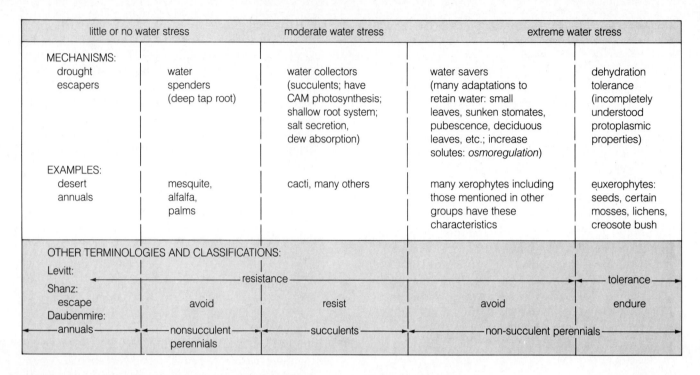

little or no water stress		moderate water stress		extreme water stress	
MECHANISMS: drought escapers	water spenders (deep tap root)	water collectors (succulents; have CAM photosynthesis; shallow root system; salt secretion, dew absorption)	water savers (many adaptations to retain water: small leaves, sunken stomates, pubescence, deciduous leaves, etc.; increase solutes: *osmoregulation*)	dehydration tolerance (incompletely understood protoplasmic properties)
EXAMPLES: desert annuals	mesquite, alfalfa, palms	cacti, many others	many xerophytes including those mentioned in other groups have these characteristics	euxerophytes: seeds, certain mosses, lichens, creosote bush

OTHER TERMINOLOGIES AND CLASSIFICATIONS:

Levitt: ←————————————— resistance —————————————→ ←— tolerance —→

Shanz:
escape

| | avoid | resist | avoid | endure |

Daubenmire:

←— annuals —→ ←— nonsucculent perennials —→ ←— succulents —→ ←————— non-succulent perennials —————→

Figure 24-5 Approach to a classification of plant responses to water stress. Some traditional approaches are also shown.

extend their roots down along tight fissures in solid limestone to depths of 30 m or more, dissolving the rock as they go.)

The so-called **desert ephemerals** are annual plants that escape the drought by existing only as dormant seeds during the dry season. When enough rain falls to wet the soil to a considerable depth, these seeds often germinate, perhaps in response to the leaching away of germination inhibitors (Chapter 21). Many of these plants grow to maturity and set at least one seed per plant before all soil moisture has been exhausted. They are eminently well suited to dry regions and thus are xerophytes in the true sense of the word, yet their active and metabolizing protoplasm is never exposed to extremely negative water potentials and is not drought hardy.

Succulent species such as the cacti, century plant (*Agave americana*), and various other crassulacean-acid-metabolism (CAM) plants (Section 10.6) are *water savers*. They resist the drought by storing water in their succulent tissues. Enough water is stored, and its rate of loss is so extremely low (because of an exceptionally thick cuticle and stomatal closure during the daytime) that they can exist for long periods without added moisture. E. T. MacDougal and E. S. Spaulding reported in 1910 that a stem of *Ibervillea sonorae* (a desert succulent from Mexico; in the cucumber family, Cucurbitaceae) stored "dry" in a museum formed new growth every summer for eight consecutive summers, decreasing in weight only from 7.5 to 3.5 kg! Since their protoplasm is not sub-

jected to extremely negative water potentials, succulents are drought avoiders but not truly drought tolerant. The water potential in their tissues is often about −1.0 MPa. Some of the succulents, especially the cacti, have extensive shallow root systems that are capable of absorbing surface moisture after a storm. Of course, the moisture is then stored in their succulent tissues.

Many other desert plants also have several adaptations that reduce water loss, although they don't actually store water in succulent tissues. For example, it is common for desert shrubs and other plants to have small leaf blades. This increases heat transfer by convection, lowering leaf temperature and thus reducing transpiration (Section 3.6). Such leaves will still be as warm as the air temperature, but air temperatures are seldom fatal. Other adaptations that apparently reduce transpiration include sunken stomates, shedding of leaves during dry periods, and heavy pubescence on leaf surfaces (Ehleringer et al., 1976). Although these modifications may indeed reduce the loss of water, they never completely prevent it and are by themselves insufficient protection against extreme drought.

As we shall see later, one of the important damaging effects of water stress is the concentration of salts within the cells of plants subjected to water stress. These salts can damage the enzymes that control metabolism and that are thus essential to life. An important adaptation found in many organisms that are subjected to water stress is the accumulation of certain organic compounds such as sucrose, the amino acid proline, and several others that lower the osmotic potential and thus the water potential in cells without limiting enzyme function. Such compounds appear in the cells of many xerophytes as water stress increases, and the resulting drop in osmotic potential is called **osmotic adjustment** or **osmoregulation** (Morgan, 1984).

Perhaps most impressive among the xerophytes are those that simply endure the drought. That is, they lose exceptionally large quantities of water so that their protoplasm is subjected to extremely negative water potentials, yet they are not killed. Such **euxerophytes** (true xerophytes) exhibit tolerance or hardiness rather than mere avoidance. Plants that only avoid drought are of great interest to ecologists, but they do not challenge our understanding as the euxerophytes do. Incidentally, many of the characteristics of drought avoiders, such as small leaves and sunken stomates, also occur in the drought endurers. Yet in the euxerophytes the ultimate weapon against drought is the ability to endure it, to be drought tolerant.

The ability of some euxerophytes to endure drought is phenomenal. Water content of the creosote bush (*Larrea divaricata*), a desert shrub of both North and South America, drops to only 30 percent of the final fresh weight before leaves die. With most plants, levels of 50 to 75 percent are lethal. Some of the most spectacular euxerophytes (called **poikilohydric** by some ecologists) are mosses and ferns—plants that we normally associate with wet environments. Their ability to dry out and then become metabolically active immediately upon rehydration apparently depends upon special features not common to other plants. Examples are *Selaginella lepidophylla* (the resurrection plant), certain grasses, and *Polypodium* (a fern).

Much work on desert xerophytes has been done in the Negev Desert of Israel. Researchers (Evenari et al., 1975) there use a somewhat more complex classification scheme than that developed by Shantz, but the same principles apply. They have studied algae, lichens, and mosses that can tolerate extreme and prolonged desiccation, as well as extreme cold and heat when they are dry. They can take up water directly and instantaneously from dew, rain, or even a moist atmosphere (some when relative humidity is only 80 percent or above), and such water absorption leads to an instantaneous switching on of metabolic activity. (Air at 80 percent RH and 20°C has a water potential of −30 MPa.)

Among higher plants, Israeli workers have identified most of the features just discussed, plus a few others. An important adaptation of their desert plants, for example, is that of **heteroblasty,** or the property of producing morphologically and physiologically different seeds from the same plant. Such seeds have different germination requirements, so only a few seeds in a given crop germinate at any given time. The risk of seedling survival is thus distributed over a variety of environmental conditions and sometimes over several years. These workers have also studied stomatal regulatory mechanisms in desert plants. In the most interesting examples, when water stress was low in the leaf, stomates opened when the temperature was raised; when water stress was higher and the temperature was again raised, stomates closed.

In the Negev Desert, plants often utilize dew. Other plants may also use dew, but the extent to which this contributes to plant growth remains somewhat controversial (Rundel, 1982). We consider later an interesting example of salt secretion on leaves. The salt absorbs moisture from the air, and this water is then absorbed into the leaves.

Incidentally, insect physiologists (Edney, 1975) have been interested for several years in the ability of certain insects to absorb water from an atmosphere with a relative humidity as low as 50 percent—and thus a water potential of almost −100 MPa! Liquid within the insect's body might have a water potential of only −1.0 to −2.0 MPa and would be in equilibrium with an atmosphere of about 99 percent relative humidity. How absorption occurs remains a mystery. We need to know much more about the fine structure and function of insect cuticles and about the properties of insect epidermal cell membranes.

There are numerous other adaptations of desert plants that cannot be discussed for lack of space. For example, the allelopathics (Section 23.6) are often produced by desert plants, restricting the germination or growth of competing plants, which reduces competition for water.

Water Stress in Mesophytes Desert xerophytes are certainly of interest to physiological ecologists, but much of the work on water stress has been done by agriculturists. Water is virtually everywhere a limiting factor in agriculture, either because of unexpected dry periods or normally low rainfall, which makes regular irrigation necessary. If one must irrigate to obtain a crop, how much and how often should water be applied to the field? The answer, which could result in the expenditure of many millions of dollars in a given state and a given year, may be strongly influenced by research on stress physiology. Here we are seldom concerned with the severe stresses endured by desert plants; rather, we are interested in the extent to which withholding relatively small amounts of water might influence crop yield.

Research has been extensive, continuing for well over a century. Thousands of papers on plant responses to drought have now been published. There are many reviews and volumes that report on sym-

Table 24-1 Generalized Sensitivity to Water Stress of Plant Processes or Parameters[a]

Process or Parameter Affected	Sensitivity to Stress (Very Sensitive → Relatively Insensitive) — Tissue ψ Required to Affect Process[b] (0 bar, −10 bars, −20 bars)	Remarks
Cell growth	————- - - (0 to ~−5 bars)	Fast-growing tissue
Wall synthesis	———— (0 to ~−5 bars)	Fast-growing tissue
Protein synthesis	——— (0 to ~−5 bars)	Etiolated leaves
Protochlorophyll formation	——— (0 to ~−5 bars)	
Nitrate reductase level	——— (0 to ~−5 bars)	
ABA accumulation	- - - ——— (~−2 to −10 bars)	
Cytokinin level	——— (~−4 to −10 bars)	
Stomatal opening	- - - - ——— - - - - - - (0 to −20 bars)	Depends on species
CO$_2$ assimilation	- - - - ——— - - - - - - (0 to −20 bars)	Depends on species
Respiration	- - - ——— (~−5 to −13 bars)	
Proline accumulation	- - - ——— (~−8 to −16 bars)	
Sugar accumulation	——— (~−10 to −18 bars)	

[a]Length of the horizontal lines represents the range of stress levels within which a process first becomes affected. Dashed lines signify deductions based on more tenuous data.

[b]With ψ of well-watered plants under mild evaporative demand as the reference point.

From Hsiao, 1973.

posia available for further study—most published since the second edition of this text (e.g., Bewley, 1979; Bewley and Krochko, 1982; Bradford and Hsiao, 1982; Hanson and Hitz, 1982; Kozlowski, 1968 to 1978; Kramer, 1983; Levitt, 1980; Morgan, 1984; Mussell and Staples, 1979; Tranquillini, 1982; and Turner and Kramer, 1980).

Theodore Hsiao (1973; Bradford and Hsiao, 1982) has been especially active in this field. His 1973 summary table (Table 24-1), which remains valid, outlines the sequence of events that occurs when water stress develops rather gradually in a plant, as water is withheld from a plant growing in a substantial volume of soil. It is important to realize that the later events are almost undoubtedly indirect responses to one or more of the early events rather than to water stress itself.

Cellular growth appears to be the most sensitive to water stress (Fig. 24-6). Decreasing the external water potential by only −0.1 MPa or less results in a perceptible decrease in cellular growth (which is irreversible cell enlargement). Hsiao suggests that this sensitivity is responsible for the common observation that many plants grow mainly at night when water stress is lowest. (But temperature and endogenous rhythms could also be involved.) In any case, the response of cellular growth to water stress appears as

a slowing of shoot and root growth. This is usually followed closely by a reduction in cell-wall synthesis. Protein synthesis in the cell may be almost equally sensitive to water stress. These responses are observed only in tissues that are normally growing rapidly (synthesizing cell-wall polysaccharides and protein as well as expanding). It has long been observed that cell-wall synthesis depends upon cell growth (Section 15.2). The effects on protein synthesis are apparently controlled at the translational level, the level of ribosome activity.

At slightly more negative water potentials, protochlorophyll formation is inhibited, although this observation is based on only a few studies. Many studies indicate that activities of certain enzymes, especially nitrate reductase, phenylalanine ammonia lyase (PAL), and a few others, decrease quite sharply as water stress increases. A few enzymes, such as α-amylase and ribonuclease, show increased activities. It was thought that such hydrolytic enzymes might break down starches and other materials to make the osmotic potential more negative, thereby resisting the drought (osmotic adjustment), but careful studies don't always support this idea. Nitrogen fixation and reduction also drop with water stress, consistent with the observed drop in nitrate-reductase activity. At levels of stress that cause observable changes in enzyme activities, cell division is also inhibited. And stomates begin to close, leading to a reduction in transpiration and photosynthesis.

At about this level of stress, abscisic acid (ABA) begins to increase markedly in leaf tissues and, to a lesser extent, in other tissues including roots (reviewed by Bradford and Hsiao, 1982; Salisbury, 1985; Walton, 1980). Increases of as much as 40- fold have been observed in leaves. This leads to stomatal closure and reduced transpiration, as we discussed in Section 3.4. In addition, ABA inhibits shoot growth, further conserving water, and root growth appears (in some studies) to be promoted, which could increase the water supply. There is also evidence that suitably low concentrations of ABA increase the rate of water conductance through roots, which would reduce the water stress in the shoots. Most of these adaptations involving ABA are best observed in mesophytes; xerophytes often have other adaptations (Kriedemann and Loveys, 1974).

There is evidence that ABA normally plays a role in the resistance of mesophytes to water stress. Most studies have been done with drought-sensitive and drought-resistant cultivars of crop plants (e.g., Quarrie, 1980). Often, resistant cultivars have higher levels of ABA when they are exposed to stress, and sensitive cultivars can be phenotypically converted to resistant types by application of ABA. But there always seem to be exceptions. A yellow lupine (*Lupinus luteus*), for example, is remarkably insensitive to applied ABA.

What is the mechanism of plant response to water stress by increased ABA in its tissues? Evidence suggests that lowered cell turgor is the trigger for ABA production (reviewed by Bradford and Hsiao, 1982), but how turgor might control ABA synthesis remains unknown.

It is interesting that ABA increases in leaves in response to several kinds of stress, including nutrient deficiency or toxicity, salinity, chilling, and water-logging. Clearly, reduced cell turgor and water potential are not involved in all of these. It may well be that ABA is a kind of universal stress hormone, its production controlled or triggered by several mechanisms. In all the cases, it seems to reduce growth and metabolism and thus conserve resources, which will then be available during recovery if and when the stress is removed.

An increased production of ethylene may be an even more sensitive indicator of water stress than increasing ABA levels. Ethylene is also produced in response to various kinds of stress besides drought, including excess water, plant pathogens, air pollution, root pruning, transplanting, and handling (Section 17.2).

Although stresses are relatively mild at $\psi = -0.3$ to -0.8 MPa, interactions and indirect responses begin to be the rule. Cytokinins decrease in leaves of

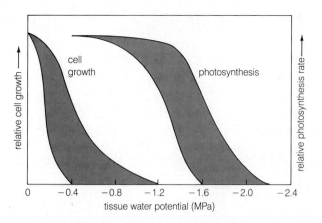

Figure 24-6 Cell growth and photosynthesis as a function of decreasing tissue water potentials. Shaded areas include ranges of response as observed with several species in different experiments. Cell growth (leaf enlargement, for example) is much more sensitive to decreasing water potential than is photosynthesis. (See Boyer, 1970, and Acevedo et al., 1971.)

some species at about these levels. At slightly more negative water potentials, the amino acid proline begins to increase sharply, sometimes building up to levels of 1 percent of the tissue dry weight! Increases of 10- to 100-fold are common. Other amino acids and amides, especially betaine, also accumulate when the stress is prolonged and depending on species. Proline arises *de novo* from glutamic acid and ultimately probably from carbohydrates. These compounds contribute to osmotic adjustment.

At higher levels of stress ($\psi = -1.0$ to -2.0 MPa), respiration, translocation of assimilates, and CO_2 assimilation drop to levels near zero. Hydrolytic enzyme activity increases considerably, and ion transport can be slowed. Actually, in many species respiration often increases, not really dropping off until water stresses of -5.0 MPa are reached.

Plants usually recover if watered when stresses are -1.0 to -2.0 MPa, meaning that, in spite of the severity of the water stress, the biological strain was elastic—or at least somewhat elastic, since growth and photosynthesis in young leaves frequently does not reach the original rate for several days, and old leaves are often shed. Clearly, since growth is especially sensitive to water stress, yields can be noticeably decreased even with moderate drought. Cells are smaller and leaves develop less during water stress, resulting in reduced area for photosynthesis.

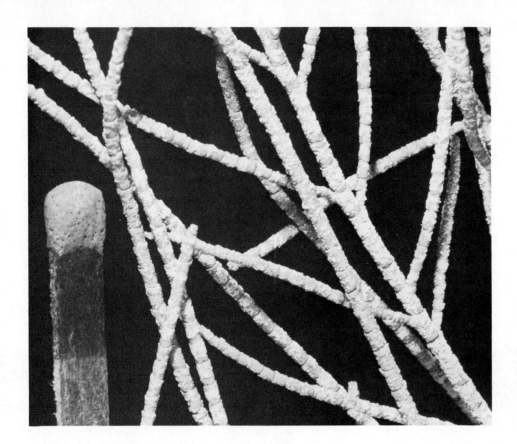

Figure 24-7 Tamarisk leaves (*Tamarix pentandra*) collected at Barstow in the Mojave Desert, California, showing heavy incrustations of salt. The paper match indicates the scale. (Photograph by F. B. Salisbury.)

Furthermore, plants may be especially sensitive to drought during certain stages, such as tassel formation in maize. Ultimately then, in the sense of final yield, biological strains are really plastic even when the water stress is only moderate.

Salt Stress A common and important stress factor in deserts is the presence of high salt concentrations in the soil. Soil salinity also restricts growth in many temperate regions besides deserts (Greenway and Munns, 1980). Millions of acres have gone out of production as salt from irrigation water accumulates in the soil (see Murray Nabor's personal essay in Chapter 5). A plant faces two problems in such areas, one of obtaining water from a soil of negative osmotic potential and another of dealing with the high concentrations of potentially toxic sodium, carbonate, and chloride ions. Some crop plants (e.g., beets, tomatoes, rye) are much more salt tolerant than others (e.g., onions, peas).

In the study of salt tolerance, the **facultative halophytes** are particularly interesting. Several such species grow best where salt levels in the soil are high, as in deserts or in soils saturated with brackish waters on the sea coasts or close to the shores of extremely salty waters such as the Great Salt Lake, where the salt content may be saturated at levels as high as 27 percent by weight. They also grow in nonsalty soils. The following genera include good examples: *Allenrolfea* (iodinebush), *Salicornia* (pickleweed or samphire), and *Limonium* (sea lavender, marsh rosemary). *Atriplex* (shadscale) and *Sarcobatus* (black greasewood) also grow in somewhat less salty soils, and certain bacteria and blue-green algae (cyanobacteria) live in the waters of the Great Salt Lake. In general, prokaryotes are the most resistant to environmental stress factors.

Barbour et al. (1980) review the literature that suggests that there are no **obligate halophytes,** plants that cannot grow unless the soil is salty. So far, all halophytes studied have been found growing naturally in non-salty soils and will grow well when planted in non-salty soils. Normally, they are not abundant in non-salty soils because they cannot compete with the glycophytes that normally grow there. As we discuss below, however, members of the genus *Halobacterium* (prokaryotes, again) accumulate large amounts of salt into their cells and cannot survive except in salty environments.

In terrestrial halophytes, the osmotic potential of leaf cell sap is invariably highly negative. Tissues from actively growing *Atriplex* species, having no special cold hardiness, for example, freeze only when temperatures drop below −14°C, implying that their osmotic potentials are as low as about −17.0 MPa. This contrasts with a normal −1.0 to −3.0 MPa in most plants. In some cases, the xylem sap does not have a highly negative osmotic potential but may be almost pure water. To obtain water from the surrounding soil, water potential within the xylem sap must then be greatly lowered by tension. This was demonstrated by Scholander and his co-workers for mangrove trees (see Fig. 4-17).

Some halophytes are referred to as **salt accumulators.** In these species the osmotic potential continues to become more negative throughout the growing season as salt is absorbed. Even in these plants, however, the soil solution is not taken directly into the plant. It is easy to calculate, based upon quantities of water transpired by the plant, that if the complete soil solution were absorbed, the plant would contain 10 to 100 times as much salt as is actually observed. Instead, water moves into the plant osmotically and not simply in bulk flow. The endodermal layer in the roots probably provides the osmotic barrier.

Halophytes in which the salt concentration within the plant does not increase during the growing season are known as **salt regulators.** Often salt does enter the plant, but the leaves swell by absorbing water, so concentrations do not increase. This leads to the development of **succulence** (a high volume/surface ratio), a common morphological feature of halophytes. Sometimes excess salt is exuded on the surface of the leaves, helping to maintain a constant salt concentration within the tissue (Fig. 24-7). In certain halophytes there are readily observable salt glands on the leaves, sometimes consisting of only two cells (Fig. 24-8). Although Na^+ ions are essential for some salt-tolerant species, it is probable that sodium pumps in cell membranes actively transport much of the ion out of the cytoplasm of both root and leaf cells, inwardly to the central vacuoles and outwardly to the extracellular spaces.

Nolana mollis, a dominant, succulent shrub of the Atacamba Desert of northern Chile grows where rainfall is less than 25 mm y^{-1}, although high fog and a relative humidity around 80 percent are common. The plant is almost always wet to the touch. Mooney et al. (1980) found that salt glands on the leaves secrete salt (mostly NaCl) that absorbs water hygroscopically from the atmosphere. If the leaves are washed with distilled water and blotted dry, they remain dry until they have had a chance to secrete more salt. If filter paper is soaked in solution collected from the leaves (or in concentrated NaCl solution) and then dried in an oven, it also absorbs water from the moist atmosphere and becomes wet to the touch. Can the plant use the water absorbed this way on its leaves? Mooney and his co-workers suggest two pathways of absorption: directly into the leaves, or through the roots after the salty solution has dripped onto the ground. Either pathway would re-

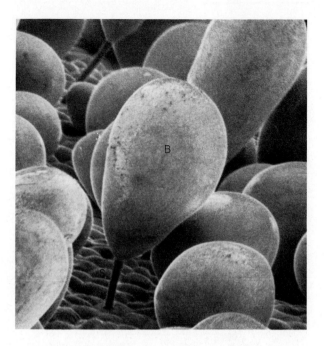

Figure 24-8 A scanning electron micrograph of salt bladders on the leaf of saltbush (*Atriplex spongiosa*). This species of saltbush is indigenous to Australia and is salt-tolerant because it has developed a mechanism to control the Na^+ and Cl^- concentrations of its tissues by accumulating salt in epidermal bladders on the surface of the aerial parts of the plant. Salt from the leaf tissues is transferred through the small stalk cell (S) and into the balloonlike bladder cell (B). As the leaf ages, the salt concentration in the cell increases, and eventually the cell bursts or falls off the leaf, releasing the salt outside the leaf. ×450. (From John Troughton and Lesley A. Donaldson, 1972.)

quire the expenditure of metabolic energy and by mechanisms that are not known to exist in plants, although they do exist in insects and arachnids (although they are not understood). The researchers calculate that ample respiratory energy is available, if a mechanism exists for its use.

Actually, large quantities of both organic and inorganic materials are leached from the leaves of many plants, both halophytes and glycophytes. Some of the leaching brought about by washing the leaves is caused by removal of materials within the tissues as well as washing off materials that have been exuded at the surface. In any case, materials that are washed from the leaves to the soil are recycled back to the plant and to other plants. Species that absorb large amounts of salt and then lose it either by leaching or by dropping their salt-filled leaves to the soil may considerably increase the salinity of the surface soil. As in other environments, desert plants sometimes profoundly influence the soil upon which they grow. Fireman and Hayward (1952) found, for example, that greasewood (*Sarcobatus vermiculatus*) in the Escalante Desert of Utah brought salts from

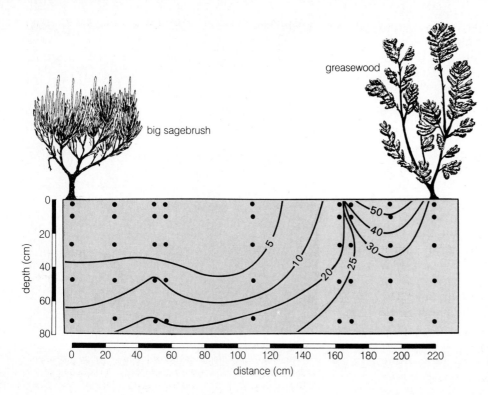

greasewood

big sagebrush

depth (cm)

distance (cm)

Figure 24-9 Salt concentrations in the soil (numbers in lines represent percent exchangeable sodium) as a function of position under a sagebrush plant (*Artemisia tridentata*) and under a greasewood plant (*Sarcobatus vermiculatus*) in the Escalante Desert of southern Utah. Note high sodium under greasewood. Dots are sampling points. (Data from Fireman and Hayward, 1952.)

depths, depositing them on the surface (Fig. 24-9). Probably the salts were contained in the leaves and released upon their fall and decay. The result was high salt concentrations beneath the greasewood plants, especially when compared with soil beneath big sagebrush (*Artemesia tridentata*) plants that did not redistribute salt in this manner. Clearly, these two species have different physiologies that are of ecological significance.

Frequently, halophytes synthesize large quantities of the amino acid proline (Fig. 24-10) as well as other amino acids and such other compounds as galactosyl glycerol and organic acids (Hellebust, 1976). As we shall see in a later section, these compounds function in osmotic adjustment.

Another potential problem for plants growing on saline soils is obtaining enough potassium. This is because sodium ions compete with the uptake of K^+ by a low affinity mechanism (Chapter 6), and K^+ is commonly present in such soils in much lower concentrations than is Na^+. In this respect, the presence of Ca^{2+} appears to be crucial. If sufficient calcium is present, a high-affinity uptake system having preference for transport of K^+ can operate well, and the plants can then obtain sufficient potassium and restrict sodium (LaHaye and Epstein, 1969). It is possible that fertilization of some saline soils with Ca^{2+} might increase their agricultural productivity. A favorable effect of Ca^{2+} on soil structure could also be important. Gypsum ($CaSO_4$) is sometimes used, providing both Ca^{2+} and some acidity, which helps in leaching out the Na^+. Elemental sulfur is also some-

times applied. It oxidizes to produce sulfuric acid, which aids in leaching. Sulfuric acid itself has been applied with some success.

Frost and Freezing Injury Although productivity of world ecosystems is probably limited more by water than by any other environmental factor, low temperature is probably most limiting to plant distribution (Parker, 1963). To grow even in subtropical regions subject occasionally to near-freezing temperatures, plants must be capable of some acclimation to low temperatures. Plants that grow in polar regions must tolerate extremely low temperatures, and only a few species can achieve this. Frost and low temperature damage in crop plants is an important hazard in most agricultural regions of the world.

It is "logical" that death of a plant results from water expansion upon freezing and subsequent disruption of cell walls and other anatomical features. Careful examination during the early decades of the 19th century showed, however, that plants actually contract rather than expand upon freezing. This is because ice crystals grow into the extracellular air spaces. Furthermore, although it must occur, ice is almost never observed within the living cells of tissues that have frozen naturally. (It is, however, observed in the dead xylem cells of trees in winter; Section 4.5.) Nor is damage to cells other than collapse observed. There is no rupture of cell walls or even cell membranes, although there is ample evidence that membranes are damaged. With rapid cooling of tissue in the laboratory (e.g., 20°C in 1 h), ice can be seen within the cells, and damage to cellular components can be observed.

Such rapid cooling does occur in nature. Acclimated American arborvitae (*Thuja occidentalis*) tissues were capable of withstanding temperatures to −85°C when cooled slowly, but southwest-facing foliage was injured when the temperature dropped 10°C per minute from 2 to −8°C at sunset. Such changes duplicated in the laboratory also injured plants. Injury symptoms could not be duplicated by any form of desiccation, and it was concluded that the winter burn was caused by the rapid temperature drop (White and Weiser, 1964).

Typically, ice crystals begin to form in the extracellular spaces, and water from within the cells diffuses out and condenses on the growing ice crystal, which may become several thousand times as large as an individual cell. In frost-hardy plants, when these ice crystals melt, the water goes back into the cells, and they resume their metabolism. In nonacclimated plants, metabolism cannot be resumed, and the water does not reenter the cells completely. There are significant differences in protoplasm between hardy and sensitive plants. Tolerant plants frequently contain higher concentrations of solutes, and their protoplasm is more elastic and remains more elastic during freezing. In recent years, the role of the membrane has been increasingly emphasized (Steponkus, 1981, 1984). Yet the molecular basis of cold tolerance is not understood, although it seems closely related to water stress.

Hardening (Acclimation) Plants exposed to low water potentials, high light levels, and such other factors as high-phosphorus and low-nitrogen fertilization become drought tolerant or hardy compared to plants of the same species not treated in this way. That is, they become **acclimated** to drought, a process of considerable importance to agriculture. This is a good example of a conditioning effect.

Actively growing plants, especially herbaceous species, are damaged or killed by temperatures of only −1 to −5°C, but many of these plants can be acclimated to survive winter temperatures of −25°C or lower. In regions where air temperatures drop below this, many plants have underground meristems protected from extreme air temperatures by soil or snow. They avoid or escape cold rather than being cold hardy (tolerant). Most of the species that survive freezing temperatures do so by tolerating some ice formation in their tissues. Generally, hardier plants can survive with more of their water frozen than less hardy plants. But there are apparently several mechanisms of hardiness (see excellent discussion in Burke et al., 1976; Levitt, 1980; Steponkus, 1981, 1984).

In practical terms, minor increases in hardiness could have a major impact on world food production.

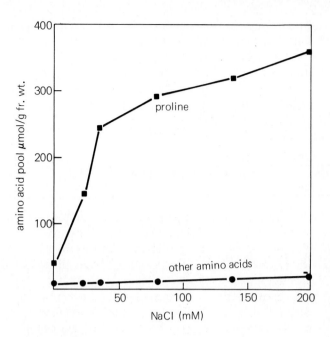

Figure 24-10 Amino acid and proline accumulation in *Triglochin maritima* grown at different salinities. Cuttings of *T. maritima* were grown in a nonsaline medium for two weeks before being transferred to saline media. Shoot tissues were harvested for analysis after 10 days of saline treatment. The extent of proline accumulation in this specialized halophyte is extreme compared with most plants, but accumulation of proline or other compatible solutes as drought or salt stress increases is common. Values in stressed mesophyte leaf tissues seldom exceed 200 μmol (g dry weight)$^{-1}$. (Note that values in this figure are for fresh weight; from Stewart and Lee, 1974.)

Winter wheats and winter rye yield 25 to 40 percent more than comparable spring varieties, for example, because they make better use of spring rains. If the winter wheats and rye could be made 2°C more cold hardy, they could replace much of the large areas of spring wheats and rye in North America and the U.S.S.R.

Frost hardiness typically develops during exposure to relatively low temperatures (e.g., 5°C) for several days. Temperatures down to −3°C are sometimes required for maximum acclimation (Larcher and Bauer, 1981; Wiser, 1970). Short days also promote acclimation in several species, and there are indications that a stimulus may move from leaf tissue to the stems. The development of frost hardiness is a metabolic process requiring an energy source. Apparently this can be provided by light and photosynthesis. Factors that promote more rapid growth inhibit acclimation: high nitrogen in the soil, pruning, irrigation, and so on. In general, nongrowing or slowly growing plants are more resistant to several environmental extremes, including air pollution. Water-stressed plants are more resistant to air pollution, partially because their stomates are closed.

Salt hardiness can also be increased somewhat

by exposure to saline conditions. Salt hardening is minimal, however, compared to drought or cold acclimation. Clearly, hardening against drought, freezing, and high salt is often a matter of hardening against water stress, but as we shall see in the next section, there are many complications.

24.4 Mechanisms of Plant Response to Water and Related Stresses

How does a euxerophyte differ from an ordinary mesophyte? Or how does a halophyte differ from a glycophyte? What is special about plants that can survive extremely low temperatures? Many different proposals have been presented to account for tolerance and acclimation, and a few of these may be common to the three kinds of stress (all related to water stress) that we are discussing here. For example, protoplasmic viscosity usually increases with high water stress, often to the point at which the protoplasm becomes brittle. Euxerophytes maintain protoplasmic plasticity much better at a given water stress than do mesophytes, and the hydrolytic activity (breakdown of starch, protein, and so on) is also less noticeable in euxerophytes. To a certain extent, these features also appear in plants capable of surviving extremely low temperature, but halophytes often approach the problem in somewhat different ways. Here are some of the possibilities.

Proposals About Responses of Mesophytes to Mild Water Stress How are mesophytes damaged by mild water stress? At least five possibilities have been proposed. First, it is known that water activity (indicating its ability to enter into chemical reactions) is a function of water potential and is thus lowered by water stress. Nevertheless, water activity is closely related to water concentration, and as we have seen (Chapter 2), water concentration changes only slightly as water potential changes considerably. At the maximum stress levels of interest to agriculturists ($\psi = -1.0$ to -2.0 MPa), water activity is lowered only slightly—probably not enough to be of any real consequence in chemical reactions. Second, solutes increase in concentration as water is lost. This could be important, but it is probably not very important under mild water stress, simply because the concentration changes would amount to only a few percent. Third, water stress might result in special changes in membranes. Such effects have indeed been demonstrated, but comparable effects can be caused by other factors without noticeable plant response, so it does not seem likely that this is an important aspect of plant response to water stress. Fourth, water stress might upset the hydration of

macromolecules, the "ice" structure of the water molecules surrounding enzymes, nucleic acids, and so on. If this water of hydration is upset, function would also be influenced. Levitt (1962) has suggested that dehydration of key enzymes would cause disulfide bonds within proteins to break and reform, sometimes reforming between adjacent molecules, leading to enzyme denaturation when molecules are rehydrated. But again, it has been calculated that mild water stress would not have much influence on the structure of water of hydration, which involves only a few percent of the water in a cell anyway. Amazingly enough, studies have shown that considerable water can be lost from a cell before enzyme function is noticeably influenced. But there could be exceptional enzymes that have not been studied.

Fifth, even the mildest water stress may profoundly change the turgor pressure within plant cells. Pressure changes of this magnitude (P = 0.1 to 1.0 MPa) probably have little effect upon enzyme activities (judging by observed responses as well as thermodynamic principles), but such changes could be the stimulus to which some special response mechanism in the cell reacts in transducing water stress to the observed cellular responses. In the large-celled marine green alga, *Valonia*, ion uptake decreases with slight increases in cellular turgor pressure. Such responses have not been observed in most higher-plant cells, although red beet tissue responds this way to decreased turgor. The observation may serve as a model for what might be taking place (see discussion in Hellebust, 1976). We have already noted that ABA is apparently produced in response to decreasing leaf-cell turgor (Section 24.3), and we have emphasized that cell expansion, meaning plant growth in general, is highly sensitive to water stress, probably via decreasing cell turgor (Fig. 24-6).

Deep Supercooling Most deciduous forest species and fruit trees avoid freezing in some of their tissues by **deep supercooling** to temperatures as low as about −40°C (lowest on record is −47°C). Ice forms in the bark and buds of such acclimated plants when temperatures are only a few degrees below the freezing point of pure water, but the ice crystals form in the spaces between cells, as in herbaceous and other drought-hardy plants; tissues of those species are not damaged by this. Xylem tissue in hardwoods is too compact to permit the formation of such ice crystals, however. When freezing does occur, xylem ray parenchyma cells are killed, the wood becomes dark and discolored, and vessels become filled with gummy occlusions (Section 4.6). Wood-rotting organisms often invade such injured trees. But the cell sap in xylem rays of most temperate woody species is capable of being supercooled without freezing. Pure water can be supercooled to −38°C if ice nucleation is prevented, and the solute-containing water in these acclimated xylem cells is apparently protected by the plasmalemma from such nucleation; the cells can often be deep supercooled to about −40°C or sometimes below. This is observed by measuring stem temperatures with a small thermocouple as temperature is lowered (Fig. 24-11). When the xylem ray cells finally freeze, the released heat of fusion causes a sharp temperature rise, as when osmotic potentials are determined cryoscopically (see Fig. 2-10). Apparently, deep supercooling is a survival mechanism for some plant tissues. When freezing does occur, ice probably forms within the living cells, and they are killed. Such species do not grow where winter temperatures drop below about −40°C (Fig. 24-12).

Extremely hardy woody plants (e.g., birches, alders, quaking aspen, willows) native to the boreal forests of North America and Asia do not deep supercool. The extracellular freezing process is similar to that found in less-hardy herbaceous plants, but ultimate hardiness is much greater. Large ice crystals form, drawing water from the cells until all but the water of hydration (bound water) is removed. Studies extended over many years have failed to show any relationship between the extent of hardiness and the amount of bound water in such plants, but their hardiness is apparently a function of their cells' ability to withstand extreme dehydration. In any case, such dormant, winter-hardy, woody plants (typically softwoods) can readily survive the liquid nitrogen temperature of −196°C. These same plants, when actively growing, may be killed by −3°C! Thus cold acclimation is much more spectacular than drought or salt hardening.

There is apparently no lower temperature limit for survival of spores, seeds, and even lichens and certain mosses in the dry condition. Such test objects have been held within a fraction of a degree of absolute zero for several hours with no apparent damage. Even active tissue may survive these low temperatures if it is experimentally cooled so rapidly that intracellular water freezes into extremely small crystals that don't damage the cytoplasm.

What are the lower temperature limits for active growth as compared with survival? Several higher plants are able to grow and even flower under the snow, where temperature is close to 0°C or sometimes below (Richardson and Salisbury, 1977). Such plants include native species such as the snow buttercup (*Ranunculus adoneus*), often called **geophytes** or **spring ephemerals,** as well as winter cereals (e.g., winter wheat and winter rye) and ornamentals (crocuses, snowdrops, tulips, hyacinths, and daffodils—the last three not necessarily actually growing under snow). They grow slowly during the winter and thus often have a significant head start when the snow melts. The native species, especially, then grow

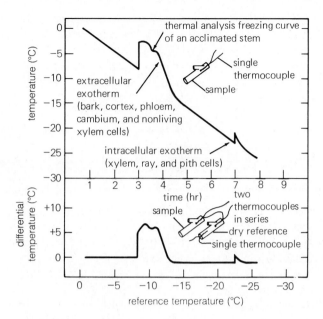

Figure 24-11 Thermal analysis and differential thermal analysis as methods to observe deep supercooling in stem samples. In both experiments, the sample is cooled over a period of time, and the sample temperature (top) or the differential temperature between the sample and a dry reference (bottom) is monitored. The peaks or exotherms show release of the heat of fusion and thus indicate freezing points. The first peak (left, warmest temperature) indicates freezing of extracellular water; the second peak (right, coldest temperature) presumably indicates freezing of water in cells. Cell freezing typically leads to the death of the cells. (From Burke et al., 1976.)

rapidly and flower before later species overtop them or before trees above them leaf out. Lichens can photosynthesize at −20°C and below, and certain bacteria can grow at temperatures on the order of −22°C. Snow algae can also grow at the freezing point of pure water and below (Aragno, 1981). The lower limits for active growth of organisms have not been determined, and there is much opportunity for research.*

Plants with no appreciable level of hardiness may escape minor frost by supercooling by 3 to 10°C. When they do freeze, it is because of various things that provide nuclei for ice crystals. Sometimes these occur within the tissues; other times they are on the surface of the leaves. Specific species of bacteria (e.g., *Pseudomonas syringae* and *Erwinia herbicola*) have been

---

*One of us (Salisbury) attended a NASA-supported Conference on Environmental Extremes held in San Diego, California, February 10-11, 1966. There it was stated by experts in the field that certain bacteria grow at −22°C as we have noted, that molds have exhibited active growth in cold storage lockers at −38°C, and that spores were formed by these organisms at −47°C! We are presently unable to document these claims, and they should be viewed with suspicion (but see Allen, 1965).

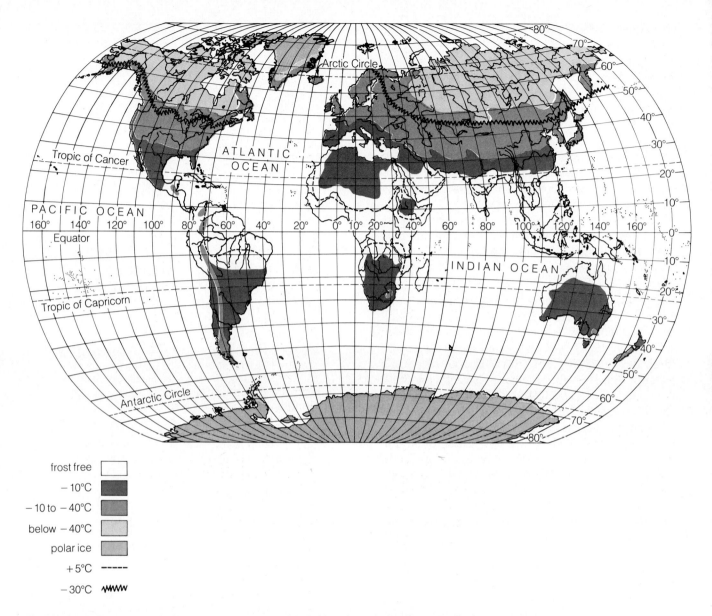

frost free ☐
− 10°C ▨ (dark)
− 10 to − 40°C ▨ (medium)
below − 40°C ▨ (light)
polar ice ▨
+5°C - - - - -
−30°C ⋏⋏⋎⋎⋏

Figure 24-12 Tentative map of low-temperature thresholds limiting plant distribution on the Earth. The lines represent average annual minimum temperatures rather than the lowest minimum temperatures ever experienced (or experienced once or twice in each century). When such unusual periods of cold occur, at least a few plants in the population usually survive, but populations cannot continue to survive if they cannot tolerate average minimum temperatures. (From Larcher and Bauer, 1981; used by permission.)

found, for example, that initiate ice formation at relatively warm temperatures. Spraying plants in the field with suspensions of these bacteria causes the plants to be killed by light frosts, whereas spraying with other bacterial species has no effect (Anderson et al., 1982; Lindow, 1983).

Compatible Solutes: Osmotic Adjustment (Osmoregulation) We have noted in several places that certain substances such as proline, betaine, and various carbohydrates build up in cells that are subject to

drought, freezing, or high salt concentrations. This is called **osmotic adjustment** or **osmoregulation** (Flowers et al., 1977; Jefferies, 1981; Morgan, 1984; Turner and Jones, 1980). Consider the situation with high salt. Because there are excessive dissolved salts in the soil solution (or sea water in the case of mangroves and other plants that grow in similar situations), the osmotic potential is negative enough to cause water to diffuse out of the tissues into the surrounding solutions—unless the water potential in the tissues is at least as negative. Actually, if the tissues are to absorb water and survive, their water potential must be more negative than that of the surrounding solution. Clearly, one way to overcome this problem would be for the cells simply to accumulate salt to the same or higher concentrations as those outside the plant. Why doesn't this happen? (We noted that those plants that do accumulate salt usually dilute it by becoming succulent or accumulate the salt into their vacuoles, where it could still dehydrate the cytoplasm osmotically.)

In all eukaryotes so far studied, salts such as NaCl denature the enzymes and thus cannot be tolerated in the cytoplasm itself. When they occur in halophyte cells, they are on the order of 10 times more concentrated in the vacuole than in the cytosol. Apparently, many plants that tolerate the various kinds of water stress do so by synthesizing compounds in their cytoplasm that can exist at high concentrations without denaturing the enzymes essential for the metabolic processes of life. The organic compounds that can be tolerated have been referred to as **compatible solutes** (or **compatible osmotica**). Paul H. Yancey and his co-workers (1982) point out that the number of compatible solutes discovered among the five kingdoms of living organisms is relatively limited. Apparently, only a few compounds can exist at relatively high concentrations in the cytoplasm without damaging the enzymes there. Table 24-2 lists a few species that exhibit different degrees of resistance to water stress (produced by drought, cold, or salt) along with the compatible solutes found in their cells. Fig. 24-13 shows the chemical structures of a few of these compounds.

The degree of osmotic adjustment is a function of the degree of outside water stress caused by salt in the surrounding medium (Fig. 24-10), by drying soil, and sometimes by frost hardening. The regulation or adjustment occurs in halophytes, xerophytes, and mesophytes. How does the regulation occur? Current research (an active field in stress physiology) is concerned with this question, but complete answers remain in the future. We have already emphasized the role of cell turgor in synthesis of ABA and in controlling the rate of cell growth as it is influenced by water stress. Most workers in the field suspect that changes in cell turgor also activate and control the degree of compatible-solute synthesis. In the marine algae, turgor pressure remains constant over a wide range of external salinities, for example, indicating that it must be regulated and suggesting that cells are sensitive to changes in turgor. Perhaps the physical properties of the plasmalemma change as the force with which it is pressed against the wall changes. There is also an interesting observation that putrescine, a polyamine, increases over 60-fold in oat leaf cells within 6 h in response to osmotic stress (sorbitol and other osmotica dissolved in the media

Table 24-2   Examples of Some Organisms and the Compatible Solutes That Increase in Their Cells During Osmotic Adjustment

Organism	Compatible Solute
BACTERIA	
Various halophiles and nonhalophiles (e.g., *Klebsiella, Salmonella, Streptococcus*)	amino acids (glutamate, proline, etc.)
Halobacterium salinarium (halophile; an archaebacterium; no osmotic adjustment.)	NaCl
FUNGI	
Chaetomium globosum (a terrestrial form)	polyhydric alcohols (mannitol, arabitol, glycerol)
Saccharomyces rouxii (an osmophilic form)	arabitol
MICROALGAE	
Chlorella pyrenoidosa (freshwater)	sucrose
Dunaliella spp. (marine and halophilic)	amino acids, glycerol
Scenedesmus obliquus (freshwater)	carbohydrate (sucrose + rafinose, glucose, fructose)
ANGIOSPERMS	
glycophytes: *Chloris gayana, Hordeum vulgare* (barley)	betaine and proline
halophytes: *Aster tripolium, Mesembryanthemum nodiflorum, Salicornia fruticosa, Triglochin maritima*	proline
halophytes: *Atriplex spongiosa, Spartina townsendii, Suaeda monoica*	betaine

Mostly from Flowers et al., 1977; and Yancey et al., 1982.

on which leaf segments were floated; Flores and Galston, 1982). Yet concentrations of putrescine at their maximum are still far too low for the compound to be acting as a compatible solute. Could the change be part of the transduction between cell turgor and the synthesis of such solutes? Future research is required to find out.

Incidentally, roles in addition to that of compatible solute have been suggested for some of the compounds in Table 24-2. Proline, for example, might serve in nitrogen storage and nitrogen transport as well as being a compatible solute.

There are a few organisms that seem to be the exception to the rule about compatible solutes; actually, they illustrate the importance of the rule. There are certain species of *Halobacterium* (Table 24-2) that

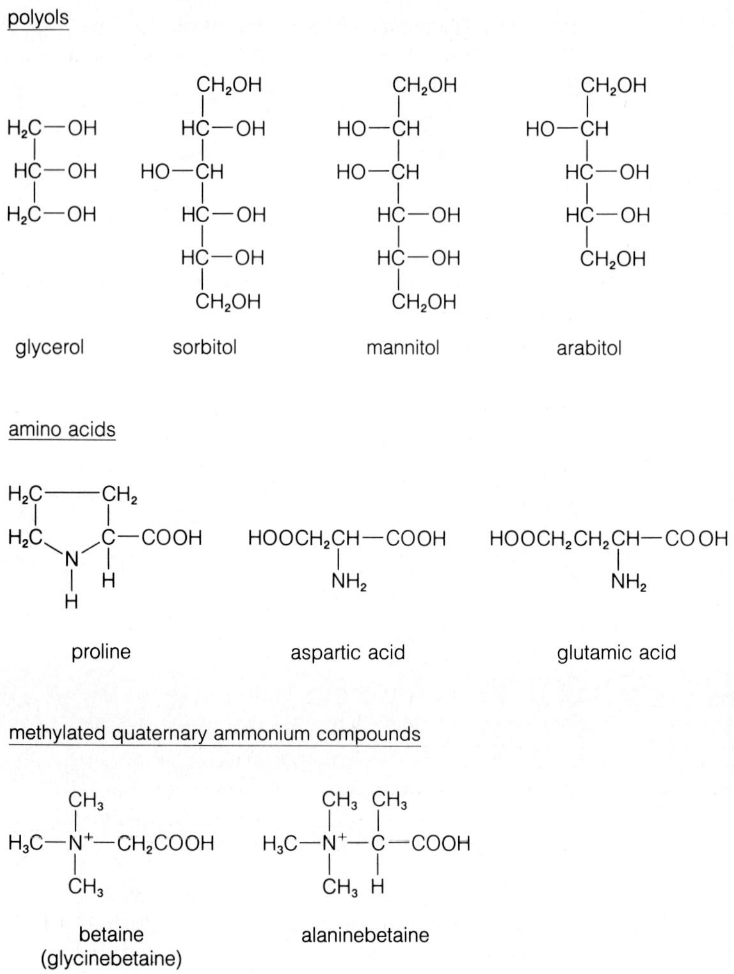

polyols

glycerol   sorbitol   mannitol   arabitol

amino acids

proline   aspartic acid   glutamic acid

methylated quaternary ammonium compounds

betaine
(glycinebetaine)   alaninebetaine

Figure 24-13 Molecular structures of some compatible solutes found in stressed plants and other organisms. (Compare Table 24-2.)

do accumulate large quantities of sodium chloride in their cells. These organisms are extremely primitive in various ways, and it has been suggested by some that they should be assigned to a sixth kingdom, the Archebacteria. Some of these organisms are unable to grow and reproduce unless they are in highly concentrated salt solutions such as the Dead Sea or the Great Salt Lake. Whereas the enzymes of eukaryotic halophytes are identical, or nearly so, to their counterparts among nonhalophytic eukaryotes, the enzymes of halobacteria have been extensively modified. They only maintain their metabolic activities when they exist in strong salt solutions, being denatured by more dilute solutions. Thus, in an evolutionary sense, it seems that the halobacteria have followed the extremely difficult pathway of modifying hundreds to thousands of enzymes so that they can tolerate high salt concentrations, while the other halophytes and salt-tolerant eukaryotes have taken the genetically much simpler approach of producing compatible solutes that do not harm cytoplasmic en-

zymes but provide a water potential balance between the cytoplasm and the surrounding environment of high water stress (Yancey et al., 1982).

24.5 Chilling Injury

Tropical or subtropical plants normally grown in southern regions of North America are sometimes damaged by frost or even by temperatures slightly above freezing (as in the winter of 1983-84). Such crops include citrus, cotton, maize, rice, sorghum, soybean, sugarcane, and sweet potato. Certain tropical fruits such as bananas are damaged even by a few hours below 13°C. (Never put bananas in the refrigerator!) Timing is crucial: Rice plants exposed to temperatures below 16°C at the time of pollen-mother cell division will not produce a crop. It has been estimated that the world rice production would decrease 40 percent if world mean temperature dropped only 0.5 to 1.0°C.

As in our discussion of water stress effects, we are faced with the question of mild chilling effects (such as those that damage bananas), as contrasted to the death caused by severe freezing. Many mechanisms have been proposed to account for chilling injury. Since chilling disrupts all the metabolic and physiological processes in plants, it seemed almost futile to look for a single key reaction that might be responsible. Nevertheless, it is possible that just such a key response might have been identified (Graham and Patterson, 1982; Lyons, 1973). As the temperature is lowered in chilling-sensitive plants, lipids in cellular membranes solidify (crystallize) at a critical temperature that is determined by the ratio of saturated to unsaturated fatty acids. This critical temperature for a phase transition from liquid to crystalline often proves to be equivalent to the temperature that causes chilling damage. As indicated, this can be as high as 10 to 13°C in sensitive species of tropical origin. Development of frost tolerance in chilling and frost-sensitive plants apparently involves changes in this ratio. An increase in the proportion of unsaturated fatty acids or in the quantity of sterols results in the membrane remaining functional at lower temperatures.

The lipid model suggests that the membrane normally exists in a liquid-crystalline condition. In this state, its enzymes have their optimal activity, and its permeability is thus under control. Below the critical temperature, the membrane exists in a solid-gel state. This change in state should bring about a contraction resulting in cracks or channels that lead to increased permeability. This would lead to the upset solute balances (ions and other solutes leaking from chill-damaged cells or mitochondria) that are observed. Enzyme activities would also be upset, leading to imbalances with non-membrane-bound enzyme systems. Thus, metabolites such as those produced in glycolysis would be expected to accumulate, because they could not be acted upon by the mitochondrial enzyme systems. Such accumulation had indeed been observed. Little ATP would be formed, because of the importance of membranes in its formation and because of these imbalances between the mitochondria and the glycolytic systems. Similar events would probably take place between the chloroplast and the cytoplasm around it.

If the temperature is raised soon enough, the membranes return to the liquid-crystalline state (since this phase transition is completely reversible), and the cell recovers. If metabolite buildup and solute leakage are allowed to occur to any great extent, however, then cells are injured or killed.

It has been observed that some cultivars are more sensitive to chilling than others, although their fatty acid ratios appear to be the same. These differences could be caused by different sensitivities to the accumulated metabolites, rather than to the initial effects on the membranes. Chilling effects can often be avoided if tissues are exposed to high temperatures for brief intervals between chilling periods, provided that initial chilling was not too prolonged. In terms of the hypothesis given, this would allow the metabolism of accumulating metabolites, so that toxic levels are not allowed to develop.

In summary, there is much support for the lipid-membrane (with associated proteins) hypothesis for chilling injury, and the basic idea might even be extended to gain at least a partial understanding of frost hardening (Harvey et al., 1982; Steponkus, 1984). But there are still problems to be solved, mostly lack of correlation between lipid compositions and chilling sensitivity in some studies (summarized by Graham and Patterson, 1982).

24.6 High-Temperature Stress

Elevated temperatures typically accompany drought conditions and are an important environmental stress factor in themselves. This is especially true for euxerophytes that are hardly cooled by transpiration.

The upper-temperature limits permitting survival have long been of interest to biologists. Plants typically die when exposed to temperatures of 44 to 50°C, but some can tolerate higher temperatures. Stem tissues near the soil line of plants in the desert, for example, may reach levels considerably above this, and *Tidestromia oblongifolia* photosynthesizes optimally in Death Valley, California, at these temperatures (see Figs. 11-4 and 11-12).

Prokaryotes are able to live at the highest temperatures (reviewed by Aragno, 1981; Brock, 1978; Kappen, 1981; and Steponkus, 1981). In thermal springs of Yellowstone National Park, bacteria grow at the boiling point of water, which is about 90°C at that elevation, but several bacteria have been found in sea-level, boiling springs in New Zealand and other areas, where the boiling point is as high as 101°C. Some exist a few degrees above this temperature at greater depths in the water, where higher pressures result in higher boiling points. Bacteria have been found growing at the bottoms of oil wells at temperatures above 100°C and greatly elevated pressures. These organisms will repeat this performance in the laboratory but will not grow at such high temperatures when pressure is reduced.

All these observations pale almost to insignificance in light of a recent discovery of J. A. Baross and J. W. Deming (1983; see also comments by Walsby, 1983). Baross et al. (1982) had found bacteria in hot water rising from sulphide-encrusted vents located along tectonic rifts and ridges on the deep ocean floor. Some of these communities occurred at temperatures exceeding 360°C! Now Baross and Deming (1983) have been able to grow these bacteria at 250°C in a titanium syringe pressurized to 265 atmospheres (26.85 MPa) and containing enriched sea water. The bacteria doubled in number every 40 minutes, increasing a hundred-fold in a few hours. The workers say they have not yet determined the upper temperature limits for growth of these organisms. Much remains to be learned. The bacteria have membranes based on a different design from other organisms: long-chain ether lipids that span from one surface to the other, forming thermostable structures. It will be interesting to learn about the structures of their proteins and nucleic acids. The organisms are probably members of the Archaebacteria, which have apparently had a long separate evolution from other bacteria. Do they represent the first living organisms on earth, originating when the earth was very warm?

Eukaryotic organisms have not been found growing at temperatures above 56 to 60°C (temperatures of hot springs that support green algae), and the upper limit for animals seems to be about 45 to 51°C. Photosynthesis apparently does not occur even in blue-green algae (prokaryotes) at temperatures above 70 to 72°C. Several dry spores and seeds of higher plants will survive temperatures well above 100°C, but these do not actively grow at such temperatures. In general, dry and dormant structures withstand various stresses well.

Plants that are hardy to high temperatures exhibit high levels of water of hydration and high protoplasmic viscosity, characteristics that are also exhibited by euxerophytes. High-temperature plants also are able to synthesize at high rates when temperatures become elevated, allowing synthetic rates to equal breakdown rates, thus avoiding ammonia poisoning. Plants can be acclimated somewhat to high temperatures, but this is minimal compared to acclimation to drought or to freezing temperatures.

24.7 Acidic Soils

Plants are found growing on soils in a pH range of at least 3 to 9, and the extremes provide another stress to which some species are adapted. Cranberries, for example, grow on acid bogs, while certain desert species normally grow only on high pH soils. In general, we know far too little about why some plants are native to low pH soils and others to soils with higher pH values. Certainly one of the reasons is competition. If we use hydroponic techniques to study the growth of various species apparently preferring different pH levels, we usually find that they do reasonably well over a fairly wide pH range. But in nature, even a slight advantage of one species over another can eventually lead to elimination of the less well-adapted one.

Soil factors closely correlated with pH are probably more important than the concentration of H^+ ions *per se*. For example, high rainfall leads to leaching of calcium and formation of acidic soils, so calcium is usually low in acidic soils and abundant in soils of high pH (calcareous soils). Moderate concentrations of this element favor development of root nodules on many legumes (Chapter 13), so nitrogen-fixing legumes will grow better on soils rich in calcium than on most acidic soils. The less-abundant calcium in acidic soils may also limit plant growth simply because H^+ is much more toxic to roots in the absence of calcium. One of the beneficial effects of liming acid soils no doubt derives from this fact. (**Liming** involves addition of calcium in various forms, often in mixtures: CaO, which is lime or burned lime; $Ca(OH)_2$, which is water-slaked or hydrated lime; or $CaCO_3$, which is limestone, dolomite, or air-slaked lime.)

The pH also strongly influences the solubility of certain elements in the soil and the rate at which they are absorbed by plants. Iron, zinc, copper, and manganese are less soluble in alkaline than acidic soils because they precipitate as hydroxides at high pH. Iron-deficiency chlorosis is thus common on soils in the western United States, which are often alkaline. Phosphate, absorbed largely as the monovalent $H_2PO_4^-$ ion, is more readily absorbed from nutrient solutions having pH values of 5.5 to 6.5 than at lower or higher pH values. In soils of high pH, more of the phosphate is present as the less readily absorbed divalent HPO_4^{2-} ion. Furthermore, much of this is usually present as insoluble calcium phosphates. In soils of low pH, where $H_2PO_4^-$ should predominate, the frequent high concentrations of aluminum ions cause its precipitation as aluminum phosphate.

The relatively high concentrations of available aluminum in many acidic soils (below about pH 4.7) can inhibit growth of some species, not only because of detrimental effects on phosphate availability but apparently also by inhibiting absorption of iron and by direct toxic effects on plant metabolism. Some species (e.g., azaleas) not only tolerate these high aluminum concentrations but thrive on such soils. Still other species tolerate amounts of various heavy metals that are toxic to most plants. An example is bentgrass (*Agrostis tenuis*) that grows in Wales and Scotland on mine tailings having unusually large amounts of lead, zinc, copper, and nickel. This grass does not exclude such toxic metals but somehow accumulates them without being injured appreciably. We don't understand the tolerance mechanism, although it was suggested that specific chelating agents (e.g., in root cell walls) form strong complexes with the metal ions and prevent their reaction with sensitive protoplasmic constituents such as enzymes. Secretion of these metals into the vacuoles would also decrease their toxic effects. There has been much recent study of these matters.

24.8 Other Stresses

Although we have examined a number of important environmental stress factors, there are others that might still be discussed. For example, certain species of plants not only survive but flourish on soils derived from serpentine rock (Kruckeberg, 1954; Whittaker, 1954) or highly acidic material that has been derived from rock that has been strongly modified by percolation of hot water in ancient hot springs (Salisbury, 1964). The serpentine soils are highly deficient in calcium and apparently have toxic amounts of other elements. Somehow serpentine species have adjusted to these stresses. Material derived from hydrothermally altered rock has most of its phosphate tied up in forms unavailable to most plants.

There is little available nitrogen, while iron, aluminum, and calcium occur in superabundant amounts. Such conditions are not only stressful but fatal to many species; yet, again, a few species have adjusted to these stress factors. Often, plants from both serpentine and hydrothermally altered material can also grow well on more normal soils. Much needs to be learned about such situations.

We're becoming acutely aware of the importance of atmospheric pollutants as stress factors. Because of the implications for agriculture as well as the health and productivity of the world's forests and other ecosystems, these stress factors have been the objects of much research in recent years. Often they form valid and important topics of plant physiology that we will arbitrarily not discuss, mostly because of space limitations.

In spite of the broad spectrum of environmental stress factors, it is well to end this chapter by returning to the importance of water in the life of most plants. Water stress is either of primary importance or is a contributing factor to reduced growth and yields (biological strains) of plants growing on much of the earth's surface.

Appendix A

The Système Internationale: The Use of SI Units in Plant Physiology

At the Eleventh General Conference on Weights and Measures, which met in October, 1960 in Paris, the metric system of units was given the name *International System of Units* with the abbreviation SI (Système International) in all languages. The system was designed to simplify the metric system then in use and to unify the application of units in all the sciences and other human endeavors. At subsequent meetings, held every three or four years, further refinements were made. For example, the Fourteenth General Conference (1971) adopted the mole (symbol mol) as an SI base unit and the pascal (symbol Pa) as a unit of pressure equal to a newton per square meter.

Table A-1 The Seven Base Units

Quantity	Unit	Symbol
Length	meter	m
Mass (not weight)	kilogram	kg
Time	second	s
Electric current	ampere	A
Temperature	kelvin	K (not °K)
Luminous intensity	candela	cd
Amount of substance	mole (Avogadro's number)	mol

Table A-2 Derived Units of Interest to Plant Physiologists

Quantity	Unit	Symbol	Definition
Force	newton	N	$1 \ kg \cdot m \ s^{-2}$
Energy, work, heat.	joule	J	$1 \ N \cdot m$
Power	watt	W	$1 \ J \ s^{-1}$
Pressure	pascal	Pa	$1 \ N \ m^{-2}$
Area	square meter	m^2	$m \cdot m$
Volume	cubic meter	m^3	$m \cdot m \cdot m$
Velocity	meter per second	$m \ s^{-1}$	$m \ s^{-1}$
Frequency	hertz	Hz	$1 \ cycle \ s^{-1}$
Voltage	volt	V	$1 \ W \ A^{-1}$
Electric resistance	ohm	Ω	$1 \ V \ A^{-1}$
Electric conductance	siemens	S	$1 \ A \ V^{-1}$
Concentration	mole per cubic meter	$mol \ m^{-3}$	$mol \ m^{-3}$
Irradiance (energy)	watt per square meter	$W \ m^{-2}$	$1 \ J \ s^{-1} \ m^{-2}$
Irradiance (photons)	moles of photons per square meter per second	$mol \ m^{-2} \ s^{-1}$	$mol \ m^{-2} \ s^{-1}$
Spectral irradiance	moles of photons per square meter per second per nanometer	$mol \ m^{-2} \ s^{-1} \ nm^{-1}$	$mol \ m^{-2} \ s^{-1} \ nm^{-1}$

Plant physiologists and other scientists have attempted to apply the system. In some cases, this has meant giving up long-accepted and familiar units; in other cases, the units to be used seemed unreasonable and even illogical, although with the passing of time, several SI units that at first seemed unacceptable now seem much less so. They are being applied by an increasing number of plant physiologists. Indeed, studies have shown that greater familiarity with the system leads to more frequent use.

Our goal in this book is to reflect the current use of units in the science of plant physiology rather than to push the use of SI units beyond their present application in the field. Our criterion for use of units in this text is their use in the most current literature of plant physiology. As it turns out, this means that most of the SI units are used in this text. In this appendix, we present a number of tables that describe the SI units, including the few exceptions still adhered to by plant physiologists.

Table A-3 Preferred and Nonpreferred prefixes (Multiples and Submultiples)

Preferred		Preferred	
kilo	k (10^3)	milli	m (10^{-3})
mega	M (10^6)	micro	μ (10^{-6})
giga	G (10^9)	nano	n (10^{-9})
tera	T (10^{12})	pico	p (10^{-12})
peta	P (10^{15})	femto	f (10^{-15})
exa	E (10^{18})	atto	a (10^{-18})

Nonpreferred		Nonpreferred	
hecto	h (10^2)	centi	c (10^{-2})
deka	da (10)	deci	d (10^{-1})

Table A-4 Abbreviated Summary of SI Style Notes

The denominator

1. The denominator should not be a multiple or submultiple of an SI unit (e.g., μN m^{-2} is acceptable, but N μm^{-2} is not).

2. Use of two solidi (/) in the same expression is not recommended; negative superscripts avoid this problem: J K^{-1} mol^{-1} (not J/K/mol); J/K mol is acceptable, since all symbols to the right of the solidus belong to the denominator.

Names of units

1. The name of a unit begins in lowercase, except at the beginning of a sentence.

2. Apply only one prefix to a unit name. The prefix and unit name are joined without a hyphen or space between.

3. If a compound unit involving division is spelled out, the word *per* is used. Only one *per* is permitted in a written unit name.

4. If a compound unit involving multiplication is spelled out, the use of a hyphen is usually unnecessary; but it can be used for clarity.

Symbols

1. Symbols are used when units are used in conjunction with numerals.

2. Symbols are never made plural (i.e., by addition of s).

3. A symbol is not followed by a period except at the end of a sentence.

4. Symbols for units named after individuals have the first letter capitalized (but the name of the unit is written in lower case; see above).

5. Symbols for prefixes greater than kilo are capitalized; all others are lower case.

6. Use numerical superscripts (^{2} and ^{3}) to indicate squares and cubes; do not use sq., cu., or c.

7. Exponents also apply to the prefix attached to a unit name; the multiple or submultiple unit is treated as a single entity.

8. Never begin a sentence with a symbol.

9. Compound symbols formed by multiplication contain a product dot ($\cdot$) to indicate multiplication.

10. Do not mix symbols and spelled-out unit names. (We sometimes break this rule when new units are introduced.)

Numerals

1. A space is left between the last digit of a numeral and symbol.

2. The period is used as a decimal marker.

3. A space is used instead of a comma to group numerals into 3-digit groups.

4. Decimal fractions are generally preferred to common fractions.

5. Decimal values less than one have a zero to the left of the decimal.

6. Multiples and submultiples are generally selected so that the numeral coefficient has a value between 0.1 and 1000. For comparison, similar quantities should use the same unit even if the values fall outside this range.

7. With numerals, do not substitute the product dot ($\cdot$) for a multiplication sign ($\times$).

8. Use numeric digits, instead of spelled-out words, for all unit coefficients.

Table A-5 Some Discarded Metric Units

Discarded Metric Unit	Acceptable SI Unit
micron (μ)	micrometer (μm)
millimicron (mμ)	nanometer (nm)
Ångstrom (Å)	0.1 nanometer (nm)
bar (bar)	0.1 megapascal (MPa); 100 kilopascal (kPa)
calorie (cal)	4.184 joule (J)
degree centigrade (°C)	degree Celsius (°C)
liter (l)	0.001 meter^3
einstein (E)	mole of quanta (mol)
parts per million (ppm)	other approaches: g m^{-3}, mol m^{-3}, etc.

Table A-6 Exceptions or Special Cases for the Plant Sciences

1. The degree Celsius is acceptable as well as the kelvin.

2. Minutes (min), hours (h), and days (d) can be used.

3. The liter is still often used (abbreviation is l; journal Plant Physiology suggests L or spell out to avoid confusion with the numeral one).

4. Molar concentrations are used (mol liter^{-1} = M); use millimolar (mM) instead of $\times 10^{-3}$ M.

5. The hectare is sometimes used (equals 10 000 m^2).

6. Sometimes it is impossible to avoid using units with prefixes in denominators (as W nm^{-1} when speaking of light energy per nanometer of wavelength).

7. The candela (one of the seven basic units: luminous intensity) is not used in plant science, since it relates to sensitivity of the human eye, as explained in Appendix B.

8. Centimeters are still widely used, but these are being replaced with millimeter (e.g., in figures for precipitation).

Appendix B

Radiant Energy: Some Definitions

Plants are strongly influenced by the radiant energy in their environment. Hence, a plant physiologist must understand the nature of radiant energy and how it might interact with a plant. The principles of radiation and how radiation interacts with matter are taught in physics and physical chemistry classes. Thus the following information is presented with two purposes in mind: to provide a brief review of the topic and to provide an accessible reference source for ideas or terms once understood but no longer remembered clearly. The format is a series of definitions, which could have been listed alphabetically. But assuming that you will want to review the entire topic in a logical fashion, we have arranged the definitions so that one builds upon another. Individual terms can be found by scanning the list.

Radiant energy (Radiation) A form of energy that is emitted or propagated through space or some material medium. It is said to be *electromagnetic* and is propagated in the form of pulsations or waves. Certain concepts and equations appropriately describe the wave nature of radiant energy, but this energy also behaves like a stream of particles. These particles without rest mass can also be described by certain equations and by reference to certain manifestations. The equations even relate the wave idea to the particle concept, but we do not yet fully understand radiant energy. The term is sometimes extended (perhaps incorrectly) to include streams of subatomic or atomic particles that do have mass, such as electrons, positrons, or the atomic nuclei that make up primary cosmic rays. Radiant energy that we can see is *light*.

The wave nature of radiant energy Several phenomena, including diffraction, interference, and polarization (mentioned later) suggest that radiant energy is propagated in the form of waves. Since familiar waves (e.g., water or sound waves) are propagated through a medium, it was postulated that radiant energy is also propagated through a medium, called the *ether*. Careful experiments around the turn of the century designed to

prove the existence of the ether failed, and the concept has been rejected. Nevertheless, the wave nature of radiant energy continues to be apparent, even though it apparently needs no medium for its propagation.

Frequency (ν) The number of wave crests (peaks in energy) passing a given point in a given interval of time. Frequency is usually given in terms of energy crests (vibrations or waves) per second (s^{-1}).* Green light has a frequency of about 6×10^{14} pulsations s^{-1}; radio waves about 10^4 to 10^{11} s^{-1}.

Velocity (c) The distance traveled by a peak of radiant energy in some specified interval of time. The velocity of all forms of radiant energy is the same in a vacuum and is equal to 3.00×10^{10} cm s^{-1} (300000 km s^{-1} or 186000 miles s^{-1}). It is virtually identical to this value in air but is slower in media such as water (2.25×10^{10} cm s^{-1}) or crown glass (1.98×10^{10} cm s^{-1}).

Wavelength (λ) The distance between waves or crests of energy in electromagnetic radiation. The wavelength is equal to the velocity divided by the frequency: $\lambda = c/\nu$. Likewise, the frequency is equal to the velocity divided by the wavelength: $\nu = c/\lambda$. Wavelengths of radiant energy vary from much shorter than the diameter of an atom to several kilometers in length (see *Electromagnetic spectrum*, below). Green light has a wavelength of about 500 nm or 5×10^{-5} cm; radio waves about 10^{-1} to 10^6 cm.

Wave number ($\bar{\nu}$) A convenient term in some applications and equal to the reciprocal of the wavelength measured in centimeters: $\bar{\nu} = 1/\lambda$. From the foregoing equations, it is apparent that the wave number is equal

---

*There are three ways of writing units when some occupy the position of a denominator; all three are equivalent: cm per s, cm/s, and cm s^{-1}. The last is coming into wide use, partially because it makes it easier to cancel units in equations such as those that follow (e.g., see *Mole of quanta*).

to the frequency divided by the velocity (in cm s^{-1}) of radiant energy: $\bar{\nu} = \nu/c$. Thus wave number is simply another way of expressing frequency. Green light has a wave number of about 2×10^4 cm^{-1}; radio waves about 10 to 10^{-6} cm^{-1}.

Refraction The change in direction (bending) that takes place when a ray of radiant energy passes from one medium into another in which its velocity is different. Refraction at the surface between glass and air makes the construction of lenses possible. Light is refracted within leaves as it passes from air into a cell wall or the cytoplasm; it may be refracted several times within a leaf. Since different wavelengths are refracted to different degrees, wavelengths are separated into a **spectrum** when they pass through a prism (Fig. B-1).

Diffraction and interference Diffraction includes those phenomena produced by the spreading of waves around and past obstacles that are comparable in size to the wavelength. Interference phenomena are caused by reinforcement when energy crests (waves) are superimposed upon each other (are in phase) or by the opposite effect, which occurs when waves are out of phase, canceling or damping each other. Thus, as waves are diffracted, they may reinforce or cancel each other by interference, producing a rainbow effect (separating the various wavelengths). Two devices often used by plant physiologists operate according to these principles: (1) The **diffraction grating,** which consists of fine lines ruled very close together on a transparent surface, separates a mixture of wavelengths into a spectrum similar to that produced by a prism; (2) **interference filters,** which have a thin layer of a reflective medium on a glass surface, the layer being of such thickness that one wavelength (or multiples thereof) is strongly reinforced by passing through the filter, while other wavelengths are canceled.

Polarization Light waves normally vibrate in many directions at right angles to the direction of propagation. When light is polarized, the wave is made to vibrate in more or less one direction; it vibrates in a plane, so it is said to be *plane polarized.* Light becomes polarized when it is passed through certain substances or is reflected. Many molecules important in plants and other living things will, in solution, rotate the plane of polarization of a beam of polarized light. These **optically active molecules** typically contain at least one asymmetric carbon atom (an atom with four different groups attached to it).

The particulate nature of radiant energy Radiant energy exists in units that cannot be further subdivided. In the photoelectric effect, for example, one electron may be ejected from a surface upon the absorption of one *particle* or *packet* of radiant energy. The particulate nature of radiant energy is described by certain equations, many of which include terms for the frequency or the wavelength—terms derived from the wave concepts of radiant energy.

Quantum or photon Terms often used interchangeably for the particles of energy in electromagnetic radiation. (The *quantum* is the unit quantity of energy in the quantum theory, while the *photon* is a quantum of the electromagnetic field.)

Quantum energy (E) The energy (E) of a quantum or photon is equivalent to the frequency (ν) times Planck's constant (h): $E = h\nu$. Thus the energy of a photon is directly proportional to the frequency of the radiation; higher frequencies have more energetic photons. Since the frequency is equal to the velocity divided by the wavelength, the energy of a photon will be inversely proportional to the wavelength: $E = hc/\lambda$. Longer wavelengths (lower frequencies) have less energetic photons. These equations are useful in calculating the energy relations of photosynthesis and other plant processes that depend on light energy.

Planck's constant (h) A universal constant of nature relating the energy of a photon to the frequency of the oscillator that emitted it (i.e., the frequency of the radiant energy). Its dimensions are energy times time per quantum or photon. It is equal to 6.6255×10^{-34} J $\cdot$ S photon^{-1}, 1.58×10^{-34} cal $\cdot$ s photon^{-1}, or 6.6255×10^{-27} erg $\cdot$ s photon^{-1}.

Quantum yield (ϕ) An expression of efficiency when the absorption of a photon by a molecule results in some photochemical reaction. The quantum yield (ϕ) is equal to the ratio of the number of molecules reacted (M) to the number of photons absorbed (Q): $\phi = M/Q$. Quantum yields for photosynthesis and for interconversion of the two forms of phytochrome (see Chapters 9 and 19) are widely studied values.

Mole of quanta or einstein A number of quanta or photons equal to Avogadro's number: 6.02×10^{23} quanta mol^{-1}. Since the einstein is not an SI unit, the *mole of quanta* is becoming increasingly common; see photosynthetic photon flux density (PPFD), below. The energy (E) in a mole of red light ($\lambda = 660$ nm or 6.6×10^{-5} cm, $\nu = 4.545 \times 10^{14}$ s^{-1}) can be calculated as follows:

$$E = \frac{hc}{\lambda} = \frac{(6.6255 \times 10^{-34}\,\text{J} \cdot \text{s photon}^{-1})\,(3.0 \times 10^{10}\,\text{cm s}^{-1})}{(6.6 \times 10^{-5}\,\text{cm})}$$

or

$$E = h\nu = (6.6255 \times 10^{-34}\,\text{J} \cdot \text{s photon}^{-1})\,(4.545 \times 10^{14}\,\text{s}^{-1})$$

$$E = 3.01 \times 10^{-19}\,\text{J photon}^{-1}$$

$$E\,\text{mol}^{-1} = (3.01 \times 10^{-19}\,\text{J photon}^{-1})$$

$$(6.02 \times 10^{23}\,\text{photon mol}^{-1})$$

$$= 181{,}000\,\text{J mol}^{-1}\ (43{,}000\,\text{cal mol}^{-1})$$

$$= 181\,\text{kJ mol}^{-1}$$

Blue light ($\lambda = 4.50 \times 10^{-5}$ cm, $\nu = 6.67 \times 10^{14}$ s^{-1})

has a frequency 6.67/4.545 times that of red, so its energy per photon is 1.467 times that of red = 4.42×10^{-19} J photon^{-1} or 266 kJ mol^{-1} (63,100 cal mol^{-1}).

The electromagnetic spectrum   The known distribution of electromagnetic energies arranged according to either wavelengths, frequencies, or photon energies (Fig. B-2). At one end of the spectrum is radiant energy of extremely short wavelengths and, consequently, extremely high frequencies and energetic photons. At this end of the spectrum are cosmic rays. (Primary cosmic rays are atomic nuclei, about 87 percent protons, and thus not photons. They have quantum energies, however, so they can be placed on the spectrum. Secondary cosmic rays do include highly energetic photons.) Slightly longer wavelengths (lower frequencies, less energetic photons) are gamma rays, which overlap broadly with the x-ray part of the spectrum. Ultraviolet radiation is expressed by wavelengths slightly shorter than those in the visible or *light* part of the spectrum, and infrared radiation has wavelengths longer than those of visible light. Radiowaves are longer still. The entire spectrum extends over at least 20 orders of magnitude with the visible portion being a part of only one order of magnitude.

Light   The visible portion of the spectrum. The term is sometimes incorrectly used to include the ultraviolet and infrared portions as well.

Color   The appearance of objects as determined by the response of the eye to the wavelengths of light coming from these objects. Short wavelengths produce the sensation we call violet or blue; the longest wavelengths produce the sensation of red. Colors of things are due to *pigments*.

Light sources   Since plant physiologists deal continually with responses of plants to light, it is important to know something about the possible sources of the light to which the plants are exposed; for example, in special growth chambers. Spectral distributions for several light sources are shown in Fig. B-3.

Incandescent light sources   Light sources such as the sun, the hot filament in an incandescent lamp, or the **plasma** (heated gas) of an arc lamp emit radiant energy in the visible spectrum because of their high temperatures. Large portions of the energy from these sources are in the infrared part of the spectrum. The hotter the incandescent source, the more the peak of the spectrum is shifted toward blue wavelengths (see Wien's Law, below). Spectra from incandescent sources are continuous rather than consisting of lines.

Fluorescent lamps   Lamps, often used in plant growth chambers, that produce light by fluorescence (see discussion below). The spectrum from such lamps consists of individual lines superimposed on a continuous spectrum. It is usually especially rich in blue but can be enriched in red wavelengths. Special fluorescent tubes have been developed for plant growth. They are especially rich in blue and red wavelengths (absorbed by chlorophyll), but experiments with these lamps give mixed results; often plants grow just as well under ordinary fluorescent lamps.

Sodium and mercury vapor lamps   Lamps in which an electric current passing through hot vapor causes light emission at specific wavelengths: orange for the sodium vapor and blue and green for the mercury vapor lamps. These lamps are called **high intensity discharge (HID) lamps.** In one highly efficient version, they contain metal salts of halide elements such as fluorine, chlorine, bromine, or iodine; these **metal halide lamps** produce a much broader spectrum than do the mercury and sodium vapor lamps, although the spectrum consists mostly of lines (is not continuous). By suitable combinations of the halides, an intensely white light can be produced.

In recent years there have been many studies in which plants are grown with the various lamps. There is considerable difference in response depending on species; but, amazingly, some species (for example, wheat) grow extremely well under HID lamps—even the low-pressure sodium-vapor lamps. The light from these lamps consists of a single wavelength at 589 nm, which in no way resembles the absorption spectrum of the leaf. Nevertheless, several species grow to maturity and produce seed and fruits when illuminated only with light from low-pressure sodium-vapor lamps (Cathy and Campbell, 1980).*

High-pressure sodium vapor lamps, like the low-pressure sodium lamps, are predominately orange but produce a considerably broader spectrum than do the low-pressure lamps with their single wavelength. In recent years, a number of plant-growth laboratories have experimented with different combinations of these lamps and have found that many species grow well under a mixture of high-pressure sodium and metal-halide lamps. These lamps are much more efficient than incandescent lamps and thus require much less power to operate at a given irradiance level, although they are expensive to install and to replace.

Irradiance   **Radiant flux** is the radiant energy falling upon a surface in an interval of time (e.g., J s^{-1}). **Irradiance** is the *radiant flux density* received on a unit surface (e.g., J s^{-1} m^{-2}). Since 1 W = 1 J s^{-1}, J s^{-1} m^{-2} = W m^{-2}. In the past, ergs m^{-2} s^{-1} have been used for low irradiances, but ergs are not SI units. Calories (cal = 4.18400 joules)[†] have also been used, although use of the calorie is now discouraged. Irradiance can also be expressed on a photon basis as mol m^{-2} s^{-1} (einsteins m^{-2} s^{-1} and μmol m^{-2} s^{-1} are commonly used). Irradiance,

---

*Cathy, Henry M. and Lowell E. Campbell. 1980. "Light and Lighting Systems for Horticultural Plants." Horticultural Reviews 2:491–537.

[†]The calorie defined by the U. S. National Bureau of Standards equals exactly 4.18400 joules. The calorie as defined in slightly different ways has slightly different values.

the correct photometric term, is often called *intensity* by plant physiologists. Use of the term **intensity** should be restricted to the light emission of the *source*. (E.g., we say that the sun or a lamp has such and such *intensity*.)

In the literature of plant physiology, irradiance has also been given in units of illumination rather than total energy or photons. Such units (e.g., 10.76 lux = 1.0 foot-candle, ft-c) are defined in terms of the sensitivity of the human eye. Plants, however, respond to the spectrum in ways quite different from the human eye, so such a measurement has no value unless exact information is given about the light source. Because plants respond to some wavelengths of light more than others (Chapters 9 & 19), even a measurement given in energy or photon units has little if any value when the spectral distribution is not also given or implied by describing the source.

Photosynthetically active radiation (PAR) Ecologists and others frequently give their measurements of irradiance in energy units for wavelengths from 400 to 700 nm, the wavelengths most active in photosynthesis. Appropriate SI units for PAR are watts meter^{-2} (W m^{-2}).

Photosynthetic photon flux density (PPFD) In a photochemical process such as photosynthesis, the end product depends upon the number of quanta absorbed rather than the total light energy absorbed. A single red photon has the same effect in photosynthesis as a single blue photon, for example, although the blue photon has more energy. Hence, in the recent literature it has become common to refer to the number of photons per unit area per unit time. Instruments that respond only to light between the wavelengths of 400 to 700 nm are often used; these may be suitably filtered and calibrated to read directly in moles (or einsteins) m^{-2} s^{-1}. During recent years and continuing to the present, not all plant physiologists have been aware of these conventions, so both W m^{-2} and mol m^{-2} s^{-1} have been referred to as PAR. PPFD always refers to moles of quanta m^{-2} s^{-1} in the wavelength region from 400 to 700 nm.

Pigment Any substance that absorbs light energy. If all the visible spectrum is absorbed, the substance appears black to the human eye; if all but the wavelengths in the green part of the spectrum are absorbed, the substance appears green.

Ground state and excited state (energy level) When atoms, ions, or molecules absorb radiant energy, they are raised to a higher **energy level.** Before absorbing any energy, they are said to be in the **ground state;** after, they are in an **excited state.** The actual change in energy content may be achieved by changes in the vibration or rotation of the atom, ion, or molecule or by changes in the electronic configuration of the atoms involved. Typically, an electron is moved a greater distance from the nucleus as energy is absorbed; it is said to be moved to a higher energy level. Due to the wave motions of the electrons orbiting the nuclei of atoms, there is not a continuous series of energy levels; rather, electrons can

exist only at certain discrete distances (energy levels) from the nucleus. To move an electron to the next higher level requires, then, a discrete amount of energy. A pigment absorbs only those photons that have exactly the required quantity of energy to bring about the change in electronic configuration. Actually there is a range of photon energies that can be absorbed in any given case because some translational, vibrational, or rotational energy changes can occur in addition to the electronic changes.

Fluorescence and phosphorescence An atom, ion, or molecule in an excited state may lose its excitation energy in any of three ways: First, it may be immediately lost as heat; that is, totally converted to translational, vibrational, or rotational energy. Second, it may be partially lost as heat with the remainder emitted as visible light of any wavelength longer (a photon of lower energy) than that absorbed. If this occurs within 10^{-9} to 10^{-5} s after absorption of the original photon, it is called **fluorescence.** If the delay is longer than that (10^{-4} to 10 s or more), it is called **phosphorescence.** Third, the energy may be used to cause a chemical reaction such as in photosynthesis.

Black body A surface that absorbs all the radiation falling upon it is a **black body.** Usually the term is used in reference to some portion of the spectrum under consideration. One may speak of a black body with reference to visible light and/or infrared radiation, for example. Carbon black or black velvet provide surfaces that approximate a black body. A still more perfect approach to true black body conditions would be a small opening in the surface of a large sphere lined with carbon black. Obviously, only a minute portion of the radiation entering this opening would ever leave through the opening.

Absorptivity coefficient This is a decimal fraction expressing the portion of impinging radiation that is absorbed. A leaf, for example, has an absorptivity coefficient of about 0.98 in the far-infrared portion of the spectrum.

Transmission and reflection Radiant energy that is not absorbed is either transmitted or reflected. Transmission or reflection are usually expressed as decimal fractions or as percentages.

Transmittance (T) or percent transmission This is the fraction of light transmitted by a substance. It is expressed as a decimal fraction, $T = I/I_o$, or a percentage ($I/I_o \times 100$), where I_o = irradiance of *incident* radiant energy and I = irradiance of *transmitted* radiant energy.

Absorbance (A) Formerly called *optical density.* The logarithm of the reciprocal of transmittance (T):

$$A = \log \frac{1}{T} = -\log T = \log \frac{I_o}{I}$$

Absorbance is often proportional to concentration of a

pigment in a transparent solution, according to the following laws:

Beer's law Each molecule of a dissolved pigment absorbs the same fraction of light incident upon it. Thus, in a nonabsorbing medium, the light absorbed should be proportional to dissolved pigment concentration. The law often holds for dilute solutions but fails as the light-absorbing properties of pigment molecules change at higher concentrations.

Lambert's law Each layer of equal thickness absorbs an equal fraction of the light that traverses it. This idea can be combined with Beer's law as the **Beer–Lambert law:** *The fraction of incident radiation absorbed is proportional to the number of absorbing molecules in its path.*

Extinction coefficient (ε) The Beer–Lambert law can be stated mathematically as follows:

$$A = \log \frac{I_o}{I} = \varepsilon \, cl$$

where ε = **extinction coefficient**
$\quad c$ = concentration of the pigment solute
$\quad l$ = length of the path of light (e.g., through a special quartz cell) in centimeters
$\quad$ (A, I_o, and I were defined earlier)

The extinction coefficient is a constant for a given pigment in dilute solution and can be determined by solving the equation just given:

$$\varepsilon = \frac{A}{cl}$$

If the concentration of solute is in moles liter^{-1} (molarity), ε is called the **molar extinction coefficient** (with units of liters mol^{-1} cm^{-1}). If concentration is known only in grams liter^{-1}, ε is the **specific extinction coefficient** (usually with the symbol a_s). The extinction coefficient is a characteristic of a given absorbing molecule in a given solvent with light of a specified wavelength. It is independent of concentration only when the Beer–Lambert law holds. The more intensely colored a pigment is at a given concentration, the larger its extinction coefficient.

Stefan–Boltzmann law All objects above the absolute zero emit radiant energy. The quantity (Q) emitted is a function of the fourth power of the absolute temperature (T) of the emitting surface, according to the Stefan–Boltzmann law:

$$Q = e\delta T^4$$

where Q = the quantity of energy radiated (in watts, using δ as below)
$\quad e$ = the emissivity (about 0.98 for leaves at growing temperatures)
$\quad \delta$ = the Stefan–Boltzmann constant (5.670×10^{-8} W m^{-2} K^{-4}, or 8.132×10^{-11} cal cm^{-2} min^{-1} K^{-4})
$\quad T$ = the absolute temperature in K (°C + 273)

The fourth power of absolute temperature in the expression means that the emission of radiant energy will increase greatly as temperature increases. Although the normal range of temperatures encountered by plants is narrow on the absolute temperature scale, the fourth power function means that energy radiated by bodies in this narrow range nevertheless varies considerably (Table B-1).

Net radiation The difference between the radiation absorbed by an object and that emitted is the net radiation. A leaf, for example, emits radiation according to the Stefan–Boltzmann law. At normal temperatures, most of this is in the far-infrared portion of the spectrum. Such emission leads to cooling. If the leaf is being illuminated by sunlight, however, it absorbs a portion of the sunlight (according to its absorptivity coefficient), which warms the leaf. Whether the leaf increases or decreases in temperature will depend upon whether more or less radiation is being absorbed or emitted—and also upon other mechanisms (convection, transpiration, and so on) that add heat to or remove heat from the leaf.

Wien's law Not only is the quantity of radiant energy emitted by an object a function of its temperature, but the quality is also influenced by temperature. With increasing temperature, the peak of emitted radiant energy (λ_{max}) shifts toward the shorter wavelengths. This peak multiplied by the absolute temperature (T) is equal to a constant, Wien's displacement constant (w): $\lambda_{max} T = w$. this is Wien's law, and it is illustrated for a wide range of temperatures in Fig. B-4. At room temperatures, objects emit maximally in the far-infrared part of the spectrum (λ_{max} = approximately 10 μm or 10,000 nm). At temperatures approximating those of an incandescent filament, the peak of emission is in the near-infrared (λ_{max} = about 1 μm); and at the temperature of the sun or other stars, the emission peak is in the visible part of the spectrum (λ_{max} = about 0.50 μm).

Color temperature As expressed by Wien's law, the emission peak of objects is a function of absolute temperature. Thus the spectral distribution of the light emitted by incandescent sources (such as an incandescent filament or the surface of a star) can be indicated in terms of absolute temperature. Epsilon Orionis and Sirius, with surface temperatures of 28,000 K and 13,600 K, are blue-white stars; the sun (5800 K) has an emission peak in the green-yellow part of the spectrum; and Betelgeuse (3600 K) is a red star. Color temperatures are widely used in photography; sensitivity of film must be balanced to the color temperature of the light source, for example.

Emissivity The curves shown in Fig. B-4 are for perfect black-body radiators. Actually, such an ideal is seldom achieved. In practice, just as objects fail to absorb some wavelengths, they also fail to emit some wavelengths; the curve for emission is the same as the curve for absorption. Thus, since a leaf has an absorptivity coefficient of about 0.98 in the far-infrared part of the spectrum, it also has an emissivity of about 0.98, with most of the radiant energy being emitted in that part of the spectrum. Incidentally, the atmosphere is far from a perfect black body, even in the far-infrared, as indicated by the solar spectrum in Fig. B-3. The "dips" in the solar spectrum represent absorption—and thus also emission—bands or lines caused by various atmospheric constituents, especially water and carbon dioxide. This must be taken into account in calculating the thermal radiation coming from the atmosphere and the thermal radiation absorbed by the atmosphere after being emitted by objects on the ground.

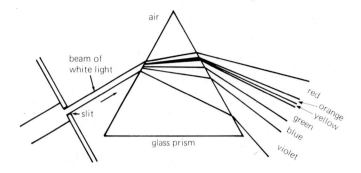

Figure B-1 White light is dispersed into its component colors by refraction when passed through a glass prism.

Table B-1 Radiation Emitted from Black-Body Surfaces at Various Temperatures

°C	K	T^4	$Q = J\,m^{-2}\,s^{-1}$	% of Q at 0°C
0	273	5.55×10^9	3.15×10^3	100
20	293	7.37×10^9	4.18×10^3	133
30	303	8.43×10^9	4.78×10^3	152
5477[a]	5750[a]	1.09×10^{15}	6.20×10^8	19,700,000

[a]Average surface temperature of the sun

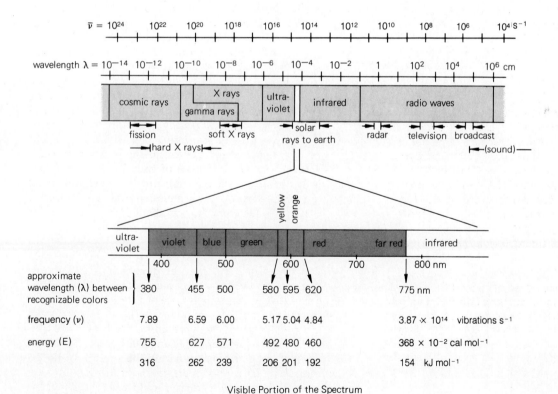

Visible Portion of the Spectrum

Figure B-2 The electromagnetic spectrum, using both wave number ($\bar{\nu}$) and wavelength (λ) in cm. Various portions of the spectrum are shown, and the visible portion is expanded to indicate the region that appears to the human eye to have various colors.

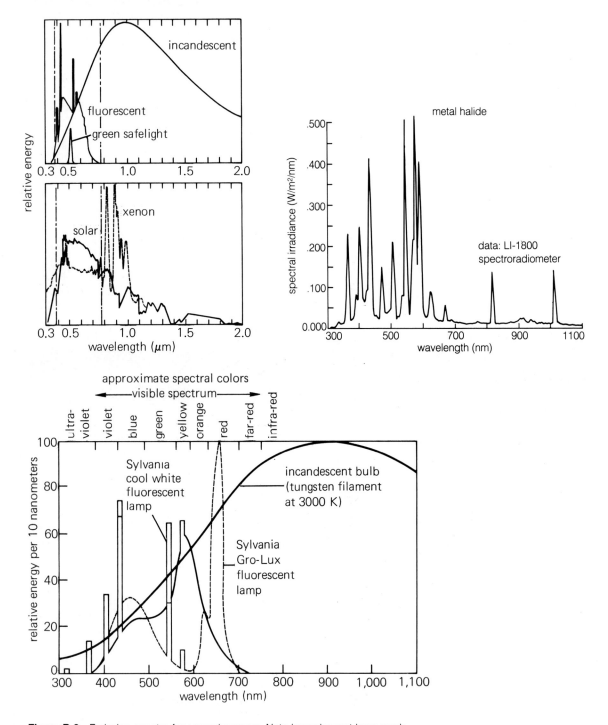

Figure B-3 Emission spectra for several sources. Note incandescent lamp peak at about 1.0 μm, mercury emission lines in fluorescent lamp spectra; infrared peaks (0.85 to 1.05 μm) from the xenon lamp; and solar peak in the midpart of the visible (indicated by dashed vertical lines).

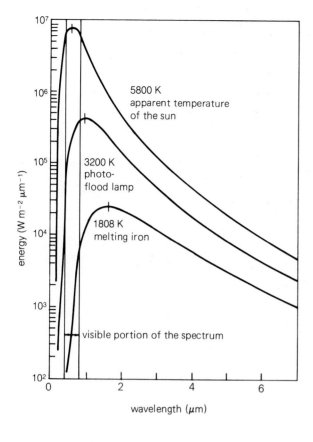

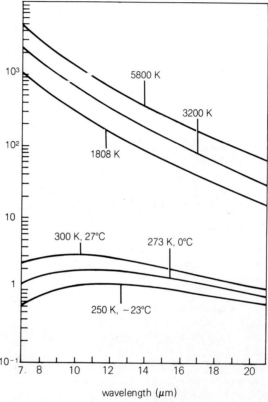

Figure B-4 Black body emission spectra compared over a wide range of energy emission and wavelength. The spectra would apply for any perfect black body radiator. Note shift in the peaks toward longer wavelengths (Wien's law), flattening of the curves, and decrease in total energy (Stefan–Boltzmann law) as temperatures decrease.

Appendix C

Gene Duplication and Protein Synthesis:
Terms and Concepts

Many times in this text, we have spoken of "transcription," "translation," and related processes, assuming that you, the reader, have some background in the concepts and terms of contemporary molecular biology. This field is discussed in textbooks of basic botany and biology, subjects you have surely studied by now. Nevertheless, this appendix is presented as a brief review and reference source to refresh your memory. Read it from beginning to end if you feel that you need a review, or scan the bold-faced terms if you want to use this appendix as a glossary.

A.1 The Central Dogma

The so-called **central dogma** of molecular biology is concerned with protein synthesis and the transfer of **information** (the **sequences** of nucleotides in nucleic acids and amino acids in proteins), first from cell generation to cell generation and then from **genes** (the carriers of heredity; actually, the carriers of sequence information) to the proteins, including the **enzymes** (which control metabolic activity and thus life). **Proteins** consist of **amino acids** arranged in specified sequences, and these sequences determine the biological activity of the proteins. The **nucleic acids,** including **DNA (deoxyribonucleic acid)** and **RNA (ribonucleic acid),** are chains of **nucleotides,** each of which consists of a nitrogen base (related to the alkaloids, page 286, and the cytokinins, pages 330–331) attached to a five-carbon sugar in the ring form (**ribose** or **deoxyribose),** which in turn is attached to a phosphate group. (A nucleotide, adenosine monophosphate, AMP, is shown as part of the NADP$^+$ molecule on page 189.) DNA occurs in the nucleus (and in mitochondria and chromosomes) and is the genetic material. One kind of RNA is formed in the nucleus, copying the sequence of nucleotides from DNA. This messenger-RNA then moves out of the nucleus into the cytoplasm, where other proteins and RNA molecules, partially in ribosomes, translate the nucleotide sequence into an amino acid sequence. This translation is protein synthesis.

A.2 The Double Helix

In DNA, there is always as many molecules of **adenine** (a **purine** nucleotide) as there are molecules of **thymine** (a **pyrimidine;** sometimes replaced by **5-methylcytosine**), and there are as many molecules of **guanine** (purine), as there are **cytosine** (pyrimidine). The nucleotides are connected (through their phosphate and deoxyribose molecules) to form long chains, and each DNA molecule consists of two such chains that are

wound together like two parallel banisters of a spiral staircase to form a **double helix.** The structures of the nucleotides are such that an adenine in one chain is always paired with a thymine in the opposite chain, and a guanine is always paired with a cytosine. That is, a purine is always paired with a specific pyrimidine in the opposite chain. This arrangement of paired nucleotides is called **complementary bonding.**

This structure of paired purines and pyrimidines in opposite helical (spiral) chains of DNA immediately suggests how information (the sequence of nucleotides in the chains) can be transmitted from one cellular generation to another. In the presence of suitable enzymes (called **DNA polymerases**), the hydrogen bonds holding the two chains of the helix together are broken so that the chains can separate from each other (probably only a small portion at a time), and free nucleotides (each with three phosphates as in ATP) can pair with their complementary bases (nucleotides) in each of the two chains. The free nucleotides are then attached to each other to form complementary chains with the exact nucleotide sequence that existed in the chains that they are replacing. Thus two double helices identical to the original are formed. As the nucleotides attach to each other to form the new chains, two phosphate groups are released as pyrophosphates, and these are immediately broken down to form phosphate, making the process irreversible.

A.3 Transcription: Copying DNA to Make RNA

Transcription is the copying process in which the sequence of nucleotides in DNA specifies a complementary sequence of nucleotides in an RNA molecule. This process also requires specific enzymes. Only one of the two chains in a DNA molecule carries the coded directions for synthesizing a protein; it is sometimes called the **sense chain.** In transcription, the two DNA chains unwind and separate, and the sense chain serves as a **template** for RNA synthesis. The RNA chain is assembled on the DNA template as free nucleotide bases link to it according to the rules for complementary pairing. The nucleotides in RNA are slightly different from those in DNA: Uracil (pyrimidine) occurs in RNA instead of thymine or 5-methylcytosine. When the triple-phosphorylated bases are paired with their complementary bases in the DNA sense chain, **RNA polymerase** links them together, again releasing pyrophosphate, which is broken down to form phosphate. The RNA that is synthesized in this transcription process is called **messenger RNA (mRNA),** since it carries the message of nucleotide sequence from the nucleus where transcription occurs to the cytoplasm, where protein synthesis occurs.

The segment of DNA that is transcribed is a **gene.** The DNA of a **chromosome** is continuous (each chromosome contains thousands of connected genes), so there may be specific initiator and terminator nucleotide sequences to mark the beginning and end of a gene. These sequences are called **punctuation marks.** Or only a single gene segment may unwind and become available for transcription at any one time. It is now known that DNA often contains segments of a single nucleotide pair repeated over and over again, but cytoplasmic mRNA does not contain comparable sequences, so it appears that the mRNA has been **edited** or **processed** from the time it is transcribed until it reaches the cytoplasm.

About 5 percent of the RNA made in the nucleus is mRNA, but about 10 to 15 percent is transfer RNA (see below), and 70 to 80 percent is **ribosomal RNA.** Each **ribosome** is made of two subunits, each with its characteristic RNAs and proteins. (In bacteria, one subunit has one large RNA of about 1500 nucleotides plus about 20 different proteins; the other subunit has one RNA with about 100 nucleotides plus about 35 proteins. Eukaryotic ribosomes are similar but larger.) Ribosomal precursors are synthesized in the **nucleolus** and assembled in the cytoplasm. At least some of the RNA is transcribed from **chromatin** (nucleoprotein that forms chromosomes) in a portion of the nucleolus called the **nucleolar organizer.**

A.4 Translation: Protein Synthesis in the Cytoplasm

In protein synthesis, the nucleotide sequence consisting of four kinds of nucleotides in mRNA undergoes a **translation** to the amino acid sequence of protein (20 kinds of amino acids). To begin with, free amino acids **(AA)** enter into a two-step process called **amino acid activation.** In the first step, an amino acid interacts with ATP in an enzyme-catalyzed reaction that removes pyrophosphate (immediately hydrolyzed to phosphate) from the ATP and attaches the amino acid to AMP. The AA-AMP complex remains attached to the enzyme that catalyzed the reaction. In the second step, the amino acid is transferred from AMP to an appropriate molecule of **transfer RNA (tRNA),** forming an **AA-tRNA** complex that is released from the enzyme that formed it. Much of the original energy of the ATP is transferred to the AA-tRNA complex; thus it is an "activation" of the amino acid. There is a specific enzyme that activates each amino acid, and there is also a specific tRNA that becomes attached to each amino acid.

A.5 The Genetic Code

Part of each tRNA molecule consists of three nucleotides that form an **anticodon.** This sequence is positioned in the tRNA molecule so that it can form complementary bonds with a three-nucleotide sequence, called a **codon,** in the mRNA strand. This makes it possible for the tRNA molecules, each carrying its specific amino acid, to line up along the mRNA strand so that the amino acids will be in the proper sequence as specified by the sequence of nucleotides, first in DNA, and then in tRNA.

Although only 20 amino acids are coded for and take part in protein synthesis, at least 60 tRNAs occur in the cytoplasm of eukaryotes. Thus a given amino acid can be carried to the mRNA at the ribosomes (where protein synthesis takes place) by more than one kind of tRNA. Note that the actual act of translation is the linking of specific amino acids to specific tRNA molecules. Translation is accomplished by each specific enzyme that recognizes a given amino acid and its appropriate tRNA, which has the anticodon sequence of nucleotides that is appropriate for that amino acid.

The sequences of nucleotides in the codons and anticodons form the **genetic code.** There are 64 (4 × 4 × 4) possible three-nucleotide sequences based on the four kinds of nucleotides in nucleic acids, so there are 64 possible codons or anticodons. But there are only 20 amino acids that enter protein at the ribosomes. Hence the genetic code is **redundant,** meaning that more than one codon codes for a given amino acid. The genetic code has been broken, meaning that we know which amino acids are specified by which codons (see basic textbooks for tables of the genetic code).

A.6 The Steps of Protein Synthesis

Having established the ground work, let us consider the actual steps in protein synthesis. The mRNA is transcribed in the nucleus, copying the sequence of nucleotides from DNA. We can think of the nucleotide sequences in both DNA and mRNA as consisting of three-nucleotide sequences called codons, although there is no break between codons in DNA or mRNA; the codons are determined by beginning at the same end and *always* reading three nucleotides at a time.

After transcription in the nucleus, the mRNA enters the cytoplasm and becomes attached to a ribosome. In the surrounding medium, there are tRNA molecules, each kind with a specific amino acid attached to it. Among these, at least one (and probably three or four, all attached to the same kind of amino acid) will have an anticodon that **recognizes** (has complementary nucleotides) the first codon in the mRNA chain of nucleotides (counting from the proper end of the chain). There will also be tRNAs with anticodons to recognize the next codon in the RNA chain, and so on. The two amino acids attached to the first two tRNA molecules to bond with the mRNA chain will be brought into close proximity, and a peptide bond will form between them, catalyzed by an enzyme that is part of the ribosome itself. Formation of the bond involves separating the amino acid from the first tRNA and transferring it to the amino acid carried by the second tRNA molecule to become attached to the mRNA strand at the ribosome. The dipeptide formed in this way is then attached (transferred) through a peptide bond to the amino acid attached to the third tRNA molecule, and so on, until a chain of amino acids has been formed. When it is large enough (some arbitrary size), this peptide chain can be called a protein.

Thus proteins form as the ribosome moves along the mRNA chain, exposing codons along the chain that can be recognized by the anticodons of tRNA molecules with their amino acids. Peptide bonds form between amino acids that are brought together in this way. The energy to form the peptide bonds comes from the energy that is released when each amino acid is separated from its tRNA. Note that a single mRNA can interact with several ribosomes at a time, forming a **polyribosome (polysome).** In addition to the components of protein synthesis that we have been discussing, the process also requires the presence of enzymes that allow protein synthesis to begin, to continue, and to terminate with release of the finished polypeptide from the ribosome.

Special codons are involved in the initiation and termination of protein synthesis. In bacteria, one of the two codons AUG (adenine-uracil-guanine) or GUG always occurs as the first word in the coded message for protein synthesis. These codons have been called **initiator codons.** Unless one of them is present as the first word in the message, protein synthesis cannot proceed. Since they code for methionine and valine, these amino acids often initiate polypeptide chains in bacteria—but sometimes they are removed after the chain forms (edited). Likewise, one of the three **terminator codons** (UAA, UAG, or UGA) must be the last word in the message or protein synthesis cannot be successfully completed. These terminator codons specify no amino acids and evidently have no corresponding tRNA molecules with complementary anticodon sites. As the ribosome reaches one of these terminator codons, no AA-tRNA binds to the mRNA. In some manner the terminator induces release of the completed polypeptide from the ribosome.

The steps in protein synthesis have been worked out almost entirely with cell-free systems isolated from bacteria, especially *Escherichia coli.* Although it is not known how completely some of the details of the mechanism apply to eukaryotic cells, the evidence is good that the basic process outlined here takes place in much the same way in higher organisms. The genetic code, for example, is known to be the same in many eukaryotic organisms.

A.7 The Regulation of Protein Synthesis

Why are certain genes being transcribed and their proteins being synthesized in any given cell at any given time? We don't know the answer yet, but we do know much about what regulates protein synthesis in certain cells. In bacteria, the **operon hypothesis** was developed to help explain the regulation of gene activity. Much evidence now supports this model, but we will not discuss it here (see basic texts).

In eukaryotic organisms, regulation is thought to occur at the level of DNA transcription and is believed to involve the two major groups of proteins that are associated with the DNA in chromatin: the histones and the nonhistones. The **histones** are positively charged and thus highly basic, opposite to most proteins, which are negatively charged. Thus histones easily attach to the negative (acidic) DNA molecules. There are at least five kinds of histone. One has 102 amino acids that are virtually identical in sequence from one end of the evolutionary scale to the other (bacteria to cattle to green plants!). They are also remarkably constant in proportional abundance to DNA from one cell type to another. All this implies that the histones must play some highly essential role in gene regulation or gene function.

In general, histones seem to repress transcription. In other words, they turn genes off. But they are too similar, both in chemical character and amount, in different tissues to account for the different proteins formed by different tissues. What, then, turns genes on in various parts of an organism? The answer appears to lie in the **nonhistone proteins.** These are a much more diverse group of proteins, presumably with a wide range of functions. Moreover, they have the important characteristic, for a regulator candidate, of being different in different cell types. Not all the data are in, but the role of the nonhistone proteins seems to be to interact with the histones in a way that allows specific parts of the DNA to transcribe mRNA. But where do they come from, and why are there different ones in different tissues? The intense research now going on in this area could soon bring answers to such questions.

Literature Cited

Prologue

Albersheim, Peter. 1975. The walls of growing plant cells. Scientific American 232(4):81–95.

Avers, Charlotte J. 1981. Cell Biology, Second Edition. Van Nostrand/Reinhold, New York.

Avers, Charlotte J. 1982. Basic Cell Biology, Second Edition. PWS Pubs. (Willard Grant Press), Boston, Massachusetts.

Berns, M. W. 1984. Cells, Second Edition. Modern Biology. Saunders, New York.

Bracegirdle, Brian and Miles, Patricia H. 1973. An Atlas of Plant Structure Vol. 1 & 2. Heinemann Educational Books, London.

Capaldi, Roderick A. 1974. A Dynamic Model of Cell Membranes. Scientific American 230(3): 26–33.

Carpita, Nicholas C. 1982. Limiting diameters of pores and the surface structure of plant cell walls. Science 218:813–814.

Carpita, Nicolas, Dario Sabularse, David Montezinos, and Deborah P. Delmer. 1979. Determination of pore size of cell walls of living plant cells. Science 205:1144–1147.

Clarkson, David T. ed. 1973. Ion Transport and Cell Structure in Plants. John Wiley & Sons, New York.

Clegg, James S. 1983. What is the cytosol? Trends in Biochemical Sciences 8:436–437.

de Duve, Christian. 1983. Microbodies in the living cell. Scientific American 248(5):74–84.

Dustin, Pierre. 1980. Microtubules. Scientific American. 243(2):66–76.

Fawcett, Don W. 1982. The Cell, Second Edition. W. B. Saunders, Philadelphia, Pennsylvania.

Grivell, Leslie A. 1983. Mitochondrial DNA. Scientific American 248(3):78–89.

Gunning, B. E. S. and R. L. Overall. 1982. Plasmodesmata and cell-to-cell transport in plants. BioScience 33:260–265.

Gunning, B. E. S. and R. L. Overall. 1983. Plasmodesmata and cell-to-cell transport in plants. BioScience 33(4):260–265.

Hall, J. L. and T. J. Flowers. 1982. Plant Cell Structure and Metabolism. Second Edition. Longman Group Limited, London.

Holtzman, E. and Novikoff, A. B. 1984. Cells and Organelles. Third Edition. Saunders, P. Philadelphia, Pennsylvania.

Jensen, William A., ed. 1970. The Plant Cell. Wadsworth Publishing Co., Inc., Belmont, California.

Lazarides, E. and Jean Paul Revel. 1979. The Molecular Basis of Cell Movement. Scientific American 240(5):100–113.

Ledbetter, Myron C. 1965. Fine structure of the cytoplasm in relation to the plant cell wall. Journal of Agriculture & Food Chemistry 13:405–407.

Ledbetter, Myron C. and Keith R. Porter. 1970. Introduction to the Fine Structure of Plant Cells. Springer-Verlag, New York.

Mazia, Daniel. 1974. The Cell Cycle. Scientific American 230(1):54–64.

Moor, H. 1959. Platin-Kohle-Abdruck-Technik angewandt auf den Feinbau der Milchröhren. (Platinum-carbon-impression technique applied to the fine structure of lacticifers.) J. Ultrastructure Research 2:393–422.

Morre, James D. and Daniel D. Jones. 1967. Golgi apparatus mediated polysaccharide secretion by outer root cap cells of Zea mays. 1. Kinetics and secretory pathway. Plants (Berl.) 74:286–301.

Palevitz, B. A. 1982. The Stomatal Complex as a Model of Cytoskeletal Participation in Cell Differentiation. Pages 345–376 in The Cytoskeleton in Plant Growth and Development, Clive W. Lloyd (ed.), Academic Press, Inc., New York.

Porter, Keith R. and Jonathan B. Tucker. 1981. The ground substance of the living cell. Scientific American 244(3):56–67.

Powell, Andrew W., Geoffrey W. Peace, Antoni R. Slabas, and Clive W. Lloyd. 1982. The detergent-resistant cytoskeleton of higher plant protoplasts contains nucleus-associated fibrillar bundles in addition to microtubules. Journal of Cell Science 56:319–335.

Robards, A. W., ed. 1974. Dynamic Aspects of Plant Ultrastructure. McGraw-Hill, London, New York, Toronto.

Roberts, Lorin W. 1976. Cytodifferentiation in Plants. Cambridge University Press, Cambridge, New York, London.

Satir, B. H., ed. 1983. Modern Cell Biology, Vol. 1. Alan R. Liss, Inc., New York.

Schopf, J. William. 1978. The Evolution of the Earliest Cells. Scientific American 239(3):111–138.

Schwartz, Lazar M. and Miguel M. Azar (eds.). 1981. Advanced Cell Biology. Van Nostrand Reinhold, NY.

Shulman, R. G. 1983. NMR Spectroscopy of Living Cells. Scientific American 248(1):89–93.

Smith, H. 1977. The Molecular Biology of Plant Cells. Botanical Monographs Vol. 14. University of California Press, Berkeley and Los Angeles.

Troughton, John and Lesley A. Donaldson. 1972. Probing Plant Structure. A. H. & A. W. Reed Ltd., Auckland, New Zealand.

Troughton, J. H. and F. B. Sampson. 1973. Plants. A scanning electron microscope survey. John Wiley & Sons Australasia Pty Ltd., New York.

Valentine, James W. 1978. The evolution of multicellular plants and animals. Scientific American 239(3):141–160.

Wiebe, Herman H. 1978. The significance of plant vacuoles. BioScience 28:327–331.

Wiebe, Herman H. 1978. What is a plant? The significance of vacuoles and cell walls. Pages 289–392 in Frank B. Salisbury and Cleon W. Ross, Plant Physiology, Second Edition. Wadsworth Publishing Co., Belmont, California.

Woese, Carl R. 1981. Archaebacteria. Scientific American 244(6):98–122.

Wolfe, S. L. 1983. Introduction to Cell Biology. Wadsworth Pub. Co., Belmont, California.

Chapter 1

Campbell, Gaylon S. 1977. An Introduction to Environmental Biophysics. Springer-Verlag, New York, Berlin, Heidelberg.

Chang, Raymond. 1981. Physical Chemistry with Application to Biological Systems, Second Edition. Macmillan Pub. Co., Inc., New York.

Eisenberg, David and Donald M. Crothers. 1979. Physical Chemistry with Application to the Life Sciences. Benjamin/Cummings, Menlo Park, California.

Fast, J. D. 1962. Entropy. McGraw-Hill, New York, Toronto, London.

Fay, James A. 1965. Molecular Thermodynamics. Addison-Wesley, Massachusetts.

Hatsopoulos, George N. and Joseph H. Keenon. 1965. Principles of General Thermodynamics. John Wiley & Sons, New York.

Klein, J. J. 1970. Maxwell, his demon and the second law of thermodynamics. American Scientist 58(1):89.

Kramer, Paul J. 1983. Water Relations of Plants. Academic Press, Santa Clara, California.

Ling, G. N. 1967. Effects of temperature on the state of water in the living cell. Pages 5–14 in Thermobiology. A. H. Rose (ed.), Academic Press, Inc., New York.

Meyer, B. S. 1938. The water relations of plant cells. Bot. Rev. 4:531–547.

Milburn, John A. 1979. Water Flow in Plants. Longman Group Limited, London. (Published in the U.S. by Longman, Inc., New York.)

Moore, Walter J. 1972. Physical Chemistry, Fourth Edition. Prentice Hall, Inc., Englewood Cliffs, New Jersey.

Morowitz, H. J. 1970. Entropy for Biologists. Academic Press, New York.

Nobel, Park S. 1983. Biophysical Plant Physiology and Ecology. W. H. Freeman and Co., San Francisco.

Noy-Meir, I. and B. Z. Ginzburg. 1967. An analysis of the water potential isotherm in plant tissue. I. The theory. Aust. J. Biol. Sci. 20:695–721.

Passioura, J. B. 1982. Water in the soil-plant-atmosphere continuum. Pages 5–33 in Encyclopedia of Plant Physiology New Series Vol. 12B, Physiological Plant Ecology. II. O. L. Lange, P. S. Nobel, C. B. Osmond, and H. Ziegler (eds.), Springer-Verlag, New York, Berlin, Heidelberg.

Renner, O. 1915. Theoretisches und Experimentelles zur Kohäsions Theorie der Wasserbewegung. (Theoretical and experimental contributions to the cohesion theory of water movement.) Jb. wiss. Bot. 56:617–667.

Slatyer, Ralph O. 1967. Plant-Water Relationships. Academic Press, London and New York.

Slatyer, R. O. and Taylor, S. A. 1960. Terminology in plant-soil-water relations. Nature 187:922–924.

Spanner, D. C. 1964. Introduction to thermodynamics. Academic Press, London, New York. (Written by a plant physiologist.)

Taylor, S. A. and R. O. Slatyer. 1962. Proposals for a unified terminology in studies of plant-soil-water relations. Arid Zone Res. 16:339–349.

Tyree, M. T. and P. G. Jarvis. 1982. Water in tissues and cells. Pages 35–77 in O. L. Lange, P. S. Nobel, C. B. Osmond, and H. Ziegler (eds.), Enc. of Plant Physiol. New Series Vol. 12B; Physiological plant Ecology. Springer-Verlag, New York, Berlin, Heidelberg.

Van Wylen, Gordon J. and Richard E. Sonntag. 1976. Fundamentals of Classical Thermodynamics, Second Edition (revised print). John Wiley & Sons, Inc., New York.

Chapter 2

Campbell, Gaylon S. and G. S. Harris. 1975. Effect of soil water potential on soil moisture absorption, transpiration rate, plant water potential and growth for Artemisia tridentata. US/IBP Desert Biome Res. Mem. 75-44. Utah State University, Logan, Utah.

Carpita, Nicholas C. 1982. Limiting diameters of pores and the surface structure of plant cell walls. Science 218:813–814.

Chardakov, V. S. 1948. New field method for the determination of the suction pressure of plants. Dokl. Akad. Nauk SSSR 60:169–172.

Cline, Richard G., and Gaylen S. Campbell. 1976. Seasonal and diurnal water relations of selected forest species. Ecology 57:367–373.

Grant, R. F., M. J. Savage, and J. D. Lea. 1981. Comparison of hydraulic press, thermocouple psychrometer, and pressure chamber for the measurement of total and osmotic leaf water potential in soybeans. South African Journal of Science 77:398–400.

Green, P. B. and R. W. Stanton. 1967. Turgor pressure: Direct manometric measurement in single cells of Nitella. Science 155:1675–1676.

Harris, J. A. 1934. The physico-chemical properties of plant saps in relation to phytogeography. Data on native vegetation in its natural environment. University of Minnesota Press, Minneapolis.

Hellkvist, J., G. P. Richards, and P. G. Jarvis. 1974. Vertical gradients of water potential and tissue water relations in sitka spruce trees measured with the pressure chamber. Journal of Applied Ecology 11:637–667.

Höffler, K. 1920. Ein Schema fur die osmotische Leistung der Pflanzenzelle (A diagram for the osmotic performance of plant cells). Berichte der Deutschen Botanischen Gesellschaft 38:288–298.

Hüsken, Dieter, Ernst Steudle, and Ulrich Zimmermann. 1978. Pressure probe technique for measuring water relations of cells in higher plants. Plant Physiology 61:158–163.

Jensen, William A. and Frank B. Salisbury. 1984. Botany, 2nd Edition. Wadsworth Pub. Co., Belmont, California.

Knipling, E. B. and P. J. Kramer. 1967. Comparison of the dye method with the thermocouple psychrometer for measuring leaf water potentials. Plant Physiology 42:1315–1320.

Kramer, Paul J. 1983. Water Relations of Plants. Academic Press, Santa Clara, California.

Kramer, Paul J., Edward B. Knipling, and Lee N. Miller. 1966. Terminology of cell-water relations. Science 153:889–890.

Lockhart, James A. 1959. A new method for the determination of osmotic pressure. American Journal of Botany 46:704–708.

Markhart, Albert H. III., Nasser Sionit, and James N. Siedow. 1980. Cell wall dilution: an explanation of apparent negative turgor potentials. Canadian Journal of Botany 59:1722–1725.

McClendon, John H. 1982. Water relations curves for plant cells: toward a realistic Höfler diagram for textbooks. What's New in Plant Physiology 13(5):17–20.

Meyer, B. S., and D. B. Anderson. 1952. Plant Physiology, Second Edition, D. Van Nostrand Co., Inc., Princeton, New Jersey.

Millar, B. D. 1982. Accuracy and usefulness of phychrometer and pressure chamber for evaluating water potentials of Pinus radiata needles. Aust. Jour. Plant Physiol. 9:499–507.

Passioura, J. B. 1980. The meaning of matric potential. Journal of Experimental Botany 31(123): 1161–1169.

Passioura, J. B. 1982. Water in the soil-plant-atmosphere continuum. Pages 5–33 in Encyclopedia of Plant Physiology New Series Vol. 12B, Physiological Plant Ecology. II. O. L. Lange, P. S. Nobel, C. B. Osmond, and H. Ziegler (eds.), Springer-Verlag, New York, Berlin, Heidelberg.

Pfeffer, Wilhelm Friedrich Phillip. 1877. Osmotische Untersuchungen. Studien Zur Zell Mechanik. (Investigations of osmosis. Studies of Cell Mechanics.) W. Engelmann Pub., Leipzig.

Pisek, Arthur. 1956. Der Wasserhaushalt der Meso und Hygrophyten (The water relations of meso and hydrophytes). Encyclopedia of Plant Physiology 3:825–853.

Ray, P. M. 1960. On the theory of osmotic water movement. Plant Physiology 35:783–795.

Renner, O. 1915. Theoretisches und Experimentelles zur Kohäsions Theorie der Wasserbewegung.
(Theoretical and experimental contributions to the cohesion theory of water movement.) Jb. wiss. Bot. 56:617–667.

Ross, Cleon W. and David L. Rayle. 1982. Evaluation of H[+] secretion relative to Zeatin-induced growth of detached cucumber cotyledons. Plant Physiology 70:1470–1474.

Scholander, P. F., H. T. Hammel, E. D. Bradstreet, and E. A. Hemmingsen. 1965. Sap pressure in vascular plants. Science 148:339–346.

Slatyer, R. O. and S. A. Taylor. 1960. Terminology in plant-soil-water relations. Nature. 187:922–924.

Spanner, D. C. 1951. The Peltier effect and its use in the measurement of suction pressure. Journal of Experimental Botany 11:145–168.

Taylor, S. A. and Slatyer, R. O. 1962. Proposals for a unified terminology in studies of plant-soil-water relations. Arid Zone Res. 16:339–349.

Tyree, M. T. and J. Dainty. 1973. The water relations of hemlock (Tsuga canadensis). II. The kinetics of water exchange between the symplast and apoplast. Canadian Journal of Botany 51:1481–1489.

Tyree, M. T. and P. G. Jarvis. 1982. Water in Tissues and Cells. Pages 35–78 in Enc. Plant Phys. 12B, Physiological Plant Ecology II. Lange, O. L., P. S. Nobel, C. B. Osmond and H. Zeigler (eds.), Springer-Verlag, Berlin, Heidelberg, New York.

Tyree, M. T. and H. T. Hammel. 1972. The measurement of the turgor pressure and the water relations of plants by the pressure-bomb technique. Journal of Experimental Botany 23(74):267–282.

Waring, R. H. and B. D. Cleary. 1967. Plant moisture stress: Evaluation by pressure bomb. Science 155:1248–1254.

West, D. W. and D. F. Gaff. 1971. An error in the calibration of xylem-water water potential against leaf-water potential. Journal of Experimental Botany 22(71):342–346.

Wiebe, Herman H. 1966. Matric potential of several plant tissues and biocolloids. Plant Physiology 41:1439–1442.

Chapter 3

Ackerson, Robert C. 1982. Synthesis and movement of abscisic acid in water-stressed cotton leaves. Plant Physiology 69:609–613.

Aylor, Donald E., Jean-Yves Parlange, and A. D. Krikorian. 1973. Stomatal mechanics. American Journal of Botany 60:163–171.

Bengtson, Curt, Brigitta Klockare, Rune Klockare, Stig Larsson, and Christer Sundqvist. 1978. The after-effect of water stress on chlorophyll formation during greening and the levels of abscisic acid and proline in dark wheat seedlings. Physiologie Plantarum 43:205–212.

Cowan, I. R., J. A. Raven, W. Hartung, and G. D. Farquhar. 1982. A possible role for abscisic acid in coupling stomatal conductance and photosynthetic carbon metabolism in leaves. Aust. J. Plant Physiol. 9:489–498.

Dacey, John W. H. 1980. Internal wind in water lilies: an adaptation for life in anaerobic sediments. Science 210:1017–1019.

Dacey, John W. H. 1981. Pressurized ventilation in the yellow water lily. Ecology 62:1137–1147.

Dacey, John W. H. and Michael J. Klug. 1982. Tracer studies of gas circulation in Nuphar: [18]O and [14]CO_2 transport. Physiologia Plantarum 56:361–366.

Dacey, John W. H. and Michael J. Klug. 1982. Ventilation by floating leaves in Nuphar. American Journal of Botany 69:999–1003.

Drake, B. G., K. Raschke, and Frank B. Salisbury. 1970. Temperatures and transpiration resistances of Xanthium leaves as affected by air temperatures, humidity, and wind speed. Plant Physiology 46:324–330.

Drake, B. G. and Frank B. Salisbury. 1972. After effects of low and high temperature pretreatment on leaf resistance, transpiration, and leaf temperature in Xanthium. Plant Physiology 50:572–575.

Edwards, Mary and Hans Meidner. 1979. Direct measurements of turgor pressure potentials, IV. Naturally occurring pressures in guard cells and their relation to solute and matric potentials in the epidermis. Journal of Experimental Botany 30(17):829–837.

Gates, David M. 1968. Transpiration and leaf temperature. Annual Review of Plant Physiology 19:211–238.

Gates, David M. 1971. The flow of energy in the biosphere. Scientific American 225(3):88–100.

Gates, David M., Harry J. Keegan, John C. Schleter, Victor R. Weidner. 1965. Spectral properties of plants. Applied Optics 4:11–20.

Hanks, R. J., ed. 1982. Predicting Crop Production as Related to Drought Stress under Irrigation. Utah Agriculture Experiment Station Research Report 65:367.

Hanks, R. J. 1983. Yield and water-use relationships: an overview. Pages 393–411 in Limitations to Efficient Water Use in Crop Production (Chapter 9A). ASA-CSSA-SSSA, 677 South Segoe Road, Madison, Wisconsin.

Henson, I. E. 1981. Changes in abscisic acid content during stomatal closure in pearl millet (Pennisetum americanum (L.) Leeke). Plant Science Letters 21:121–127.

Humble, G. D. and K. Raschke. 1971. Stomatal opening quantitatively related to potassium transport: Evidence from electron probe analysis. Plant Physiology 48:447–453.

Janac, J., J. Catsky, and P. G. Jarvis. 1971. Infra-red gas analyzers and other physical analyzers. Pages 111–193 in Photosynthetic Plant Production: Manual of Methods, Sestak, Z., J. Catsky, and P. G. Jarvis (eds.), Dr. W. Junk (publ.), The Hague.

Jensen, William A. and Frank B. Salisbury. 1984. Botany, Second Edition. Wadsworth Publishing Co., Belmont, California.

Lancaster, J. E. and J. D. Mann. 1977. Extrastomatal control of transpiration in leaves of yellow lupin (Lupinus luteus var. Weiko III) after drought or abscisic acid treatment. Journal of Experimental Botany 28:1373–1379.

Long, S. P. 1982. Measurement of photosynthetic gas exchange. Pages 25–34 in Techniques in Bio-productivity and Photosynthesis, Coombs, J., and D. O. Hall (eds.), Pergamon Press, Oxford.

Nobel, Park S. 1974. Biophysical Plant Physiology. W. H. Freeman and Co., San Francisco, California.

Nobel, Park S. 1980. Leaf anatomy and water use efficiency. Pages 43–55 in Neil C. Turner and Paul J. Kramer (eds.), Adaptation of Plants to Water and High Temperature Stress. John Wiley & Sons, New York, Chichester, Brisbane, Toronto.

Nobel, Park S. 1983. Biophysical Plant Physiology and Ecology. W. H. Freeman and Co., San Francisco, California.

Outlaw, William H., Jr. 1982. Carbon metabolism in guard cells. In Leroy L. Creasy and Geza Hrazdina (eds.), Cellular and Subcellular Localization in Plant Metabolism. Plenum Pub. Corp., New York.

Outlaw, William H., Jr. 1983. Current concepts on the role of potassium in stomatal movements. Physiologia Plantarum 59:302–311.

Outlaw, William H., Jr., J. Manchester and V. E. Zenger. 1982. Potassium involvement not demonstrated in stomatal movements of Paphiopedilum. Qualified confirmation of the Nelson-Mayo report. Canadian Journal of Botany 60: 240–244.

Permadasa, M. A. 1981. Photocontrol of stomatal movements. Biology Review 56:551–588.

Radin, John W. and Robert C. Ackerson. 1982. Does abscisic acid control stomatal closure during water stress? What's New in Plant Physiology 13(3):9–12.

Raschke, K. 1976. How stomata resolve the dilemma of opposing priorities. Phil. Trans. R. Soc. Lond. B. 273:551–560.

Raschke, K. and G. D. Humble. 1973. No uptake of anions required by opening stomata of Vicia faba: Guard cells release hydrogen ions. Planta (Berlin) 115:47–57.

Raschke, K. and Jan A. D. Zeevaart. 1976. Abscisic acid content, transpiration, and stomatal conductance as related to leaf age in plants of Xanthium strumarium L. Plant Physiology 58:169–174.

Salisbury, Frank B. 1979. Temperature. Pages 75–116 in T. W. Tibbitts, and T. T. Kozlowski (eds.), Controlled Environment Guidelines for Plant Research. Academic Press, New York.

Salisbury, Frank B. and George G. Spomer. 1964. Leaf temperatures of alpine plants in the field. Planta 60:497–505.

Sinclair, T. R., C. B. Tanner and J. M. Bennett. 1984. Water-use efficiency in crop production. BioScience 34(1):36–40.

Thom, A. S. 1968. The exchange of momentum, mass, and heat between an artificial leaf and the airflow in a wind tunnel. Quart. J. Roy. Met. Soc. 94:44–55.

Willmer, C. M., J. C. Rutter and H. Meidner. 1983. Potassium involvement in stomatal movements of Paphiopedilum. Journal of Experimental Botany 34(142):507–513.

Wilson, J. A., A. B. Ogunkanmi and T. A. Mansfield. 1978. Effects of external potassium supply on stomatal closure induced by abscisic acid. Plant, Cell and Environment 1:199–201.

Zeiger, E. and P. K. Helpler. 1977. Light and stomatal function: Blue light stimulates swelling of guard cell protoplasts. Science 196:887–889.

Ziegenspeck, H. 1955. Die Farbenmikrophotographie, ein Hilfsmittel zum objektiven Nachweis submikroscopeischer Strukturelemente. Die Radiomicellierung und Filierung der Schleisszellen von Ophioderma pendulum. Photographie und Wissenschaft. (Color photomicrography, an aid to the objective study of submicroscopic structural elements. Radialmicellation and filiation of guard cell of Ophioderma pendulum.) 4:19–22.

Chapter 4

Apfel, R. E. 1972. The tensile strength of liquids. Scientific American, 227(6):58–71.

Askenasy, E. 1897. Beiträge zur Erklärung des Saftstrigens. (Contributions to a Clarification of the Ascent of Sap). Naturhisto.-Med. Ver. Heidelberg. 5:429–448.

Begg, J. E. and N. C. Turner. 1970. Water potential gradients in field tobacco. Plant Physiology 46:343–346.

Briggs, Lyman. J. 1950. Limiting negative pressure of water. Journal of Applied Physics, 21:721–722.

Comstock, G. G. and W. A. Cote, Jr. 1968. Factors affecting permeability and pit aspiration in coniferous sapwood. Wood Science and Technology 2:279–291.

Cooke, R. and I. D. Kuntz. 1974. The properties of water in biological systems. Annual Review of Biophysics and Bioengineering, 3:95–126.

Crafts, A. S. and T. C. Broyer. 1938. Migration of salts and water into xylem of the roots of higher plants. American Journal of Botany, 25:529–535.

Dixon, Henry H. 1914. Transpiration and the Ascent of Sap in Plants. Macmillan and Co., Ltd., London.

Dixon, Henry H. and J. Joly. 1894. Notes. Annals of Botany Lond., 8:468–470.

Dixon, Henry H. and J. Joly. 1895. On the ascent of sap. Roy. Soc. London Phil. Trans., B 186: 563–576.

Esau, Katherine. 1977. Anatomy of Seed Plants, Second Edition. John Wiley & Sons, New York.

Greenridge, K. N. H. 1958. Rates and patterns of moisture movement in trees. Pages 43–69 in Thimann, K. V. (ed.), The physiology of forest trees. Ronald, New York.

Guinness Book of World Records, 1983. Tallest Tree. 97–106.

Hammel, H. T. 1967. Freezing of xylem sap without cavitation. Plant Physiology, 42:55–66.

Hartig, Th. 1878. Anatomie und Physiologie der Holzpflanzen. (Anatomy and Physiology of Woody Plants.) Springer, Berlin.

Hayward, A. T. J. 1971. Negative pressure in liquids: Can it be harnessed to serve man? American Scientist, 59:434–443.

Heine, R. W. and D. J. Farr. 1973. Comparison of heat-pulse and radioisotope tracer method for determining sap-flow in stem setments of poplar. Journal of Experimental Botany, 24:649–654.

Huber, B. and E. Schmidt. 1936. Weitere thermoelektrische Untersuchungen Über den Transpirationsstrom der Bäume. (Further thermoelectric investigations on the transpiration stream in trees.) Tharandt Forst Jb 87:369–412.

Jensen, William A. and Frank B. Salisbury. 1984. Botany, 2nd Edition. Wadsworth Publishing Co. Belmont, California.

Legge, N. J. 1980. Aspects of transpiration in mountain ash Eucalyptus regnans F. Muell. PhD Thesis, LaTrobe Univ. Melbourne.

Lybeck, B. R. 1959. Winter freezing in relation to the rise of sap in tall trees. Plant Physiology 34: 482–486.

MacKay, J. F. G. and P. E. Weatherby. 1973. The effects of transverse cuts through the stems of transpiring woody plants on water transport and stress in the leaves. Journal of Experimental Botany, 24:15–28.

Milburn, J. A. and R. P. C. Johnson. 1966. The conduction of sap. II. Detection of vibrations produced by sap cavitation in Ricinus xylem. Planta 69:43–52.

Morohashi, Y. and M. Shimorkoriyama. 1971. Apparent free space of germinating pea seeds. Physiologie Plantarum 25:341–345.

Münch, E. 1930. Die Stoffbewegungen in der Pflanze. (Translocation in Plants.) Fischer, Jena.

Nobel, Park S. 1983. Biophysical Plant Physiology and Ecology. W. H. Freeman and Co., San Francisco.

Oertli, J. J. 1971. The stability of water under tension in the xylem. Z. Pflanzenphysiol. 65:195–209.

O'Leary, J. W. 1970. Can there be a positive water potential in plants? BioScience, 20:858–859.

Pickard, William F. 1981. The ascent of sap in plants. Progress in Biophysics and Molecular Biology 37:181–229.

Plumb, R. C. and W. B. Bridgman. 1972. Ascent of sap in trees. Science, 176:1129–1131.

Preston, R. D. 1952. Movement of water in higher plants. Pages 257–321 in Frey-Wyssling (ed.), Deformation and flow in biological systems. Elsevier North-Holland, Amsterdam.

Rein, H. 1928. Die Thermo-Stromuhr. Ein Verfahren welches mit etwa ±10 Prozent Genauigkeit die unblutige langdauerde Messung der mittleren Durchflußmengen an gleichzeitigen Gefäßen gestattet. (The thermostream clock. A procedure that allows an unbloody, long-duration measurement of the average flow quantities in simultaneous vessels with an accuracy of ±10 percent.) Z Biol. 87:394–418.

Renner, O. 1911. Experimentelle Beiträge zur Zenntnis der Wasserbewegung. (Experimental contributions to the knowledge of water movement.) Flora, 103:171.

Renner, O. 1915. Theoretisches und Experimentelles zur Kohäsionstheorie der Wasserbewegung. (Theoretical and experimental contribution to the cohesions theory of water movement.) Jahrbuch fur Wissenschaftliche Botanik, 56:617–667.

Scholander, Per F. 1968. How mangroves desalinate sea water. Physiology Plantarum 21:251–268.

Scholander, Per F., H. T. Hammel, E. D. Bradstreet, and E. A. Hemmingsen. 1965. Sap pressure in vascular plants. Science, 148:339–346.

Scholander, P. F., E. Hemmingsen, and W. Garey. 1961. Cohesive lift of sap in the rattan vine. Science, 134:1835–1838.

Spomer, G. G. 1968. Sensors monitor tension in transpiration streams of trees. Science, 161: 484–485.

Stone, J. F. and G. A. Shirazi. 1975. On the heat-pulse method for the measurement of apparent sap velocity in stems. Planta, 122:166–177.

Strasburger, E. 1893. Über das Saftsteigen. (About the ascent of sap.) S. Fischer, Jena.

Sucoff, E. 1969. Freezing of conifer xylem and the cohesion-tension theory. Physiologia Plantarum 22:424–431.

Troughton, John and Lesley A. Donaldson. 1972. Probing Plant Structure. McGraw-Hill Book Co., New York.

Wiebe, H. H., R. W. Brown, T. W. Daniel, and E. Campbell. 1970. Water potential measurements in trees. BioScience 20:225–226.

Zimmermann, M. H. 1983. Xylem Structure and the Ascent of Sap. Springer-Verlag, Berlin, Heidelberg, New York.

Chapter 5

Alexander, G. V. and L. T. McAnulty. 1983. Multielement analysis of plant-related tissues and fluids by optical emission spectrometry. Journal of Plant Nutrition 3:51–59.

Arnon, D. I. and P. R. Stout. 1939. The essentiality of certain elements in minute quantity for plants with special reference to copper. Plant Physiology 14:371–375.

Asher, C. J. and D. G. Edwards. 1983. Modern solution culture techniques. Pages 94–119 in Läuchli, A. and Bieleski, R. L. (eds.), 1983a.

Baker, A. J. M. 1981. Accumulators and excludersstrategies in the response of plants to heavy metals. Journal of Plant Nutrition 3:643–654.

Bollard, E. G. 1983. Involvement of unusual elements in plant growth and nutrition. Pages 695–744 in Läuchli, A. and Bieleski, R. L. (eds.), 1983b.

Bouma, D. 1983. Diagnosis of mineral deficiencies using plant tests. Pages 120–146 in Läuchli, A. and R. L. Bieleski (eds.), 1983a.

Brown, J. C. 1978. Mechanism of iron uptake by plants. Plant, Cell and Environment 1:249–257.

Brown, J. C. 1979. Genetic improvement and nutrient uptake in plants. BioScience 29:289–292.

Brown, T. A. and A. Shrift. 1982. Selenium: toxicity and tolerance in higher plants. Biological Reviews 57:59–84.

Brownell, P. F. 1979. Sodium as an essential micronutrient element for plants and its possible role in metabolism. Advances in Botanical Research 7:117–224.

Brownell, P. F. and C. J. Crossland. 1974. Growth responses to sodium by *Bryophyllum tubiflorum* under conditions inducing crassulacean acid metabolism. Plant Physiology 54:416–417.

Clark, R. B. 1982. Iron deficiency in plants grown in the Great Plains of the U.S. Journal of Plant Nutrition 5:251–268.

Chapin, F. S., III. 1980. The mineral nutrition of wild plants. Annual Review of Ecology and Systematics 11:233–260.

Cheung, W. Y. 1982. Calmodulin. Scientific American 246(6):62–70.

Christiansen, M. N. and C. F. Lewis, eds. 1982. Breeding Plants for Less Favorable Environments. John Wiley & Sons, Inc., New York.

Clarkson, D. T. and J. B. Hanson. 1980. The mineral nutrition of higher plants. Annual Review of Plant Physiology 31:239–298.

Cooper, A. 1979. The ABC of NFT (nutrient film technique). Grower Books, London.

Davideseu, D. and V. Davideseu. 1982. Evaluation of fertility by plant and soil analysis. Abascus Press, Kent, England.

Dugger, W. M. 1983. Boron in plant metabolism. Pages 626–650 in Läuchli, A. and R. L. Bieleski, eds., Encyclopedia of Plant Physiology, New Series, Vol. 15, Inorganic Plant Nutrition. Springer-Verlag, Berlin.

Emery, T. 1982. Iron metabolism in humans and plants. American Scientist 70:626–632.

Epstein, E. 1972. Mineral Nutrition of Plants: Principles and Perspectives. John Wiley & Sons, Inc., New York.

Epstein, E. and A. Läuchli. 1983. Pages 196–205 in Kommedahl, T. and P. N. Williams (eds.), Challenging Problems in Plant Health. American Phytopathological Society, St. Paul, Minn.

Flowers, T. J. and A. Läuchli. 1983. Sodium versus potassium: substitution and compartmentation. Pages 651–681 in Läuchli, A. and R. L. Bieleski (eds.), 1983b.

Gauch, H. G. 1972. Inorganic Plant Nutrition. Dowden, Hutchinson and Ross, Inc., New York.

Graves, C. J. 1983. Nutrient film technique. In: Janick, J. (ed.), Horticultural Reviews, Vol. 5. AVI Publishing Co., Inc., Westport, Connecticut.

Hewitt, E. J. 1966. Sand and Water Culture Methods Used in the Study of Plant Nutrition. Second Edition. Technical Communication no. 22 (Revised). Commonwealth Agricultural Bureaux, Farnham Royal, Bucks, England.

Hewitt, E. J. and T. A. Smith. 1975. Plant Mineral Nutrition. John Wiley & Sons, Inc., New York.

Hodson, M. J., A. G. Sangster, and D. W. Parry. 1982. Silicon deposition in the inflorescence bristles and microhairs of *Setaria italica* (L.) Beauv. Annals of Botany 50:843–850.

Jones, J. B., Jr. 1982. Hydroponics: its history and use in plant nutrition studies. Journal of Plant Nutrition 5:1003–1030.

Jones, L. H. and K. A. Handreck. 1967. Silica in soils, plants and animals. Advances in Agronomy 19:107–149.

Kabata-Pendias, A. and H. Pendias. 1984. Trace elements in soils and plants. CRC Press, Inc., Boca-Raton, Florida.

Kerr, P. S., D. G. Blevins, B. J. Rapp, and D. D. Randall. 1983. Soybean leaf urease: comparison with seed urease. Physiologia Plantarum 57:339–345.

Läuchli, A. and R. L. Bieleski, eds. 1983a. Encyclopedia of Plant Physiology, New Series, Vol. 15 A, Inorganic Plant Nutrition. Springer-Verlag, Berlin.

———. 1983b. Encyclopedia of Plant Physiology, New Series, Vol. 15B, Inorganic Plant Nutrition. Springer-Verlag, Berlin.

Lewis, J. and B. E. F. Reimann. 1969. Silicon and plant growth. Annual Review of Plant Physiology 20:289–304.

Lindsay, W. L. 1974. Role of chelation in micronutrient availability. Pages 507–524 in Carson, E. W. (ed.), The Plant Root and Its Environment. Univ. of Virginia Press, Charlottesville.

———. 1979. Chemical Equilibria in Soils. John Wiley & Sons, Inc., New York.

Lindsay, W. L. and P. Schwab. 1982. The chemistry of iron in soils and its availability to plants. Journal of Plant Nutrition 5:821–840.

Longeragan, J. F. and K. Snowball. 1969. Calcium requirements of plants. Australian Journal of Agricultural Research 20:465–478.

Lovatt, C. J. 1985. Evolution of the xylem resulted in a requirement for boron in the apical meristems of vascular plants. New Phytologist (in press).

Lovatt, C. J. and W. M. Dugger, 1984. Biochemistry of boron. In Frieden, E. (ed.), Biochemistry of the Ultra-Trace Elements, Plenum Press, New York (in press).

Marmé, D. 1983. Calcium transport and function. Pages 599–625 in Läuchli, A. and R. L. Bieleski, eds., 1983b.

Mengel, K. and E. A. Kirkby. 1982. Principles of Plant Nutrition, Third Edition. International Potash Institute, Worblaufen-Bern, Switzerland.

———. 1980. Potassium in crop production. Advances in Agronomy 33:59–111.

Mertz, W. 1981. The essential trace elements. Science 213:1332–1338.

Miller, W. J. and M. N. Neathery. 1977. Newly recognized trace mineral elements and their role in animal nutrition. BioScience 27:674–679.

Moorby, J. and R. T. Besford. 1983. Mineral nutrition and growth. Pages 481–527 in Läuchli, A. and R. L. Bieleski (eds.), 1983b.

Nabors, M. W. 1983. Increasing the salt and drought tolerance of crop plants. Pages 165–184 in Randall, D. D. (ed.), Current Topics in Plant Biochemistry and Physiology, Vol. 2, University of Missouri, Columbia.

Olsen, R. A., J. C. Brown, and J. N. Bennett. 1982. Reduction of Fe^{+3} as it relates to Fe chlorosis. Journal of Plant Nutrition 5:433–445.

Pilbeam, D. J. and E. A. Kirkby. 1983. The physiological role of boron in plants. Journal of Plant Nutrition 6:563–582.

Plant and Soil, vol. 72, 1983. This entire issue contains articles largely related to genetic variability in mineral nutrition.

Poston, J. M. 1978. Coenzyme B_{12}-dependent enzymes in potatoes: leucine2,3-aminomutase and methylmalonyl-CoA mutase. Phytochemistry 17:401–402.

Powell, P. E., P. J. Szaniszlo, G. R. Cline, and C. P. P. Reid. 1982. Hydroxamate siderophores in the iron nutrition of plants. Journal of Plant Nutrition 5:653–673.

Quispel, A. 1983. Dinitrogen-fixing symbioses with legumes, non-legume angiosperms and associative symbioses. Pages 286–329 in Läuchli, A. and R. L. Bieleski (eds.), 1983a.

Raven, J. A. 1980. Short- and long-distance transport of boric acid in plants. New Phytologist 84:231–249.

Rees, T. A. V. and I. A. Bekheet. 1982. The role of nickel in urea assimilation by algae. Planta 156:385–387.

Reisenauer, H. M. 1966. Mineral nutrients in soil solution. Pages 507–508 in Altman, P. L. and D. S. Dittmer(eds.), Environmental Biology. Federation of American Societies for Experimental Biology, Bethesda, Maryland.

Robb, D. A. and W. S. Peirpoint, eds. 1983. Metals and Micronutrients. Uptake and Utilization by Plants. Academic Press, Inc., New York.

Römheld, V. and H. Marschner. 1983. Mechanism of iron uptake by peanut plants. I. Fe^{+3} reduction, chelate splitting, and release of phenolics. Plant Physiology 71:949–954.

Rovira, A. D., G. D. Bowen, and R. C. Foster. 1983. The significance of rhizosphere microflora and mycorrhizas in plant nutrition. Pages 61–93 in Läuchli, A. and R. L. Bieleski (eds.), 1983a.

Rubin, J. L., C. G. Gaines, and R. A. Jensen. 1982. Enzymological basis for herbicidal action of glyphosate. Plant Physiology 70:833–839.

Ryan, D. F. and F. H. Bormann. 1982. Nutrient resorption in northern hardwood forests. BioScience 32:29–32.

Sandman, G. and P. Böger. 1983. The enzymological function of heavy metals and their role in electron transfer processes of plants. Pages 563–596 in Läuchli, A. and R. L. Bieleski (eds.), 1983b.

San Pietro, A., ed. 1982. Biosaline Research: A Look to the Future. Plenum Press, New York.

Schwab, A. P. and W. L. Lindsay. 1983. Effect of redox on the solubility and availability of iron. Soil Science Society of America 47:201–205.

Seckback, J. 1982. Ferreting out the secrets of plant ferretin: a review. Journal of Plant Nutrition 5:369–394.

Shear, C. B. and M. Faust. 1980. Nutritional ranges in deciduous tree fruits and nuts. Horticultural Reviews 2:142–163.

Soltanpour, P. N., J. B. Jones, Jr., and S. M. Workmann. 1982. Optical emission spectrometry. Pages 29–65 in Methods of Soil Analysis, Part 2, Chemical and Microbiological Properties (Agronomy Monograph no. 9).

Sprague, H. B., ed. 1964. Hunger Signs in Crops. A Symposium. David McKay Co., Inc., New York.

Stadtman, T. C. 1980. Selenium-dependent enzymes. Annual Review of Biochemistry 49:93–110.

Sutcliffe, J. F. and D. A. Baker. 1981. Plants and Mineral Salts, Second Edition. Edward Arnold Publishers Ltd., London.

Terry, N. 1977. Photosynthesis, growth, and the role of chloride. Plant Physiology 60:69–75.

Titus, J. S. and S. M. Kang. 1982. Nitrogen metabolism, translocation, and recycling in apple trees. Horticultural Reviews 4:204–246.

Uren, N. C. 1981. Chemical reduction of an insoluble higher oxide of manganese by plant roots. Journal of Plant Nutrition 4:65–71.

Vallee, B. L. 1976. Zinc biochemistry: a perspective. Trends in Biochemical Sciences 1:88–91.

Vose, P. B. 1982. Iron nutrition in plants: a world overview. Journal of Plant Nutrition 5:233–249.

Welch, R. M. 1981. The biological significance of nickel. Journal of Plant Nutrition 3:345–356.

Werner, D. and R. Roth. 1983. Silica metabolism. Pages 682–694 in Läuchli, A. and R. L. Bieleski, eds., 1983b.

Wilcox, G. E. 1982. The future of hydroponics as a research and plant production method. Journal of Plant Nutrition 5:1031–1038.

Wilson, S. B. and D. T. D. Nicholas. 1967. A cobalt requirement for nonnodulated legumes and for wheat. Phytochemistry 6:1057–1066.

Wolf, B. 1982. A comprehensive system of leaf analyses and its use for diagnosing crop nutrient status. Communications in Soil Science and Plant Analysis 13:1035–1059.

Woolhouse, H. W. 1983. Toxicity and tolerance in the responses of plants to metals. Pages 245–300 in Lange, O. L., P. S. Nobel, C. B. Osmond, and H. Ziegler (eds.), Encyclopedia of Plant Physiology, New Series, Vol. 12C, Physiological Plant Ecology III, Responses to the Chemical and Biological Environment. Springer-Verlag, Berlin.

Chapter 6

Anderson, W. P. 1975. Ion transport through roots. Pages 437–463 in Torrey, J. G. and D. T. Clarkson, (eds.), The Development and Function of Roots. Academic Press, Inc., London.

Atkinson, D. 1980. The distribution and effectiveness of the roots of tree crops. Horticultural Reviews 2:424–490.

Baker, D. A. 1978. Proton co-transport of organic solutes by plant cells. New Phytologist 81:485–497.

Baker, D. A. and J. L. Hall, eds. 1975. Ion Transport in Plant Cells and Tissues. North-Holland Publishing Company, Amsterdam.

Berhowitz, R. L. and R. L. Travis. 1982. A comparative evaluation of the level of concanavalin A binding by enriched plasma membrane fractions from developing soybean roots. Plant Physiology 69:379–384.

Borstlap, A. C. 1983. The use of model-fitting in the interpretation of 'dual' uptake isotherms. Plant, Cell and Environment 6:407–416.

Bowles, D. J. 1982. Membrane glycoproteins. Pages 584–600 in Loewus, F. A. and W. Tanner, (eds.), Plant Carbohydrates I. Intracellular Carbohydrates, Encyclopedia of Plant Physiology, New Series, Vol. 13A. Springer-Verlag, Berlin.

Bowling, D. J. J. 1981. Release of ions to the xylem in roots. Physiologia Plantarum 53:392–397.

Briskin, D. P. and R. T. Leonard. 1982. Partial characterization of a phosphorylated intermediate associated with the plasma membrane ATPase of corn roots. Proceedings of the National Academy of Sciences USA. 79:6922–6926.

Brummer, B. and R. W. Parish. 1983. Identification of specific proteins and glycoproteins associated with membrane fractions isolated from Zea mays L. coleoptiles. Planta 157:446–453.

Caldwell, M. and L. B. Camp. 1974. Below-ground productivity of two cool desert communities. Oecologia 17:123–130.

Carpita, N. C., D. Sabularse, D. M. Montezinas, and D. P. Delmer. 1979. Determination of the pore size of the cell walls of living plant cells. Science 205:1144–1147.

Chapin, F. S., III. 1980. The mineral nutrition of wild plants. Annual Review of Ecology and Systematics 11:233–260.

Clarkson, D. T. 1974. Ion Transport and Cell Structure in Plants. McGraw-Hill Book Company, New York.

———. 1981. Nutrient interception and transport by root systems. Pages 307–330 in Johnson, C. B., (ed.), Physiological Processes Limiting Plant Productivity. Butterworths, London.

Clarkson, D. T. and J. B. Hanson. 1980. The mineral nutrition of higher plants. Annual Review of Plant Physiology 31:239–298.

Clarkson, D. T. and A. W. Robards. 1975. The endodermis, its structural development and physiological role. Pages 415–436 in Torrey, J. G. and D. T. Clarkson (eds.), The Development and Function of Roots. Academic Press, Inc., London.

Cram, J. 1983. Characteristics of sulfate transport across plasmalemma and tonoplast of carrot root cells. Plant Physiology 72:204–211.

Demel, R. A. and B. DeKruyff. 1976. The function of sterols in membranes. Biochimica et Biophysica Acta 457:109–132.

Dittmer, H. J. 1949. Root hair variation in plant species. American Journal of Botany 36:152–155.

Drew, M. C. 1975. Comparison of the effects of a localized supply of phosphate, nitrate, ammonium and potassium on the growth of the seminal root system, and the shoot, in barley. New Phytologist 75:479–490.

———. 1979. Properties of roots which influence rates of absorption. Pages 21–38 in Harley, J. L. and R. S. Russell (eds.), The Soil-Root Interface. Academic Press, Inc., New York.

Drew, M. C. and L. R. Baker. 1978. Nutrient supply and the growth of the seminal root system in barley. Journal of Experimental Botany 29:435–451.

Elbein, A. D. 1982. Glycolipids and other glycosides. Pages 601–631 in Loewus, F. A. and W. Tanner (eds.), Plant Carbohydrates I, Intracellular Carbohydrates, Encyclopedia of Plant Physiology, New Series, Vol. 13A. Springer-Verlag, Berlin.

Epstein, E. 1972. Mineral Nutrition of Plants: Principles and Perspectives. John Wiley & Sons, Inc., New York.

———. 1976. Kinetics of ion transport and the carrier concept. Pages 70–94 in Lüttge, U. and M. G. Pitman (eds.), Encyclopedia of Plant Physiology, New Series, Vol. 2, Part B. Springer-Verlag, Berlin.

———. 1977. The Role of Roots in the Chemical Economy of Life on Earth. BioScience 27:783–787.

Esau, K. 1971. Plant Anatomy, Third Edition. John Wiley & Sons, Inc., New York.

Gerdemann, J. W. 1975. Vesicular-arbuscula mycorrhizae. Pages 575–591 in Torrey, J. G. and D. T. Clarkson, 1975 (eds.), The Development and Function of Roots. Academic Press, Inc., New York.

Giaquinta, R. T. 1983. Phloem loading of sucrose. Annual Review of Plant Physiology 34:347–387.

Giese, A. C. 1968. Cell Physiology, Third Edition. W. B. Saunders Co., Philadelphia, Pennsylvania.

Glass, A. D. M. 1983. Regulation of ion transport. Annual Review of Plant Physiology 34:311–326.

Gunning, B. E. S. and R. L. Overall. 1983. Plasmodesmata and cell-to-cell transport in plants. BioScience 33:260–265.

Hall, J. L. and D. A. Baker. Cell membranes. Pages 39–77 in Baker, D. A. and J. L. Hall, (eds.), Ion Transport in Plant Cells and Tissues. North-Holland Publishing Company, Amsterdam.

Harley, J. L. and R. S. Russell, eds. 1979. The Soil-Root Interface. Academic Press, Inc., New York.

Hayman, D. S. 1980. Mycorrhiza and crop production. Nature 287:487–488.

Higinbotham, N. 1973. Electropotentials of plant cells. Annual Review of Plant Physiology 24:25–46.

Higinbotham, N., B. Etherton, and R. J. Foster. 1967. Mineral ion contents and cell transmembrane electropotentials of pea and oat seedling tissue. Plant Physiology 42:37–46.

Hodges, T. K. 1973. Ion absorption by plant roots. Advances in Agronomy 25:163–207.

Hodges, T. K. 1976. ATPases associated with membranes of plant cells. Pages 260–283 in Lüttge, U. and M. G. Pitman (eds.), 1976a.

Houslay, M. D. and K. K. Stanley. 1982. Dynamics of Biological Membranes. John Wiley & Sons, Inc., New York.

Itoh, S. and S. A. Barber. 1983. Phosphorus uptake by six plant species as related to root hairs. Agronomy Journal 75:457–461.

Jensen, W. A., B. Heinrich, D. B. Wake, and M. H. Wake. 1979. Biology. Wadsworth Publishing Co., Belmont, California.

Jeschke, W. D., W. Stelter, B. Reising, and R. Behl. 1983. Vacuolar Na/K exchange, its occurrence in root cells of Hordeum, Atriplex, and Zea and its significance for K/Na discrimination in roots. Journal of Experimental Botany 34:964–979.

Kochian, L. V. and W. L. Lucas. 1982. Potassium transport in corn roots. I. Resolution of kinetics into a saturable and linear component. Plant Physiology 70:1723–1731.

———. 1983. Potassium transport in corn roots. II. The significance of the root periphery. Plant Physiology 73:208–215.

———. 1984. The significance of the linear component for K+ influx by corn roots. In Cram, W. J., R. Rybova, and K. Janacek (eds.), Plant Membrane Transport. Wiley-Interscience (in press).

Kramer, P. J. and T. T. Kozlowski. 1979. Physiology of Woody Plants. Academic Press, Inc., New York.

Kyte, J. 1981. Molecular considerations relevant to the mechanism of active transport. Nature 292:201–204.

Larkin, P. J. 1980. Plant protoplast agglutination and immobilization. Pages 135–160 in Loewus, F. A. and C. A. Ryan (eds.), Recent Advances in Phytochemistry, Vol. 15. Plenum Press, New York.

Lass, B. and C. I. Ullrich-Eberius. 1984. Evidence for proton/sulfate cotransport and its kinetics in Lemna gibba G1. Planta 161:53–60.

Läuchli, A. 1976. Apoplasmic transport in tissues. Pages 3–34 in Lüttge, U. and M. G. Pitman (eds.), 1976b.

Lee, R. B. 1982. Selectivity and kinetics of ion uptake by barley plants following nutrient deficiency. Annals of Botany 50:429–449.

Leigh, R. A. 1983. Minireview: Methods, progress and potential for the use of isolated vacuoles in studies of solute transport in higher plant cells. Physiologia Plantarum 57:390–396.

Lüttge, U. and M. G. Pitman, eds. 1976a. Transport in Plants II, Part A, Cells, Encyclopedia of Plant Physiology, New Series, Vol. 2. Springer-Verlag, Berlin.

———. 1976b. Transport in Plants II, Part B, Tissues and Organs, Encyclopedia of Plant Physiology, New Series, Vol. 2. Springer-Verlag, Berlin.

Lüttge, U. and E. Schnepf. 1976. Organic substances. Pages 244–277 in Lüttge, U. and M. G. Pitman (eds.), 1976b.

Marks, G. C. and T. T. Kozlowski, eds. 1973. Ectomycorrhizae, Their Ecology and Physiology. Academic Press, Inc., New York.

Marmé, D., E. Marrè, and R. Hertel, eds. 1982. Plasmalemma and Tonoplast: Their Functions in the Plant Cell. Elsevier Biomedical Press, Amsterdam.

Maronek, D. M., J. W. Hendrix, and J. Kiernan. 1981. Mycorrhizal fungi and their importance in horticultural crop production. Horticultural Reviews 3:172–213.

Marty, F. and D. Branton. 1980. Analytical characterization of beetroot vacuole membrane. Journal of Cell Biology 87:72–83.

Marx, D. H. and N. C. Schenck. 1983. Potential of mycorrhizal symbiosis in agricultural and forest productivity. Pages 334–347 in Kommedahl, T. and P. H. William (eds.), Challenging Problems in Plant Health. American Phytopathological Society, St. Paul, Minnesota.

Mosse, B. 1975. A microbiologist's view of anatomy. Pages 39–65 in Walker, N. (ed.), Soil Microbiology. John Wiley & Sons, Inc., New York.

Mosse, B., D. P. Stribley, and F. Letacon. 1981. Ecology of mycorrhizae and fungi. Pages 137–210 in Alexander, M. (ed.), Advances in Microbial Ecology, Vol. 5. Plenum Press, New York.

Nissen, P. 1974. Uptake mechanisms: Inorganic and organic. Annual Review of Plant Physiology 25:53–79.

Nissen, P. and O. Nissen. 1983. Validity of the multiphasic concept of ion absorption in plants. Physiologia Plantarum 57:47–56.

Perry, T. O. 1981. Tree roots—where they grow: implications and practical significance. Pages 39–48 in New Horizons, a publication of the Horticultural Research Institute, Washington, D. C.

Pitman, M. G. 1976. Ion uptake by plant roots. Pages 95–128 in Lüttge, U. and M. G. Pitman (eds.), 1976b.

———. 1977. Ion transport into the xylem. Annual Review of Plant Physiology 28:71–88.

Pitman, M. G., W. P. Anderson and U. Lüttge. 1976. Transport processes in roots. Pages 57–69 in Lüttge, U. and M. G. Pitman (eds.), 1976b.

Polley, L. D. and J. W. Hopkins. 1979. Rubidium (potassium) uptake by Arabidopsis. A comparison of uptake by cells in suspension culture and by roots of intact seedlings. Plant Physiology 64:374–378.

Pressman, B. C. 1976. Biological applications of ionophores. Annual Review of Biochemistry 45:501–530.

Raven, J. A. and R. A. Smith. 1974. Significance of hydrogen ion transport in plant cells. Canadian Journal of Botany 52:1035–1048.

Reinhold, L. and A. Kaplan. 1984. Membrane transport of sugars and amino acids. Annual Review of Plant Physiology 35:45–83.

Reisenauer, H. M. 1966. Mineral nutrients in soil solution. Pages 507–508 in Altman, P. L. and D. S. Dittmer, (eds.), Environmental Biology. Federation of American Society for Experimental Biology, Bethesda, Maryland.

Reynolds, E. K. C. 1975. Tree rootlets and their distribution. Pages 163–177 in Torrey, J. G. and D. T. Clarkson (eds.), The Development and Function of Roots. Academic Press, Inc., New York.

Ritchie, G. A. and J. R. Dunlap. 1980. Root growth potential: its development and expression in forest tree seedlings. New Zealand Journal of Forestry Science 10:218–248.

Robards, A. W. 1975. Plasmodesmata. Annual Review of Plant Physiology 26:13–29.

Ruehle, J. L. and D. H. Marx. 1979. Fiber, food, fuel and fungal symbionts. Science 206:419–422.

Russell, S. 1977. Plant Root Systems. McGraw-Hill Book Co. Ltd.

Sabater, B. 1982. A mechanism for multiphasic uptake of solutes in plants. Physiologia Plantarum 55:121–128.

Safir, G. R. 1980. Vesicular-arbuscular mycorrhizae and crop production. Pages 231–251 in Carlson, P. S. (ed.), The Biology of Crop Production. Academic Press, Inc., New York.

Sanders, F. E., B. Mosse, and P. B. Tinker, eds. 1976. Endomycorrhizas. Academic Press, Inc., New York.

Schuurmans, Stekhoven, F. and S. L. Borting. 1981. Transport adenosine triphosphatases: properties and functions. Physiological Reviews 61:1–76.

Singer, S. L. and G. L. Nicolson. 1972. The fluid mosaic model of the structure of cell membranes. Science 175:720–731.

Smith, F. A. and J. A. Raven. 1976. H$^+$ transport and regulation of cell pH. Pages 317–346 in Lüttge, U. and M. G. Pitman (eds.), 1976a.

Smith, S. S. E. 1980. Mycorrhizas of autotrophic higher plants. Biological Reviews 55:475–510.

Spanswick, R. M. 1976. Symplasmic transport in tissues. Pages 35–53 in Lüttge, U. and M. G. Pitman (eds.), 1976b.

Stocking, C. R. and U. Heber, eds. 1976. Transport in plants III, Intracellular Interactions and Transport Processes, Encyclopedia of Plant Physiology, New Series, Vol. 3, Springer-Verlag, Berlin.

Sutcliff, J. F. 1976. Regulation in the whole plant. Pages 394–417 in Lüttge, U. and M. G. Pitman, (eds.), 1976b.

Sutton, R. F. 1980. Root system morphogenesis. New Zealand Journal of Forestry Science 10:264–292.

Sze, Heven. 1980. Nigericin-stimulated ATPase activity in microsomal vesicles of tobacco callus. Proceedings of the National Academy of Science USA 77:5904–5908.

Tanford, C. 1983. Mechanism of free energy coupling in active transport. Annual Review of Biochemistry 52:379–409.

Troughton, A. 1981. Length and life of grass roots. Grass and Forage Science 36:117–120.

Vermeer, J. and M. E. McCully. 1982. The rhizosphere in Zea: new insight into its structure and development. Planta 156:45–61.

Wagner, G. J. 1982. Compartmentation in plant cells: the role of the vacuole. Pages 1–45 in Creasy, L. L. and G. Hrazdina (eds.), Cellular and Subcellular Organization Plant Metabolism. Plenum Press, New York.

Weaver, J. E. and E. Zink. 1946. Length of life of roots of ten species of perennial range and pasture grasses. Plant Physiology 24:201–217.

Whatley, M. H. and L. Sequeira. 1981. Bacterial attachment to cell walls. Pages 213–240 in Loewus, F. A. and C. A. Ryan (eds.), Recent Advances in Phytochemistry, Vol. 15, The Phytochemistry of Cell Recognition and Cell Surface Interactions. Plenum Press, New York.

Wiebe, H. 1978. The significance of plant vacuoles. BioScience 28:327–331.

Yoshida, S., M. Uemura, T. Niki, A. Sakai, and L. V. Gusta. 1983. Partition of membrane particles in aqueous two-polymer phase system and its practical use for purification of plasma membranes from plants. Plant Physiology 72:105–114.

Chapter 7

Arnold, W. N. 1968. The selection of sucrose as the translocate of higher plants. Journal of Theoretical Biology 21:13–20.

Biddulph, O. and R. Cory. 1957. An analysis of translocation in the phloem of the bean plant using THO, ^{32}P, and ^{14}CO$_2$. Plant Physiology 32:608–619.

Botha, C. E. J., R. F. Evert, and R. D. Walmsley. 1975. Observations of the penetration of the phloem in leaves of Nerium oleander (Linn.) by stylets of the aphid, Aphis meril (B. de F.). Protoplasma 86:309–319.

Christy, A. Lawrence and Donald B. Fisher. 1978. Kinetics of ^{14}C-photosynthate translocation in morning glory vines. Plant Physiology 61:283–290.

Crafts, A. S. and O. Lorenz. 1944. Fruit growth and food transport in Cucurbits. Plant Physiology 19:131–138.

Cronshaw, J. 1981. Phloem structure and function. Annual Review of Plant Physiology 32:465–484.

de Vries, H. 1885. Über die Bedeutung der Circulation und der Rotation des Protoplasma für das Stofftransport in der Pflanze (On the significance of the circulation and rotation of protoplasm for translocation in plants). Botanische Zeitung 43:1–26.

Eisinger, W. 1983. Regulation of pea internode expansion by ethylene. Annual Review of Plant Physiology 34:225–240.

Esau, K. 1966. Plant Anatomy. John Wiley & Sons, Inc. New York.

Esau, K. 1977. Anatomy of Seed Plants, Second Edition. John Wiley & Sons, Inc. New York.

Eschrich, W. 1967. Beidirektionelle Translokation in Siebröhren (Bidirectional translocation in sieve tubes). Planta 73:37–49.

Fahn, A. 1982. Plant Anatomy, Third Edition. Pergamon Press, Oxford, New York.

Fensom, D. S. 1972. A theory of translocation in phloem of Heracleum by contractile protein microfibrillar material. Canadian Journal of Botany 50:479–497.

Field, R. J. and A. J. Peel. 1971. The movement of growth regulators and herbicides into the sieve elements of willow. New Phytologist 70:997–1003.

Fisher, D. B. 1975. Structure of functional soybean sieve elements. Plant Physiology 56:555–569.

Geiger, D. R., R. T. Giaquinta, S. A. Sovonick, and R. J. Fellows. 1973. Solute distribution in sugar beet leaves in relation to phloem loading and translocation. Plant Physiology 52:585–589.

Geiger, D. R. and S. A. Sovonick. 1976. II. Effects of temperature, anoxia and other metabolic inhibitors on translocation. Pages 480, 504. In Zimmermann, M. H. and J. A. Milburn (eds.), Transport in plants. In Phloem transport. In A. Pirson and M. H. Zimmerman (eds.), Encyclopedia of Plant Physiology Vol. 1. Springer-Verlag, Berlin.

Geiger, D. R., S. A. Sovonick, T. L. Shock, and R. J. Fellows. 1974. Role of free space in translocation in sugar beet. Plant Physiology 54:892–898.

Giaquinta, R. 1976. Evidence for phloem loading from the apoplast: chemical modification of membrane sulfhydryl groups. Plant Physiology 57:872–875.

Giaquinta, R. T. and D. R. Geiger. 1972. Mechanisms of inhibition of translocation by localized chilling. Plant Physiology 51:372–377.

Giaquinta, R. T. 1983. Phloem loading of sucrose. Annual Review of Plant Physiology 34:347–387.

Gifford, R. M. and L. T. Evans. 1981. Photosynthesis, carbon partitioning, and yield. Annual Review of Plant Physiology 32:485–509.

Gunning, Brian E. S. 1977. Transfer cells and their roles in transport of solutes in plants. Scientific Progress Oxf. 64:539–568.

Gunning, Brian E. S., J. S. Pate, F. R. Minchin, and I. Marks. 1974. Quantitative aspects of transfer cell structure in relation to vein loading in leaves and solute transport in legume nodules, Pages 87–126 in Sleigh, M. A., D. H. Jennings (eds.), Transport at the Cellular Level, Cambridge Press.

Gunning, B. E. S. and M. W. Steer. 1975. Pages 208–209 in Ultrastructure and the biology of plant cells, Edward Arnold Publishers, Ltd.

Hammel, H. T. 1968. Measurement of turgor pressure and its gradient in the phloem of oak. Plant Physiology 43:1042–1048.

Hampson, S. E., R. S. Loomis, and D. W. Rains. 1978. Regulation of sugar uptake in hypocotyls of cotton. Plant Physiology 62:851–855.

Heyser, W. 1980. Phloem loading in the maize leaf. Berichte der Deutsche Botanische Gesellschaft 93:221–224.

Housley, T. L. and D. B. Fisher. 1977. Estimation of osmotic gradients in soybean sieve tubes by quantitative microautoradiography. Qualified support for the Münch hypothesis. Plant Physiology 59:701–706.

Kennedy, J. S. and T. E. Mittler. 1953. A method for obtaining phloem sap via the mouth-parts of aphids. Nature 171:528.

Komor, E., M. Rotter, J. Waldhauser, E. Martin, and B. H. Cho. 1980. Sucrose proton symport for phloem loading in the Ricinus seedling. Berichte der Deutsche Botanischen Ges. 93:211–219.

Läuchli, A. 1972. Translocation of inorganic solutes. Annual Review of Plant Physiology 23:197–218.

McReady, C. C. 1966. Translocation of growth regulators. Annual Review of Plant Physiology 17:283–294.

Minchin, P. E. H. and J. H. Troughton. 1980. Quantitative interpretation of phloem translocation data. Plant Physiology 31:191–215.

Münch, E. 1927. Versuche über den Saftkreislauf (Experiments on the circulation of sap). Berichte der Deutsche Botanische Gesellschaft 45:340–356.

Münch, E. 1930. Die Stoffbewegungen in der Pflanze (Translocation in Plants) Fischer, Jena.

Passioura, J. B. and A. E. Ashford. 1974. Rapid translocation in the phloem of wheat roots. Australian Journal of Plant Physiology 1:521–527.

Pate, J. S. 1975. Exchange of solutes between phloem and xylem and circulation in the whole plant. Pages 451–473. In M. H. Zimmermann and J. A. Milburn (eds.), Transport in Plants. I. Phloem transport. In A. Pirson and M. H. Zimmermann, Encyclopedia of Plant Physiology New Series, Vol. 1, Springer-Verlag, Berlin, Heidelberg, New York.

Pate, J. S. 1980. Transport and partitioning of nitrogenous solutes. Annual Review of Plant Physiology 31:313–340.

Pate, J. S., D. B. Layzell, and D. L. McNeil. 1979. Modeling the transport and utilization of C and N in a nodulated legume. Plant Physiology 63:730–738.

Pate, J. S., P. J. Sharkey, and C. A. Atkins. 1977. Nutrition of a developing legume fruit. Functional economy in terms of carbon, nitrogen, and water. Plant Physiology 59:506–510.

Patrick, J. W. 1979. An assessment of auxin-promoted transport in decapitated stems and whole shoots of Phaseolus vulgaris L. Planta 146:107–112.

Peel, A. J. 1974. Transport of nutrients in plants. John Wiley & Sons, Inc., New York.

Peterson, C. A. and H. B. Currier. 1969. An investigation of bidirectional translocation in the phloem. Physiologia Plantarum 22:1238–1250.

Phillis, E. and T. G. Mason. 1933. Studies on the transport of carbohydrates in the cotton plant. III. The polar distribution of sugar in the foliage leaves. Annals of Botany 47:585–634.

Reid, M. S. and R. L. Bieleski. 1974. Sugar changes during fruit ripening—Whither sorbitol? Mechanisms of regulation of plant growth. Pages 823–830 in R. L. Bieleski, A. R. Ferguson, and M. M. Cresswell (eds.), Bulletin 12, The Royal Society of New Zealand, Wellington.

Roeckle, B. 1949. Nachweise eines Konzentrationshubs zwischen Palisadenzellen und Siebröhren (Proof for a concentration buildup between palisade cells and sieve tubes). Planta 36:530–550.

Rogers, S. and A. J. Peel. 1975. Some evidence for the existence of turgor pressure gradients in the sieve tubes of willow. Planta 126:259–267.

Sauter, J. J., W. Iten and M. H. Zimmermann. 1973. Studies of the release of sugar into the vessels of sugar maple (Acer saccharum). Canadian Journal of Botany 51:1–8.

Sij, J. W. and C. A. Swanson. 1973. Effects of petiole anoxia on phloem transport in squash. Plant Physiology 51:368–371.

Smith, J. C. Andrew, and J. A. Milburn. 1980. Phloem turgor and the regulation of sucrose loading in Ricinus communis L. Planta 148:41–48.

Thaine, R., S. L. Ovenden, and J. S. Turner. 1959. Translocation of labelled assimilates in the soybean. Australian Journal of Biological Sciences 12:349–372.

Troughton, J. H., J. Moorby, and B. G. Currie. 1974. Investigations of carbon transport in plants. I. The use of carbon-11 to estimate various parameters of the translocation process. Journal of Experimental Botany 25:684–694.

Watson, B. T. 1975. The influence of low temperature on the rate of translocation in the phloem of *Salix viminalis* L. Annals of Botany 39:889–900.

Wiebe, H. H. 1962. Physiological response of plants to drought. Utah Science 23:70–71.

Wright, J. P. and D. B. Fisher. 1980. Direct measurement of sieve tube turgor pressure using severed aphid stylets. Plant Physiology 65:1133–1135.

Ziegler, H. 1975. Nature of transported substances. Pages 59–94 *in* Zimmermann, M. H. and J. A. Milburn (eds.), Transport in plants. I. Phloem transport. *In* A. Pirson and M. H. Zimmermann (eds.), Enc. of Plant Physiology New Series, Vol. 1, Springer-Verlag, Berlin.

Zimmermann, M. H. 1961. Movement of organic substances in trees. Science 133:73–79.

Zimmermann, M. H. and J. A. Milburn, eds. 1975. Transport in plants. I. Phloem transport. *In* A. Pirson and M. H. Zimmermann (eds.), Enc. of Plant Physiology New Series, Vol. 1, Springer-Verlag, Berlin.

Zimmermann, M. H. and H. Ziegler. 1975. Appendix III: List of sugars and sugar alcohols in sieve-tube exudates. Pages 480–504 *in* Zimmermann, M. H. and J. A. Milburn (eds.), Transport in plants. I. Phloem transport. *In* A. Pirson and M. H. Zimmermann (eds.), Enc. of Plant Physiology Vol. 1. Springer-Verlag, Berlin.

Chapter 8

Adams, C. A. and R. W. Rinne. 1981. The occurrence and significance of dispensable proteins in plants. New Phytologist 89:1–14.

Anfinsen, C. B. 1959. The Molecular Basis of Evolution. John Wiley & Sons, Inc. New York.

Boulter, D. and B. Parthier, eds. 1982. Nucleic Acids and Proteins in Plants. I. Structure, Biochemistry and Physiology of Proteins. Enc. of Plant Physiology, New Series, Vol. 14, Part A. Springer-Verlag, Berlin.

Burke, J. J., J. N. Siedow, and D. E. Moreland. 1982. Succinate dehydrogenase. A partial purification from mung bean hypocotyls and soybean cotyledons. Plant Physiology 70:1577–1581.

Cammack, R., D. Hall, and K. Rao. 1971. Ferredoxins: Are they living fossils? New Scientist 23:696–698.

Creasy, L. L. and G. Hrazdina, eds. 1982. Cellular and Subcellular Localization in Plant Metabolism. Recent Advances in Phytochemistry, Vol. 16. Plenum Press, New York.

Davies, D. D. 1977. Regulation of enzyme levels and activity. Pages 306–328 *in* Smith, H. (ed.), The Molecular Biology of Plant Cells, Botanical Monographs, Vol. 14. Univ. of California Press, Berkeley.

Dickerson, R. E. and I. Geis. 1969. The Structure and Action of Proteins. W. A. Benjamin, Inc., Menlo Park, California.

Engel, P. C. 1977. Enzyme Kinetics. John Wiley & Sons, Inc. New York.

Gottlieb, L. D. 1971. Gel electrophoresis: new approach to the study of evolution. BioScience 21:939–944.

Hall, J. L. and A. L. Moore. 1983. Isolation of Membranes and Organelles from Plant Cells. Academic Press, Inc., New York.

Harpstead, D. D. 1971. High lysine corn. Scientific American 225(2):34–42.

IUB recommendations on enzyme nomenclature. 1979. Recommendations of nomenclature committee of the international union of biochemistry on the nomenclature and classification of enzymes. Academic Press, Inc., New York.

Jensen, U. and D. E. Fairbrothers, eds. 1983. Proteins and Nucleic Acids in Plant Systematics. Springer-Verlag, Berlin.

Johnson, V. A. and C. L. Lay. 1974. Genetic improvement of plant protein. Journal of Agricultural and Food Chemistry 22:558–566.

Koshland, D. E., Jr. 1973. Protein shape and biological control. Scientific American 229(4):52–64.

———. 1984. Control of enzyme activity and metabolic pathways. Trends in Biochemical Sciences 9:155–159.

Lemport, D. A. 1980. Structure and function of plant glycoproteins. Pages 501–541 *in* Preiss, J. (ed.), The Biochemistry of Plants, Vol. 3, Carbohydrates: Structure and Function. Academic Press, Inc., New York.

Larkins, B. A. 1981. Seed storage proteins: characterization and biosynthesis. Pages 449–489 *in* Marcus, A. (ed.), The Biochemistry of Plants, Vol. 6, Proteins and Nucleic Acids. Academic Press, Inc., New York.

Markert, C. L., ed. 1975. Isozymes II, Physiological Function. Academic Press, Inc., New York.

McNaughton, S. J. 1972. Enzyme thermal adaptations. The evolution of homeostasis in plants. American Naturalist 106:165–172.

Orr, M. L. and B. K. Watt. 1957. Amino acid content of foods. United States Department of Agriculture Home Economics Research Report 4:1–82.

Randall, D. D. and P. M. Rubin. 1977. Plant pyruvate dehydrogenase complex. II. ATP-dependent inactivation and phosphorylation. Plant Physiology 59:1–3.

Ricard, J. 1980. Enzyme flexibility as a molecular basis for metabolic control. Pages 31–80 *in* Davies, D. D. (ed.), The Biochemistry of Plants, Vol. 2, Metabolism and Respiration. Academic Press, Inc., New York.

Ross, C. W. 1981. Biosynthesis of nucleotides. Pages 169–205 *in* Marcus, A. (ed.), The Biochemistry of Plants, Vol. 6, Proteins and Nucleic Acids. Academic Press, Inc., New York.

Schulz, G. E. and R. H. Schirmer. 1979. Principles of Protein Structure. Springer-Verlag, New York.

Simon, J. P. 1979. Differences in thermal properties of NAD malate dehydrogenase in genotypes of *Lathyrus japonicus* Willd. (Leguminosae) from maritime and continental sites. Plant, Cell and Environment 2:23–33.

Smith, H., ed. 1977. Regulation of enzyme synthesis and activity in higher plants. Academic Press, Inc., London.

Spiker, S. 1975. An evolutionary comparison of plant histones. Biochimica et Biophysica Acta 400:461–467.

Tanksley, S. D. and T. J. Orton, eds. 1983. Isozymes in Plant Genetics and Breeding, Parts A and B, Elsevier Science Publishers, Amsterdam.

Ting, I. P., I. Führ, R. Curry, and W. C. Zschoche. 1975. Malate dehydrogenase isozymes in plants: Preparation, properties, and biological significance. Pages 369–383 *in* Markert, C. L. (ed.), 1975.

Trewavas, A. 1976. Post-transitional modification of proteins by phosphorylation. Annual Review of Plant Physiology 27:349–374.

Walsh, C. T. 1984. Enzyme mechanisms. Trends in Biochemical Sciences 9:159–162.

Wolfe, S. L. 1972. Biology of the Cell. Wadsworth Publishing Company, Belmont, California.

Chapter 9

Allen, J. F. 1983. Protein phosphorylation—carburettor of photosynthesis? Trends in Biochemical Sciences 8:369–373.

Allen, J. F., J. Bennett, K. E. Steinback, and C. Arntzen. 1981. Chloroplast protein phosphorylation couples plastoquinone redox state to distribution of excitation energy between photosystems. Nature 291:25–29.

Amesz, T. 1983. The role of manganese in photosynthetic oxygen evolution. Biochimica et Biophysica Acta 726:1–12.

Amzel, L. M. and P. L. Pedersen. 1983. Proton ATPases: structure and mechanism. Annual Review of Biochemistry 52:801–824.

Anderson, J. M. 1982. Distribution of the cytochromes of spinach chloroplasts between the appressed membranes of grana stacks and stroma-exposed thylakoid regions. FEBS Letters 138:62–66.

Anderson, J. M. and A. Melis. Localization of different photosystems in separate regions of chloroplast membranes. Proceedings of the National Academy of Sciences USA 80:745–749.

Avron, M. 1981. Photosynthetic electron transport and photophosphorylation. Pages 163–191 *in* Hatch, M. D. and N. K. Boardman (eds.), 1981.

Balegh, S. E. and O. Biddulph. 1970. The photosynthetic action spectrum of the bean plant. Plant Physiology 46:1–5.

Barber, J. 1982. Influence of surface charges on thylakoid structure and function. Annual Review of Plant Physiology 23:261–295.

Barber, J., ed. 1982. Topics in Photosynthesis, Vol. 4, Electron Transport and Photophosphorylation. Elsevier Biomedical Press, Amsterdam.

Barber, J. 1983. Photosynthetic electron transport in relation to thylakoid membrane composition and organization. Plant, Cell and Environment (Commissioned Review). 6:311–322.

———. 1984. Has the mangano-protein of the water splitting reaction of photosynthesis been isolated? Trends in Biochemical Sciences 9:79–80.

Black, C. C. 1977. How herbicides work—the bipyridilium herbicides. Weeds Today 9:12–15.

Brown, J. S. 1972. Forms of chlorophyll *in vivo*. Annual Review of Plant Physiology 23:73–86.

Clayton, R. K. 1981. Photosynthesis: Physical Mechanisms and Chemical Patterns. Cambridge University Press, Cambridge, U. K.

Cogdell, R. J. 1983. Photosynthetic reaction centers. Annual Review of Plant Physiology 34:21–45.

Cramer, W. A. and J. Whitmarsh. 1977. Photosynthetic cytochromes. Annual Review of Plant Physiology 28:133–172.

Ehleringer, J. and O. Björkman. 1977. Quantum yields for CO_2 uptake in C_3 and C_4 plants: dependence on temperature, CO_2, and O_2 concentrations. Plant Physiology 59:86–90.

Ehleringer, J. R. and R. W. Pearcy. 1983. Variation in quantum yield for CO_2 uptake among C_3 and C_4 plants. Plant Physiology 73:555–559.

Fong, F. K., ed. 1982. Light Reaction Path of Photosynthesis. Springer-Verlag, Berlin.

Glazer, A. N. 1982. Phycobilisomes: structure and dynamics. Annual Review of Microbiology 36:173–198.

Goedheer, J. C. 1979. Carotenoids in the photosynthetic apparatus. Berichte der Deutschen Botan. Gesamte 92:427–436.

Goodwin, T. W. 1980. The Biochemistry of the Carotenoids. Vol. 1: Plants. Second Edition. Methuen, Inc., New York.

Govindjee, ed. 1982. Photosynthesis, Vol. 1. Energy Conversion by Plants and Bacteria. Academic Press, Inc., New York.

Haaker, H. and C. Veeger. 1984. Enzymology of nitrogen fixation. Trends in Biochemical Sciences 9:188–192.

Halliwell, B. 1981. Chloroplast Metabolism—The Structure and Function of Chloroplasts in Green Leaf Cells. Clarendon Press, Oxford.

Hammes, G. 1983. Mechanism of ATP synthesis and coupled proton transport: studies with purified chloroplast coupling factor. Trends in Biochemical Sciences 8:131–136.

Hatch, M. D. and N. K. Boardman, eds. 1981. Photosynthesis. Vol. 8, The Biochemistry of Plants. Academic Press, Inc., New York.

Hiller, R. G. and D. J. Goodchild. 1981. Thylakoid membrane and pigment organization. Pages 1–49 *in* Hatch, M. D. and N. K. Boardman (eds.), 1981.

Hinckle, P. C. and R. E. McCarty. 1978. How cells make ATP. Scientific American 238(3):104–123.

Hirschberg, J. and L. McIntosh. 1983. Molecular basis of herbicide resistance in *Amaranthus hybridus*. Science 222:1346–1349.

Jagendorf, A. T. 1967. The chemiosmotic hypothesis of photophosphorylation. Pages 69–78 *in* San Pietro, A., F. A. Greer, and T. J. Army (eds.), Harvesting the Sun, Photosynthesis in Plant Life. Academic Press, Inc., New York.

Junge, W. 1982. Electrogenic reactions and proton pumping in green plant photosynthesis. Current Topics in Membranes and Transport 16:431–465.

Klimov, V. V. and A. A. Krasnovskii. 1981. Pheophytin as the primary electron acceptor in photosystem 2 reaction centers. (Review.) Photosynthetica 15:592–609.

Kok, B. 1976. Photosynthesis: the path of energy. Pages 845–885 in Bonner, J. and J. E. Varner (eds.), Plant Biochemistry, Third Edition. Academic Press, Inc., New York.

Kyle, D. J., L. A. Staehelin, and C. J. Arntzen. 1983. Lateral mobility of the light-harvesting complex in chloroplast membranes controls excitation energy distribution in higher plants. Archives of Biochemistry and Biophysics 222:527–541.

Malkin, R. 1982. Photosystem I. Annual Review of Plant Physiology 33:455–479.

Melis, A. and J. S. Brown. 1980. Stoichiometry of system I and system II reaction centers and of plastoquinone in different photosynthetic membranes. Proceedings of the National Academy of Sciences USA 77:4712–4716.

Mitchell, P. 1966. Chemical coupling in oxidative and photosynthetic phosphorylation. Biological Reviews 41:445–502.

Osborne, B. A. and M. K. Garrett. 1983. Quantum yields for CO_2 uptake in some diploid and tetraploid plant species. Plant, Cell and Environment 6:135–144.

Possingham, J. V. 1980. Plastid replication and development in the life cycle of higher plants. Annual Review of Plant Physiology 31:113–129.

Staehelin, L. A. and C. J. Arntzen. 1983. Regulation of chloroplast membrane function: Protein phosphorylation changes the spatial organization of membrane components. (Review.) Journal of Cell Biology 97:1327–1337.

Stemler, A. and R. Radmer. 1975. Source of photosynthetic oxygen in bicarbonate-stimulated Hill reaction. Science 190:457–458.

Strottman, H. and J. Schumann. 1983. Structure, function and regulation of chloroplast ATPase. (Minireview) Physiologia Plantarum 57:375–382.

Trebst, A. 1978. Plastoquinones in photosynthesis. Philosophical Transactions of the Royal Society of London (Series B) 284:591–599.

Trebst, A. and M. Avron, eds. 1977. Photosynthesis I. Photosynthetic Electron Transport and Photophosphorylation, Vol. V, Encyclopedia of Plant Physiology, New Series. Springer-Verlag, Berlin.

Virgin, H. I. and H. Egnéus. 1983. Control of plastid development in higher plants. Pages 289–311 in Shropshire, W., Jr. and H. Mohr (eds.), Photomorphogenesis. Encyclopedia of Plant Physiology, New Series, Vol. 16A. Springer-Verlag, Berlin.

Chapter 10

Akazawa, T. 1976. Polysaccharides. Pages 381–403 in Bonner, J. and J. E. Varner (eds.), Plant Biochemistry. Academic Press, Inc., New York.

———. 1979. Ribulose-1,5-bisphosphate carboxylase. Pages 208–229 in Gibbs, M. and E. Latzko (eds.), 1979.

Anderson, L. E., A. R. Ashton, A. H. Mohamed, and R. Scheibe. 1982. Light/dark modulation of enzyme activity in photosynthesis. BioScience 32:103–107.

Avigad, G. 1982. Sucrose and other disaccharides. Pages 217–347 in Loewus, F. A. and W. Tanner (eds.), 1982.

Banks, W. and D. D. Muir. 1980. Structure and chemistry of the starch granule. Pages 321–369 in Preiss, J. (ed.), 1980.

Bassham, J. A. 1965. Photosynthesis: the path of carbon. Pages 875–902 in Bonner, J. and J. E. Varner (eds.), Plant Biochemistry. Second Edition. Academic Press, Inc., New York.

———. 1979. The reductive pentose phosphate cycle and its regulation. Pages 9–30 in Gibbs, M. and E. Latzko (eds.), 1979.

Beck, E. 1979. Glycolate synthesis. Pages 327–337 in Gibbs, M. and E. Latzko (eds.), 1979.

Bewley, J. D., ed. 1981. Nitrogen and Carbon Metabolism. Development in Plant and Soil Sciences 3. Martinus Nijhoff/Dr. W. Junk, the Hague, Boston.

Black, C. C., W. H. Campbell, T. M. Chen, and P. Dettrich. 1973. The monocotyledons: their evolution and comparative biology. III. Pathways of carbon metabolism related to net carbon dioxide assimilation by mono cotyledons. Quarterly Review of Biology 48:299–313.

Black, C. C., N. W. Carnal, and W. H. Kenyon. 1982. Compartmentation and the regulation of CAM. Pages 51–68 in Ting, I. P. and M. Gibbs (eds.), 1982.

Buchanan, B. B. 1980. Role of light in the regulation of chloroplast enzymes. Annual Review of Plant Physiology 31:341–374.

———. 1984. The ferredoxin/thioredoxin system: a key element in the regulatory function of light in photosynthesis. BioScience 34:378–383.

Campbell, W. H. and C. C. Black. 1982. Cellular aspects of C_4 leaf metabolism. Pages 223–248 in Creasy, L. L. and G. Hrazdina (eds.), 1982.

Chapman, K. S. R. and M. D. Hatch. 1981. Regulation of C_4 photosynthesis: mechanism of activation and inactivation of extracted pyruvate, inorganic phosphate dikinase in relation to dark/light regulation. Archives of Biochemistry and Biophysics 210:82–89.

Chollet, R. 1977. The biochemistry of photorespiration. Trends in Biochemical Sciences 2:155–158.

Creasy, L. L. and G. Hrazdina, eds. 1982. Cellular and Subcellular Localization in Plant Metabolism. Recent Advances in Phytochemistry, Vol. 16. Plenum Press, New York.

de Duve, C. 1983. Microbodies in the living cell. Scientific American 248(5):74–84.

Downton, W. J. S. 1975. The occurrence of C-4 photosynthesis among plants. Photosynthetica 9:96–105.

Edwards, G. E. and S. C. Huber. C_4 pathway. Pages 237–281 in Hatch, M. D. and N. K. Boardman (eds.), 1981.

Edwards, G, and D. A. Walker. 1983. C_3:C_4 Mechanisms, and Cellular and Environmental Regulation of Photosynthesis. Blackwell Scientific Publications, Oxford.

Ellis, R. J. 1979. The most abundant protein in the world. Trends in Biochemical Sciences 4:241–244.

Gibbs, M. and E. Latzko. 1979. Photosynthesis II, Encyclopedia of Plant Physiology, New Series, Vol. 6. Springer-Verlag, Berlin.

Gibson, A. C. 1982. The anatomy of succulence. Pages 1–17 in Ting, I. P. and M. Gibbs (eds.), 1982.

Govindjee, ed. 1983. Photosynthesis, Vol. II: Development, Carbon Metabolism and Plant Productivity. Academic Press, Inc., New York.

Graham, D. and E. A. Chapman. 1979. Interactions between photosynthesis and respiration in higher plants. Pages 150–162 in Gibbs, M. and E. Latzko (eds.), 1979.

Halliwell, B. 1981. Chloroplast Metabolism. Clarendon Press, Oxford, England.

Hatch, M. D. 1977. C_4 pathway photosynthesis: mechanism and physiological function. Trends in Biochemical Sciences 2:199–201.

Hatch, M. D. and N. K. Boardman, eds. 1981. Photosynthesis. The Biochemistry of Plants, Vol. 8. Academic Press, Inc., New York.

Hatch, M. D. and C. B. Osmond. 1976. Compartmentation and transport in C_4 photosynthesis. Pages 144–184 in Stocking, C. R. and U. Heber (eds.), 1976.

Heber, U. and D. A. Walker. 1979. The chloroplast envelope—barrier or bridge? Trends in Biochemical Sciences 4:252–256.

Heldt, H. W. 1976. Metabolite carriers of chloroplasts. Pages 137–143 in Stocking, C. R. and U. Heber (eds.), 1976.

———. 1979. Light-dependent changes of stromal H^+ and Mg^{+2} concentrations controlling CO_2 fixation. Pages 202–207 in Gibbs, M. and E. Latzko (eds.), 1979.

Hesla, B. I., L. L. Tieszen, and S. K. Imbamba. 1982. A systematic survey of C_3 and C_4 photosynthesis in the Cyperaceae of Kenya, East Africa. Photosynthetica 16:196–205.

Holmgren, A. 1981. Thioredoxin: structure and functions. Trends in Biochemical Sciences 6:26–29.

Holtum, J. A. M. and K. Winter. 1982. Activity of enzymes of carbon metabolism during the induction of crassulacean acid metabolism in Mesembryanthemum crystallinum L. Planta 155:8–16.

Huang, A. H. C., R. N. Trelease, and T. S. Moore, Jr. 1983. Plant Peroxisomes. Academic Press, Inc., New York.

Huber, S. C. 1982. Photosynthetic carbon metabolism in chloroplasts. Pages 151–184 in Creasy, L. L. and G. Hrazdina (eds.), 1982.

Jenner, C. F. 1982. Storage of starch. Pages 700–747 in Loewus, F. A. and W. Tanner (eds.), 1982.

Jensen, R. G. 1980. Biochemistry of the chloroplast. Pages 273–313 in Tolbert, N. E. (ed.), The Biochemistry of Plants, Vol. 1. Academic Press, Inc., New York.

Jordan, D. B. and W. L. Ogrun. 1981. Species variation in the specificity of ribulose bisphosphate carboxylase/oxygenase. Nature 291:513–515.

Keeley, J. E. and B. A. Morton. 1982. Distribution of diurnal acid metabolism in submerged aquatic plants outside the genus Isoetes. Photosynthetica 16:546–553.

Kluge, M. 1979. The flow of carbon in crassulacean acid metabolism (CAM). Pages 113–125 in Gibbs, M. and E. Latzko (eds.), 1979.

Kluge, M. and I. P. Ting. 1978. Crassulacean Acid Metabolism: Analysis of an Ecological Adaptation. Springer-Verlag, New York.

Krenzer, E. G., Jr., D. N. Moss, and R. K. Crookston. 1975. Carbon dioxide compensation points of flowering plants. Plant Physiology 56:194–206.

Laetsch, W. M. 1974. The C-4 syndrome: a structural analysis. Annual Review of Plant Physiology 25:27–52.

Loewus, F. A. and W. Tanner, eds. 1982. Encyclopedia of Plant Physiology, New Series, Vol. 13A, Plant Carbohydrates I: Intracellular Carbohydrates. Springer-Verlag, Berlin.

Meier, H. and J. S. G. Reid. 1982. Reserve polysaccharides other than starch in higher plants. Pages 418–471 in Loewus, F. A. and W. Tanner (eds.), 1982.

Miziorko, H. M. and G. H. Lorimer. 1983. Ribulose-bisphosphate carboxylase-oxygenase. Annual Review of Biochemistry 52:507–535.

Nakamoto, H. and G. E. Edwards. 1983. Control of the activation/inactivation of pyruvate, Pi dikinase from the C_4 plant maize by the adenylate energy charge, pyruvate, and analogs of pyruvate. Biochemical and Biophysical Research Communications 115:673–679.

Neales, T. F., A. A. Patterson, and V. J. Hartney. 1968. Physiological adaptation to drought in the carbon assimilation and water loss of xerophytes. Nature 219:469–472.

O'Leary, M. H. 1982. Phosphoenolpyruvate carboxylase: an enzymologist's view. Annual Review of Plant Physiology 33:297–315.

Osmond, C. B. and J. A. Holtum. 1981. Pages 329–374 in Hatch, M. D. and N. K. Boardman (eds.), 1981.

Pearcy, R. W. 1983. The light environment and growth of C_3 and C_4 tree species in the understory of a Hawaiian forest. Oecologia 58:19–25.

Perchorowicz, J. T. and R. G. Jensen. 1983. Photosynthesis and activation of ribulose bisphosphate carboxylase in wheat seedlings. Regulation by CO_2 and O_2. Plant Physiology 71:955–960.

Preiss, J., ed. 1980. Carbohydrates: Structure and Function, Vol. 3, The Biochemistry of Plants. Academic Press, Inc., New York.

Preiss, J. 1982a. Regulation of the biosynthesis and degradation of starch. Annual Review of Plant Physiology 33:431–454.

———. 1982b. Biosynthesis of starch and its regulation. Pages 397–417 in Loewus, F. A. and W. Tanner (eds.), 1982.

———. 1984. Starch, sucrose biosynthesis and partition of carbon in plants are regulated by orthophosphate and triose-phosphates. Trends in Biochemical Sciences 9:24–27.

Preiss, J. and C. Levi. 1979. Metabolism of starch in leaves. Pages 282–312 in Gibbs, M. and E. Latzko (eds.), 1979.

Rathnam, C. K. M. 1978. C$_4$ photosynthesis: the path of carbon in bundle sheath cells. Scientific Progress, Oxford 65:409–435.

Rathnam, C. K. M. and R. Chollett. 1980. Photosynthetic carbon metabolism in C$_4$ plants and C$_3$-C$_4$ intermediate species. Pages 1–48 in Reinhold, L., J. B. Harborne, and T. Swain (eds.), Progress in Phytochemistry, Vol. 6. Pergamon Press, Oxford.

Rathnam, C. K. M. and R. Chollett. 1980. Regulation of photorespiration. Current Advances in Plant Science 38:1–19.

Ray, T. B. and C. C. Black. 1979. The C$_4$ pathway and its regulation. Pages 77–101 in Gibbs, M. and E. Latzko (eds.), 1979.

Robinson, S. P. and D. A. Walker. 1981. Photosynthetic carbon reduction cycle. Pages 193–236 in Hatch, M. D. and N. K. Boardman (eds.), 1981.

Sarojini, G. and D. J. Oliver. 1983. Extraction and partial characterization of the glycine decarboxylase multienzyme complex from pea leaf mitochondria. Plant Physiology 72:194–199.

Steward, F. C. and R. G. S. Bidwell, eds. 1983. Plant Physiology, a Treatise. Vol. VII, Energy and Carbon Metabolism. Academic Press, Inc., New York.

Stocking, C. R. and U. Heber. 1976. Transport in Plants III, Encyclopedia of Plant Physiology, New Series, Vol. 3. Springer-Verlag, Berlin.

Szarek, S. R. and I. P. Ting. 1977. The occurrence of crassulacean acid metabolism among plants. Photosynthetica 11:330–342.

Ting, I. P. and M. Gibbs, eds., 1982. Crassulacean Acid Metabolism. Waverly Press, Baltimore, Maryland.

Tolbert, N. E. 1979. Glycolate metabolism by higher plants and algae. Pages 338–352 in Gibbs, M. and E. Latzko (eds.), 1979.

———. 1981. Metabolic pathways in peroxisomes and glyoxysomes. Annual Review of Biochemistry 50:133–157.

Wagner, J. and W. Larcher. 1981. Dependence of CO$_2$ gas exchange and acid metabolism of the alpine CAM plant Sempervivum montanumon on temperature and light. Oecologia 50:88–93.

Walker, D. A. 1976. Plastids and intracellular transport. Pages 85–136 in Stocking, C. R. and U. Heber (eds.), 1976.

Waller, S. S. and J. K. Lewis. 1979. Occurrence of C$_3$ and C$_4$ photosynthetic pathways in North American grasses. Journal of Range Management 32:12–28.

Whittingham, C. P., A. J. Keys, and I. F. Bird. 1979. The enzymology of sucrose synthesis in leaves. Pages 313–326 in Gibbs, M. and E. Latzko (eds.), 1979.

Wildner, G. F. 1981. Ribulose-1,5-bisphosphate carboxylase-oxygenase: aspects and prospects. Physiologia Plantarum 52:385–389.

Winter, K. and J. H. Troughton. 1978. Photosynthetic pathways in plants of coastal and inland habitats of Israel and the Sinai. Flora 167:1–34.

Chapter 11

Allen, E. J. and R. K. Scott. 1980. An analysis of growth of the potato crop. Journal of Agricultural Science, Cambridge 94:583–606.

Austin, R. B., J. Bingham, R. D. Blackwell, L. T. Evans, M. A. Ford, C. L. Morgan, and M. Taylor. 1980. Genetic improvements in winter wheat yields since 1900 and associated physiological changes. Journal of Agricultural Science, Cambridge 94:675–689.

Azcón-Bieto, J. 1983. Inhibition of photosynthesis by carbohydrates in wheat leaves. Plant Physiology 73:681–686.

Baker, D. N. 1965. Effects of certain environmental factors on net assimilation in cotton. Crop Science 5:53–56.

Baker, D. N. and R. B. Musgrave. 1964. Photosynthesis under field conditions. V. Further plant chamber studies on the effects of light on corn (Zea mays L.). Crop Science 4:127–131.

Bauer, A. and P. Martha. 1981. The CO$_2$ compensation point of C-3 plants—a re-examination. I. Interspecific variability. Zeitschrift für Pflanzenphysiologie 103:445–450.

Berry, J. A. 1975. Adaptation of photosynthetic processes to stress. Science 188:644–650.

Berry, J. A. and O. Björkman. 1980. Photosynthetic response and adaptation to temperature in higher plants. Annual Review of Plant Physiology 31:491–543.

Björkman, O. 1980. The response of photosynthesis to temperature. Pages 273–301 in Grace, J., E. D. Ford, and P. G. Jarvis (eds.), Plants and Their Atmospheric Environment. Blackwell Scientific Publications, Oxford.

———. 1981. Responses to different quantum flux densities. Pages 57–107 in Lange, P. S. Nobel, C. B. Osmond, and H. Ziegler (eds.), Encyclopedia of Plant Physiology, New Series, Vol. 12A, Physiological Plant Ecology I. Springer-Verlag, Berlin.

Björkman, O. and P. Holmgren, 1963. Adaptability of the photosynthetic apparatus to light intensity in ecotypes from exposed and shaded habitats. Physiologia Plantarum 16:889–914.

Black, C. C. 1973. Photosynthetic carbon fixation in relation to net CO$_2$ uptake. Annual Review of Plant Physiology 24:253–286.

Boardman, N. K. 1977. Comparative photosynthesis of sun and shade plants. Annual Review of Plant Physiology 28:355–377.

Brown, R. H. 1978. A difference in N use efficiency in C$_3$ and C$_4$ plants and its implications in adaptation and evolution. Crop Science 18:93–98.

Cardwell, V. B. 1982. Fifty years of Minnesota corn production: sources of yield increase. Agronomy Journal 74:984–990.

Chang, J-H. 1981. Corn yield in relation to photoperiod, night temperature, and solar radiation. Agricultural Meteorology 24:253–262.

Corré, N. C. 1983. Growth and morphogenesis of sun and shade plants I. The influence of light intensity. Acta Botanica Neerlandica 32:49–62.

Edwards, G. E., S. B. Ku, and J. G. Foster. 1983. Physiological constraints to maximum yield potential. Pages 105–109 in Kommedahl, T. and P. H. Williams (eds.), Challenging Problems in Plant Health. The American Phytopathological Society, St. Paul, Minn.

Ehleringer, J. and R. W. Pearcy. 1983. Variation in quantum yield for CO$_2$ uptake among C$_3$ and C$_4$ plants. Plant Physiology 73:555–559.

Evans, J. R. 1983. Nitrogen and photosynthesis in the flag leaf of wheat (Triticum aestivum L.). Plant Physiology 72:297–302.

Evans, L. T. 1980. The natural history of crop yield. American Scientist 68:388–397.

Gaastra, P. 1959. Photosynthesis of crop plants as influenced by light, carbon dioxide, temperature, and stomatal diffusion resistance. Mededelinger van de Landbouwhogeschool Te Wagenigen 59:1–68.

Gifford, R. M. and L. T. Evans. 1981. Photosynthesis, carbon partitioning, and yield. Annual Review of Plant Physiology 32:485–509.

Gifford, R. M., J. H. Thorne, W. D. Hitz, and R. T. Giaquinta. 1984. Crop productivity and photoassimilate partitioning. Science 225:801–808.

Good, N. E. and D. H. Bell. 1980. Photosynthesis, plant poductivity, and crop yield. Pages 3–51 in Carlson, P. S. (ed.), The Biology of Crop Productivity. Academic Press, Inc., New York.

Govindjee, ed. 1983. Photosynthesis, Vol. II: Development, Carbon metabolism and Plant Productivity. Academic Press, Inc., New York.

Hadley, N. F. and S. R. Szarek. 1981. Productivity of desert ecosystems. BioScience 31:747–753.

Hall, N. P. and A. J. Keys. 1983. Temperature dependence of the enzymic carboxylation and oxygenation of ribulose 1,5-bisphosphate in relation to effects of temperature on photosynthesis. Plant Physiology 72:945–948.

Hanover, J. W. 1980. Control of tree growth. BioScience 30:756–762.

Hanson, H. C. 1917. Leaf structure as related to environment. American Journal of Botany 4:533–560.

Hargrove, T. R. and V. L. Cabanilla. 1979. The impact of semidwarf varieties on Asian rice-breeding programs. BioScience 29:731–735.

Herold, A. 1980. Regulation of photosynthesis by sink activity—the missing link. New Phytologist 86:131–144.

Hesbeth, J. D. 1963. Limitations to photosynthesis responsible for differences among species. Crop Science 3:493–496.

Hicklenton, P. R. and P. A. Jolliffe. 1980. Alterations in the physiology of CO$_2$ exchange in tomato plants grown in CO$_2$-enriched atmosphere. Canadian Journal of Botany 58:2181–2189.

Johnson, C. B., ed. 1981. Physiological Processes Limiting Plant Productivity. Butterworths, London.

Kommedahl, T. and P. H. Williams, eds. 1981. Challenging Problems in Plant Health. The American Phytopathological Society, St. Paul, Minnesota.

Lange, O. L., P. S. Nobel, C. B. Osmond, and H. Ziegler. 1981. Physiological Plant Ecology I. Encyclopedia of Plant Physiology, New Series, Vol. 12A. Springer-Verlag, Berlin.

Larcher, W. 1969. The effect of environmental and physiological variables on the carbon dioxide gas exchange of trees. Photosynthetica 3:167–198.

Leopold, A. C. and P. E. Kriedemann. 1975. Plant Growth and Development. McGraw-Hill Book Company, New York.

Long, S. P. 1983. C$_4$ photosynthesis at low temperatures. (Commissioned Review.) Plant, Cell and Environment 6:345–363.

Marzola, D. L. and D. P. Bartholomew. 1979. Photosynthetic pathway and biomass energy production. Science 205:555–559.

McClendon, J. H. and G. G. McMillen. 1982. The control of leaf morphology and the tolerance of shade by woody plants. Botanical Gazette 143:79–83.

McCree, K. J. 1981. Photosynthetically active radiation. Pages 41–55 in Lange, O. L., P. S. Nobel, C. B. Osmond, and H. Ziegler (eds.), Encyclopedia of Plant Physiology, New Series, Vol. 12A, Physiological Plant Ecology I. Springer-Verlag, Berlin.

Moss, D. N. and L. H. Smith. 1972. A simple classroom demonstration of differences in photosynthetic capacity among species. Journal of Agronomic Education 1:16–17.

Oquist, G. 1983. Effects of low temperature on photosynthesis. Plant, Cell and Environment 6:281–300.

Osborne, B. A. and M. K. Garrett. 1983. Quantum yields for CO$_2$ uptake in some diploid and tetraploid plant species. Plant, Cell and Environment 6:135–144.

Osmond, C. B., O. Björkman, and D. J. Anderson, eds. 1980. Physiological Processes in Plant Ecology. Springer-Verlag, Berlin.

Pearcy, R. W. and J. Ehleringer. 1984. Comparative ecophysiology of C$_3$ and C$_4$ plants. (A Review.) Plant, Cell and Environment 7:1–13.

Puckridge, D. W. 1968. Photosynthesis of wheat under field conditions. I. The interaction of photosynthetic organs. Australian Journal of Agricultural Research 19:711–719.

Radmer, R. and B. Kok. 1977. Photosynthesis: limited yields, unlimited dreams. BioScience 27:599–605.

Revelle, R. 1982. Carbon dioxide and world climate. Scientific American 247(2):35–43.

Rycroft, M. J. 1982. Analysing atmospheric carbon dioxide levels. Nature 295:190–191.

Šesták, Z. 1981. Leaf ontogeny and photosynthesis. Pages 147–158 in Johnson, C. B. (ed.), Physiological processes Limiting Plant Productivity. Butterworths, London.

Somerville, C. R. and W. L. Ogren. 1982. Genetic modification of photorespiration. Trends in Biochemical Sciences 7:171–174.

Stuiver, M. 1978. Atmospheric carbon dioxide and carbon reservoir changes. Science 199:253–258.

Thomas, M. D. and G. R. Hill. 1949. Photosynthesis under field conditions. Pages 19–52 in Franck, J. and W. E. Loomis (eds.), Photosynthesis in Plants. Iowa State University Press, Ames.

Warlaw, I. F. 1980. Translocation and source-sink relationships. Pages 297–399 in Carlson, P. S. (ed.), The Biology of Crop Productivity, Academic Press, Inc., New York.

Whittaker, R. H. 1975. Communities and Ecosystems. Second Edition. Macmillan, Inc., New York.

Wittwer, S. H. 1980. The shape of things to come. Pages 413–459 in Carlson, P. S. (ed.), The Biology of Crop Productivity. Academic Press, Inc., New York.

Zelitch, I. 1982. The close relationship between net photosynthesis and crop yield. BioScience 32:796–802.

Chapter 12

Alpi, A. and H. Beevers. 1983. Effects of O_2 concentration on rice seedlings. Plant Physiology 71:30–34.

ap Rees, T. 1980. Assessment of the contribution of metabolic pathways to plant respiration. Pages 1–29 in Davies, D. D. (ed.), 1980.

Atkinson, D. E. 1968. The energy charge of the adenylate pool as a regulatory parameter. Interaction with feedback modifiers. Biochemistry 7:4030–4034.

———. 1977. Cellular Energy Metabolism and its Regulation. Academic Press, Inc., New York.

Atwell, B. J., I. Waters, and H. Greenway. 1982. The effect of oxygen and turbulence on elongation of coleoptiles of submergence-tolerant and -intolerant rice cultivars. Journal of Experimental Botany 33:1030–1044.

Bajrachara, D., H. Falk, and P. Schopfer. 1976. Phytochrome-mediated development of mitochondria in the cotyledons of mustard (Sinapis alba L.) seedlings. Planta 131:253–261.

Barclay, A. M. and R. M. M. Crawford. 1982. Plant growth and survival under strict anaerobiosis. Journal of Experimental Botany 33:541–549.

Beevers, H. 1974. Conceptual developments in metabolic control, 1924–1974. Plant Physiology 54:437–442.

Biale, J. B. and R. E. Young. 1981. Respiration and ripening of fruits—retrospect and prospect. Pages 1–39 in Friend, J. and M. J. C. Rhodes (eds.), Advances in the Biochemistry of Fruits and Vegetables. Academic Press, Inc., London.

Bradford, K. J. and S. F. Yang. 1981. Physiological responses of plants to waterlogging. Horticultural Science 16:25–30.

Brunori, M. and M. T. Wilson. 1982. Cytochrome oxidase. Trends in Biochemical Sciences 7:295–299.

Buchanan, B. B. 1984. The ferredoxin/thioredoxin system: a key element in the regulatory function of light in photosynthesis. BioScience 34:378–383.

Campbell, R. and M. C. Drew. 1983. Electron microscopy of gas space (aerenchyma) formation in adventitious roots of Zea mays L. subjected to oxygen shortage. Planta 157:350–357.

Carnal, N. W. and C. C. Black. 1983. Phosphofructokinase activities in photosynthetic organisms. The occurrence of pyrophosphate-dependent 6-phosphofructokinase in plants. Plant Physiology 71:150–155.

Casey, R. P. and A. Azzi. 1983. An evaluation of the evidence for H^+ pumping by reconstituted cytochrome c oxidase in the light of recent criticism. FEBS Letters 154:237–342.

Chang, C. W. 1982. Enzymic degradation of starch in cotton leaves. Phytochemistry 21:1263–1269.

Cori, C. F. 1983. Embden and the glycolytic pathway. Trends in Biochemical Sciences 8:257–259.

Crawford, R. M. M. 1982. Physiological responses to flooding. Pages 453–477 in Lange, O. L., P. S. Nobel, C. B. Osmond, and H. Ziegler (eds.), Encyclopedia of Plant Physiology, New Series, Vol. 12B. Physiological Plant Ecology II. Water Relations and Carbon Assimilation. Springer-Verlag, Berlin.

Davies, D. D., ed. 1980. Metabolism and Respiration. The Biochemistry of Plants, Vol. 2. Metabolism and Respiration. Academic Press, Inc., New York.

Davies, D. D. 1980. Anaerobic metabolism and the production of organic acids. Pages 581–611 in Davies, D. D. (ed.), 1980.

Day, D. A., G. P. Arron, and G. G. Laties. 1980. Nature and control of respiratory pathways in plants: the interaction of cyanide-resistant respiration with the cyanide-sensitive pathway. Pages 198–241 in Davies, D. D. (ed.), 1980.

Drew, M. C. 1979. Plant responses to anaerobic conditions in soil and solution culture. Current Advances in Plant Science 36:1–14.

Dunn, G. 1974. A model for starch breakdown in higher plants. Phytochemistry 13:1341–1346.

Elthon, T. E. and C. R. Steward. 1983. A chemiosmotic model for plant mitochondria. BioScience 33:687–692.

Erecińska, M. and D. V. Wilson. 1982. Topical review. Regulation of cellular energy metabolism. Journal of Membrane Biology 70:1–14.

Gill, C. J. 1970. The flooding tolerance of woody plants—a review. Forest Abstracts 31:671–678.

Goodwin, T. W. and E. I. Mercer. 1983. Introduction to Plant Biochemistry. Second Edition. Pergamon Press, Oxford.

Halmer, P. and J. D. Bewley. 1982. Control by external and internal factors over the mobilization of reserve carbohydrates in higher plants. Pages 748–793 in Loewus, F. A. and W. Tanner (eds.), Encyclopedia of Plant Physiology, New Series, Vol. 13A, Plant Carbohydrates I: Intracellular Carbohydrates. Springer-Verlag, Berlin.

Hanson, J. B. and D. A. Day. 1980. Plant mitochondria. Pages 315–358 in Tolbert, N. E. (ed.), The Biochemistry of Plants, Vol. 1, The Plant Cell. Academic Press, Inc., New York.

Henry, M. F. and E. J. Nyns. 1975. Cyanide-insensitive respiration. An alternative mitochondrial pathway. Sub-Cellular Biochemistry 4:1–65.

Ishizaki, Y., H. Taniguchi, Y. Maruyama, and M. Nakamura. 1983. Debranching enzymes of potato tubers (Solanum tuberosum L.). I. Purification and some properties of potato isoamylase. Agricultural and Biological Chemistry 47:771–779.

James, W. O. 1963. Plant Physiology. Sixth Edition. Oxford University Press, London.

Jenner, C. F. 1982. Storage of Starch. Pages 700–747 in Loewus, F. A. and W. Tanner (eds.), Encyclopedia of Plant Physiology, New Series, Vol. 13A, Plant Carbohydrates I: Intracellular Carbohydrates. Springer-Verlag, Berlin.

Jensen, W. A. and F. B. Salisbury. 1974. Botany: An Ecological Approach. Wadsworth Publishing Company, Inc., Belmont, California.

Joly, C. A. and R. M. M. Crawford. 1982. Variation in tolerance and metabolic responses to flooding in some tropical trees. Journal of Experimental Botany 135:799–809.

Kawase, M. 1981. Anatomical and morphological adaptation of plants to waterlogging. Horticultural Science 16:30–34.

Kennedy, R. A., M. E. Rumpho, and D. Yanderzee. 1983. Germination of Echinochloa crus-galli (Barnyard grass) seeds under anaerobic conditions. Plant Physiology 72:787–794.

Kozlowski, T. T. 1984. Flooding and Plant Growth. Academic Press, Inc., New York.

———. 1984. Plant responses to flooding of soil. BioScience 34:162–167.

Lambers, H. 1980. The physiological significance of cyanide-resistant respiration in higher plants. Plant, Cell and Environment 3:293–302.

Laties, G. G. 1982. The cyanide-resistant, alternative path in higher plant respiration. Annual Review of Plant Physiology 33:519–555.

Lee, D. C. 1977. Plant mitochondria. Pages 105–135 in Smith, H. (ed.), The Molecular Biology of Plant Cells, Botanical Monographs, Vol. 14. Univ. of Calif. Press, Berkeley.

Lendzian, K. J. 1980. Modulation of glucose-6-phosphate dehydrogenase by NADPH, $NADP^+$, and dithiothreitol at variable $NADPH/NADP^+$ ratios in an illuminated reconstituted spinach (Spinacia oleraceae) chloroplast system. Planta 148:1–6.

Lipmann, F. 1975. Reminiscences of Embden's formulation of the Embden-Meyerhof cycle. Molecular and Cellular Biochemistry 6:171–175.

Malone, C., D. E. Koeppe, and R. J. Miller. 1974. Corn mitochondria swelling and contraction—an alternate interpretation. Plant Physiology 53:918–927.

Meeuse, B. J. D. 1975. Thermogenic respiration in aroids. Annual Review of Plant Physiology 26:117–126.

Meier, H. and J. S. G. Reid. 1982. Reserve polysaccharides other than starch in higher plants. Pages 418–471 in Loewus, F. A. and W. Tanner (eds.), Encyclopedia of Plant Physiology, New Series, Vol. 13A, Plant Carbohydrates I: Intracellular Carbohydrates. Springer-Verlag, Berlin.

Mitchell, P. 1979. Keilin's respiratory chain concept and its chemiosmotic consequences. Science 206:1148–1159.

Nash, D. and J. T. Wiskich. 1983. Properties of substantially chlorophyll-free pea leaf mitochondria prepared by sucrose density gradient separation. Plant Physiology 71:627–634.

O'Brien, T. P. and M. E. McCully. 1969. Plant Structure and Development. Collier-Macmillan, Limited, London.

Öpik, H. 1974. Mitochondria. Pages 52–83 in Robards, A. W. (ed.), Dynamic Aspects of Plant Ultrastructure. McGraw-Hill, New York.

Palmer, J. M. 1976. The organization and regulation of electron transport in plant mitochondria. Annual Review of Plant Physiology 27:133–157.

———. 1979. The uniqueness of plant mitochondria. Biochemical Society Transactions 7:246–252.

———. 1984. Physiology and Biochemistry of Plant Respiration. Cambridge University Press, Cambridge, United Kingdom.

Palmer, J. M. and I. M. Moller. 1982. Regulation of NAD(P)H dehydrogenases in plant mitochondria. Trends in Biochemical Sciences 7:258–261.

Penning deVries, F. W. T. 1975. The cost of maintenance processes in plant cells. Annals of Botany 39:77–92.

Penning deVries, F. W. T., A. H. M. Brunsting, and H. H. van Laar. 1974. Products, requirements and efficiency of biosynthesis: a quantitative approach. Journal of Theoretical Biology 45:339–377.

Philipson, J. J. and M. P. Coutts. 1980. The tolerance of tree roots to waterlogging. IV. Oxygen transport in woody roots of sitka spruce and lodgepole pine. New Phytologist 85:489–494.

Pradet, A. and P. Raymond. 1983. Adenine nucleotide ratios and adenylate energy charge in energy metabolism. Annual Review of Plant Physiology 34:199–224.

Preiss, J. 1982. Regulation of the biosynthesis and degradation of starch. Annual Review of Plant Physiology 33:431–454.

———. 1984. Starch, sucrose biosynthesis and partition of carbon in plants are regulated by orthophosphate and triose-phosphates. Trends in Biochemical Sciences 9:24–27.

Raison, J. K. 1980. Effect of low temperature on respiration. Pages 613–626 in Davies, D. D. (ed.), 1980.

Rhodes, M. J. C. 1980. Respiration and senescence of plant organs. Pages 419–462 in Davies, D. D. (ed.), 1980.

Rumpho, M. E. and R. A. Kennedy. 1983. Anaerobiosis in Echinochloa crus-galli (barnyard grass) seedlings. Intermediary metabolism and ethanol tolerance. Plant Physiology 72:44–49.

Siedow, J. W. 1982. The nature of the cyanide-resistant pathway in plant mitochondria. Recent Advances in Phytochemistry 16:47–83.

Smith, A. M., G. Kalsi, and H. W. Woolhouse. 1984. Products of fermentation in the roots of alders (*Alnus* Mill.). Planta 160:272–275.

Solomos, T. 1977. Cyanide-resistant respiration in higher plants. Annual Review of Plant Physiology 28:279–297.

Steup, M., H. Robenek, and M. Melkonian. 1983. In-vitro degradation of starch granules isolated from spinach chloroplasts. Planta 158:428–436.

Storey, B. T. 1980. Electron transport and energy coupling in plant mitochondria. Pages 125–195 *in* Davies, D. D. (ed.), 1980.

Turner, J. F. and D. H. Turner. 1980. The regulation of glycolysis and the pentose phosphate pathway. Pages 279–316 *in* Davies, D. D. (ed.), 1980.

Wikström, M. 1984. Pumping of protons from the mitochondrial matrix by cytochrome oxidase. Nature 308:558–560.

Wiskich, J. T. 1980. Control of the Krebs cycle. Pages 244–278 *in* Davies, D. D. (ed.), 1980.

Wu, M., D. A. Smyth, and C. C. Black, Jr. 1983. Fructose 2,6-bisphosphate and the regulation of pyrophosphate-dependent phosphofructokinase activity in germinating pea seeds. Plant Physiology 73:188–191.

Chapter 13

Abrol, Y. P., S. K. Sawhney, and M. S. Naik. 1983. Light and dark assimilation of nitrate in plants. Plant, Cell and Environment 6:595–599.

Allen, O. N. and E. K. Allen. 1981. The Leguminosae. A Source Book of Characteristics, Uses and Nodulation. Univ. of Wisconsin Press, Madison.

Anderson, J. W. 1980. Assimilation of inorganic sulfate into cysteine. Pages 203–225 *in* Miflin, B. J. (ed.), 1980a.

Anderson, K. T. Shanmugam, S. T. Lim and eight other authors. 1980. Genetic engineering in agriculture with emphasis on nitrogen fixation. Trends in Biochemical Sciences 5:35–39.

Appelby, C. A. 1984. Leghemoglobin and *Rhizobium* respiration. Annual Review of Plant Physiology 35:443–478.

Aslam, M. and R. C. Huffaker. 1984. Dependency of nitrate reduction on soluble carbohydrates in primary leaves of barley under aerobic conditions. Plant Physiology 75:623–628.

Barker, A. V. and H. A. Mills. 1980. Ammonium and nitrate nutrition of horticultural crops. Horticultural Reviews 2:395–423.

Barton, K. A. and W. J. Brill. 1983. Prospects in plant genetic engineering. Science 219:671–676.

Beevers, Leonard and R. H. Hageman. 1980. Nitrate and nitrite reduction. Pages 116–168 *in* Miflin, B. J. (ed.) 1980a.

Bell, E. A. 1980. Non-protein amino acids in plants. Pages 403–423 *in* Bell, E. A. and B. V. Charlwood (eds.), Encyclopedia of Plant Physiology, New Series, Vol. 8, Secondary Plant Products. Springer-Verlag, Berlin.

———. 1981. The non-protein amino acids occurring in plants. Progress in Phytochemistry 7:171–196.

Bergersen, F. J. 1982. Root Nodules of Legumes: Structure and Functions. John Wiley & Sons, Inc., New York.

Bergersen, F. J. and D. J. Goodchild. 1973. Aeration pathways in soybean root nodules. Australian Journal of Biological Science 26:729–740.

Bonas, U. K. Schmitz, H. Rennenberg, and H. Bergmann. 1982. Phloem transport of sulfur in *Ricinus*. Planta 155:82–88.

Bond, G. 1963. *In* Nutman, P. S. and B. Mosse (eds.), Symbiotic Associations, Thirteenth Symposium of the Society for General Microbiology.

———. 1976. The results of the IBP survey of root-nodule formation in nonleguminous angiosperms. Pages 443–474 *in* Nutman, P. S. (ed.), Symbiotic Nitrogen Fixation in Plants. International Biological Programme, Vol. 7. Cambridge University Press, Cambridge.

Bray, C. M. 1983. Nitrogen Metabolism in Plants, Longman Group Limited, London.

Brimblecombe, P. and D. H. Stedman. 1982. Historical evidence for a dramatic increase in the nitrate component of acid rain. Nature 298:460–462.

Carroll, B. J. and P. M. Gresshoff. 1983. Nitrate inhibition of nodulation and nitrogen fixation in white clover. Zeitschrift für Pflanzenphysiologie 110:77–88.

Davies, D. D. 1973. Control of and by pH. Symposia for the Society for Experimental Biology 27:513–529.

Dawson, J. O. 1983. Dinitrogen fixation in forest ecosystems. Canadian Journal of Microbiology 29:979–992.

Dry, I., W. Wallace, and D. J. D. Nicholas. 1981. Role of ATP in nitrite reduction in roots of wheat and pea. Planta 152:234–238.

Duxbury, J. M., D. R. Bouldin, R. E. Terry, and R. L. Tate III. 1982. Emissions of nitrous oxide from soils. Nature 298:462–464.

Eisbrenner, G. and H. J. Evans. 1983. Aspects of hydrogen metabolism in nitrogen-fixing legumes and other plant-microbe associations. Annual Review of Plant Physiology 34:105–136.

Freney, J. R. and I. E. Galbally, eds. 1982. Cycling of Carbon Nitrogen, Sulfur and Phosphorus in Terrestrial and Aquatic Ecosystems. Springer-Verlag, New York.

Giovanelli, J., S. H. Mudd, and A. H. Datko. 1980. Sulfur amino acids in plants. Pages 454–506 *in* Miflin, B. J. (ed.), 1980a.

Givan, C. V. 1979. Metabolic detoxification of ammonia in tissues of higher plants. Phytochemistry 18:375–382.

———. 1980. Aminotransferases in higher plants. Pages 329–358 *in* Miflin, B. J. (ed.), 1980a.

Goodchild, D. J. and F. J. Bergersen. 1966. Electron microscopy of the infection and subsequent development of soybean nodule cells. Journal of Bacteriology 92:204–213.

Gordon, J. C. and J. O. Dawson. 1979. Potential uses of nitrogen fixing trees and shrubs in commercial forestry. Botanical Gazette 140:588–590.

Granstedt, R. C. and R. C. Huffaker. 1982. Identification of the leaf vacuole as a major nitrate storage pool. Plant Physiology 70:410–413.

Guerrero, M. G., J. M. Vega, and M. Lasada. 1981. The assimilatory nitrate-reducing system and its regulation. Annual Review of Plant Physiology 32:169–204.

Gutschick, V. P. 1978. Energy and nitrogen fixation. BioScience 28:571–575.

Haynes, R. J. and K. M. Goh. 1978. Ammonium and nitrate nutrition of plants. Biological Reviews 53:465–510.

Hewitt, E. J. and C. S. Gundry. 1970. The molybdenum requirement of plants in relation to nitrogen supply. Journal of Horticultural Science 45:351–358.

Higgins, T. J. V. 1984. Synthesis and regulation of major proteins in seeds. Annual Review of Plant Physiology 35:191–221.

Hubbell, D. H. 1981. Legume infection by Rhizobium: a conceptual approach. BioScience 31:832–837.

Huffaker, R. C. 1982. Biochemistry and physiology of leaf proteins. Pages 370–400 *in* Boulter, D. B. and B. Parthier (eds.), Encyclopedia of Plant Physiology, New Series, Vol. 14A, Nucleic Acids and Proteins, Part 1. Springer-Verlag, Berlin.

LaRue, T. A. and T. G. Patterson. 1981. How much nitrogen do legumes fix? Pages 15–38 *in* Brady, N. C. (ed.), Advances in Agronomy, Vol. 34. Academic Press, Inc., New York.

Lea, P. J. and K. W. Joy. 1983. Amino acid interconversion in germinating seeds. Recent Advances in Phytochemistry 17:12–35.

Lea, P. J. and B. J. Miflin. 1980. Transport and metabolism of asparagine and other nitrogen compounds within the plant. Pages 569–608 in Miflin, B. J. (ed.), 1980a.

Lott, J. N. A. 1980. Protein bodies. Pages 589–623 *in* Tolbert, N. E. (ed.), The Biochemistry of Plants, Vol. 1, The Plant Cell. Academic Press, Inc., New York.

Maga, J. A. 1982. Phytate: its chemistry, occurrence, food interactions, nutritional significance, and methods of analysis. Journal of Agricultural and Food Chemistry 30:1–9.

Miflin, B. J., ed. 1980a. The Biochemistry of Plants: Amino Acids and Derivatives, Vol. 5. Academic Press, Inc., New York.

———. 1980b. Nitrogen metabolism and amino acid biosynthesis in crop plants. Pages 255–296 *in* Carlson, P. S. (ed.), The Biology of Crop Productivity. Academic Press, Inc., New York.

Miflin, B. J. and P. J. Lea. 1980. Ammonia assimilation. Pages 169–202 *in* Miflin, B. J. (ed.), 1980a.

Naik, M. S., Y. P. Abrol, T. V. R. Nair, and C. S. Ramaras. 1982. Nitrate assimilation—its regulation and relationship to reduced nitrogen in higher plants. Phytochemistry 21:495–504.

Ninomiya, Y. and S. Sato. 1984. A ferredoxin-like electron carrier from non-green cultured tobacco cells. Plant and Cell Physiology 25:453–458.

Noel, K. D., M. Carneol, and W. J. Brill. 1982. Nodule protein synthesis and nitrogenase activity of soybeans exposed to fixed nitrogen. Plant Physiology 70:1236–1241.

Oberleas, D. 1983. Phytate content in cereals and legumes and methods of determination. Cereal Foods World 28:352–357.

Pate, J. S. 1973. Uptake, assimilation and transport of nitrogen compounds by plants. Soil Biology and Biochemistry 5:109–119.

———. 1980. Transport and partitioning of nitrogenous solutes. Annual Review of Plant Physiology 31:313–340.

Pate, J. S. and C. A. Atkins. 1983. Xylem and phloem transport and the functional economy of carbon and nitrogen of a legume leaf. Plant Physiology 71:835–840.

Pernollet, Jean-Claude. 1978. Protein bodies of seeds: ultrastructure, biochemistry, biosynthesis and degradation. Phytochemistry 17:1473–1480.

Peters, G. A. 1978. Blue-green algae and algal associations. BioScience 28:580–585.

Pierce, M. and W. D. Bauer. 1983. A rapid regulatory response governing nodulation in soybean. Plant Physiology 73:286–290.

Postgate, J. R. and F. C. Cannon. 1981. The molecular and genetic manipulation of nitrogen fixation. Philosophical Transactions of the Royal Society of London, Series B 292:589–599.

Raven, J. A. and F. A. Smith. 1976. Nitrogen assimilation and transport in vascular land plants in relation to intracellular pH regulation. New Phytologist 76:415–431.

Rennenberg, H. 1984. The fate of excess sulfur in higher plants. Annual Review of Plant Physiology 35:121–153.

Reynolds, P. H. S., D. G. Blevins, M. J. Boland, K. R. Schubert, and D. D. Randall. 1982. Enzymes of ammonia assimilation in legume nodules: a comparison between ureide- and amide-transporting plants. Physiologia Plantarum 55:255–260.

Reynolds, P. H. S., M. J. Boland, D. G. Blevins, D. D. Randell, and K. R. Schubert. 1982. Ureide biogenesis in leguminous plants. Trends in Biochemical Sciences 7:366–368.

Rice, E. L. 1974. Allelopathy. Academic Press, Inc., New York.

Roberts, G. P. and W. J. Brill. 1981. Genetics and regulation of nitrogen fixation. Annual Review of Microbiology 35:207–235.

Robertson, J. G. and K. J. F. Farnden. 1980. Ultrastructure and metabolism of the developing legume root nodule. Pages 65–115 *in* Miflin, B. J. (ed.), 1980a.

Rognes, S. E. 1980. Anion regulation of lupin asparagine synthetase: chloride activation of the glutamine-utilizing reactions. Phytochemistry 19:2287–2293.

Runge, M. 1983. Physiology and ecology of nitrogen nutrition. Pages 163–200 *in* Lange, O. L., P. S. Nobel, C. B. Osmond, and H. Ziegler (eds.), Physiological Plant Ecology III, Responses to the Chemical and Biological Environment, Encyclopedia of Plant Physiology, New Series, Vol. 12C. Springer-Verlag, Berlin.

Schank, S. C., R. L. Smith, and R. C. Littell. 1983. Establishment of associative N₂-fixing systems. Soil and Crop Science Society of Florida Proceedings 42:113–117.

Schmidt, A. 1979. Photosynthetic assimilation of sulfur compounds. Pages 481–496 in Gibbs, M. and E. Latzko (eds.), Encyclopedia of Plant Physiology, New Series, Vol. 6. Springer-Verlag, New York.

Simpson, R. J., H. Hambers, and M. J. Dalling. 1982. Translocation of nitrogen in a vegetative wheat plant (*Triticum aestivum*). Physiologia Plantarum 56:11–17.

Sinclair, T. R. and C. T. de Witt. 1975. Photosynthate and nitrogen requirement for seed production by various crops. Science 189:565–567.

Somers, D. A., T. M. Kuo, A. Kleinhofs, R. L. Warner, and A. Oaks. 1983. Synthesis and degradation of barley nitrate reductase. Plant Physiology 72:949–952.

Srivastava, H. S. 1980. Regulation of nitrate reductase activity in higher plants. Phytochemistry 19:725–733.

Steward, F. C. and R. G. S. Bidwell, eds. 1983. Plant Physiology: A Treatise. Vol. 8, Nitrogen Metabolism. Academic Press, Inc., New York.

Stewart, G. R., A. F. Mann and P. A. Fentom. 1980. Enzymes of glutamate formation: glutamate dehydrogenase, glutamine synthetase, and glutamate synthase. Pages 271–328 in Miflin, B. J. (ed.), 1980a.

Stewart, W. D. P. 1982. Nitrogen fixation—its current relevance and future potential. Israel J. Botany 31:5–44.

Stewart, W. D. P. and J. R. Gallon, eds. 1980. Nitrogen Fixation. Academic Press, Inc., New York.

Thomas, R. J. and L. E. Schrader. 1981. Ureide metabolism in higher plants. Phytochemistry 20:361–371.

Titus, J. S. and S. Kang. 1982. Nitrogen metabolism, translocation, and recycling in apple trees. Horticultural Reviews 4:204–246.

Torrey, J. G. 1978. Nitrogen fixation by actinomycete-nodulated angiosperms. BioScience 28:586–592.

Urquhart, A. A. and K. W. Joy. 1982. Transport, metabolism, and redistribution of xylem-borne amino acids in developing pea shoots. Plant Physiology 69:1226–1232.

Van Berkum, P. and B. B. Bohlool. 1980. Evaluation of nitrogen fixation by bacteria in association with roots of tropical grasses. Microbiological Reviews 44:491–517.

Vance, C. P. and G. H. Heichel. 1981. Nitrate assimilation during vegetative regrowth of alfalfa. Plant Physiology 68:1052–1056.

Vincent, J. M., ed. 1982. Nitrogen Fixation in Legumes. Academic Press, Inc., New York.

Vose, P. B. and A. P. Ruschel, eds. 1981. Associative N₂-Fixation. CRC Press, Inc., Baco Raton, Florida.

Wallsgrove, R. M., A. J. Keys, P. J. Lea, and B. J. Miflin. 1983. Photosynthesis, photorespiration and nitrogen metabolism. Plant, Cell and Environment 6:301–309.

Weber, E. and D. Neumann. 1980. Protein bodies, storage organelles in plant seeds. Biochemie und Physiologie der Pflanzen 175:279–306.

Weiland, R. T., C. A. Stutte, and P. R. F. Silva. 1982. Nitrogen volatilization from plant foliage. Arkansas Agricultural Experiment Station Report Series 266:3–40.

Wetselaar, R. and G. D. Farquhar. 1980. Nitrogen losses from tops of plants. Advances in Agronomy 33:263–302.

Yates, M. G. 1980. Biochemistry of nitrogen fixation. Pages 1–64 in Miflin, B. J. (ed.), 1980a.

Chapter 14

Asen, S., R. N. Stewart, and K. E. Norris. 1975. Anthocyanin, flavonol copigments, and *p*H responsible for larkspur flower color. Phytochemistry 14:2677–2682.

Barbour, M. G., J. H. Burk, and W. Pitts. 1980. Terrestrial Plant Ecology. The Benjamin/Cummings Publ. Co., Inc., Menlo Park, California.

Beevers, H. 1979. Microbodies in higher plants. Annual Review of Plant Physiology 30:159–193.

———. 1980. The role of the glyoxylate cycle. Pages 117–130 in Stumpf, P. K. (ed.), 1980a.

Bell, E. A. and B. V. Charlwood, eds. 1980. Secondary Plant Products. Encyclopedia of Plant Physiology, New Series, Vol. 8. Springer-Verlag, Berlin.

Bowers, W. S., T. Ohta, J. S. Cleere, P. A. Marsella. 1976. Discovery of insect anti-juvenile hormones in plants. Science 193:542–547.

Brown, S. A. 1981. Coumarins. Pages 269–300 in Conn, E. E. (ed.), 1981.

Buchanan, R. A., F. H. Otey, and M. O. Bagby. 1980. Botanochemicals. Pages 1–22 in Swain, T. and R. Kleiman (eds.), Recent Advances in Phytochemistry, Vol. 14, The Resource Potential in Phytochemistry. Plenum Press, New York.

Chang, F. and C. L. Hsü. 1981. Preocene II affects sex attractancy in medfly. BioScience 31:676–677.

Cohen, J. D. and W. J. Meudt. 1983. Investigations on the mechanism of the brassinosteroid reponse. I. Indole-3-acetic acid metabolism and transport. Plant Physiology 72:691–694.

Conn, E. E., ed. 1981. The Biochemistry of Plants, Volume 7, Secondary Plant Products. Academic Press, Inc., New York.

Cutler, D. F., K. L. Alvin, and C. E. Price. 1980. The Plant Cuticle. Academic Press, Inc., New York.

Darvill, A. G. and P. Albersheim. 1984. Phytoalexins and their elicitors—a defense against microbial infection in plants. Annual Review of Plant Physiology 35:243–275.

Deverall, B. J. 1977. Deference Mechanism of Plants. Cambridge University Press, Cambridge.

Dodson, C. H., R. L. Dressler, H. G. Hills, R. M. Adams, and N. H. Williams. 1969. Biologically active compounds in orchid fragrance. Science 164:1243–1249.

Douce, R. and J. Joyard. 1980. Plant galactolipids. Pages 321–362 in Stumpf, P. K. (ed.), 1980a.

Dressler, R. L. 1982. Biology of the orchid bees (Euglossini). Annual Review of Ecology and Systematics 13:373–394.

Earle, F. R. and Q. Jones. 1962. Analyses of seed samples from 113 plant families. Economic Botany 16:221–250.

Edwards, P. J. and S. D. Wratten. 1980. Ecology of Insect-Plant Relations. Edward Arnold, University Park Press, Baltimore, Maryland.

Evans, R. C., D. T. Tingey, M. L. Gumpertz, and W. F. Burns. 1982. Estimates of isoprene and monoterpene emission rates in plants. Botanical Gazette 143:304–310.

Friend, T. 1981. Plant phenolics, lignification and plant disease. Pages 197–261 in Swain, T., T. B. Harborne, and C. F. Van Summer (eds.), Recent Advances in Phytochemistry, Vol. 7. Pergamon Press, New York.

Galliard, T. 1980. Degradation of acyl lipids: hydrolytic and oxidative enzymes. Pages 85–116 in Stumpf, P. K. (ed.), 1980a.

Gerhardt, B. 1983. Localization of β-oxidation enzymes in peroxisomes isolated from nonfatty plant tissues. Planta 159:238–246.

Geuns, J. M. C. 1978. Steroid hormones and plant growth and development. Phytochemistry 17:1–14.

———. 1982. Plant steroid hormones—what are they and what do they do? Trends in Biochemical Sciences 7:7–9.

Goodwin, T. W. 1980. Biosynthesis of sterols. Pages 485–507 in Stumpf, P. K. (ed.), 1980a.

Gould, J. M. 1983. Probing the structure and dynamics of lignin *in situ*. What's New in Plant Physiology 14:5–8.

Grisebach, H. 1981. Lignins. Pages 457–478 in Conn, E. E. (ed.), 1981.

Grove, M. D. and eight other authors. 1979. Brassinolide, a plant growth-promoting steroid isolated from *Brassica napus* pollen. Nature 281:216–217.

Grunwald, C. 1980. Steroids. Pages 221–256 in Bell, E. A. and B. V. Charlwood (eds.), Encyclopedia of Plant Physiology, Vol. 8, Secondary Plant Products. Springer-Verlag, Berlin.

Gurr, M. I. 1980. The biosynthesis of triacylglycerols. Pages 205–248 in Stumpf, P. K. (ed.), 1980a.

Hahlbrock, K. 1981. Flavonoids. Pages 425–456 in Conn, E. E. (ed.), 1981.

Harborne, J. B. 1976. Functions of flavonoids in plants. Pages 1056–1095 in Harborne, J. B., T. J. Mabry, and H. Mabry (eds.), Chemistry and Biochemistry of Plant Pigments, Vol. 1. Academic Press, Inc., New York.

———, ed. 1978. Biochemical Aspects of Plant and Animal Coevolution. Academic Press, Inc., New York.

———. 1979. Flavonoid Pigments. Pages 619–656 in Rosenthal, G. A. and D. H. Janzen (eds.), 1979.

———. 1982. Introduction to Ecological Biochemistry. Academic Press, Inc., New York.

Harborne, J. B. and T. J. Mabry, eds. 1982. The Flavonoids: Advances in Research 1975–1981. Methuen, Inc., New York.

Harwood, J. L. 1980. Plant acyl lipids: structure, distribution, and analysis. Pages 2–55 in Stumpf, P. K. (ed.), 1980a.

Haslam, E. 1981. Vegetable tannins. Pages 527–556 in Conn, E. E. (ed.), 1981.

Heftmann, E. 1975. Functions of steroids in plants. Phytochemistry 14:891–901.

———. 1983. Biogenesis of steroids in the Solanaceae. Phytochemistry 22:1843–1860.

Hewitt, S., J. R. Hillman, and B. A. Knights. 1980. Steroidal oestrogens and plant growth and development. New Phytologist 85:329–350.

Holloway, P. J. 1980a. Structure and histochemistry of plant cuticular membranes: an overview. Pages 1–32 in Cutler, D. F., K. L. Alvin, and C. E. Price (eds.), The Plant Cuticle. Academic Press, Inc., New York.

———. 1980b. The chemical constitution of plant cutins. Pages 45–86 in Cutler, D. F., K. L. Alvin, and C. E. Price (eds.), The Plant Cuticle. Academic Press, Inc., New York.

Hoshino, T., U. Matsumoto, and T. Goto. 1981. Self-association of some anthocyanins in neutral aqueous solution. Phytochemistry 20:1971–1976.

Hrazdina, G., R. Alscher-Herman, and V. M. Kish. 1980. Subcellular localization of flavonoid synthesizing enzymes in *Pisum, Phaseolus, Brassica* and *Spinacia* cultivars. Phytochemistry 19:1355–1359.

Huang, A. H. C., R. N. Trelease, and T. S. Moore, Jr., eds. 1983. Microbodies in Plants. Academic Press, Inc., New York.

Jones, D. H. 1984. Phenylalanine ammonia lyase: regulation of its induction, and its role in plant development. Phytochemistry 23:1349–1359.

Juniper, B. E. and C. E. Jeffree. 1982. Plant Surfaces. Edward Arnold/University Park Press, Baltimore, Maryland.

Keeler, R. F. 1975. Toxins and teratogens of higher plants. Lloydia 38:56–86.

Kolattukudy, P. E. 1980a. Cutin, suberin, and waxes. Pages 571–645 in Stumpf, P. K. (ed.), 1980a.

———. 1980b. Biopolyester membranes of plants: cutin and suberin. Science 208:990–1000.

———. 1981. Structure, biosynthesis, and biodegradation of cutin and suberin. Annual Review of Plant Physiology 32:539–567.

Lewis, W. H. and M. P. F. Elvin-Lewis. 1977. Medical Botany: Plants Affecting Man's Health. John Wiley & Sons, Inc., New York.

Lin, Y.-H., R. A. Moreau, and A. H. C. Huang. 1982. Involvement of glyoxysomal lipase in the hydrolysis of storage triacylglycerols in the cotyledons of soybean seedlings. Plant Physiology 70:108–112.

Loomis, W. D. and R. Croteau. 1980. Biochemistry of terpenoids. Pages 363–418 in Stumpf, P. K. (ed.), 1980a.

Lord, J. M. and L. M. Roberts. 1980. Formation of glyoxysomes. Trends in Biochemical Sciences 5:271–274.

Lovett, J. V. 1982. Allelopathy and self-defense in plants. Australian Weeds 21:33–36.

Mabry, T. J. 1980. Betalins. Pages 513–533 in Bell, E. A. and B. V. Charlwood (eds.), Secondary Plant Products, Vol. 8, Encyclopedia of Plant Physiology, New Series. Springer-Verlag, Berlin.

Mabry, T. J. and J. E. Gill. 1979. Sesquiterpene lactones and other terpenoids. Pages 502–538 in Rosenthal, G. A. and D. H. Janzen (eds.), 1979.

Macey, M. and P. K. Stumpf. 1983. β-oxidation enzymes in microbodies from tubers of *Helianthus tuberosus*. Plant Science Letters 28:207–212.

Mäder, M. and V. Amberg-Fisher. 1982. Role of peroxidase in lignification of tobacco cells. I. Oxidation of nicotinamide adenine dinucleotide and formation of hydrogen peroxide by cell wall peroxidases. Plant Physiology 70:1128–1131.

Mandava, N. B., J. M. Sasse, and J. H. Yopp. 1981. Brassinolide, a growth-promoting steroidal lactone. II. Activity in selected gibberellin and cytokinin bioassays. Physiologia Plantarum 53:453–461.

McClure, J. W. 1979. The physiology of phenolic compounds in plants. Pages 525–556 in Swain, T., T. B. Harborne, and C. F. Van Summer (eds.), Recent Advances in Phytochemistry, Vol. 12, Biochemistry of Plant Phenolics. Pergamon Press, New York.

McMorris, T. C. 1978. Antheridiol and the oogoniols, steroid hormones which control sexual reproduction in *Achlya*. Philosophical Transactions of the Royal Society of London, Series B 284:459–470.

Mettler, I. J. and H. Beevers. 1980. Oxidation of NADH in glyoxysomes by a malate-aspartate shuttle. Plant Physiology 66:555–560.

Moore, T. S., Jr. 1982. Phospholipid biosynthesis. Annual Review of Plant Physiology 33:235–259.

Mudd, J. B. 1980. Phospholipid biosynthesis. Pages 249–282 in Stumpf, P. K. (ed.), 1980a.

———. 1980b. Sterol interconversions. Pages 509–534 in Stumpf, P. K. (ed.), 1980a.

Muller, C. H. The role of chemical inhibition (allelopathy) in vegetational composition. Bulletin of the Torrey Botanical Club 93:332–351.

Muller, C. H. and C. L. Chou. 1972. Phytotoxins: an ecological phase of phytochemistry. Pages 201–216 in Harborne, J. B. (ed.), Phytochemical Ecology. Academic Press, Inc., New York.

Nes, W. R. 1974. Role of sterols in membranes. Lipids 9:596–612.

Pathak, M. D. and R. C. Saxena. 1976. Insect resistance in crop plants. Current Advances in Plant Science 27:1233–1252.

Piattelli, M. 1981. The betalains: Structure, biosynthesis, and chemical taxonomy. Pages 557–626 in Conn, E. E. (ed.), 1981.

Price, C. E. 1980. A review of the factors influencing the penetration of pesticides through plant leaves. Pages 237–252 in Cutler, D. F., K. L. Alvin, and C. E. Price (eds.), The Plant Cuticle. Academic Press, Inc., New York.

Putnam, A. R. and R. M. Heisey. 1983. Allelopathy: chemical interactions between plants. What's New in Plant Physiology 14:21–24.

Rice, E. L. 1979. Allelopathy—an update. Botanical Review 45:15–109.

———. 1984. Allelopathy. Second Edition. Academic Press, Inc., New York.

Robinson, T. 1979. The evolutionary ecology of alkaloids. Pages 413–448 in Rosenthal, G. A. and D. H. Janzen (eds.), 1979.

———. 1980. The Organic Constituents of Higher Plants. Fourth Edition. Cordus Press, North Amherst, Massachusetts.

———. 1982. The Biochemistry of Alkaloids. Second Edition. Springer-Verlag, Berlin.

Roeske, C. N., J. N. Seiber, L. P. Brower, and C. M. Moffitt. 1976. Milkweed cardenolides and their comparative processing by monarch butterflies. Pages 93–167 in Wallace, J. W. and R. L. Mansell (eds.), Recent Advances in Phytochemistry, Vol. 10, Biochemical Interaction Between Plants and Insects. Plenum Press, New York.

Rosenthal, G. A. and D. H. Janzen. 1979. Herbivores: Their Interaction with Secondary Plant Metabolites. Academic Press, Inc., New York.

Seigler, D. S. 1981. Secondary metabolites and plant systematics. Pages 139–176 in Conn, E. E. (ed.), 1981.

Shutt, D. A. 1976. The effects of plant oestrogens on animal reproduction. Endeavour 35:110–113.

Sinclair, T. R. and C. T. deWit. 1975. Photosynthate and nitrogen requirements for seed production by various crops. Science 189:565–567.

Sláma, K. 1979. Insect hormones and antihormones in plants. Pages 683–700 in Rosenthal, G. A. and D. H. Janzen (eds.), 1979.

———. 1980. Animal hormones and antihormones in plants (a review). Biochemisches und Physiologie Pflanzen 175:177–193.

Sorokin, H. 1967. The spherosomes and the reserve fat in plant cells. American Journal of Botany 54:1008–1016.

Spurgeon, S. L. and J. W. Porter. 1980. Biochemistry of terpenoids. Pages 419–483 in Stumpf, P. K. (ed.), 1980a.

Stoessl, A. 1980. Phytoalexins—a biogenetic perspective. Phytopath. Zeitschrift 99:251–272.

Street, H. E. and H. Öpik. 1970. The Physiology of Flowering Plants: Their Growth and Development. American Elsevier Publishing Co., Inc., New York.

Stumpf, P. K., ed. 1980a. Lipids, Structure and Function. The Biochemistry of Plants, Vol. 4. Academic Press, Inc., New York.

Stumpf, P. K. 1980b. Biosynthesis of saturated and unsaturated fatty acids. Pages 177–204 in Stumpf, P. K. (ed.), 1980a.

———. 1983. Plant fatty acid biosynthesis—present status, future prospects. What's New in Plant Physiology 14:25–28.

Swain, T. 1979. Tannins and lignins. Pages 657–682 in Rosenthal, G. A. and D. H. Janzen (eds.), 1979.

Takeno, K. and R. P. Pharis. 1982. Brassino steroid-induced bending of the leaf lamina of dwarf rice seedlings: an auxin-mediated phenomenon. Plant and Cell Physiology 23:1275–1281.

Thomson, W. W., J. B. Mudd, and M. Gibbs, eds. 1983. Biosynthesis and Function of Plant Lipids. American Society of Plant Physiologists, Rockville, Maryland.

Tolbert, N. E. 1981. Metabolic pathways in peroxisomes and glyoxysomes. Annual Review of Biochemistry 50:133–157.

Trelease, R. N. 1984. Biogenesis of glyoxysomes. Annual Review of Plant Physiology 35:321–347.

Troughton, J. and L. A. Donaldson. 1972. Probing Plant Structure. McGraw-Hill Book Co., New York.

Vickery, B. and M. L. Vickery. 1981. Secondary Plant Metabolism. University Park Press, Baltimore, Maryland.

Waller, G. R. and O. C. Dermer. 1981. Enzymology and alkaloid metabolism in plants and microorganisms. Pages 317–402 in Conn, E. E. (ed.), 1981.

Wanner, G., H. Formanek, and R. R. Theimer. 1981. The ontogeny of lipid bodies (spherosomes) in plant cells. Planta 151:109–123.

Went, F. W. 1974. Reflections and speculations. Annual Review of Plant Physiology 25:1–26.

Whittaker, R. H. and P. P. Feeny. 1971. Allelochemics: chemical interactions between species. Science 171:757–770.

Wong, E. 1976. Biosynthesis of flavonoids. Pages 464–526 in Goodwin, T. W. (ed.), Chemistry and Biochemistry of Plant Pigments, Vol. 1. Academic Press, Inc., New York.

Yatsu, L. Y., T. J. Jacks, and T. P. Hensarling. 1971. Isolation of spherosomes (oleosomes) from onion, cabbage, and cottonseed tissue. Plant Physiology 48:675–682.

Chapter 15

Barlow, P. W. 1975. The root cap. Pages 21–54 in Torrey, J. G. and D. T. Clarkston (eds.), The Development and Function of Roots. Academic Press, Inc., New York.

Beck, William A. 1941. Production of solutes in growing epidermal cells. Plant Physiology 16:637–641.

Berlyn, G. P. 1972. Seed germination and morphogenesis. Pages 223–312 in T. T. Kozlowski (ed.), Seed Biology, Vol. 1, Academic Press, Inc., New York.

Bewley, J. Derek and M. Black. 1978. Physiology and Biochemistry of Seeds in Relation to Germination. Vol. 1, Development, Germination, and Growth. Springer-Verlag, Berlin, New York.

Cleland, R. 1971. Cell mass extension. Annual Review of Plant Physiology 22:197–222.

Clowes, F. A. L. 1975. The quiescent center. Pages 3–19 in Torrey, J. G. and D. T. Clarkston (eds.), The Development and Function of Roots. Academic Press, Inc., New York.

Coombe, B. G. 1976. The development of fleshy fruits. Annual Review of Plant Physiology 26:207–228.

Cutter, E. G. 1969. Plant Anatomy: Experiment and Interpretation Part I, Cells and Tissues. Addison-Wesley Pub. Co., Inc., Reading, Massachusetts.

Dale, John E. 1982. The Growth of Leaves. Edward Arnold, London.

Dure, L. S. III. 1975. Seed formation. Annual Review of Plant Physiology 26:259–278.

Erickson, Ralph O. 1976. Modeling of plant growth. Annual Review of Plant Physiology 27:407–434.

Erickson, Ralph O. and David R. Goddard. 1951. An analysis of root growth in cellular and biochemical terms. Tenth Growth Symposium. Growth 17 (Supp.):89–116.

Erickson, Ralph O. and Katerine B. Sax. 1956. Elemental growth rate of the primary root of *Zea mays*. Proc. Amer. Phil. Soc. 100:487–498.

Erickson, Ralph O. and Wendy K. Silk. 1980. The kinematics of plant growth. Scientific American 242(5):134–151.

Esau, K. 1977. Anatomy of Seed Plants, 2nd Ed. John Wiley & Sons, Inc., New York.

Fahn, A. 1982. Plant Anatomy, 3rd Ed. Pergamon Press, Oxford, New York.

Feldman, Lewis J. 1984. Regulation of root development. Annual Review of Plant Physiology 35:223–242.

Foster, R. C. 1982. The fine structure of epidermal cell mucilages of roots. New Phytologist 91:727–740.

Green, P. B., R. O. Erickson and J. Buggy. 1971. Metabolic and physical control of cell elongation rate. *In vivo* studies in *Nitella*. Plant Physiology 47:423–430.

Hepler, P. K. and B. A. Palevitz. 1974. Microtubules and microfilaments. Annual Review of Plant Physiology 25:309–362.

Hepler, P. K. 1976. Plant Microtubules. Pages 147–187 in Bonner, J. and J. E. Varner (eds.), Plant Biochemistry. Academic Press, Inc., New York.

Hsiao, T. C. 1973. Plant responses to water stress. Annual Review of Plant Physiology 24:519–570.

Hulme, A. C. 1970. The Biochemistry of Fruits and Their Products. Vol. 1. Academic Press, Inc., New York.

Jaffe, Mordecai J. 1980. Morphogenetic responses of plants to mechanical stimuli or stress. BioScience 30:239–243.

Jensen, William A. and Frank B. Salisbury. 1984. Botany, 2nd Ed. Wadsworth Pub. Co., Belmont, California.

John, P. C. L., ed. 1982. The Cell Cycle. Society for Experimental Biology Seminar Series, 10. Cambridge University Press, New York.

Kende, Hans and Bruno Baumgartner. 1974. Regulation of aging in flowers of *Ipomoea tricolor* by ethylene. Planta 116:279–289.

Kramer, Paul J. 1943. Amount and duration of growth of various species of tree seedlings. Plant Physiology 18:239–251.

Kramer, Paul J. and T. T. Kozlowski. 1960. Physiology of Trees. McGraw-Hill Book Co., New York.

Leopold, A. C. and P. E. Kreidemann. 1975. Plant Growth and Development. McGraw-Hill Book Co., New York.

Lockhart, James. 1965. Cell Extension. Pages 826–849 in Bonner, J. A. and J. B. Varner (eds.), Plant Biochemistry. Academic Press, Inc., New York.

Lloyd, Clive W. 1983. Helical microtubular arrays in onion root hairs. Nature 305:311–313.

McNeil, D. L. 1976. The basis of osmotic pressure maintenance during expansion growth in *Helianthus annuus* hypocotyls. Australian Journal of Plant Physiology 3:311–324.

Nursten, H. E. 1970. Volatile compounds: The aroma of fruits. Pages 239–268 *in* Hulme, A. C. (ed.), The Biochemistry of Fruits and Their Products, Vol. I. Academic Press, Inc., New York.

Perry, T. O. 1971. Dormancy of trees in winter. Science 17:29–36.

Ray, P. M., P. B. Green and R. Cleland. 1972. Role of turgor in plant cell growth. Nature 239:163–164.

Rhodes, M. J. C. 1980. The maturation and ripening of fruits. Pages 157–206 *in* Kenneth V. Thimann (ed.), Senescence in Plants. CRC Press, Inc., Boca Raton, Florida.

Richards, F. C. 1969. The quantitative analysis of growth. Pages 2–76 *in* Stewart, F. C. (ed.), Plant Physiology, Vol. 5A, Analysis of growth: Behavior of Plants and Their Organs. Academic Press, Inc., New York.

Richards, F. C. and W. W. Schwabe. 1969. Pages 79–116 *in* Stewart, F. C. (ed.), Plant Physiology, Vol. 5A, Analysis of Growth: Behavior of Plants and Their Organs. Academic Press, Inc., New York.

Sachs, R. M. 1965. Stem elongation. Annual Review of Plant Physiology 16:73–96.

Silk, Wendy Kuhn. 1980. Growth rate patterns which produce curvature and implications for the physiology of the blue light response. Pages 643–655 *in* Senger, H. (ed.), The Blue Light Syndrome. Springer-Verlag, Berlin, Heidelberg.

Silk, Wendy Kuhn. 1984. Quantitative Descriptions of Development. Annual Review of Plant Physiology 35:479–518.

Silk, Wendy Kuhn and Ralph O. Erickson. 1978. Kinematics of hypocotyl curvature. American Journal of Botany 65:310–319.

Sinnott, E. W. 1960. Plant Morphogenesis. McGraw-Hill Book Co., New York.

Spencer, D. and T. J. V. Higgins. 1982. Seed maturation and deposition of storage proteins. Pages 306–336 *in* Smith, H. and D. Grierson (eds.), The Molecular Biology of Plant Development. University of California Press, Berkeley and Los Angeles, California.

Taiz, Lincoln. 1984. Plant cell expansion: regulation of cell wall mechanical properties. Annual Review of Plant Physiology 35:585–622.

Troughton, John and Lesley A. Donaldson. 1972. Probing Plant Structure. McGraw-Hill Book Co., New York.

Went, Fritz W. 1957. The Experimental Control of Plant Growth. The Ronald Press, New York.

Whaley, W. Gordon. 1961. Growth as a general process. *In* Ruhland, W. (ed.), Encyclopedia of Plant Physiology. Growth and Growth Substances. 14:71–112.

Zimmermann, M. H. and C. L. Brown. 1971. Trees—Structure and Function. Springer-Verlag, Berlin.

Chapter 16

Adams, P. A., P. B. Kaufman, and H. Ikuma. 1973. Effects of gibberellic acid and sucrose on the growth of oat (*Avena*) stem segments. Plant Physiology 51:1102–1108.

Adams, P. A., M. J. Montague, M. Tepfer, D. L. Rayle, H. Ikuma, and P. B. Kaufman. 1975. Effect of gibberellic acid on the plasticity and elasticity of *Avena* stem segments. Plant Physiology 56: 757–760.

Akozawa, T. and S. Miyata. 1982. Biosynthesis and secretion of α-amylase and other hydrolases in germinating cereal seeds. Pages 40–78 *in* Campbell, P. N. and R. D. Marshall, (eds.), Essays in Biochemistry, Vol. 18. The Biochemical Society.

Atzorn, R. and E. W. Weiler. 1983. The role of endogenous gibberellins in the formation of α-amylase by aleurone layers of germinating barley caryopses. Planta 159:289–299.

Bandurski, R. S. and H. M. Nonhebel. 1984. Auxins. Pages 1–20 *in* Wilkins, M. B. (ed.), 1984.

Batra, M. W., K. L. Edwards, and T. K. Scott. 1975. Auxin transport in roots. Pages 300–325 *in* Torrey, J. G. and D. T. Clarkson (eds.), The Development and Function of Roots. Academic Press, Inc., New York.

Bernal-Lugo, I., R. N. Beachy, and J. E. Varner. 1981. The response of barley aleurone layers to gibberellic acid includes the transcription of new sequences. Biochemical and Biophysical Research Communications 102:617–623.

Brenner, M. L. 1981. Modern methods for plant growth substance analysis. Annual Review of Plant Physiology 32:511–538.

Broughton, W. J. and A. J. McComb. 1971. Changes in the pattern of enzyme development in gibberellin-treated pea internodes. Annals of Botany 35:213–228.

Carpita, N. C. and K. M. Tarmann. 1982. Promotion of hypocotyl elongation in loblolly pine (*Pinus taeda* L.) by indole-3-acetic acid. Physiologia Plantarum 55:149–154.

Chapman, J. M. and H. V. Davies. 1983. Control of the breakdown of food reserves in germinating dicotyledonous seeds—a reassessment. Annals of Botany 52:593–595.

Cleland, R. E. 1980. Auxin and H$^+$-excretion: the state of our knowledge. Pages 71–78 *in* Skoog, F. (ed.), Plant Growth Substances, 1979. Academic Press, Inc., New York.

Cohen, J. D. and R. S. Bandurski. 1982. Chemistry and physiology of the bound auxins. Annual Review of Plant Physiology 33:403–454.

Crozier, A. 1981. Aspects of the metabolism and physiology of gibberellins. Pages 33–149 *in* Woolhouse, H. W. (ed.), Advances in Botanical Research. Academic Press, Inc., New York.

———, ed. 1983. The Biochemistry and Physiology of Gibberellins, Vols. 1 and 2. Praeger Publishers, New York.

Crozier, A. and J. R. Hillman, eds. 1984. The Biosynthesis and Metabolism of Plant Hormones. Society of Experimental Biology Seminar Series, Vol. 23. Cambridge University Press, London.

Dodds, J. H. and M. A. Hall. 1980. Plant hormone receptors. Scientific Progress, Oxford 66:513–535.

Evans, M. L. 1974. Rapid responses to plant hormones. Annual Review of Plant Physiology 25:195–223.

Galston, A. W. 1964. The Life of the Green Plant. Prentice-Hall, Inc., Englewood Cliffs, New Jersey.

Gibbons, G. C. 1981. On the relative role of the scutellum and aleurone in the production of hydrolases during germination of barley. Carlsberg Research Communications 46:215–225.

Glasziou, K. T. 1969. Control of enzyme formation and inactivation in plants. Annual Review of Plant Physiology 20:63–88.

Goldsmith, M. H. M. 1977. The polar transport of auxin. Annual Review of Plant Physiology 28:439–478.

———. 1982. A saturable site responsible for polar transport of indole-3-acetic acid in sections of maize coleoptiles. Planta 155:68–75.

Gove, J. P. and M. C. Hoyle. 1975. The isozymic similarity of indoleacetic acid oxidase to peroxidase in birch and horseradish. Plant Physiology 56:684–687.

Hassig, B. E. 1974. Origins of adventitious roots. New Zealand Journal of Forestry Science 4:229–310.

Hedden, P. 1979. Aspects of gibberellin chemistry. Pages 19–56 *in* Mandava, N. B. (ed.), 1979.

Hill, T. A. 1980. Endogenous Plant Growth Substances. Second Edition. Edward Arnold, London.

Hillman, J. R., ed. 1978. Isolation of Plant Growth Substances. Cambridge University Press, Cambridge, England.

Hillman, J. R. 1984. Apical dominance. Pages 127–148 *in* Wilkins, M. B. (ed.), 1984.

Ilan, I. and S. Gepstein. 1981. Hormonal regulation of food reserve breakdown in germinating dicotyledonous seeds. Israel Journal of Botany 29:193–206.

Jackson, D. L. and J. A. McWha. 1984. Grain or coleoptile tip as the source of IAA in cereal shoots? Plant, Cell and Environment 7:15–21.

Jacobs, M. and S. F. Gilbert. 1983. Basal localization of the presumptive auxin transport carrier in pea stem cells. Science 220:1297–1300.

Jacobs, W. P. 1979. Plant Hormones and Plant Development. Cambridge University Press, Cambridge, England.

Jones, R. L. 1980. The physiology of gibberellin-induced elongation. Pages 188–195 *in* Skoog, F. (ed.), 1980.

Jones, R. L. and J. MacMillan. 1984. Gibberellins. Pages 21–52 *in* Wilkins, M. B. (ed.), 1984.

Jones, R. L. and I. D. J. Phillips, 1964. Agar diffusion technique for estimating gibberellin production by plant organs. Nature 204:497–499.

Key, J. L. 1969. Hormones and nucleic acid metabolism. Annual Review of Plant Physiology 20: 449–474.

Klingman, G., F. M. Ashton, and L. J. Noordhoff. 1982. Weed Science. Principles and Practices, 2nd Ed. John Wiley & Sons, Inc., Somerset, New Jersey.

Kormondy, E. J., T. F. Sherman, F. B. Salisbury, N. T. Spratt, Jr., and G. McCain. 1977. Biology. The Integrity of Organisms. Wadsworth Publishing Co., Inc., Belmont, California.

Letham, D. S., P. B. Goodwin, and T. J. V. Higgins. 1978. Phytohormones and Related Compounds—a Comprehensive Treatise. Vol. I, The Biochemistry of Phytohormones and Related Compounds. Vol. II, Phytohormones and the Development of Higher Plants. Elsevier/North Holland, Inc., New York.

Liu, P. B. W. and J. B. Loy. 1976. Action of gibberellic acid on cell proliferation in the subapical shoot meristem of watermelon seedlings. American Journal of Botany 63:700–704.

MacMillan, J., ed. 1980. Hormonal Regulation of Development. I. Molecular Aspects of Plant Hormones. Encyclopedia of Plant Physiology, New Series, Vol. 9. Springer-Verlag, Berlin.

Mandava, N. B., ed. 1979. Plant Growth Substances. American Chemical Society, Washington, D. C.

Mariott, K. M. and D. H. Northcote. 1975. The induction of enzyme activity in the endosperm of germinating castor-bean seeds. The Biochemical Journal 152:65–70.

Marré, E. 1979. Fusicoccin: a tool in plant physiology. Annual Review of Plant Physiology 30:273–312.

McComb, A. J. and W. J. Broughton. 1972. Metabolic changes in internodes of dwarf pea plants treated with gibberellic acid. Pages 407–413 *in* Carr, D. J. (ed.), Plant Growth Substances 1970. Springer-Verlag, Berlin.

Mitchell, J. W., D. P. Skaggs, and W. P. Anderson. 1951. Plant growth-stimulating hormones in immature bean seeds. Science 114:159–161.

Moloney, M. M., M. C. Elliott, and R. E. Cleland. 1981. Acid growth effects in maize roots: evidence for a link between auxin-economy and proton extrusion in the control of root growth. Planta 152:285–291.

Moore, T. C. 1979. Biochemistry and Physiology of Plant Hormones. Springer-Verlag, Berlin.

Moreland, D. E. 1980. Mechanisms of action of herbicides. Annual Review of Plant Physiology 31:597–638.

Morgan, P. W. and J. I. Durham. 1983. Strategies for extracting, purifying, and assaying auxins from plant tissues. Botanical Gazette 144:20–31.

Mozer, J. 1980. Partial purification and characterization of the mRNA for α-amylase from barley aleurone layers. Plant Physiology 65:834–837.

Mulkey, T. J., K. M. Kuzmanoff, and M. L. Evans. 1982. Promotion of growth and shift in the auxin dose/response relationship in maize roots treated with the ethylene biosynthesis inhibitors aminoethoxyvinylglycine and cobalt. Plant Science Letters 25:43–48.

Nickell, L. G. 1978. Controlling biological behavior with chemicals. Chemical and Engineering News, Oct. 9.

——. 1979. Controlling biological behavior of plants with synthetic plant growth regulating chemicals. Pages 263–279 in Mandava, N. B. (ed.), 1979.

——. 1982. Plant Growth Regulators. Agricultural Uses. Springer-Verlag, Berlin.

——, ed. 1983. Plant Growth Regulating Chemicals. Vols. I and II. CRC Press, Inc., Boca Raton, Florida.

Oertli, J. J. 1975. Effects of external solute supply on cell elongation in barley coleoptiles. Zeitschrift für Pflanzenphysiologie 74:440–450.

Pengelly, W. L. and J. G. Torrey. 1982. The relationship between growth and indole-3-acetic acid content of roots of Pisum sativum L. Botanical Gazette 143:195–200.

Pharis, R. P. and C. C. Kuo. 1977. Physiology of gibberellins in conifers. Canadian Journal of Forest Research 7:299–325.

Phillips, I. D. J. 1975. Apical dominance. Annual Review of Plant Physiology 26:341–367.

Phinney, B. O. 1979. Gibberellin biosynthesis in the fungus Gibberella fujikuroi and in higher plants. Pages 57–78 in Mandava, N. B. (ed.), 1979.

——. 1983. The history of the gibberellins. Pages 19–52 in Crozier, A. (ed.), The Biochemistry and Physiology of the Gibberellins, Vol. 1. Praeger Publishers, New York.

——. 1984. Gibberellin A₁, dwarfism and the control of shoot elongation in higher plants. Pages 17–41 in Crozier, A. and J. R. Hillman (eds.), The Biosynthesis and Metabolism of Plant Hormones. Society for Experimental Biology Seminar Series, Vol. 23. Cambridge University Press, London.

Phinney, B. O. and C. Spray. 1982. Chemical genetics and the gibberellin pathway. Pages 101–110 in Wareing, P. F. (ed.), 1982.

Railton, I. D. 1982. Gibberellin metabolism in plants. Cell Biology International Reports 6:319–337.

Rayle, D. L. and R. Cleland. 1979. Control of plant cell enlargement by hydrogen ions. Current Topics in Developmental Biology 11:187–214.

Reeve, D. R. and A. Crozier. 1974. An assessment of gibberellin structure-activity relationships. Journal of Experimental Botany 25:431–445.

——. 1980. Quantitative analysis of plant hormones. Pages 203–280 in MacMillan, J. (ed.), 1980.

Rubenstein, B. and M. A. Nagao. 1976. Lateral bud outgrowth and its control by the apex. The Botanical Review 42:83–113.

Rubery, P. H. 1981. Auxin receptors. Annual Review of Plant Physiology 32:569–596.

Sachs, R. M. 1965. Stem elongation. Annual Review of Plant Physiology 16:73–96.

Salisbury, F. B. and R. V. Parke. 1964. Vascular Plants: Form and Function. Wadsworth Publishing Co., Inc., Belmont, California.

Scott, T. K. 1972. Auxins and roots. Annual Review of Plant Physiology 23:235–258.

Sembdner, G., D. Gross, H.-W. Liebisch, and G. Schneider. 1980. Biosynthesis and metabolism of plant hormones. Pages 281–444 in MacMillan, J. (ed.), 1980.

Skoog, F., ed. 1980. Plant Growth Substances 1979. Springer-Verlag, Berlin.

Stevenson, T. T. and R. E. Cleland. 1981. Osmoregulation in the Avena coleoptile in relation to auxin and growth. Plant Physiology 67:749–753.

Stuart, D. A. and R. L. Jones. 1977. The roles of extensibility and turgor in gibberellin- and dark-stimulated growth. Plant Physiology 59:61–68.

——. 1978. The role of acidification in gibberellic acid- and fusicoccin-induced elongation growth of lettuce hypocotyl sections. Planta 142:135–145.

Taiz, L. 1984. Plant cell expansion: regulation of mechanical properties. Annual Review of Plant Physiology 35:585–657.

Tanimoto, E. and Y. Masuda. 1971. Role of the epidermis in auxin-induced elongation of light-grown pea stem segments. Plant and Cell Physiology 12:663–673.

Terry, M., D. McGraw, and R. L. Jones. 1982. Effect of IAA on growth and soluble cell wall polysaccharides centrifuged from pine hypocotyl sections. Plant Physiology 69:323–326.

Theologis, A. and P. M. Ray. 1982. Early auxin-regulated polyadenylylated mRNA sequences in pea stem tissue. Proceedings of the National Academy of Sciences USA.

Thimann, K. V. 1980. The development of plant hormone research in the last 60 years. Pages 15–33 in Skoog, F. (ed.), 1980.

Trewavas, A. J. 1981. How do plant growth substances work? Plant, Cell and Environment 4:203–228.

——. 1982. Growth substance sensitivity: the limiting factor for plant development. Physiologia Plantarum 55:60–72.

Vanderhoef, L. N. 1980. Auxin-regulated elongation: a summary hypothesis. Pages 90–96 in Skoog, F. (ed.), 1980.

Vanderhoef, L. N., C. Stahl, N. Siegel, and R. Zeigler. 1973. The inhibition by cytokinin of auxin promoted elongation in excised soybean hypocotyl. Physiologia Plantarum 29:22–27.

Wareing, P. F., ed. 1982. Plant Growth Substances 1982. Academic Press, Inc., London.

Weiler, E. W. 1984. Immunoassay of plant growth regulators. Annual Review of Plant Physiology 35:85–95.

Wightman, F. and D. L. Lighty. 1982. Identification of phenylacetic acid as a natural auxin in the shoots of higher plants. Physiologia Plantarum 55:17–24.

Wightman, F., E. A. Schneider, and K. V. Thimann. 1980. Hormonal factors controlling the initiation and development of lateral roots. II. Effects of exogenous growth factors on lateral root formation in pea roots. Physiologia Plantarum 49:304–314.

Wilkins, M. B., ed. 1984. Advanced Plant Physiology. Pitman Publishing Limited, London.

Yokota, T., N. Murofushi, and N. Takahashi. 1980. Extraction, purification, and identification. Pages 113–201 in MacMillan, J. (ed.), 1980.

Zurfluh, L. L. and T. J. Guilfoyle. 1982a. Auxin-induced changes in the population of translatable messenger RNA in elongating sections of soybean hypocotyl. Plant Physiology 69:332–337.

——. 1982b. Auxin-induced changes in the population of translatable messenger RNA in elongating maize coleoptile sections. Planta 156:525–527.

Chapter 17

Abeles, F. B. 1973. Ethylene in Plant Biology. Academic Press, Inc., New York.

Adams, D. D. and S. F. Yang. 1979. Ethylene biosynthesis: identification of 1-aminocyclopropane-1-carboxylic acid as an intermediate in the conversion of methionine to ethylene. Proceedings of the National Academy of Sciences USA 76:170–174.

Addicott, F. T., ed. 1982. Abscission. University of California Press, Berkeley.

——, ed. 1983. Abscisic Acid. Praeger Publishers, New York.

Addicott, F. T. and Carns, H. R. 1983. History and introduction. Pages 1–21 in Addicott, F. T. (ed.), 1983.

Barton, K. A. and W. J. Brill. 1983. Prospects in plant genetic engineering. Science 219:671–676.

Bewley, J. D. and M. Black. 1982. Physiology and Biochemistry of Seeds, Vol. 2. Springer-Verlag, Berlin.

Bewli, I. S. and F. H. Witham. 1976. Characterization of kinetin-induced water uptake by detached radish cotyledons. Botanical Gazette 137:58–64.

Beyer, E. M., Jr. 1976. A potent inhibitor of ethylene action in plants. Plant Physiology 58:268–271.

Beyer, E. M., Jr., P. W. Morgan, and S. F. Yang. 1984. Ethylene. Pages 111–126 in Wilkins, M. B. (ed.), 1984.

Biale, J. B. and R. E. Young. 1981. Respiration and ripening of fruits—retrospect and prospect. Pages 1–39 in Friend, J. and M. J. C. Rhodes (eds.), Advances in the Biochemistry of Fruits and Vegetables. Academic Press, Inc., London.

Black, M. 1983. Abscisic acid in seed germination and dormancy. Pages 331–363 in Addicott, F. T. (ed.), 1983.

Black, M. and J. E. Shuttleworth. 1974. The role of the cotyledons in the photocontrol of hypocotyl extension in Cucumis sativus L. Planta 117:57–66.

Bradford, K. 1983. Involvement of plant growth substances in the alteration of leaf gas exchange of flooded tomato plants. Plant Physiology 73:480–483.

Bradford, K. and S. F. Yang. 1981. Physiological responses of plants to waterlogging. HortScience 16:25–30.

Burg, S. P. 1973. Ethylene in plant growth. Proceedings of the National Academy of Sciences USA 70:591–597.

Camp, P. J. and J. L. Wickliff. 1981. Light or ethylene treatments induce transverse cell enlargement in etiolated maize mesocotyls. Plant Physiology 67:125–128.

Carmi, A. and J. van Staden. 1983. Role of roots in regulating the growth rate and cytokinin content in leaves. Plant Physiology 73:76–78.

Chaleff, R. S. 1983. Isolation of agronomically useful mutants from plant cell cultures. Science 219:676–682.

Chen, C. M. and D. K. Melitz. 1979. Cytokinin biosynthesis in a cell-free system from cytokinin-autotrophic tobacco tissue cultures. FEBS Letters 107:15–20.

Chen, C. M. and B. Petschow. 1978. Cytokinin biosynthesis in cultured rootless tobacco plants. Plant Physiology 62:861–865.

Coombe, B. G. 1976. The development of fleshy fruits. Annual Review of Plant Physiology 27:207–228.

Davies, D. R. 1981. Cell and tissue culture: potentials for plant breeding. Philosophical Transactions of the Royal Society of London, Series B 292:547–556.

Davies, W. J. and T. A. Mansfield. 1983. The role of abscisic acid in drought avoidance. Pages 237–268 in Addicott, F. T. (ed.), 1983.

Dilley, D. R. 1978. Approaches to maintenance of postharvest integrity. Journal of Food Biochemistry 2:235–242.

Downs, R. J. 1962. Photocontrol of growth and dormancy in woody plants. Pages 133–148 in Kozlowski, T. (ed.), Tree Growth. Ronald Press.

Downs, R. J. and H. W. Siegelman. 1963. Photocontrol of anthocyanin synthesis in milo seedlings. Plant Physiology 38:25–30.

Eisinger, W. 1983. Regulation of pea internode expansion by ethylene. Annual Review of Plant Physiology 34:225–240.

Ernst, D., W. Schäfer, and D. Oesterhelt. 1983. Isolation and identification of a new, naturally occurring cytokinin (6-benzylaminopurineriboside) from an anise cell culture (Pimpinella anisum L.). Planta 159:222–225.

Fosket, D. E. 1977. The regulation of the plant cell cycle by cytokinin. Pages 62–91 in Rost, T. L. and E. M. Gifford, Jr. (eds.), Mechanisms and Control of Cell Division. Dowden, Hutchinson, and Ross, Inc., Stroudsburg, Pennsylvania.

Fosket, D. E., L. C. Morejohn, and K. E. Westerling. 1981. Control of growth by cytokinin: an examination of tubulin synthesis during cytokinin-induced growth in cultured cells of Paul's scarlet rose. Pages 193–211 in Guern, J. and C. Péaud-Lenoël (eds.), 1981.

Galston, A. W. 1983. Polyamines as modulators of plant development. BioScience 33:382–388.

Gersani, M. and H. Kende. 1982. Studies on cytokinin-stimulated translocation in isolated bean leaves. Journal of Plant Growth Regulation 1:161–171.

Gorham, J. 1977. Lunularic acid and related compounds in liverworts, algae, and Hydrangea. Phytochemistry 16:249–253.

Greene, E. M. 1980. Cytokinin production by microorganisms. The Botanical Review 46:25–74.

Guern, J. and C. Péaud-Lenoël, eds. 1981. Metabolism and Molecular Activities of Cytokinins. Springer-Verlag, Berlin.

Halevy, A. H. and S. Mayak. 1981. Senescence and postharvest physiology of cut flowers—part 2. Horticultural Reviews 3:59–143.

Hartman, K. M. 1967. Ein wirkungsspektrum der photomorphogenese unter hochenergiebedingungen und seine interpretation auf der basis des phytochroms. Zeitschrift für Naturforschung 22b:1172–1175.

Hasegawa, K. and T. Hashimoto. 1975. Variation in abscisic acid and batasin content of yam bulbils—effects of stratification and light exposure. Journal of Experimental Botany 26:757–764.

Hoffman, N. E. and S. F. Yang. 1980. Changes of 1-aminocyclopropane-1-carboxylic acid content in ripening fruits in relation to their ethylene production rates. Journal of the American Society for Horticultural Science 105:492–495.

Holmes, M. G. and H. Smith. 1975. The function of phytochrome in plants growing in natural environments. Nature 254:512–514.

Horgan, R. 1984. Cytokinins. Pages 53–75 in Wilkins, M. B. (ed.), 1984.

Huff, A. K. and C. W. Ross. 1975. Promotion of radish cotyledon enlargement and reducing sugar content by zeatin and red light. Plant Physiology 56:429–433.

Hutton, M. J. and J. van Staden. 1983. Transport and metabolism of 8(^{14}C) zeatin applied to leaves of Citrus sinensis. Zeitschrift für Pflanzenphysiologie 111:75–83.

Imaseki, H., Y. Yoshii, and I. Todaka. 1982. Regulation of auxin-induced ethylene biosynthesis in plants. Pages 259–268 in Wareing, P. F. (ed.), 1982.

Jackson, M. B. 1982. Ethylene as a growth hormone under flooded conditions. Pages 291–301 in Wareing, P. F. (ed.), 1982.

Kawase, M. 1981. Anatomical and morphological adaptation of plants to waterlogging. HortScience 16:30–34.

Klein, W. H., L. Price, and K. Mitrakos. 1963. Light stimulated starch degradation in plastids and leaf morphogenesis. Photochemistry and Photobiology 2:233–240.

Kozlowski, T. T., ed. 1973. The Shedding of Plant Parts. Academic Press, Inc., New York.

Kushad, M. M., D. G. Richardson, and A. J. Ferro. 1983. Intermediates in the recycling of 5-methylthioribose to methionine in fruits. Plant Physiology 73:257–261.

Laloue, M. and C. Pethe. 1982. Dynamics of cytokinin metabolism in tobacco cells. Pages 185–195 in Wareing, P. F. (ed.), 1982.

Laughlin, R. G., R. L. Munyon, S. K. Ries, and V. F. Wert. 1983. Growth enhancement of plants by fentomole doses of colloidally dispersed triacontanol. Science 219:1219–1221.

Leopold, A. C. and P. E. Kriedemann. 1975. Plant Growth and Development. McGraw-Hill, Inc., New York.

Letham, D. S. 1971. Regulators of cell division in plant tissues. XII. A cytokinin bioassay using excised radish cotyledons. Physiologia Plantarum 25:391–396.

———. 1974. Regulators of cell division in plant tissues. XX. The cytokinins of coconut milk. Physiologia Plantarum 32:66–70.

Letham, D. S. and L. M. S. Palni. 1983. The biosynthesis and metabolism of cytokinins. Annual Review of Plant Physiology 34:163–197.

Lew, R. and H. Tsuji. 1982. Effect of benzyladenine treatment duration on delta-aminolevulinic acid accumulation in the dark, chlorophyll lag phase abolition, and long-term chlorophyll production in excised cotyledons of dark-grown cucumber seedlings. Plant Physiology 690663–667.

Lieberman, M. 1979. Biosynthesis and action of ethylene. Annual Review of Plant Physiology 30:533–589.

Loy, J. B. 1980. Promotion of hypocotyl elongation in watermelon seedlings by 6-benzyladenine. Journal of Experimental Botany 31:743–750.

Matsubara, S. 1980. Structure-activity relationships of cytokinins. Phytochemistry 19:2239–2253.

Milborrow, B. V. 1967. The identification of (+) abscicin II ((+)-dormin) in plants and measurement of its concentrations. Planta 76:93–113.

———. 1983. Pathways to and from abscisic acid. Pages 79–111 in Addicott, F. T. (ed.), 1983.

Mohr, H. 1957. Der Einfluss monochromatischer Strahlung auf das Längenwachstum des hypocotyls und auf die anthocyanbildung bei Keimlinger von Sinapis alba. Planta 49:389–405.

Murai, N., F. Skoog, M. E. Doyle, and R. S. Hanson. 1980. Relationships between cytokinin production, presence of plasmids, and fasciation caused by strains of Corynebacterium fascians. Proceedings of the National Academy of Sciences USA 77:619–623.

Nabors, M. W. 1976. Using spontaneously occurring and induced mutations to obtain agriculturally useful plants. BioScience 26:761–768.

———. 1983. Increasing the salt and drought tolerance of crop plants. Pages 165–184 in Randall, D. D., D. G. Blevins, R. L. Larson, and B. J. Rapp (eds.), Current Topics in Plant Biochemistry and Physiology, Vol. 2. University of Missouri, Columbia, Missouri.

Narain, A. and M. M. Laloraya. 1974. Cucumber cotyledon expansion as a bioassay for cytokinins. Zeitschrift für Pflanzenphysiologie 71:313–322.

Ng, P. P., A. L. J. Cole, P. E. Jameson, and J. A. Mcwha. 1982. Cytokinin production by ectomycorrhizal fungi. New Phytologist 91:57–62.

Nickell, L. G. 1979. Controlling biological behavior of plants with synthetic plant growth regulating chemicals. Pages 263–279 in Mandava, N. B. (ed.), Plant Growth Substances. American Chemical Society, Washington, D. C.

Osborne, D. J. 1982. The ethylene regulation of cell growth in specific target tissues of plants. Pages 279–290 in Wareing, P. F. (ed.), 1982.

Palni, L. M. S., R. Horgan, N. M. Darrall, T. Stuchbury, and P. F. Wareing. Cytokinin biosynthesis in crown-gall tissue of Vinca rosea. The significance of nucleotides. Planta 159:50–59.

Parker, M. W., S. B. Hendricks, H. A. Borthwick, and F. W. Went. 1949. Spectral sensitivities for leaf and stem growth of etiolated pea seedlings and their similarity to action spectra for photoperiodism. American Journal of Botany 36:194–204.

Parthier, B. 1979. The role of phytohormones (cytokinins) in chloroplast development (a review). Biochemie und Physiologie der Pflanzen 174:173–214.

Patrick, J. W. 1976. Hormone-directed transport of metabolites. Pages 433–446 in Wardlaw, I. F. and J. Passioura (eds.), Transport and Transfer Processes in Plants. Academic Press, Inc., New York.

Pillay, I. and I. D. Railton. 1983. Complete release of axillary buds from apical dominance in intact, light-grown seedlings of Pisum sativum L. following a single application of cytokinin. Plant Physiology 71:972–974.

Preger, R. and S. Gepstein. 1984. Carbon dioxide-independent and -dependent components of light inhibition of the conversion of 1-aminocyclopropane-1-carboxylic acid to ethylene in oat leaves. Physiologia Plantarum 60:187–191.

Radin, J. W. and R. C. Ackerson. 1982. Does abscisic acid control stomatal closure during water stress? What's New in Plant Physiology 13(3):9–12.

Raskin, I. and H. Kende. 1984. Regulation of growth in stem sections of deep-water rice. Planta 160:66–72.

Rayle, D. L., C. W. Ross, and N. Robinson. 1982. Estimation of osmotic parameters accompanying zeatin-induced growth of detached cucumber cotyledons. Plant Physiology 70:1634–1636.

Ries, S. and V. Wert. 1982. Rapid effects of triacontanol in vivo and in vitro. Journal of Plant Growth Regulation 1:117–127.

Ross, C. W. and D. L. Rayle. 1982. Evaluation of H$^+$ secretion relative to zeatin-induced growth of detached cucumber cotyledons. Plant Physiology 70:1470–1474.

Sacher, J. A. 1983. Abscisic acid in leaf senescence. Pages 479–522 in Addicott, F. T. (ed.), 1983.

Sexton, R. and J. A. Roberts. 1982. Cell biology of abscission. Annual Review of Plant Physiology 33:133–162.

Sexton, R. and H. W. Woolhouse. 1984. Senescence and abscission. Pages 469–497 in Wilkins, M. B. (ed.), 1984.

Shepard, J. F., D. Bidney, T. Barsby, and R. Remble. 1983. Genetic transfer in plants through interspecific protoplast fusion. Science 219:683–688.

Siegelman, H. W. and S. B. Hendricks. 1957. Photocontrol of anthocyanin formation in turnip and red cabbage seedlings. Plant Physiology 32:393–398.

———. 1958. Photocontrol of anthocyanin synthesis in apple skin. Plant Physiology 33:185–190.

Skene, K. G. M. 1975. Cytokinin production by roots as a factor in the control of plant growth. Pages 365–396 in Torrey, J. G. and D. T. Clarkson (eds.), The Development and Function of Roots. Academic Press, Inc., New York.

Skoog, F. and R. Y. Schmitz. 1979. Biochemistry and physiology of cytokinins. Pages 335–413 in Litwack, G. (ed.), Biochemical Actions of Hormones. Vol. VI. Academic Press, Inc., San Francisco.

Stewart, R. N., M. Lieberman, and A. T. Kunishi. 1974. Effects of ethylene and gibberellic acid on cellular growth and development in apical and subapical regions of etiolated pea seedlings. Plant Physiology 54:1–3.

Stoddart, J. L. and H. Thomas. 1982. Leaf senescence. Pages 592–636 in Boulter, D. and B. Parthier (eds.), Encyclopedia of Plant Physiology, New Series, Vol. 14A, Nucleic Acids and Proteins in Plants I. Springer-Verlag, Berlin.

Tabor, C. W. and H. Tabor. 1984. Polyamines. Annual Review of Biochemistry 53:749–790.

Thimann, K. V., ed. 1980. Senescence in Plants. CRC Press, Inc., Boca Raton, Florida.

Thimann, K. V., S. O. Satler, and V. Trippi. 1982. Further extension of the syndrome of leaf senescence. Pages 539–548 in Wareing, P. F. (ed.), 1982.

Thomas, J., C. W. Ross, C. J. Chastain, N. Koomanoff, J. E. Hendrix, and E. van Volkenburgh. 1981. Cytokinin-induced wall extensibility in excised cotyledons of radish and cucumber. Plant Physiology 68:107–110.

Tillberg, E. 1983. Levels of endogenous abscisic acid in achenes of Rosa rugosa during dormancy release and germination. Physiologia Plantarum 58:243–248.

Torrey, J. G. 1976. Root hormones and plant growth. Annual Review of Plant Physiology 27:435–459.

Vanderhoef, L. N., C. A. Stahl, N. Siegel, and R. Zeigler. 1973. The inhibition by cytokinin of auxin-promoted elongation in excised soybean hypocotyl. Physiologia Plantarum 29:22–27.

Van Staden, J. 1983. Seeds and cytokinins (minireview). Physiologia Plantarum 58:340–346.

Varner, J. E. and D. T. Ho. 1976. Hormones. Pages 713–770 in Bonner, J. and J. E. Varner (eds.), Plant Biochemistry. Academic Press, Inc., New York.

Walton, D. C. 1980/1981. Does ABA play a role in seed germination? Israel Journal of Botany 29:168–180.

———. 1980. Biochemistry and physiology of abscisic acid. Annual Review of Plant Physiology 31:453–489.

Wareing, P. F., ed. 1982. Plant Growth Substances 1982. Academic Press, Inc., London.

Wilkins, M. B., ed. 1984. Advanced Plant Physiology. Pitman Publishing Limited, London.

Withrow, R. B., W. H. Klein, and V. Elstad. 1957. Action spectra of photomorphogenic induction and its photoinactivation. Plant Physiology 32:453–462.

Woolhouse, H. W. 1978. Senescence processes in the life cycle of flowering plants. BioScience 28:25–31.

Wright, S. T. C. 1966. Growth and cellular differentiation in the wheat coleoptile (*Triticum vulgare*). II. Factors influencing the growth response to gibberellic acid, kinetin, and indole-3-acetic acid. Journal of Experimental Botany 17:165–176.

Yang, S. F., H. E. Hoffman, T. McKeon, J. Riov, C. H. Kao, and K. H. Yung. 1982. Mechanism and regulation of ethylene biosynthesis. Pages 239–248 in Wareing, P. F. (ed.), 1982.

Zavala, M. E. and D. L. Brandon. 1983. Localization of a phytochrome using immunocytochemistry. Journal of Cell Biology 97:1235–1239.

Chapter 18

Audus, L. J. 1979. Plant geosensors. Journal of Experimental Botany 30:1051–1073.

Ball, N. G. 1969. Nastic responses. Pages 277–300 in M. B. Wilkins (ed.), Physiology of plant growth and development. McGraw-Hill Book Company, New York.

Bandurski, Robert S., A. Schulze, P. Dayanandan, and P. B. Kaufman. 1984. Response to gravity by *Zea mays* seedlings. I. Time course of the response. Plant Physiology 74:284–288.

Behrens, H. M., M. H. Weisenseel, and A. Sievers. 1982. Rapid changes in the pattern of electric current around the root tip of *Lepidium sativum* L. following gravistimulation. Plant Physiology 70:1079–1083.

Bewley, J. D. and M. Black. 1982. Physiology and Biochemistry of Seeds in Relation to Germination. In Two Volumes. Springer-Verlag, Berlin, Heidelberg, New York.

Björkman, Olle and S. B. Powles. 1981. Leaf movement in the shade species *Oxalis oregana*. I. Response to light level and light quality. Annual Report of the Director, Dept. of Plant Biology, Stanford, California. Carnegie Institution of Washington Year Book, 80:59–62. (See also following article, Powles and Björkman, pages 63–66 and report in Science News (1981) 120:392.)

Blaauw, A. H. 1909. Die Perzeption des Lichtes (The perception of light). Rec. Trav. Botica Neerlandica. 5:209–272.

Blaauw, O. H. and G. Blaauw-Jansen. 1970a. The phototropic responses of *Avena* coleoptiles. Acta Botica Neerlandica 19:755–763.

Blaauw, O. H. and G. Blaauw-Jansen. 1970b. Third positive (c-type) phototropism in the *Avena* coleoptile. Acta Botica Neerlandica 19:764–776.

Brain, Robert D., J. A. Freeberg, C. V. Weiss, and Winslow R. Briggs. 1977. Blue light-induced absorbance changes in membrane fractions from corn and *Neurospora*. Plant Physiology 59:948–952.

Brauner, L. and A. Hager. 1958. Versuche Zur Analyse Der Geotropischen Perzeption (Experiments to analyze geotropic perception). Planta, Bd. 51, S. 115–147.

Briggs, Winslow R. 1963. Mediation of phototropic responses of corn coleoptiles by lateral transport of auxin. Plant Physiology 38:237–247.

Briggs, Winslow W. and M. Iino. 1983. Blue-light-absorbing photoreceptors in plants. Philosophical Transactions of the Royal Society of London B303:347–359.

Chilton, Mary-Dell. 1983. A vector for introducing new genes into plants. Scientific American 248(6):51–59.

Cholodny, N. 1926. Beiträge zur Analyse der geotropischen Reaktion. Jahrb. Wiss. (Contributions to the analysis of the geotropic reactions.) Botany 65:447–459.

Ciesielski, Theophil. 1872. Untershungen über die Abwärtskrümmung der Wurzel. (Investigations on the downward bending of roots.) Beitraege Zur Biologie de Pflanzen 1:1–30. (The Darwins cited Ciesielski's inaugural dissertation, Abwärtskrümmung der Wurzel, Breslau, 1871.)

Clifford, Paul E., D. S. Fensom, B. I. Munt, and W. D. McDowell, 1982. Lateral stress initiates bending responses in dandelion peduncles: a clue to geotropism? Canadian Journal of Botany 50:2671–2673.

Clifford, Paul E., D. M. Reid, and R. P. Pharis. 1983. Endogenous ethylene does not initiate but may modify geobending—A role for ethylene in autotropism. Plant, Cell and Environment 6:433–436.

Curry, G. M. 1969. Phototropism. Pages 241–273 in The Physiology of Plant Growth and Development. Wilkins, M. B. (ed.), McGraw-Hill, New York.

Darwin, Charles, assisted by Francis Darwin. 1881. The Power of Movement in Plants. Appleton-Century-Crofts, Inc., New York ("Authorized Edition," 1896, D. Appleton and Company, New York; reprinted by De Capo Press, New York, 1966).

Davis, Eric and Anne Schuster. 1981. Intercellular communication in plants: evidence for a rapidly generated, bidirectionally transmitted wound signal. Proceedings of the National Academy of Sciences 78:2422–2426.

Dayanandan, P., F. V. Hebard, and P. B. Kaufman. 1976. Cell elongation in the grass pulvinus in response to geotropic stimulation and auxin application. Planta (Berlin) 131:245–252.

Dayanandan, P., V. H. Frederick, V. D. Baldwin, and P. B. Kaufman. 1977. Structure of gravity-sensitive sheath and internodal pulvini in grass shoots. American Journal of Botany 64(10):1189–1199.

Dennison, David S. 1979. Phototropism In W. Haupt and M. E. Feinbeib (eds.), Physiology of Movements, Vol. 7 of A. Pirson and M. H. Zimmerman (eds.), Encyclopedia of Plant Physiology (New Series), Springer-Verlag, Berlin, Heidelberg, New York. 7:506–508.

Dennison, David S. 1984. Phototropism. Pages 149–162 in M. B. Wilkins (ed.), Advanced Plant Physiology. Pitman Pub. Limited, London and Marshfield, Massachusetts.

Digby, John and Richard D. Firn. 1976. A critical assessment of the Cholodny-Went theory of shoot geotropism. Current Advances in Plant Science 8:953–960.

Dolk, H. E. 1974. Geotropie en groestrof. (Geotropism and growth substances.) Dissertation, Utrecht.

du Buy, H. G. and E. Nuernbergk. 1934. Phototropismus und Wachstum der Pflanzen. (Photogropism and growth of plants.) II. Ergeb. Biol. 10:207–322.

Ehleringer, J. and I. Forseth. 1980. Solar tracking by plants. Science 210:1094–1098.

El-Antably, H. M. M. 1975. Redistribution of endogenous indole-acetic acid, abscisic acid and gibberellins in geotropically stimulated *Ribes nigrum* roots. Zeitschrift für Pflanzenphysiologie 76:400–410.

Everett, M. 1974. Dose-response curves for radish seedling phototropism. Plant Physiology 54:222–225.

Feldman, L. J. 1981. Light-induced inhibitors from intact and cultured caps of *Zea* roots. Planta 153:471–475.

Firn, Richard D. and John Digby. 1980. The establishment of tropic curvatures in plants. Annual Review of Plant Physiology 31:131–148.

Franssen, J. M. and J. Bruinsma. 1981. Relationships between xanthoxin, phototropism, and elongation growth in the sunflower seedling *Helianthus annuus* L. Planta 151:365–370.

Franssen, J. M., S. A. Cooke, John Digby, and Richard D. Firn. 1981. Measurements of differential growth causing phototropic curvature of coleoptiles and hypocotyls. Z. Pflanzenphysiol. 103:207–216.

Franssen, J. M., Richard D. Firn, and John Digby. 1982. The role of the apex in the phototropic curvature of *Avena* coleoptiles: positive curvature under conditions of continuous illumination. Planta 155:281–286.

Gould, F. W. 1968. Grass systematics. McGraw-Hill Book Company, New York.

Haberlandt, F. 1902. Über die Statolithefunction der Stärkekörner. (About the statolith function of starch grains.) Berichte der Deutschen Botanisches Gesellschaft 20:189–195.

Hamner, P. A., C. A. Mitchell, and T. C. Weiler, 1974. Height control in greenhouse chrysanthemum by mechanical stress. Hortscience. 9:474–475

Haupt, Wolfgang and M. D. Feinleib. 1979. Introduction. In W. Haupt and M. E. Feinleib (eds.), Physiology of Movements, Vol. 7 of A. Pirson and M. H. Zimmermann (eds.), Encyclopedia of Plant Physiology (New Series), Springer-Verlag, Berlin, Heidelberg, New York. 7:1–8.

Heslop-Harrison, J. 1976. Carnivorous plants a century after Darwin. Endeavour 35:114–122.

Heslop-Harrison, Yokande. 1978. Carnivorous Plants. Scientific American 238(2):104–115.

Hillman, S. K. and M. B. Wilkins. 1982. Gravity perception in decapped roots of *Zea mays*. Planta 155:267–271.

Inversen, T. H. 1969. Elimination of geotropic responsiveness in roots of cress (*Lepidium sativum*) by removal of statolith starch. Physiologia Plantarum 22:1251–1262.

Iversen, T. H. 1974. The roles of statoliths, auxin transport, and auxin metabolism in root geotropism. K. norske Vidensk, Selsk, Mus. Miscellanea 15:1–216.

Jackson, M. B. and P. W. Barlow. 1981. Root geotropism and the role of growth regulators from the cap: a re-examination. Plant, Cell and Environment 4:107–123.

Jaffe, M. J. 1973. Thigmomorphogenesis: The response of plant growth and development to mechanical stimulation. Planta 114:143–157.

Jaffe, M. J. 1976. Thigmomorphogenesis: a detailed characterization of the response of beans (*Phaseolus vulgaris* L.) to mechanical stimulation. Zeitschrift für Pflanzenphysiologie 77:437–453.

Jaffe, M. J. 1980. Morphogenetic responses of plants to mechanical stimuli or stress. BioScience 30(4):239–243.

Jaffe, M. J. and A. W. Galston. 1968. The physiology of tendrils. Annual Review of Plant Physiology 19:417–434.

Jensen, William A. and Frank B. Salisbury. 1972. Botany: An Ecological Approach. Wadsworth Publishing Company, Inc., Belmont, California.

Johnsson, Anders. 1971. Aspects on gravity-induced movements in plants. Quarterly Reviews of Biophysics 2-4:277–320.

Juniper, Barrie E. 1976. Geotropism. Annual Review of Plant Physiology. 27:385–406.

Juniper, Barrie E., S. Groves, B. Landua-Schachar, and L. J. Audus. 1966. Root cap and the perception of gravity. Nature (London) 209:93–94.

Lee, J. S., T. J. Mulkey, and M. L. Evans. 1983. Reversible loss of gravitropic sensitivity in maize roots after tip application of calcium chelators. Science 220:1375–1376.

MacDonald, Ian R., James W. Hart, and Dennis C. Gordon. 1983. Analysis of growth during geotropic curvature in seedlings hypocotyls. Plant, Cell and Environment 6:401–406.

Mandoli, Dina F. and Winslow R. Briggs. 1982. Optical properties of etiolated plant tissues. Proceedings of the National Academy of Sciences USA. 79:1902–1906.

Mandoli, Dina F. and Winslow R. Briggs. 1983. Physiology and optics of plant tissues. What's New in Plant Physiology 14:13–16.

Mandoli, Dina F. and Winslow R. Briggs. 1984. Fiber-optic plant tissues: spectral dependence in dark-grown green tissues. Photochemistry and Photobiology 39(3):419–424.

Mertens, Rüdiger and Elmar W. Weiler. 1983. Kinetic studies on the redistribution of endogenous growth regulators in gravireacting plant organs. Planta 158:339–348.

Meyer, A. M. 1969. Versuche zue Trennung von 1. positiver und negativer Krümmung der *Avena* koleoptile. (Experiments to separate first positive and negative curvature of the *Avena* coleoptile). Zeitschrift für Pflanzenphysiologie 60:135–146.

Meyer, B. S. and D. B. Anderson. 1952. Plant Physiology, Second Edition. D. Van Nostrand Company, Inc., New York.

Mitchell, Cary A. 1977. Influence of mechanical stress on auxin-stimulated growth of excised pea stem sections. Physiologia Plantarum 41:129–134.

Mitchell, Cary A., Candace J. Severson, John A. Wott, and P. Allen Hammer. 1975. Seismomorphogenic regulation of plant growth. Journal of American Society of Horticultural Science 100(2):161–165.

Mueller, Wesley J., Frank B. Salisbury, and P. Thomas Blotter. 1985. Gravitropism in higher plant shoots. II. Dimensional and pressure changes during stem bending. Plant Physiology (in press).

Mulkey, Timothy J. and Michael L. Evans. 1981. Geotropism in corn roots: Evidence for its mediation by differential acid efflux. Science 212:70–71.

Mulkey, Timothy J. and Michael L. Evans. 1982. Suppression of asymmetric acid efflux and gravitropism in maize roots treated with auxin transport inhibitors or sodium orthovanadata. J. Plant Growth Regul. 1:259–265.

Mulkey, Timothy J., Konrad M. Kuzmanoff, and Michael L. Evans. 1981. The agar-dye method for visualizing acid efflux patterns during tropistic curvatures. What's New in Plant Physiology 12:9–12.

Munoz, V. and W. L. Butler. 1975. Photoreceptor pigments for blue light in Neurospora crassa. Plant Physiology 55:421–426.

Nemec, B. 1901. Über die Wahrnehmung des Schwerkraftreizes bei den Pflanzen. (About the perception of gravity by plants.) Jahrb. Wiss. Bot. 36:80–178.

Pickard, B. G. 1973. Action potentials in higher plants. Botanical Review 39:172–201.

Pilet, P.-E. 1977. Root georeaction: gravity effect on hormone balance. In Life Sciences Research in Space. Burke, W. R., T. D. Guyenne (eds.), E. S. A. Scientific and Techical Publications Branch, Noordwijk, The Netherlands.

Roblin, G. 1982. Movements and bioelectrical events induced by photostimulation in the primary pulvinus of Mimosa pudica. Zeitschrift für Pflanzenphysiologie 106:299–303.

Salisbury, Frank B. 1963. The Flowering Process. Pergamon Press, Oxford, London, New York, Paris.

Salisbury, Frank B., Helga Hilscher, and Linda Gillespie. 1984. The mechanics of gravitropic bending in dicot stems. Plant Physiology 75(1):179.

Salisbury, Frank B., Julianne E. Sliwinski, Wesley J. Mueller, and Chauncy S. Harris. 1982a. How stems bend up. Utah Science 43:42–49.

Salisbury, Frank B. and Ray M. Wheeler. 1981. Interpreting plant responses to clinostating. Plant Physiology 67:677–685.

Salisbury, Frank B., Ray M. Wheeler, Julianne E. Sliwinski, and Wesley J. Mueller. 1982b. Plants, gravity, and mechanical stresses. Utah Science 43:14–21.

Samejima, Michikazu and Takao Sibaoka. 1980. Changes in the extracellular ion concentration in the main pulvinus of Mimosa pudica during rapid movement and recovery. Plant and Cell Physiology 21:467–479.

Sangwan, R. S. and B. Norreel. 1975. Induction of plants from pollen grains of Petunia cultured in vitro. Nature 257:222–224.

Satter, Ruth L. and Arthur W. Galston. 1981. Mechanisms of control of leaf movements. Annual Reviews of Plant Physiology. 32:83–110.

Satter, R. L., D. D. Sabnis, and A. W. Galston. 1970. Phytochrome controlled nyctinasty in Albizzi julibrissin. I. Anatomy and fine structure of the pulvinule. American Journal of Botany 57:374–381.

Schopfer, P. 1984. Photomorphogenesis. Pages 380–407 in M. B. Wilkins (ed.), Advanced Plant Physiology. Pitman Pub. Limited, London.

Schrempf, M., Ruth L. Satter, and Arthur W. Galston. 1976. Potassium-linked chloride fluxes during rhythmic leaf movement of Albizzia julibrissin. Plant Physiology 58:190–192.

Schwartz, Amnon and Dov Koller. 1978. Phototropic response to vectorial light in leaves of Lavatera cretica L. Plant Physiology 61:924–928.

Schwartz, Amnon and Dov Koller. 1980. Role of the cotyledons in the phototropic response of Lavatera cretica seedlings. Plant Physiology 66:82–87.

Scurfield, G. 1973. Reaction wood: Its structure and function. Science 179:647–655.

Shakel, K. A. and A. E. Hall. 1979. Reversible leaflet movements in relation to drought adaptation of cowpeas, Vigna unguiculata L. Walp. Aust. J. Plant Physiol. 6:265–276.

Shaw, S. and W. B. Wilkins. 1973. The source and lateral transport of growth inhibitors in geotropically stimulated roots of Zea mays and Pisum sativum. Planta 109:11–26.

Shen-Miller, J. and R. R. Hinchman. 1974. Gravity sensing in plants: A critique of the statolith theory. BioScience 24:643–651.

Shepard, James F. 1982. The regeneration of potato plants from leaf-cell protoplasts. Scientific American 246(5):154–166.

Sibaoka, T. 1969. Physiology of rapid movements in higher plants. Annual Review of Plant Physiology 20:165–184.

Simons, P. J. 1981. The role of electricity in plant movements. New Phytologist 87:11–37.

Sinnot, E. W. 1960. Plant morphogenesis. McGraw-Hill Book Company, New York.

Sliwinski, Julianne E. and Frank B. Salisbury. 1985. Gravitropism in higher plant shoots. III. Cell dimensions during gravitropic bending; perception of gravity. Plant Physiology (in press).

Strotmann, H. and S. Bickel-Sandkotter. 1984. Immunoassay of plant growth regulators. Annual Review of Plant Physiology 35:85–96.

Sunderland, N. 1970. Pollen plants and their significance. New Scientist 47:142–144.

Suzuke, T., N. Kondo and T. Fujii. 1979. Distribution of growth regulators in relation to the light-induced geotropic responsiveness in Zea roots. Planta 145:323–329.

Thimann, K. V. and G. M. Curry. 1960. Phototropism and phototaxis. Pages 243–309 in Comparative Biochemistry: A Comparative Treatise. Florkin, M., H. S. Mason (eds.), New York: Academic Press, Vol. I, Sources of Free Energy.

Tibbitts, T. W. and W. M. Hertzberg. 1978. Growth and epinasty of marigold plants maintained from emergence on horizontal clinostats. Plant Physiology 61:199–203.

Umrath, Karl and G. Kastberger. 1983. Action potentials of the high-speed conduction in Mimosa pudica and Neptunia plena. Phyton 23:65–78.

Van Sambeek, Jerome W. and Barbara G. Pickard. 1976. Mediation of rapid electrical, metabolic, transpirational, and photosythetic changes by factors released from wounds. III. Measurements of CO_2 and H_2O flux. Canadian Journal of Botany 54:2662–2671.

Vierstra, R. D. and K. L. Poff. 1981. Role of carotenoids in the phototropic response of corn seedlings. Plant Physiology 68:798–801.

Volkmann, D. and A. Sievers. 1979. Graviperception in multicellular organs. In W. Haupt and M. E. Feinbeib (eds.), Physiology of Movements, Vol. 7 of A. Pirson and M. H. Zimmermann (eds.), Encyclopedia of Plant Physiology (New Series). Springer-Verlag, Berlin, Heidelberg, New York. 7:573–600.

Wainwright, C. M. 1977. Sun-tracking and related leaf movements in a desert lupine (Lupinus arizonicus). American Journal of Botany 64:1032–1041.

Watanabe, S. and T. Sibaoka. 1983. Light- and auxin-induced leaflet opening in detached pinnae of Mimosa pudica. Plant and Cell Physiology 24:641–647.

Went, F. W. 1926. On growth accelerating substances in the coleoptile of Avena sativa. Proc. K. Akad. Wet. Amsterdam 30:10–19.

Wheeler, Raymond M. and Frank B. Salisbury. 1979. Water spray as a convenient means of imparting mechanical stimulation of plants. HortScience 14(3):270–271.

Wheeler, Raymond M. and Frank B. Salisbury. 1981. Gravitropism in higher plant shoots. I. A role for ethylene. Plant Physiology 67:686–690.

Wilkins, M. B. 1975. The role of the root cap in root geotropism. Current Advances in Plant Science 8:317–328.

Wilkins, M. B. 1979. Growth-control mechanisms in gravitropism. In W. Haupt and M. E. Feinleib (eds.), Physiology of Movements, Vol. 7 of A. Pirson and M. H. Zimmermann (eds.), Encyclopedia of Plant Physiology (New Series). Springer-Verlag, Berlin, Heidelberg, New York. 7:601–626.

Wilkins, Malcolm B. 1984. Gravitropism. Pages 163–185 in M. B. Wilkins (ed.), Advanced Plant Physiology. Pitman Pub. Limited, London and Marshfield, Massachusetts.

Wilkins, H. and R. L. Wain. The root cap and control of root elongation in Vea mays L. seedlings exposed to white light. Planta 121:1–8.

Williams, Stephen E. and Alan B. Bennett. 1982. Leaf closure in the Venus flytrap: an acid growth response. Science 218:1120–1122.

Wilson, Brayton F. and Robert R. Archer. 1977. Reaction wood: induction and mechanical action. Annual Review of Plant Physiology 28:23–43.

Wright, Luann Z. and David L. Rayle. 1983. Evidence for a relationship between H^+ excretion and auxin in shoot gravitropism. Plant Physiology 72:99–104.

Yin, H. C. 1938. Diaphototropic movements of the leaves of Malva neglecta. American Journal of Botany 25:1–6.

Zawadzki, Tadeusz and Kazimierz Trebacz. 1982. Action potentials in Lupinus angustifolius L. shoots. IV. Propagation of action potential in the stem after the application of mechanical block. Journal of Experimental Botany 33(132):100–110.

Zimmermann, B. D. and W. R. Briggs. 1963. Phototropic dosage-response curves for oat coleoptiles. Plant Physiology 38:248–253.

Chapter 19

Bartley, M. R. and B. Frankland. 1982. Analysis of the dual role of phytochrome in the photoinhibition of seed germination. Nature 300:750–752.

Bewley, J. D. and M. Black. 1982. Physiology and Biochemistry of Seeds, Vol. 2, Viability, Dormancy, and Environmental Control. Springer-Verlag, Berlin.

Black, M. and J. Shuttleworth. 1976. Inter-organ effects in the control of growth. Pages 317–322 in Smith, H. (ed.), Light and Plant Development. Butterworth and Co., Ltd., London.

Borthwick, H. 1972. History of phytochrome. Pages 3–23 in Mitrakos, K. and W. Shropshire, Jr. (eds.), Phytochrome. Academic Press, Inc., New York.

Briggs, W. R. 1976. H. A. Borthwick and S. B. Hendricks—pioneers of photomorphogenesis. Pages 1–6 in Smith H. (ed.), Light and Plant Development. Butterworth and Co., Ltd., London.

Briggs, W. R. and M. Iino. 1983. Blue-light-absorbing photoreceptors in plants. Philosophical Transactions of the Royal Society of London, Series B 303:347–359.

Castelfranco, P. A. and S. I. Beale. 1983. Chlorophyll biosynthesis: recent advances and areas of current interest. Annual Review of Plant Physiology 34:241–278.

Colbert, J. T., H. P. Hershey, and P. H. Quail. 1983. Autoregulatory control of translatable phytochrome messenger RNA levels. Proceedings of the National Academy of Sciences USA 80:2248–2252.

Cordonnier, M.-M., C. Smith, H. Greppin, and L. H. Pratt. 1983. Production and purification of monoclonal antibodies to Pisum and Avena phytochrome. Planta 158:369–376.

Creswell, E. G. and J. P. Grime. 1981. Induction of a light requirement during seed development and its ecological consequences. Nature 291:583–585.

Dale, J. E. 1982. The Growth of Leaves. Edward Arnold Publishers Limited, London.

De Greef, J. A. and H. Fredericq. 1983. Photomorphogenesis and hormones. Pages 401–427 in Shropshire, W., Jr. and H. Smith (eds.), 1983a.

Deregibus, V. A., R. A. Sanchez, and J. J. Casal. 1983. Effects of light quality on tiller production in Lolium spp. Plant Physiology 72:900–902.

Esashi, Y., R. Kuraishi, N. Tanaka, and S. Satoh. 1983. Transition from primary to secondary dormancy in cocklebur seeds. Plant, Cell and Environment 6:493–499.

Frankland, B. 1981. Germination in shade. Pages 187–204 in Smith, H. (ed.), 1981.

Frankland, B. and R. Taylorson. 1983. Light control of seed germination. Pages 428–456 in Shropshire, W., Jr. and H. Mohr (eds.), 1983a.

Gorski, T., K. Górska, and J. Rybicki. 1978. Studies on the germination of seeds under leaf canopy. Flora 167:289–299.

Gorton, H. L. and W. R. Briggs. 1980. Phytochrome responses to end-of-day irradiations in light-grown corn in the presence and absence of Sandoz 9789. Plant Physiology 66:1024–1026.

Gressel, J. 1980. Blue light and transcription. Pages 133–153 in Senger, H. (ed.), The Blue Light Syndrome. Springer-Verlag, Berlin.

Grime, J. P. 1979. Plant Strategies and Vegetation Processes. John Wiley, London.

———. 1981. Plant strategies in shade. Pages 159–186 in Smith, H. (ed.), 1983.

Hahlbrock, K. and H. Grisebach. 1975. Biosynthesis of flavonoids. Pages 866–915 in Harborne, J. B., T. J. Mabry, and H. Mabry (eds.), The Flavonoids. Academic Press, Inc., New York.

Haupt, W. 1966. Phototaxis in plants. International Review of Cytology 19:267–299.

Hendricks, S. B. 1960. Comparative Boichemistry of Photoreactive Systems. Academic Press, Inc., New York.

Holmes, M. G. 1983. The perception of shade. Philosophical Transactions of the Royal Society of London, Series B 303:503–521.

Inoue, Y. and K. Shibata. 1973. Light-induced chloroplast rearrangements and their action spectra as measured by absorption spectrophotometry. Planta 114:341–358.

Jabben, M. and G. F. Dietzer. 1978. A method for measuring phytochrome in plants grown in white light. Photochemistry and Photobiology 27:797–802.

Jabben, M. and M. G. Holmes. 1983. Phytochrome in light-grown plants. Pages 704–722 in Shropshire, W., Jr. and H. Mohr (eds.), 1983b.

Jensen, W. A. and F. B. Salisbury. 1972. Botany: An Ecological Approach. Wadsworth Publishing Company, Inc., Belmont, California.

Johnson, C. B. 1982. Photomorphogenesis. Pages 365–403 in Smith, H. and D. Grierson (eds.), The Molecular Biology of Plant Development. University of California Press, Berkeley.

Jose, A. M. and E. Schäfer. 1978. Distorted phytochrome action spectra in green plants. Planta 138:25–28.

Kasemir, H. 1983. Light control of chlorophyll accumulation in higher plants. Pages 662–686 in Shropshire, W., Jr. and H. Mohr (eds.), 1983b.

Kendrick, R. E. and J. de Kok. 1983. Intermediates in phytochrome transformation. Scientific Progress, Oxford 68:475–486.

Kendrick, R. E. and B. Frankland. 1983. Phytochrome and Plant Growth. Edward Arnold, University Park Press, Baltimore, Maryland.

Koller, D. 1969. The physiology of dormancy and survival of plants in desert environments. Pages 449–469 in Woolhouse, H. W. (ed.), Dormancy and Survival. Symposia for the Society of Experimental Biology, no. 23. Academic Press, Inc., New York.

Lamb, C. J. and M. A. Lawton. 1983. Photocontrol of gene expression. Pages 213–257 in Shropshire, W., Jr. and H. Mohr (eds.), 1983a.

Mancinelli, A. L. 1983. The photoregulation of anthocyanin synthesis. Pages 640–661 in Shropshire, W., Jr. and H. Mohr (eds.), 1983b.

Mancinelli, A. L. and I. Rabino. 1978. The "high irradiance responses" of plant photomorphogenesis. The Botanical Review 44:129–180.

Mandoli, D. F. and W. R. Briggs. 1981. Phytochrome control of two low-irradiance responses in etiolated oat seedlings. Plant Physiology 67:733–739.

Mitrakos, K. and W. Shropshire, Jr. (eds.) 1972. Phytochrome. Academic Press, Inc., New York.

Mohr, H. 1974. The role of phytochrome in controlling enzyme levels in plants. Pages 37–81 in Paul, J. (ed.), The Biochemistry of Cell Differentiation. M. T. P. International Review of Science, Vol. 9. Butterworth and Co., Ltd. London.

———. 1983. Pattern specification and realization in photomorphogenesis. Pages 338–357 in Shropshire, W., Jr. and H. Mohr (eds.), 1983a.

Mohr, H. and E. Schäfer. 1983. Photoperception and de-etiolation. Philosophical Transactions of the Royal Society of London, Series B 303: 489–501.

Morgan, D. C. 1981. Shadelight quality effects on plant growth. Pages 205–222 in Smith, H. (ed.), 1981.

Morgan, D. C. and H. Smith. 1976. Linear relationship between phytochrome photoequilibrium and growth in plants under simulated natural radiation. Nature 262:210–212.

———. 1979. A systematic relationship between phytochrome-controlled development and species habitat, for plants grown in simulated natural radiation. Planta 145:253–258.

———. 1981. Non-photosynthetic responses to light quality. Pages 109–134 in Lange, O. L., P. S. Nobel, C. B. Osmond, and H. Ziegler (eds.), Physiological Plant Ecology I, Encyclopedia of Plant Physiology, New Series, Vol. 12A. Springer-Verlag, Berlin.

Nabors, M. W. and A. Lang. 1971. The growth physics and water relations of red-light induced germination in lettuce seeds. I. Embryos germinating in osmoticum. Planta 101:1–25.

Newbury, H. J. 1983. Photocontrol of enzyme activation and inactivation. Philosophical Transactions of the Royal Society of London, Series B 303:433–441.

Ninnemann, H. 1980. Blue light photoreceptors. BioScience 30:166–170.

Pratt, L. H. 1982. Phytochrome: the protein moiety. Annual Review of Plant Physiology 33:557–582.

———. 1983. Assay of photomorphogenic photoreceptors. Pages 152–177 in Shropshire, W., Jr. and H. Mohr (eds.), 1983a.

Quail, P. H., J. T. Colbert, H. P. Hershey, and R. D. Vierstra. 1983. Phytochrome: molecular properties and biogenesis. Philosophical Transactions of the Royal Society of London, Series B 303:387–402.

Rollin, P. 1972. Phytochrome control of seed germination. Pages 229–254 in Mitrakos, K. and W. Shropshire, Jr. (eds.), Phytochrome. Academic Press, Inc., New York.

Rüdiger, W. 1983. Chemistry of the phytochrome photoconversions. Philosophical Transactions of the Royal Society of London, Series B 303: 377–386.

Rüdiger, W. and H. Scheer. 1983. Chromophores in photomorphogenesis. Pages 119–151 in Shropshire, W., Jr. and H. Mohr (eds.), 1983a.

Saunders, M. J., M.-M. Cordonnier, B. A. Palevitz, and L. H. Pratt. 1983. Immunofluorescence visualization of phytochrome in Pisum sativum L. epicotyls using monoclonal antibodies. Planta 159:545–553.

Schäfer, E. and W. Haupt. 1983. Blue-light effects in phytochrome-mediated responses. Pages 722–744 in Shropshire, W., Jr. and H. Mohr (eds.), 1983b.

Schäfer, E., T.-U. Hassig, and P. Schopfer. 1982. Phytochrome-controlled extension growth of Avena sativa L. seedlings. II. Fluence rate response relationships and action spectra of mesocotyl and coleoptile responses. Planta 154:231–240.

Schopfer, P., K.-H. Fidelak, and E. Schäfer. 1982. Phytochrome-controlled extension growth of Avena sativa L. seedlings. I. Kinetic characterization of mesocotyl, coleoptile, and leaf responses. Planta 154:224–230.

Senger, H., ed. 1980. The Blue Light Syndrome. Springer-Verlag, Heidelberg.

Shimazaki, Y., Y. Moriyasu, L. H. Pratt, and M. Furuya. 1981. Isolation of the red-light-absorbing form of phytochrome from light-grown pea shoots. Plant and Cell Physiology 22:1165–1173.

Shropshire, W., Jr. and H. Mohr (eds.) 1983a. Photomorphogenesis. Encyclopedia of Plant Physiology, New Series, Vol. 16A. Springer-Verlag, Berlin.

———, eds. 1983b. Photomorphogenesis Encyclopedia of Plant Physiology, New Series, Vol. 16B. Springer-Verlag, Berlin.

Small, J. G. C., C. J. P. Spruit, G. Blaauw-Jansen, and O. H. Blaauw. 1979. Action spectra for light-induced germination in dormant lettuce seeds. II. Blue region. Planta 144:133–136.

Smith, H. 1972. The photocontrol of flavonoid synthesis. Pages 433–481 in Mitrakos, K. and W. Shropshire, Jr. (eds.), 1972.

———, ed. 1981. Plants and the Daylight Spectrum. Academic Press, Inc., New York.

———. 1982. Light quality, photoperception and plant strategy. Annual Review of Plant Physiology 33:481–518.

———. 1983. Is P$_{fr}$ the active form of phytochrome? Philosophical Transactions of the Royal Society of London, Series B 303:443–452.

Smith, W. O. 1983. Phytochrome as a molecule. Pages 96–118 in Shropshire, W., Jr. and H. Mohr, (eds.), 1983a.

Thomas, B. 1981. Specific effects of blue light on plant growth and development. Pages 443–460 in Smith, H., (ed.), 1981.

Thomson, B. 1954. The effect of light on cell division and cell elongation in seedlings of oats and peas. American Journal of Botany 41:326–332.

Toole, V. K. 1973. Effects of light, temperature, and their interactions on the germination of seeds. Seed Science and Technology 1:339–396.

Van Volkenburgh, E. and R. E. Cleland. 1979. Separation of cell enlargement and division in bean leaves. Planta 146:245–247.

———. 1981. Control of light-induced bean leaf expansion: role of osmotic potential, wall yield stress, and hydraulic conductivity. Planta 153: 572–577.

Vierstra, R. D. and P. H. Quail. 1983a. Photochemistry of 124 kilodalton Avena phytochrome in vitro. Plant Physiology 72:264–267.

———. 1983b. Purification and initial characterization of 124 kilodalton phytochrome. Biochemistry 22:2498–2505.

Vierstra, R. D., M.-M. Cordonnier, L. H. Pratt, and P. H. Quail. 1984. Native phytochrome: immunoblot analysis of relative molecular mass and in-vitro proteolytic degradation for several plant species. Planta 160:521–528.

Vince-Prue, D. 1975. Photoperiodism in Plants. McGraw-Hill Book Co., New York.

Vince-Prue, D. and A. E. Canham. 1983. Horticultural significance of photomorphogenesis. Pages 517-544 in Shropshire, W., Jr. and M. Mohr (eds.), 1983b.

Wong, E. 1976. Biosynthesis of flavonoids. Pages 464–526 in Goodwin, T. W. (ed.), Chemistry and Biochemistry of Plant Pigments, Vol. 1. Academic Press, Inc., New York.

Chapter 20

Aschoff, J., K. Hoffman, H. Pohl, and R. Wever. 1975. Retrainment of circadian rhythms after phase-shifts of the Zeitgeber. Cronobiologia 2:23–78.

Behling, Ingeborg. 1929. Über das Zeitgedächtnis der Bienen. (On the time memory in honey bees.) Zeitschrift für vergleichende Physiologie 9:259–338.

Blakemore, Richard P. and Richard B. Frankel. 1981. Magnetic navigation in bacteria. Scientific American 245(6):58–65.

Brown, Frank, A., Jr. and Carol S. Chow. 1973a. Interorganismic and environmental influences through extremely weak electromagnetic fields. Biology Bulletin 144:437–461.

Brown, Frank A., Jr. and Carol S. Chow. 1973b. Lunar-correlated variations in water uptake by bean seeds. Biology Bulletin 145:265–278.

Brown, F. A., Jr. and H. M. Webb. 1948. Temperature relations of an endogenous daily rhythmicity in the fiddler crab, *Uca*. Physiologie Zoologie 21:371–381.

Bünning, Erwin. 1960. Opening Address: Biological Clocks. Cold Spring Harbor Symposia on Quantitative Biology 15:1–9.

Bünning, Erwin. 1973. The Physiological Clock. Third Edition. Academic Press, London, England.

Bünning, Erwin. 1977. Fifty years of research in the wake of Wilhelm Pfeffer. Annual review of Plant Physiology 28:1–22.

Bünning, E. and Moser, I. 1972. Influence of valinomycin on circadian leaf movements of *Phaseolus*. Proceedings of the National Academy of Sciences. USA 69:2732–2733.

Bünning, E. and K. Stern. 1930. Über die tagesperiodischen Bewegungen der Primärblätter von *Phaseolus multiflorus* II. Die Bewengungen bei Termokonstanz. (The diurnal movements of the primary leaves of *Phaseolus multiflorus*. II. Movement at constant temperature.) Berichte der Deutschen Botanischen Gesellschaft 48:227–252.

Bünning, E., K. Stern and R. Stoppel. 1930. Versuche über den Einfluss von Luftionen auf die Schlafbewegungen von *Phaseolus*. (Experiments on the influence of atmosphere ions on the sleep movements of *Phaseolus*.) Planta/Archiv für wissenschaftliche Botanik 11(1):67–74.

Darwin, Charles assisted by Francis Darwin. 1881. The Power of Movement in Plants. Appleton, New York.

De Mairan, J. 1729. Observation botanique. (Botanical observations.) Histoire de l'Academie Royale des Sciences. Paris, p. 35.

Ehret, Charles F. 1960. Action spectra and nucleic acid metabolism in circadian rhythms at the cellular level. Pages 149–158 in Cold Spring Harbor Symposia on Quantitative Biology, Vol. XXV, Biological Clocks. The Biological Laboratory, Cold Spring Harbor, New York.

Engelmann, Wolfgang and Martin Schrempf. 1980. Membrane models for circadian rhythms. Photochemical and Photobiological Reviews 5:49–86.

Enright, J. T. 1982. Sleep movements of leaves: in defense of Darvin's interpretation. Oecologia 54:253–259.

Feldman, Jerry F. 1982. Genetic approaches to circadian clocks. Annual Review of Plant Physiology 33:583–608.

Feldman, Jerry F. 1983. Genetics of circadian clocks. BioScience 33:426–431.

Galston, Arthur W. 1974. The unscientific method. Natural History, March, pp. 18–24.

Garner, W. W. and H. A. Allard. 1920. Effect of the relative length of day and night and other factors of the environment on growth and reproduction in plants. Journal of Agricultural Research 18:553–606.

Garner, W. W. and H. A. Allard. 1931. Effect of abnormally long and short alterations of light and darkness on growth and development of plants. Journal of Agricultural Research 42:629–651.

Gorton Holly L. and Ruth L. Satter, 1983. Circadian rhythmicity in leaf pulvini. BioScience 33:451–456.

Harris, Philip J. C. and M. B. Wilkins. 1978. Evidence of phytochrome involvement in the entrainment of the circadian rhythm of carbon dioxide metabolism in *Bryophyllum*. Planta 138:271–278.

Harris, Philip J. C. and M. B. Wilkins. 1978. The circadian rhythm in *Pryophyllum* leaves: phase control by radiant energy. Planta 143:323–328.

Halberg, Franz, Erna Halberg, Cyrus P. Barnum, and John J. Bittner. 1959. Physiologic 24-hour periodicity in human beings and mice, the lighting regimen and daily routine. Pages 803–878 in R. B. Withrow (ed.), Photoperiodism and Related Phenomena in Plants and Animals. American Association for the Advancement of Science, Washington DC.

Highkin, Harry R. and John B. Hanson. 1954. Possible interaction between light-dark cycles and endogenous daily rhythms on the growth of tomato plants. Plants Physiology 29:301–304.

Hillman, William S. 1976. Biological rhythms and physiological timing. Annual Review of Plant Physiology 27:159–179.

Johnson, C. H., James F. Roeber, and J. W. Hastings. 1984. Circadian changes in enzyme concentration account for rhythm of enzyme activity in *Gonyaulax*. Science 223:1428–1430.

Kleinhoonte, Anthonia. 1932. Untersuchungen über die autonomen Bewegungen der Primärblätter von *Canavalia ensiformis* (Investigations on the autonomous leaf movements of the primary leaves of *Canavalia ensiformis*). Jahrbuch für wissenschaftliche. Botanik 75:679–725.

Lörcher, L. 1957. Die wirkung vershiedener lichtqualitäten auf die endogene Tagersrhythmik von *Phaseolus* (Effect of various light qualities on the endogenous diurnal rhyths of *Phaseolus*). Zeitschrift für Botanik 46:209–241.

Luce, Gay Gaer. 1971. Biological Rhythms in Human and Animal Physiology. Dover Publications, Inc. New York.

Mackenzie, Jean. 1973. How biorhythms affect your life. Science Digest 74(2):18–22.

Menaker, Michael. 1965. Circadian rhythms and photoperiodism in *Passer domesticus*. Circadian Clocks (Proceedings of the Feldafing Summer School, 7–18 September, 1964—Jürgen Aschoff, ed. North-Holland Publishing Co., Amsterdam.

Morre-Ede, Martin C., F. M. Sulzman, and C. A. Fuller. 1982. The Clocks That Time Us. Harvard University Press, Cambridge, Massachusetts, and London, England.

Munoz, Victor and Warren L. Butler. 1975. Photoreceptor pigment for blue light in *Neurospora crassa*. Plant Physiology 55:421–426.

Palmer, John D. 1975. Biological clocks of the tidal zone. Scientific American 232(2):70–79.

Pfeffer, W. 1915. Beiträge zur Kenntnis der Entstehung der Schlafbewegungen. (Contributions to the knowledge of the origin of leaf movements). Adhandl. Math. Phys, Kl. Kön. Sächs. Ges. d. Wiss., 34:1–154.

Pittendrigh, Colin S. 1954. On temperature independence in the clock system controlling emergence time in *Drosophila*. Proceedings of the National Academy of Sciences 40:1018–1029.

Pittendrigh, Colin S. 1967. On the mechanism of entrainment of a circadian rhythm by light cycles. Pages 277–297 in J. Aschoff (ed.), Circadian Clocks. North-Holland Publishing Co., Amsterdam.

Rodgers, C. W., R. L. Sprinkle, and F. H. Lindbert. 1974. Biorhythms: three tests of the "critical days" hypothesis. International Journal of Chronobiology 2:215–310.

Salisbury, Frank B. 1963. Biological timing and hormone synthesis in flowering of *Xanthium*. Planta 59:518–534.

Salisbury, Frank B., George G. Spomer, Martha Sobral, and Richard T. Ward. 1968. Analysis of an alpine environment. Botanical Gazette 129(1):16–32.

Satter, Ruth L. and Arthur W. Galston. 1981. Mechanisms of control of leaf movement. Annual Review of Plant Physiology 32:82–110.

Saunders, D. S. 1976. The biological clock of insects. Scientific American 234(2):114–121.

Simon, Esther, Ruth L. Satter, and Arthur W. Galston. 1976. Circadian rhythmicity in excised *Samanea* Pulvini. Plant Physiology 58:421–425.

Spruyt, E., L. Maes, J. P. Verbelen, E. Moereels, and J. A. DeGreef. 1983. Circannual course of photomorphogenetic reactivity in etiolated bean seedlings. Photochemistry and Photobiology 37 (4):471–473.

Sulzman, Frank M. 1983. Primate circadian rhythms. BioScience 33:445–450.

Sweeney, Beatrice M. 1969. Rhythmic Phenomena in Plants. Academic Press, London and New York.

Sweeney, B. M. 1976. Pros and cons of the membrane model for circadian rhythms in the marine algae, *Gonyaulax* and *Acetabularia*. Pages 63–76 in Biological Rhythms in the Marine Environment. P. J. De Coursey (ed.), University of South Carolina Press.

Sweeney, Beatrice M. 1983. Biological clocks—an introduction. BioScience 33:424–425. (See other articles in the same issue.)

Thommen, G. 1973. Biorhythms: Is This Your Day? Avon Books, New York.

Thompson, Marcia J. and David W. Harsha. 1984. Our rhythms still follow the African sun. Psychology Today 18(1):50–54.

Wilkins, Malcolm B. and Philip J. C. Harris. 1975. Phytochrome and phase setting of endogenous rhythms. Pages 399–417 in H. Smith (ed.), Light and Plant Development. Butterworths, London and Boston.

Zimmer, Rose. 1962. Phasenverschiebung und andere Störolichtwirkungen auf die endogen tagesperiodischen Blütenblattbewegungen (Phase shift and other light-interruption effects on endogenous diurnal petal movements). Planta 58:283–300.

Zimmerman, Natille H., and Michael Menaker. 1975. Neural connections of sparrow pineal: role in circadian control of activity. BioScience 190:477–479.

Chapter 21

Altman, P. L. and Dorothy S. Dittmer, eds. 1962. Growth, Including Reproduction and Development. Federation of the American Society for Experimental Biology, Washington, D.C.

Arias, I., P. M. Williams, and J. W. Bradbeer. 1976. Studies in seed dormancy. IX. The role of gibberellin biosynthesis and the release of bound gibberellin in the post-chilling accumulation of gibberellin in seeds of *Corylus avellana* L. Planta 131:135–139.

Bewley, J. D. and M. Black. 1982. Physiology and Biochemistry of Seeds in Relation to Germination. In two volumes. Springer-Verlag, Berlin, Heidelberg, New York.

Caspari, E. W. and R. W. Marshak. 1965. The rise and fall of Lysenko. Science 149:275–278.

Chailakhyan, M. K. 1968. Internal factors of plant flowering. Annual Review of Plant Physiology 19:1–36.

Chouard, P. 1960. Vernalization and its relations to dormancy. Anual Review of Plant Physiology 11:191–238.

Crocker, W. and L. V. Barton. 1953. Physiology of Seeds. Chronica Botanica Co., Waltham, Massachusetts.

Dell, B. 1980. Structure and function of the strophiolar plug in seeds of *Albizia lophantha*. American Journal of Botany 67(4):556–563.

Downs, R. J. and H. A. Borthwick. 1956. Effects of photoperiod on growth of trees. Botanical Gazette 117:310–326.

Evenari, M. 1957. The physiological action and biological importance of germination inhibitors. Society of Experimental Biology Symposium 11:21–43.

Friend, D. J. C. and V. A. Helson. 1976. Thermoperiodic effects on the growth and photosynthesis of wheat and other crop plants. Botanical Gazette 137(1):75–84.

Gassner, G. 1918. Beiträge zue physiologischen Charakteristik Sommer und Winter annueller Gewächse insbesondere der Getreidepflanzen. (Contributions to the physiological characteristics of summer and winter annual growth, particularly with cereals). Zeitschrift für Botanik 10:417–430.

Hartsema, Annie M. 1961. Influence of temperatures on flower formation and flowering of bulbous and tuberous plants. Encyclopedia of Plant Physiol. 16:123–167.

Heide, O. M. 1974. Growth and dormancy in Norway spruce ecotypes (*Picea abies*). I. Interaction of photoperiod and temperature. Physiologie Plantarum 30:1–12.

Henson, Cynthis A., Larry E. Schrader and Stanley H. Duke. 1980. Effects of temperature on germination and mitochondrial dehydrogenases in two soybean (Glycine max) cultivars. Physiologia Plantarum 48:168–174.

Jann, R. C. and R. D. Amen. 1977. What is germination? Pages 7–28 in A. A. Khan (ed.), The Physiology and Biochemistry of Seed Dormancy and Germination. North-Holland Pub. Co., Amsterdam and New York.

Junttila O. 1973. The mechanism of low temperature dormancy in mature seeds of Syringa species. Physiologia Plantarum 29:256–263.

Ketring, D. L. 1973. Germination inhibitors. Seed Science Technology 1(2):305–324.

Khan, A. A. 1977. Seed dormancy: changing concepts and theories. Pages 29–50 in A. A. Khan (ed.), The Physiology and Biochemistry of Seed Dormancy and Germination. North-Holland Pub. Co., Amsterdam and New York.

Koller, D. 1957. Germination-regulating mechanisms in some desert seeds. IV: Atriplex dimorphostegia. Ecology 38:1–13.

Kramer, Paul J. and T. T. Koslowski. 1979. Physiology of Woody Plants. Academic Press, New York.

Lang, Anton. 1957. The effect of gibberellin upon flower formation. Proceedings of the National Academy of Sciences 43:709–711.

Lang, A. 1965a. Effects of some internal and external conditions on seed germination. Pages 849–893 in W. Ruhland (ed.), Encyclopedia of Plant Physiology, Vol. 15, Part 2. Springer-Verlag, Berlin.

Lang, A. 1965b. Physiology of flower iniation. Encyclopedia of Plant Physiology. W. Ruhland (ed.), Springer-Verlag, Berlin. 15(1):1380–1536.

Leopold, A. C. and Paul E. Kriedeman. 1975. Plant Growth and Development. Second Edition. McGraw-Hill Book Co., Inc., New York.

Lerman, J. C. and E. M. Cigliano. 1971. New Carbon-14 Evidence for six-hundred years old Canna pacta seed. Nature 232:568–570.

Mayer, A. M. 1974. Control of seed germination. Annual Review of Plant Physiology 25:167–193.

Mayer, A. M. and Poljakoff-Mayber. 1975. The Germination of Seeds. Second Edition. Pergamon Press, New York, London.

Melchers, G. 1937. Die Wirkung von Genen, tiefen Temperaturen und blühenden Pfropfpartnern auf die Blühreife von Hyoscymas niger L. (The effect of genes, low temperatures, and flowering graft partners on ripeness to flower of Hyoscymas niger.) Biologisches Zentralblatt 57:568–614.

Nikolaeva, M. G. 1969. Physiology of Deep Dormancy in Seeds. Translated from Russian and published for the National Science Foundation by the Israel Program for Scientific Translations.

Osborne, Daphne. 1980. Senescence in seeds. In K. V. Thimann (ed.), Senescence in Plants. CRC Press Inc., Boca Raton, Florida. Chapter 2.

Perry, T. O. and H. Hellmers. 1973. Effects of abscisic acid on growth and dormancy of two races of red maple. Botanical Gazette 134:283–289.

Phillips, I. D. J., J. Miners, and J. R. Roddick. 1980. Effects of light and photoperiodic conditions on abscisic acid in leaves and roots of Acer pseudoplatanus L. Planta 149:118–122.

Porsild, A. E., C. R. Harington, G. A. Mulligan. 1967. Lupinus articus Wats. grown from seeds of Pleistocene age. Science 158:113–114.

Priestly, David A. and Maarten A. Posthumus. 1982. Extreme longevity of lotus seeds from Plantien. Nature 299:148–149.

Purvis, O. M. 1961. The physiological analysis of vernalization. Encyclopedia of Plant Physiology 16:76–122.

Rees, A. R. 1972. The Growth of Bulbs. Academic Press, New York.

Roberts, E. H. and R. D. Smith. 1977. Dormancy and the pentose phosphate pathway. In A. A. Khan (ed.), The Physiology and Biochemistry of Seed Dormancy and Germination. North-Holland Pub. Co., Amsterdam and New York.

Salisbury, Frank B. 1963. The Flowering Process. Pergamon Press, Oxford, London, New York, Paris.

Samish. 1954. Dormancy in woody plants. Annual Review of Plant Physiology 5:183–204. (A pomologist who uses the term "rest.")

Thompson, H. C. 1953. Vernalization of growing plants. In Growth and Differentiation in Plants. W. E. Loomis (ed.), The Iowa State College Press, Ames, Iowa.

Ulrich, Albert. 1955. Influence of night temperature and nitrogen nutrition on the growth, sucrose accumulation and leaf minerals of sugar beet plants. Plant Physiology 30:250–257.

Vegis, A. 1964. Dormancy in higher plants. Annual Review of Plant Physiology 15:185–224.

Vest, E. Dean. 1972. Shadscale and fungus: desert partners. Pages 725–726 in W. A. Jensen and F. B. Salisbury. Botany, An Ecological Approach. Wadsworth Publishing Co., Belmont, California.

Villiers, T. A. 1972. Seed dormancy. Pages 219–281 in T. T. Kozlowski (ed.), Seed Biology, Vol. II. Academic Press, New York.

Walton, D. C. 1977. Abscisic acid and seed germination. Pages 145–156 in A. Khan (ed.), The Physiology and Biochemistry of Seed Dormancy and Germination. Elsevier/North Holland Biomed. Press., Amsterdam and New York.

Walton, D. C. 1980. Biochemistry and physiology of abscisic acid. Annual Review of Plant Physiology 31:453–489.

Wareing, P. F. 1956. Photoperiodism in woody plants. Annual Review of Plant Physiology 7:191–214.

Wellensiek, S. J. 1964. Dividing cells as the prerequisite for vernalization. Plant Physiology 39:832–835.

Went, F. W. 1957. The Experimental Control of Plant Growth. Chronica Botanica Co., Waltham, Massachusetts.

Went, F. W. 1974. Reflections and speculations. Annual Review of Plant Physiology 25:1–26.

Wood, W. M. L. 1953. Thermonasty in tulip and crocus flowers. Journel of Experimental Botany 4:65–77.

Chapter 22

Battle, R. W., J. K. Gaunt and D. L. Laidman. 1976. the effect of photoperiod on edogenous γ-tocopherol and plastochromanol in leaves of Xanthium strumarium L. Biochemical Society of London Transactions 4:484–486.

Battle, R. W., D. L. Laidman, and J. K. Gaunt. 1977. The relationship between floral induction and γ-tocopherol concentrations in leaves of Xanthium strumarium L. Biochemical Society of London Transactions 5:322–324.

Bernier, G., J. Kinet, and R. M. Sachs. 1981. The physiology of flowering. Vol. I. The initiation of flowers. Vol. II. Transition to reproductive growth. CRC Press, Boca Raton, Florida.

Bernier, G., M. V. S. Raju, A. Jacqmard, M. Bodson, J. M. Kinet, and A. Havelange. 1974. Release in mitosis of the meristematic G2 cells as an early and essential effect of the floral stimulus in mustard and cocklebur. Bulletin, The Royal Society of New Zealand 12:547–551.

Bjornseth, Ian Petter. 1981. Effects of natural daylength and nutrition on the cessation of cambial activity in young plants of Picea abies. Mitteilungen der Forstlichen Bundesversuchsanstalt. Wien. 167–176.

Bollig, I. 1977. Different circadian rhythms regulate photoperiodic flowering response and leaf movement in Pharbitis nil L. Choisy. Planta 135:137–142.

Bose, T. K. 1974. Effect of temperature and photoperiod on growth, flowering and seed formation in tossa jute. Indian Journal of Agricultural Science 44:32–35.

Bünning, Erwin. 1937. Die endonome Tagesrhythmik als Grundlage der photoperiodischen Reaktion (The endogenous daily rhythm as the basis of the photoperiodic reaction). Berichte der Deutschen Botanischen Gesellschaft 54:590–607.

Carr, D. J. 1967. The relationship between florigen and the flower hormones. Pages 144:305–312 in Plant Growth Regulators. J. F. Fredrick and E. M. Weyer (eds.), Annals of the New York Academy of Sciences.

Chailakhyan, Mikhail K. H. 1968. Internal factors of plant flowering. Annual Review of Plant Physiology 19:1–36.

Claes, H. and A. Lang. 1947. Die wirkung von β-indolylessigsäure und 2,3,5-trijodbenzoesäure auf die blüten bildung von Hyoscymas niger (The action of Hyoscyamus niger). Zeitschrift für Naturforschung 26:56–63.

Cleland, C. F., and A. Ajami. 1974. Identification of flower-inducing factor isolated from aphid honeydew as being salicylic acid. Plant Physiology 54:904–906. (See also 54:899–903.)

Cumming, Bruce G. 1959. Extreme sensitivity of germination and photoperiodic reaction in the genus Chenopodium (Tourn.) L. Nature 184:1044–1045.

Cumming, Bruce G. 1969. Chenopodium rubrum L. and related species. Pages 156–185 in L T. Evans (ed.), The Induction of Flowering. Macmillan of Australia.

Deitzer, G. F., R. Hayes and M. Jabben. 1979. Kinetics and time dependence of the effect of far red light on the photoperiodic induction of flowering in winter barley. Plant Physiology 64:1015–1021.

Evans, L. T., ed. 1969. The Induction of Flowering, Some Case Histories. Macmillan Company of Austrilia Pty. Ltd. Victoria, Australia.

Evans, L. T. 1975. Daylength and the Flowering of Plants. Benjamin, Menlo Park, California.

Fischer, J. 1916. Zur Frage der Kohlensäureerhährung der Pflanze. (On the question of the carbon dioxide nutrition of plants.) Gartenflora 65:232.

Gerner, W. W. and H. A. Allard. 1920. Effect of the relative length of day and night and other factors of the environment of growth and reproduction in plants. Journal of Agricultural Research 18:553–606.

Garner, W. W. and H. A. Allard. 1923. Further studies in photoperiodism, the response of plants to relative length of day and night. Journal of Agricultural Research 23:871–920.

Gibby, David D. and Frank B. Salisbury. 1971. Participation of long-day inhibition in flowering of Xanthium strumarium L. Plant Physiology 47:784–789.

Hamner, Karl C. 1969. Glycine max L. Merrill. Pages 62–89 in L. T. Evans (ed.), The Induction of Flowering. Some Case Histories. Macmillan of Australia, South Melbourne.

Hamner, K. C. and J. Bonner. 1938. Photoperiodism in relation to hormones as factors in floral initiation and development. Botanical Gazette 100:388.

Hamner, K. C. 1963. Endogenous rhythms in controlled environments. Pages 215–232 in Environmental Control of Plant Growth. L. T. Evans, (ed.), Academic Press, Inc., New York.

Havelange, A. 1980. The quantitative ultrastructure of the meristematic cells of Xanthium strumarium during the transition to flowering. American Journal of Botany 67:1171–1178.

Hay, R. K. M. and O. M. Heide. 1983. Specific photoperiodic stimulation of dry matter production in a high-latitude cultivar of Poa pratensis. Physiologia Plantarum 57:135–142.

Hillman, W. S. 1979. Photoperiodism in plants and animals. J. J. Head (ed.), Carolina Biology Readers. Carolina Biology Supply Company, Burling, North Carolina.

Hodson, H. K. and K. C. Hamner. 1970. Floral inducing extract from Xanthium. Science 167:384–385.

Jabben, Merten and Gerald F. Deitzer. 1979. Effects of the herbicide San 9789 on photomorphogenic responses. Plant Physiology 63:481–485.

Jacqmard, A., M. V. S. Raju, J. M. Kinet, and G. Bernier. 1976. The early action of the floral stimulus on mitotic activity and DNA synthesis in the apical meristem of Xanthium strumarium. American Journal of Botany 63:166–174.

Junttila, Olavi and O. M. Heide. 1981. Shoot and needle growth in *Pinus sylvestris* as related to temperature in northern Fennoscandia. Forest Science 27(3):423–430.

Kato, Y., N. Fukunharu, and R. Kobayashi. 1958. Stimulation of flower bud differentiation of conifers by gibberellin. Pages 67–68 *in* Abstracts of the Second Meeting of the Japan Gibberellin Research Association.

King, R. W. 1975. Multiple circadian rhythms regulate photoperiodic flowering responses in *Chenopodium rubrum*. Canadian Journal of Botany 53:2631–2638.

King, R. W. 1979. Photoperiodic time measurement and effects of temperature on flowering in *Chenopodium ruburm* L. Australian Journal of Plant Physiology 6:417–422.

King, R. W. and J. A. D. Zeevaart. 1973. Floral stimulus movement in *Perilla* and flower inhibition caused by noninduced leaves. Plant Physiology 51:727–738.

Klebs, G. 1918. Über die Blütenbildüng bei *Sempervivum*. (On flower formation in *Sempervivum*). Flora (Jena) 128:111–112.

Kraus, E. J. and H. R. Kraybill. 1918. Vegetation and reproduction with special reference to the tomato. Oregon Agricultural Experiment Station Bulletin 149:5.

Lane, H. C., H. W. Siegelman, W. L. Butler, and E. N. Firer. 1963. Detection of phytochrome in green plants. Plant Physiology 38:414.

Lang, Anton. 1965. Physiology flower initiation. Pages 1380–1535 *in* W. Ruhland (ed.), Encyclopedia of Plant Physiology. Vol. 15, Springer-Verlag, Berlin.

Lang, A., M. K. Chailakhyan, and I. A. Krolova. 1977. Promotion and inhibition of flower formation in a day-neutral plant in grafts with a short-day plant and a long-day plant. Proceedings of the National Academy of Science. USA 74: 2412–2416.

Lincoln, R. G., A. Cunningham, B. H. Carpenter, J. Alexander, and D. L. Mayfield. 1966. Florigenic acid from fungal cultures. Plant Physiology 41:1079–1080.

Lumsden, Peter, Brian Thomas, and Daphne Vince-Prue. 1982. Photoperiodic control of flowering in dark-grown seedlings of *Pharbitis nil Choisy*. Plant Physiology 70:277–282.

McMillan, C. 1974. Photoperiodic adaptation of *Xanthium strumarium* in Europe, Asia Minor, and northern Africa. Canadian Journal of Botany 52:1779–1791.

Mooney, H. A. and W. D. Billings. 1961. Comparative physiological ecology of arctic and alpine populations of *Oxyria digyna*. Ecological Monographs 31:1–29.

Olmsted, C. E. 1944. Growth and development in range grasses. IV. Photoperiodic responses in twelve geographic strains of side-oats grama. Botanical Gazette 106:46–74.

Papenfuss, H. D. and Frank B. Salisbury. 1967. Aspects of clock resetting in flowering of *Xanthium*. Plant Physiology 42:1562–1568.

Pharis, Richard P. and C. G. Kuo. 1977. Physiology of gibberellins in conifers. Canadian Journal of Forest Research 7(2):299–325.

Pryce, R. J. 1972. Gallic acid as a natural inhibitor of flowering in *Kalanchoe blossfeldiana*. Phytochemistry 11:1911–1918.

Rowan, W. 1925. Relation of light to bird migration and developmental changes. Nature 115:494–495.

Sachs, R. M. 1978. Nutrient diversion: an hypothesis to explain the chemical control of flowering. Hortscience 12:220–222.

Sachs, R. M. and W. P. Hackett. 1983. Souce-Sink relationships and flowering. Meudt, Werner J. (ed.), Pages 263–272 *in* Beltsville Symposia in Agricultural Research Vol. 6, Strategies of Plant Reproduction.

Salisbury, Frank B. 1955. The dual role of auxin in flowering. Plant Physiology 30:327–334.

Salisbury, Frank B. 1963. The Flowering Process. Pergamon Press, Cambridge, New York.

Salisbury, Frank B. 1963. Biological timing and hormone synthesis in flowering in *Xanthium*. Planta 59:518–534.

Salisbury, Frank B. 1965. Time measurement and the light period in flowering. Planta (Berl.) 66:1–26.

Salisbury, Frank B. 1981. The twilight effect: initiating dark measurement in photoperiodism of *Xanthium*. Plant Physiology 67:1230–1238.

Salisbury, Frank B. 1981. Response to photoperiod. Chapter 5, pages 135–167 *in* O. L. Lange, P. S. Nobel, C. B. Osmond, and H. Ziegler (eds.), Physiological Plant Ecology I. Vol. 12A. *In* A. Pirson, M. H. Zimmermann (eds.), Encyclopedia of Plant Physiology, New Series, Springer-Verlag, Berlin, Heidelberg.

Salisbury, Frank B. 1982. Photoperiodism. Horticultural Reviews 4:66–105.

Salisbury, Frank B. and James Bonner. 1956. The reactions of the photoinductive dark period. Plant Physiology 31:141–147.

Salisbury, Frank B. and Alice Denney. 1974. Noncorrelation of leaf movements and photoperiodic clocks in *Xanthium strumarium* L. *Chronobiology* Proceedings of the International Society of Chronobiology, Little Rock, Arkansas. Nov. 8–10, 1071, pages 679–696, Igaku Shoin, Ltd.

Schwabe, W. W. 1971. Physiology of vegetative reproduction and flowering. Pages 233–411 *in* F. C. Steward (ed.), Plant Physiology—A Treatise. Volume 7A. Academic Press, New York.

Takimoto, A. and K. Ikeda. 1961. Effect of twilight on photoperiodic induction in some short day plants. Plant Cell Physiology 2:213–229.

Tournois, J. 1914. Etudes sur la sexulite du houblon. (Studies on the sexuality of hops). Annals des Sciences Naturelles (Botanique) 19:49–191.

Vince-Prue, Daphne. 1975. Photoperiodism in Plants. McGraw-Hill, London.

Vince-Prue, Daphne, and K. E. Cockshull. 1981. Photoperiodism and crop production. *In* Physiological Process Limiting Plant Productivity, C. B. Johnson (ed.), pp. 175–197.

Wellensiek, S. J. 1973. Genetics and flower formation of annual *Lunaria*. Netherland Journal of Agricultural Science 21:163–166.

Zeevaart, Jan A. D. 1976. Physiology of flower formation. Annual Review of Plant Physiology 27:321–348.

Zeevaart, Jan A. D. 1976. Phytohormones and flower formation. *In* D. S. Letham et al (eds.), Plant Hormones and Related Compounds. ASP Biological and Medical Amsterdam.

Chapter 23

Barbour, M. G., J. H. Burk, and W. D. Pitts. 1980. Terrestrial Plant Ecology. Benjamin Cummings Publishing Company, Inc. Menlo Park, California.

Berger, P. J., N. C. Negus, E. H. Sanders, and P. D. Gardner. 1981. Chemical triggering of reproduction in *Microtus montanus*. Science 214:69–70.

Bernays, E. A. and S. Woodhead. 1982. Plant phenols utilized as nutrients by a phytophagous insect. Science 216:201–203.

Bhargava, S. C. 1975. Photoperiodicity and seed germination in rice. Indian Journal of Agricultural Sciences 45:447–451.

Billings, W. D. 1970. Second Edition. Plants, Man and the Ecosystem. Wadsworth Publishing Company, Inc., Belmont, California.

Björkman, O. 1968. Further studies on differentiation of photosynthetic properties in sun and shade ecotypes of *Solidago virgaurea*. Physiologia Plantarum 21:84–99.

Björkman, O. 1981. Responses to Different Quantum Flux Densities. Pages 57–107 *in* Lange, O. L., P. S. Nobel, C. B. Osmond, and H. Ziegler (eds.), Encyclopedia of Plant Physiology 12A, Physiological Plant Ecology I. Springer-Verlag, Berlin, Heidelberg, New York.

Blackman, F. F. 1905. Optima and limiting factors. Annals of Botany 14(74):281–295.

Bryant, J. P. 1981. Phytochemical deterrence of snowshoe hare browsing by adventitious shoots of four Alaskan trees. Science 213:889–890.

Caldwell, M. M. 1981. Plant response to solar ultraviolet radiation. Pages 170–194 *in* Lange, O. L., P. S. Nobel, C. B. Osmond, and H. Ziegler (eds.), Encyclopedia of Plant Physiology 12A, Physiological Plant Ecology I. Springer-Verlag, Berlin, Heidelberg, New York.

Clausen, Jens, David D. Keck, and William M. Hiesey. 1940. Experimental Studies on the Nature of Species. I. Effect of Varied Environments on Western North American Plants. Carnegie Institution of Washington, Publication No. 520. Washington D. C.

Clements, Frederic E., Emmett V. Martin, and Frances L. Long. 1950. Adaptation and Origin in the Plant World. Chronica Botanica Company, Waltham, Massachusetts.

Downs, R. J. and Borthwick, H. A. 1956. Effects of photoperiod on growth of trees. Botanical Gazette 117:310–326.

Gates, D. M. 1962. Energy exchange in the Biosphere. Harper and Row, New York.

Highkin, H. R. and A. Lang. 1966. Residual effect of germination temperature on the growth of peas. Planta 68:94–98.

Jarvis, B. B., J. O. Midiwo, and D. Tuthill. 1981. Interaction between the antibiotic Trichothecenes and the higher plant *Baccaris megapotamica*. Science 214:460–462.

Kigel, J., M. Ofir, and D. Koller. 1977. Control of the germination responses of *Amaranthus retroflexus* L. seeds by their parental photothermal environment. Journal of Experimental Botany 28: 1125–1136.

Koller, Dov. 1962. Preconditioning of germination in lettuce at time fruit ripening. American Journal of Botany 49:841–844.

Krebs, C. J. 1978. Ecology: The Experimental Analysis of Distribution and Abundance. Second Edition. Harper and Row, New York.

Lange, O. L., P. S. Nobel, C. B. Osmond, and H. Ziegler. 1981. Introduction: Perspectives in Ecological Plant Physiology. Pages 1–9 *in* Lange, O. L., P. S. Nobel, C. B. Osmond, and H. Ziegler (eds.), Encyclopedia of Plant Physiology 12A, Physiological Plant Ecology I. Springer-Verlag, Berlin, Heidelberg, New York.

Lockhart, James A. 1965. The analysis of interactions of physical and chemical factors on plant growth. Annual Review of Plant Physiology 16: 37–52.

Mayer, A. M. and Poljakoff-Mayber, A. 1975. The germination of Seeds. Second Edition, Pergamon, New York, London.

Mohr, H. 1972. Lectures on Photomorphogenesis. Springer-Verlag, Berlin, Heidelberg, New York.

Morgan, D. C. and H. Smith. 1981. Nonphotosynthetic Responses to Light Quality. Pages 109–130 *in* Lange, O. L., P. S. Nobel, C. B. Osmond, and H. Ziegler (eds.), Encyclopedia of Plant Physiology 12A, Physiological Plant Ecology I. Springer-Verlag, Berlin, Heidelberg, New York.

Pourrat, Yvonne and Roger Jacques. 1975. The influence of photoperiodic conditions received by the mother plant on morphological and physiological characteristics of *Chenopodium polyspermum* L. seeds. Plant Science Letters 4:273–279.

Richardsen, S. and Frank B. Salisbury. 1977. Plant responses to the light penetrating snow. Ecology 58:1152–1158.

Ricklefs, Robert E. 1979. Ecology. Second Edition. Chiron Press.

Salisbury, Frank B. 1975. Multiple factor effects on plants. Pages 501–520 *in* F. John Vernberg (ed.), Physiological Adaptation to the Environment. Intext Educational Publishers, New York.

Salisbury, Frank B. 1981a. Responses to photoperiod. Pages 135–167 *in* Lange, O., P. S. Nobel, C. B. Osmond, and H. Ziegler (eds.), Encyclopedia of Plant Physiology Vol 12A, Physiological Plant Ecology I. Springer-Verlag, Berlin, Heidelberg.

Salisbury, Frank B. 1981b. The twilight effect: initiating dark measurement in photoperiodism of *Xanthium*. Plant Physiology 67:1230–1238.

Salisbury, Frank B. 1982. Photoperiodism. Horticultural Reviews 4:66–105.

Schneider, D., M. Boppre, J. Zqeig, S. B. Horsley, T. W. Bell, J. Meinwald, K. Hansen, and E. W. Diehl. 1982. Scent orhan development in *Creatonotos* moths: regulation by pyrrolizidine alkaloids. Science 215:1264–1265.

Schopfer, Peter. 1969. Die Hemmung der Streckungs—wachstrums durch Phytochrom—ein Stoffaufnahme erfordernder Prozess? (Inhibition of elongation growth by phytochrome—a process requiring substrate uptake?) Planta (Berlin) 85:383–388.

Schultz, J. C. and I. T. Baldwin. 1982. Oak leaf quality declines in response to defoliation by Gypsy Moth larvae. Science 217:149–150.

Smith, Harry. 1983. Light quality, photoperception, and plant strategy. Annual Review of Plant Physiology 33:481–518.

Spomer, G. G. 1973. The concepts of "interaction" and "Operational environment" in environmental analyses. Ecology 54(1):200–204.

Swain, Tony. 1977. Secondary compounds as protective agents. Annual Review of Plant Physiology 28:479–501.

Turesson, G. 1922. The genotypic response of the plant species to the habitat. Hereditas 3:211–350.

Vince-Prue, D. 1975. Photoperiodism in plants. McGraw-Hill, London.

Whittaker, R. H. 1975 (Second Edition). Communities and Ecosystems. Macmillan Publishing Company, Inc., New York; Collier Macmillan Publishers, London.

Chapter 24

Acevedo, Edmundo, Theodore C. Hsiao, and D. W. Henderson. 1971. Immediate and subsequent growth responses of maize leaves to changes in water status. Plant Physiology 48:631–636.

Allen, T. 1965. The Quest—A Report on Extraterrestrial Life. Chilton Books.

Anderson, J. A., D. W. Buchanan, R. E. Stall, and C. B. Hall. 1982. Frost injury of tender plants increased by *Pseudomonas syringae* van Hall. Journal of the American Society of Horticultural Science 107:123–125.

Arangno, M. 1981. Responses of micoorganisms to temperature. Pages 339-369 in Encyclopedia of Plant Physiology. New Series Vol. 12A, Physiological Plant Ecology. Lange, O. L., P. S. Nobel, C. B. Osmond, and Zeigler (eds.), Springer-Verlag, Berlin, Heidelberg, New York.

Barbour, M. G., J. H. Burk, and W. D. Pitts. 1980. Terrestrial Plant Ecology. Benjamin/Cummings Publishing Company, Inc. Menlo Park, California.

Baross, J. A. and J. W. Deming. 1983. Growth of "black smoker" bacteria at temperatures of at least 250°C. Nature 303:423–426.

Baross, J. A., M. D. Lilley, and L. I. Gordon. 1982. Is the CH_4, H_2, and CO venting from submarine hydrothermal systems produced by thermophylic bacteria? Nature 298:366–368.

Bewley, Derek J. 1979. Physiological aspects of desiccation tolerance. Annual Review of Plant Physiology 30:195–238.

Bewley, J. D. and J. E. Krochko. 1982. Desiccation-tolerance. Pages 325–378 in Encyclopedia of Plant Physiology. New Series Vol. 12B; Physiological Plant Ecology II, Lange, O. L., P. S. Nobel C. B. Osmond and H. Ziegler (eds.), Springer-Verlag, Berlin, Heidelberg, New York.

Boyer, J. S. 1970. Leaf enlargement and metabolic rates in corn, soybean, and sunflower at various leaf water potentials. Plant Physiology 46:233–235.

Bradford, K. J. and T. C. Hsiao. 1982. Physiological responses to moderate water stress. In O. L. Lange, P. S. Nobel, C. B. Osmond, and H. Ziegler (eds.), Physiological Plant Ecology II, Water Relations and Carbon Assimilation. Pages 263–324 in A. Pirson, M. H. Zimmermann (eds.), Encyclopedia of Plant Physiology, New Series. Vol. 12B. Springer-Verlag, Berlin, Heidelberg, New York.

Brock, T. D. 1978. Thermophilic Microorganisms and Life at High Temperatures. Springer-Verlag, Berlin, Heidelberg, New York.

Burke, M. J., L. V. Gusta, H. A. Quamme, C. J. Weiser, and P. H. Li. 1976. Freezing and injury in plants. Annual Review of Plant Physiology 27:507–528.

Edney, E. B. 1975. Absorption of water vapor from unsaturated air. In F. John Vernberg (ed.), Physiological Adaptation to the Environment. Intext Educational Publishers, New York.

Ehleringer, J., Olle Björkman, and Harold A. Mooney. 1976. Leaf pubescence: effects on absorptance and photosynthesis in a desert shrub. Science 192:376–377.

Evenari, M., E. D. Schultze, L. Kappen, U. Buschbom, and O. L. Lange. 1975. Adaptive mechanisms in desert plants. In F. John Vernberg (ed.), Physiological Adaptation to the Environment. Intext Educational Publishers, New York.

Fireman, M. and H. E. Hayward. 1952. Indicator significance of some shrubs in the Escalante Desert, Utah. Botanical Gazette 114:143–154.

Flores, H. E. and A. W. Galston. 1982. Polyamines and plant stress: activation of putrescine biosynthesis by osmotic shock. Science 217:1259–1260.

Flowers, T. J., P. F. Troke, and A. R. Yeo. 1977. The mechanism of salt tolerance in halophytes. Annual Review of Plant Physiology 28:89–121.

Graham, Douglas and Brian D. Patterson. 1982. Responses of plants to low, nonfreezing temperatures: proteins, metabolism, and acclimation. Annual Review of Plant Physiology. 33:347–372.

Greenway, H. and Rana Munns. 1980. Mechanisms of salt tolerance in nonhalophytes. Annual Review of Plant Physiology 30:149–190.

Hanson, Andrew D. and William D. Hitz. 1982. Metabolic responses of mesophytes to plant water deficits. Annual Review of Plant Physiology 33:163–203.

Harvey, G. W., L. V. Gusta, D. C. Fork and J. A. Berry. 1982. The relation between membrane lipid phase separation and frost tolerance of cereals and other cool climate plant species. Plant, Cell and Environment 5:241–244.

Hellebust, Johan A. 1976. Osmoregulation. Annual Review of Plant Physiology 27:485–505.

Hsiao, T. C. 1973. Plant responses to water stress. Annual Review of Plant Physiology 24:519–570.

Jefferies, R. L. 1981. Osmotic adjustment and the response of halophytic plants to salinity. BioScience 31(1):42–46.

Jensen, W. A. and F. B. Salisbury. 1984. Botany, Second Edition. Wadsworth Publishing Company., Inc. Belmont, California.

Kappen, L. 1981. Ecological significance of resistance to high temperature. Pages 439–474 in Encyclopedia of Plant Physiology. New Series Vol. 12A, Physiological Plant Ecology. Lange, O. L., P. S. Nobel, C. S. Osmond, and H. Ziegler (eds.), Springer-Verlag, Berlin, Heidelberg, New York.

Kozlowski, T. T. 1968–1978. Water Deficits and Plant Growth. Vol. I. II, III, IV, V, and VI. Academic Press, New York and London.

Kramer, Paul J. 1980. Drought, stress, and the origin of adaptations. Pages 7–20 in Turner, N. C. and P. J. Kramer, Adaptation of Plants to Water and High Temperature Stress. John Wiley and Sons, New York.

Kramer, Paul J. 1983. Water Relations of Plants. Academic Press, New York and London.

Kriedemann, P. E. and B. R. Loveys. 1974. Hormonal mediation of plant responses to environmental stress. Pages 461–465 in R. L. Bieleski, A. R. Ferguson, and M. M. Creswell (eds.), Mechanisms of Regulation of Plant Growth, The Royal Society of New Zealand, Wellington.

Krucheberg, A. R. 1954. The ecology of serpentine soils. III. Plant species in relation to serpentine soils. Ecology 35:267–274.

LaHaye, P. A. and E. Epstein. 1969. Salt toleration by plants: enhancement with calcium. Science 166:395–396.

Larcher, W. and H. Bauer. 1981. Ecological significance of resistance to low temperature. Pages 403–437 in Encyclopedia of Plant Physiology. New Series Vol. 12A, Physiological Plant Ecology, Lange, O. L., P. S. Nobel, C. S. Osmond, and H. Ziegler (eds.), Springer-Verlag, Berlin, Heidelberg, New York.

Levitt, J. 1962. A sulfhydryl-disulfide hypothesis of frost injury and resistance in plants. Journal of Theoretical Biology 3:355–391.

Levitt, J. 1972. Responses of Plants to Environment Stresses. Academic Press, Inc., New York and London.

Levitt, Jacob. 1980. response of Plants to Environmental Stresses. Second Edition, Vol. I and II. Academic Press, Inc., New York and London.

Lindow, Steven. 1983. The role of bacterial ice nucleation in frost injury to plants. Annual Review of Phytopathology 21:363–384.

Lyons, J. M. 1973. Chilling injury in plants. Annual Review of Plant Physiology 24:445–466.

MacDougal, D. T. and E. S. Spaulding. 1910. The water-balance of succulent plants. Carnegie Institute Washington Publications 141:77.

Mooney, H. A., S. L. Gulmon, J. Ehleringer, and P. W. Rundel. 1980. Atmospheric water uptake by an Atacama desert schrub. Science 209:693–694.

Morgan, James M. 1984. Osmoregulation and water stress in higher plants. Annual Review of Plant Physiology 35:299–348.

Mussell, H. and R. Staples. 1979. Stress Physiology in Crop Plants. Wiley-Interscience, New York.

Parker, J. 1963. Cold resistance in woody plants. Botanical Review 29:124–201.

Quarrie, S. A. 1980. Genotypic differences in leaf water potential, abscisic acid and proline concentrations in spring wheat during drought stress. Annals of Botany 46:383–394.

Richardson, S. G. and F. B. Salisbury. 1977. Plant responses to the light penetrating snow. Ecology 58:1152–1158.

LeRudulier, D., A. R. Strom, A. M. Dandekar, L. T. Smith and R. C. Valentine. 1984. Molecular biology of osmoregulation. Science 224:1064–1068.

Rundel, P. W. 1982. Water uptake by organs other than roots. Pages 111–128 in Lange, O. L., P. S. Nobel, C. B. Osmond, and H. Ziegler (eds.), Encyclopedia of Plant Physiology. Vol. 12B, Physiology Plant Ecology II. Springer-Verlag, Berlin, Heidelberg, New York.

Salisbury, Frank B. 1964. Soil formation and vegetation on hydrothermally altered rock material in Utah. Ecology 45:1–9.

Salisbury, Frank B. and N. G. Marinos. 1985. The ecological role of plant growth substances. In R. P. Pharis and D. M. Reid (eds.), Encyclopedia of Plant Physiology. Vol. 11. Springer-Verlag, Berlin, Heidelberg, New York.

Shantz, H. C. 1927. Drought resistance and soil moisture. Ecology 8:145–157.

Spomer, G. G. 1973. The concepts of "interaction" and "operational environment" in environmental analyses. Ecology 54(1):200–204.

Steponkus, P. L. 1981. Responses to extreme temperatures. Cellular and subcelluar bases. Pages 371–402 in Encyclopedia of Plant Physiology, New Series, Vol. 12A Physiological Plant Ecology. Lange, O. L., P. S. Nobel, C. S. Osmond, and H. Ziegler (eds.), Springer-Verlag, Berlin, Heidelberg, New York.

Steponkus, Peter L. 1984. Role of the plasma membrane in freezing injury and cold acclimation. Annual Review of Plant Physiology 35:543–693.

Stewart, G. R. and J. A. Lee. 1974. The role of proline accumulation in Halophytes. Planta 120:279–289.

Tranquillini, W. Frost-drought and its ecological significance. 1982. Pages 379–400 *in* Encyclopedia of Plant Physiology. New Series Vol. 12B; Physiological Plant Ecology, Lange, O. L., P. S. Nobel, C. B. Osmond, and H. Ziegler (eds.), Springer-Verlag, Berlin, Heidelberg, New York.

Troughton, J. and L. A. Donaldson. 1972. Probling Plant Structure. McGraw-Hill Book Company, New York.

Turner, Neil C. and Madeleine M. Jones. 1980. Turgor maintenance by osmotic adjustment: A review and evaluation. Pages 87–103 *in* Turner, N. C. and Kramer, P. J. (eds.), Adaptation of Plants to Water and High Temperature Stress. Wiley-Interscience, New York.

Walsby A. E. 1983. Bacteria that grow at 250°C. Nature 303:381.

Walton, D. C. 1980. Biochemistry and physiology of abscisic acid. Annual Review of Plant Physiology 31:453–489.

Weiser, C. J. 1970. Cold resistance and injury in woody plants. Science 169:1269–1278.

White, W. C. and C. J. Weiser. 1964. The relation of tissue desiccation, extreme cold, and rapid temperature fluctations to winter inury of American Arborvitae. American Society for Horticultural Science 85:554–563.

Whittaker, R. H. 1954. The ecology of serpentine soils. IV. The vegetational response to serpentine soils. Ecology 35:278–288.

Yancey, P. H., M. E. Clark, S. C. Hand, R. D. Bowlus, and G. N. Somero. 1982. Living with water stress: evolution of osmolyte systems. Science 217:1214–1222.

Author Index

Index of Species and Subjects

Deficiency symptoms, 106
Deficient zone, 102
Defoliation, 443
Degradation of starch, 230
Delayed effects, 412
Delayed rhythm, 400
Delphinium barbeyi, 286
Delta-aminolevulinic acid (ALA), **385**
Denaturation, **173**
Denitrification, 252
Derris elliptica, 285
Desert: ephemerals, 474; pavement, 470; and photodormancy, 381; plant-energy exchange in, 73–74; as stressful environment, 469–471, *470, 471*
Desert pavement, **470**
Designed leakage, 85
Desmotubule, **120–121**
Destruction of phytochrome, 379
Determinate structure, 292
Development, 2, **290**; steps in, 292–294
Devernalization, **414**
Dextrins, **149**, 230
D-glucose, *148*
Dianthus, 274
Dicamba, 319
Dicumarol, **281**
Dielectric constants, **21**
Differentially permeable membranes, 39
Differentiation, **290**, 350–373; principles of, 373
Diffusion, 18–32; defined, **22**; and entropy, 26; model of, 24–25, *29*; rate of, 31–32; and solute absorption, 130–131
Digalactosyldiglyceride, 122
Digilanides, **275–276**
Digitalis, 275–276
Digitaria sanguinalis, 201, 305
Dihydrophaseic acid, *346*, **346**
Dihydroxyacetone phosphate, 199
Dihydrozeatin, *331*, **331**
Dilution, 38
Dioecious plants, **305**
Dionea muscipula, 354
Dioxin, 319
Directional differential growth, 355–356
Disaccharide, *148*, **148**
Distal part, 302
Disulfide bonds, **169**
Diterpenoids, **277**
Diurnal cycle, *392*, 459–461
Dixon cohesion theory, *88*, 88–94
DNA, 2; and cytokinins, 330; and low temperatures, 414; and nitrogen, 265; replication, 294–295
Dormancy, **416**; abscisic acid and, 346–347; of buds, 420–422; photodormant seeds, 380; in potato tubers, 423; and radiant energy, 465; of seeds, 418–420
Dorsal, **350**
Dose-response relations, 357–359
Douglas fir tree, 90–91, 390
Drosophila, 402
Drought, **469**, 473–475; tolerance, 111
Drought spot of apples, 112
Dry matter, **96**
Dry weight, *291*, **291**
Dual-daylength plants, 429
Dual kinetics, 129
Dusk, *399*, 437–438
Dwarf plants and gibberellins, 322–324
Dynamic equilibrium, **24**

E'₀, **192**
Ecdysones, **276**
Echinochloa: crus-galli, 247; *utilis*, 100
Ecophenes, **457**
Ecosystems: genetics and, 456–458; and photodormancy, 381; plant-energy exchange in, 73–74; primary productivity estimates, 217
Ecotypes, 456–458, **457**
Ectendotrophic, **116**
Ectomycorrhizae, **116**, *117*
EDTA, 105
Einstein, 183
Elastic biological strain, **468–469**
Elasticity of cell wall, **295**
Electrolytes, 21
Electron-transport: of mitochondria, 237–239; in photosynthesis, 187–190, *188*; summary, 239–240
Electrophoresis, **170**
Electropotentials, 131–132
Elicitors, **281**
Elongation, **293**–*294*
Embryo, **292**
Embryogenesis, **334**
Emerson enhancement effect, 185–**186**
End-product inhibition, **177**
Endodermis, **83**
Endogenous timing, **394**
Endogonacae, 117
Endomycorrhizae, **116**
Endoplasmic reticulum (ER), *4, 10,* **11**; in eukaryotic cells, 11; and fats, 271; in sieve tube, 143
Endoreduplication, **373**
Energy: of activation, **171–172**, *411*; nitrogen fertilizers and, 253; in respiration, 244–246; spillover, 191; and thermodynamics, 25; and transpiration, 66–67, 70–73
Energy charge (EC), 244, **244**
Englemann spruce, 222
Enongenous auxin, 315
Enthalpy, **27**
Entrained rhythms, 399
Entrainment, 400
Entropy: changes, 26; and thermodynamics, 25–27
Environment: and abscission, 349; defined, **449–451**; and evaporation, 54; interaction with, 2; and isozymes, 171; problems of, 449; radiation in, 458–465; and stomates, 60–61; stress physiology, 468–490; in thermodynamics, 25; types of responses to, 455–457; and water stress, 473–487
Environmental physiology, **449**
Enzymes, 2, **164–165**; allosteric enzymes, 177–178; branching enzymes, 211; chemical composition of, 166–168; denaturation, 173; factors influencing reactions, 173–177; induction, 260; major classes and subclasses, 166; mechanisms of action, 171–173; properties of 165–171; prosthetic groups, 168–169; substrate concentration, 172, *174*; temperature dilemma, 409–412, *411*; three-dimensional structures of, 169–170; and water stress, 476
Enzyme-substrate complex, *172*, **172**
Epicotyl hook, **300**, *301*
Epidermis, **83**
Epigeally, **385**
Epimerase, 242
Epinasty, *342*, **350**

Epitopes, **387**
Equilibrium: in entropy, 26; and matric potential, 53
Equisetum arvense, 101
Ergastic substances, 8
Ergosterol, **275**
Erythrose-4-phosphate, 197, 242, 390
Erythroxylon coca, 286
Escherichia coli, 3
Essential oils, 277–278
Ester linkage, 269
Estrogens, **276**
Ethephon, **344**
Ether, 403
Ethrel, **344**
Ethyl alcohol, 403
Ethylene; 339–344, *340, 341, 343*; antagonists of, 344; and auxins, 343; and bud dormancy, 422; -diaminetetraacetic acid (EDTA), 105; and flowering, 343–344, 444; and gravitropism, 367–368; and stems and roots, 342–343; synthesis, 340–342; and underground storage organs, 422; and water stress, 477
Etiolated plants, **374**; mustard seedling, *389*; and radiant energy, 464; stems, 359
Etioplasts, *338*, **338**
Eucalyptus regnans, 75
Euglena, 396
Eukaryotic cells, 2, **4–15**; components of, *5*; defined, 3; high temperature and, 489
Eutrophication, **104**
Euxerophytes, **475**
Evans' modified Shive's solution, 98–99
Evaporation, 56, 88
Evapotranspiration, **55**
Excited state of molecule, 183
Exobasidium, 336
Exogenous timing, **394**
External mantle, 116, *117*

Facilitated diffusion, **130–131**
Facultitative halophytes, **478**
Facultative requirement, **412**
Fagus sylvatica, 371
False pulvinus, **369**
Faraday constant, **132**
Fasciation disease, 335–336
Fats, *269*; converted to sugars, 271–274; formation of, 270–271; importance of, 269–270; in membrane, 122; solubility problem, 125–126
Fatty acids, **268–274**; to sugars, 271–274
Fe-EDDHA, **105**
Fe-S proteins, **186**
Feedback: and allosteric enzymes, 177–178; inhibition, 177, **177–178**; loops, *63*, **63**; and low temperature response, 425
Fermentation, **233**; diagram of, *235*; of flooded tree seeds, 248
Ferredoxin, **166**; -thioredoxin system, 207
Ferredoxin; NADP⁺ reductase, **188**
Fertilizers, 104; chelates, 104–105; nitrogen, 253
Ferulic acid, 275, **279**
Fibers, *78*, *79*
Fibrillar cytoskeleton, 10
Fick's First Law, **32**; and evaporation, 56
Field capacity, **86**
Field effects, **373**
First Law of Thermodynamics, 25
First-negative curvature, **357**
First-positive curvature, **357**

Fisher's analysis of variance, 454–455
Flagella, **4**
Flavin, 168; in phototropism, 359
Flavones, 284, *285*
Flavonoids, **283–285**; light-enhanced synthesis and, 390–391
Flavonols, 284, *285*
Flavoproteins, 237, *359*
Flooding, 473
Florigen concept, 442–444
Florigenic acid, **443**
Flow paraboloid, *81*
Flow process, 299
Flow velocities, 89
Flowers, 305–306; and ethylene, 343–344, 444; florigen concept, 442–444; and giberellins, 324; gravitropism and, 369–370; induced state and, 445–446; photo-periodism and, 405, 428–429; responses to applied hormones, 444–445; ripeness to flower, 435; *see also* Buds
Flowthrough ventilation system, 70
Fluid mosaic model, 123–124
Fluids, **22**. *See also* Liquids
Flourescence, **183**; and allophycocyanin, 193
Flourine, 101
Flux, **32**
Föhn, **469**
Foliar plantlets, *433*
Forcing, **423**, *424*
Forestry, 449
4-chloro-indoleacetic acid, **310**
Fraxinus sp., 82, 153
Free-base cytokinin, **331**
Free-running period, **395–396**
Freeze drying, **136**
Freezing: air blockage in, 94; and dormancy, 416; injury from, 480–482; and osmotic potential, *44*
Freezing point method, **44**
Fresh weight, *291*, **291**
Frost injury, 480–482
frq, 403
Fructans (fructosans), **209**; formation of, 212–213; hydrolysis of, 232
Fructose-1,6-bisphosphate, 197
Fructose-2,6-bisP, 246
Fructose-6-phosphate, 197
Fruit: and ethylene, 339; growth, 298, 306–308
Fucoxanthin, 184
Function and structure, 2–3
Furanose, 149
Fusicoccum amydali, 318

Galactolipids, 183
Gallic acid, **279**, 280–281
Gallotannins, 280–281
Gamma plantlets, **291**
Gemmae, **347**
Geophytes, **483**
Geotropism, 355
Germination 416, *419*; *See also* Seeds
Gibberellic acid, **320**
Gibberellins, 319–**329**, *320*; commercial uses of, 329; dormant seeds and, *324*; flower response, 444–445; food and mineral elements, 324–325, 328; and gravitropism, 367; mechanisms of, 328–329; metabolism of, 320–321; and photodormancy, 383; and vernalin, 414–415
Gibbs free energy, **26–27**, 36
Ginko biloba, 378